Impedance Spectroscopy

Impedance Spectroscopy Theory, Experiment, and Applications

Second Edition

Edited by

Evgenij Barsoukov

J. Ross Macdonald

WILEY-INTERSCIENCE

A John Wiley & Sons, Inc., Publication

Published by John Wiley & Sons, Inc., Hoboken, New Jersey.
Published simultaneously in Canada.

Library of Congress Cataloging-in-Publication Data:
ISBN: 0-471-64749-7

Printed in the United States of America
10 9 8 7 6 5 4

Contents

Preface

The principal audience that will benefit from this book are M.Sc. and Ph.D. students with specialization in physical chemistry, electrochemistry, or physics, as well as researchers and engineers in the field of electrochemistry, particularly in areas of semiconductors, solid electrolytes, corrosion, solid state devices, and electrochemical power sources. Impedance spectroscopy has firmly established itself as one of the most informative and irreplaceable investigation methods in these areas of research. In addition, the book provides a valuable source of information and resource for established researchers and engineers working in one or more of the above fields.

The book should enable understanding of the method of impedance spectroscopy in general, as well as detailed guidance in its application in all these areas. It is the only book in existence that brings together expert reviews of all the main areas of impedance applications. This book covers all the subjects needed by a researcher to identify whether impedance spectroscopy may be a solution to his/her particular needs and to explain how to set up experiments and how to analyze their results. It includes both theoretical considerations and the know-how needed to begin work immediately. For most subjects covered, theoretical considerations dealing with modeling, equivalent circuits, and equations in the complex domain are provided. The best measurement methods for particular systems are discussed and sources of errors are identified along with suggestions for improvement. The extensive references to scientific literature provided in the book will give a solid foundation in the state of the art, leading to fast growth from a qualified beginner to an expert.

The previous edition of this book became a standard textbook on impedance spectroscopy. This second extended edition updates the book to include the results of the last two decades of research and adds new areas where impedance spectroscopy has gained importance. Most notably, it includes completely new sections on batteries, supercapacitors, fuel cells, and photochromic materials. A new section on commercially available measurements systems reflects the reality of impedance spectroscopy as a mainstream research tool.

EVGENIJ BARSOUKOV
Dallas, Texas

Preface to the First Edition

Impedance spectroscopy (IS) appears destined to play an important role in fundamental and applied electrochemistry and materials science in the coming years. In a number of respects it is the method of choice for characterizing the electrical behavior of systems in which the overall system behavior is determined by a number of strongly coupled processes, each proceeding at a different rate. With the current availability of commercially made, high-quality impedance bridges and automatic measuring equipment covering the millihertz to megahertz frequency range, it appears certain that impedance studies will become increasingly popular as more and more electrochemists, materials scientists, and engineers understand the theoretical basis for impedance spectroscopy and gain skill in the interpretation of impedance data.

This book is intended to serve as a reference and/or textbook on the topic of impedance spectroscopy, with special emphasis on its application to solid materials. The goal was to produce a text that would be useful to both the novice and the expert in IS. To this end, the book is organized so that each individual chapter stands on its own. It is intended to be useful to the materials scientist or electrochemist, student or professional, who is planning an IS study of a solid state system and who may have had little previous experience with impedance measurements. Such a reader will find an outline of basic theory, various applications of impedance spectroscopy, and a discussion of experimental methods and data analysis, with examples and appropriate references. It is hoped that the more advanced reader will also find this book valuable as a review and summary of the literature up to the time of writing, with a discussion of current theoretical and experimental issues. A considerable amount of the material in the book is applicable not only to solid ionic systems but also to the electrical response of liquid electrolytes as well as to solid ones, to electronic as well as to ionic conductors, and even to dielectric response.

The novice should begin by reading Chapter 1, which presents a broad overview of the subject and provides the background necessary to appreciate the power of the technique. He or she might then proceed to Chapter 4, where many different applications of the technique are presented. The emphasis in this chapter is on presenting specific applications of IS rather than extensive reviews; details of how and why the technique is useful in each area are presented. To gain a fuller appreciation of IS, the reader could then proceed to Chapters 2 and 3, which present the theory and measuring and analysis techniques.

For someone already familiar with IS, this text will also be useful. For those familiar with one application of the technique the book will provide both a convenient source for the theory of IS, as well as illustrations of applications in areas possibly unfamiliar to the reader. For the theorist who has studied IS, the applications discussed in Chapter 4 pose questions the experimentalist would like answered; for the experimentalist, Chapters 2 and 3 offer different (and better!) methods to analyze IS data. All readers should benefit from the presentation of theory, experimental data, and analysis methods in one source. It is our hope that this widened perspective of the field will lead to a more enlightened and thereby broadened use of IS.

In format and approach, the present book is intended to fall somewhere between the single-author (or few-author) text and the "monograph" of many authors and as many chapters. Although the final version is the product of 10 authors' labors, considerable effort has been made to divide the writing tasks so as to produce a unified presentation with consistent notation and terminology and a minimum of repetition. To help reduce repetition, all authors had available to them copies of Sections 1.1–1.3, 2.2, and 3.2 at the beginning of their writing of the other sections. We believe that whatever repetition remains is evidence of the current importance to IS of some subjects, and we feel that the discussion of these subjects herein from several different viewpoints is worthwhile and will be helpful to the readers of the volume.

J. Ross Macdonald
Chapel Hill, North Carolina
March 1987

Contributors

Evgenij Barsoukov
Senior Applications Engineer
Portable Power Management
Texas Instruments Inc.

Nikolaos Bonanos
Senior Research Scientist
Materials Research Department
Risø National Laboratory,
P.O. Box 49, DK-4000 Roskilde
Denmark

Brian Conway
Department of Chemistry
University of Ottawa
Business Address: Marion Hall
140 Louis-Pasteur, Ottawa
On KIN 6N5

Donald Franceschetti
Department of Physics and Chemistry
The University of Memphis
Memphis, TN 38152-6670

Albert Goossens
Delft Department of Chemical Technology (DelftChemTech)
Faculty of Applied Sciences
Delft University of Technology
Julianalaan 136
2628 BL DELFT
The Netherlands

Digby Macdonald
Director, Center for Electrochemical Science and Technology
201 Steidle Building
University Park, PA 16802

J. Ross Macdonald
Department of Physics and Astronomy
University of North Carolina
Chapel Hill, NC 27599-3255, USA

Gunnar A. Niklasson
Department of Materials Science
The Ångstrom Laboratory
Uppsala University
P.O. Box 534
SE-75121 Uppsala, Sweden

Brian Sayers
Product Manager, Solartron Analytical
Unit B1, Armstrong Mall,
Southwood Business Park,
Farnborough,
Hampshire, England GU14 0NR

Norbert Wagner
Deutsches Zentrum für Luft- und Raumfahrt e.V.
(German Aerospace Center)
Institut für Technische Thermodynamik
Pfaffenwaldring 38-40
D-70569 Stuttgart

Contributors to the First Edition

Nikolaos Bonanos
Senior Research Scientist
Materials Research Department
Risø National Laboratory,
P.O. Box 49, DK-4000 Roskilde
Denmark

E.P. Butler
ALCAN International Ltd.
Banbury Laboratories
Banbury, Tennessee

Donald Franceschetti
Department of Physics
The University of Memphis
Manning Hall, Room 328
Campus Box 526670
Memphis, TN 38152-6670

William B. Johnson
W. L. Gore & Associates
201 Airport Road
P.O. Box 1488
Elkton, MD 21922-1488

Digby Macdonald
Director, Center for Electrochemical Science and Technology
201 Steidle Building
University Park, PA 16802

Dr. J. Ross Macdonald
(William R. Kenan, Jr., Professor of Physics, Emeritus)
Department of Physics and Astronomy
University of North Carolina
Chapel Hill, NC 27599-3255, USA

Michael McKubre
Director, SRI International
333 Ravenswood Ave.
Menlo Park, CA 94025-3493

Ian D. Raistrick
Los Alamos National Laboratory
Los Alamos, NM 97545

B. C. H. Steele
Department of Metallurgy and Material Science
Imperial College of Science and Technology
London, England

Wayne L. Worrell
Department of Materials
Science and Engineering
University of Pennsylvania
Philadelphia, Pennsylvania

Chapter 1

Fundamentals of Impedance Spectroscopy

J. Ross Macdonald
William B. Johnson

1.1 BACKGROUND, BASIC DEFINITIONS, AND HISTORY

1.1.1 The Importance of Interfaces

Since the end of World War II we have witnessed the development of solid state batteries as rechargeable high-power-density energy storage devices, a revolution in high-temperature electrochemical sensors in environmental, industrial, and energy efficiency control, and the introduction of fuel cells to avoid the Carnot inefficiency inherent in noncatalytic energy conversion. The trend away from corrosive aqueous solutions and toward solid state technology was inevitable in electrochemical energy engineering, if only for convenience and safety in bulk handling. As a consequence, the characterization of systems with solid–solid or solid–liquid interfaces, often involving solid ionic conductors and frequently operating well above room temperature, has become a major concern of electrochemists and materials scientists.

At an interface, physical properties—crystallographic, mechanical, compositional, and, particularly, electrical—change precipitously and heterogeneous charge distributions (polarizations) reduce the overall electrical conductivity of a system. Proliferation of interfaces is a distinguishing feature of solid state electrolytic cells, where not only is the junction between electrode and electrolyte considerably more complex than in aqueous cells, but the solid electrolyte is commonly polycrystalline. Each interface will polarize in its unique way when the system is subjected to an applied potential difference. The rate at which a polarized region will change when the applied voltage is reversed is characteristic of the type of interface: slow for chemical reactions at the triple phase contacts between atmosphere, electrode, and electrolyte, appreciably faster across grain boundaries in the polycrystalline elec-

Impedance Spectroscopy, Second Edition, edited by Evgenij Barsoukov and J. Ross Macdonald
ISBN 0-471-64749-7

trolyte. The emphasis in electrochemistry has consequently shifted from a time/concentration dependency to frequency-related phenomena, a trend toward small-signal ac studies. Electrical double layers and their inherent capacitive reactances are characterized by their relaxation times, or more realistically by the distribution of their relaxation times. The electrical response of a heterogeneous cell can vary substantially depending on the species of charge present, the microstructure of the electrolyte, and the texture and nature of the electrodes.

Impedance spectroscopy (IS) is a relatively new and powerful method of characterizing many of the electrical properties of materials and their interfaces with electronically conducting electrodes. It may be used to investigate the dynamics of bound or mobile charge in the bulk or interfacial regions of any kind of solid or liquid material: ionic, semiconducting, mixed electronic–ionic, and even insulators (dielectrics). Although we shall primarily concentrate in this monograph on solid electrolyte materials—amorphous, polycrystalline, and single crystal in form—and on solid metallic electrodes, reference will be made, where appropriate, to fused salts and aqueous electrolytes and to liquid-metal and high-molarity aqueous electrodes as well. We shall refer to the experimental cell as an electrode–material system. Similarly, although much of the present work will deal with measurements at room temperature and above, a few references to the use of IS well below room temperature will also be included. A list of abbreviations and model definitions appears at the end of this work.

In this chapter we aim to provide a working background for the practical materials scientist or engineer who wishes to apply IS as a method of analysis without needing to become a knowledgeable electrochemist. In contrast to the subsequent chapters, the emphasis here will be on practical, empirical interpretations of materials problems, based on somewhat oversimplified electrochemical models. We shall thus describe approximate methods of data analysis of IS results for simple solid-state electrolyte situations in this chapter and discuss more detailed methods and analyses later. Although we shall concentrate on intrinsically conductive systems, most of the IS measurement techniques, data presentation methods, and analysis functions and methods discussed herein apply directly to lossy dielectric materials as well.

1.1.2 The Basic Impedance Spectroscopy Experiment

Electrical measurements to evaluate the electrochemical behavior of electrode and/or electrolyte materials are usually made with cells having two identical electrodes applied to the faces of a sample in the form of a circular cylinder or rectangular parallelepiped. However, if devices such as chemical sensors or living cells are investigated, this simple symmetrical geometry is often not feasible. Vacuum, a neutral atmosphere such as argon, or an oxidizing atmosphere are variously used. The general approach is to apply an electrical stimulus (a known voltage or current) to the electrodes and observe the response (the resulting current or voltage). It is virtually always assumed that the properties of the electrode–material system are time-invariant, and it is one of the basic purposes of IS to determine these properties, their interrelations, and their dependences on such controllable variables as temperature, oxygen partial pressure, applied hydrostatic pressure, and applied static voltage or current bias.

A multitude of fundamental microscopic processes take place throughout the cell when it is electrically stimulated and, in concert, lead to the overall electrical response. They include the transport of electrons through the electronic conductors, the transfer of electrons at the electrode–electrolyte interfaces to or from charged or uncharged atomic species which originate from the cell materials and its atmospheric environment (oxidation or reduction reactions), and the flow of charged atoms or atom agglomerates via defects in the electrolyte. The flow rate of charged particles (current) depends on the ohmic resistance of the electrodes and the electrolyte and on the reaction rates at the electrode–electrolyte interfaces. The flow may be further impeded by band structure anomalies at any grain boundaries present (particularly if second phases are present in these regions) and by point defects in the bulk of all materials. We shall usually assume that the electrode–electrolyte interfaces are perfectly smooth, with a simple crystallographic orientation. In reality, of course, they are jagged, full of structural defects, electrical short and open circuits, and they often contain a host of adsorbed and included foreign chemical species that influence the local electric field.

There are three different types of electrical stimuli which are used in IS. First, in transient measurements a step function of voltage [$V(t) = V_0$ for $t > 0$, $V(t) = 0$ for $t < 0$] may be applied at $t = 0$ to the system and the resulting time-varying current $i(t)$ measured. The ratio $V_0/i(t)$, often called the indicial impedance or the time-varying resistance, measures the impedance resulting from the step function voltage perturbation at the electrochemical interface. This quantity, although easily defined, is not the usual impedance referred to in IS. Rather, such time-varying results are generally Fourier or Laplace-transformed into the frequency domain, yielding a frequency-dependent impedance. If a Fourier-transform is used, a distortion arising because of the non-periodicity of excitation should be corrected by using windowing. Such transformation is only valid when $|V_0|$ is sufficiently small that system response is linear. The advantages of this approach are that it is experimentally easily accomplished and that the independent variable, voltage, controls the rate of the electrochemical reaction at the interface. Disadvantages include the need to perform integral transformation of the results and the fact that the signal-to-noise ratio differs between different frequencies, so the impedance may not be well determined over the desired frequency range.

A second technique in IS is to apply a signal $v(t)$ composed of random (white) noise to the interface and measure the resulting current. Again, one generally Fourier-transforms the results to pass into the frequency domain and obtain an impedance. This approach offers the advantage of fast data collection because only one signal is applied to the interface for a short time. The technique has the disadvantages of requiring true white noise and then the need to carry out a Fourier analysis. Often a microcomputer is used for both the generation of white noise and the subsequent analysis. Using a sum of well-defined sine waves as excitation instead of white noise offers the advantage of a better signal-to-noise ratio for each desired frequency and the ability to analyze the linearity of system response.

The third approach, the most common and standard one, is to measure impedance by applying a single-frequency voltage or current to the interface and measuring the phase shift and amplitude, or real and imaginary parts, of the resulting current at that frequency using either analog circuit or fast Fourier transform (FFT) analysis of the response. Commercial instruments (see Section 3.2) are available which measure the impedance as a function of frequency automatically in the frequency ranges of about 1 mHz to 1 MHz and are easily interfaced to laboratory microcomputers. The advantages of this approach are the availability of these instruments and the ease of their use, as well as the fact that the experimentalist can achieve a better signal-to-noise ratio in the frequency range of most interest. In addition to these three approaches, one can combine them to generate other types of stimuli. The most important of these, ac polarography, combines the first and third techniques by simultaneously applying a linearly varying unipolar transient signal and a much smaller single-frequency sinusoidal signal (Smith [1966]).

Any intrinsic property that influences the conductivity of an electrode–materials system, or an external stimulus, can be studied by IS. The parameters derived from an IS spectrum fall generally into two categories: (a) those pertinent only to the material itself, such as conductivity, dielectric constant, mobilities of charges, equilibrium concentrations of the charged species, and bulk generation–recombination rates; and (b) those pertinent to an electrode–material interface, such as adsorption–reaction rate constants, capacitance of the interface region, and diffusion coefficient of neutral species in the electrode itself.

It is useful and not surprising that modern advances in electronic automation have included IS. Sophisticated automatic experimental equipment has been developed to measure and analyze the frequency response to a small-amplitude ac signal between about 10^{-4} and $>10^{6}$ Hz, interfacing its results to computers and their peripherals (see Section 3.1). A revolution in the automation of an otherwise difficult measuring technique has moved IS out of the academic laboratory and has begun to make it a technique of significant importance in the areas of industrial quality control of paints, emulsions, electroplating, thin-film technology, materials fabrication, mechanical performance of engines, corrosion, and so on.

Although this book has a strong physicochemical bias, the use of IS to investigate polarization across biological cell membranes has been pursued by many investigators since 1925. Details and discussion of the historical background of this important branch of IS are given in the books of Cole [1972] and Schanne and Ruiz-Ceretti [1978].

1.1.3 Response to a Small-Signal Stimulus in the Frequency Domain

A monochromatic signal $v(t) = V_m \sin(\omega t)$, involving the single frequency $\nu \equiv \omega/2\pi$, is applied to a cell and the resulting steady state current $i(t) = I_m \sin(\omega t + \theta)$ measured. Here θ is the phase difference between the voltage and the current; it is zero for purely resistive behavior. The relation between system properties and response to periodic voltage or current excitation is very complex in the time domain. In general, the solution of a system of differential equations is required. Response of capacitive and inductive elements is given as $i(t) = [dv(t)/dt]\ C$ and $v(t) = [di(t)/dt]\ L$ correspondingly, and combination of many such elements can produce an intractable complex problem.

Fortunately, the use of Fourier transformation allows one to simplify significantly the mathematical treatment of this system. The above differential equations can be transformed into $I(j\ \omega) = C \cdot \omega \cdot j \cdot V(j\ \omega)$ and $I(j\ \omega) = \cdot V(j\ \omega)/(L \cdot \omega \cdot j)$. Here $j \equiv \sqrt{-1}$, which is also often denoted in the literature as "i". For the case of sine-wave excitation as above, Fourier transforms of voltage and current $V(j\ \omega)$ and $I(j\ \omega)$ become $V_m\pi$ and $I_m\pi \cdot \exp(\theta j)$ respectively. It can be easily seen that in the frequency domain voltage/current relations can be rearranged to a form similar to Ohm's law for dc current: $I(j\ \omega) = V(j\ \omega)/Z(j\ \omega)$ where for capacitance the complex quantity $Z(j\ \omega)$ is $1/(C \cdot \omega \cdot j)$ and for inductance $Z(j\ \omega)$ is $L \cdot \omega \cdot j$. The complex quantity $Z(j\ \omega)$ is defined as the "impedance function", and its value at a particular frequency is "impedance" of the electric circuit. For simplicity, $Z(j\ \omega)$ is usually written as just $Z(\omega)$. Because of this Ohm's law-like relationship between complex current and voltage, the impedance of a circuit with multiple elements is calculated using the same rules as with multiple resistors, which greatly simplifies calculations.

Impedance may be defined not only for discrete systems but also for arbitrary distributed systems as the Fourier transform of the differential equation defining the voltage response divided by the Fourier transform of the periodic current excitation: $Z(j\ \omega) = F\{v(t)\}/F\{i(t)\}$. Here the $F\{\}$ operator denotes a Fourier transform. However, Fourier transformation only reduces differential equations to simple Ohm's law-like form under conditions of linearity, causality, and stationarity of the system; therefore impedance is properly defined only for systems satisfying these conditions.

The concept of electrical impedance was first introduced by Oliver Heaviside in the 1880s and was soon after developed in terms of vector diagrams and complex representation by A. E. Kennelly and especially C. P. Steinmetz. Impedance is a more general concept than resistance because it takes phase differences into account, and it has become a fundamental and essential concept in electrical engineering. Impedance spectroscopy is thus just a specific branch of the tree of electrical measurements. The magnitude and direction of a planar vector in a right-hand orthogonal system of axes can be expressed by the vector sum of the components a and b along the axes, that is, by the complex number $Z = a + jb$. The imaginary number $j \equiv \sqrt{-1} \equiv \exp(j\pi/2)$ indicates an anticlockwise rotation by $\pi/2$ relative to the x axis. Thus, the real part of Z, a, is in the direction of the real axis x, and the imaginary part b is along the y axis. An impedance $Z(\omega) = Z' + jZ''$ is such a vector quantity and may be plotted in the plane with either rectangular or polar coordinates, as shown in Figure 1.1.1. Here the two rectangular coordinate values are clearly

$$\mathrm{Re}(Z) \equiv Z' = |Z|\cos(\theta) \quad \text{and} \quad \mathrm{Im}(Z) \equiv Z'' = |Z|\sin(\theta) \tag{1}$$

with the phase angle

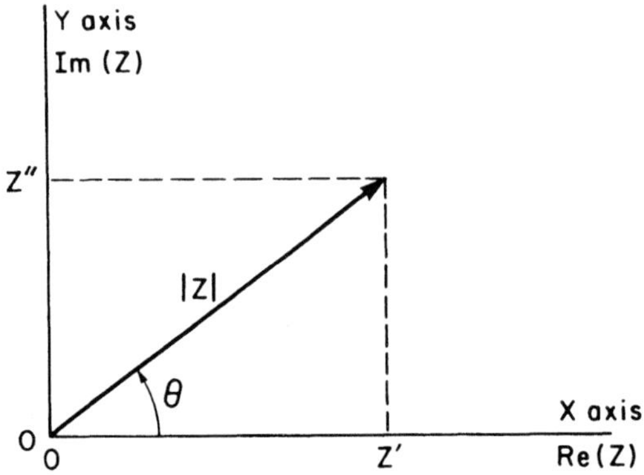

Figure 1.1.1. The impedance Z plotted as a planar vector using rectangular and polar coordinates.

$$\theta = \tan^{-1}(Z''/Z') \tag{2}$$

and the modulus

$$|Z| = \left[(Z')^2 + (Z'')^2\right]^{1/2} \tag{3}$$

This defines the Argand diagram or complex plane, widely used in both mathematics and electrical engineering. In polar form, Z may now be written as $Z(\omega) = |Z|\exp(j\theta)$, which may be converted to rectangular form through the use of the Euler relation $\exp(j\theta) = \cos(\theta) + j\sin(\theta)$. It will be noticed that the original time variations of the applied voltage and the resulting current have disappeared, and the impedance is time-invariant (provided the system itself is time-invariant).

In general, Z is frequency-dependent, as defined above. Conventional IS consists of the (nowadays often automated) measurement of Z as a function of ν or ω over a wide frequency range. It is from the resulting structure of the $Z(\omega)$ vs. ω response that one derives information about the electrical properties of the full electrode–material system.

For nonlinear systems, i.e. most real electrode–material systems, IS measurements in either the time or the frequency domain are useful and meaningful in general only for signals of magnitude such that the overall electrode–material system response is electrically linear. This requires that the response to the sum of two separate input-measuring signals applied simultaneously be the sum of the responses of the signals applied separately. A corollary is that the application of a monochromatic signal, one involving $\sin(\omega t)$, results in no, or at least negligible, generation of harmonics in the output, that is components with frequencies $n\nu$ for $n = 2, 3, \ldots$. Both solid and liquid electrochemical systems tend to show strong nonlinear behavior, especially in their interfacial response, when applied voltages or currents are large. But so long as the applied potential difference (p.d.) amplitude V_m is less than the thermal voltage, $V_T \equiv RT/F \equiv kT/e$, about 25 mV at 25°C, it can be shown that the basic differential equations which govern the response of the system become linear to an excellent approximation. Here k is Boltzmann's constant, T the absolute temperature, e the proton charge, R the gas constant, and F the faraday. Thus if the applied amplitude V_m is appreciably less than V_T, the system will respond linearly. Note that in the linear regime it is immaterial as far as the determination of $Z(\omega)$ is concerned whether a known $v(\omega t)$ is applied and the current measured or a known $i(\omega t)$ applied and the resulting voltage across the cell measured. When the system is nonlinear, this reciprocity no longer holds.

1.1.4 Impedance-Related Functions

The impedance has frequently been designated as the ac impedance or the complex impedance. Both these modifiers are redundant and should be omitted. Impedance without a modifier always means impedance applying in the frequency domain and usually measured with a monochromatic signal. Even when impedance values are derived by Fourier transformation from the time domain, the impedance is still defined for a set of individual frequencies and is thus an alternating-current impedance in character.

Impedance is by definition a complex quantity and is only real when $\theta = 0$ and thus $Z(\omega) = Z'(\omega)$, that is, for purely resistive behavior. In this case the impedance is completely frequency-independent. When Z' is found to be a variable function of frequency, the Kronig–Kramers (Hilbert integral transform) relations (Macdonald and Brachman [1956]), which holistically connect real and imaginary parts with each other, ensure that Z'' (and θ) cannot be zero over all frequencies but must vary with frequency as well. Thus it is only when $Z(\omega) = Z'$, independent of frequency, so $Z' = R$, an ordinary linear resistance, that $Z(\omega)$ is purely real.

There are several other measured or derived quantities related to impedance which often play important roles in IS. All of them may be generically called immittances. First is the admittance, $Y \equiv Z^{-1} \equiv Y' + jY''$. In the complex domain where v, i, and Z are all taken complex, we can write $v = Zi$ or alternatively $i = Yv$. It is also customary in IS to express Z and Y in terms of resistive and capacitance components as $Z = R_s(\omega) - jX_s(\omega)$ and $Y = G_p(\omega) + jB_p(\omega)$, where the reactance $X_s = [\omega C_s(\omega)]^{-1}$ and the susceptance $B_p \equiv \omega C_p(\omega)$. Here the subscripts s and p stand for "series" and "parallel."

The other two quantities are usually defined as the modulus function $M = j\omega C_c Z = M' + jM''$ and the complex dielectric constant or dielectric permittivity $\varepsilon = M^{-1} \equiv Y/(j\omega C_c) \equiv \varepsilon' - j\varepsilon''$. In these expressions $C_c \equiv \varepsilon_0 A_c/l$ is the capacitance of the empty measuring cell of electrode area A_c and electrode separation length l. The quantity ε_0 is the dielectric permittivity of free space, 8.854×10^{-12} F/m. The dielectric constant ε is often written elsewhere as ε^* or $\hat{\varepsilon}$ to denote its complex character. Here we shall reserve the superscript asterisk to denote complex conjugation; thus $Z^* = Z' - jZ''$. The interrelations between the four immittance functions are summarized in Table 1.1.1.

The modulus function $M = \varepsilon^{-1}$ was apparently first introduced by Schrama [1957] and has been used appreciably by McCrum *et al.* [1976], Macedo *et al.* [1972], and Hodge *et al.* [1975, 1976]. The use of the complex dielectric constant

Table 1.1.1. Relations Between the Four Basic Immittance Functions[a]

	M	Z	Y	ε
M	M	μZ	μY^{-1}	ε^{-1}
Z	$\mu^{-1}M$	Z	Y^{-1}	$\mu^{-1}\varepsilon^{-1}$
Y	μM^{-1}	Z^{-1}	Y	$\mu\varepsilon$
ε	M^{-1}	$\mu^{-1}Z^{-1}$	$\mu^{-1}Y$	ε

[a] $\mu \equiv j\omega C_c$, where C_c is the capacitance of the empty cell.

goes back much further but was particularly popularized by the work of Cole and Cole [1941], who were the first to plot ε in the complex plane.

Some authors have used the designation *modulus spectroscopy* to denote small-signal measurement of M vs. ν or ω. Clearly, one could also define admittance and dielectric permittivity spectroscopy. The latter is just another way of referring to ordinary dielectric constant and loss measurements. Here we shall take the general term *impedance spectroscopy* to include all these other very closely related approaches. Thus IS also stands for *immittance spectroscopy*. The measurement and use of the complex $\varepsilon(\omega)$ function is particularly appropriate for dielectric materials, those with very low or vanishing conductivity, but all four functions are valuable in IS, particularly because of their different dependence on and weighting with frequency.

1.1.5 Early History

Impedance spectroscopy is particularly characterized by the measurement and analysis of some or all of the four impedance-related functions Z, Y, M, and ε and the plotting of these functions in the complex plane. Such plotting can, as we shall see, be very helpful in interpreting the small-signal ac response of the electrode–material system being investigated. Historically, the use of Z and Y in analyzing the response of electrical circuits made up of lumped (ideal) elements (R, L, and C) goes back to the beginning of the discipline of electrical engineering. An important milestone for the analysis of real systems, that is ones distributed in space, was the plotting by Cole and Cole [1941] of ε' and ε'' for dielectric systems in the complex plane, now known as a Cole–Cole plot, an adaption at the dielectric constant level of the circle diagram of electrical engineering (Carter [1925]), exemplified by the Smith–Chart impedance diagram (Smith [1939, 1944]). Further, Z and/or Y have been widely used in theoretical treatments of semiconductor and ionic systems and devices from at least 1947 (e.g. Randles [1947], Jaffé [1952], Chang and Jaffé [1952], Macdonald [1953], and Friauf [1954]). Complex plane plots have sometimes been called Nyquist diagrams. This is a misnomer, however, since Nyquist diagrams refer to transfer function (three- or four-terminal) response, while conventional complex plane plots involve only two-terminal input immittances.

On the experimental side, one should mention the early work of Randles and Somerton [1952] on fast reactions in supported electrolytes; no complex plane plotting appeared here. But complex plane plotting of G_p/ω vs. C_p was used by Macdonald [1955] for experimental results on photoconducting alkali halide single crystals. Apparently the first plotting of impedance in the impedance plane for aqueous electrolytes was that of Sluyters [1960] (theory) and Sluyters and Oomen [1960] (experiment). The use of admittance plane plotting for accurate conductivity determination of solid electrolytes was introduced by Bauerle [1969], the first important paper to deal with IS for ionic solids directly. Since then, there have been many pertinent theoretical and experimental papers dealing with IS and complex plane plots. Many of them will be cited later, and we conclude this short survey of early history pertinent to IS with the mention of three valuable reviews: Sluyters-Rehbach and Sluyters [1970], Armstrong *et al.* [1978], and Archer and Armstrong [1980]. The first and second of these deal almost entirely with liquid electrolytes but are nevertheless somewhat pertinent to IS for solids.

1.2 ADVANTAGES AND LIMITATIONS

Although we believe that the importance of IS is demonstrated throughout this monograph by its usefulness in the various applications discussed, it is of some value to summarize the matter briefly here. IS is becoming a popular analytical tool in materials research and development because it involves a relatively simple electrical measurement that can readily be automated and whose results may often be correlated with many complex materials variables: from mass transport, rates of chemical reactions, corrosion, and dielectric properties, to defects, microstructure, and compositional influences on the conductance of solids. IS can predict aspects of the performance of chemical sensors and fuel cells, and it has been used extensively to investigate membrane behavior in living cells. It is useful as an empirical quality control procedure, yet it can contribute to the interpretation of fundamental electrochemical and electronic processes.

A flow diagram of a general characterization procedure using IS is presented in Figure 1.2.1. Here CNLS stands for complex nonlinear least squares fitting (see Section 3.3.2). Experimentally obtained impedance data for a given electrode–materials system may be analyzed by using an exact mathematical model based on a plausible physical theory that predicts theoretical impedance $Z_t(\omega)$ or by a relatively empirical equivalent circuit whose impedance predictions may be denoted by $Z_{ec}(\omega)$. In either the case of the relatively empirical equivalent circuit or of the exact mathematical model, the parameters can be estimated and the experimental $Z_e(\omega)$ data compared to either the predicted equivalent circuit impedance $Z_{ec}(\omega)$ or to the theoretical impedance $Z_t(\omega)$. Such fitting is most accurately accomplished by the CNLS method described and illustrated in Section 3.3.2.

An analysis of the charge transport processes likely to be present in an experimental cell (the physical model) will often suggest an equivalent circuit of ideal resistors and capacitors (even inductors or negative capacitors in some instances)

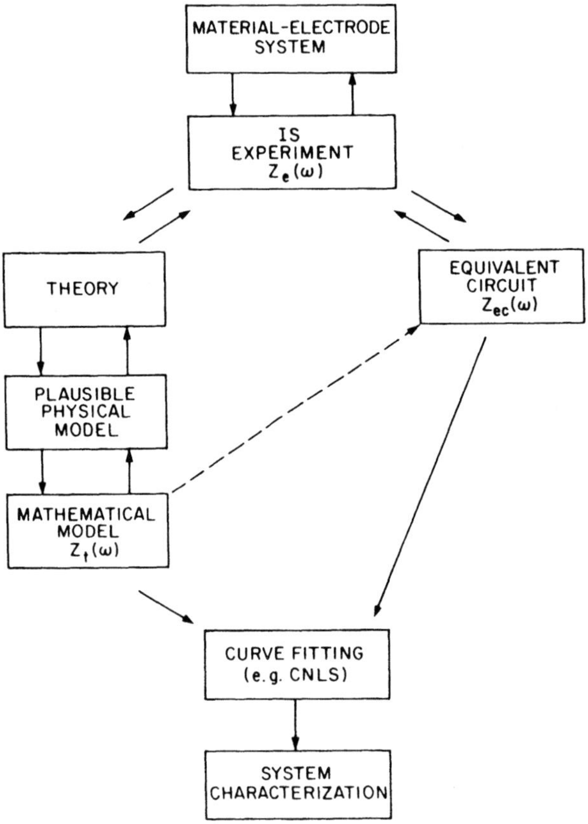

Figure 1.2.1. Flow diagram for the measurement and characterization of a material–electrode system.

and may account adequately for the observed IS response. For example Schouler *et al.* [1983] found that the effects of densification by sintering a polycrystalline electrolyte will reduce the magnitude of the resistance across the grain boundaries and simultaneously decrease the surface area associated with the interface capacitance. These components will clearly be electrically in parallel in this situation. Their combination will be in series with other similar subcircuits representing such processes as the ionization of oxygen at the electrodes.

In another example, the oxidation–reduction reaction for the Zn^{2+} couple in an aqueous solution with a dropping mercury electrode (Sluyters and Oomen [1960]) can be represented by a reaction resistance R_R, arising from the transfer of electrons between the electrode and the solution, in parallel with a capacitor C_R associated with the space charge diffuse double layer near the electrode surface. It is not difficult to calculate the theoretical impedance for such a circuit in terms of the parameters R_R and C_R. From an analysis of the parameter values in a plausible equivalent circuit as the experimental conditions are changed, the materials system can be characterized by analysis of its observed impedance response, leading to estimates of its microscopic parameters such as charge mobilities, concentrations, and electron transfer reaction rates.

The disadvantages of IS are primarily associated with possible ambiguities in interpretation. An important complication of analyses based on an equivalent circuit (e.g. Bauerle [1969]) is that ordinary ideal circuit elements represent ideal lumped-constant properties. Inevitably, all electrolytic cells are distributed in space, and their microscopic properties may be also independently distributed. Under these conditions, ideal circuit elements may be inadequate to describe the electrical response. Thus, it is often found that $Z_e(\omega)$ cannot be well approximated by the impedance of an equivalent circuit involving only a finite number of ordinary lumped-constant elements. It has been observed by many in the field that the use of distributed impedance elements [e.g. constant-phase elements (CPEs) (see Section 2.2.2.2)] in the equivalent circuit greatly aids the process of fitting observed impedance data for a cell with distributed properties.

There is a further serious potential problem with equivalent circuit analysis, not shared by the direct comparison with $Z_t(\omega)$ of a theoretical model: What specific equivalent circuit out of an infinity of possibilities should be used if one is necessary? An equivalent circuit involving three or more circuit elements can often be rearranged in various ways and still yield exactly the same $Z_{ec}(\omega)$. For the different interconnections the values of the elements will have to be different to yield the same $Z_{ec}(\omega)$ for all , but an essential ambiguity is present. An example is presented in Figure 1.2.2. In these circuits the impedance Z_i is arbitrary and may be made up of either lumped elements, distributed elements, or a combination of these types. Examples of other circuits which demonstrate this type of ambiguity will be presented in Section 2.2.2.3. Which one of two or more circuits which all yield exactly the same $Z_{ec}(\omega)$ for all ω should be used for physicochemical analysis and interpretation? This question cannot be answered for a single set of $Z_e(\omega)$ data alone. An approach to its solution can only be made by employing physical intuition and by carrying out several $Z_e(\omega)$ sets of measurements with different conditions, as discussed in Section 2.2.2.3.

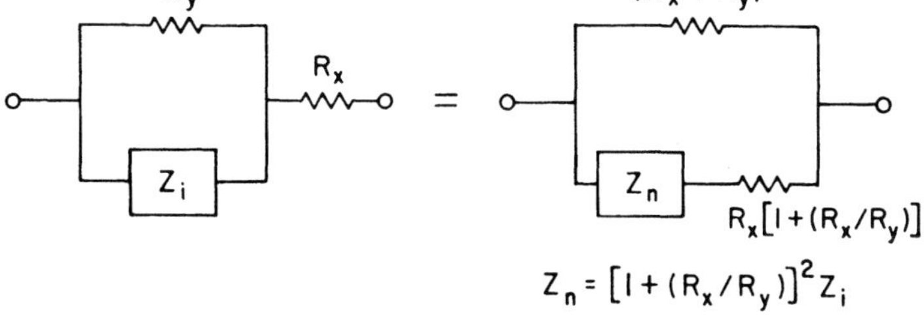

Figure 1.2.2. An example of different circuits with the same overall impedance at all frequencies.

1.2.1 Differences Between Solid State and Aqueous Electrochemistry

The electrochemist who works with aqueous electrolytes has available at least one major stratagem not accessible to those who work with solid electrolytes. Investigators interested in the interfacial behavior of a particular charged species, are usually free to add to the solution an excess of a second electrolyte, the ions of which are neither adsorbed nor react at the interface, but which by sheer numbers are able to screen the interior of the electrolyte from any electric field and cause nearly all the potential drop to occur within a few angstroms of the interface. The investigator is thus (at least by assumption) freed from having to take into account the effect of a nonuniform electric field on the transport of the electroactive species through the bulk electrolyte and need not (again by assumption) puzzle over the fraction of the applied signal which directly governs the exchange of ions or electrons between the electrode surface and the adjacent layer of electrolyte. The added electrolyte species which thus simplifies the interpretation of the experimental results is termed the *indifferent* or *supporting electrolyte*, and systems thus prepared are termed *supported systems*. Solid electrolytes must necessarily be treated as unsupported systems, even though they may display some electrical characteristics usually associated with supported ones. The distinction between unsupported and supported situations is a crucial one for the interpretation of IS results.

It is thus unfortunate that there has been a tendency among some workers in the solid electrolyte field to take over many of the relatively simple theoretical results derived for supported conditions and use them uncritically in unsupported situations, situations where the supported models and formulas rarely apply adequately. For example the expression for the Warburg impedance for a redox reaction in a supported situation is often employed in the analysis of data on unsupported situations where the parameters involved are quite different (e.g. Sections 2.2.3.2 and 2.2.3.3).

There are a few other important distinctions between solid and liquid electrolytes. While liquid electrolytes and many solid electrolytes have negligible electronic conductivity, quite a number of solid electrolytes can exhibit substantial electronic conductivity, especially for small deviations from strict stoichiometric composition. Solid electrolytes may be amorphous, polycrystalline, or single-crystal, and charges of one sign may be essentially immobile (except possibly for high temperatures and over long time spans). On the other hand, all dissociated charges in a liquid electrolyte or fused salt are mobile, although the ratio between the mobilities of positive and negative charges may differ appreciably from unity. Further, in solid electrolytes mobile ions are considered to be able to move as close to an electrode as permitted by ion-size steric considerations. But in liquid electrolytes there is usually present a compact inner or Stern layer composed of solvent molecules, for example H_2O, immediately next to the electrode. This layer may often be entirely devoid of ions and only has some in it when the ions are specifically adsorbed at the electrode or react there. Thus capacitative effects in electrode interface regions can be considerably different between solid and liquid electrolyte systems.

1.3 ELEMENTARY ANALYSIS OF IMPEDANCE SPECTRA

1.3.1 Physical Models for Equivalent Circuit Elements

A detailed physicoelectrical model of all the processes which might occur in investigations on an electrode–material system may be unavailable, premature, or perhaps too complicated to warrant its initial use. One then tries to show that the experimental impedance data $Z_e(\omega)$ may be well approximated by the impedance $Z_{ec}(\omega)$ of an equivalent circuit made up of ideal resistors, capacitors, perhaps inductances, and possibly various distributed circuit elements. In such a circuit a resistance represents a conductive path, and a given resistor in the circuit might account for the bulk conductivity of the material or even the chemical step associated with an electrode reaction (see, e.g., Randles [1947] or Armstrong *et al.* [1978]). Similarly, capacitances and inductances will be generally associated with space charge polarization regions and with specific adsorption and electrocrystallization processes at an electrode. It should be pointed out that ordinary circuit elements, such as resistors and capacitors, are always considered as lumped-constant quantities which involve ideal properties. But all real resistors are of finite size and are thus distributed in space; they therefore always involve some inductance, capacitance, and time delay of response as well as resistance. These residual properties are unimportant over wide frequency ranges and therefore usually allow a physical resistor to be well approximated in an equivalent circuit by an ideal resistance, one which exhibits only resistance over all frequencies and yields an immediate rather than a delayed response to an electrical stimulus.

The physical interpretation of the distributed elements in an equivalent circuit is somewhat more elusive. They are, however, essential in understanding and interpreting most impedance spectra. There are two types of distributions with which we need to be concerned. Both are related, but in different ways, to the finite spatial extension of any real system. The first is associated directly with nonlocal processes, such as diffusion, which can occur even in a completely homogeneous material, one whose physical properties, such as charge mobilities, are the same everywhere. The other type, exemplified by the constant-phase element (CPE), arises because microscopic material properties are themselves often distributed. For example the solid electrode–solid electrolyte interface on the microscopic level is not the often presumed smooth and uniform surface. It contains a large number of surface defects such as kinks, jags, and ledges, local charge inhomogeneities, two- and three-phase regions, adsorbed species, and variations in composition and stoichiometry. Reaction resistance and capacitance contributions differ with electrode position and vary over a certain range around a mean, but only their average effects over the entire electrode surface can be observed. The macroscopic impedance which depends, for example, on the reaction rate distribution across such an interface is measured as an average over the entire electrode. We account for such averaging in our usual one-dimensional treatments (with the dimension of interest perpendicular to the electrodes) by assuming that pertinent material properties are continuously distributed over a given range from minimum to maximum values. For example when a given time constant, associated with an interface or bulk processes, is thermally activated with a distribution of activation energies, one passes from a simple ideal resistor and capacitor in parallel or series to a distributed impedance element, for example the CPE, which exhibits more complicated frequency response than a simple undistributed RC time constant process (Macdonald [1984, 1985*a*, *c*, *d*], McCann and Badwal [1982]).

Similar property distributions occur throughout the frequency spectrum. The classical example for dielectric liquids at high frequencies is the bulk relaxation of dipoles present in a pseudoviscous liquid. Such behavior was represented by Cole and Cole [1941] by a modification of the Debye expression for the complex dielectric constant and was the first distribution involving the important constant phase element, the CPE, defined in Section 2.1.2.3. In normalized form the complex dielectric constant for the Cole–Cole distribution may be written

$$(\varepsilon - \varepsilon_\infty)/(\varepsilon_s - \varepsilon_\infty) = \left[1 + (j\omega\tau_0)^{1-\alpha}\right]^{-1} \tag{1}$$

where ε is the dielectric constant, ε_s and ε_∞ are the static and high-frequency limiting dielectric constants, τ_0 the mean relaxation time, and α a parameter describing the width of the material property distribution (in this case a distribution of dielectric relaxation times in frequency space).

1.3.2 Simple RC Circuits

Figure 1.3.1 shows two RC circuits common in IS and typical Z and Y complex plane responses for them. The response of Figure 1.3.1*a* is often present (if not always measured) in IS results for solids and liquids. Any electrode–material system in a measuring cell has a geometrical capacitance $C_g \equiv C_\infty = C_1$ and a bulk resistance $R_b \equiv R_\infty = R_1$ in parallel with it. These elements lead to the time constant $\tau_D = R_\infty C_\infty$, the dielectric relaxation time of the basic material. Usually, τ_D is the smallest time constant of interest in IS experiments. It is often so small ($<10^{-7}$ s) that for the highest angular frequency applied, ω_{max}, the condition $\omega_{max}\tau_D << 1$ is satisfied and little or nothing of the impedance plane curve of Figure 1.3.1*b* is seen. It should be noted, however, that lowering the temperature will often increase τ_D and bring the bulk arc within the range of measurement. Since the peak frequency of the complete semicircle of Figure 1.3.1*b*, ω_p, satisfies $\omega_p\tau_D = 1$, it is only when $\omega_{max}\tau_D >> 1$ that nearly the full curve of Figure 1.3.1*b* is obtained. Although the bulk resistance is often not appreciably distributed, particularly for single crystals, when it is actually distributed the response of the circuit often leads to a repressed semicircle in the Z plane, one whose center lies below the real axis instead of to a full semicircle with its center on the real axis. Since this distributed element situation is frequently found for processes in the $\omega << \tau_D^{-1}$ frequency range, however, we shall examine in detail one simple representation of it shortly.

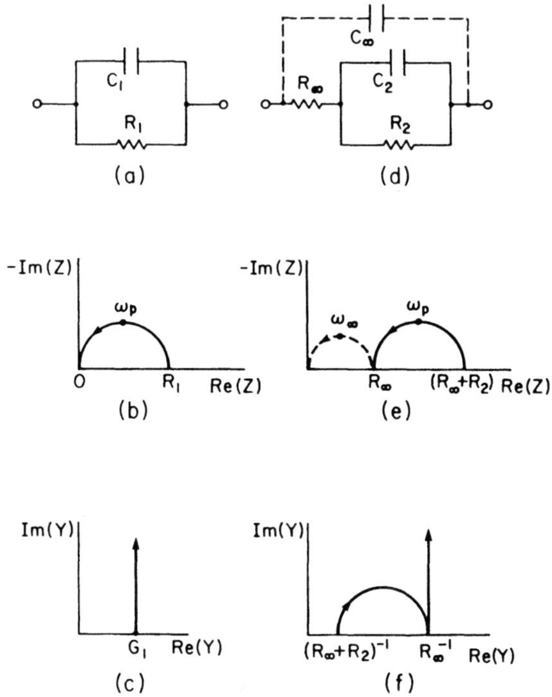

Figure 1.3.1. Figures 1.3.1*a* and *d* show two common RC circuits. Parts *b* and *e* show their impedance plane plots and *c* and *f* their admittance plane plots. Arrows indicate the direction of increasing frequency.

Besides $R_1 = R_\infty$ and $C_1 = C_\infty$, one often finds parallel R_1, C_1 response associated with a heterogeneous electrode reaction. For such a case we would set $R_1 = R_R$ and $C_1 = C_R$, where R_R is a reaction resistance and C_R is the diffuse double-layer capacitance of the polarization region near the electrode in simplest cases. The circuit of Figure 1.3.1*d* combines the above possibilities when $R_2 = R_R$ and $C_2 = C_R$. The results shown in Figure 1.3.1*e* and *f* are appropriate for the well-separated time constants, $R_\infty R_\infty << R_2 C_\infty$. It is also possible that a parallel RC combination can arise from specific adsorption at an electrode, possibly associated with delayed reaction processes. The response arising from R_∞ and C_∞ in Figure 1.3.1*e* is shown dotted to remind one that it often occurs in too high a frequency region to be easily observed. Incidentally, we shall always assume that the capacitance and resistance of leads to the measuring cell have been subtracted out (e.g. by using the results of a preliminary calibration of the system with the cell empty or shorted) so that we always deal only with the response of the material–electrode system alone.

In the complex plane plots, the arrows show the direction of increasing frequency. Further, $G_1 \equiv R_1^{-1}$, $G_\infty \equiv R_\infty^{-1}$, $G_2 \equiv R_2^{-1}$. Because IS results usually involve capacitance and rarely involve inductance, it has become customary to plot impedance in the –Im (Z), Re (Z) plane rather than the Im (Z), Re (Z) plane, thereby ensuring that the vast majority of all curves fall in the first quadrant, as in Figure 1.3.1*b*. This procedure is also equivalent to plotting $Z^* = Z' - iZ''$ rather than Z, so we can alternatively label the ordinate Im (Z^*) instead of –Im (Z). Both choices will be used in the rest of this work.

The admittance of the parallel RC circuit of Figure 1.3.1*a* is just the sum of the admittances of the two elements, that is,

$$Y_a = G_1 + j\omega C_1 \tag{2}$$

It immediately follows that

$$Z_a = Y_a^{-1} = R_1/(R_1 Y_a) = R_1/[1 + j\omega R_1 C_1] \tag{3}$$

This result can be rationalized by multiplying by $[1 - j\omega R_1 C_1]$, the complex conjugate of $[1 + j\omega R_1 C_1]$, in both numerator and denominator. The response of the Figure 1.3.1*a* circuit is particularly simple when it is plotted in the Y plane, as in Figure 1.3.1*c*. To obtain the overall admittance of the Figure 1.3.1*d* circuit, it is simplest to add R_∞ to the expression for Z_a above with $R_1 \rightarrow R_2$ and $C_1 \rightarrow C_2$, convert the result to an admittance by inversion, and then add the $j\omega C_\infty$ admittance. The result is

$$Y_d = j\omega C_\infty + [1 + j\omega R_2 C_2]/[(R_2 + R_\infty) + j\omega C_2 R_2 R_\infty] \tag{4}$$

Although complex plane data plots, such as those in Figures 1.3.1*b*, *c*, *e* and *f* in which frequency is an implicit variable, can show response patterns which are often very useful in identifying the physicochemical processes involved in the electrical response of the electrode–material system, the absence of explicit frequency dependence information is frequently a considerable drawback. Even when frequency values are shown explicitly in such two-dimensional (2-D) plots, it is usually found that with either equal intervals in frequency or equal frequency ratios, the frequency points fall very nonlinearly along the curves. The availability of computerized plotting procedures makes the plotting of all relevant information in a single graph relatively simple. For example three-dimensional (3-D) perspective plotting, as introduced by Macdonald, Schoonman, and Lehnen [1981], displays the frequency dependence along a new log (ν) axis perpendicular to the complex plane (see Section 3.3). For multi-time-constant response in particular, this method is particularly appropriate. The full response information can alternately be plotted with orthographic rather than perspective viewing.

1.3.3 Analysis of Single Impedance Arcs

Analysis of experimental data that yield a full semicircular arc in the complex plane, such as that in Figure 1.3.1*b*, can provide estimates of the parameters R_1 and C_1 and hence lead to quantitative estimates of conductivity, faradic reaction rates, relaxation times, and interfacial capacitance (see detailed discussion in Section 2.2.3.3). In practice, however, experimental data are only rarely found to yield a full semicircle with its center on the real axis of the complex plane. There are three common perturbations which may still lead to at least part of a semicircular arc in the complex plane:

1. The arc does not pass through the origin, either because there are other arcs appearing at higher frequencies and/or because $R_\infty > 0$.
2. The center of an experimental arc is frequently displaced below the real axis because of the presence of distributed elements in the material–electrode system. Similar displacements may also be observed in any of the other complex planes plots (Y, M, or $\in$). The relaxation time τ is then not single-valued but is distributed continuously or discretely around a mean, $\tau_m = \omega_m^{-1}$. The angle θ by which such a semicircular arc is depressed below the real axis is related to the width of the relaxation time distribution and as such is an important parameter.
3. Arcs can be substantially distorted by other relaxations whose mean time constants are within two orders of magnitude or less of that for the arc under consideration. Many of the spectra shown in following chapters involve overlapping arcs.

We shall begin by considering simple approximate analysis methods of data yielding a single, possibly depressed, arc. Suppose that IS data plotted in the impedance plane (actually the Z^* plane) show typical depressed circular arc behavior, such as that depicted in Figure 1.3.2. Here we have included R_∞ but shall initially ignore any effect of C_∞. We have defined some new quantities in this figure which will be used in the analysis to yield estimates of the parameters R_∞, $R_R \equiv R_0 - R$, τ_R and the fractional exponent ψ_{ZC}, parameters which fully characterize the data when they are well represented by the distributed-element ZARC impedance expression (see Section 2.2.2.2),

$$Z - R_\infty \equiv Z_{\text{ZARC}} \equiv (R_0 - R_\infty) I_Z \tag{5}$$

where

$$I_Z \equiv \left[1 + (j\omega\tau_R)^{\psi_{ZC}}\right]^{-1} \equiv \left[1 + (js)^{\psi_{ZC}}\right]^{-1} \tag{6}$$

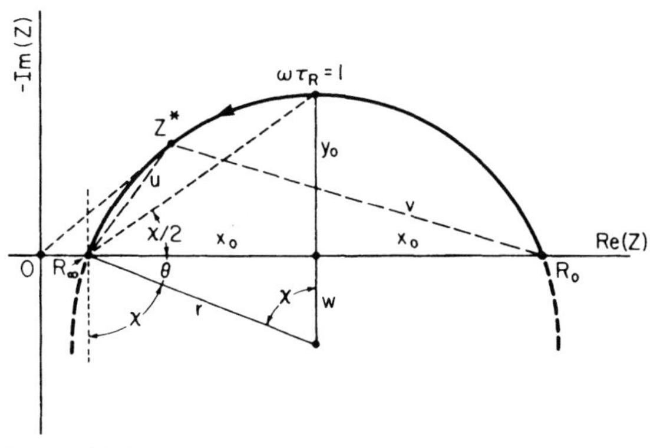

Figure 1.3.2. Impedance plane plot for a depressed circular arc showing definitions of quantities used in its analysis.

Here $s \equiv \omega\tau_R$ is a normalized frequency variable, and I_Z is the normalized, dimensionless form of Z_{ZARC}. Notice that it is exactly the same as the similarly normalized Cole–Cole dielectric response function of Eq. (1) when we set $\psi_{ZC} = 1 - \alpha$. We can also alternatively write the ZARC impedance as the combination of the resistance R_R in parallel with the CPE impedance Z_{CPE} (see Section 2.2.2.2). The CPE admittance is (Macdonald [1984])

$$Y_{\text{CPE}} = Z_{CPE}^{-1} \equiv A_0(j\omega)^{\psi_{zc}} \equiv (j\omega\tau_R)^{\psi_{zc}} \tag{7}$$

Then Eq. (5) may be expressed as

$$Z_{\text{ZARC}} = R_R \Big/ \left[1 + B_0(j\omega)^{\psi_{zc}}\right] \tag{8}$$

where $B_0 \equiv \tau_R^{\psi_{zc}} \equiv R_R A_0$. The fractional exponent ψ_{ZC} satisfies $0 \le \psi_{ZC} \le 1$.

Let us start by considering two easy-to-use approximate methods of estimating the parameters, methods often adequate for initial approximate characterization of the response. The estimates obtained by these approaches may also be used as initial values for the more complicated and much more accurate CNLS method described and illustrated in Section 3.3.2. Note that the single $R_R C_R$ situation, that where $\theta = 0$ and $\psi_{ZC} = 1$, is included in the analysis described below.

From the figure, $-Z''$ reaches its maximum value, y_0, when $\omega = \omega_m = \tau_R^{-1}$ and thus $s = 1$. At this point the half-width of the arc on the real axis is $Z' - R_\infty = x_0 \equiv R_R/2$. Now from the data, the complex plane plot, and estimated values of x_0, y_0, and ω_m, one can immediately obtain estimates of R_∞, R_0, R_R, and τ_R. In order to obtain θ, one must, of course, find the direction of the circle center. The easiest graphical method is to draw on the Z^* plane plot several lines perpendicular to the semicircle; the center will be defined by their intersection. Two other more accurate approaches will be described below. Incidentally, when there is more than one arc present and there is some overlap which distorts the right, lower-frequency side of the arc, the present methods can still be used without appreciable loss of accuracy provided overlap distortion is only significant for $\omega < \omega_m$, that is, on the right side of the center of the left arc. Then all parameters should be estimated from the left side of the arc, that is, for $\omega \ge \omega_m$. A similar approach may be used when data are available only for $\omega \le \omega_m$. From Figure 1.3.2 and Eq. (5) we readily find that $\theta = \pi/2 - x \equiv (\pi/2)(1 - \psi_{ZC})$; thus when $\psi_{ZC} = 1$ there is no depression and one has simple single-time-constant ($\tau_R \equiv R_R C_R$) Debye response with $A_0 \equiv C_R$. When $\psi_{ZC} < 1$, $\tau_R = (R_R A_0)^{1/\omega_{zc}}$, but an ideal C_R capacitor cannot be directly defined, reflecting the distributed nature of the response.

The rest of the analysis proceeds as follows. First, one may obtain an estimate of ψ_{ZC} from the θ value using $\psi_{zc} = 1 - 2\theta/\pi$. But a superior alternative to first obtaining θ by finding the circle center approximately is to use the values of x_0 and y_0 defined on the figure. For simplicity, it will be convenient to define

$$q \equiv (\omega\tau_R)^{\psi_{zc}} \equiv (s)^{\psi_{zc}} \tag{9}$$

$$\chi \equiv (\pi/2) - \theta \equiv (\pi/2)\psi_{zc} \tag{10}$$

and note that

$$x_0 \equiv (R_0 - R_\infty)/2 \equiv R_R/2 \tag{11}$$

We may now rewrite Eq. (6) for I_Z as

$$I_Z(q,\chi) = \frac{[1+q\cos(\chi)] - jq\sin(\chi)}{1+2q\cos(\chi)+q^2} \tag{12}$$

For $q = 1$, the peak point, one finds

$$I_Z(1,\chi) = 0.5[1 - j\tan(\chi/2)] \tag{13}$$

Let us further define for later use the quantity

$$\psi_J \equiv \tan(\chi/2) = \tan(\pi\psi_{ZC}/4) \tag{14}$$

Now in general from Eq. (12) we may write

$$-I_Z''/I_Z' = q\sin(\chi)/[1+q\cos(\chi)] \tag{15}$$

which becomes, for $q = 1$,

$$-I_Z''/I_Z'\big|_{q=1} = y_0/x_0 = \tan(\chi/2) \equiv \psi_J \tag{16}$$

Thus from knowledge of y_0 and x_0 one can immediately calculate χ, ψ_J, ψ_{ZC}, and θ For completeness, it is worth giving expressions for w and r which follow from the figure. One finds

$$w = x_0\,\mathrm{ctn}(\chi) = x_0\tan(\theta) = x_0[1-\psi_J^2)/2\psi_J] \tag{17}$$

and

$$r = y_0 + w = x_0\csc(\chi) = x_0\sec(\theta) = x_0[(1+\psi_J^2)/2\psi_J] \tag{18}$$

A further method of obtaining ψ_{ZC} and θ is to first estimate R_∞ and plot $(Z - R_\infty)^{-1}$ in the Y plane. Then a spur inclined at the angle $[(\pi/2) - \theta] = \chi$ will appear whose $\omega \to 0$ intercept is $(R_0 - R_\infty)^{-1}$. A good estimate of ψ_{ZC} may be obtained from the χ value when the spur is indeed a straight line. Now at $\omega = \omega_m$, it turns out that $B_0\omega_m^{\psi_{ZC}} = 1$. Thus one may obtain an estimate of B_0 from $\omega_m^{-\psi_{ZC}}$. Then $\tau_R = B_0^{1/\psi_{ZC}} = \omega_m^{-1}$ and $A_0 = R_R^{-1}B_0$. Thus all the parameters of interest have then been estimated.

The above simple methods of estimating ψ_{ZC} depend only on the determination of x_0 and y_0 from the impedance complex plane arc or on the use of a few points in the admittance plane. Although they are often adequate for initial investigation, it is worth mentioning a relatively simple alternative procedure which can be used to test the appropriateness of Eqs (5) and (6) and obtain the parameter estimates of interest. Consider the point Z^* on the arc of Figure 1.3.2, a point marking a specific value of Z. It follows from the figure and Eq. (5) that $Z^* - R_\infty = (R_0 - R_\infty)I_Z^* \equiv u$ and $R_0 - Z^* = (R_0 - R_\infty)\,(1 - I_Z^*) \equiv v$. Therefore,

$$\ln|v/u| = \ln\left|\left(I_Z^*\right)^{-1}\right| = \ln(q) = \psi_{ZC}[\ln(\omega) + \ln(\tau_R)] \tag{19}$$

If one assumes that R_0 and R_∞ may be determined adequately from the complex plane plot, not always a valid assumption, then v and u may be calculated from experimental Z data for a variety of frequencies. A plot of $\ln|v/u|$ vs. $\ln(\omega)$ will yield a straight line with a slope of ψ_{ZC} and an intercept of ψ_{ZC} ln(τ_R) provided Eq. (19) holds. Ordinary linear least squares fitting may then be used to obtain estimates of ψ_{ZC} and ln (τ_R).

Although a more complicated nonlinear least squares procedure has been described by Tsai and Whitmore [1982] which allows analysis of two arcs with some overlap, approximate analysis of two or more arcs without much overlap does not require this approach and CNLS fitting is more appropriate for one or more arcs with or without appreciable overlap when accurate results are needed. In this section we have discussed some simple methods of obtaining approximate estimates of some equivalent circuit parameters, particularly those related to the common symmetrical depressed arc, the ZARC. An important aspect of material–electrode characterization is the identification of derived parameters with specific physicochemical processes in the system. This matter is discussed in detail in Sections 2.2 and 3.3 and will not be repeated here. Until such identification has been made, however, one cannot relate the parameter estimates, such as R_R, C_R, and ψ_{ZC}, to specific microscopic quantities of interest such as mobilities, reaction rates, and activation energies. It is this final step, however, yielding estimates of parameters immediately involved in the elemental processes occurring in the electrode–material system, which is the heart of characterization and an important part of IS.

1.4 SELECTED APPLICATIONS OF IS

In this section two applications will be presented which illustrate the power of the IS technique when it is applied to two very diverse areas, aqueous electrochemistry and fast ion transport in solids. These particular examples were chosen because of their historical importance and because the analysis in each case is particularly simple. Additional techniques and applications of IS to more complicated systems will be presented in Chapter 4 as well as throughout the text.

The first experimental use of complex plane analysis in aqueous electrochemistry was performed in 1960 (Sluyters and Oomen [1960]). This study is a classic illustration of the ability of impedance spectroscopy to establish kinetic parameters in an aqueous electrochemical system. Using a standard hanging mercury drop cell, the impedance response of the $Zn(Hg)/Zn^{2+}$ couple in a 1M $NaClO_4$ + 10^{-3}M $HClO_4$ electrolyte was examined at 298 K. For this couple, the reaction rate is such that in the frequency range of 20 Hz to 20 kHz the kinetics of charge transfer is slower than ion diffusion in the electrolyte. The results (Figure 1.4.1) show a single semicircle characteristic of kinetic control by an electrochemical charge transfer step at the electrode–electrolyte interface. The physical model appropriate to this system is the same as that presented in Figure 1.3.1*d*. The semicircle beginning at the origin in Figure 1.3.1*e* is not observed in Figure 1.4.1 because the frequency range was limited to below 20 kHz. Thus, in Figure 1.4.1, R_∞ is the solution resistance, R_2 is the charge transfer resistance. The double-layer capacitance, C_2 can be obtained by analysis of Z'' frequency dependence.

By solving the standard current–potential equation for an electrochemical reaction (see, for example, Bard and Faulkner [1980]) under the conditions of kinetic

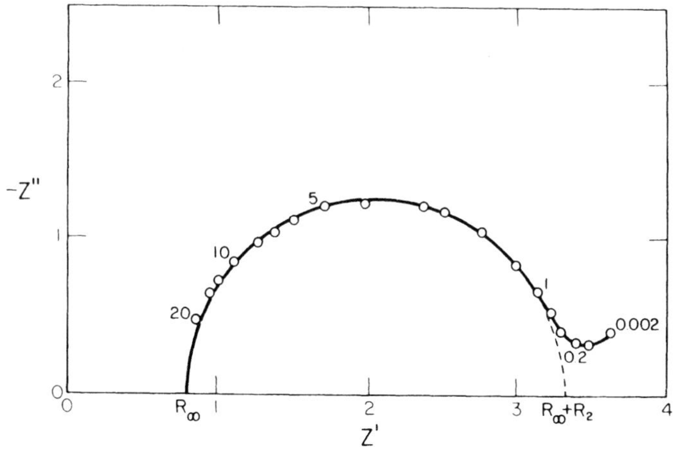

Figure 1.4.1. The impedance results of a $Zn(Hg)/Zn^{2+}$ couple in 1M $NaClO_4$ + 10^{-3}M $HClO_4$ with $C_{Zn} = 8 \times 10^{-6}$ moles/cm^3 and $C_{Zn^{2+}} = 8 \times 10^{-6}$. The numbers represent the frequency in kilohertz; the axes are in arbitrary scale units. (Sluyters and Oomen [1960])

Table 1.4.1. Calculation of Rate Constant of $Zn(Hg)/Zn^{2+}$ Couple

$C_{Zn} = C_{Zn^{2+}}$ (moles/cm^3)	R_2 (Ω-cm^2)	$R_2 \times C_{Zn}$ (moles-Ω/cm)	k (cm/sec)[a]
2×10^{-6}	10.17	20.3×10^{-6}	
4	4.95	19.8	
5	4.26	21.3	
8	2.41	19.3	$3.26 \times 10^{-3} \pm 3.6\%$
10	2.13	21.3	
16	1.27	20.3	
16	1.28	20.5	

[a] Calculated from the average value of $R_2 \times C_{Zn} = 20.4 \times 10^{-6}$ by $k = (R_2 C_{Zn} n^2 F^2)^{-1} RT$ according to Eq. (1).

Source: Sluyters and Oomen [1960].

control (i.e. the rate of charge transfer is much slower than diffusive processes in the system), the value of R_2 can be evaluated. For a known concentration of Zn at the amalgam–electrolyte interface, $C_{Zn(Hg)}$, and a known concentration of Zn^{2+} at the electrolyte–electrode interface, $C_{Zn^{2+}}$, the value of R_2 is given by Eq. (1):

$$R_2 = \frac{RT}{n^2 F^2 k (C_{Zn^{2+}})^{\alpha} (C_{Zn(Hg)})^{1-\alpha}} \qquad (1)$$

where n is the number of electrons transferred, F is Faraday's constant, k is the rate constant for the electrochemical charge transfer reaction, α is the electrochemical transfer coefficient, R is the ideal gas constant, and T is the absolute temperature. When the concentration of Zn in the amalgam is equal to the concentration of Zn ions in the solution, then the rate constant k can be determined. Results at several different equal concentrations of Zn and Zn^{2+} (Table 1.4.1) gave a mean value of $k = 3.26 \times 10^{-3}$ cm/s. By using different concentrations of Zn and Zn^{2+} the transfer coefficient α (Tables 1.4.2 and 1.4.3) was found to be 0.70. In addition, the value of the double-layer capacitance could be easily determined in each of the experiments.

In a similar experiment, the Hg/Hg^{2+} reaction in 1M $HClO_4$ has also been investigated (Sluyters and Oomen [1960]) using IS in the frequency range of 20 Hz to 20 kHz and for concentrations between 2×10^{-6} and 10×10^{-6} moles/cm^3 Hg^{2+}. The results (Figure 1.4.2) show linear behavior in the complex plane with an angle of 45° to the real axis. Such a response is indicative of a distributed element as discussed in the previous section. In this case, the system is under diffusion control as the kinetics of the charge transfer at the electrode–electrolyte interface is much faster than the diffusion of the Hg^{2+} ions in the solution. Solution of the diffusion equation with the appropriate boundary conditions under a small ac perturbation gives the diffusional contribution to the impedance in the complex plane as (see Chapter 2 for a detailed discussion)

$$W = \sigma\omega^{-1/2} - j\sigma\omega^{-1/2} \tag{2}$$

where the impedance, W, is generally called the Warburg impedance, ω is the angular frequency, j is equal to $(-1)^{1/2}$ and σ is a constant given by

$$\sigma = \frac{RT}{n^2F^2\sqrt{2}}\left[\frac{1}{C_{\mathrm{Hg}^{2+}}\left(D_{\mathrm{Hg}^{2+}}\right)^{1/2}} + \frac{1}{\left[C_{\mathrm{Hg}}\left(D_{\mathrm{Hg}}\right)^{1/2}\right]}\right] \tag{3}$$

Table 1.4.2. Calculation of Transfer Coefficient α of $Zn(Hg)/Zn^{2+}$ Couple

C_{Zn} (moles/cm^3)	$C_{Zn^{2+}}$ (moles/cm^3)	R_2 (Ω-cm^2)	$\log R_2$	$-\log C_{Zn^{2+}}$	α^a
16×10^{-6}	16×10^{-6}	1.28	0.107	4.796	
16	8	2.00	0.301	5.097	0.70
16	4	3.29	0.517	5.398	
16	2	5.37	0.730	5.699	

[a] From slope of $-\log C_{Zn^{2+}}$ vs. $\log R_2$ plot.
Source: Sluyters and Oomen [1960].

Table 1.4.3. Calculation of Transfer Coefficient $1 - \alpha$ of $Zn(Hg)/Zn^{2+}$ Couple

$C_{Zn^{2+}}$ (moles/cm^3)	C_{Zn} (moles/cm^3)	R_2 (Ω-cm^2)	$\log R_2$	$-\log C_{Zn}$	$1 - \alpha^a$
16×10^{-6}	16×10^{-6}	1.28	0.107	4.796	
16×10^{-6}	8	1.56	0.193	5.097	0.29
16×10^{-6}	4	1.93	0.286	5.398	

[a] From slope of $-\log C_{Zn}$ vs. $\log R_2$ plot.
Source: Sluyters and Oomen [1960].

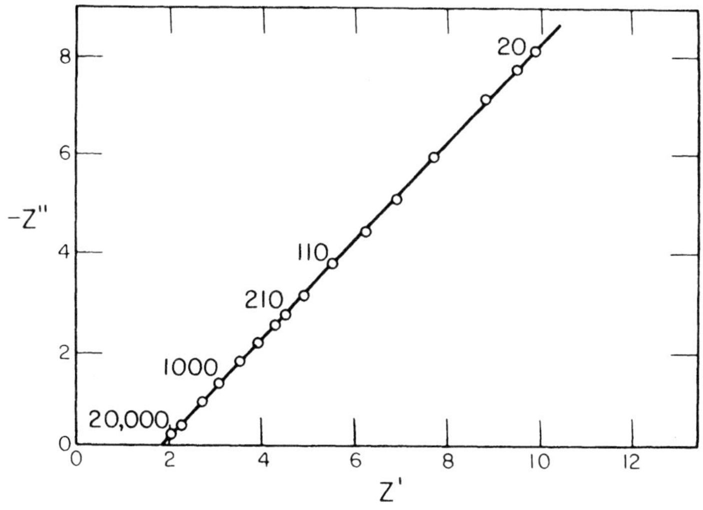

Figure 1.4.2. The impedance results of a Hg_2^{2+}/Hg couple in 1M $HClO_4$ electrolyte with $C_{Hg^{2+}} = 2 \times 10^{-6}$ moles/cm^3. The numbers represent the frequency in kilohertz; the axes are in arbitrary scale units. (Sluyters and Oomen [1960])

where $D_{Hg^{2+}}$ and D_{Hg} are the diffusivity of mercurous ions in solution and mercury in amalgam, respectively, and the other terms are defined as above. This impedance is to be added (see Sluyters [1960]) and the discussion in Chapter 2) in series with R_2 of Figure 1.3.1*d*. When the impedance of this circuit is plotted in the complex plane, one obtains a semicircle combined with a straight line at an angle of 45° to the real axis. The line, when extended to the real axis, has an intercept of $R_\infty + R_2 - 2\sigma C_{dl}$. If $2\sigma C_{dl}$ is small, as in the present case, the semicircle is suppressed and the product of the imaginary part of W, Im (W) and $\omega^{1/2}$ will be equal to σ at all frequencies.

The experimental results in Figure 1.4.2 are thus consistent with a system under diffusion control. The diffusivity of Hg_2^{2+} ions in solution can be easily calculated (Table 1.4.4) at several different concentrations of Hg_2^{2+} in the solution from the value of σ. No further information can be obtained from this data because the time constant associated with the kinetics is too fast to be measured at frequencies below 20 kHz.

The frequency range chosen in the above experiments was dictated by the limited electronics available in 1960 and the cumbersome experimental approach associated with it, which required that the impedance be measured independently at each frequency. The introduction of automated impedance analysis instruments removes this restriction and allows the experimenter to choose the most appropriate frequency range for a given experiment. This choice should be determined by the nature of the interfaces in the experiment and the time constants that are associated with them. For example corrosion studies, which often involve a slow aqueous diffusion process, generally have relatively large time constants (on the order of

Table 1.4.4. Calculation of Diffusion Coefficient of Hg^{2+} in 1N $HClO_4$

$C_{Hg_2^{2+}}$ (moles/cm^3)	σ^a (Ω-sec$^{-1/2}$ cm^2)	$D^b_{Hg_2^{2+}}$ (cm^2/sec)	$R_\infty + R_2$ (Ω-cm^2)
10×10^{-6}	2.09	0.241×10^{-5}	0.190
5	4.10	0.251	0.188
4	4.99	0.264	0.188
3	6.60	0.268	0.195
2	9.73	0.277	0.193

[a] $\sigma = \mathrm{Im}\,(W)^{1/2}_{\omega}$ was found to be independent of frequency within 2%.

[b] $D_{Hg^{2+}} = [RT(\sigma n^2 F^2 \sqrt{2}\ C_{Hg^{2+}})^{-1}]^2$ according to Eq. (3) with $1/[C_{Hg}(D_{Hg})^{-1/2}] << 1/[C_{Hg^{2+}}(D_{Hg^{2+}})^{1/2}]$, as is the case here with a pure Hg electrode.

Source: Sluyters and Oomen [1960].

0.1–10 s), and thus most impedance studies of corroding systems use frequencies between a few millihertz and 100 kHz. On the other hand, studies of solid ionic conductors require higher frequencies to measure the time constant associated with ionic motion (milli- to microseconds), which is generally smaller than those found in aqueous diffusion processes. Thus, frequencies between a few hertz and 15 MHz are most appropriate here.

That is not to say that the frequency range should always be restricted based upon predetermined expectations. In the above studies, a wider frequency range would probably have allowed a determination of additional information. For the Zn/Zn^{2+} couple, lower frequencies would have allowed the measurement of the diffusivity of zinc ions in the solution. For the study of the Hg/Hg^{2+} couple, the kinetics of the electrochemical reaction at the interface could have been explored by using higher frequencies. Nevertheless, an understanding of the relationship between the time constant in an experiment and the frequencies with which to measure it provides an intelligent starting point in the choice of the most appropriate frequency range.

A second example which illustrates the utility of IS to solid state chemists is the application of impedance analysis to zirconia–yttria solid electrolytes (Bauerle [1969]). At elevated temperatures solid solution zirconia–yttria compounds are known to be oxygen-ion conductors which function by transport of oxygen ions through vacancies introduced by the dopant yttria. By examining cells of the form

$$\mathrm{Pt}, \mathrm{O_2}|(\mathrm{ZrO_2})_{0.9}(\mathrm{Y_2O_3})_{0.1}|\mathrm{O_2}, \mathrm{Pt} \tag{4}$$

using IS, admittance plots were obtained (Figure 1.4.3*a*). The equivalent circuit proposed to fit this data is shown in Figure 1.4.3*b*. By a careful examination of the effect of the electrode-area-to-sample-length ratio, and by measuring the dc conductivity of the samples, the high-frequency semicircle (the one on the right in Figure 1.4.3*a*) was ascribed to bulk electrolyte behavior, while the low-frequency semicircle (on

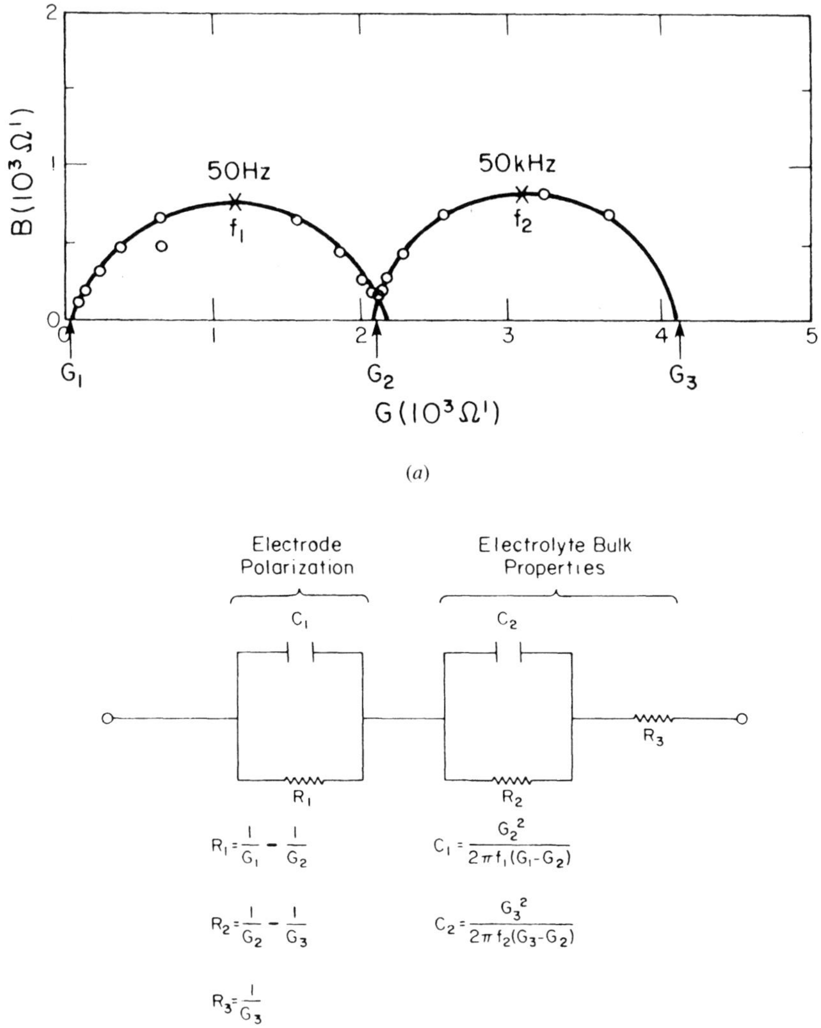

Figure 1.4.3. (*a*) Admittance behavior of the electrochemical cell given in TABLE1.4.4 at 873 K for a specimen with naturally porous electrodes (sputtered Pt). (*b*) The equivalent circuit for the behavior in part *a* showing the two impedance elements associated with each semicircle. (Bauerle [1969])

the left in Figure 1.4.3*a*) corresponded to the electrode polarization. In the terminology of Figure 1.4.3*b*, R_1 and C_1 correspond to electrode polarization phenomena, while R_2, R_3, and C_2 describe processes which occur in the bulk of the electrolyte specimen. Furthermore, by varying temperature, oxygen partial pressure, and electrode preparation, the role of each component in the overall conduction mechanism was determined. In particular, R_1 represents an effective resistance for the electrode reaction

$$\frac{1}{2}O_{2(g)} + 2e^- = O^{2-}(\text{electrolyte}) \tag{5}$$

where C_1 is the double-layer capacitance of the electrode; R_2 is a "constriction" or intergranular resistance corresponding to resistance of conduction across two different grains, primarily due to impurities located there; C_2 is the capacity across the intergranular region; and R_3 is the resistance to conduction within the grains. Electron microprobe studies supported the theory of impurities at the grain boundary. Thus, in a system as electrochemically complex at this, with many different effects interacting, one can still obtain fundamental information about processes occurring at each interface and in the bulk specimen.

This second study illustrates a very important point about IS. Although it is an extremely powerful technique in its own right, the analysis of complicated systems must be correlated with other experimental information to verify that the chosen circuit is physically reasonable. Furthermore, agreement between independently determined experimental values and those determined in a fitting procedure of the complex plane results can only strengthen the IS results and thus should never be overlooked.

Chapter 2

Theory

Ian D. Raistrick
Donald R. Franceschetti
J. Ross Macdonald

2.1 THE ELECTRICAL ANALOGS OF PHYSICAL AND CHEMICAL PROCESSES

2.1.1 Introduction

One of the most attractive aspects of impedance spectroscopy as a tool for investigating the electrical and electrochemical properties of materials and systems is the direct connection that often exists between the behavior of a real system and that of an idealized model circuit consisting of discrete electrical components. The investigator typically compares or fits the impedance data to an equivalent circuit, which is representative of the physical processes taking place in the system under investigation. The main objective of the present section is to define and discuss the analogies between circuit elements and electrochemical processes, so that the results of data fitting can be more easily converted into physical understanding. That such a close connection exists between electrochemistry and the behavior of idealized circuit elements is not surprising, since the fundamental laws which connect charge and potential and which define the properties of linear systems are unchanged in passing from electronic to ionic materials.

There are, however, dangers in the indiscriminate use of analogies to describe electrochemical systems. The first point to be made is that equivalent circuits are seldom unique. Only the simplest circuits can be said to be unambiguous in their description of experimental data; in complex situations, choices based upon other physical data are often necessary. It should also be remembered that electrolytes, interfaces, and so on are only *approximately* modeled by idealized circuit elements over a limited range of experimental conditions. One general condition, which will be assumed through much of this volume, is that we are dealing with small signals; that is, linear behavior is implied. The impedance is supposed to be independent of the amplitude of the applied signal, or at least to approach a constant finite limit as

Impedance Spectroscopy, Second Edition, edited by Evgenij Barsoukov and J. Ross Macdonald
ISBN 0-471-64749-7

the amplitude of the signal is decreased. Electrochemical systems, of course, can be highly nonlinear, and response to large signals includes rectification and higher harmonic generation. In Section 2.1.4 we discuss the linearization of interfacial kinetics to produce a charge transfer resistance. Interfacial capacitances are also voltage-dependent and mass transport will also be nonlinear if diffusion coefficients or thermodynamic terms, present in the diffusion expression, are a function of concentration (see Section 2.1.3.2). The use of small signals, however, is in general a distinct advantage of the impedance approach as compared to cyclic voltammetry, for example, where the wealth of information contained in a single experiment may prove too difficult to deconvolute. Usually, the voltage dependence of the electrochemical parameters is rather small, and a linear expansion of the ac current, in terms of the variation of the perturbed concentrations and so on is well justified. Higher-order effects are not discussed in this chapter, but a discussion of electrochemical applications may be found in the review by Sluyters-Rehbach and Sluyters [1984].

Two other limitations on the exact correspondence between equivalent circuits and electrochemical systems are also addressed. The first of these is the effect of geometry on the current distribution. The effect of this on the frequency dispersion of the impedance is only beginning to be explored by those interested in impedance methods, and will prove to be important as the technique is extended to more complex geometries and small structures. A number of problems where current distribution is undoubtedly important, for example in the behavior of polycrystalline solid electrolytes and the effect of roughness on interfacial impedance, have recently been considered by Fleig and Maier [1996].

A further limitation is the often observed anomalous frequency dependence of both bulk and interface parameters. Several electrochemical properties, for example conductivity and interface capacitance, are predicted to be independent of frequency, whereas, in fact, they often show significant deviations from this behavior. This type of phenomenon has achieved recognition only since the application of ac techniques to a wide variety of problems, since a small degree of frequency dispersion is difficult to recognize in transient (time domain) experiments. Although good parameterization of this frequency dispersion has been achieved, and certain general or "universal" forms suggested, a universally accepted microscopic description has not yet emerged. Some of the aspects of this phenomenology are discussed in Sections 2.1.2.3 and 2.1.2.7, and we have brought together some of the various attempts which have appeared in a wide variety of fields to deal with this problem.

The general approach adopted in this section is to treat bulk and interfacial phenomena separately. First the electrical properties of homogeneous phases are discussed. There are two aspects to this treatment, relating respectively to dielectric relaxation and long-range dc conductivity.

Although the well-established measurement of dielectric loss is not, in its narrowest sense, strictly impedance spectroscopy, a discussion of relaxation behavior is central to the family of techniques that use the interaction of a time-varying electromagnetic signal with a material to deduce microscopic detail. The generalization of the treatment of systems with a single relaxation time (Debye behavior) to those with multiple relaxations or distributions of relaxation times is discussed in Section

2.1.2.3. Recently, the application of impedance methods to disordered, condensed phases, such as organic polymers and glasses, warrants a general appreciation of the concepts involved. Dielectric loss measurements are also important and are used extensively to study the energetics of relaxations of complex ionic defects, such as those found in the fluorite family of materials.

The determination of dc ionic conductivity is perhaps the most widespread and also the simplest application of impedance spectroscopy. By using ac methods, electrode polarization can be *correctly* eliminated from an electrochemical system, and other sources of spurious frequency dispersion, such as grain boundary effects, may also be removed under certain circumstances. Electrodes may be inert foreign metals, thus eliminating the need for demonstrating the reversibility of parent metal electrodes.

Conductivity is of course closely related to diffusion in a concentration gradient, and impedance spectroscopy has been used to determine diffusion coefficients in a variety of electrochemical systems, including membranes, thin oxide films, and alloys. In materials exhibiting a degree of disorder, perhaps in the hopping distance or in the depths of the potential wells, *simple* random walk treatments of the statistics are no longer adequate; some modern approaches to such problems are introduced in Section 2.1.2.7.

The above mentioned sections deal with bulk phenomena. The other important area about which impedance spectroscopy gives important information is that of the electrochemical interface. This is usually a junction between an electronic and an ionic conductor; electrochemical devices utilize the charge transfer that occurs at this interface. The kinetics of this process as well as the electrical nature of the interface region are discussed in Section 2.1.4.

The emphasis of this section is on solid systems; therefore, several important aspects of electrical response appropriate to liquid electrochemistry are either neglected or given little emphasis. Examples are the omission of convection in the treatment of mass transport and the related neglect of ac impedance at a rotating disk electrode. Similarly, porous electrodes are not discussed, although related "rough" electrodes are briefly considered. Complex electrochemical mechanisms at solid–solid interfaces have been hardly mentioned in the literature; the treatment of the topic here reflects that. However, some attempt has been made to give a sufficiently general approach to interface kinetics and the development of expressions for the faradic impedance so that solid state scientists may be aware of the advanced state of development of the theory used by aqueous electrochemists.

2.1.2 The Electrical Properties of Bulk Homogeneous Phases

2.1.2.1 Introduction

In this section we are concerned with the electrical response of solids with a uniform composition. Mass and charge transport in the presence of concentration gradients are discussed in Section 2.1.3.

On the time scale of interest to electrochemists (i.e. greater than ~1 μs) an electric field can interact with a solid in two principal ways. These are, respectively, the reorientation of defects having electric dipole moments (usually complex defects) and the translational motion of charge carriers (usually simple defects such as vacancies, ionic interstitials, and defect electronic species).

The first interaction leads to a displacement current,

$$i = d\mathbf{D}/dt \tag{1}$$

where **D** is the electric displacement (defined as the total charge density on the electrodes),

$$\mathbf{D} = \varepsilon_0\mathbf{E} + \mathbf{P} \tag{2}$$

where **E** is the electric field, ε_0 is the permittivity of free space, and **P** is the polarization of the dielectric material.

The second type of interaction leads to a purely real (dc) conductivity σ,

$$i = \sigma\mathbf{E} \tag{3}$$

We therefore deal first with the phenomenon of dielectric relaxation in materials with a single time constant and an absence of conductivity and later with materials that show long-range conductivity. As we are primarily concerned with developing the electrical analogs of these processes, little consideration is given to the characteristics of individual defects or materials. Similarly, the thermodynamics of the formation of defects, which determines their concentration, is also ignored.

2.1.2.2 Dielectric Relaxation in Materials with a Single Time Constant

When an electric field **E** is applied to an insulating material, the resulting polarization **P** may be divided into two parts according to the time constant of the response:

1. An almost instantaneous polarization due to the displacement of the electrons with respect to the nuclei. This defines the high-frequency dielectric constant ε_∞ related to the refractive index.

$$\varepsilon_\infty - 1 = \mathbf{P}_\infty/\mathbf{E}\varepsilon_0 \tag{4}$$

The time constant of this process is about 10^{-16} s and therefore occurs in the UV region of the electromagnetic spectrum. Ionic vibrations have a time constant which usually occurs in the infrared and are also therefore instantaneous as far as electrochemical experiments are concerned.

2. A time-dependent polarization $\mathbf{P}'(t)$ due to the orientation of dipoles in the electric field. If the field remains in place for an infinitely long time, the resulting total polarization $\mathbf{P}_s$ defines the static dielectric constant ε_s:

$$\varepsilon_s - 1 = \mathbf{P}_s/\mathbf{E}\varepsilon_0 \tag{5}$$

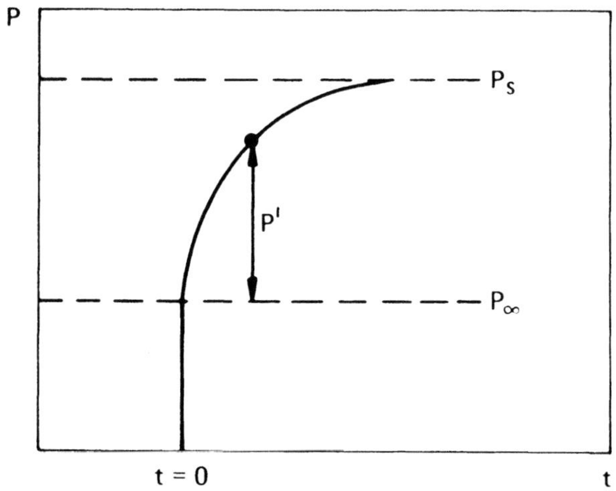

Figure 2.1.1. Time dependence of the polarization P after the application of an electric field to an insulator at $t = 0$.

$$\mathbf{P}_s = \mathbf{P}_\infty + \mathbf{P}'(t = \infty) \tag{6}$$

The simplest assumption allowing calculation of the properties of such a system is that $\mathbf{P}'(t)$ is governed by first-order kinetics, that is, a single-relaxation time τ, such that

$$\tau d\mathbf{P}'(t)/dt = \mathbf{P}_s - \mathbf{P} \tag{7}$$

In other words, the rate at which $\mathbf{P}$ approaches $\mathbf{P}_s$ is proportional to the difference between them. Referring to Figure 2.1.1, on application of a unit step voltage $u_0(t)$

$$\mathbf{P} = \mathbf{P}_\infty u_0(t) + \mathbf{P}'(t) \tag{8}$$

If we take the Laplace transforms of the last two equations and solve for $\{\mathbf{P}\}$, we obtain

$$\{\mathbf{P}\} = \frac{\mathbf{P}_\infty}{(p+\omega_0)} + \frac{\omega_0 \mathbf{P}_s}{p(p+\omega_0)} \tag{9}$$

where $\{\mathbf{P}\}$ is the Laplace transform of the polarization and $\omega_0 = \tau^{-1}$; p is the complex frequency variable.

The current density is obtained using the relation

$$\{i\} = p\{\mathbf{P}\} - \mathbf{P}(t = 0) \tag{10}$$

or by differentiating in the time domain, to give

$$\{i\} = \mathbf{P}_\infty + (\mathbf{P}_s - \mathbf{P}_\infty)\frac{\omega_0}{(p+\omega_0)} \tag{11}$$

and therefore

$$i(t) = \mathbf{P}_\infty \delta(t) + (\mathbf{P}_s - \mathbf{P}_\infty)\tau^{-1} \exp(-t/\tau) \tag{12}$$

This is the same result as that obtained for the circuit of Figure 2.1.2, with the identities

$$\begin{aligned} \tau &= RC_2 \\ C_2 &= (\varepsilon_s - \varepsilon_\infty)\varepsilon_0 \\ C_1 &= \varepsilon_\infty \varepsilon_0 \end{aligned} \tag{13}$$

The admittance due the relaxation process is, since $\{\mathbf{E}\} = 1/p$:

$$Y = \frac{\{i\}}{\{\mathbf{E}\}} = \varepsilon_\infty \varepsilon_0 p + (\varepsilon_s \varepsilon_\infty)\varepsilon_0 \frac{\omega_0 p}{(p + \omega_0)} \tag{14}$$

or, separating the real and imaginary parts ($p = j\omega$),

$$Y = \frac{\omega^2 RC_2}{1 + \omega^2 R^2 C_2^2} + j\frac{\omega C_2}{1 + \omega^2 R^2 C_2^2} + j\omega C_1 \tag{15}$$

The expression may be rewritten in terms of the complex dielectric constant $\varepsilon = Y/j\omega\varepsilon_0$:

$$\varepsilon - \varepsilon_\infty = \frac{\varepsilon_s - \varepsilon_\infty}{1 + \omega^2\tau^2} - j\frac{\omega\tau(\varepsilon_s - \varepsilon_\infty)}{1 + \omega^2\tau^2} \tag{16}$$

The real and imaginary parts of this expression are the Debye relaxation relations, which have remained the basic model of dielectric relaxation since their inception (Debye [1929]). They are plotted against one another in Figure 2.1.3 and in Figure 2.1.4 are separately plotted against the normalized frequency ω/ω_0. The dielectric loss peak ε'', which corresponds to the real part of the admittance, has been widely used in solid state measurement for the characterization of relaxation processes. As will be seen later, the equivalent circuit of Figure 2.1.2 is also used in the interpretation of ac impedance data for solid electrolyte systems even though the physical phenomena describing the relaxation processes, that is conductivity and space charge accumulation and depletion, are quite different.

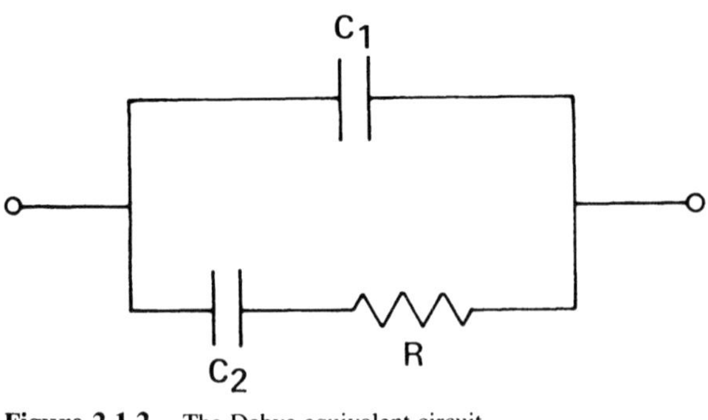

Figure 2.1.2. The Debye equivalent circuit.

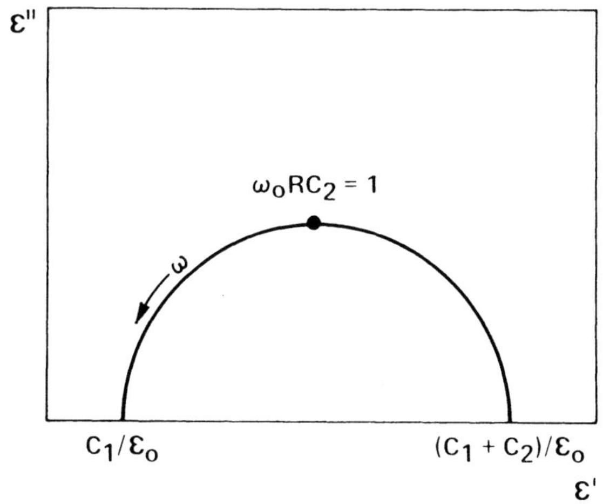

Figure 2.1.3. Complex plane plot of the frequency dependence of the complex permittivity modeled by the circuit of Figure 2.1.2.

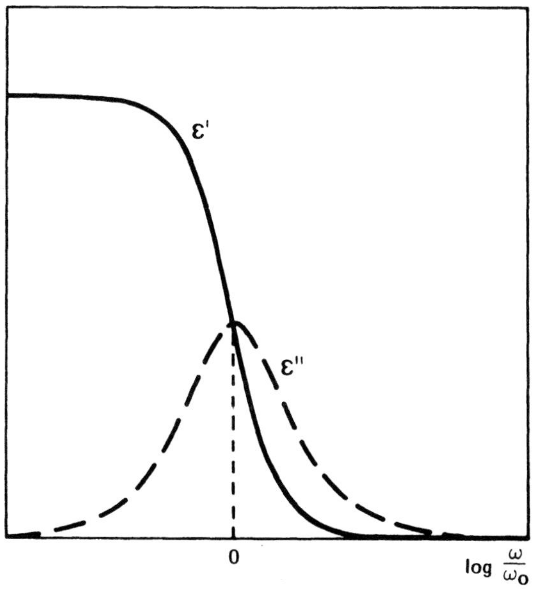

Figure 2.1.4. Real and imaginary parts of the complex permittivity as a function of normalized radial frequency.

The principal difference between a dielectric loss experiment and an impedance spectrum is that the former usually utilizes temperature as the independent variable and measurements are made at several fixed frequencies. A typical example of the use of dielectric loss measurements to obtain data about the relaxations of defects in crystalline solids is the paper by Wapenaar *et al.* [1982], who studied LaF_3-doped

BaF_2. Two main types of relaxation are found in this material, corresponding to dipole moments caused by association of substitutional lanthanum with an interstitial fluorine along the <100> and <111> crystal axes, respectively. Loss peaks are seen at low levels of doping corresponding to both defects. Calculation of the respective dipole moments allows calculation of the concentration of defects from the strengths of the losses (i.e. from the associated values of C_2).

In practice, very few systems obey the Debye equations with accuracy; an extensive literature exists on the real properties of dielectric materials. It is often found, especially in disordered materials (e.g. glasses and amorphous thin films), that the Debye peak is considerably broadened over its theoretical half-width of $\log[(2+\sqrt{3})/(2-\sqrt{3})]$ decades. A number of empirical relaxation functions have been proposed to parameterize the observations, usually interpreted in terms of a distribution of relaxation times. This theme is discussed in the next section.

2.1.2.3 Distributions of Relaxation Times

The Debye dispersion relations were derived above for a process with a single relaxation time. Figure 2.1.4 showed that the dielectric loss function ε'' corresponding to this situation is symmetric about a central frequency, with a characteristic shape and width. The current flow in such a material, after the imposition of a voltage step function, decays exponentially with time. In view of the observations mentioned at the end of the previous section, attempts have been made to extend the Debye model by including processes with more than one relaxation time. By choosing a distribution of relaxation times with appropriate strengths and frequencies, it should prove possible to parameterize the broad response of many dielectric materials. The distribution of relaxation times approach has also been suggested as the origin of the "constant-phase elements" that are often seen in impedance studies of solid electrolytes and the solid–solid interface. In this section, some of the main features of this line of reasoning are presented. The dielectric literature in this area is extensive; no attempt is made here to be comprehensive.

Assuming linear relaxation processes of the type modeled by the series—RC—branch, the principle of superposition allows the dielectric function $\varepsilon^* - \varepsilon_\infty$ to be generalized:

$$\varepsilon - \varepsilon_\infty = \int_0^\infty \frac{(\varepsilon_s - \varepsilon_\infty)G(\tau)d\tau}{1+p\tau} \tag{17}$$

where p is the complex frequency variable and $G(\tau)$ represents a distribution of relaxation times. The distribution should be normalizable,

$$\int_0^\infty G(\tau)d\tau = 1 \tag{18}$$

and should have upper and lower limits. Here $G(\tau)$ represents the fraction of the total dispersion $(\varepsilon_s - \varepsilon_\infty)$, which is contributed by processes having relaxation times between τ and $\tau + d\tau$.

Now ε may be divided into real and imaginary parts corresponding to the frequency $j\omega$,

$$\varepsilon' - \varepsilon_\infty = \int_0^\infty \frac{G(\tau)(\varepsilon_s - \varepsilon_\infty)d\tau}{1+(\omega\tau)^2} \tag{19}$$

$$\varepsilon'' = \int_0^\infty \frac{\omega\tau G(\tau)(\varepsilon_s - \varepsilon_\infty)d\tau}{1+(\omega\tau)^2} \tag{20}$$

Kirkwood and Fuoss [1941] first showed that $G(\tau)$ could be recovered by integration from a set of ε'' values. A general treatment has been given by Macdonald and Brachman [1956], who provided a useful set of relations between the various functions used to describe networks and systems as well as between responses to various types of input.

Using the notation of those authors, the network function is defined as

$$Q(p) = \int_0^\infty \frac{G(\tau)d\tau}{1+p\tau} \tag{21}$$

where $Q(p)$ corresponds to $(\varepsilon - \varepsilon_\infty)/(\varepsilon_s - \varepsilon_\infty)$.

The admittance is related to $Q(p)$:

$$Y(p) = pQ(p) \tag{22}$$

The step function response $A(t)$ and the impulse response $B(t)$ are

$$A(t) = L^{-1}[Q(p)] \tag{23}$$

$$B(t) = L^{-1}[Y(p)] \tag{24}$$

where L^{-1} is the inverse Laplace transform operator.

It was shown that $G(\tau)$ was derivable from these quantities through the relations

$$\tau G(t) = D(\lambda) = L^{-1}L^{-1}[Q(p)] \tag{25}$$

or

$$D(\lambda) = L^{-1}A(t) \tag{26}$$

where $D(\lambda)$ is a distribution function of the new variable $\lambda = \tau^{-1}$. Other relationships may also be derived. The authors give useful examples of various types of network functions and derived distributions. The simplest is

$$G(\tau) = \delta(\tau - \tau_0) \tag{27}$$

which corresponds to a single relaxation time and leads to the simple Debye dispersion equations. Rewriting $G(\tau)$ in terms of λ,

$$D(\lambda) = \tau G(\tau) = \tau\delta(\tau - \tau_0) \tag{28}$$

and therefore

$$D(\lambda) = \lambda^{-1}\delta(\lambda^{-1} - \lambda_0^{-1}) \tag{29}$$

$$= \lambda_0\delta(\lambda - \lambda_0)$$

Hence,

$$A(t) = L\{D(\lambda)\} \tag{30}$$
$$= \lambda_0 \exp(-\lambda_0 t) \tag{31}$$
$$= (1/\tau_0)\exp(-t/\tau_0) \tag{32}$$

and

$$Q(p) = 1/(1+\tau_0 p) \tag{33}$$

Observed relaxation times may occur over many orders of magnitude, and it seems reasonable that such a range of variation would, for a thermally activated process of the type

$$\tau = \tau^* \exp(E^*/kT) \tag{34}$$

correspond to a distribution of activation energies rather than to a distribution of τ^*. We may therefore define a distribution $K(E^*)$ such that

$$K(E^*)dE^* = G(\tau)d\tau \tag{35}$$

Evidently, from Eq. (34)

$$K(E^*) = (\tau/kT)G(\tau) \tag{36}$$

Macdonald [1962] has pointed out that if $K(E^*)$ is independent of T, then $G(\tau)$ cannot be so independent. Both the midpoint τ_0 and the width of the distribution will change with temperature. Not all of the $G(\tau)$ proposed in the literature are consistent with this postulate. See Section 2.2.3.5 for further discussion of activation energy distributions.

Van Weperen *et al.* [1977] noted that in fluorite-structure materials the dielectric and ionic thermocurrent peaks broadened with increasing concentration of dipoles (for example Johnson *et al.* [1969]) and developed a theory of dipole–dipole interactions which predicted an almost Gaussian distribution of activation energies:

$$K(E^*) = \frac{1}{\sigma\sqrt{(2\pi)}}\exp\left[-\frac{(E^* - E_0^*)^2}{2\sigma^2}\right] \tag{37}$$

The corresponding distribution of τ is lognormal, the Wagner distribution

$$G(\tau) = \frac{b}{\tau\sqrt{\pi}}\exp\left[-b^2\left(\ln\frac{\tau}{\tau_0}\right)^2\right] \tag{38}$$

where $\sigma = kT/b\sqrt{2}$; σ^2 is the variance of the $K(E^*)$. If $K(E^*)$ is to be invariant with T, then b should be proportional to T.

The importance of this distribution, apart from being well defined in a physical sense, is its behavior for large σ, that is wide distributions. As b becomes small, $G(\tau)$ becomes proportional to $1/\tau$, and $A(t)$, the current response to a unit step function, becomes proportional to $1/t$. The power spectrum may mimic $1/f$ behavior over several decades, and the dielectric function will show a very gradual frequency dispersion.

One of the most widely used distributions is that proposed by Cole and Cole [1941] to describe the occurrence of depressed semicircular arcs in the $\varepsilon'' - \varepsilon'$ plots obtained for a wide variety of polar liquids and solids. The dielectric constant behavior was described by the equation

$$\varepsilon - \varepsilon_\infty = \frac{(\varepsilon_s - \varepsilon_\infty)}{\left[1 + (j\omega\tau_0)^{1-\alpha}\right]} \tag{39}$$

where $\alpha\pi/2$ is the angle between the real axis and the line to the center of the circle from the high-frequency intercept. Now ε may be separated into real and imaginary parts

$$\frac{\varepsilon' - \varepsilon_\infty}{\varepsilon_s - \varepsilon_\infty} = \frac{1}{2}\left[1 - \frac{\sinh(1-\alpha)x}{\cosh(1-\alpha)x + \cos\alpha\pi/2}\right] \tag{40}$$

$$\frac{\varepsilon''}{\varepsilon_s - \varepsilon_\infty} = \frac{1}{2}\frac{\cos\alpha\pi/2}{\cosh(1-\alpha)x + \sin\alpha\pi/2} \tag{41}$$

where $x = \log(\omega\tau_0)$. These expressions reduce of the Debye relationships for $\alpha \to 0$. The derived distribution function of time constants is

$$G(\tau) = \frac{1}{2\pi\tau}\frac{\sin\alpha\pi}{\cosh(1-\alpha)\log(\tau/\tau_0) - \cos\alpha\pi} \tag{42}$$

from which the distribution of activation energies would be, using Eq. (36),

$$K(E^*) = \frac{1}{2\pi kT}\frac{\sin\alpha\pi}{\cosh(1-\alpha)(E^* - E_0^*)/kT - \cos\alpha\pi} \tag{43}$$

Unlike the Wagner distribution, $K(E^*)$ cannot be rendered temperature-independent except in limiting cases where $\alpha \to 0$ or 1. The Cole–Cole distribution, like the log-normal distribution, is symmetrical with respect to a central frequency or relaxation time. The distribution of time constants is plotted as a function of the variable $s \equiv \log(\tau/\tau_0)$ in Figure 2.1.5, and the complex plane plots for various values of α are given in Figure 2.1.6.

As was pointed out by Cole and Cole, dielectric response corresponding to the function of Eq. (39) may be decomposed into the circuit shown in Figure 2.1.7, which contains a constant-phase element (CPE). The CPE is an empirical impedance function of the type

$$Z_{\mathrm{CPE}} = A(j\omega)^{-\alpha} \tag{44}$$

which has proved of considerable value in data fitting. The admittance of this circuit may be expressed as

$$Y = j\omega C_1 + \frac{j\omega C_2}{\left[1 + C_2 A(j\omega)^{1-\alpha}\right]} \tag{45}$$

Dividing by $j\omega\varepsilon_0$ and comparing with Eq. (39), we find that

$$A = \frac{\tau_0^{1-\alpha}}{(\varepsilon_s - \varepsilon_\infty)\varepsilon_0} \tag{46}$$

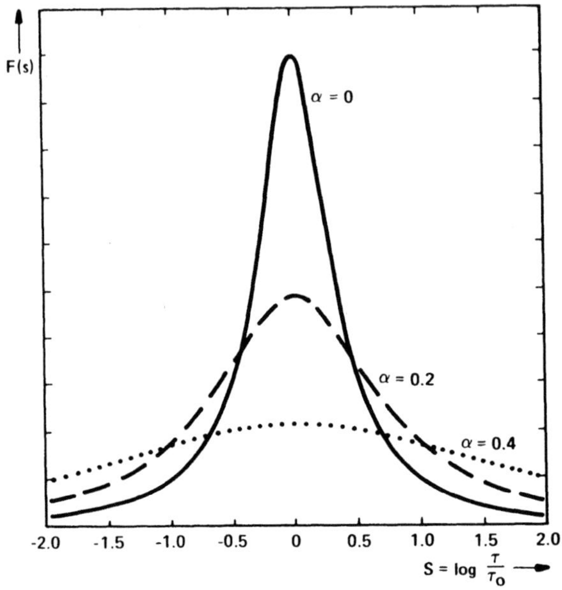

Figure 2.1.5. The distribution function $F(s)$ associated with the Cole–Cole distribution of relaxation times [Eq. (42)] for different values of α.

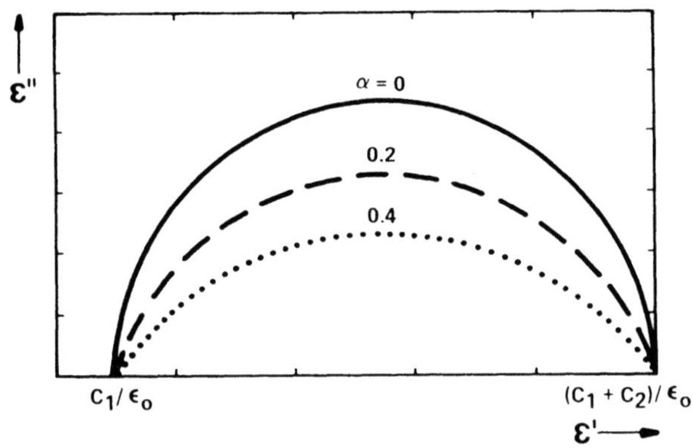

Figure 2.1.6. Complex permittivity associated with the Cole–Cole expression [Eq. (39)].

It is interesting to enquire about the distribution of relaxation times implied by the presence of Z_{CPE} alone. If

$$Z_{\text{CPE}} = Ap^{-\alpha} \tag{47}$$

then the admittance is

$$Y(p) = A^{-1}p^{\alpha} \tag{48}$$

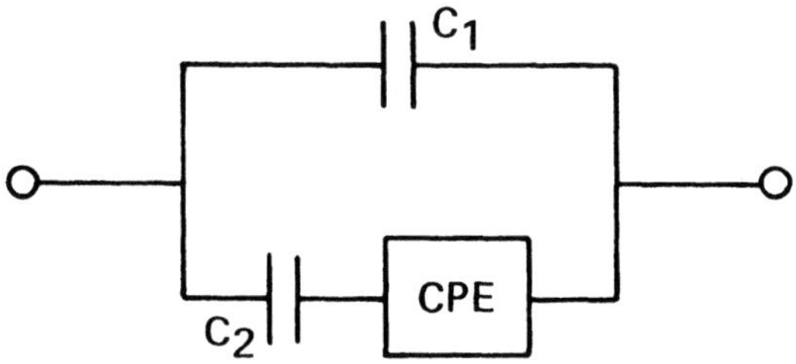

Figure 2.1.7. The analog of Figure 2.1.2, which models the electrical response associated with the Cole–Cole expression.

Therefore,

$$A(t) = L^{-1}\{A^{-1}p^{\alpha-1}\} \tag{49}$$

$$= \frac{A^{-1}t^{-\alpha}}{\Gamma(1-\alpha)} \tag{50}$$

where Γ is the gamma function. And

$$D(\lambda) = \frac{A^{-1}\lambda^{\alpha-1}}{\Gamma(1-\alpha)\Gamma(\alpha)} \tag{51}$$

or

$$G(\tau) = \frac{\sin \alpha\pi}{\pi} \cdot A^{-1}\tau^{-\alpha} \tag{52}$$

Thus, the distribution of relaxation times is proportional to $1/\tau^{\alpha}$. It has often been pointed out that this distribution is nonnormalizable. Physical acceptability may be restored by truncating the distribution at upper and lower limits of τ. The resulting distribution has been discussed by Matsumoto and Higasi [1962].

Assuming an expression for τ of the form given by Eq. (34), we find a distribution of activation energies:

$$K(E^*) \propto \exp[(1-\alpha)E^*/kT] \tag{53}$$

The important point to be made here is that the assumption of (1) an exponential distribution of activation energies and (2) an exponential form for τ lead directly to CPE behavior. The exponential distribution of activation energies has been further discussed by Macdonald [1963]. See also Section 2.2.3.5.

Two other distribution functions are due to Kirkwood and Fuoss [1941] and Davidson and Cole [1951] (see also Davidson [1961]).

The first of these is symmetric, and is again based on an extension of the Debye theory functions. In the Debye theory

$$\varepsilon''/\varepsilon''_{\max} = \operatorname{sech} x \tag{54}$$

where $x = \log \omega/\omega_0$. Instead of this, Kirkwood and Fuoss wrote

$$\varepsilon''/\varepsilon''_{\max} = \operatorname{sech} \alpha x \tag{55}$$

which leads to a distribution of the form

$$G(s) = \frac{2}{\pi} \cdot \frac{\cos(\alpha\pi/2)\ \cosh \alpha s}{\cos^2(\alpha\pi/2) + \sinh^2 \alpha s} \tag{56}$$

where s is again equivalent to $\log(\tau/\tau_0)$.

The Davidson–Cole equation

$$\frac{\varepsilon - \varepsilon_\infty}{\varepsilon_s - \varepsilon_\infty} = \frac{1}{(1 + j\omega\tau_0)^\beta} \tag{57}$$

leads to a skewed arc in the $\varepsilon'' - \varepsilon'$ plane. It is a semicircle at low frequency, but asymptotic to $\beta\pi/2$ at high frequencies (Figure 2.1.8).

The real and imaginary parts are

$$\varepsilon' - \varepsilon_\infty = (\varepsilon_s - \varepsilon_\infty) \cos \beta y (\cos y)^\beta \tag{58}$$

$$\varepsilon'' = (\varepsilon_s - \varepsilon_\infty) \sin \beta y (\cos y)^\beta \tag{59}$$

where $y = \tan^{-1} \omega\tau_0$.

The current response to the application of a step function potential difference is

$$i(t) = \mathbf{P}_\infty \delta(t) + \frac{\mathbf{P}_s - \mathbf{P}_\infty}{\tau_0 \Gamma(\beta)} \left(\frac{t}{\tau_0}\right)^{\beta-1} \exp-\left(\frac{t}{\tau_0}\right) \tag{60}$$

which may be compared with the equivalent Cole–Cole expression

$$i(t) = \mathbf{P}_\infty \delta(t) + \frac{\mathbf{P}_s - \mathbf{P}_\infty}{\tau_0 \Gamma(1+\alpha)} \left(\frac{t}{\tau_0}\right)^{(1-\alpha)} \tag{61}$$

and the Debye expression [Eq. (12)].

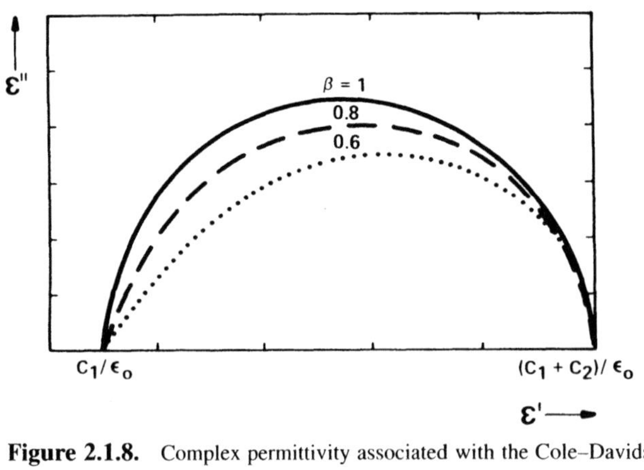

Figure 2.1.8. Complex permittivity associated with the Cole–Davidson expression [Eq. (57)].

The distribution of relaxation times is highly asymmetric:

$$\begin{aligned} G(\tau) &= \frac{\sin\beta\pi}{\pi}\left(\frac{\tau}{\tau_0-\tau}\right)^{\beta}, \quad && \tau<\tau_0 \\ &= 0 && \tau>\tau_0 \end{aligned} \tag{62}$$

Thus, the distribution ends abruptly at its most probable value.

A further generalization of the Debye approach was made by Williams and Watts [1970] and Williams *et al.* [1970] who introduced the use of the following fractional exponential form to describe the decay of polarization after the removal of a constant field.

$$\phi(t) = \mathbf{P}(t)/(\mathbf{P}_s - \mathbf{P}_\infty) = \exp-(t/\tau)^{\alpha}, \quad 0<\alpha<1 \tag{63}$$

neglecting the instantaneous drop in polarization. The equivalent Debye expression has $\alpha = 1$.

Thus,

$$A(t) = -d\phi/dt = (\alpha t^{\alpha-1}/t^{\alpha})\exp-(t/\tau)^{\alpha} \tag{64}$$

The Laplace transform of $A(t)$, that is $Q(p)$, may be evaluated by series expansions, although care is needed because of slow convergence for certain ranges of α (Williams *et al.* [1970]). For the particular case of $\alpha = \frac{1}{2}$, an analytical expression is easily derived:

$$Q(p) = \frac{\sqrt{\pi}}{2\sqrt{\tau_0}}\exp\frac{1}{\tau_0}\operatorname{erfc}\frac{1}{\sqrt{(\tau_0 p)}} \tag{65}$$

The inverse Laplace transform of functions of this type, which gives the distribution of relaxation times, is given by Montroll and Bendler [1984]. A simple expression for $\alpha = \frac{1}{2}$ may be derived:

$$G(\tau) = \frac{1}{2\sqrt{(\pi\tau\tau_0)}}\exp\left[-(\tau/4\tau_0)\right] \tag{66}$$

Like the Cole–Davidson function, the Williams–Watts approach gives asymmetric plots in the complex ε plane. A detailed comparison of these two forms has been made by Lindsey and Patterson [1980].

Interest in the Williams–Watts approach has arisen, not only because of its empirical success in fitting dielectric data, but also because of its relation to certain types of diffusion and random walk problems. The mechanistic relation between diffusion and relaxation was introduced by Glarum [1960], who suggested a process in which a mobile defect enabled a “frozen in” dipole to relax. Further aspects of random walk processes and their relation to CEPs and other empirical functions are discussed in a later section.

2.1.2.4 Conductivity and Diffusion in Electrolytes (Ibl [1983a], Newman [1973])

In the previous sections the expressions for the admittance of materials were developed on the assumption that they had no dc conductivity. The real part of the admittance arose from the dissipative process of dipole reorientation. Energy was absorbed by the system when the orientation of dipoles was changed with respect to the electric field vector.

Dissipation may also occur by mass transport of particles in the bulk of the phase. Work must be done against the frictional forces of the medium through which the particle moves. In solids, migration and diffusion are usually important; in liquids and membranes, hydrodynamic mass transport must also be considered.

A convenient starting point for discussion of transport properties in electrolytes is a consideration of the physical laws which connect charge and electric potential. In a medium of uniform dielectric constant we may write Poisson's equation, which connects the gradient of the electric field with the charge density

$$\nabla^2\Phi = -\rho/\varepsilon_s\varepsilon_0 \tag{67}$$

Here Φ is the electric potential. The charge density ρ is equal to the sum over the local concentrations of species multiplied by their charges.

$$\rho = F\Sigma z_i c_i \tag{68}$$

Because of the magnitude of the constant F/ε_0, very large electric fields result from very small deviations from electroneutrality. It is, therefore, a very good approximation to write for the interior of an electrolyte

$$\Sigma z_i c_i = 0 \tag{69}$$

provided the separation of the electrodes is not too small. In other words, except for the smallest systems, the bulk of the electrolyte is electrically neutral. This situation is generally otherwise at interfaces where there exists the possibility of large electric fields. It also follows that, in general, Laplace's equation

$$\nabla^2\Phi = 0 \tag{70}$$

is a good approximation in the electrolyte bulk.

A second basic equation expresses the conservation of mass in the system

$$\partial c_i/\partial t = -\nabla \cdot j_i + R_i \tag{71}$$

This equation states that the rate of accumulation of a species i in a given volume element is equal to the negative of the divergence of the flux plus any terms that lead to the production or deletion of i, such as chemical reactions or recombination in the bulk of the material.

Third, we can write an equation for the electric current density in terms of the fluxes of changed species

$$i = F\Sigma z_i j_i \tag{72}$$

We now need an expression for the flux of species i in terms of the forces acting on the particles. The assumption that is usually made is that a particle has a characteristic mobility u_i which is the proportionality constant between its velocity and the force causing it to move. The driving force is supposed to be the gradient in electrochemical potential η_i of the species, so that when the mobility is multiplied by the driving force and the concentration, we obtain the flux of i:

$$j_i = -c_i u_i \nabla \eta_i \tag{73}$$

The problem with this equation lies in the formulation of the force term. In general, a particle may move in response to gradients in the electrochemical potentials of other species, leading to cross-terms in the flux equation. In principle, the presence of cross-terms will occur whenever a component is present whose chemical potential may vary independently of that of species i. Thus, the motion of i may depend not only on $\nabla \eta_i$ but also on $\nabla \eta_j$ if η_j is independent of η_i (i.e. it is not coupled through a Gibbs–Duhem relation). The flux equation may therefore be generalized as

$$j_i = -c_i u_i \left[\nabla \eta_i + \sum_j \alpha_{ij} \nabla \eta_j \right] \tag{74}$$

where the α_{ij} are the coefficients expressing the influence of $\nabla \eta_i$ on i.

It may be shown that this equation is equivalent to the phenomenological equations derived from irreversible thermodynamics, as well as the multicomponent diffusion equations derived from the Stefan–Maxwell equations, which were first used to describe diffusion in multicomponent gases.

Further development of transport theory involves solution of Eqs (71) and (73), subject to the appropriate initial and boundary conditions to give currents, concentration profiles, and so on.

The simplest approach, often adopted in practice, particularly in solution electrochemistry, may be termed dilute solution approximation. We can write, for the gradient in electrochemical potential for a dilute solution,

$$\nabla \eta_i = RT \nabla c_i / c_i + z_i F \nabla \Phi \tag{75}$$

Thus,

$$j_i = -RTu_i \nabla c_i - z_i F c_i u_i \nabla \Phi \tag{76}$$

The quantity RTu_i is called the diffusion coefficient (Nernst–Einstein relation)

$$j_i = -D_i \nabla c_i - z_i F c_i u_i \nabla \Phi \tag{77}$$

and the current density is given by the expression

$$i = -F \sum z_i \nabla c_i D_i - F^2 \sum z_i^2 c_i u_i \nabla \Phi \tag{78}$$

Substituting Eq. (77) into Eq. (71), we obtain

$$\partial c_i / \partial t = z_i F \nabla \cdot (u_i c_i \nabla \Phi) + \nabla \cdot (D_i \nabla c_i) + R_i \tag{79}$$

or, in one dimension,

$$\partial c_i/\partial t = z_i F u_i \frac{\partial c_i}{\partial x}\mathbf{E} + D_i\frac{\partial^2 c_i}{\partial x^2} + R_i \tag{80}$$

if we assume $\partial\mathbf{E}/\partial x = 0$, that is, electroneutrality. This is the classical Nernst–Planck equation.

Instead of assuming dilute or ideal behavior, it is possible to write

$$\nabla\eta_i = \nabla\mu_i + z_i F\nabla\Phi \tag{81}$$

$$j_i = -c_i u_i \nabla\mu_i - z_i F c_i u_i \nabla\Phi \tag{82}$$

$$= -D_k c_i \nabla\mu_i / RT - z_i F c_i u_i \nabla\Phi \tag{83}$$

Here D_k is known as the component diffusion coefficient. The importance of this definition lies in the fact that Nernst–Einstein proportionality between a diffusion coefficient and a mobility has been retained, even though the condition of ideality has been relaxed. This is important since the apparent violation of the Nernst–Einstein equation in nonideal solutions is not a failure of the proportionality between mobility and mean displacement; it is a weakness in the method of formulating the driving force for diffusion in terms of a concentration gradient (Fick's law) rather than in terms of an activity or chemical potential gradient.

2.1.2.5 Conductivity and Diffusion—a Statistical Description

In the previous section, the Nernst–Planck equation was developed from the macroscopic flux density and mass conservation equations [(71) and (73), respectively]. The same equation can also be derived by statistical methods, which describe the probability of finding a particle within a volume region at a time t, given an initial distribution and a set of jump probabilities. For the simplest case, in one dimension, with equal probabilities of the particle making a jump to the right or to the left, the time evolution of an initial delta function in concentration at $x = 0$ is Gaussian:

$$n(x,t)\Delta x = \frac{1}{2\sqrt{(\pi D t)}}\exp(-x^2/4Dt)\Delta x \tag{84}$$

where D, the diffusion coefficient, is equal to $\nu l^2/2$, where ν is the number of steps of length l the particle makes per unit time. A further assumption involved in the use of the statistical arguments which lead to this equation is that the jumps are statistically independent. $n(x, t)$ is plotted for several different times in Figure 2.1.9. The total area under the curve is constant.

The mean displacement of the particles is zero:

$$\langle x\rangle = \int x n(x,t)dx = 0 \tag{85}$$

as long as the jump probabilities are symmetrical. The second moment is

$$\langle x\rangle^2 = \int x^2 n(x,t)dx = 2Dt \tag{86}$$

The probability density $n(x, t)$, given by Eq. (84), is a solution of the diffusion equation

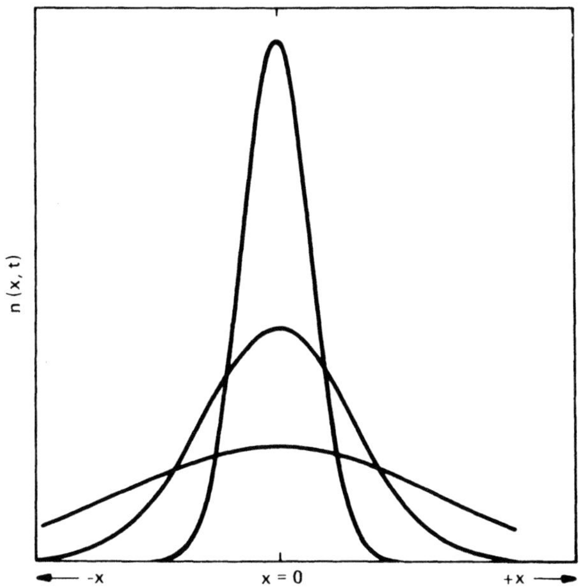

Figure 2.1.9. The Gaussian or normal distribution function $n(x, t)$ plotted as a function of distance from the origin for different times.

$$\partial n(x,\ t)/\partial t = D\partial^2 n(x,\ t)/\partial x^2 \tag{87}$$

This is identical to the Nernst–Planck Eq. (80) in the absence of an electric field term and a generation–recombination term.

If the jump probabilities are not symmetrical—for example, in the presence of an electric field—then $\langle x \rangle$ is no longer equal to zero, and the probability distribution is

$$n(x,\ t)\Delta t = \frac{1}{2\sqrt{(\pi D t)}} \exp\left[(-x - \langle v \rangle t)^2 t / 4Dt\right]\Delta x \tag{88}$$

Here $\langle v \rangle$ is the mean drift velocity $\langle x \rangle / t$. The mean drift velocity per unit field is the drift mobility b_i, and the conductivity σ_i is defined by

$$\sigma_i = b_i c_i z_i F \tag{89}$$

$$= \langle v \rangle c_i z_i F / \mathbf{E} \tag{90}$$

Note that b_i is equal to $z_i F u_i$.

Now $n(x, t)$ from Eq. (88) is the solution to Eq. (80) with R_i set to zero and describes a propagating Gaussian packet. The ratio of the dispersion $\sqrt{\left(\langle x^2 \rangle - \langle x \rangle^2\right)}$ to distance traveled is inversely proportional to the square root of time.

A more general approach via the master equation, leading to Fokker–Planck equations, may also be followed and may be found in texts on statistical physics (see, for example, Reichl [1980]).

The type of diffusion discussed here may be termed "*normal*" or *Gaussian diffusion*. It arises simply from the statistics of a process with two possible outcomes, which is attempted a very large number of times. In Section 2.1.2.7, the statistical basis of diffusion is enlarged to include random walks in continuous rather than discrete time, and also situations where different distributions of jump distances occur.

2.1.2.6 Migration in the Absence of Concentration Gradients

Under certain circumstances, the passage of electric current through an electrolyte does not lead to a concentration gradient, and Eq. (78) becomes

$$i = -F^2 \Sigma c_i z_i^2 u_i \nabla \Phi \tag{91}$$

The term $F^2\Sigma c_i z_i^2 u_i$ is called the conductivity σ, and under these conditions Ohm's law is obeyed by the electrolyte. Examples of this behavior are found where only one electrolyte species is mobile, for example in a solid electrolyte, with reversible electrodes (see Section 2.1.3) or at high frequencies where several carriers may move, but where neither electrode nor concentration polarizations have time to build up.

For a simple hopping conductivity process, in the absence of long-range interactions, the conductivity is expected to be independent of frequency. Here, a single particle is presumed to move along an infinite lattice of identical potential wells (Figure 2.1.10*a*). This might be contrasted with the case of a single particle hopping backward and forward in double well, where the low-frequency conductivity is zero, and a Debye-like transition region is followed by a constant high-frequency conductivity (Figure 2.1.10*b*).

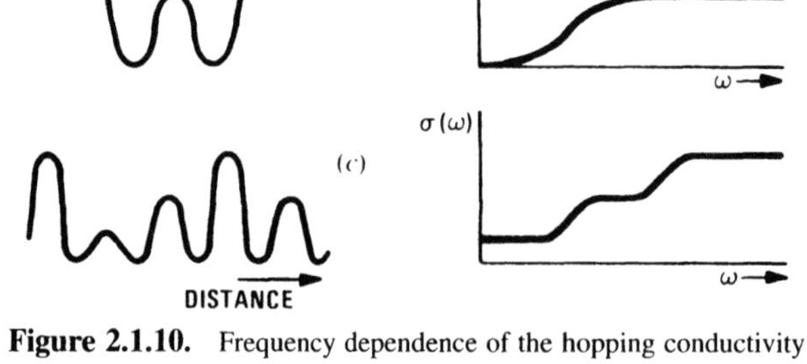

Figure 2.1.10. Frequency dependence of the hopping conductivity for different potential energy profiles: (*a*) Periodic constant activation energy, (*b*) a single bi-well, and (*c*) a potential profile with multiple activation energies.

Apart from the most dilute solutions, however, we do expect interactions between unassociated defect species, and in general this will lead to a frequency-dependent conductivity. This has been recognized for many years by electrochemists (Onsager [1926, 1927], Debye and Falkenhagen [1928]). The form of the frequency dependence, however, is of considerable interest.

Following the work of Jonscher (see, for example, Jonscher [1977, 1980]), who showed that a large number of different types of conductors exhibited a frequency dispersion of the CPE type, the presence of interactions has been invoked to explain the observed frequency dispersion in the conductivity of a number of solid electrolytes by Almond, West, and coworkers (Almond and West [1983a, b]), Almond *et al.* [1982, 1983, 1984]. These authors expressed the real part of the ionic conductivity of a number of materials by an expression of the type

$$\sigma(\omega) = \sigma_0 + A\omega^n \tag{92}$$

where σ_0 is a "dc" or frequency-independent part, and the second term is of the CPE type. Making use of Jonscher's empirical expressions, Eq. (92) was rewritten:

$$\sigma(\omega) = K\omega_p + K\omega_p^{1-n}\omega^n \tag{93}$$

where ω_p is the hopping frequency and K depends on the concentration of the mobile charge carriers. The high-temperature limiting value for ω_p should be equal to the attempt frequency, which is independently accessible using IR spectroscopy. For the case of sodium beta alumina, good agreement was found. Once ω_p is known, then the carrier concentration and activation entropy can also be deduced. For the case of beta alumina the hopping rate calculation has been confirmed by mechanical relaxation measurements. In a number of materials, however, including β''-alumina, the low-frequency region is not independent of frequency and a second CPE term must be included. The work of Almond and West is also discussed beginning at Eq. (78) of Section 2.2.3.5.

According to Jonscher, the origin of the frequency dependence of the conductivity was due to relaxation of the ionic atmosphere after the movement of the particle. This idea, and the earlier concepts of Debye, Onsager, and Falkenhagen, have been developed into a quantitative model suitable for solids by Funke [1986]. It is assumed that immediately after an ion hops to a new site (a new minimum in lattice potential energy) it is still displaced from the true minimum in potential energy, which includes a contribution from other mobile defects. At long times the defect cloud relaxes, until the true minimum coincides with the lattice site. The model predicts upper and lower frequency-limiting conductivities and a region in between of power law (CPE) behavior.

In general, both conductivity and dipolar relaxation processes may be present in the same material, and the total conductivity is given by

$$\sigma_\Sigma = \sigma + j\varepsilon_0\varepsilon\omega \tag{94}$$

Where ε is given by Eq. (16).

The equivalent circuit for such a combination of processes is shown in Figure 2.1.11. An—RC—series combination will be present for each relaxation process

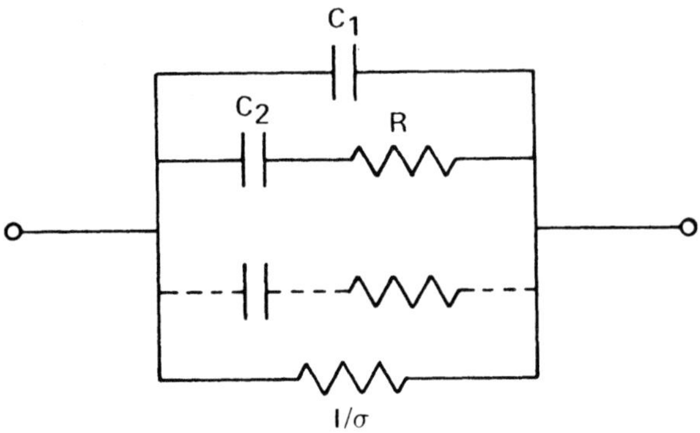

Figure 2.1.11. An equivalent circuit which models the behavior of a material which has both long-range conductivity (σ) and a number of discrete relaxation processes.

present in the material. The dielectric loss peaks will be superimposed on a background loss due to the long-range conductivity process. Due to interactions, for example, dipole–dipole, or lattice relaxations, as discussed above, a distribution of relaxation times is to be expected, and Wapenaar and Schoonman [1981] have included Cole–Cole branches (series—CZ_{CPE}—combinations) rather than Debye branches to fit data on $Ba_{1-x}La_xF_{2+x}$ fluorite structure solid solutions. The impedance spectra of doped tysonite materials have been investigated by Roos *et al.* [1984], who again found excellent agreement between data and a circuit which included Cole–Cole branches. However, the activation energies were not consistent with a simple dielectric relaxation, and a model in which conducting species move between inequivalent sites has been developed (Franceschetti and Shipe [1984]). The motion of the defect species may involve several distinguishable jump processes, each governed by a different activation energy. The interpretation of the relaxation branches in the circuit is that they describe a polarization arising from the inequalities in jump probabilities.

The important point to be emphasized here is that, although for a single jump frequency, in the absence of interactions, no particular structure is expected in the conductivity or dielectric constant at that frequency, a more complex model which incorporates several jump frequencies (Figure 2.1.10*c*) indicates that frequency dispersion is expected in this range. Although the model of Franceschetti and Shipe was developed for the tysonite structure, a more general treatment for a small number of sublattices has been independently given (Wong and Brodwin [1980]) and confirms the main conclusions. As the number of possible jump frequencies increases, it is apparent that generalization of the model will eventually describe conductivity and diffusion in disordered materials, that is, in materials in which the jump probabilities are described by random variables.

2.1.2.7 Transport in Disordered Media

In a previous section reference was made to the random walk problem (Montroll and Schlesinger [1984], Weiss and Rubin [1983]) and its application to diffusion in solids. Implicit in these methods are the assumptions that particles hop with a fixed jump distance (for example between neighboring sites on a lattice) and, less obviously, that jumps take place at fixed equal intervals of time (discrete time random walks). In addition, the processes are Markovian, that is the particles are without memory: the probability of a given jump is independent of the previous history of the particle. These assumptions force normal or Gaussian diffusion. Thus, the diffusion coefficient and conductivity are independent of time.

In recent years, more complex types of transport processes have been investigated and, from the point of view of solid state science, considerable interest is attached to the study of transport in disordered materials. In glasses, for example, a distribution of jump distances and activation energies are expected for ionic transport. In crystalline materials, the best ionic conductors are those that exhibit considerable disorder of the mobile ion sublattice. At interfaces, minority carrier diffusion and discharge (for example electrons and holes) will take place in a random environment of mobile ions. In polycrystalline materials the lattice structure and transport processes are expected to be strongly perturbed near a grain boundary.

In general, the study of transport processes in disordered media has its widest application to electronic materials, such as amorphous semiconductors, and very little attention has been given to its application to ionic conductors. The purpose of this section is to discuss briefly the effect of disorder on diffusion process and to point out the principles involved in some of the newly developing approaches. One of the important conclusions to be drawn is that frequency-dependent transport properties are predicted to be of the form exhibited by the CPE if certain statistical properties of the distribution functions associated with time or distance are fulfilled. If these functions exhibit anomalously long tails, such that certain moments are not finite, then power law frequency dispersion of the transport properties is observed. However, if these moments are finite, then Gaussian diffusion, at least as limiting behavior, is inevitable.

Although the general problem of a random walk on a random lattice is difficult, there has been considerable success in approaching this problem from the point of view of the continuous time random walk, which was first proposed by Montroll and Weiss [1965]. In this approach, the walk is supposed to take place on a regular lattice; disorder is introduced by defining a distribution of waiting or residence times for a particle on a site. In a disordered material there will be a distribution of energy barriers. It therefore seems reasonable that a particle in a deep well will spend more time there than will a particle in a shallow well. The waiting time distribution function $\psi(t)$ therefore describes the probability for an event to happen at a time t after a previous event. The original $\psi(t)$ of Montroll and Weiss was generalized to a position-dependent $\psi(r, t)$ by Scher and Lax [1973a]. The application of the approach to electronic transport in amorphous semiconductors has been discussed in several papers (Scher and Montroll [1975], Pfister and Scher [1978], Scher and Lax [1973b]).

There is a qualitative difference in transport properties depending on the nature of $\psi(t)$. If $\psi(t)$ is such that the time between hops has a finite first moment, that is, a mean residence time $\langle t \rangle$ can be defined, then classical diffusion is observed. An example would be

$$\psi(t) = \lambda \exp(-\lambda t) \tag{95}$$

where the mean waiting time is $1/\lambda$. If, on the other hand, $\int t\psi dt$ diverges, then non-Gaussian or "dispersive" transport is seen. Of particular interest in this respect are $\psi(t)$ with long time tails

$$\psi(t) = \alpha A t^{-1-\alpha} / \Gamma(1-\alpha), \quad 0 < \alpha < 1 \tag{96}$$

In other words, the hopping probability is a slowly decaying function of time. Under these conditions, the dispersion of the concentration, $\langle x^2(t) \rangle$, becomes proportional to t^{α}, and the diffusion coefficient

$$D(t) = (1/2\Delta) d\langle x^2 \rangle / dt \tag{97}$$

and conductivity become time-dependent, with a power law dependence on frequency and time. Here Δ is the dimensionality of the system.

The physical origin of a power law distribution function for waiting times might arise from an exponential distribution of activation energies. Suppose, the distribution function of activation energies was of the form

$$K(E^*) = K_0 \exp(-E^*)/E_0^* \tag{98}$$

Then, if the waiting time were proportional to the exponential of the activation energy, the distribution of waiting times would have a power law dependence on time, as required by Eq. (96).

The concepts of the continuous time random walk (CTRW) approach have been applied to ionic conductivity in glasses by Abelard and Baumard [1984]. In an alkali silicate glass, it is usually assumed that only a small number of the alkali metal ions are mobile, and the remainder are associated with nonbridging oxygens. These latter form dipoles which may reorient in the presence of an electric field. Interpretation of the complex impedance or dielectric constant of these glasses therefore is usually made in terms of a distribution of these relaxation times, in the manner discussed in the previous section. Abelard and Baumard, however, suggest that a more appropriate approach is to consider all alkali ions mobile, but with a distribution of activation energies associated with the potential wells in which they are situated. There is, therefore, a distribution of waiting times which leads to the observed frequency dependence of the real part of the conductivity.

An interesting extension of the dispersive transport model is its application to dielectric relaxation. As mentioned earlier, Glarum [1960] proposed that "frozen-in" polarizations could be relaxed when a defect (e.g. a mobile charge carrier) approached them. Bordewijk [1975] extended the model and showed that in one-dimensional transport a Williams–Watts dielectric relaxation function with $\alpha = \frac{1}{2}$

resulted, but normal Debye relaxation was predicted in three dimensions ($\alpha = 1$). Schlesinger and Montroll [1984] have shown that if diffusion of the mobile defect is restricted to a CTRW with a long-time-tailed $\psi(t)$, then a Williams–Watts dielectric function for the relaxation of the dipoles is expected. The relaxation process is treated as a diffusion controlled chemical reaction.

The essentially different nature of transport processes with $\psi(t) \propto t^{-(1-\alpha)}$ should be stressed. Processes with this type of waiting time distribution function show an absence of scale. They exhibit very sporadic behavior. Long dormancies are followed by bursts of activity. They have been described as *fractal time* processes (Schlesinger [1984]). *Fractal space* processes, in which the absence of scale is present in the spatial aspects of the transport, are considered later in this section.

A different approach to transport in disordered systems has been developed by considering the excitation dynamics of random one-dimensional chains (Alexander *et al.* [1981]). Such a system may be represented by a master equation of the form

$$dP_n/dt = w_{n,n-1}(P_{n-1} - P_n) + W_{n,n+1}(P_{n+1} - P_n) \tag{99}$$

where the P's are the amplitude of the excitations (site occupancies) and the W's are the transition probabilities between the nodes or sites, n and so on. This equation is obviously a discrete form of the diffusion equation, with the W's stochastic variables described by a distribution function. The electrical analog to this equation is the random transmission line, described by the equation

$$C_n dP_n/dt = W_{n,n-1}(P_{n-1} - P_n) + W_{n,n+1}(P_{n+1} - P_n) \tag{100}$$

Here the C_n are the random capacitances, the W_n are the random conductances, and the P_n are the node potentials.

Alexander *et al.* [1981] have obtained solutions to this type of equation for various types of distribution functions of W for an initial delta function input in P. In particular, they considered distribution functions for which a mean transition rate $\langle W \rangle$ could be defined, and functions which were of the form

$$\rho(W) = \rho_0(T)W^{-\alpha(T)} \tag{101}$$

where no mean transition rate exists. This is similar to the distribution function of waiting times for a CTRW defined in Eq. (96), and the arguments suggesting its use are essentially the same. The transition rate is an exponential function of activation energies, and the activation energies are supposed to be exponentially distributed, leading to a power law form for the distribution of transition rates. A similar argument can be used if a distribution of jump distances is assumed, that is configurational disorder rather than randomness in the activation energies is assumed.

As for the case of the CTRW method, qualitatively different solutions are obtained depending on whether a mean transition rate can be defined or not. In the former case, the system behaves as if it were ordered with a single transition rate [$\rho(W) = \delta(W - W_{av})$], even though the W's are random variables. These systems exhibit a frequency-independent low-frequency conductivity.

For power law distributions, however, the low-frequency conductivity tends to the form

$$\sigma(\omega) \propto (-j\omega)^{\alpha/(2-\alpha)} \tag{102}$$

as $\omega \to 0$. There may also be a situation in which a crossover between the two distributions occurs as a function of time, in which case the mean square particle displacement is given by

$$\langle x^2 \rangle = 2D_0 t + Bt^{1-s} \tag{103}$$

and the real part of the conductivity is

$$\sigma'(\omega) = \sigma(0) + A\omega^s \tag{104}$$

Thus, at low frequencies a constant conductivity would be seen, but at higher frequencies a power law contribution enters. In these equations, D_0 is the limiting diffusion coefficient and A and B are thermally activated constants.

Experimental observations of the frequency dependence of the conductivity in the one-dimensional ionic conductor potassium hollandite (Bernasconi *et al.* [1979]) show a pronounced power law dependence, as predicted by the model. It was proposed that the transport process in this material was limited by random barrier heights caused by the presence of impurities.

In the previous paragraphs it was pointed out that a discrete time random walk, or a CTRW with a finite first moment for the waiting time distribution, on a lattice with a fixed jump distance led to a Gaussian diffusion process with a probability density given by Eq. (84). The spatial Fourier transform of this equation is

$$\mathbf{n}(q,t) = \exp(-Dtq^2) \tag{105}$$

Disorder was introduced into this system by postulating a distribution of waiting times. A complementary extension of the theory may be made by considering a distribution of jump distances. It may be shown that, as a consequence of the central limit theorem, provided the single-step probability density function has a finite second moment, Gaussian diffusion is guaranteed. If this condition is not satisfied, however, then Eq. (105) must be replaced by

$$\mathbf{n}(q,t) = \exp\left(-At\,|q|^{\mu}\right) \tag{106}$$

where μ lies between 0 and 2. This distribution function is known as a Levy or stable distribution. This distribution is a solution of the equation

$$\partial\mathbf{n}(q,t)/\partial t = -A|q|^{\mu}\,\mathbf{n}(q,t) \tag{107}$$

which is, of course, the Fourier transform of the diffusion equation when $\mu = 2$. A number of authors have considered the type of random walk process, defined by the Levy distribution (Hughes *et al.* [1981]). A particle executes a walk which may be transient and clustered. In other words, not all regions of space are visited by the walker, and a hierarchy of clusters is developed. The clusters may be self-similar, and Mandelbrot [1983] has stressed the fractal nature of walks with this distribution.

Other workers have also considered transport on a self-similar geometry, through the connection with percolation. Close to a percolation threshold, the conductivity and dielectric constant behave with a power law exponent in the concentration of one of the components, and the percolating cluster at the threshold has been identified as a fractal object. The temporal behavior of the diffusion process close to the percolation threshold has also been considered (Gefen *et al.* [1983]) and, using scaling arguments, it has been shown that the mean square displacement is

$$\left\langle r(t)^2 \right\rangle = at^{\theta} \tag{108}$$

where a and θ are constants, leading to a time- and frequency-dependent diffusion coefficient and conductivity. Experimental verification of this power law dependence has been obtained for two-dimensional percolation in thin gold films (Laibowitz and Gefen [1984]).

An alternative approach, used to describe the properties of ionically conducting glasses, is conceptually closely related to the earlier discussion of transport properties in materials with a small number of sublattices for the conducting species.

It has been recognized for a considerable time that if the translational invariance of the conductivity activation energy barriers is lost, then the dielectric and conductivity properties become frequency-dependent (Figure 2.1.10*c*). For a material with no dipolar relaxation processes (i.e. ε_s is not a function of time), but with a conductivity σ_0, then Eq. (94) becomes

$$Y = \sigma_0 + j\omega\varepsilon_s\varepsilon_0 \tag{109}$$

The equivalent circuit is simply a parallel—RC—combination, and thus

$$\varepsilon = \varepsilon_s - \frac{j\sigma_0}{\omega\varepsilon_0} \tag{110}$$

Macedo, Moynihan, and Bose [1972] defined the *conductivity relaxation time*

$$\tau_\sigma = \varepsilon_0\varepsilon_s/\sigma_0 \tag{111}$$

Hence

$$\varepsilon = \varepsilon_s - \frac{j\varepsilon_s}{\omega\tau_\sigma} \tag{112}$$

Macedo and others (Hodge *et al.* [1975, 1976]) have stressed the electric modulus formalism ($M^* = 1/\varepsilon^*$) for dealing with conducting materials, for the reason that it emphasizes bulk properties at the expense of interfacial polarization. Equation (112) transforms to

$$M = M_s \frac{j\omega\tau_\sigma}{1 + j\omega\tau_\sigma} \tag{113}$$

where $M_s = 1/\varepsilon_s$.

For a material with a single relaxation time τ_σ, a plot of M'' vs. $\log(f)$ shows a maximum, in just the same way that ε''shows a maximum for a dielectric relaxation process. Glassy conductors however, often show broad and asymmetric modulus spectra, and, in complete analogy to the discussion of Section 2.1.2.3, Macedo *et al.*

[1972] introduced a distribution function of conductivity relaxation times $G(\tau_\sigma)$ such that

$$M = M_s \int_0^\infty G(\tau_\sigma) \frac{j\omega\tau_\sigma}{1 + j\omega\tau_\sigma} d\tau_\sigma \tag{114}$$

They were able to fit experimental modulus data for a calcium–potassium nitrate melt and a lithium aluminosilicate glass using a double lognormal distribution function.

The decay function for the electric field after the imposition of a charge on the electrodes

$$\phi'(t) = \mathbf{E}(t)/\mathbf{E}(t = 0) \tag{115}$$

may also be defined, in analogy to the decay of polarization function [see Eq. (63)]. For a single relaxation time,

$$\phi'(t) = \exp - t/\tau_\sigma \tag{116}$$

The modulus is related to ϕ' through the expression

$$\frac{M}{M_s} = 1 - \left\{ \left(-\frac{d\phi'}{dt} \right) \right\} \tag{117}$$

where, as usual, { } denote a Laplace transform.

Moynihan *et al.* [1973] used the Williams–Watts form of ϕ',

$$\phi'(t) = \exp\left[-(t/\tau_\sigma)^\beta \right], \quad 0 < \beta \le 1 \tag{118}$$

to obtain a much better fit to the same glass data mentioned above. The same function has also been used to analyze modulus data for lithium phosphate glasses (Martin and Angell [1986]), where it was found that the parameter β was largely independent of temperature but the distribution broadened with increasing alkali content. The inappropriateness of the Moynihan Williams-Watts approach has, however, been emphasized in Macdonald [1996, 2004] and in Chapter 4.2.

2.1.3 Mass and Charge Transport in the Presence of Concentration Gradients

2.1.3.1 Diffusion

In the absence of an electric field and terms in R_i, Eq. (30) reduces to

$$\partial c_i / \partial t = -\nabla \cdot j_i \tag{119}$$

$$= \nabla \cdot (D_i \nabla c_i) \tag{120}$$

In one dimension, for constant D_i,

$$\partial c_i / \partial t = D_i \partial^2 c_i / \partial x^2 \tag{121}$$

This equation has been introduced from two points of view. In the macroscopic approach it was assumed that the flux or diffusion current is proportional to a concentration gradient or a chemical potential gradient and also satisfies a continuity equation. The generalizations of and justifications for this approach lie at the basis of nonequilibrium thermodynamics; as such, they are independent of the atomistic nature of the processes involved.

On the other hand, in the atomistic approach, the time-dependent configurations of the system are determined from the probabilities of the elementary atomic process. The random walk approach calculates the probability of finding the system in a certain state after a certain time given an initial distribution of particles. It is then possible to show that the distributions are solutions of the diffusion equation.

It was shown in Section 2.1.2.4 that the general flux equations (e.g. the Nernst–Planck equation) contain, in addition to the diffusion terms, a contribution from migration, that is the movement of charged particles under the influence of an electric field. Under certain circumstances it is quite possible to carry out experiments in which the field is negligibly small compared to the concentration or activity driving force.

In aqueous electrochemistry this situation is usually achieved by use of a supporting electrolyte. This is an inactive salt that is added to the solution in high concentration to increase the conductivity enough that the migration term in Eq. (78) or Eq. (82) becomes very small. In solid state electrochemistry it is difficult to achieve the same effect in such a simple way. The movement of a minority charge carrier, either electronic or ionic, in a good solid electrolyte is an analogous situation. This result is exploited in the Wagner [1933] asymmetric polarization experiment in which the partial conductivity of electronic species in a solid electrolyte is measured assuming that the driving force for electronic conductivity is an activity gradient rather than an electric field. In the next section, another example, that of chemical diffusion in a majority electronic carrier, is discussed in more detail. It is worth mentioning that local electric fields arising from the coupled motion of two charged species through an approximate electroneutrality condition are not necessarily absent from the equations of this section. This local field may be present and profoundly affect the diffusion of species, without any net average field across the bulk of a sample leading to a migration process. In this section, it is assumed that a field which leads to a migration current is absent, but local fields, such as those present in neutral electrolyte diffusion or ambipolar diffusion, may be present.

The purposes of this section is to discuss the electrical analogs of diffusion processes in the absence of migration and to present suitable electrical equivalent circuits for analysis of data obtained under these circumstances.

From the point of view of impedance spectroscopy, solutions of the diffusion equation are required in the frequency domain. The Laplace transform of Eq. (121) is an ordinary differential equation

$$p\{c\} - c(t=0) = Dd^2\{c\}/dx^2 \tag{122}$$

where p is the complex frequency variable

$$p = \sigma + j\omega \tag{123}$$

This transformation of a partial differential equation into an ordinary differential equation illustrates a general advantage of working in the frequency domain. Solutions are of the form

$$\{\Delta c\} = A\exp(-\alpha x) + B\exp(\alpha x) \tag{124}$$

where $\{\Delta c\}$ is the Laplace transform of the excess concentration

$$\Delta c = c(x,t) - c(x,0) \tag{125}$$

and $\alpha = \sqrt{(p/D)}$. Here A and B are constants to be determined by the boundary conditions. Experimentally, one boundary is usually the interface between the electrode and the electrolyte (x = 0). Consider the case of semiinfinite diffusion into the electrode:

$$\Delta c \rightarrow 0 \text{ as } x \rightarrow \infty$$

and therefore $B = 0$.

At $x = 0$ (the electrode–electrolyte interface) the solution is

$$\{\Delta c\}_{x=0} = \{\Delta i\}/zF\sqrt{(pD)} \tag{126}$$

where Δi is the ac current, which is equal to $-zFDd\Delta c/dx$. To calculate the impedance we need a relationship between $\Delta c_{x=0}$ and Δv, the ac component of the voltage. For small perturbations around equilibrium, we may write

$$\Delta v/\Delta c = (dE/dc) \tag{127}$$

where (dE/dc) represents the change in electrode potential with concentration, which may be developed from a model (e.g. ideal solution assumptions) or from a separate thermodynamic measurement. For an ideal solution $dE/dc = RT/zFc$. For small-signal conditions, the perturbation may also be expanded around a steady state dc potential, in which case the surface concentrations due to the dc current must also be calculated from a steady state flux equation.

Taking the Laplace transform of Eq. (127) and substituting into Eq. (126) gives

$$Z(p) = \{\Delta v_{x=0}\}/\{\Delta i\} = (dE/dc)/zF\sqrt{(pD)} \tag{128}$$

Setting $\sigma = 0$ and separating the real and imaginary parts gives

$$Z(j\omega) = (dE/dc)(\omega^{-0.5} - j\omega^{-0.5})/zF\sqrt{(2D)} \tag{129}$$

The complex impedance is therefore inversely proportional to the square root of frequency. In the complex plane it is a straight line inclined at $\pi/4$ to the real axis.

Equation (121) has analogies in both heat conduction and electrical circuit theory. Consider the semiinfinite transmission line composed only of resistors and capacitors (Figure 2.1.12). If r is the resistance per unit length and c is the capacitance per unit length, then

$$I = -(\partial V/\partial x)/r \tag{130}$$

$$\partial V/\partial x = -(\partial I/\partial x)/c \tag{131}$$

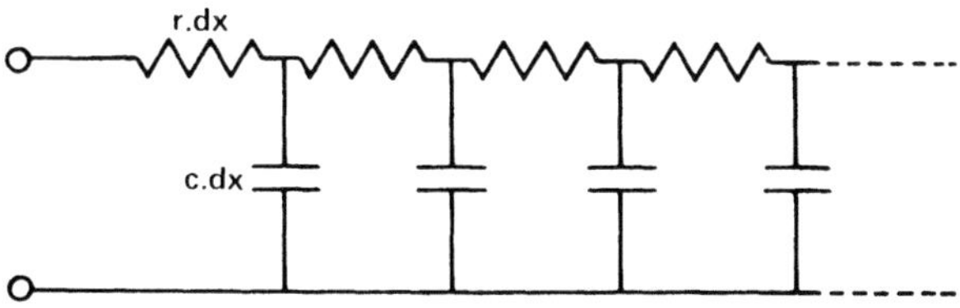

Figure 2.1.12. A resistive–capacitive transmission line which describes the behavior of a semiinfinite diffusion process.

Differentiating Eq. (130) and combining the two equations gives

$$\partial V/\partial t = (\partial^2 V/\partial x^2)/rc \tag{132}$$

The analogy with diffusion may be made more specific if we compare the appropriate driving forces and fluxes. The electric potential difference V in the transmission line case is analogous to the electrochemical potential difference in the case of diffusion. Thus,

$$\frac{\partial V}{\partial x} \equiv \frac{RT}{zF} \cdot \frac{1}{c} \cdot \frac{\partial c}{\partial x} \tag{133}$$

for an ideal solution (the case for which Fick's law is least ambiguously valid). The reciprocal of the resistance per unit length is analogous to cDF^2z^2/RT, and the capacitance per unit length is analogous to z^2F^2c/RT.

Thus the reciprocal of the rc product plays the role of the diffusion coefficient. The impedance of the transmission line is

$$Z_R = \sqrt{(r/pc)} \tag{134}$$

which is exactly the same form as Eq. (129) if appropriate substitutions are made. For a nonideal solution, RT/zFc may be replaced by (dE/dc).

So far, only semiinfinite boundary conditions have been considered. For many problems, however, thin samples dictate the use of finite-length boundary conditions. A reflective boundary $dc/dx = 0$ has been considered by Ho *et al.* [1980] and the impedance derived for this case:

$$Z(j\omega) = \frac{dE}{dc} \cdot \frac{1}{zF} \cdot \frac{\cosh l\sqrt{(j\omega/D)}}{\sqrt{(j\omega D)}} \tag{135}$$

The equivalent circuit analog of this situation is a finite-length transmission line terminated with an open circuit. A constant activity or concentration is also a common condition for the interface removed from $x = 0$. In this case the finite-length transmission line would be terminated in a resistance, and the impedance is given by the expression

$$Z(j\omega) = \frac{dE}{dc} \cdot \frac{1}{zF} \cdot \frac{\tanh l\sqrt{(j\omega/D)}}{\sqrt{(j\omega D)}} \tag{136}$$

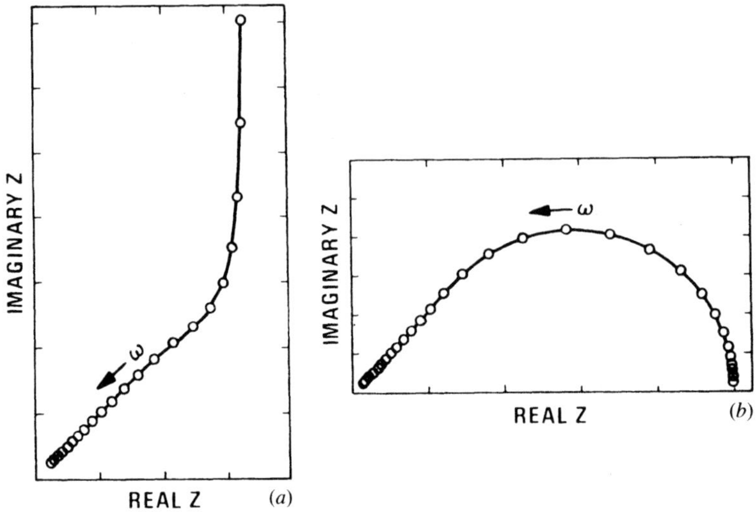

Figure 2.1.13. Complex plane representations of the impedance due to a finite-length diffusion process with (*a*) reflective, (*b*) transmissive boundary conditions at $x = 1$.

The complex plane representations of these two impedance behaviors are shown in Figure 2.1.13.

In this section the following principal assumptions were made. First, it was assumed that the surface concentrations and potentials were given by their equilibrium or dc steady state values. In other words, there was supposed to be no barrier preventing or slowing down the transfer of matter across the electrode–electrolyte interface. In general, of course, this will not be true, and the impedance associated with the interface forms a very important aspect of impedance spectroscopy as applied to electrochemical situations. This is in contrast to applications where the only interest lies in bulk effects. The interfacial impedance, due to both the storage and the dissipation of energy, will be addressed in Section 2.1.4.

Second, although the form of the diffusion equation was derived from ideal solution theory, it will be seen in the next section that the form of the equation may be retained, even though the dilute solution assumptions are relaxed, through definition of a chemical diffusion coefficient.

2.1.3.2 Mixed Electronic–Ionic Conductors

It was seen in the previous paragraphs that the presence of a supporting electrolyte leads to a situation in which a charged species may diffuse in an essentially field-free environment. Under these circumstances, Fick's law of diffusion may be solved for the appropriate boundary conditions, and the electrical response of the system may be modeled by a transmission line composed of distributed elements. Although such situations are very common in aqueous electrochemistry, the analogous situation in solids, where a minority ion diffuses in a solid supporting electrolyte, occurs infrequently.

More interesting is the commonly encountered situation where an ion diffuses in a majority electronic conductor. Thus, diffusion in metallic and semiconducting alloys or of inserted species in transition metal oxides and chalcogenides fall into this category. Many electrode reactions are of this type. Lithium diffusion in β-LiAl and other alloys is of interest in negative electrode reactions for advanced lithium batteries; hydrogen and lithium diffusion in oxides (e.g. V_6O_{13}) and sulfides (e.g. TiS_2) are of importance as positive electrode reactions for batteries and electrochromic devices.

In materials of this type the diffusion process may be regarded as involving a neutral atomic species, or as a coupled process in which an ionic and electronic species move together. In the simplest case, where the electronic partial conductivity is much greater than the ionic, the flux equation for a neutral species may be written

$$j = -c^* u \frac{d\mu^*}{dx} = -c^* uRT \frac{d \log \alpha^*}{dx} = -c^* D_k \frac{d \log \alpha^*}{dx} \tag{137}$$

Here we cannot assume ideal or dilute solution behavior since the mobile species activity may vary widely especially if the electron activity changes rapidly over the stoichiometric domain of the phase of interest. After rearrangement we obtain the equation

$$j = -D_k \frac{d \log \alpha^*}{d \log c^*} \cdot \frac{dc^*}{dx} = -D_k \left[1 + \frac{d \log \gamma^*}{d \log c^*} \right] \frac{dc^*}{dx} \tag{138}$$

which is equivalent to Fick's law if we write

$$D_c = D_k \left[1 + \frac{d \log \gamma^*}{d \log c^*} \right] \tag{139}$$

Here u is the mobility, μ the chemical potential, and γ the activity coefficient of the mobile species. The "*" denotes that the relations are written for a neutral species. Now D_k is the component diffusion coefficient which, as pointed out in Section 2.1.2.4, obeys the Nernst–Einstein relation

$$D_k = uRT \tag{140}$$

for all situations, irrespective of ideality assumptions.

The term in square brackets in Eq. (138) expresses the variation of activity coefficient of the *neutral* species with concentration. Thus, in addition to the statistical contribution to diffusion, expressed by the familiar gradient-in-concentration term, there is a chemical driving force due to the variation of free energy with composition and hence position. The term in square brackets is known as the thermodynamic enhancement factor and was identified by Darken [1948]. The diffusion coefficient D_c is known as the chemical diffusion coefficient, and its use is appropriate whenever diffusion takes place in an appreciable concentration gradient and when ideal solution laws cannot be applied to the solute. The concept was extended by C.

Wagner [1953], and a general treatment has been given by Weppner and Huggins [1977]. This general approach involves elimination of the field term from simultaneous equations of the type

$$j_i = -RTu_i\left[\frac{d\log a_i}{d\log c_i}\frac{dc_i}{dx} + \frac{z_i F c_i}{RT}\frac{d\Phi}{dx}\right] \tag{141}$$

written for ionic and electronic species. The result is a general equation

$$j_i = -D_{k_i}\left|(1-t_i)\frac{d\log a_i^*}{d\log c_i^*} - \sum_{j\neq i,e,h} t_j \frac{z_i}{z_j}\frac{d\log a_j^*}{d\log c_i^*}\right|\frac{dc_i}{dx} \tag{142}$$

valid for general transference numbers and thermodynamic parameters. For small-signal conditions (constant enhancement factor over the concentration range of the experiment), the diffusion equations are still of the Fick's law type and therefore lead to $\sqrt{\omega}$ dependence of the admittance (Warburg behavior).

In recent years, cells of the type

$$\mathrm{Li}|\mathrm{Li}^{+}\mathrm{electrolyte}|\mathrm{Li}_y M$$

(where M is a mixed conducting host material for the inserted lithium) have been used to investigate the diffusion of Li in a number of alloys and oxides using ac impedance methods. The boundary condition at the electrolyte–electrode interface is a sinusoidally varying chemical potential of (neutral) lithium. It is important to recognize that the potential difference applied across a cell determines the activity of the electroactive species at the point at which the conductivity of the system changes from being predominantly ionic to being predominantly electronic. In this kind of experiment the thermodynamic enhancement factor is conveniently determined *in situ* by measuring the dependence of equilibrium cell potential on electrode composition.

For thin samples, the second boundary condition may be modified to include either transmissive or reflective interfaces, as discussed in Section 2.1.3.1.

2.1.3.3 Concentration Polarization

The situation is often encountered where, upon the passage of current through an electrochemical cell, only one of the mobile species is discharged at the electrodes. Examples are (a) the use of a liquid or polymeric electrolyte, where both ions are mobile, and yet where only one is able to participate in the electrode reaction; and (b) a mixed conducting solid in which current is passed by electrons, but in which cations also have a significant transport number.

Consider a system consisting of a binary, unsupported electrolyte between electrodes which are reversible only to the cation. The cell is initially at equilibrium (no net currents are passing). At very short times after the imposition of a potential difference, the concentrations of all species in the bulk of the electrolyte are uniform and the ions move in response to the applied field. The current is determined by the uniform electrolyte conductivity.

$$i = -F^2 \sum c_i z_i^2 u_i \frac{d\Phi}{dx} \tag{143}$$

At long times, on the other hand, the flux of the blocked anion falls to zero, and a constant flux of cations passes through the system. In order to maintain electroneutrality there must also be a gradient in anion concentration and hence in electric potential, which just balances the gradient in anion chemical potential.

$$\frac{d\Phi}{dx} = -\frac{1}{z_- F} \frac{d\mu_-}{dx} \tag{144}$$

Thus, there is effectively a gradient in the concentration of neutral species across the cell, and therefore we must include in the total potential difference a Nernstian term which is equal to the potential difference that would exist immediately after the interruption of current flow but before the reestablishment of uniform concentration profiles. The other contribution to the potential difference, that which is due to the flow of current itself, is a term arising from the gradient in conductivity due to the variations in concentration. This of course arises from differences in the mobility of the two species. It must be distinguished from the ohmic term present at very short times (high frequencies) due to the initially uniform conductivity of the electrolyte. The concentration polarization is therefore the additional polarization which is present due to concentration gradients caused by the current flow; this is compared to the ohmic polarization that would be present if the current flow (and distribution) were the same but the concentration gradients were absent.

Substitution of the condition (144) into the flux equation for cations

$$j_+ = -c_+ u_+ RT \frac{d \log c_+}{dx} - c_+ u_+ F \frac{d\Phi}{dx} \tag{145}$$

(assuming ideal solutions) gives the steady state cation current:

$$i_+ = -2u_+ RT \frac{dc^*}{dx} \tag{146}$$

The ohmic potential difference may then be found by integration of Eq. (144) across the cell using this flux equation. Since the current is constant, the concentration profile must also be uniform. The Nernstian term may be included as the potential of a concentration cell with the same concentration profile. It is possible to show that the ratio of the steady-state resistance to the high-frequency resistance depends on the transference numbers of the ions.

At large potentials this model predicts a limiting current density at which point the concentration at one of the electrodes has fallen to zero. For example in a solid or polymeric electrolyte, with plane parallel electrodes and an initial uniform concentration of c,

$$i_L = \frac{4c^* D_+ F}{l} \tag{147}$$

where l is the thickness of the electrolyte.

It is therefore apparent that in passing from high to low frequency in a system of this kind, there is an additional impedance due to concentration polarization. Macdonald and Hull [1984] considered this effect on the electrical response of this type of system. Under many circumstances, the presence of concentration polarization might be confused with an interface impedance. At different ratios of mobilities of anions and cations, either diffusion-like response (finite-length transmission line behavior) or parallel capacitative–resistive behavior may appear. Ac impedance methods have been used to determine ionic transference numbers in polymeric electrolytes using this principle (Sorensen and Jacobsen [1982]).

2.1.4 Interfaces and Boundary Conditions

2.1.4.1 Reversible and Irreversible Interfaces

Although it is quite reasonable to discuss the bulk properties of homogeneous phases in isolation, it is seldom possible in electrochemical situations to neglect the interfaces, since potentials and fluxes are usually measured or defined at junctions between ionically and electronically conducting phases. In general, two extreme types of interface are recognized.

The first type is an interface which is reversible to the species under consideration. The term *reversible* implies certain thermodynamic and kinetic properties. Thermodynamically, it means that an equilibrium relation of the type

$$ {}^{1}\Delta^{2}\eta_i = 0 \tag{148}$$

may be written for the ith species, which applies to points immediately on each side of the interface in phases 1 and 2. Here, η is the electrochemical potential. Thus, a clean interface between a parent metal M and a binary M^+ conducting solid electrolyte is thermodynamically reversible; the activities of M, M^+, and e^- are fixed by equilibrium at the interface.

Kinetically, the term is less well defined and depends more explicitly on the nature of the experiment. In practice, it means that the exchange current density (the microscopic flux crossing the interface equally in both directions at equilibrium) is very much greater than the net current density crossing the interface during the experiment or the measuring process. At appreciable current densities, however, the net current density may eventually exceed the exchange current density; interface kinetics then become important.

In the electrochemical literature it is useful to refer to a reversible interface or interfacial reaction as one whose potential is determined only by the thermodynamic potentials of the various electroactive species at the electrode surface. In other words, it is only necessary to take into account mass transport to and from the interface, and not the inherent heterogeneous kinetics of the interfacial reaction itself, when discussing the rate of the charge transfer reaction. This nomenclature has two principal disadvantages. First, it neglects the fact that mass transport to the interface, whether migration or diffusion, is inherently an irreversible or dissipative process in a thermodynamic sense. Second, it neglects the time dependence of the system. At short times the rate may be largely determined by interfacial reaction rates; at long times it may be determined by mass transport processes. This is particularly clear when ac experiments are performed; steady states may be achieved in the frequency domain that correspond to transient conditions in the time domain.

An interface may be reversible to one species, but blocking to others. In addition, in multicomponent systems, a reversible electrode may not necessarily define the thermodynamic potentials of all components present at the interface.

An electrode which is reversible to electrons but irreversible to ions is a common situation in both aqueous and solid state electrochemistry. For determinations of *ionic* conductivity in electrolytes, this type of electrode has proved useful, because the concentrations of majority ionic species do not depend critically on the imposition of a well-defined thermodynamic activity of the electroactive neutral species. Measurements with two irreversible electrodes of a nonreactive metal are then permissible; numerous examples are found in the solid-electrolyte literature. Minority electronic transport however, typically depends very strongly on the activity of neutral components, and care must be taken to utilize thermodynamically meaningful experiments to determine minority conductivities. Asymmetric cells using one reversible electrode and one irreversible electrode are then appropriate, but have actually been little explored using ac impedance methods.

A real electrode with some degree of reversibility will therefore allow a steady state current to pass; in the sense that such a current obeys Faraday's laws, it is termed a *faradic current*. A completely polarizable electrode passes no faradic current. In transient or ac experiment however, a polarizable electrode and a reversible electrode both pass a nonfaradic current, corresponding to charging or discharging of the interface capacitance and perhaps changes in the nature and concentration of any adsorbed species. The distinction between the two types of current is important in developing expressions for the impedance of the electrode–electrolyte interface.

2.1.4.2 Polarizable Electrodes

Within the voltage limits set by the thermodynamic stability range of the electrolyte, foreign metal electrodes may sometimes be regarded as ideally polarizable or blocking. The metal electrodes must not react with the electrolyte, and for the moment adsorption and underpotential deposition will be neglected. From an electrochemical point of view, this is the simplest type of interface and has furnished much of the information we have about the electrified interface.

Depending on the initial positions of the Fermi levels of the electrolyte and electrode, a small amount of charge flows in one direction or the other and a field is created on the electrolyte side of the contact. The mobile charges in the electrolyte distribute themselves over this field; the charge density of the metal is confined to the surface of the electrode. The excess charge density at any point within the electrolyte is given by Boltzmann statistics

$$\rho(x) = \Sigma z_i F c_i$$
$$= \Sigma z_i F c_i^0 \exp\left(\frac{-z_i F \Phi}{RT}\right) \tag{149}$$

and the relationship between charge and potential is given by the Poisson equation, and thus

$$\frac{d^2\Phi}{dx^2} = \frac{-1}{\varepsilon\varepsilon_0} \Sigma z_i F c_i^0 \exp\left(\frac{-z_i F \Phi}{RT}\right) \tag{150}$$

Solution of this equation leads to the space charge (diffuse double-layer) capacitance. For a symmetrical ($z_+ = z_-$) electrolyte,

$$C_d = \left[\frac{2z^2 F^2 \varepsilon\varepsilon_0 c_i^0}{RT}\right]^{1/2} \cosh\left(\frac{zF\Phi_0}{2RT}\right) \tag{151}$$

where Φ_0 is the potential at $x = 0$ (measured relative to $\Phi = 0$ at $x = \infty$). The diffuse double layer therefore behaves as a simple parallel plate capacitor. The perturbation in concentration due to the electric field extends into the electrolyte a distance on the order of the Debye length L_D:

$$L_D = \left[\frac{RT\varepsilon\varepsilon_0}{2z^2 F^2 c_i^0}\right]^{1/2} \tag{152}$$

Thus, the higher the concentration of the electrolyte, the thinner the diffuse double layer.

In addition to the use of Boltzmann statistics, the model has assumed:

a. Point charges, and hence no limit on the distance of closest approach to the interface.

b. A uniform dielectric constant.

c. A sharp boundary between the metal and the electrolyte; that is, the electronic wave functions do not extend beyond the geometrical plane of the interface.

d. No screening effects such as those found in the Debye–Hückel theory of electrolytes.

At some potential, the situation will occur where there is no excess charge on either side of the interface and the concentration profiles are flat. This point is known as the potential of zero charge.

Up to now, no real distinction has been made between solid and liquid electrolytes. In an extrinsically conducting solid, the complementary charge carrier will be absent; therefore, it will not be included in the distribution. However, the theory is basically equally applicable to solids, molten salts, and polar solvent electrolytes.

In the presence of a polar solvent molecule such as water, considerable attention has been focused on the role of the solvent. Since the dipole moment is free to rotate in the presence of an electric field, it is reasonable that in a layer of water close to the interface there will be a net dipolar orientation and the water will not exhibit its normal dielectric constant. In addition, hydrated ions will not be able to approach indefinitely close to the interface. Thus, up to a monolayer of charge will exist at a distance of closest approach to the electrode; this distance is determined by the size of the (hydrated) ion. Beyond this inner layer, the diffuse layer will extend back into the solution. The interface therefore behaves as two capacitors in series: an inner (Stern) layer and an outer (Gouy–Chapman) layer. The model takes into account both the dipolar nature of the solvent and some of the finite-size effects.

In a solid, of course, there is no solvent. However, we still expect an inner-layer capacitance, since there is still a finite distance of closest approach of the ions to the interface. Therefore,

$$C_i = \varepsilon\varepsilon_0 / d \tag{153}$$

where C_i is the capacitance per unit area. If d is of the order of a few angstroms and ε is of order 1, then C_i should lie between 1 and $10\,\mu\text{F/cm}^2$. We also expect it to be independent, or a slowly varying function, of interfacial potential difference. Since C_d, however, depends exponentially on voltage, we expect it to become large quickly, and therefore C_i will dominate the interface capacitance except when it is close to the potential of zero charge.

There are few experimental studies of the solid electrolyte–solid electrode interface carried out in such a way that meaningful potential capacitance data can be obtained. This would involve the asymmetric cell type of arrangement; for example,

M/M^+ electrolyte/inert metal

where the thermodynamic quantities of the electrolyte are fixed at the inert metal–electrolyte interface by the application of a potential difference. Some studies of this type have been performed, but there seems to be few instances of the observation of a well-defined diffuse double-layer capacitance. Most of the experiments were, however, performed on highly conducting materials, which would be expected to have very thin diffuse double layers.

A possible exception to this is the study of the graphite–AgBr interface (Kimura *et al.* [1975]), where a broad minimum in capacitance was found at potentials somewhat positive of the Ag–AgBr electrode potential at temperatures between 219 and 395°C. The minimum was somewhat broader than expected from the theoretical model, but of the correct order of magnitude. On the other hand, a similar experiment by Armstrong and Mason [1973] showed no particular minimum in capacitance at a similar temperature.

More often, the double-layer capacitance for the silver conductors seems to show a small potential dependence, more easily interpreted, at least qualitatively, in terms of an inner-layer phenomenon.

The experimental study of the solid–solid interface is complicated by a further problem. It is often (perhaps usually) observed that, instead of a purely capacitative behavior, the interface shows significant frequency dispersion. Several authors have found excellent agreement of this behavior with the dispersion shown by the constant-phase element (Bottelberghs and Broers [1976], Raistrick *et al.* [1977]). Although the amount of frequency dispersion is influenced by electrode roughness and other aspects of the quality of the interface (i.e. nonuniform current distribution), these are evidently not the only contributions to the observation of CPE behavior. Although no well-defined microscopic theory of the CPE has emerged, this empirically important aspect of both interface and bulk behavior is discussed further in this section and in Section 2.2. As is true for the case of the potential dependence of the capacitance, there have been too few studies of the frequency dispersion of the interface. It should be mentioned that the microscopically smooth liquid metal–aqueous electrolyte interfaces apparently do not show frequency dispersion of the capacitance if the systems are quite pure.

2.1.4.3 Adsorption at the Electrode–Electrolyte Interface

In the previous section the distance of closest approach of ions to a planar electrode–electrolyte interface was discussed. In solid-electrolyte systems, this distance is assumed to be approximately the radius of the mobile ion. In the presence of a polar solvent the hydration sheath of the ion and the solvent layer adjacent to the metal are also important. The only forces acting on the interface have been assumed to be electrostatic in origin. These forces orient the solvent dipoles and determine the distribution of ions with distance from the interface.

It is possible, however, that an ion can interact chemically with the electrode material. If this happens the ion may break through the solvent layers or, as in the case of the solid, become displaced from a normal lattice site. This possibility is known as specific adsorption. In aqueous electrochemistry the locus of the centers of the specifically adsorbed ions is known as the inner Helmholtz plane. Neutral molecules may also adsorb and hence affect the faradic current, for example by blockage of the reaction sites. Neutral molecule effects have not been studied in the case of solid systems and will therefore not be considered further.

In order to include adsorption in a discussion of the electrical response, it is necessary to know the relationship between the surface concentration of the adsorbed species and the concentration in the electrolyte just outside the double layer. This last concentration can then be related to the bulk or average concentration through appropriate diffusion equations.

For a neutral molecule, potential dependence will still be expected, since at large potential differences the force acting on the dipole of a polar solvent will be sufficient to compete with all but the strongest adsorption bonds.

A simple isotherm, due originally to Langmuir, assumes that the free energy of adsorption ΔG_i^0 is the same all over the surface and that interactions between adsorbed species are neglected. Under these conditions, the surface concentration Γ_i is related to the surface concentration at full coverage Γ_0 by the expression

$$\frac{\Gamma}{\Gamma_0 - \Gamma} = \frac{\theta}{1-\theta} = a_i^b \exp\left(\frac{-\Delta G_i^0}{RT}\right) \exp\left(\frac{-\Phi z_i F}{RT}\right) \tag{154}$$

where a_i^b is the bulk activity of i.

The capacitance associated with the adsorption can be obtained by differentiation of the charge due to the adsorbed species:

$$q = \theta q_i \tag{155}$$

where q_i is the charge corresponding to one monolayer:

$$C = dq/d\Phi = (dq/d\theta)(d\theta/d\Phi) = q_i(z_i F/RT)\theta(1-\theta) \tag{156}$$

Various attempts have been made to include interactions between adsorbed species. As pointed out by Conway *et al.* [1984], the correct way to handle interactions is to include the appropriate pairwise or long-range interaction term into the partition function, which allows calculation of the Helmholtz free energy and the chemical potential. These quantities are a function of θ due to (a) the configurational term, as included in the Langmuir case; and (b) the interaction or deviation from ideality.

As an example, Frumkin's isotherm may be derived by assuming a pairwise interaction of the form

$$U(\theta) = r\theta^2/2 \tag{157}$$

where r is positive for a repulsive interaction and negative for an attractive force. This leads to

$$\frac{\theta}{1-\theta} = a_i^b \exp\left(\frac{-\Delta G_i^0 - r\theta}{RT}\right)\exp\left(\frac{-\Phi z F}{RT}\right) \tag{158}$$

This yields a capacitance of the form

$$C(\theta) = q_i\left(\frac{z_i F}{RT}\right)\cdot\frac{\theta(1-\theta)}{1+r\theta(1-\theta)} \tag{159}$$

Comparing this expression with Eq. (156) indicates that the new capacitance expression can be expressed as a "Langmuir" capacitance in series with an "interaction" capacitance:

$$C(\theta) = \left(C_L^{-1} + C_I^{-1}\right)^{-1} \tag{160}$$

Other expressions for different forms of the interaction term have been given by Conway *et al.* [1984].

The Temkin isotherm attempts to account for heterogeneity of the electrode surface by making the energy of adsorption vary linearly with coverage, which gives

$$\exp(r\theta) = K a_i^b \exp(z_i F\Phi/RT) \tag{161}$$

and

$$C(\theta) = q_i(z_i F/RT)\cdot 1/r \tag{162}$$

The rates of adsorption are usually rapid and hence the kinetics are determined by other electrochemical or chemical steps and mass transport. Armstrong has pointed out that in solid-electrolyte systems, where the interfacial potential difference cannot be varied independently of the concentration of the mobile species, the adsorption of that species cannot be controlled by a diffusional process.

Raleigh [1976] has put forward a model of competitive chemisorption of anions and cations in silver halides that leads to a broad maximum in capacitance at the potential of zero charge, in agreement with observations on some of these compounds. This approach is greatly extended in Macdonald *et al.* [1980].

The kinetics of complex electrochemical reactions in the presence of adsorbed intermediates and its effect on the impedance of the interface is discussed in Section 2.1.4.4.

2.1.4.4 Charge Transfer at the Electrode–Electrolyte Interface

The rate of heterogeneous charge transfer reaction

$$O + ne = R \tag{163}$$

is given by the expression

$$-i_F = nF[k_f c_O - k_b c_R] \tag{164}$$

where i_F is the faradic current density, $k_{f,b}$ are the forward and reverse rate constants, and $c_{O,R}$ are the concentrations of the reactants and products at the interface at time t.

The current, in general, is composed of a steady state or dc part determined by the mean dc potential E and the mean dc concentrations at the interface, c_O and c_R, and an ac part, Δi_F, determined by the ac perturbing potential ΔE and the fluctuating concentrations Δc_i. The faradic impedance is given by the ratio of the Laplace transforms of the ac parts of the voltage and current

$$Z_F = \{\Delta E\}/\{\Delta i_F\} \tag{165}$$

Because charge transfer is involved, the presence of an electric field at the interface affects the energies of the various species differently as they approach the interfacial region. In other words, the activation energy barrier for the reaction depends on the potential difference across the interface. It is convenient to express the potential dependence of the rate constants in the following manner:

$$k_f = k_0 \exp -\alpha(E - E^0)nF/RT \tag{166}$$

$$k_b = k_0 \exp(1-\alpha)(E - E^0)nF/RT \tag{167}$$

where k_0 is the rate constant at the formal electrode potential E^0 and α is the apparent cathodic transfer coefficient. Hence

$$\frac{k_f}{k_b} = \exp \frac{nF}{RT}(E - E^0) \tag{168}$$

Generally, we can express Δi_F as an expansion of the ac parts of the concentrations and electrode potential,

$$\Delta i_F = \Sigma\left(\frac{\partial i_F}{\partial c_i}\right)\Delta c_i + \left(\frac{\partial i_F}{\partial E}\right)\Delta E + \text{higher-order terms} \tag{169}$$

Neglecting all but the first-order terms (linearization) and solving for ΔE,

$$-\Delta E = \frac{1}{(\partial i_F/\partial E)}\left[\Sigma\left(\frac{\partial i_F}{\partial c_i}\right)\Delta c_i - \Delta i_F\right] \tag{170}$$

and hence

$$Z_F = \frac{1}{(\partial i_F/\partial E)}\left[1 - \Sigma\left(\frac{\partial i_F}{\partial c_i}\right)\frac{\{\Delta c_i\}}{\{\Delta i_F\}}\right] \tag{171}$$

The first term is the *charge transfer resistance*; the second term contains the influence of the ac part of the mass transport on the impedance. Here $\{\Delta c_i\}/\{\Delta i_F\}$ can be expressed as a solution of the diffusion equation. For example, for semiinfinite diffusion to a plane, we can use Eq. (126):

$$\frac{\{\Delta c_i\}}{\{\Delta i_F\}} = \frac{1}{nF\sqrt{(pD_i)}} \tag{172}$$

The coefficients in parentheses may be evaluated from the rate expressions discussed above,

$$\left(\frac{\partial i_F}{\partial E}\right) = k_f \frac{n^2F^2}{RT}\left[\alpha c_O + (1-\alpha)c_R \exp\frac{nF}{RT}(E - E^0)\right] \tag{173}$$

and

$$-\left(\frac{\partial i_F}{\partial c_O}\right) = nFk_f, \quad -\left(\frac{\partial i_F}{\partial c_R}\right) = -nFk_f \exp\frac{nF}{RT}(E - E^0) \tag{174}$$

Here c_O and c_R are determined by the solution of the appropriate dc mass transport equations. The coefficients may then be substituted into Eq. (166) to give the overall faradic impedance.

At the equilibrium potential E_r, the net current is zero; therefore, c_O and c_R are equal to their bulk values c_O^* and c_R^* and are related through the Nernst equation

$$\frac{c_O^*}{c_R^*} = \exp\frac{nF}{RT}(E_r - E^0) \tag{175}$$

Under these circumstances, the coefficient $(\partial i_F/\partial E)$ simplifies to give

$$\frac{\partial i_F}{\partial E} = \frac{n^2F^2}{RT}k^0 \exp\left[-\frac{\alpha nF}{RT}(E_r - E^0)c_O^*\right] \tag{176}$$

and the charge transfer resistance is

$$r_{ct} = RT/nFi_0 \tag{177}$$

where the exchange current density is

$$\begin{aligned} i_0 &= nFk^0 c_O^* \exp[-\alpha(nF/RT)(E_r - E^0)] \\ &= nFk^0 c_O^{*(1-\alpha)} c_R^{*a} \end{aligned} \tag{178}$$

When mass transport to the electrode is unimportant, substitution of Eq. (178) into Eq. (164) gives the Butler–Volmer equation

$$-i_F = i_0\left[\exp-\frac{nF}{RT}\alpha(E - E_r) - \exp\frac{nF}{RT}(1-\alpha)(E - E_r)\right] \tag{179}$$

When E is sufficiently far removed from E_r, the current in one direction may be neglected, leading to the Tafel relation

$$(E - E_r) = a + b \log i \tag{180}$$

It should be emphasized that the development of the expressions for charge transfer kinetics given here is not completely general. It rests on the assumptions of absolute rate theory. More general treatments have been given in the literature where no *a priori* assumption of the form of the dependence of the rate constants on potential is made (Birke [1971], Holub *et al.* [1967]). A point which arises from these more general treatments is worth pursuing here. For the case of semiinfinite diffusion to a planar interface, the faradic impedance may be written in the form

$$Z_F = r_{ct} + (\sigma_0 + \sigma_R)(1/\sqrt{\omega})(1 - j) \tag{181}$$

where r_{ct} is given by the inverse of Eq. (168), and σ_0 and σ_R are of the form

$$\sigma_O = \frac{RT}{n^2 F^2 \sqrt{2}} \cdot \frac{1/\sqrt{D_0}}{\alpha c_O + (1-\alpha) c_R \exp(nF/RT)(E - E^0)} \tag{182}$$

and

$$\sigma_R = \frac{RT}{n^2 F^2 \sqrt{2}} \cdot \frac{(1/\sqrt{D_R}) \exp(nF/RT)(E - E^0)}{\alpha c_O + (1-\alpha) c_R \exp(nF/RT)(E - E^0)} \tag{183}$$

The terms in $\sigma_i/\sqrt{\omega}$ correspond to the normal Warburg impedance; they do not contain the heterogeneous rate constants. The more general treatments, however, indicate that the Warburg impedance does in general contain coefficients that depend on the rate constants and their potential dependence. It is only on the basis of absolute rate theory that these coefficients cancel out of the final expression.

The complete equivalent circuit for a single-step charge transfer reaction in the presence of diffusion is given in Figure 2.1.14. The electrolyte resistance and double-layer capacitance have also been added to this figure.

A second aspect of the theory developed in this section is the assumption that the faradic current is decoupled from the nonfaradic current. In other words, the impedance due to the double-layer capacitance is included afterward and placed in parallel to the faradic impedance, since

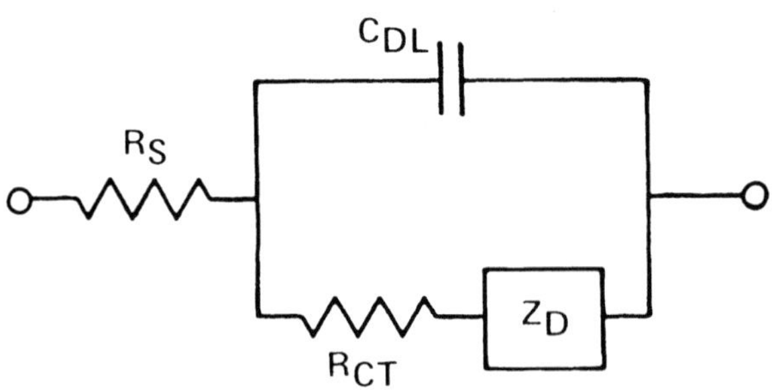

Figure 2.1.14. The Randles equivalent circuit, which describes the response of a single-step charge-transfer process with diffusion of reactants and/or products to the interface.

$$i = i_F + i_{NF} \tag{184}$$

In general, however,

$$i_{NF} = \left(\frac{dq}{dt}\right) = \left(\frac{\partial q}{\partial E}\right)\left(\frac{\partial E}{\partial t}\right) + \Sigma\left(\frac{\partial q}{\partial c_i}\right)\frac{\partial \Delta c_i}{\partial t} \tag{185}$$

where the summation extends over all species, including O and R. Thus, the nonfaradic component is coupled to the faradic current unless experimental steps are taken to decouple them. This is usually achieved by making the concentration of electroactive ions very small compared to the inactive charge carriers, which do most of the double-layer charging.

In solid electrolytes, however, the unsupported electroactive species is often the sole charge carrier. It is thus impossible to change the interfacial potential difference without changing the concentration of ions in the double-layer region. This means of course, that the normal Warburg impedance is not seen, but it also means that there is a coupling between the faradic current and the double-layer charging.

This has been recognized by Armstrong [1974], who has proposed the rate equation

$$i = i_0\left[1 + \frac{C_{dl}(E - E_r)}{|q_-| - \Delta E C_{dl}}\right]\exp\frac{nF}{RT}\alpha(E - E_r) \tag{186}$$

for metal deposition from a solid electrolyte. Here, ΔE is the difference between E and the potential of zero charge, q_- is the charge density of anions in the inner layer, and C_{dl} is taken independent of voltage.

The treatment given above for a single-step charge-transfer reaction may be readily extended to more complex reaction schemes. For a multistep reaction, the partial currents of the individual steps must be appropriately coupled and the mass transport relations defined for each step.

For example, for a surface-adsorbed species, intermediate in a two-step reaction, there will be an additional relationship of the type

$$d\Gamma/dt = \Delta i_F^1/n_1 F - \Delta i_F^2/n_2 F \tag{187}$$

where Γ is the surface concentration of the adsorbed intermediate produced by reaction 1 and removed by reaction 2. This case is of considerable importance in aqueous electrochemistry. The form of the impedance and the expected equivalent circuits have been discussed by Gerischer and Mehl [1955], Armstrong and Henderson [1972], Grahame [1952], and Epelboin and Keddam [1970]. In the absence of mass transport control, the equivalent circuit is of the form shown in Figure 2.1.15. Here R_{CT} is a charge transfer resistance, and R_{ADS} and C_{ADS} are components which contain the contribution of the surface concentration (coverge) of the adsorbed intermediate and the rate of adsorption or desorption, respectively. Under certain circumstances R_{ADS} and C_{ADS} can become negative, leading to the appearance of inductive behavior in the impedance spectrum.

As applications of impedance spectroscopy to very complex reactions in the solid state have not yet been made, further development of the theory of the faradic

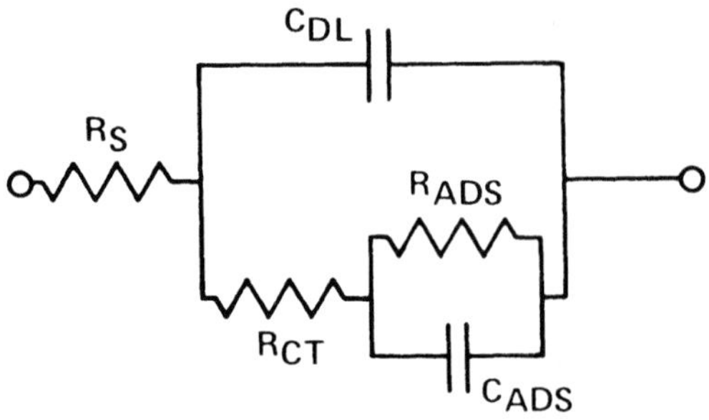

Figure 2.1.15. An equivalent circuit which describes the electrical response of an electrochemical reaction with a strongly adsorbed intermediate.

impedance seems unwarranted here. The linear operator approach to calculating the impedance of the systems is due to Rangarajan [1974] and is well described in the review of Sluyters-Rehbach and Sluyters [1984].

2.1.5 Grain Boundary Effects

It was suggested earlier that the electrical analog of an isotropic, homogeneous, ionically conducting solid is a pure resistance in parallel with a high-frequency ideal capacitor. This model assumes the absence of electrode polarization and of relaxation processes within the crystal that would lead to additional parallel branches in the equivalent circuit. This model is generally accepted, and several studies of single-crystal materials have demonstrated its validity.

Many solids are, however, studied in polycrystalline form, either because they are only available as such or because this is the manner in which they will be utilized. Polycrystalline materials usually have less than theoretical density (voidage) and misorientated grains (important in anisotropic materials). In the simplest case, these effects would lead to purely geometric reductions in the conductivity with respect to the single crystal. In addition, impurities may be present as a second phase at the grain boundaries. Because of the importance of ac impedance spectroscopy as a tool for measuring ionic conductivity, there have been several studies of the effect of polycrystallinity on the impedance of solid electrolytes.

The problem was first attacked in a modern manner by Bauerle [1969], whose paper was the first application of impedance spectroscopy to solid electrolytes. He studied both high-purity and "impure" polycrystalline zirconia. Bauerle found that the presence of a second phase at the grain boundaries in dense material led to the introduction of a second time constant in the equivalent circuit. This additional impedance was absent in the very-high-purity material. Bauerle envisaged the ionically insulating second phase as introducing a constriction in the area of contact between the grains of the highly conducting phase. Beekmans and Heyne [1976] found similar behavior in calcia-stabilized zirconia and suggested as well that a distribution of time constants for the grain boundary behavior was appropriate, rather than the single—RC—time constant as was suggested by Bauerle.

Later, it became apparent that a second phase need not be present in polycrystalline materials for the grain boundaries to make a contribution to the impedance of the system. There have been several studies of the impedance of polycrystalline sodium β-alumina, a very nonisotropic solid electrolyte. Hooper [1977] systematically studied the relationship between single-crystal and polycrystalline material and showed that it was possible to extract "true" bulk values from polycrystalline samples. This intragrain conductivity had the same activation energy as the single-crystal material; there still was, however, a relatively small difference between the absolute conductivity values, probably mostly due to the geometric effects introduced by the anisotropy of the material, and preferential orientation in the pressed samples. Grain boundary (intergrain) conductivity had a greater activation energy and disappeared at high temperatures.

There have also been several studies of the more isotropically conducting materials based on the Li_4SiO_4 and γ-II Li_3PO_4 ("LISICON") structures (Ho [1980] and Bruce and West [1983]). Ho varied the density of polycrystalline $Li_{4+y}Si_{1-y}Al_yO_4$ from nearly the theoretical density down to about 60% of theoretical density. At the highest densities, only a single circular arc was seen in the impedance plots, but at lower densities two arcs became apparent. The resistance associated with the lower-frequency arc exhibited a higher activation energy than that associated with the higher-frequency arc, which was attributed to intergrain impedance. As with Hooper's study of β-alumina, this contribution disappeared at higher temperatures. Ho's study and the 1976 work of Raistrick *et al.* [1976] on other polycrystalline alumino silicates noted, however, that except for the very densest of materials, polycrystalline samples always showed some anomalous frequency dispersion. The circuit element now often known as a constant-phase element (CPE) was introduced to fit the data

$$Y^*_{\text{CPE}} = A(j\omega)^{\alpha} \tag{188}$$

Each of the circular arcs found in the polycrystalline materials was of the form shown in Figure 2.1.16. It seems that polycrystallinity introduces anomalous frequency dispersion into the bulk impedance behavior before a second and separate

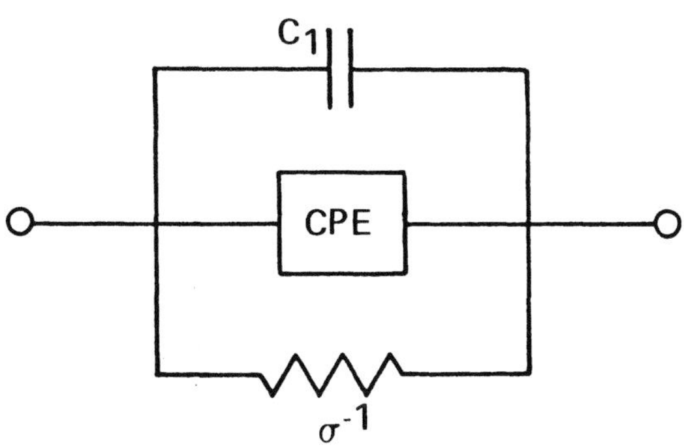

Figure 2.1.16. An equivalent circuit which describes the electrical response of polycrystalline solid electrolytes.

contribution from an intergrain impedance appears. A study by Bruce and West [1983] of polycrystalline LISICON essentially reached the same conclusions. Unlike Ho, however, the same activation energy was found for both the inter- and intragrain resistances. This suggests that essentially the same physical processes are involved and that the authors attributed the intergrain impedance to a constriction effect not unlike that proposed by Bauerle. Bruce and West, however, attributed the constriction to the smaller area of contact between grains rather than to the presence of an ionically insulating second phase. A recent investigation of the effect of a single-grain boundary on the response of an otherwise single-crystal CaO-doped CeO_2 specimen also found two semicircles in the impedance plane (El Adham and Hammou [1983]). Depending on the presence or absence of CaO enrichment at the grain boundary, a different activation energy was present for the resistance associated with the boundary. A detailed modeling of the properties of a constriction resistance of this type should be possible, but does not yet appear to have been carried out. It is also possible that close to a grain boundary, the transport properties of the crystal are controlled by imperfections, expected to be present there in higher concentration that in the center of a grain, leading to an additional contribution to the intergrain impedance. This idea is very close to the observation of conductivity enhancement due to heterogeneous doping of solid electrolytes. Here, a second insoluble, nonconducting phase is introduced into the solid electrolyte as finely dispersed particles. The internal space charge created at the phase boundaries may lead to a significant increase in the concentration of mobile defects. A detailed impedance study of such a system would be of considerable interest, but does not yet appear to have been carried out.

2.1.6 Current Distribution, Porous and Rough Electrodes—the Effect of Geometry

2.1.6.1 Current Distribution Problems (Ibl [1983b])

It was pointed out in Section 2.1.2.4 that in most electrochemical systems, including those situations in which there is a concentration gradient but in which regions close to interfaces between different phases are excluded, electroneutrality is a reasonable assumption. Under these circumstances, the potential variation is given by Laplace's equation

$$\nabla^2\Phi = 0 \tag{189}$$

In principle, this equation may be solved subject to the following boundary conditions:

$$\Phi = \text{constant (conducting boundary)}$$
$$\partial\Phi/\partial n = \text{constant (insulating boundary)}$$

where n is the normal to the boundary.

The potential and current distributions derived from these boundary conditions are called the *primary distribution*. They depend only on the geometry of the system. For very simple geometries, analytical solutions to Laplace's equation have been found. In recent years, numerical solutions have often become the preferred method, and both finite-difference and finite-element methods, as well as techniques based on Green's function methods, are valuable. Equation (189) corresponds to steady temperature in heat conduction problems; a useful discussion is found in Carslaw and Jaeger [1959].

Often, in solid state experiments, the most common experimental arrangement is the most satisfactory from a primary current distribution point of view. The electrodes completely cover the ends of the electrolyte and there is no spreading of the current lines. The current distribution should, however, be considered whenever more complex geometries are involved or the placement of a reference electrode is in question.

The real importance of current distribution problems in impedance measurements, however, lies in the fact that the distribution is frequency-dependent. This arises because of the influence of interfacial polarization combined with the geometrical aspects of the arrangement.

The electrode–electrolyte interface is not an equipotential surface. This is because the interfacial potential difference is typically a function of the local current density. Thus, even for a purely resistive interface impedance—independent of local current density—there is a smoothing effect on the current distribution in the system. The greater the current density at a particular point, the greater the potential drop across the interface, which in turn tends to lower the local current density. The tendency is therefore to make a uniform current distribution. The magnitude of the smoothing depends on the relative magnitudes of the interface and bulk impedances, as well as the geometry of the system. The current distribution in the presence of interfacial polarization (but neglecting mass transport effects) is called the *secondary distribution*.

In general, interfacial impedance is partly capacitative as well as resistive in nature. At high frequencies, the capacitance short-circuits the interface, and the primary distribution is observed for the ac part of the current. As the frequency is lowered, the interface impedance increases, causing a changeover to the secondary distribution. Of necessity, this effect leads to a frequency dependence of the equivalent circuit parameters which describe the system. Of course, if the primary distribution is uniform, there will be no frequency dispersion arising from this source.

The question of the frequency dependence of the current distribution and its effect on the measured impedance of a solid state electrochemical system has been hardly considered, although it is important in discussing the impedance of, for example, porous gas electrodes on anion conductors, of rough electrodes (discussed below), and also perhaps of polycrystalline materials. In aqueous electrochemical situations the effects has been considered with respect to the rotating disk electrode, where there may be severe current distribution problems.

2.1.6.2 Rough and Porous Electrodes

It is recognized that porosity or roughness of the electrode surface could be expected to lead to a frequency dispersion of the interfacial impedance even in the absence of detailed considerations of the current distribution problems as outlined above.

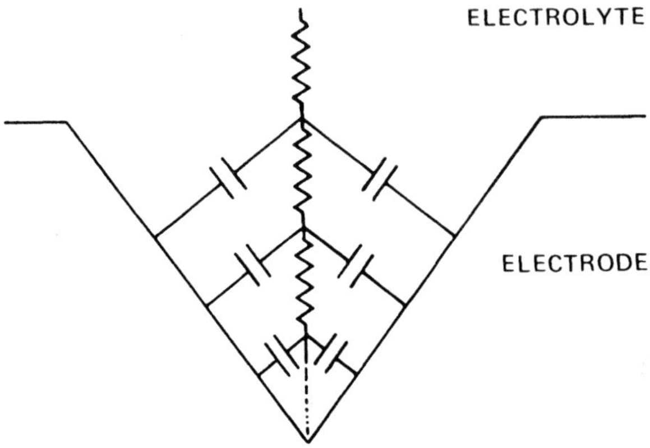

Figure 2.1.17. Transmission line model for a V-shaped groove in an electrode surface.

A simple approach to the problem of porous and rough interfaces is based on the use of transmission line analogies (de Levie [1967]). Consider a cylindrical pore in a conducting electrode. If the series resistance of the electrolyte per unit length is r, and the interfacial capacitance per unit length is c, then the pore behaves as a transmission line and has an impedance given by Eq. (135). This approach can be extended to more complex situations which include pores of finite depth, nonuniform pores, and situations where the interfacial capacitance is replaced by a complex faradic admittance, corresponding to an electrochemical reaction taking place down the depth of the pore. It is also possible to include finite electrode resistance.

In order to describe a rough electrode, de Levie [1965] suggested a model based on the V-shaped groove shown in Figure 2.1.17. It was assumed that the double-layer capacitance was uniform over the true surface of the electrode and that the current lines were perpendicular to the *macroscopic* surface of the electrode. The impedance of the groove is then analogous to a transmission line in which the components are a function of the distance into the groove. The transmission line equation may be readily modified for the present case, where r and c are functions of x, which is the distance down the line:

$$I = -[1/r(x)]\partial V/\partial x \tag{190}$$

$$\partial V/\partial t = -[1/c(x)]\partial I/\partial x \tag{191}$$

On combining these equations, we obtain

$$-c(x)\frac{\partial V}{\partial t} = \frac{1}{r(x)^2}\cdot\left[\frac{\partial V}{\partial x}\frac{\partial r(x)}{\partial x} - r(x)\frac{\partial^2 V}{\partial x^2}\right] \tag{192}$$

The Laplace transform of this equation is, like the diffusion equation, an ordinary differential equation

$$r^2 \frac{d^2\{V\}}{dx^2} - r\frac{dr}{dx} \cdot \frac{d\{V\}}{dx} - r^3 cp\{V\} = 0 \tag{193}$$

For r and c simple functions of x, the equation becomes a modified Bessel equation and may be solved analytically. For example, in the case of the groove geometry, as considered by de Levie,

$$Z = (\rho/\tan\beta)I_0(\lambda)/\lambda I_1(\lambda) \tag{194}$$

where Z is the impedance, 2β is the angle at the apex of the groove, ρ is the specific solution resistivity, and λ is $2\sqrt{(\rho l p\kappa/\sin\beta)}$; κ is the interface capacitance per unit area. In the limit of low and high frequency, the phase angle of the impedance changes from $\pi/2$ to $\pi/4$.

Using this approach, a number of different geometries can be analyzed (Keiser et al. [1976]). The interface impedance, considered in the example above to be a pure capacitance, could be generalized to include both real and imaginary components, for example a Warburg impedance. In general, however, ρ and κ would become functions of distance into the groove if significant diffusional effects were included in the calculation, and the diffusion layer thickness, relative to the thickness of the surface features, becomes important.

As pointed out by de Levie, however, the most important weakness in the model is the assumption that the current distribution is normal to the macroscopic surface, that is a neglect of the true current distribution. For a rough surface, the lines of electric force do not converge evenly on the surface. The double layer will therefore be charged unevenly, and the admittance will be time and frequency dependence.

The tangential components of the interface charging were recognized and included in a qualitative model by Scheider [1975], who suggested the use of branched transmission lines to model the effects of uneven surface topology. The suggestion was significant in that transmission lines of this type do represent a circuit which, unlike the unbranched transmission line of de Levie, agrees with experimental observations of the impedance at rough electrodes.

The basic type of line suggested by Scheider is shown in Figure 2.1.18 for a single type of branching. In general, both the series (as shown) and the parallel components of a simple—RC—line may be replaced by other transmission lines, which may themselves be branched. The degree of branching may be unlimited. Let the series impedance per unit length be z and the parallel admittance per unit length be y (Figure 2.1.19). In addition, let all the z and all the y be independent of distance down the line. Using the continued-fraction approach to write the total impedance Z_T,

$$Z_T = \cfrac{1}{y + \cfrac{1}{z + \cfrac{1}{y + \cdots}}} \tag{195}$$

For an infinite line,

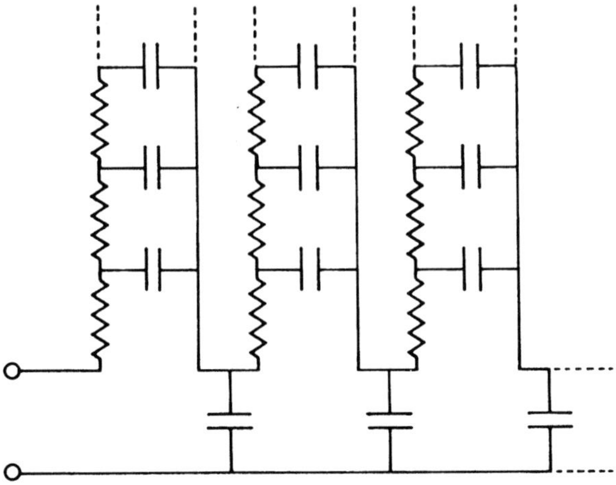

Figure 2.1.18. A branched transmission line circuit which shows CPE behavior.

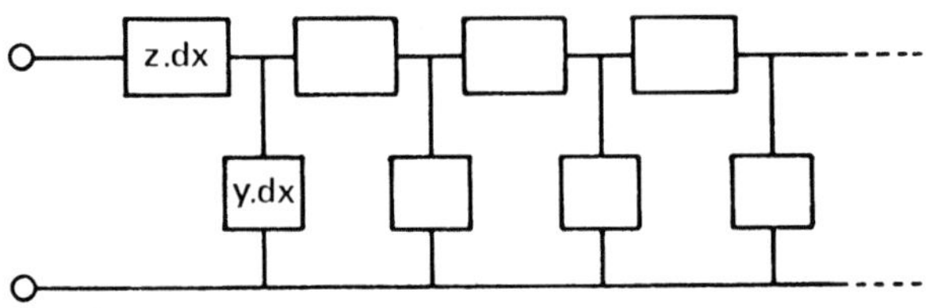

Figure 2.1.19. A generalized transmission line where z and y are respectively the series impedance and interfacial admittance per unit length.

$$Z_T = \frac{1}{y + \dfrac{1}{z + Z_T}} \tag{196}$$

or

$$Z_T = -\frac{z}{2} + \sqrt{\frac{z^2}{4} + \frac{z}{y}} \tag{197}$$

In the limit of $dx \to 0$, $z, y \to 0$ but z/y is finite. Therefore,

$$Z_T = \sqrt{\frac{z}{y}} \tag{198}$$

Suppose, if z represents a transmission line $z = A(j\omega)^{-1/2}$, and y is a pure capacitance $y = j\omega C$, then

$$Z_T = \sqrt{\left(\frac{A}{C}\right)(j\omega)^{-3/2}} \tag{199}$$

Evidently, a first-order branching of the series component leads to an impedance with a phase angle of (3/4) ($\pi/2$).

In fact, the value of the frequency exponents is limited only by the degree of branching of the circuit. The interval between frequency exponents is

$$[1/2]^{\theta+1}$$

where θ is the order of the branching. The frequency dependence is determined by the branching type and not by the magnitude of the components.

The important feature of these lines is that they produce a constant phase angle, like a Warburg impedance, but with the phase angle not restricted to $\pi/4$. This is exactly the behavior often found at the electrode–electrolyte interface and has been termed a *constant-phase element* (CPE). It appears to be true that roughness is an important contributing factor to the observed frequency dispersion. Scheider's model, however, remains qualitative, and the microscopic link between the topology and the circuit is absent.

In general, a transmission line with nonuniform components, such as that described by Eq. (192), does not lead to CPE behavior. Schrama [1957] has shown that for lines with a particular type of nonuniformity, CPE behavior is predicted. This relationship is, for a discrete—RC—line,

$$R_k = \frac{2\Gamma(1-\alpha)}{\Gamma(\alpha)} \cdot \frac{\Gamma(\alpha+\kappa)}{\Gamma(1-\alpha+\kappa)} \cdot h^{\alpha} \tag{200}$$

$$C_k = (2\kappa+1)\frac{\Gamma(\alpha)}{\Gamma(1-\alpha)} \cdot \frac{\Gamma(1-\alpha+\kappa)}{\Gamma(1+\alpha+\kappa)} \cdot h^{1-\alpha} \tag{201}$$

where h is a positive real number. Schrama suggested an interpretation in terms of nonuniform diffusion coefficients and driving forces.

It has been shown (Liu [1985]) that at a fractal interface, a nonuniform transmission line will model the electrical response. The fractal geometry assumed was that of the triadic bar of Cantor, illustrated schematically in Figure 2.1.20*a*. The

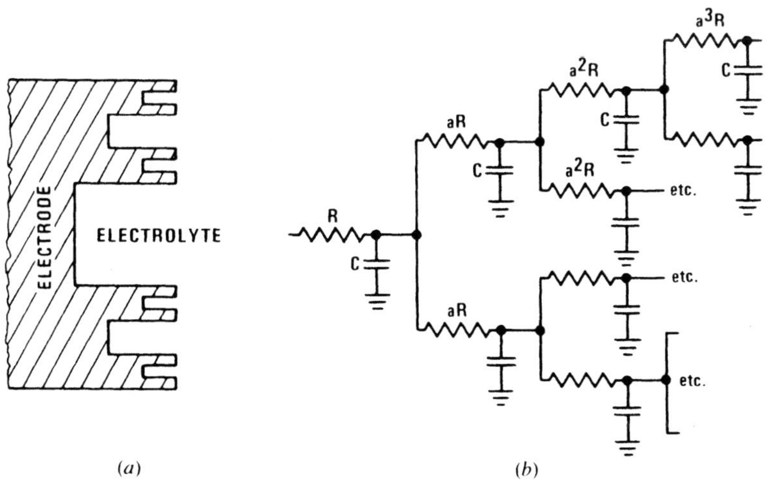

Figure 2.1.20. (*a*) Formal model for a fractal electrode–electrolyte interface, and (*b*) an equivalent circuit which models the double-layer charging behavior.

equivalent circuit corresponding to such an interface is shown in Figure 2.1.20*b*. It is evident that the circuit may be folded over into an—RC—transmission line with the resistance and capacitance per unit length related to one another in a specific manner.

The possible relationship between CPE behavior and a fractal interface geometry have also been emphasized by Le Mehaute [1984] and Le Mehaute and Crepy [1983].

This section was written by Ian D. Raistrick and updated by J. R. Macdonald and D. R. Franceschetti. The U.S. government retains a royalty-free license to reproduce or reuse this material for in-house or governmental purpose.

2.2 PHYSICAL AND ELECTROCHEMICAL MODELS

Donald R. Franceschetti
J. Ross Macdonald

2.2.1 The Modeling of Electrochemical Systems

Numerous theoretical models have been developed to explain and predict the behavior of electrochemical systems and to guide the design of systems with desired characteristics. The models which have been developed thus far fall generally into three broad categories, corresponding to three different levels of description of the system. From a practical standpoint, no one group of models is more important than another: the models which may be helpful to a materials scientist in fabricating a new solid electrolyte may provide no assistance at all to the engineer attempting to design a portable power source which meets rigid cost and performance specifications.

At the most fundamental level of description are atomistic or microscopic models which attempt to provide an accurate description of the motions of individual charge-carrying particles in the system. At the least detailed level are the equivalent circuit models, in which hypothetical electrical circuits, consisting of elements with well-defined electrical properties, are used to describe the response of the system to a range of possible signals. Such models are of special interest in impedance spectroscopy, since the frequency response behavior of linear electrical circuits is now extremely well understood. An important variant of the equivalent circuit is the equivalent electrical network, composed of simple circuit elements of differential size. An introduction to equivalent circuit models is provided in Section 2.2.2 below. The electrical network approach is discussed in 2.2.3.4. The level intermediate between equivalent circuits or networks and microscopic models might be termed the *continuum level*, as the bulk regions of the electrodes and electrolyte are regarded as continuous media. The transport of mobile species is thus assumed to be governed by differential equations, and the transfer of charge across interfaces obeys rate laws which serve as boundary conditions for these equations. The parameters of the continuum model can be regarded as averages of the parameters appearing in an atomistic model. As can be seen in Section 2.2.3, the continuum parameters can usually be related to the parameters of an equivalent circuit model, and the analysis of the continuum model can thus guide in the construction of an appropriate equivalent circuit.

2.2.2 Equivalent Circuits

2.2.2.1 *Unification of Immittance Responses*

For a long time, dimensionless normalization has been used in the dielectric constant measurement area of IS. As in Eq. (1) in Section 1.3, one supposes that there exists a low-frequency limiting value of the dielectric constant, ε_s, and a high-frequency limiting value ε_∞. In the latter case there may be even higher-frequency dispersions than that whose limit is ε_∞, but in ordinary IS it is usually sufficient to establish (or assume!) that $\varepsilon = \varepsilon_\infty$ over a wide range of high frequencies, and primary attention is directed to the response in the frequency region between $\varepsilon = \varepsilon_s$ and ε_∞. Then the normalized response may be written

$$k \equiv \frac{\varepsilon - \varepsilon_\infty}{\varepsilon_S - \varepsilon_\infty} \tag{1}$$

where $\kappa \rightarrow 1$ as $\nu \rightarrow 0$ and $\kappa \rightarrow 0$ as $\nu \rightarrow \infty$ (or to the region where $\varepsilon \cong \varepsilon_\infty$). The function κ is a normalized immittance, defined at the dielectric constant level.

For solid electrolytes one usually is concerned with intrinsically conducting systems rather than with intrinsically nonconducting (dielectric) ones. It is then appropriate and usual to consider basic system response at the impedance rather than the complex dielectric constant level. Then if one assumes that the overall impedance of the system, Z_z, approaches R_0 at sufficiently low frequencies and R_∞ at sufficiently high ones, one can form the normalized dimensionless quantity

$$I_Z \equiv \frac{Z_Z - R_\infty}{R_0 - R_\infty} \tag{2}$$

in analogy to Eq. (1). We have given a specific expression for I_z in Eq. (6) of Section 1.3. Again, $I_z \rightarrow 1$ as $\nu \rightarrow 0$ and $I_z \rightarrow 0$ as $\nu \rightarrow \infty$.

Since Eqs (1) and (2) are similar in form, we can combine them in the single expression

$$I_k = \frac{U_k - U_{k\infty}}{U_{k0} - U_{k\infty}} \tag{3}$$

where $k = \varepsilon$ or Z, $U_\varepsilon = \varepsilon^* = \varepsilon' + j\varepsilon''$, and $U_Z = Z = Z' + jZ''$. As before, U_{k0} and $U_{k\infty}$ are, respectively, the low-and highs-frequency limiting real values of U_k. We actually use ε^*, the complex conjugate of ε, rather than ε here so that the imaginary parts of I_ε and I_z may be defined with the same sign.

Now, as already mentioned in Section 1.3, the I_Z of that section's Eq. (6) is of just the same form as the well-known Cole–Cole dielectric dispersion response function (Cole and Cole [1941]). In its normalized form, the same I_k function can thus apply at either the impedance or the complex dielectric constant level. We may generalize this result (J. R Macdonald [1985a,c,d]) by asserting that *any* IS response

Table 2.2.1. Relations Between the General, Unified Immittance Function I and Specific System Functions

Conductive System ($k = Z$)	General System Normalized	Dielectric System ($k = \varepsilon$)
M	μI_k	Y
Z	I_k	ε
Y	I_k^{-1}	M
ε	$(\mu I_k)^{-1}$	Z

Note: Here $\mu \equiv j\omega C_c$.

function that can be normalized as in Eq. (3) may be used at either the complex dielectric constant level or at the impedance level. It is very important to note that when the same function (with possibly different parameter values) is applied at both the complex dielectric constant and the impedance levels, it defines different systems at these levels. This matter is discussed in more detail in J. R. Macdonald [1985c]. Thus a theoretical derivation of response at one level automatically yields response of the same kind for the other level but applying to a different type of system. When the form of Eq. (3) is used to discuss bulk properties we will use I_ρ for the complex, frequency-dependent resistivity and I_σ for the complex frequency-dependent conductivity.

The above results allow us to use a single general $I(\omega)$ function to represent normalized response at either the ε or Z system levels. When the k subscript of I_k is omitted, it will be understood to be general in this sense. The use of the general normalized immittance response function I allows one to subsume two kinds of systems and response with a single function and will be so employed in the next section. Table 2.2.1 shows how I_k, for $k = Z$ or ε, is related for the various immittance levels to the specific conductive and dielectric systems functions. Here, as before, $\mu \equiv j\omega C_c$. Alternatively, to maintain dimensionless quantities at all levels, one might replace μ by $j\omega\tau \equiv js$, where τ is a specific relaxation time and $s \equiv \omega\tau$. Note that all functions are simply related to I. Let us illustrate these relations with a specific example. Take, for concreteness, $I_k = F_k(\psi_{kc}, s)$, the specific response function of Eq. (6) in Section 1.3. Then we can write

$$\varepsilon_\varepsilon = \varepsilon_\infty + (\varepsilon_S - \varepsilon_\infty)F_\varepsilon(\psi_{\varepsilon C}, s) \tag{4}$$

$$Y_\varepsilon = i\omega C_c \varepsilon_\varepsilon \tag{5}$$

$$Z_\varepsilon = \{i\omega C_C[\varepsilon_\infty + (\varepsilon_S - \varepsilon_\infty)F_\varepsilon(\psi_{\varepsilon C}, s)]\}^{-1} \tag{6}$$

and

$$Z_Z = R_\infty + (R_0 - R_\infty)F_Z(\psi_{ZC}, s) \tag{7}$$

$$Y_Z = Z_Z^{-1} \tag{8}$$

$$\varepsilon_Z = \{i\omega C_c[R_\infty + (R_0 - R_\infty)F_Z(\psi_{ZC}, s)]\}^{-1} \tag{9}$$

We have used specific subscripts here to designate which type of system is involved. Now when one compares, say, Eqs (6) and (7), both at the impedance level but for different types of systems, or Eqs (4) and (9), both at the complex dielectric constant level, one sees that although the normalized expressions for I_z and I_ε are of exactly the same form, Z_ε and Z_Z as well as ε_ε and ε_Z yield very different frequency response. The main unification produced by the introduction of the general $I \equiv I_k$ is that of allowing a single function to represent typical normalized response of either a conductive ($k = Z$) or a dielectric ($k = \varepsilon$) system.

2.2.2.2 *Distributed Circuit Elements*

Diffusion-Related Elements. Although we usually employ ideal resistors, capacitors, and inductances in an equivalent circuit, actual real elements only approximate ideality over a limited frequency range. Thus an actual resistor always exhibits some capacitance and inductance as well and, in fact, acts somewhat like a transmission line, so that its response to an electrical stimulus (output) is always delayed compared to its input. All real elements are actually distributed because they extend over a finite region of space rather than being localized at a point. Nevertheless, for equivalent circuits which are not applied at very high frequencies (say over 10^7 or 10^8 Hz), it will usually be an adequate approximation to incorporate some ideal, lumped-constant resistors, capacitors, and possibly inductances.

But an electrolytic cell or dielectric test sample is always finite in extent, and its electrical response often exhibits two generic types of distributed response, requiring the appearance of distributed elements in the equivalent circuit used to fit IS data. The first type, that discussed above, appears just because of the finite extent of the system, even when all system properties are homogeneous and space-invariant. Diffusion can lead to a distributed circuit element (the analog of a finite-length transmission line) of this type. When a circuit element is distributed, it is found that its impedance cannot be exactly expressed as the combination of a finite number of ideal circuit elements, except possibly in certain limiting cases.

The second generic type of distributed response is quite different from the first, although it is also associated with finite extension in space. In all ordinary IS experiments one uses electrodes of macroscopic dimensions. Therefore, the total macroscopic current flowing in response to an applied static potential difference is the sum of a very large number of microscopic current filaments originating and ending at the electrodes. If the electrodes are rough and/or the bulk properties of the material are inhomogeneous, the individual contributions to the total current will all be different, and the distribution in electrode surface or bulk properties will lead to a distributed resistance (many different elemental resistances) or conductance.

The situation is even more complicated when small-signal frequency and time dependence is considered. Consider a material involving ion-hopping conduction. The immediate microscopic surroundings of different ions may be different at a given instant either because of inhomogeneous material properties or because the dynamic relaxation of the positions of atoms surrounding an ion has progressed a different amount for different ions. The result may be described in terms of a distribution of relaxation times, which, for example, might be associated with a distribution of hopping-barrier-height activation energies. Such a distribution of relaxation times will lead to frequency-dependent effects which may, at least approximately, often be described through the use of certain simple distributed circuit examples.

The first distributed element introduced into electrochemistry was the infinite-length Warburg [1899] impedance, often termed *the* Warburg impedance by those possibly unaware of the more general finite-length Warburg solution. The infinite-length Warburg impedance is obtained from the solution of Fick's second law, the diffusion equation, for one-dimensional diffusion of a particle in a semiinfinite space, a situation mathematically analogous to wave transmission on a semiinfinite distributed RC transmission line (e.g. Franceschetti and Macdonald [1979c]). Diffusion of atomic oxygen in an infinitely thick electrode might be described by this impedance, an impedance which we shall designate $Z_{W\infty}$ (see below). But real physical situations never involve infinite lengths (although this limit may sometimes be a useful one to consider). The solution for the diffusion of particles in a finite-length region (equivalent to a finite-length, shorted transmission line) appears first to have been presented by Llopis and Colon [1958], for the *supported* situation, where the finite length considered was the thickness of the Nernst diffusion layer, appropriate for a stirred electrolyte or a rotating electrode. But particles diffusing in an electrode of thickness l_e or in an electrolytic cell of unstirred liquid or in solid material are free to move through the entire available region l_e or l. Thus it is reasonable to take the finite-length region where diffusion occurs as l or l_e in cases of present interest. General Warburg response for charge motion in a finite-length region of an *unsupported* electrolyte appears in the first exact solution of this problem (Macdonald [1953]). It was identified and discussed in later work (Macdonald [1971a,b, 1974a,b] and Macdonald and Franceschetti [1978]). These results, particularly appropriate for solid electrolytes, will be discussed later in Section 2.2.3.3. Here it is sufficient to give the expression for Z_W for an uncharged particle diffusing in a finite-length region of length l_e, which might be the thickness of an electrode (Franceschetti and Macdonald [1979c]), and show how it reduces to $Z_{W\infty}$ as $l_e \to \infty$. The result may be written

$$Z_W \equiv R_{D0}\left[\tanh\left(\sqrt{js}\right/\sqrt{js}\right)\right] \tag{10}$$

where $s \equiv l_e^2(\omega/D)$ and D is the diffusion coefficient of the diffusing particle. Here the diffusion resistance R_{D0} is the $\omega \to 0$ limit of $Z_W(\omega)$. It may be expressed in a form involving various rate constants if so desired (Macdonald and Franceschetti [1978], Franceschetti [1981]). Series expansion readily shows that when $s << 3$, Z_W is well approximated by R_{D0} in parallel with a capacitance C_{D0}, where

$$C_{D0} \equiv l_e^2 / 3DR_{D0} \tag{11}$$

When plotted in the complex plane, Z_W leads to an initial straight-line region with $\theta = 45°$: it reaches a peak value of $-Z''_w = 0.417\ R_{D0}$ at $s = 2.53$ and then begins to decrease toward the real axis, finally approaching it vertically, as required by the limiting R_{D0} and C_{D0} in parallel

When $s >> 3$, the tanh term approaches unity and Z_W approaches $Z_{W\infty}$, given by

$$Z_{W\infty} = R_{D0}/\sqrt{js} = (R_{D0}/l_e)(2\omega/D)^{-1/2}(1-j) \tag{12}$$

clearly showing the 45° response of $Z_{W\infty}$. Let us define the frequency-dependent diffusion length as $l_D \equiv \sqrt{D/\omega}$; then $s \equiv (l_e/l_D)^2$. It is obvious that when $l_D << l_e$, for example at high frequencies, $Z_{W\infty}$ response is found: the diffusion length is then much less than the entire region available for diffusion. But when l_D begins to approach l_e, $Z_{W\infty}$ response is no longer appropriate since diffusion begins to be limited, and $Z_{W\infty}$ must then be replaced with Z_W. In fact, it is always reasonable and appropriate to use Z_W. The quantity Z_W will always be referred as the Warburg or diffusion impedance in this work. Finally, the resistance R_{D0} is proportional to l_e and is thus extensive. Equation (11) also shows that C_{D0} is also extensive (but proportional to l_e rather than to l_e^{-1} as in an ordinary plane parallel capacitance). Thus Warburg response becomes extensive and depends on electrode separation at sufficiently low frequencies. But as Eq. (12) shows, $Z_{W\infty}$ is entirely intensive since (R_{D0}/l_e) is itself intensive. One way of identifying Warburg response is to validate the transition from intensive to extensive behavior using measurements with two or more different values of l_e.

The Warburg impedance Z_W is the diffusion analog of the impedance of a finite-length, uniformly distributed RC transmission line (see Figure 2.2.1) with a short at the far end, equivalent in the diffusion case to unhindered disappearance of the diffusing particles at $x = l_e$. But this special situation, while common, by no means includes all cases of interest. An expression for the impedance Z_D of the finite-length diffusion problem with more general conditions at the far end has been presented by Franceschetti and Macdonald [1979c]. It is recommended that the general Z_D be used initially in an equivalent circuit representation and for fitting unless and until it can be established that $Z_D \cong Z_W$.

Although we shall not discuss the general $Z_D(\omega)$ further here, there is one additional specific case which follows from it and deserves mention. Suppose that the finite-length transmission line analog is open-circuited (see Franceschetti and Macdonald [1979c]). Then no direct current can flow in the actual system, as it could with Z_W (but not $Z_{W\infty}$), and the concentration of the diffusing particle increases at the

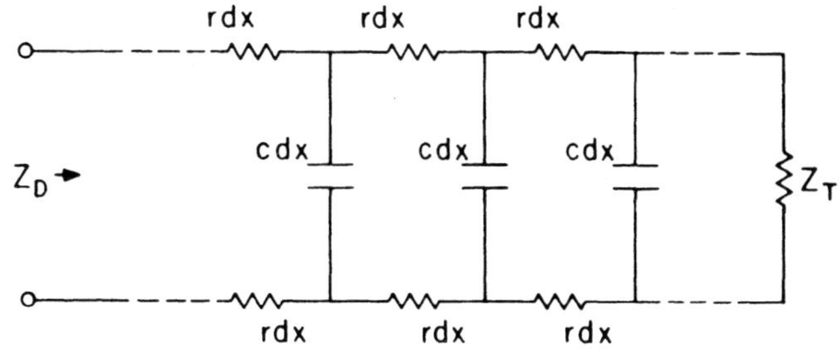

Figure 2.2.1. Uniform continuous transmission line involving series resistance of $2r$ per unit length and shunt capacitance of c per unit length, terminated by an impedance Z_T. When $Z_T = 0$, $Z_D = Z_W$, and when $Z_T = \infty$, $Z_D = Z_{DOC}$.

far end of the diffusion region where complete blocking occurs. The final low-frequency behavior of the open-circuit impedance, Z_{DOC} is thus capacitative. Its representation at the dielectric constant level is

$$\varepsilon \equiv Y/j\varepsilon C_c = (C_{DOC}/C_c)[\tanh\sqrt{js}/\sqrt{js}] \tag{13}$$

Thus, the process leads to the limiting capacitance C_{DOC} as $\omega \to 0$: further, the frequency response at this level is exactly the same as that for Z_W at the impedance level [see Eq. (10)] and thus involves an initial straight line at $\theta = 45°$. At the impedance level, $Z_{DOC} \equiv (j\omega C_c\varepsilon)^{-1}$, is given by

$$Z_{DOC} = (\tau_{DOC}/C_{DOC})[\text{ctnh}(\sqrt{js})/\sqrt{js}] \tag{14}$$

where we have written $s \equiv \omega\tau_{DOC}$ so that $\tau_{DOC} \equiv l_e^2/D$. The appearance of response of the present type has been found in electrochromic thin films (see Glarum and Marshall [1980], Ho *et al.* [1980], Franceschetti and Macdonald [1982]). Equations (10) and (13) show immediately that the general I_k function associated with shorted or open-circuited diffusion is just

$$I_k = \tanh\sqrt{js}/\sqrt{js} \tag{15}$$

A distributed element which shares some features with Z_W, Eq. (14), is the Gerischer impedance,

$$Z_G(\omega) = \frac{Z_{G0}}{\sqrt{k + j\omega}} \tag{16}$$

which arises when an electroactive species undergoes a chemical reaction in the bulk (Gerischer [1951], Sluyters-Rehbach and Sluyters [1970], [1984], Boukamp and Bouwmeester [1984]). The Gerischer function is indistinguishable from the finite-length Warburg in the high frequency limit, and like it yields purely resistive behavior in the low frequency limit, but differs from it in its detailed frequency dependence, yielding, for example, an arc closer to the real axis in the Z plane. Like the infinite length Warburg, the Gerischer impedance is representable by a semi-infinite transmission line with a series resistance per unit length but with a parallel resistance added to the transmission line capacitance per unit length. A related result describing the diffusion of an electroactive species which is trapped at fixed sites was developed by Franceschetti [1984].

The somewhat similar function

$$Z_L(\omega) = \frac{Z_{L0}}{\sqrt{k + j\omega}}\tanh(\sqrt{k + j\omega}) \tag{17}$$

was found by Lorenz [1954] for an electrode reaction in which atoms are added to a ledge on a crystalline electrode. This function was proposed by Franceschetti and Macdonald [1979a] based on a treatment by Wang and Nowick [1979] involving oxygen diffusion along strip-like electrode contacts with an oxygen conducting electrolyte. Franceschetti and Ross [1989] extended this result to circular electrode contacts, obtaining an impedance function in terms of cylindrical Bessel functions of $[k + j\omega]^{1/2}$.

The Constant Phase Element and Its Simple Combinations. Although Warburg and open-ended diffusion effects frequently appear in supported situations and sometimes in unsupported ones and exhibit characteristic $\theta = 45°$ lines in the Z^* or ε plane, one often finds approximate straight-line behavior over a limited frequency range with $\theta \neq 45°$ (e.g. McCann and Badwal [1982]). Then the frequency response of Z' and Z'' is no longer proportional to $\omega^{-1/2}$ but to some other power of ω. To describe such response it is convenient to write, as in Eq. (7) in Section 1.3, at the admittance level,

$$Y_{CPE} = A_0(j\omega)^{\psi} = A_0\omega^{\psi}[\cos(\psi\pi/2) + j\sin(\psi\pi/2)] \tag{18}$$

where A_0 and ψ are frequency-independent parameters which usually depend on temperature, and $0 \leq \psi \leq 1$. This admittance has been designated the constant-phase element (CPE) admittance because its characteristic feature, and that of Z_{CPE} as well, is a phase angle independent of frequency. Although a slightly more general form $a_0(j\omega\tau)^{\psi}$ may be written, the combination $(a_0\tau^{\psi})$ cannot be resolved into its parts using single-temperature frequency response measurements and fitting. The parameter A_0 will be intensive for interface processes and may be extensive for bulk ones. Unlike the finite-length Warburg impedance, the CPE exhibits no transition from intensive to extensive behavior as the frequency decreases. Note that a resistance R_∞ in series with $Z_{\text{CPE}} = Y_{\text{CPE}}^{-1}$ yields an inclined spur (straight line) in the Z plane, with an $\omega \to \infty$ intercept of R_∞.

The importance of constant phase response was probably first emphasized by Fricke [1932]; the CPE was explicitly mentioned by Cole and Cole [1941], and its importance and ubiquity have been independently emphasized in recent times by Jonscher [1974, 1975a,b, 1980, 1983]. Some discussion of its history, relation to physical processes, and applicability has been given by Macdonald [1984]. Note that it describes an ideal capacitor for $\psi = 1$ and an ideal resistor for $\psi = 0$. It is generally thought to arise, when $\psi \neq 0$ or 1, from the presence of inhomogeneities in the electrode–material system, and it can be described in terms of a (nonnormalizable) distribution of relaxation times (Macdonald and Brachman [1956]), or it may arise from nonuniform diffusion whose electrical analog is an inhomogeneously distributed RC transmission line (Schrama [1957]).

Although CPE-like response appears in the majority of experimental data on solid and liquid electrolytes, it is always well approximated only over a finite range of frequency. In fact, the CPE cannot be applied for all frequencies and becomes physically unrealizable for sufficiently low or high frequencies (Macdonald [1984, 1985b,c,d]). Although many response theories lead to the CPE type of response for a finite frequency range, they must deviate from such response at the frequency extremes in order to yield realistic, physically realizable response. Because of the lack of full physical realizability, the CPE, as in Eq. (18), cannot be normalized in the usual I_k fashion. For example, the $\omega \to 0$ limits of neither ε_{CPE} nor Z_{CPE} exist. With this understood, we shall nevertheless write a unified expression for the CPE, taking it to represent just a dimensionless form of either ε or Z. Then we have

$$I_{\text{CPEk}} = B_k(j\omega)^{-\psi_k} \equiv (js)^{-\psi_k} \tag{19}$$

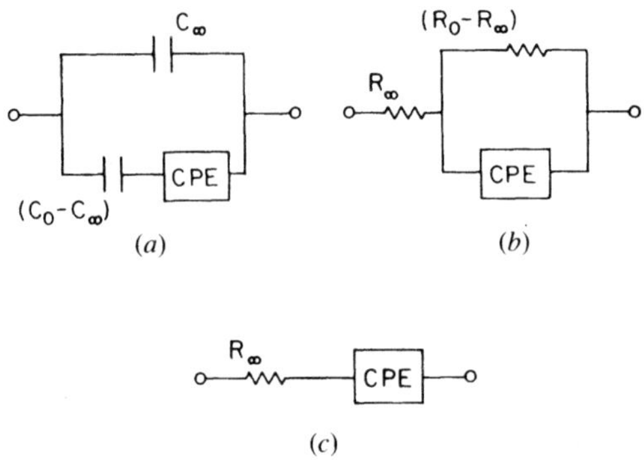

Figure 2.2.2. Three compound circuits involving the CPE: (a) the εARC, (b) the ZARC, and (c) the YARC.

where B_k is a frequency-independent constant and the second form is less general than the first (Macdonald [1984]). For $k = Z$, one usually sets $\psi_Z = n$ and for $k = \varepsilon$, $\psi_\varepsilon = 1 - n$, where as usual $0 \leq n \leq 1$. These choices ensure that at the admittance level, for either $k = \varepsilon$ or $k = Z$, the fractional exponent ψ in Eq. (18) is just n.

There are three important subcircuits, shown in Figure 2.2.2, which consist of the CPE in conjunction with other circuit elements. Although they can always be treated by considering the CPE contribution separately, their wide use as combined elements and their historical importance justify their separate discussion as compound circuit elements and our assignment of specific designations to them. Further, it has not usually been recognized how such compound expressions can involve individual CPE's. The first subcircuit, that in Figure 2.2.2*a*, the εARC, yields a depressed symmetrical semicircular arc in the complex ε plane; the second, shown in Figure 2.2.2*b*, the ZARC or ZC, yields an arc in the *Z* plane; and the third, the YARC, leads to an arc in the *Y* plane. The depression of the arc of course depends on the CPE parameter ψ or n. We have already discussed in Section 1.3 the ZARC and simple ways of analyzing data exhibiting it. The same methods apply to depressed arcs in any complex plane.

The circuit shown in Figure 2.2.2*a* is just that originally proposed by Cole and Cole [1941] for dielectric systems. It yields capacitances C_0 in the $\omega \to 0$ limit and C_∞ in the $\omega \to \infty$ limit. When one uses the Eq. (18) definition of the CPE admittance, it leads to the following expression for ε:

$$\varepsilon = \varepsilon_\varepsilon = \varepsilon_\infty + \frac{\varepsilon_S - \varepsilon_\infty}{1 + (j\omega\tau)^{\psi_\varepsilon}} \tag{20}$$

where $\varepsilon_\infty \equiv C_\infty/C_c$, $\varepsilon_s \equiv C_0/C_c$, $\tau \equiv [(C_0 - C_\infty)/A_0]^{1/\psi_\varepsilon}$, and $\psi_\varepsilon \equiv 1 - \psi$. Cole and Cole related Eq. (20) to a particular distribution of relaxation times, and it has been widely used with $\psi_\varepsilon \equiv 1 - \alpha$ for the interpretation of IS results for dielectric and low-conductivity liquid and solid materials. Clearly when $0 < \psi_\varepsilon < 1$, Eq. (20) leads to a depressed arc in the complex ε plane. For $\psi_\varepsilon = 1$, it yields a single-relaxation-constant Debye curve, a full semicircle.

In terms of the original CPE parameters of Eq. (18), the εARC admittance of Figure 2.2.2*a* may be written as

$$Y_\varepsilon = Y_{\varepsilon\mathrm{ARC}} = j\omega\left[C_\infty + \frac{C_0 - C_\infty}{1 + [(C_0 - C_\infty)/A_0](j\omega)^{1-\psi}}\right] \tag{21}$$

For simplicity we shall usually ignore C_∞ in the definition of εARC; it may readily be included when needed. The εARC circuit element will, in fact, often appear in solid electrolyte equivalent circuits with C_∞ zero (or appearing elsewhere in the overall circuit). Then it is clear that the εARC function of Eq. (21) may be considered to represent a distributed (complex) capacitor. When $\psi = 0$, it involves an ordinary capacitor C_0 and resistor A_0^{-1} in series, and when $\psi = 1$, it involves C_0 in series with the capacitor A_0. These results are consistent with the behavior of the CPE alone. The CPE reduces to an ideal capacitor for $\psi = 1$ and to a resistor for $\psi = 0$. An example of the use of the εARC function for the analysis of a polycrystalline sample is provided by the work of Casciola and Fabiani [1983].

It was independently suggested some time ago (Ravaine and Souquet [1973], Sandifer and Buck [1974], and Macdonald [1976b]) that the following impedance form might be used to describe the depressed arcs which often appear when impedance data on solids is plotted in the Z^* plane, namely,

$$Z_Z = Z_{Z\mathrm{ARC}} = \frac{R_0}{1 + (j\omega\tau)^{1-\alpha}} \tag{22}$$

where no R_∞ is included and $0 \le \alpha \le 1$. When R_∞ is included, one can write

$$Z_Z = R_\infty + \frac{R_0 - R_\infty}{1 + (j\omega\tau)^{\psi_Z}} \tag{23}$$

which is just the impedance of the Figure 2.2.2*b* circuit with $\psi = \psi_Z$. This is an exact analog, at the impedance level, of the Cole–Cole complex dielectric constant expression of Eq. (20). Although the two forms may be described in terms of the same formal distribution of relaxation times, this distribution applies at different response levels for the two cases and thus describes quite different system behavior (Macdonald and Brachman [1956]), Macdonald [1985*a–c*]). Now it is clear that Eq. (22), which applies when $R_\infty = 0$ or is neglected, may be rewritten as

$$Z_Z = \frac{R_0}{1 + A(j\omega)^{\psi_Z}} \tag{24}$$

involving the parameter $A \equiv \tau^{\psi_Z}$. Now the parallel combination of a resistance R_0 and a CPE with parameter A_0, as in Figure 2.2.2*b*, yields just

$$Z_Z = \frac{R_0}{1 + A_0 R_0 (j\omega)^{\psi_Z}} \tag{25}$$

equivalent to the result in Eq. (24) if $A = A_0R_0$ or $\tau = (A_0R_0)^{1/\psi_Z}$.

The foregoing results and the definition of I_k of Eq. (3) lead immediately to

$$I_k = \left[1 + (js)^{\psi_k}\right]^{-1} \tag{26}$$

where, as usual, $s \equiv \omega\tau$ (see the discussion in the last section). Since when $k = Z$ one obtains the ZARC function and, when $k = \varepsilon$, the Cole–Cole equation, we suggest that the general normalized response function of Eq. (26) be designated the ZC function. Again it is most appropriate to take $\psi_Z = \psi = \alpha$ and $\psi_\varepsilon = 1 - \alpha$. Although CNLS fitting of data with either a CPE and R in parallel, as in Eq. (25), or with the unified expression of Eq. (26), involving τ, will yield exactly the same fit, the two approaches involve different parameterizations (R_0, A_0, and ψ_Z or R_0, τ, and ψ_Z). One or the other will generally yield smaller estimated standard errors for A_0 or τ and less correlation of one of these quantities with the other parameters. That choice should be used. Analysis of Na β-alumina data (Macdonald and Cook [1985]) gave better results, for example, with the τ parameterization.

It has already been mentioned that the series combination of a resistance R_∞ and a CPE as in Figure 2.2.2*c* leads to a depressed arc in the Y plane. Since such arcs are also often encountered experimentally, it is reasonable to define them as YARCs, for which the admittance may be written as

$$Y_{\mathrm{YARC}} = \frac{G_\infty}{1 + (R_\infty A_0)^{-1}(\mathrm{I}\,j\omega)^{-\psi}} \tag{27}$$

where $G_\infty \equiv 1/R_\infty$. The similarity to Eq. (25) is obvious, although frequency increases along the ZARC and the YARC in opposite directions, as usual for Z- and Y-plane plots. Although it is possible to define a dimensionless function like I_k which can represent either Y-system or M-system response, just as I_k represents either conductive or dielectric system response, the matter will not be pursued here. In fact, a single dimensionless function with superscript $k = \varepsilon$, Y, Z, or M may be used to represent response of any of the four different immittance-level systems if the normalization is properly defined at each level.

It is worth mentioning that although Eqs (20)–(27) may be interpreted as involving nonuniform diffusion (Schrama [1957]) either in bulk or at an interface, another allied but somewhat different approach which also leads to Eq. (22) has been proposed (Le Mehaute and Crepy [1983]) without reference to its earlier history and use. This theory involves mass transfer at a fractal interface, one with apparent fractal dimensionality d, with $d = \psi^{-1} = (1 - \alpha)^{-1}$. A more solidly based treatment of a fractal interface has been published by Liu [1985]. Unfortunately, neither of these approaches provides a quantitative interpretation in terms of microscopic parameters of why ψ, determined from data fitting on solids or liquids, often depends appreciably on temperature.

The above results show that the $\psi_\varepsilon = 1 - \alpha$ parameter which appears in the εARC Cole–Cole function, Eq. (20), associated with a CPE and ideal capacitor in series, and the ψ's appearing in the ZARC and YARC functions, Eqs (25) and (27), associated with a CPE and resistor in parallel or in series, may all be interpreted as the ψ of a CPE. The ψ values estimated from fitting with these forms are thus comparable. Although the CPE has sometimes been found in equivalent circuit data fitting to appear separately and not directly in any of the above compound forms (e.g. Macdonald, Hooper, and Lehnen [1982]), its presence as a direct part of the εARC, ZARC, and YARC functions, ones which have long been used in the inter-

Table 2.2.2. Some Distributed Elements and Their Descriptions at the Impedance Level

Symbol	Name and Description	Defining Equation
Z_W	Finite-length Warburg diffusion	(10)
$Z_{W\infty}$	Infinite-length Warburg diffusion	(12)
Z_D	Diffusion with general boundary conditions	—
Z_{DOC}	Open-circuit (blocked) diffusion	(14)
Z_{CPE}	Constant-phase element	(18), (19)
$Z_{\varepsilon\text{ARC}}$	Depressed semicircle in complex dielectric constant plane (see Fig. 2.2.2*a*)	(20), (21)
Z_{ZARC}	Depressed semicircle in impedance plane (see Fig. 2.2.2*b*)	(22), (23)
Z_{ZC}	ZC element, general form of εARC and ZARC	(26)
Z_{YARC}	Depressed semicircle in admittance plane (see Fig. 2.2.2*c*)	(27)

pretation of a wide variety of IS data on dielectric and conduction materials, underlines the wide usefulness of the CPE.

As already mentioned, the lack of any physically based relation for the temperature dependence of the CPE and CPE-like fractional exponent ψ (or n or α) is an important weakness in the theories which lead to frequency response with such exponents. A new theory which involves a distribution of activation energies and does predict temperature dependence for ψ often in agreement with experiment (Macdonald [1985a,c,d]) will be discussed in Section 2.2.3.5, along with some empirical frequency response relations suggested by Jonscher. We summarize the various distributed elements (simple and compound) discussed in this section in Table 2.2.2.

2.2.2.3 Ambiguous Circuits

Let us now further consider the inherent ambiguity of equivalent circuit fitting. One example of two different equivalent circuits having the same overall impedance at all frequencies has already been presented in Figure 1.2.2. Incidentally, if we change all the resistors in both circuits to capacitors, we have another instance of the same kind of ambiguity. Another series of circuits which may all have the same impedance is shown in Figure 2.2.3 (see Franceschetti and Macdonald [1977]). Here we have again given the actual relations between the various components. Some adsorption models (see later) yield inductive-type behavior and a resulting arc which falls below the real axis in the Z^* plane. Sometimes the apparent inductance can be very large. But it is only an apparent inductance since real inductance requires storage of energy in a magnetic field and there is no appreciable ac magnetic field energy present in low-current IS measurements. The actual situation involves an inductive type of phase shift, but rather than represent it by the inductive circuits of Figure 2.2.3*a* and *b*, which give a somewhat misleading picture of the process, we recommend following earlier work (e.g. Franceschetti and Macdonald [1977]) and using circuit (c), which involves both a negative differential capacitor and a negative dif-

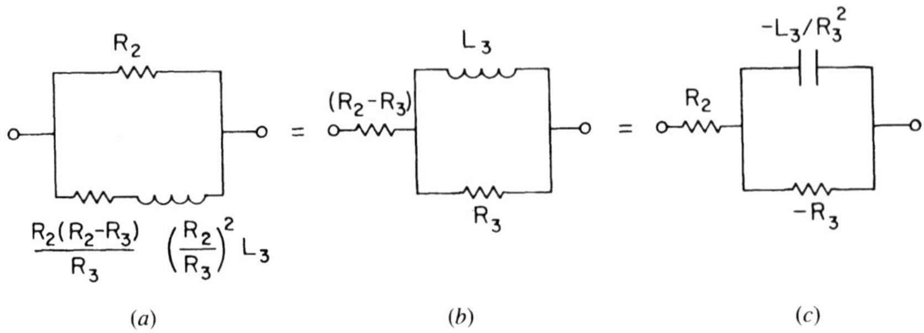

Figure 2.2.3. Three circuits having the same impedance at all frequencies.

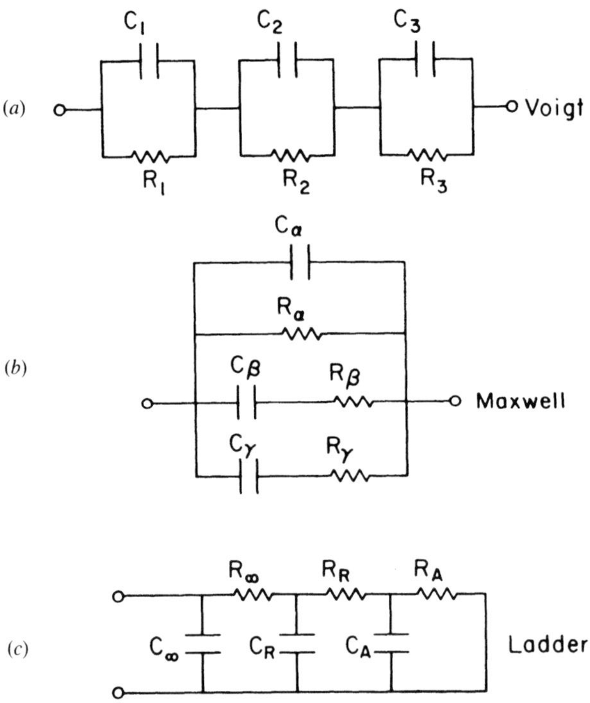

Figure 2.2.4. Three further circuits which can have the same impedance at all frequencies when the parameters of the circuit are properly interrelated.

ferential resistor. Since adsorption often can be represented electrically by a positive resistor and capacitor in the (c) type of circuit, continuity is served by allowing both these elements to become negative when appropriate. It is then unnecessary to pass from an ordinary RC circuit to a LC one as adsorption changes; instead the *R* and *C* can just go from positive values to negative ones.

Figure 2.2.4 presents three more electrical circuits often encountered in IS work. They exhibit three time constants ($N = 3$) and can all yield the same impedance for all frequencies when their elements are properly related. All three circuits yield three distinct arcs in the Z^* plane when the three time constants are well separated. Starting from the Voigt circuit with only two time constants, one can find relatively simple algebraic formulas yielding expressions for the elements in the other two $N = 2$ circuits which ensure that the impedance is the same (Novoseleskii *et al.* [1972]), but such simple relations do not always exist when one starts with another of the circuits or when there are three or more time constants present. In practice, however, the detailed relations between the elements are not particularly important when CNLS fitting procedures are available. First, parameter estimates for any of the circuits may be obtained by such fitting and those for the different circuits compared. Incidentally, the degree of fit is completely independent of which of the three circuits is employed. Second, when the time constants are separated by factors of 100 or more, the R's and C's of, say, the top and bottom circuits closely approach each other. It is particularly when this condition is not satisfied, however, that CNLS fitting is necessary to resolve the overlapping arcs in the Z^* plane.

But CNLS fitting is not always available or may not be justified for preliminary fits. When $N = 2$ in the circuits of Figure 2.2.4, several of the relations between the circuit elements of the three types of circuits can prove very useful in graphical fits. These relationships are summarized in the appendix at the end of this section. Once the parameters of a particular $N = 2$ circuit have been graphically estimated from impedance spectra, estimates of the parameters of the other two circuits may be obtained using these relations, avoiding the need for graphical fitting of the other circuits. Indeed, the equivalence relations are the easiest way to obtain the parameters of the ladder network (Figure 2.2.4*c*), which cannot be well estimated from either an impedance or admittance plot but require a laborious process involving more than one type of plot (de Levie and Vukadin [1975]). It is simple to fit to the $N = 2$ circuit of Figure 2.2.4*a* or *b* and transform to the ladder representation. Further, the equivalence relations may be useful when no IS data are available but circuit element estimates are, as in published work of others.

Since all three of the Figure 2.2.4 circuits are equivalent as far as fitting is concerned, how does one choose between them, particularly in cases when element estimates for the different circuits are quite different? First, one may use continuity and knowledge of the physical processes involved, as in the above brief discussion of inductive-like effects in adsorption. Second, one may be able to compare the circuits with the predictions of a physical model—one which yields simpler expressions for the elements of one of the circuits than for the others. This has been done for the unsupported conduction case (Franceschetti and Macdonald [1977]) and the work showed that in the case of charge of a single-sign mobile the ladder circuit was much superior to the others (see Section 2.2.2.3).

Finally, one should apply the criterion of simplicity. Given equally good fits, the circuit with the smallest number of elements should be used. Second, when electrode separation l, temperature T, or possibly oxygen partial pressure $p(O_2)$ (e.g. Verkerk and Burggraaf [1983], Badwal [1984]) is changed, one expects some or all of the fitting parameters to change. But that circuit in which the changes are least, simplest, and/or closest to theoretical expectations should certainly generally be preferred. By carrying out CNLS fitting with several different but plausible circuits, such as those in Figure 2.2.4, for various l, T, and/or $p(O_2)$ conditions, one can often reach an unambiguous choice of the "best" fitting circuit (out of those considered) to use. Since various processes occur in various, often widely separated, different frequency regions, it should be emphasized that IS measurements must include such regions to allow identification and analysis of the individual processes present. Generally, then, as wide a frequency span as possible should be covered by the experimental measurements.

Appendix. This appendix summarizes the relations between the elements of the three circuits of Figure 2.2.4 when $N = 2$. Here the subscripts a and b are used in place of the α and β of Figure 2.2.4.

(a) Voigt → Maxwell

$$C_a = \frac{C_1 C_2}{C_1 + C_2} \tag{A1}$$

$$C_b = \frac{(R_1 C_1 - R_2 C_2)^2}{(C_1 + C_2)(R_1 + R_2)^2} \tag{A2}$$

$$R_a = R_1 + R_2 \tag{A3}$$

$$R_b = \frac{R_1 R_2 (R_1 + R_2)(C_1 + C_2)^2}{(R_1 C_1 - R_2 C_2)^2} \tag{A4}$$

(b) Maxwell → Voigt

$$C_{1,2} = 2C_a\left(1 \mp \frac{R_b/R_a - C_a/C_b + 1}{k^{1/2}}\right)^{-1} \tag{A5}$$

$$R_{1,2} = \frac{R_a}{2}\left(1 \pm \frac{C_a/C_b - R_b/R_a + 1}{k^{1/2}}\right) \tag{A6}$$

where

$$k = \left(\frac{C_a}{C_b} + \frac{R_b}{R_a} + 1\right)^2 - 4\frac{C_a R_b}{C_b R_a} \tag{A7}$$

Here $R_{1,2}$ and $C_{1,2}$ are defined such that $R_1 C_1 > R_2 C_2$.

(c) Maxwell → Ladder

$$C_\infty = C_a \tag{A8}$$

$$R_\infty = \frac{R_a R_b}{R_a + R_b} \tag{A9}$$

$$C_R = C_b\left(\frac{R_a + R_b}{R_a}\right)^2 \tag{A10}$$

$$R_R = \frac{R_a^2}{R_a + R_b} \tag{A11}$$

(d) Ladder → Maxwell

$$C_a = C_\infty \tag{A12}$$

$$R_a = R_\infty + R_R \tag{A13}$$

$$C_b = \left(\frac{R_R}{R_\infty + R_R} \right)^2 C_R \tag{A14}$$

$$R_b = \frac{R_\infty}{R_R}(R_\infty + R_R) \tag{A15}$$

2.2.3 Modeling Results

2.2.3.1 Introduction

In any modeling situation one must first specify the physical conditions considered. Is the system in equilibrium or in a steady state? What species of mobile and immobile charges are present? Is the material between the electrodes homogeneous (liquid or single-crystal) or not (amorphous or polycrystalline)? In the polycrystalline case, what boundary conditions should be used at the interfaces between crystallites? In all cases, what kind of electrodes are assumed and thus what are the appropriate boundary conditions at the electrode–material interfaces?

Perhaps the most general problem one would like to solve in the present area, a sufficiently general situation that would include almost all simpler ones of interest, is the following: a biased situation with applied dc p.d. of arbitrary size and a small-signal ac p.d. also applied; an arbitrary number of charged species (but not exceeding, say, six) with arbitrary mobilities and bulk concentrations present; general interactions (e.g. generation–recombination) possible between the various positive and negative charged species; arbitrary (i.e. general) blocking–nonblocking, reaction–adsorption conditions for each of the mobile species at the electrode interfaces; and separate treatments of homogeneous and polycrystalline situations. One would like to calculate the direct current I over a wide range of applied potential difference and, at any given applied steady state potential difference, calculate the impedance as a function of frequency.

Unfortunately, this general problem, which usually involves an inhomogeneous distribution of charge within the material, has not been solved. The situation is highly nonlinear and, although the many coupled differential equations and boundary conditions which could be used to specify it mathematically could, in principle, be solved with a large computer, only purely numerical results depending on a very large number of input parameter values (e.g. mobilities, equilibrium concentrations) would be obtained. Of course even such a general, and almost useless, solution would still be approximate since the equations used would themselves still be approximations to the actual physical situation.

Thus far, only much simpler idealizations of the general problem have been solved, and in Section 2.2.3.3 we shall discuss some of their results. When the simpler solutions are thought to be adequate, they may be used to analyze experimental data and obtain estimates of such interesting quantities as electrode charge-transfer reaction rate. Many of the simpler solutions can be represented exactly or approximately by an equivalent circuit, but some yield only a complicated expression for $Z(\omega)$ which cannot always be so represented in a useful manner.

Solutions for unbiased, flat-band situations (i.e. where there are no intrinsic space charge layers at the boundaries) are simplest, and only these will be discussed below except when otherwise noted. We shall present a brief discussion of the supported situation, primarily appropriate for liquids and mixed conductors, and devote more space to results for unsupported materials, since most solid electrolytes involve unsupported ionic conduction under conditions of primary interest. For simplicity, theoretical results for flux, currents, impedances, and other circuit elements will be given in specific form, per unit of electrode area A_c, so this area will not appear directly in the formulas.

The possible behaviors of an electrode–electrolyte interface are variously discussed in the literature in terms of polarizability, blocking or nonblocking character, and reversibility, with usage differing somewhat from one author to another. For clarity and precision we shall use the term *polarizability* to denote the electrical behavior of the electrode–electrolyte interface and the terms *blocking* (or *nonblocking*) and *reversibility* to describe the electrochemical character of the interface. An electrode–electrolyte interface is nonpolarizable if the potential drop across the interface is independent of the current through the interface. It is partially polarizable if the interfacial potential difference is dependent on the current and completely polarizable if it completely prohibits the flow of (faradic) current. An interface is blocking with respect to a given charge-carrying species in the electrolyte if that species cannot cross the interface or exchange charge (in the form of electrons) with the electrode; otherwise it is nonblocking with respect to the given species. A nonblocking interface is generally thermodynamically reversible since, in thermal equilibrium, the electrochemical potential of the species involved in the interfacial charge transfer will obey an equilibrium relation. The interface is kinetically reversible if the rate of the electrode charge-transfer reaction is rapid enough that the equilibrium relation is maintained in the immediate vicinity of the interface as current passes through the system.

The polarization of an electrode–electrolyte interface can result either from the slowness of the electrode reaction, as in the case of nonblocking but kinetically nonreversible electrodes, or from any factor which limits the transport of any of the species participating in the electrode reaction, for example slow diffusion of the reactant or product species away from the interface or the generation or consumption of one of the species by a slow chemical reaction in the electrolyte.

We shall be concerned primarily with the behavior of ionic charge carriers at the interface, and the electrode–electrolyte combinations to be encountered will fall into two general groups: parent-atom electrodes and redox electrodes. In parent-atom electrodes, charge can cross the interface in ionic form. Electrodes of this type include parent-metal electrodes such as Ag in the solid state cell Ag|AgCl|Ag, in which the electrode serves both as a source of ions and as an electronic conductor, and parent-nonmetal electrodes, as in the cell Br_2(Pt)|AgBr|Br_2(Pt), in which an inert metal phase must be present to serve as the electronic conductor. In redox electrodes, charge crosses the interface in the form of electrons and the reaction may be written in the form

$$\mathrm{Red}^{(z-n_e)+} \underset{k_b}{\overset{k_f}{\rightleftarrows}} \mathrm{Ox}^{z+} + n_e e^- \tag{28}$$

where k_f and k_b are forward and reverse reaction rate constants, respectively. Both Ox and Red are usually soluble in the electrolyte, but if $z = n_e$, the Red species is uncharged, and if it is a gas, it may evolve at an electrode and/or diffuse into the electrode, especially if the electrode is somewhat porous. The admittance behavior of more complex electrode reactions (in aqueous electrolytes) than those mentioned above has been discussed by Seralathan and de Levie [1987] and is not considered herein.

Finally, while still dealing with interface effects, it is worth stating one of the most important equations of reaction rate electrochemistry, the Butler–Volmer equation (see Vetter [1967], Franceschetti [1982]). Written in terms of flux for a simple redox reaction, it is

$$\begin{aligned} J \equiv (I/nF) = \{ & k_f^0 c_{\mathrm{Red}}^H \exp[(nF/RT)\alpha\eta_{MH}] \\ & - k_b^0 c_{\mathrm{Ox}}^H \exp[-(nF/RT)(1-\alpha)\eta_{MH}]\} \end{aligned} \tag{29}$$

Here n is the number of moles of electrons involved in the reaction; the rate constants have been assumed to be thermally activated; and k_f^0 and k_b^0 are potential-independent rate constant parameters. The potential-dependent concentrations c_{Red} and c_{Ox} are evaluated at their points of closest approach to the electrode, taken here as the outer Helmholtz plane. Further, α is here a dimensionless symmetry factor often assumed to be 0.5 (the symmetrical barrier case), and η_{MH} is the charge transfer overvoltage effective in driving the reaction away from equilibrium (for $\eta_{MH} = 0$, $J = 0$). The Butler–Volmer equation is usually a good approximation for both biased and unbiased conditions.

2.2.3.2 Supported Situations

Half-Cells. The concept of a supported electrolyte has proven quite valuable in solution electrochemistry by allowing great theoretical simplification at (usually) only a small cost in accuracy. The several (often implicit) assumptions made in treating the electrolyte in a given cell as supported, however, deserve careful attention as they generally do not apply in the case of solid state electrochemical systems. It should also be noted that it is usually possible in solution electrochemistry to use a large, essentially kinetically reversible counterelectrode so that all but a negligible fraction of the applied potential difference falls across the electrode–electrolyte interface of interest. In its simplest form, the supported approach assumes that all the potential difference in the system falls across the compact double layer—approximately one solvent molecule diameter in thickness—at this electrode, and the approach of the electroactive species to the boundary of the compact layer, the outer Helmholtz plane, occurs purely by diffusion. Corrections for the buildup of space charge near the interface (the diffuse double layer) and the ohmic drop across the electrolyte are then made in piecewise fashion as needed. In solid state electrochemistry, one is usually concerned with measurements on a cell with an unsupported electrolyte and identical plane parallel electrodes. The impedance of such a cell will be twice that of its two half-cells. Although most solid electrolytes are essentially unsupported for temperatures of interest, Archer and Armstrong [1980] have suggested that when a solid electrolyte contains immobile anions and two different species of cations of similar mobilities, if one of the ionic species is present in much higher bulk concentration than the other, it can act as support for the other species of mobile ion—a supported case. Also, mixed electronic–ionic conduction can sometimes lead to supported conditions. Thus the supported situation may even be of some direct interest for solid electrolytes.

A symmetrical full cell with identical electrodes, no intrinsic space charge layers, and zero potential drop across the compact double layers is mathematically equivalent to two half cells in series since the concentrations of the diffusing species will be zero in the center. In this one case, the normalized impedance of the half-cell is the same as that of the full cell. For identical and kinetically reversible redox electrodes and both oxidized and reduced species mobile in the supported electrolyte, a complete solution, neglecting compact layer capacitance, has been obtained by Sluyters [1963]. This solution, derived and presented as real functions, is highly complicated in appearance. Macdonald [1971a] pointed out that the Sluyters result simplifies greatly when the impedance is written in complex variable form. Franceschetti [1987] treated the same situation allowing for a finite charge transfer rate and a compact layer capacitance, and in the process found Sluyters result as the sum of two finite-length Warburg elements. In the same paper, Franceschetti found an impedance for a supported system with two dissimilar electrodes of the parent-metal type (perhaps a pure metal and an amalgam). The resulting equivalent circuit requires an inductive element and three lengths of transmission line, one of which includes an inductive series reactance.

Let us now consider only infinitesimal deviations from zero bias equilibrium conditions and assume that the equilibrium distribution of all charges is constant throughout the material, so for equilibrium there is no polarization and electroneutrality applies everywhere. Solution of Fick's laws of diffusion under supported small-signal ac conditions for a simple one-step reaction (single reacting charged species) (Randles [1947], Sluyters-Rehbach and Sluyters [1970], Armstrong *et al.* [1978], Franceschetti [1982]) leads to the equivalent circuit of Figure 2.2.5. This

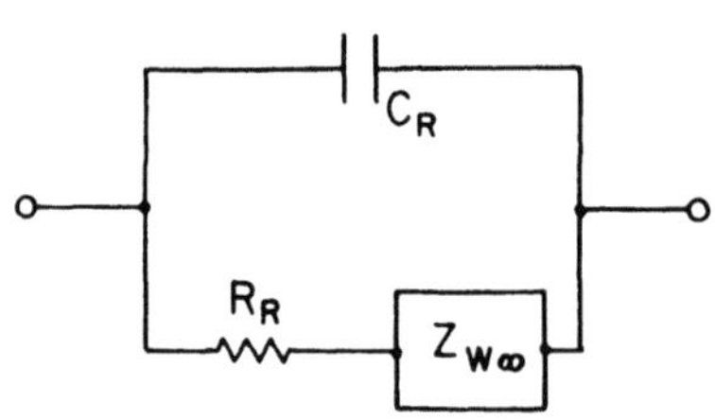

Figure 2.2.5. Equivalent circuit for a single electrode and its interface under supported conditions.

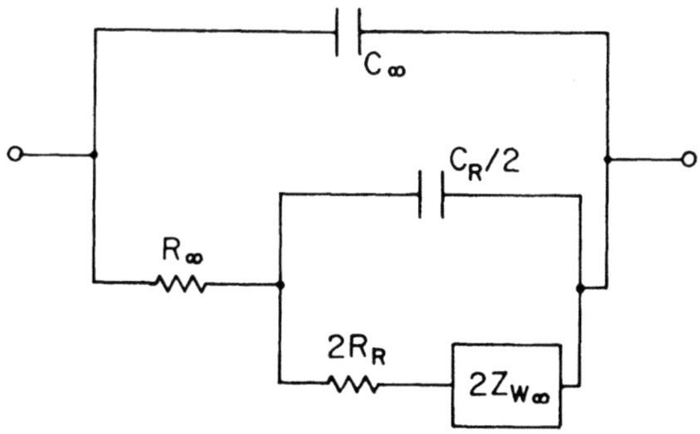

Figure 2.2.6. Equivalent circuit for a full cell with two identical electrodes under supported conditions. This circuit only applies when $Z_W \cong Z_{W\infty}$.

circuit may be taken to apply to a half-cell of infinite extent to the right of the electrode. It thus does not include any bulk or solution resistance R_∞ which would depend on the finite extent of an actual cell. It does include an infinite-length Warburg impedance, the charge-transfer reaction resistance R_R, and the capacitance associated with it, C_R, the diffuse double-layer capacitance, sometimes denoted C_{dl}. This circuit, with the addition of a bulk or solution resistance in series with it, is customarily known as the *Randles circuit*. Appreciable discussion of the circuit from an electrochemical viewpoint appears in Macdonald [1971a].

Full-Cell Results. We can generalize the circuit of Figure 2.2.5 to a full-cell situation with identical electrodes as long as Z_W is well approximated by $Z_{W\infty}$, that is as long as the diffusion length is much smaller than l. One then obtains two contributions to the impedance of the Figure 2.2.5 type, one associated with each electrode. Since the electrodes are taken identical, the two intensive impedances are identical and may be combined to yield a result of twice the individual impedances. When we additionally add a geometric capacitance $C_g \equiv C_\infty$ and the bulk or solution resistance R_∞, we obtain the circuit of Figure 2.2.6.

Although we will always refer to actual circuit elements in an equivalent circuit, we shall, for simplicity, give expressions for these elements per unit area. We shall not, however, usually distinguish between a quantity and its per-unit-area specific form, so, for example, a capacitance-unit area will still be referred to as a *capacitance*. If the separation of the electrodes is l, the geometrical capacitance (per unit area) is given by

$$C_\infty = \varepsilon\varepsilon_0/l \tag{30}$$

where ε is the effective dielectric constant of the electrolyte. It is customary to omit C_∞ in supported situations since IS measurements on liquids rarely extend to high enough frequencies for it to affect the overall impedance appreciably. This is not always the case for solid electrolytes where impedance contributions from other elements may be high, especially at low temperatures.

Now ionic conduction in a full cell actually occurs in a region of length $l_{\text{eff}} = 1 - 2l_H$, where l_H is the effective thickness of the inner region, next to the electrode, into which ions cannot fully penetrate. Because of the finite size of ions, the minimum steric but not necessarily electrical value of l_H is an ionic radius. Except for unrealistically thin cells, the distinction between l_{eff} and l is not important for most circuit elements and will usually be neglected hereafter. It should be mentioned, however, that in the study of thin (sometimes monomolecular) membranes in the biological field, using high-molarity liquid electrolyte electrodes, the distinction may be important. We may now write the expression for R_∞ as

$$R_\infty = (l/F)\left(\sum_{i=1}^{m} z_i \mu_i c_i^0\right)^{-1} \tag{31}$$

where z_i and μ_i are the valence numbers and mobilities of the m charged species of bulk concentration c_i^0 present in the electrolyte. We have followed past work in assigning the ∞ subscripts to C_∞ and R_∞ herein since these elements lead to the semicircle in the complex impedance plane which occurs at higher frequencies than do any other impedance plane structures. This semicircle peaks at $\omega = \tau_D^{-1}$, where $\tau_D = \mathrm{R}_\infty \mathrm{C}_\infty$ is the dielectric relaxation time of the material, an intensive quantity.

An important quantity in solid and liquid electrolytes is the Debye length L_D, given by

$$L_D = \left[(\varepsilon\varepsilon_0 RT)^{-1} F^2 \sum_{i=1}^{m'} z_i^2 c_i^0\right]^{-1/2} \tag{32}$$

where the sum includes mobile charge species only. The Debye length is a measure of the distance in the electrolyte over which a small perturbation in potential or electric field decays. Such a perturbation creates a region of space charge where electroneutrality no longer holds. This region extends only over a few Debye lengths.

The Gouy–Chapman diffuse double-layer differential capacitance C_R which is associated with the charge-transfer reaction resistance R_R appearing in Figure 2.2.6 is given by

$$C_R = \varepsilon\varepsilon_0 / L_D \tag{33}$$

in the absence of bias. An expression taking dc bias into account appears in Macdonald [1954], and a further generalization taking finite-ion-size effects into account as well by means of a lattice gas treatment is presented in Macdonald, Franceschetti, and Lehnen [1980] (see also Franceschetti [1982]).

The supported electrolyte half-cell reaction resistance R_R may be written in the general form

$$R_R = (RT/n^2F^2)(k_i^0 c_i^0)^{-1} \tag{34}$$

where k_i^0 and c_i^0 depend on the specific type of reaction considered. For the parent metal electrode situation, $M|M^{z+}$, where M denotes the metal electrode, $k_i^0 = k_b^0$, a potential-independent Butler–Volmer-type reaction rate parameter, and $c_i^0 = c_{M+}^0$, the bulk equilibrium concentration of the reacting cation. Similarly for a redox situation one finds $k_i^0 c_i^0 = (I_0/nF) = k_f^0 c_{\text{Red}}^0 = k_b^0 c_{\text{Ox}}^0$, where I_0 is the exchange current, a measure of the rates at which oxidation and reduction processes occur in equilibrium (Vetter [1967], Gabrielli [1981]). It is the common magnitude of the equal and opposite electrical currents associated with oxidation and with reduction (no net current in equilibrium). A more complicated expression for R_R and for infinite-length Warburg impedance under steady state conditions where the dc bias is nonzero has been given by Sluyters-Rehbach and Sluyters [1970].

Diffusion Effects. Next, consider small-signal unbiased diffusion effects. We initially discuss the parent-ion equal electrode situation for simplicity. The problem with applying the circuit of Figure 2.2.6 to the supported full-cell situation at low frequencies is that C_R remains an intensive quantity associated with an interface region but $2Z_{W\infty} \to Z_D$, some new diffusion impedance, and becomes extensive at low enough frequencies. It is then not correct to continue to allow the element $C_R/2$ to bridge the series combination of $2R_\infty$ and $2Z_{W\infty}$. In fact, at sufficiently low frequencies where the diffusion effects are extensive, there can only be a single diffusion impedance Z_D for the entire cell. At sufficiently high frequencies Z_D will be well approximated by $Z_{W\infty}$, but the full cell then exhibits an effective diffusion impedance of $2Z_{W\infty}$, not $Z_{W\infty}$, as shown in Figure 2.2.6. There is thus a transition region in frequency as the frequency is decreased in which the effective diffusion impedance goes from $2Z_{W\infty}$ to $Z_D = Z_W$, *not* $2Z_W$, and the $C_R/2$ connection shown in Figure 2.2.6 is also no longer entirely applicable. Let us therefore first consider sufficiently low frequencies that the extensive Z_D is present and the admittance of $C_R/2$ is negligible. We can then consider Z_D alone and its transition to $Z_{W\infty}$. We shall start with the expressions (Franceschetti [1981]) appropriate for Z_D in the parent-metal electrode and redox cases.

In the parent-metal electrode case take the diffusion coefficient of the metal ion M^{z+} as D_M. Then one finds

$$Z_D = Z_W = \left[\frac{RT}{(zF)^2}\right]\left[\frac{l}{D_M c^0_{M+}}\right]\left[\frac{\tanh\sqrt{j\omega l^2/D_M}}{\sqrt{j\omega l^2/D_M}}\right] \tag{35}$$

which should be compared with the result in Eq. (10). Now when $(\omega l^2/D_M) >> 3$, $Z_W \to Z_{W\infty}$ and one obtains

$$Z_{W\infty} = \frac{RT/(zF^2)}{(c^0_{M+})(j\omega D_M)^{1/2}} \tag{36}$$

The situation is somewhat more complicated in the redox case because of the presence of the two charged species in the electrolyte [unless $(z - n_e) = 0$]. Then one finds

$$Z_D = Z_{W,\text{Ox}}\text{Ox} + Z_{W,\text{Red}}$$

where

$$Z_{W,\text{Ox}} = \left[\frac{RT}{(nF)^2}\right]\left[\frac{l}{D_{\text{Ox}} c^0_{\text{Ox}}}\right]\left[\frac{\tanh\sqrt{j\omega l^2/D_{\text{Ox}}}}{\sqrt{j\omega l^2/D_{\text{Ox}}}}\right] \tag{37}$$

$$Z_{W,\mathrm{Red}} = \left[\frac{RT}{(nF)^2}\right]\left[\frac{l}{D_{\mathrm{Red}}c^0_{\mathrm{Red}}}\right]\left[\frac{\tanh\sqrt{j\omega l^2/D_{\mathrm{Red}}}}{\sqrt{j\omega l^2/D_{\mathrm{Red}}}}\right] \tag{38}$$

and D_{Ox} and D_{Red} are the relevant diffusion coefficients and all concentrations are those in the bulk. When both tanh terms are well approximated by unity, the expression for Z_D reduces to

$$\begin{aligned} Z_D &= Z_{W,\mathrm{Ox}}\mathrm{Ox} + Z_{W,\mathrm{Red}} \\ &\equiv Z_{W\infty} = \left[\frac{RT}{(nF)^2}\right]\left[\left(D_{\mathrm{Ox}}\left(c^0_{\mathrm{Ox}}\right)^2\right)^{-1/2} + \left(D_{\mathrm{Red}}\left(c^0_{\mathrm{Red}}\right)^2\right)^{-1/2}\right](j\omega)^{-1/2} \end{aligned} \tag{39}$$

the classical result (Armstrong *et al.* [1978], Gabrielli [1981]). The above results show that we may expect to find two finite-length Warburgs in series (and generally displaced in frequency) at sufficiently low frequencies. It is worth again emphasizing that although it will be the $Z_{W\infty}$ of Eq. (39) which appears in the half-cell circuit, it is $2Z_{W\infty}$ which appears for the full cell under supported conditions at sufficiently high frequencies.

We have attempted to give supported results in a form appropriate for comparison with unsupported ones by considering full-cell conditions. The transition problems discussed above only occur for unstirred (liquid) electrolytes or for solid electrolytes. When a stirred solution or rotating electrode with laminar flow is employed, the l which appears in Z_D and Z_W expressions is replaced by δ_N, where δ_N is the thickness of the Nernst diffusion layer. It decreases as the frequency of rotation of a rotating electrode increases and the experiment is always carried out for conditions where $\delta_N << l$.

2.2.3.3 Unsupported Situations: Theoretical Models

Introduction. We shall discuss results for unsupported situations under two categories: (1) those that follow directly or indirectly from exact solutions of the small-signal differential equations of charge motion in the material–electrode system; and (2) those which largely arise from empirical analysis of data and often use such ubiquitous distributed elements as CPEs. The first category deals with more idealized situations than the second but generally leads to more detailed results and to more specific relations between macroscopic equivalent circuit elements and microscopic processes occurring in the system. At the present early stage of theoretical analysis of real systems, both approaches have important roles to play.

Most theories of charge transport and interfacial charge transfer in unsupported situations involve a model which assumes a homogeneous material, for example a single crystal. Here we shall initially discuss the electrical response following from the application of a small-signal ac potential difference to homogeneous materials without applied dc bias or built-in Frenkel space charge layers and with identical plane parallel electrodes—the idealized full-cell situation. Theoretical results are only available so far for conditions where there may be a single species of mobile positive charge and a single species of mobile negative charge present with electroneutrality in the bulk. Results for polycrystalline materials and for homogeneous ones with Frenkel layers or applied dc bias will be discussed later on.

We shall start with a discussion of the exact results obtained from the solution of the most general model yet considered, but one which is still appreciably idealized (Macdonald and Franceschetti [1978]). Then work relaxing some of the idealizations will be discussed. Some of the present results have been included in the solid electrolyte reviews of Archer and Armstrong [1980] and by Franceschetti [1982]. The model of Macdonald and Franceschetti involves mobile positive and negative charges which may arise from three sources: the partial or full dissociation of (a) neutral intrinsic centers, (b) neutral donor centers, and (c) neutral acceptor centers. The model is general enough to include disordered sublattice materials and single crystals with Schottky or Frenkel disorder. Arbitrary amounts of generation–recombination (G/R) are allowed. After dissociation of a neutral center, the resulting positive and negative charges are taken to have arbitrary mobilities; so, for example, a donor center might dissociate to yield an immobile positive charge and a mobile negative charge. We shall denote the mobility ratio for negative and positive charges as $\pi_m \equiv \mu_n/\mu_p$. Although the present model also allows arbitrary valence numbers for the mobile charged species, we shall primarily restrict attention here to the usual uni-univalent case.

Boundary Conditions: Adsorption–Reaction Effects. The Macdonald–Franceschetti model involves relatively general boundary conditions at the electrodes and so includes the possibility of charge transfer reactions and specific adsorption. Because of its generality, however, the model prediction for $Z_t(\omega)$ is very complicated and, in general, cannot be well represented by even a complicated equivalent circuit. The $Z_t(\omega)$ expression, may, however, be used directly in CNLS fitting. Here, for simplicity, we shall consider only those specific situations where an approximate equivalent circuit is applicable. Idealizations involved in the model include the usual assumption of diffusion coefficients independent of field and position, the use of the simplified Chang–Jaffé [1952] boundary conditions, and the omission of all inner layer and finite-ion-size effects. Some rectification of the latter two idealizations will be discussed later.

The Chang–Jaffé boundary conditions involve the physical assumption that the current arising from the reaction of a charge carrier of a given species is proportional to the excess concentration of that species at the interface, that is for, say, a negatively charged species,

$$I_n = -z_n e k_n (n - n^0) \tag{40}$$

where z_n is the valence number of the charge carrier, k_n is a reaction rate parameter at the reaction plane, and n^0 is the bulk concentration of the species. The Chang–Jaffé conditions, as compared to the Butler–Volmer equation, are unrealistic in two important respects: there is a complete neglect of the finite size of the charge carriers (i.e. the compact double layer), and it is assumed that the charge transfer rate does not depend at all on the local concentration of the electrode reaction products. These deficiencies are not, however, nearly as limiting as one might at first expect. The neglect of the compact double layer introduces only a small error (which becomes zero at zero frequency) in many solid state situations (Macdonald [1974b], Franceschetti and Macdonald [1977]). The accumulation of the electrode reaction product can be neglected when (a) the product species is a metal atom which is rapidly incorporated into a parent-metal electrode, (b) the product is a gas atom which equilibrates very rapidly with the ambient atmosphere, or (c) the product species is soluble in the electrolyte or electrode and diffuses away from the interface very quickly. If the accumulation of the product species is not eliminated by one of these processes, but the transport of the product is governed by diffusion and therefore is independent of the electric field, it may be incorporated into the Chang–Jaffé boundary condition through the artifice of a complex, frequency-dependent reaction rate constant as described below.

It proves convenient in the theoretical work to use the dimensionless Chang–Jaffé rate parameters

$$\rho_n = (l/2)(k_n/D_n) \tag{41}$$

and

$$\rho_p = (l/2)(k_p/D_p) \tag{42}$$

where the k's are effective rate constants and the D's the diffusion coefficients of the negative and positive species. These parameters have in some previous work been given in terms of the alternate equivalent quantities $\rho_n = \rho_2 = r_2/2 = r_n/2$ and $\rho_p = \rho_1 = r_1/2 = r_p/2$. Clearly when $\rho_n = 0$ the electrode is completely blocking for the negative species and for $\rho_n = \infty$ it is completely nonblocking and nonpolarized.

Now since, to a good approximation, specific adsorption of an ion at an electrode and then a reaction of the adsorbed ion to form a neutral species occur at very nearly the same point in space, one might expect that these sequential interface processes would be largely decoupled from bulk and double-layer effects which occur elsewhere in the system. Some time ago Lànyi [1975] introduced the concept of frequency-dependent complex rate constants, and they have been found very useful in allowing reaction–adsorption effects to be included in a very simple way (Macdonald [1976a], Franceschetti and Macdonald [1977]). In essence, if a $Z_t(\omega)$ solution has been found for a certain situation involving the presence of real, frequency-independent ρ_n and ρ_p boundary parameters, one only needs to change them to complex frequency-dependent quantities to automatically include adsorption effects. No other parts of the solution are affected. As an example, suppose that a negative carrier is adsorbed and the adsorbed species then reacts to form a neutral species whose concentration remains, or is held, constant. One finds that the real ρ_n originally present in the solution need only be replaced by

$$\rho_n = \frac{\rho_{n0} + j(\omega\tau_D)\xi_{na}\rho_{n\infty}}{1 + j(\omega\tau_D)\xi_{na}} \tag{43}$$

where ρ_{n0} and $\rho_{n\infty}$ are the $\omega \to 0$ and $\omega \to \infty$ limits of ρ_n: $\xi_{na} \equiv \tau_{na}/\tau_D$; and τ_{na} is the adsorption relaxation time. In this case $\rho_{n\infty}$ is the rate constant for the first step in the adsorption–reaction sequence in which charge is exchanged between the electrolyte and the adsorbed layer, and ρ_{n0} is a function of both $\rho_{n\infty}$ and the rate constant for the second step in the process, in which charge is exchanged between the adsorbed layer and the electrode. For the case of pure adsorption, ρ_{n0} is zero and ρ_n becomes zero at $\omega \to 0$. In the limit in which the second (reaction) step is much faster than the initial (adsorption) step, the adsorbed layer becomes inconsequential, $\tau_{na} \to 0$, $\rho_{n0} \to \rho_{n\infty}$, and $\rho_{n\infty}$ becomes real and frequency-independent. The quantities ρ_{n0} and $\rho_{n\infty}$ or, equivalently, k_{n0} and $k_{n\infty}$, may be expressed in terms of partial derivatives with respect to various surface concentrations of the small-signal boundary conditions written in terms of current (Franceschetti and Macdonald [1977]). Derivative definitions of this form which depend on Taylor series expansions, appropriate for small-signal conditions, were developed earlier by Armstrong and Henderson [1972], for example (see also Armstrong *et al.* [1978]).

The rate-limiting diffusion of an electrode reaction can also be incorporated into the Chang–Jaffé boundary conditions by a similar approach. In this case, assuming, for example, diffusion through a semiinfinite electrode, the result obtained is

$$\rho_n = \frac{\rho_{n\infty}\sqrt{i\omega D}}{\rho_n' + \sqrt{i\omega D}} \tag{44}$$

where, as before, $\rho_{n\infty}$ is the $\omega \to \infty$ limit of ρ_n, ρ_n' is a rate parameter for the inverse electrode reaction, and D is the diffusion coefficient of the reaction product. For sufficiently large D, diffusion becomes undetectable and $\rho_n = \rho_{n\infty}$. Various adsorption–reaction–diffusion sequences have been considered by Franceschetti and Macdonald ([1979c, 1982]) and Franceschetti ([1982, 1984]).

DC Response. Before passing to the uni-univalent case which we will consider in detail, let us consider the full dc resistance of the system for arbitrary valence numbers, but only for two (or possibly one) species of mobile charge with equilibrium bulk concentrations n^0 and p^0, valence numbers z_n and z_p, and electrical mobilities μ_n and μ_p. Then electroneutrality in the bulk leads to $z_n n^0 = z_p p^0$. The bulk conductance G_∞ may be expressed as

$$G_\infty = R_\infty^{-1} = G_{\infty n} + G_{\infty p} \tag{45}$$

where

$$G_{\infty n} = (F/l)(z_n \mu_n n^0) \tag{46}$$

and

$$G_{\infty p} = (F/l)(z_p \mu_p p^0) \tag{47}$$

Let us further define the conductivity fractions (or bulk transport numbers) $\varepsilon_n \equiv G_{\infty n}/G_\infty$ and $\varepsilon_p \equiv G_{\infty p}/G_\infty$. These quantities may be written in the simple forms

$$\varepsilon_n = \left(1 + \pi_m^{-1}\right)^{-1} \tag{48}$$

and

$$\varepsilon_p = (1 + \pi_m)^{-1} \tag{49}$$

under intrinsic conditions, the only case to be considered in detail here. Finally, define the Debye length when only one species of charge is mobile as L_{D1}. The important quantities $M \equiv (l/2)/L_D$ and $M_1 \equiv (l/2)/L_{D1}$ then measure the number of Debye lengths in a half-cell (half a symmetrical cell of full electrode separation l). In the present case where L_D and M refer to a single species of positive and a single species of negative charge mobile, $L_{D1} = \sqrt{2} L_D$.

Let us (apparently arbitrarily) now define the small-signal half-cell adsorption–reaction impedances associated with the positively and negatively charged species as

$$Z_{Rn} \equiv RT / z_n^2 F^2 k_n n^0 \tag{50}$$

and

$$Z_{Rp} \equiv RT / z_p^2 F^2 k_p n^0 \tag{51}$$

where the k's may be complex. Using the Einstein relation $D_j = (RT/F)\,(\mu_j/z_j)$ for the $j = n$ and p species, one readily finds that ρ_j may be written as $(l/2)\,(F/RT)\,(z_j k_j/\mu_j)$, so if ρ_j is complex, so is k_j. Note that the $\omega \to 0$ limits of Z_{Rn} and Z_{Rp} are

$$R_{\theta n} \equiv RT / z_n^2 F^2 k_n n^0 \tag{52}$$

and

$$R_{\theta p} \equiv RT / z_p^2 F^2 k_{p0} p^0 \tag{53}$$

where k_{n0} and k_{p0} are related as above to ρ_{n0} and ρ_{p0}.

Now it is often useful to consider normalized quantities in theoretical analysis or even in an equivalent circuit or 3-D plot. We shall, when desirable, normalize impedances with the bulk resistance R_∞, so $Z_N \equiv Z/R_\infty$, and capacitances with C_∞, so $C_N \equiv C/C_\infty$. The normalized expressions for some of the circuit elements defined above simplify considerably and are

$$G_{\infty nN} \equiv G_{\infty n} / G_\infty = \varepsilon_n \tag{54}$$

$$G_{\infty pN} \equiv G_{\infty p} / G_\infty = \varepsilon_p \tag{55}$$

$$G_{\theta nN} \equiv R_\infty / R_{\theta n} = 2\varepsilon_n \rho_{n0} \tag{56}$$

and

$$G_{\theta pN} \equiv R_\infty / R_{\theta p} = 2\varepsilon_p \rho_{p0} \tag{57}$$

We are now finally in a good position to write down the expression for the full-cell complete dc resistance following from the present model, $R_D \equiv Z_t(\omega \to 0)$. We write it here as R_D or R_{DN} rather than R_0 to agree with earlier usage. The exact R_D result applies for arbitrary valences for the mobile charged species, arbitrary mobilities, intrinsic, extrinsic, or intrinsic and extrinsic conduction, and any dissociation–recombination conditions. In unnormalized form it is just

$$R_D \equiv G_D^{-1} = (G_n + G_p)^{-1} \tag{58}$$

where

$$G_n \equiv (R_{\infty n} + 2R_{\theta n})^{-1} \tag{59}$$

and

$$G_p \equiv (R_{\infty p} + 2R_{\theta p})^{-1} \tag{60}$$

These results show that the total dc conductance is made up of a branch G_n, involving negative charge carrier effects only, in parallel with a similar branch involving only positive carrier effects. Each individual branch involves a bulk resistive contribution and two equal adsorption–reaction resistances, one associated with each electrode. The expression for R_D in normalized form, R_{DN} (e.g. Franceschetti and Macdonald [1977]), is even simpler, namely,

$$R_{DN} = \left(\frac{\varepsilon_n}{1+\rho_{n0}^{-1}} + \frac{\varepsilon_p}{1+\rho_{p0}^{-1}} \right)^{-1} \tag{61}$$

Note that when $\rho_{n0} = \rho_{p0} = \infty$, one obtains $R_{DN} = (\varepsilon_n + \varepsilon_p)^{-1} \equiv 1$; so $R_D = R_\infty$, ohmic behavior and thus not very interesting. Of course in the completely blocking $\rho_{n0} = \rho_{p0} = 0$ case, $R_{DN} = \infty$. To set a scale, it is interesting to note that when $\rho_{n0} = \rho_{p0} = 1$, $R_{DN} = 2$ and adsorption–reaction effects have contributed an additional R_∞ resistance to R_D. In general, when $\rho_{p0} = \rho_{n0} \equiv \rho_e$, then $R_{DN} = 1 + \rho_e^{-1}$, a result entirely independent of mobilities and π_m, except indirectly through R_∞ and ρ_e themselves.

Adsorption–Reaction and Reaction–Diffusion Predictions. Next, in order to investigate adsorption and reaction effects more fully, let us consider Z_{Rn} and Z_{Rp} in normalized form. It is straightforward to show that

$$Z_{RnN} = (2\varepsilon_n \rho_n)^{-1} = \rho_{n0} R_{\theta nN} / \rho_n \tag{62}$$

and

$$Z_{RpN} = (2\varepsilon_p \rho_p)^{-1} = \rho_{p0} R_{\theta pN} / \rho_p \tag{63}$$

If we now substitute the complex ρ_n from Eq. (43) into Eq. (62) and a similar expression for ρ_p into Eq. (63), we readily find that the resulting impedances each lead to a simple ladder network whose hierarchical form is consonant with the sequential processes: adsorption then reaction. But for the full cell there are two identical interface impedances in series. The circuit for a half-cell with total impedance Z_{Rn} is shown in Figure 2.2.7*a*. The full-cell impedance is just $2Z_{Rn}$. The normalized elements of Figure 2.2.7*a* are readily found to be given by

$$R_{RnN} = (\varepsilon_n \rho_{n\infty})^{-1} \tag{64}$$

$$R_{AnN} = \rho_{nm} / \varepsilon_n \rho_{n0} \rho_{n\infty} \tag{65}$$

and

$$C_{AnN} = \xi_n \varepsilon_n \rho_{n\infty}^2 / \rho_{nm} \tag{66}$$

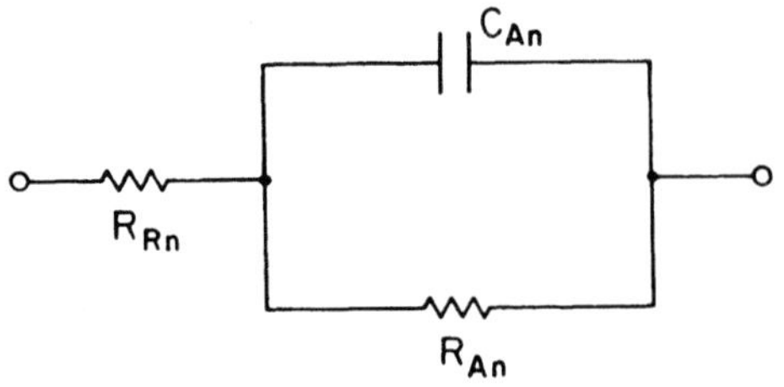

(a)

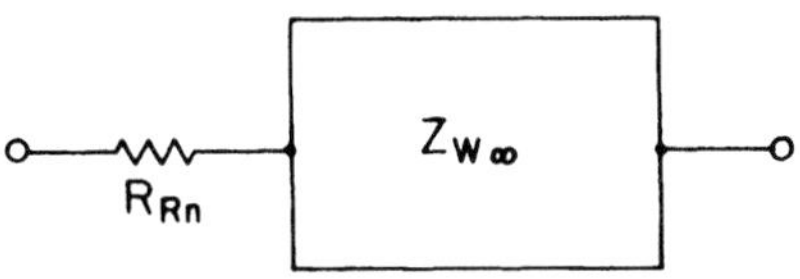

(b)

Figure 2.2.7. (a) Equivalent circuit for two identical simple electrode adsorption–reaction processes in series, one at each electrode, with negative charge carriers reacting. Unsupported conditions. (b) Equivalent circuit for two identical reaction–diffusion processes with negative charge carriers reacting. Unsupported conditions.

where $\rho_{nm} \equiv \rho_{n\infty} - \rho_{n0}$. All the elements in the circuit of Figure 2.2.7 are intensive, as they should be for interface effects. The normalized dc resistance of the circuit is just

$$2Z_{RnN}(\omega \to 0) = R_{RnN} + R_{AnN} = (\varepsilon_n \rho_{n0})^{-1} \equiv 2R_{\theta nN} \tag{67}$$

which is as expected. Note that since ρ_{nm} may be either positive or negative, R_{An} and C_{An}, specific adsorption elements, have the same sign and also may be positive or negative, in agreement with earlier discussion. In the absence of the adsorption step, $2Z_{Rn} = R_{Rn}$, the reaction resistance (for two electrodes), since R_{An} is then zero and C_{An} infinite. See Figure 4.4.25 for some of the complex plane shapes which follow from the present approach.

In like manner, if we substitute the complex ρ_n from Eq. (44) into Eq. (62), we obtain the circuit shown in Figure 2.2.7*b*, appropriate for a reaction–diffusion sequence without an intermediate adsorption stage. Here R_{RnN} is given by Eq. (64) as before and

$$Z_{WN} = \rho_n' / \varepsilon_n \rho_{n\infty} \sqrt{i\omega D} \tag{68}$$

Theoretical Results for Various Cases of Interest. Thus far we have only considered some exact $\omega \to 0$ results and typical adsorption–reaction interface frequency response for a half-cell or full cell. Let us now turn to further predictions of the complete full-cell model (Macdonald and Franceschetti [1978]), predictions derived from its specific analytical results in several simplified cases and from a large amount of CNLS fitting of various equivalent circuits to the exact model predictions (e.g. Franceschetti and Macdonald [1977], Macdonald, Franceschetti, and Meaudre [1977], Macdonald and Franceschetti [1979a], Macdonald and Hull [1984]). For simplicity, we consider only the uni-univalent case ($z_n = z_p = 1$), intrinsic conduction, and $M >> 1$. The latter condition excludes the behavior of very thin layers and membranes, but their response has been discussed in the literature. Let us define cases of interest by their $[\rho_p, \rho_n, \pi_m]$ values. Actual values of ρ_p and ρ_n cited in this way will always be real, but when the symbols are used, they may include complex cases. Because the electrodes are taken to be identical, ρ_n and ρ_p values apply to both electrodes.

Little has been done on the $[\rho_e, \rho_e, \pi_m]$ case, that where the normalized reaction rates (but not necessarily k_n and k_p) are equal and nonzero. Although this is a situation of small experimental interest except for $\rho_e \cong 0$, a formal expression for its admittance has been given (Macdonald and Franceschetti [1978]) but does not lead to a useful approximate equivalent circuit representation except when $\pi_m = 1$. When $\rho_e = 0$ as well, there is no finite-length Warburg present. Because of its complexity, model predictions for the general $[\rho_p, \rho_n, \pi_m]$ case with both ρ_p and ρ_n values nonzero and noninfinite have been little explored, although it has been found that for a nonzero ρ_n value say, and even for $\pi_m = 1$, as ρ_p increases from zero toward ρ_n a finite-length Warburg arc which appears in the impedance plane rapidly decreases in size (Macdonald [1975]). Note that when a diffusion arc is present, one finds (Macdonald [1974b]) that $C_p(\omega)$, the total frequency-dependent parallel capacitance associated with Y, exhibits $\omega^{-3/2}$ and $\omega^{-1/2}$ dependence, quite different from ω^{-2} simple Debye response, yet frequently observed experimentally.

The situation of most experimental interest, especially for solid electrolytes, is defined by $[0, \rho_n, \pi_m]$ or, equivalently, $[\rho_p, 0, \pi_m]$. Thus only one species of charge carrier discharges, but both positive and negative ones may be mobile. The equivalent circuit we believe to be most appropriate in this case is presented in Figure 2.2.8*b*. First, we see the usual elements C_∞, R_∞, R_R, C_A, and R_A already discussed. For this $[0, \rho_n, \pi_m]$ situation, R_R, R_A, and C_A are given by just the unnormalized forms of R_{RnN}, R_{AnN}, and C_{AnN} presented in Eqs (64)–(66). The additional elements Z_{De}, Z_W, C_R, and $R_{R\infty}$ which appear in the circuit require discussion. But first it should be emphasized that much theoretical analysis and fitting of theory to different equivalent circuits makes it quite clear that the hierarchical ladder network form of this circuit is far more appropriate than either the series Voigt or the parallel Maxwell forms. The ladder network, which leads to a continued fraction expression for the total impedance, ensures that the $R_\infty C_\infty$ arc in the impedance plane will occur at the highest frequencies, followed (with or without overlap) by a $R_R C_R$ arc and then $R_A C_A$ arc as the frequency decreases. The Voigt circuit, for example, imposes no such requirements. The diffusion arc(s) may actually occur in any frequency range for the present circuit but is usually found at the right of the diagram, the lowest frequency region.

The element Z_{De} is a general diffusion impedance added to account for possible diffusion of uncharged reactants such as oxygen atoms (but not parent-electrode atoms) in the electrodes (for the present full-cell situation it accounts, of course, for diffusion in both electrodes). When $\rho_n \equiv 0$ it should not appear. In most cases of

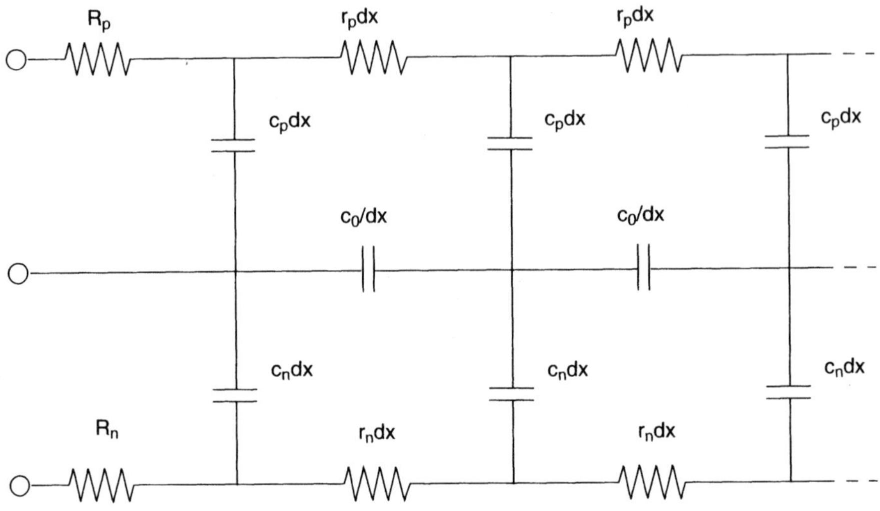

(*a*)

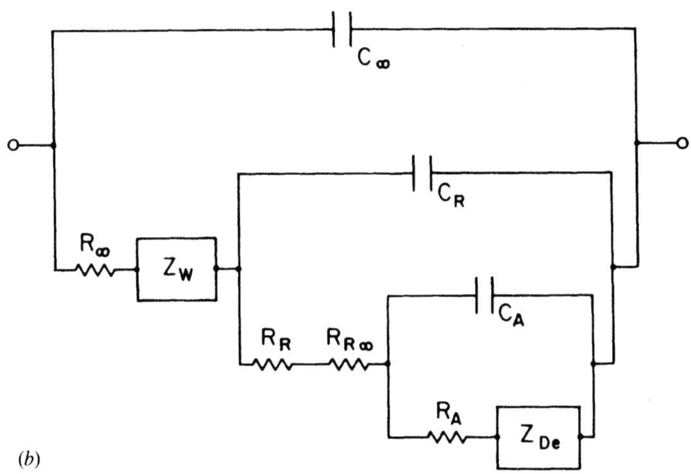

(*b*)

Figure 2.2.8. (a) Transmission line representation of Nernst–Planck Poisson equation system for a binary electrolyte. R_p and R_n are charge transfer resistances for positive and negative charge species at the electrode, respectively. (b) General approximate equivalent circuit (full-cell, unsupported) for the [0, ρ_n, π_m] cases applying to a homogeneous liquid or solid material.

interest, it will probably be best represented by a finite-length Warburg impedance, so $Z_{De} = Z_{We}$. One might at first be surprised to find Z_{De} in series with the adsorption related resistance R_A rather than in series with C_∞ and R_∞. Although the diffusion process in the electrode does occur after the charge carrier has been transported through the electrolyte, it has only a negligible effect on the potential difference across the electrode, which is essentially zero for a metallic electrode. Rather, it contributes to the interfacial impedance by hindering the discharge of the adsorbed species. If adsorption does not occur to any significant extent then $C_A \to 0$, $R_A \to 0$, and Z_{De} is in series with R_R.

The next new element is a finite-length Warburg impedance Z_W, associated with diffusion of charged particles within the electrolyte. It is primarily present in low-frequency regions where the more mobile or more abundant charged species have time in a half-cycle to rearrange positions so as to screen the less mobile or less abundant charges from the electric field, leaving diffusion as the primary conduction method for such low frequencies. Since this diffusional process occurs in the electrolyte bulk, Z_W is placed in series with R_∞. For the present situation, the diffusion impedance is well approximated by

$$Z_{WN} = Z_W / R_\infty = \pi_m^{-1}\left[\tanh(j\Omega H_N^2)^{1/2} / (j\Omega H_N^2)^{1/2}\right] \tag{69}$$

except when $\rho_n = 0$ and $\pi_m = 1$ simultaneously, a situation where $Z_W = 0$. Here $\Omega \equiv \omega\tau_D \equiv \omega R_\infty C_\infty$. The quantity H_N is found to be

$$H_N = (M_e/2)(\pi_m^{-1} + 2 + \pi_m)^{1/2} \tag{70}$$

We defined $M_e \equiv l/2L_{De}$, where L_{De} in the present case is given by

$$L_{De} = \left[\varepsilon\varepsilon_0 RT/8\pi F^2 q n^0\right]^{1/2} \tag{71}$$

and q takes intrinsic G/R into account and is actually frequency-dependent and complex in the exact theory. Here it will be sufficient to take $q = 1$ when charges of both sign are mobile and usually take $q = 0.5$ when only one species is mobile (see later discussion). Then L_{De} equals either L_D or L_{D1}. Note that when $\pi_m = 1$, $H_N = M_e$. Alternatively, when $\pi_m \to 0$ because $\mu_n \to 0$, there will be no dc path through the circuit and $Z_W \to \infty$.

The appearance of a Warburg impedance—generally associated with a diffusion process—arising from the motion of charged particles in an unsupported system has been the source of some confusion. A discussion and comparison of the various sources of finite-length Warburg impedances in both supported and unsupported systems can be found in Franceschetti *et al.* [1991]. The essential point is that the appearance of Warburg behavior in itself is not necessarily a signature of mobile neutral species coupled to the electrode reaction.

It should be mentioned that when the product of a reaction at the electrode is neutral, it may possibly diffuse back into the electrolyte as well as into the electrode (Franceschetti [1981]). Then another Z_W appears in series with R_R. How can one distinguish between up to three different Z_W's, all effectively in series with R_∞? By changes in the electrode thickness l_e and the separation of the electrodes l, one should be able to identify a given Z_W arc as arising from diffusion in one or the other region. One can then decide whether a Z_W which depends on l involves charged or uncharged species by changing (if possible) the equilibrium concentration of the neutral species in the electrolyte, either directly or by changing the composition or pressure of the ambient atmosphere. Of course, in most experimental situations only a single Z_W arc appears in the very-low-frequency region (or measurements do not extend to low enough frequencies to show others). As we shall see subsequently, a single Z_W can only arise from neutral-species diffusion if $\mu_p = 0$ and $\mu_n \neq 0$, so $\pi_m = \infty$, the one-mobile case. Similar results appear for $\mu_n = 0$ and $\mu_p \neq 0$.

The next element requiring discussion is C_R, the reaction capacitance, arising from the series combination of equal diffuse double-layer capacitance effects at each electrode. It is usually very well approximated by

$$C_{RN} \equiv C_R / C_\infty = M_e \operatorname{ctnh}(M_e) - 1 \tag{72}$$

essentially equal to M_e in the present $M >> 1$ case, but note that $C_R + C_\infty = C_\infty M_e$ ctnh (M_e). When $M_e >> 1$, the usual situation, this full-cell result is just half of the conventional diffuse double-layer capacitance, an intensive quantity, given in Eq. (33) for the supported half-cell situation.

Now what about the remaining element, $R_{R\infty}$? In the present case, Eq. (61) leads to $R_{DN} = \varepsilon_n^{-1}[1 + \rho_{n0}^{-1}] = 1 + \pi_m^{-1} + [\varepsilon_n\rho_{n0}]^{-1}$, an exact result. But on omitting Z_{De}, the circuit of Figure 2.2.8*b* leads to $R_{DN} = R_{\infty N} + Z_{WN0} + (R_{RN} + R_{AN}) + R_{R\infty N} = 1 + \pi_m^{-1} + (\varepsilon_n\rho_{n0})^{-1} + R_{R\infty N}$. Thus $R_{R\infty}$ must actually be zero, at least in the $\omega \rightarrow 0$ limit, unless expressions for one or more of the other parameters are incorrect in this limit. Macdonald and Hull [1984] found that even when $R_A = 0$ and $R_R = 0$ (taking $\rho_n = \rho_{n\infty} = \infty$) a circuit similar to the present one with the present Z_W could be best fitted to the exact $[0, \infty, \pi_m]$ case with a nonzero $R_{R\infty}$ approximately given by

$$R_{R\infty N} = 2M^{-1}[\pi_m^{-1} - 1] \tag{73}$$

for $\pi_m \leq 1$ and by zero for $\pi_m \geq 1$. Thus even in the absence of a normal reaction resistance, CNLS fitting of exact data leads to a nonzero apparent reaction resistance. For large M it is only of importance when π_m is very small (a high-resistance case), since when $R_{R\infty} < 0.01\ R_\infty$ its effect will be essentially negligible and difficult to resolve even with CNLS fitting. Note also that $R_{R\infty N}$ will always be appreciably smaller than $Z_{WN0} = \pi_m^{-1}$ for $M >> 1$. Nevertheless, the presence of this element in the circuit and the natural tendency to consider the measured $R_R + R_{R\infty}$ as "the" reaction resistance can lead to incorrect estimates of the rate parameter unless the presence of $R_{R\infty}$ is explicitly recognized (Macdonald and Hull [1984]). The fact that $R_{R\infty}$ should not actually appear at $\omega = 0$, yet is needed in the fitting circuit, is an indication of some inappropriateness in the fitting circuit itself. But for the present it seems the best circuit available.

Let us continue to ignore Z_{De} and investigate two simpler cases. First consider the important completely blocking case $[0, 0, \pi_m]$ where $R_R = \infty$. There is still a Warburg impedance present in general, but it can only contribute to making the normally vertical spur present at low frequencies in the impedance plane and associated with complete blocking show less than vertical behavior over a finite frequency range. But an inadequacy of the present expression for Z_W appears when $\pi_m = 1$; then the exact solution leads to no Z_W but Eq. (69) still yields a nonzero Z_W. The special [0, 0, 1] case must therefore be handled separately until a more complete expression for Z_W is found or unless direct CNLS comparison between data and model predictions is employed.

The Case of Charge of Only a Single-Sign Mobile. The remaining one-mobile case, $[0, \rho_n, \infty]$, is of particular interest for solid electrolytes with only a single-species-of-charge (here negative-charge) mobile. This situation is the most usual one for solid electrolytes, although it should be realized that it is always something of an approximation. At nonzero temperature both positive- and negative-charge species present in a solid material or a fused salt are mobile, although their mobility ratio, $\pi_m \equiv \mu_n/\mu_p$, may be either very large or very small. The relatively immobile species may have so low a mobility at a given temperature that motion of this species is negligible during a half-cycle of the lowest frequency applied. Then the one-mobile approximation will be a good one. Although we have taken ρ_p as zero in the above case designation, its value is immaterial since the positive charges are taken immobile ($\mu_p = 0$) and cannot react at an electrode. In the present case all the ε_n's which appear in the defining equations are unity and all π_m^{-1}'s zero. Therefore, as Eq. (69) shows, $Z_W = 0$, and no charged-particle Warburg arc is present, and the only Warburg diffusion response possible must arise from diffusion of neutral particles in the electrodes or the bulk of the material. The exact theoretical results show that in the present case the circuit of Figure 2.2.8*b* (with $R_{R\infty} = 0$) is completely applicable with all frequency-independent elements given exactly by their values following from the foregoing expressions except that for C_{RN}. When $Z_{De} = 0$ and the time constants are well separated, so that $R_A C_A >> R_R C_R >> R_\infty C_\infty$, the circuit of Figure 2.2.8*b* leads to just three distinct arcs in the Z^* plane.

It is in the partially dissociated one-mobile case that G/R can play a role of some importance (e.g. Macdonald [1953], Macdonald and Franceschetti [1978]). We have already mentioned that the L_{De} which appears in the equation for C_{RN} should usually be taken as L_{D1} ($q = 0.5$) in the one-mobile case. This choice is particularly relevant for fully dissociated charges, such as might arise from the complete ionization of immobile donors. But in the partly dissociated situation, appropriate for intrinsic conduction, G/R can lead to an effective mobility for the immobile charge species (except at dc). Then, over some region of frequency, as ν decreases L_{De} changes from L_{D1} to L_D because of the frequency dependence of q. Thus, although the formal expression for C_{RN} given above remains valid for the one-mobile case, one must consider the physical situation to decide whether to use L_{D1} or L_D in M_e. Alternatively, it is more accurate to use the full frequency-dependent expression for q as a part of the definition of C_{RN} in this case. Then CNLS fitting can, in principle, lead to information about the degree of dissociation of intrinsic centers and the associated G/R parameters. Such a procedure would only be justified, however, for excellent data. It was originally thought (Macdonald [1976b]) that G/R might lead to a separate semicircle in the impedance plane, but later work (Macdonald *et al.* [1977]) suggests that it does not for ionic conduction. Thus its effects for the one-mobile case are entirely restricted to C_R only and are relatively small even there. Note that in the two-mobile case, as long as the mobilities are not greatly different, effective mobilization, arising from G/R, of the species with the smaller mobility will still lead to negligible effects.

Some Results for More General and Realistic Situations. Next let us consider the removal of some of the approximations inherent in the foregoing model. For the small-signal flat-band case, it turns out that the half-cell reaction resistance in the supported case, Eq. (34), derived using the Butler–Volmer equation, and the half-cell reaction resistance in the unsupported case, $R_{\theta n}$, Eq. (52), which followed from use of the Chang–Jaffé boundary equations, are essentially identical (Macdonald [1974a,b]). Furthermore, Franceschetti and Macdonald [1977] and Macdonald and Franceschetti [1979a] later showed that the calculation of the reaction resistance in either the unsupported or supported case gave the same result (because of compensating errors) whether Chang–Jaffé or Butler–Volmer equations were employed, provided, however, that any inner- or compact-layer capacitance C_c present was much larger than the diffuse double-layer capacitance. In addition, a method of transforming an unsupported small-signal impedance solution based on Chang–Jaffé boundary conditions to one employing Butler–Volmer, or even more general boundary equations, was developed. This method obviates the difficult task of solving the small-signal equations *ab initio* with the new boundary conditions.

There are plausible physical reasons to prefer Butler–Volmer to Chang–Jaffé conditions, especially when a compact layer is present, since the Butler–Volmer equations can account for the p.d. across this layer. Macdonald and Franceschetti [1979a] therefore studied how, for the $[0, \rho_{n0}, \infty]$ case without adsorption, C_R and R_R are changed from the results given here to new values when Butler–Volmer, or even more general boundary conditions, are used instead of Chang–Jaffé conditions and a compact layer of arbitrary constant capacitance was assumed present as well. Theoretical results were given and CNLS fitting of such results to an equivalent circuit were carried out in order to find the simplest adequate modifications needed. This approach not only allowed $\omega \to 0$ modifications to appear but yielded information on changes in the interface impedance over all ω values of interest arising from the presence of C_c and the more general boundary conditions.

Results found from the above approach were surprisingly simple. The present expression for $R_{\theta N}$ or $R_{\theta n N}$ (with $\varepsilon_n = 1$) was shown to hold exactly in the Butler–Volmer case, independent of the size of C_{cN}. A simple expression for a new effective full-cell C_R, say C_{Re}, was found in the Butler–Volmer case, namely,

$$C_{ReN} = C_{RN} - (C_{RN} + 1 + \rho_{n0})^2 / (C_{RN} + 1 + C_{cN}) \tag{74}$$

Now in the usual $M >> 1$ case where $C_{RN} >> 1 + \rho_{n0}$, this result reduces to just $C_{ReN}^{-1} \cong C_{RN}^{-1} + C_{cN}^{-1}$, a series combination of the original C_R and the compact-layer capacitance C_c. However, this is just the $\omega \to 0$ result always used in the supported case! Although these results were derived for the $[0, \rho_{n0}, \infty]$ case without adsorption, they should hold quite adequately for the nonadsorption $[0, \rho_{n0}, \pi_m]$ case as well.

When adsorption is present and the effects of a compact layer are included as well, it has been shown (Macdonald *et al.* [1980]) that in the $[0, 0, \infty]$ case the $\omega \to 0$ expression for the total differential capacitance is more complicated than just C_{Re} and C_A in parallel. One needs first to separate C_c into two series parts so $C_c^{-1} = C_\alpha^{-1} + C_\beta^{-1}$, where C_α is the capacitance between the electrode and the charge centroids of the adsorbed ions (at the inner Helmholtz plane), and C_β is that from this plane to the outer Helmholtz plane, where the diffuse layer of charge begins. Then one obtains the circuit of Figure 2.2.9, which reduces to the above result for

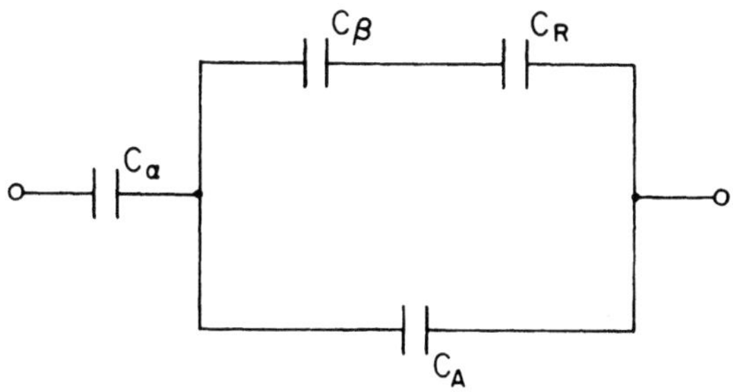

Figure 2.2.9. Circuit for the total interface differential capacitance in the (0, 0, ∞) case without adsorption for $\omega \to 0$.

C_{Re} when $C_\alpha = \infty$. When $C_\alpha < \infty$, it is not clear how R_R should be added to this circuit since it should bridge C_α and still be in series with C_A. For most solid electrolyte situations, however, it will usually be an adequate approximation to take $C_\alpha = \infty$ in fact and to put R_R in series with C_A, returning to the usual form of the interface part of the circuit (Figure 2.2.8*b*). First, there is the probability of the electron wave function spilling out from the surface of a metal electrode and reducing the effective thickness of C_α toward zero (Kornyshev *et al.* [1982]). Second, for solid materials there will be no inner uncharged layer of solvent material, as in liquid electrolyte situations. This means that the plane marking the beginning of the diffuse layer is nearly as close to the electrode as the plane where adsorption occurs; when $C_{\alpha N} \to \infty$ or it is very large, $C_{\beta N}$ will also be extremely large and may often be neglected compared to C_R, so $C_{Re} \cong C_R$. Incidentally, in the presence of dc bias, C_R (and C_A) are both functions of the effective overpotential and can increase greatly over their flat-band values under some conditions. They are limited in maximum value, however, because of the finite size of ions (Macdonald *et al.* [1980]). When one finds experimentally that C_{Re} is nearly independent of applied dc bias and temperature, it is likely that it is dominated by C_c rather than C_R.

Now let us briefly consider some results for the non-flat-band case without adsorption (Franceschetti and Macdonald [1979a,b, 1980]). Both transient response and biased small-signal frequency response results have been obtained using computer simulation, that is numerical solution of coupled sets of partial or ordinary second-order differential equations describing the model. Chang–Jaffé and Butler–Volmer boundary conditions were both employed. Here we shall discuss only the frequency response results. First, the response of cells with (0, 0, π_m) conditions at the left electrode and (∞, ∞, π_m) conditions at the right one was considered, leading to essentially half-cell conditions. The dc bias was assumed to arise from either built-in Frenkel space charge regions or an actual applied p.d. No direct current flowed in these completely blocking situations. Second, results were obtained for the full-cell system with (0, 0, π_m) conditions at each electrode. Many complex plane Z and Y plots were presented to show how arcs and spurs varied with applied bias. More importantly, it was found that the equivalent circuit of Figure 2.2.8*b* with $Z_{De} = 0$, $\rho_{n0} = 0$, and no adsorption applied quite well, not only for zero bias (flat-band) but for either polarity of applied bias up to the maximum magnitude applied of about 15 (RT/F), sufficient bias to make the system very nonlinear. There are no R_A and R_R elements present under these conditions, but Z_W did appear for $\pi_m \neq 1$ and for $\pi_m \neq \infty$ conditions. It is impractical to summarize here all the results found and reported, but the dependences of C_R and the components of Z_W generally varied with bias in reasonable and expected ways. The bulk parameters R_∞ and C_∞ showed negligible variation with bias.

Later, large-bias frequency response results were obtained for one-mobile partially blocking situations with no adsorption and with either $(\rho_{p0}, -, 0)$ for the left electrode and $(\infty, -, 0)$ for the right (half-cell conditions) or $(\rho_{p0}, -, 0)$ for both. Then a direct current can flow, and steady state current–voltage curves for Chang–Jaffé conditions were compared with those for Butler–Volmer ones. Appreciable differences occurred for biases bigger in magnitude than (RT/F). As expected, no diffusion effects were present in the response. The circuit of Figure 2.2.8*b* was again found adequate to describe the response, here with $Z_{De} = 0$, $R_A = 0$, and $Z_W = 0$. For Chang–Jaffé conditions $R_{\infty N}$ remained very close to its expected unity value and $C_{\infty N}$ was held fixed at unity, but for Butler–Volmer conditions $R_{\infty N}$ and $C_{\infty N}$ were somewhat bias-dependent and differed from unity. The dependences of R_R and C_R on bias were in accord with predictions based on the buildup of charge accumulation or depletion regions near the partly blocking electrodes, and it was found that C_{RN} was smaller in the Butler–Volmer case, because of compact-layer effects, than in the Chang–Jaffé one. For the same reason R_{RN} was less bias-dependent for Butler–Volmer than for Chang–Jaffé.

The foregoing results show the wide scope of the Figure 2.2.8*b* general circuit. It applies with good approximation for both flat-band small-signal conditions and under equilibrium or nonequilibrium biased conditions provided its elements are properly interpreted to account for the presence or absence of a compact layer and the appropriate type of boundary conditions.

Thus far we have dealt with either half-cells, where the right half-cell boundary involves nonpolarizing, ohmic boundary conditions (∞, ∞, π_m) and the left involves conditions such as $(0, \rho_n, \pi_m)$ or with full cells with identical boundary conditions at each electrode. Another interesting full-cell case is that of crossed reactions where the left electrode involves $(0, \rho_n, \pi_m)$ and the right $(\rho_p, 0, \pi_m)$. Different electrodes are used, so charge of one sign reacts at one electrode and that of the other sign at the other electrode. This double-injection model cannot pass dc and has been analyzed by Glarum and Marshall [1980]. Although their solution is rather complicated, it reduces under conditions of interest $(\pi_m << 1)$ to just the Z_{DOC} impedance of Section 2.2.2.2, that for an open-circuited transmission line. As already noted, it leads to an ordinary finite-length Warburg arc in the ε, not the Z, complex plane. Glarum and Marshall have used this result with some success in analyzing data for iridium oxide thin films (see also Franceschetti and Macdonald [1982]). It is therefore likely that a modification of the general Figure 2.2.8*b* circuit, useful in some $R_D = 0$ situations, would be to replace Z_W by Z_{DOC}. In fact, the most general modification would be to replace Z_W by Z_D, allowing the possibility of any kind of uniform-transmission-line-like behavior.

2.2.3.4 Unsupported Situations: Equivalent Network Models

Sah [1970] introduced the use of networks of electrical elements of infinitesimal size to describe charge carrier motion and generation/recombination in semiconductors. Barker [1975] noted that the Nernst–Planck–Poisson equation system for an unsupported binary electrolyte could be represented by a three-rail transmission line (Figure 2.2.8*a*), in which a central conductor with a fixed capacitive reactance per unit length is connected by shunt capacitances to two resistive rails representing the individual ion conductivities. Electrical potentials measured between points on the central rail correspond to electrostatic potential differences between the corresponding points in the cell while potentials computed for the resistive rails correspond to differences in electrochemical potential. This idea was further developed by Brumleve and Buck [1978], and by Franceschetti [1994] who noted that nothing in principle prevents extension of the model to two or three dimensional systems.

Most recently, the electrical network formalism has been used extensively by Jamnik, Maier and their collaborators (Jamnik *et al.* [1999], Jamnik and Maier [2001], Jamnik [2003]), whose work has highlighted the utility of the models for qualitative reasoning about phenomena occurring in systems with a mixed ionic and electronic conducting electrolyte (MIEC) and electrodes nearly blocking for one species. These workers highlight the fact that the capacitors connecting the ionic rails to the central rail representing bulk dielectric behavior are in fact chemical, rather than electrostatic, capacitances related to the derivative of chemical potential with respect to ionic concentration. In modeling mixed conductors with a large enough electrode separation to have bulk electroneutrality, Jamnik *et al.* [1999] and Jamnik and Maier [2001] obtain informative approximate results by eliminating the central or displacement rail in Figure 2.2.8*a* and replacing it with a geometric capacitance in parallel with the transmission line. In particular, they are able to model the transition between a "nearly single-carrier system" not exhibiting a Warburg region and a two carrier system, exhibiting a Warburg. Jamnik [2003] describes the use of the model to describe multiphase systems and to model grain boundary effects.

While electrical networks offer a number of useful features, as a guide to qualitative reasoning about electrochemical systems and as a means of shedding light on the origin of the elements appearing in "lumped" circuits, it should be remembered that they are simply a graphical representation for what amounts to a finite difference representation of the underlying Nernst–Planck, Poisson, Maxwell and Diffusion equations. Flieg and Maier [1997] have made use of this direct computational approach to study the effects of partial contact with a real electrode while Jamnik *et al.* [2000] have used it to study the effective rate constant for structured arrays of contact points or strips.

2.2.3.5 Unsupported Situations: Empirical and Semiempirical Models

In this section, we shall first discuss some ways in which the theoretical model results and equivalent circuits of the last section may be modified to attempt to account for less ideal conditions than assumed in the theory, conditions often appreciably closer to those found in real material–electrode systems. Then we shall discuss empirical and semiempirical models which may be useful as elements in equivalent circuits used for fitting real IS data.

Possible Circuit and Model Modifications. Further modifications of the Figure 2.2.8*b* equivalent circuit are often necessary, especially for polycrystalline material. One frequently finds experimentally that one or more of the $R_\infty C_\infty$, $R_R C_R$, or $R_A C_A$ semicircles in the complex plane are depressed so their centers lie below the real axis. It is more probable for the $R_R C_R$ arc to show such depression than the bulk $R_\infty C_\infty$ one for single-crystal material. Such depression may be interpreted in terms of a distribution of relaxation times, possibly arising in the case of the $R_R C_R$ arc from electrode surface roughness and/or porosity (e.g. de Levie [1967], Franklin [1975]). Although the exact small-signal solution with identical electrodes actually leads to some arc depression when π_m is very different from unity (Macdonald [1974b]), the amount of depression possible from widely different positive and negative charge mobilities is insufficient to explain most experimental depressions.

In the absence of a fully adequate microscopic theory leading to arc depression in the impedance plane, it has become customary to use the ZARC function defined in Section 2.2.2.2 to describe the depression analytically. This function involves either a resistance and a CPE in parallel or a unified impedance as in Eqs (23) and (25). For describing depressed arcs in the ε plane, the εARC (Cole–Cole) function also defined in Section 2.2.2.2 has long been used. Note that the CPE which appears in both the ZARC and the εARC may be associated with a non-uniform transmission line and nonuniform diffusion in a region of infinite extent, but the CPE has not been generalized so far to the finite-length diffusion regime. When an adequate expression is available, it will represent a more complex process than ordinary finite-length (uniform) diffusion, for example, Z_W, a subset of such generalized CPE response. Although this process would be appropriate for distributed (nonuniform) bulk response, since it would change from intensive to extensive behavior with decreasing frequency as Z_W does, the ordinary CPE, which can be taken either intensive or extensive, seems more appropriate for intensive interface processes. Whenever a straight-line spur in the impedance of plane is found with an angle from the real axis different from $\pi/4$, the CPE should replace Z_W.

Although there is no complete derivation of a generalized CPE yet available which arises from nonuniform diffusion (NUD) in a finite-length region, one may heuristically modify the CPE and Warburg diffusion expressions in such a way as to generalize them both. The result is

$$Z_{\mathrm{NUD}} = R_0 \tanh\left[A_0 R_0 (j\omega)^{\psi}\right] / \left[A_0 R_0 (j\omega)^{\psi}\right] \tag{75}$$

where R_0, which might be R_∞ for bulk behavior, is taken extensive and we require $0 < \psi < 1$. For $\psi = 0.5$, this expression reduces to just Z_W when one takes $A_0 = l_e/R_0\sqrt{D}$. For $A_0 R_0 \omega^{\psi} >> 1$, $Z_{\mathrm{NUD}} \cong Z_{\mathrm{CPE}}$. Further, for any ψ, $Z_{\mathrm{NUD}} \rightarrow R_0$, an extensive quantity, when $A_0 R_0 \omega^{\psi} << 1$, and, for $\psi < 1$, to $[A_0(j\omega)^{\psi}]^{-1}$ for $A_0 R_0 \omega^{\psi} >> 1$. This result is intensive, as it should be if A_0 is taken intensive, as it is in the Warburg limit. Note that while both the ZARC impedance and Z_{NUD} involve $R_0 A_0$, they will have somewhat different shapes at the lowest frequencies where they approach the real axis. Further, the present expression for Z_{NUD} is just an empirical stopgap result and is probably most useful for $0 < \psi \leq 0.6$. We shall denote this heuristic generalization as the generalized finite-length Warburg model (GFW).

The fact that the current ungeneralized CPE has no dc path (while a generalized one would) can lead to problems in using the CPE in hierarchical circuits such as that of Figure 2.2.8*b*. Whenever it seems appropriate to replace an ordinary capacitor in the circuit by a CPE, no problem arises. But one cannot replace a resistor needed as part of a dc path by a CPE and still maintain the dc path; all one can do is put a CPE in parallel with the resistor, producing a ZARC function, or perhaps to use Z_{NUD}. But there is a problem in using such elements in hierarchical circuits. Although the real and imaginary parts of, say, a ZARC function could be separated, and the imaginary part used in place of an ideal capacitor in an hierarchical circuit and the real part in place of an ideal resistor, there is no physical justification for such separation.

In the polycrystalline case, one must consider the processes which occur within an individual single-crystal grain and what happens at the grain boundaries, taking into account that there is almost certainly a distribution of grain sizes and orientations present. Since the response is a three-dimensional average of the response of a great many interacting grains, one expects that the bulk response for composite materials will both be more complicated than the $R_\infty C_\infty$ semicircle expected for a perfect homogeneous material and may often be described by a distribution of relaxation times, either discrete or continuous. Since one usually finds that it is experimentally impossible to distinguish results arising from a continuous distribution and its approximation by, say, 10 or more discrete relaxations, it is often easiest to use the continuous distributions, since many less parameters need be specified. For the distorted and displaced arcs which are usually seen in the impedance plane for composite materials such as ceramics, it of course makes no difference in which order the elements representing this overall bulk response appear in the equivalent circuit (Voigt, not hierarchical connection), and it has also been customary to try to represent the response by one or more ZARC functions in series. No charged-particle Warburg response will be present if charge of only a single sign is mobile.

Although is does not seem reasonable to build a hierarchical ladder network circuit using the separated real and imaginary parts of a distributed element such as the ZARC, we can still achieve considerable generality and flexibility if we form a circuit using *only* unified distributed elements as in the three-level circuit of Figure 2.2.10. Here DE represents a general distributed element, one like the ZARC which can well approximate either an ideal resistor or an ideal capacitor in limiting cases of its fractional frequency dependence exponent ψ. Thus, in Figure 2.2.10, the odd-numbered DEs could, in the limit, be taken as capacitors, and the even-numbered ones as resistors. In practical cases, however, one would often find it necessary to choose some of the DEs as nonideal distributed elements. Note that if the electrodes were nonblocking, one would need to ensure a dc path through the circuit by, for example, taking the even-numbered DEs as resistors, ZARCs, or some other unitary or composite nonblocking distributed element.

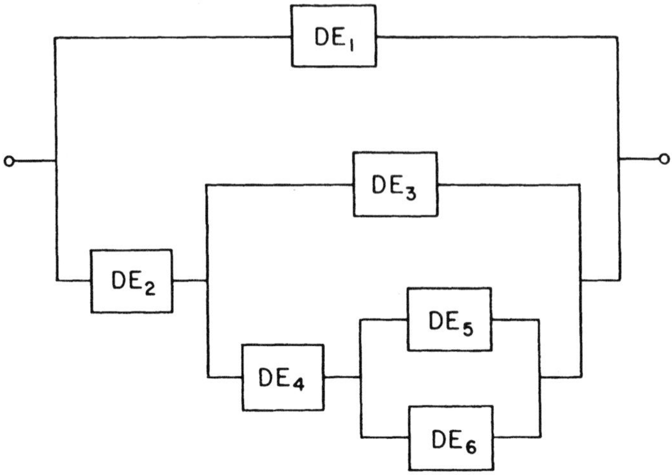

Figure 2.2.10. General equivalent circuit showing hierarchical structure and involving general distributed elements.

Further Empirical and Semiempirical Models. Although various empirical distributed-element models have already been discussed, particularly in Section 2.2.2.2, the subject is by no means exhausted. Here we briefly mention and discuss various old and new elements which may sometimes be of use in a fitting circuit such as that of Figure 2.2.10. Complex plane plots of IS data by no means always yield perfect or depressed semicircular arcs; often the arc is unsymmetric and cannot be well approximated by the ZC. An unsymmetrical impedance plane arc usually exhibits a peak at low frequencies and CPE-like response at sufficiently high frequencies. The reverse behavior is not, however, unknown (Badwal [1984]). An expression originally proposed in the dielectric field by Davidson and Cole [1951] yields ordinary asymmetric behavior. Its I_k generalization is

$$I_k = [1 + js]^{-\psi_k} \tag{76}$$

with $s \equiv \omega\tau$ and $0 \le \psi_k \le 1$, and it reduces to symmetric Debye response for $\psi_k = 1$. This model will be denoted by DC. A further empirical approach, due to Kohlrausch [1854] and Williams and Watts [1970], yields transient response of the fractional exponential form $\exp\ [-(t/\tau_0)^{\psi}$, with $0 < \psi \le 1$. It yields frequency-domain response generally rather similar to that of Eq. (76) and such response will be denoted in this section by WW. The LEVM/LEVMW CNLS computer program, available for free downloading from J. R. Macdonald and Solartron Inc. (Macdonald and Solartron Inc. [2003]) allows accurate calculation of such response for both simulation and data fitting. See the discussion in section 4.2.2.3.

Jonscher [1974, 1975a,b, 1980, 1983], in an extensive series of papers, working primarily in the dielectric area, independently emphasized the importance and ubiquity of constant phase response, and proposed and demonstrated the utility of three different empirical frequency response functions in IS data fitting. These three equations, termed *universal dielectric response* by Jonscher, were originally expressed in terms of the imaginary part of the complex dielectric susceptibility, χ''. They may alternatively be expressed, of course, in terms of ε'' or Y'. Further, they may all be generalized to the I''_k ($k = \varepsilon$ or Z) representation. Finally, it has been found (Macdonald [1985d]) that two of them may be written in full complex form, not just as I''_k.

The three generalized Jonscher equations may be expressed as

$$I_k = B_{k0}(j\omega)^{-\psi_k} \tag{77}$$

$$I_k = B_{k1}\left[(j\omega\tau_{k1})^{-\psi_{k1}} + (j\omega\tau_{k2})^{-\psi_{k2}}\right] \tag{78}$$

and

$$I'_k = -B_{k2}\left[(\omega/\omega_{kp})^{-\psi_{k3}} + (\omega/\omega_{kp})^{-\psi_{k4}}\right]^{-1} \tag{79}$$

It is clear that Eq. (77) is just the CPE and Eq. (78) is a combination of two CPEs (in parallel for $k = \varepsilon$ and in series for $k = Z$). Of course, the ω_{kp} of Eq. (79), which denotes a peak frequency, could be replaced by $\tau_{kp} \equiv \tau_{kp}^{-1}$. The possible range of all the exponents is (0, 1). In Jonscher's $k = \varepsilon$ "universal dielectric response" case, one has $\psi_\varepsilon = 1 - n$, $\psi_{\varepsilon 1} = 1 - n_1$, $\psi_{\varepsilon 2} = 1 - n_2$, $\psi_{\varepsilon 3} = m$, and $\psi_{\varepsilon 4} = 1 - n$. If we further choose $\psi_Z = n$, $\psi_{Z1} = n_1$, and $\psi_{Z2} = n_2$, Eqs (77) and (78) yield the same frequency dependence exponents at the admittance level when a single term dominates. For example, Eq. (77) yields $Y_\varepsilon \propto (j\omega)^n$ and $Y_Z \propto (j\omega)^n$ for the above choices. We shall term Eq. (78) the *generalized second Jonscher equation* (GJ2) and Eq. (79) the *generalized third Jonscher equation* (GJ3).

The minus sign in Eq. (79) arises because we have defined I_k with a plus sign as $I'_k = I'_k + iI''_k$. When $\omega = \omega_{kp}$ in Eq. (79), $|I''_k|$ reaches a maximum. Further, when $\psi_{k3} = \psi_{k4}$, this equation reduces to the long-known Fuoss–Kirkwood [1941] form, yielding a symmetrical curve for $-I''$ vs. $\log(\omega/\omega_{kp})$. We shall denote this special form of the GJ3 as the GFKJ equation. Although no fully complex general expression consistent with Eq. (79) is available, when $\psi_k \equiv \psi_{k3} = \psi_{k4}$ complex forms have been given for various fractional values of ψ_k. Jonscher and his collaborators have shown that the χ'' forms of Eqs (77)–(79) can fit a great deal of dielectric and conductive system data. Unfortunately, the fits never used CNLS, and no ordinary nonlinear least squares fits of χ'' giving fitted parameter estimates and standard deviation estimates have been presented. Further discussion of "universal dielectric response" appears in Macdonald [1985d].

Some time ago Almond, West, and Grant [1982], Bruce, West, and Almond [1982], and Almond and West [1983b] specialized the χ'' form of Jonscher's Eq. (78) for hopping conduction situations to obtain

$$\sigma(\omega) = K(\omega_p + \omega_p^{1-n}\omega^n) \tag{80}$$

where $\sigma(\omega)$ is the ac conductivity, K is a temperature-dependent constant, and ω_p was identified as the thermally activated ionic hopping frequency ν_H. Now Eq. (80) may be rewritten at the Y level as

$$Y'(\omega) = G_0\left[1 + (\omega/\omega_p)^n\right] \tag{81}$$

Next, generalization of Eq. (81) to the complex plane yields

$$Y(\omega) = G_0\left[1 + (j\omega\tau_0)^n\right] \tag{82}$$

which is fully consistent with Eq. (81) when

$$\omega_p = \left\{\tau_0[\cos(n\pi/2)]^{1/n}\right\}^{-1} \tag{83}$$

or, equivalently,

$$\tau_0 \equiv \omega_0^{-1} = [\cos(n\pi/2)]^{-1/n}/\omega_p \tag{84}$$

Finally, Eq. (82) yields

$$Z = R_0/\left[1 + (j\omega\tau_0)^n\right] \tag{85}$$

where $R_0 = G_0^{-1}$. This expression is just the long-known ZARC [compare Eq. (22)]. Thus, it appears that the principal new element in the Almond–West work is the identification of ω_p as the hopping frequency. This interesting suggestion has been examined at some length recently (Macdonald and Cook [1985]), with the conclusion that the case is not proven so far. If, in fact, the hopping frequency ν_H is directly involved in the ZARC when it is applied to hopping conduction situations, it seems most plausible that $\nu_H = \omega_0 = \tau_0^{-1}$, or perhaps $\omega_0/2\pi$, rather than ω_p. Equation (83) shows that $\omega_p \to \infty$ as $n \to 1$, an unlikely result and one avoided by the choice of ω_0 instead.

Several of the empirical model responses discussed above have been given a more theoretical basis (see Macdonald [1985c,d] for references), but they still suffer from two important weaknesses. They do not generally lead to physically realistic response at both high- and low-frequency extremes, and they do not lead to any predictions for possible temperature dependence of the fractional exponent(s) ψ. A semiempirical theory whose frequency response results are briefly discussed below does, however, avoid these weaknesses. Since any real material will have a largest (τ_∞) and a smallest (τ_0) response time, response at longer (shorter) times than these will be determined by these limiting responses (for a single type of physical process). But such single-time-constant behavior leads to frequency response proportional to ω for $\omega << \tau_\infty^{-1}$ and to ω^{-1} for $\omega >> \tau_0^{-1}$. Although simple Debye behavior with $\tau_0 = \tau_\infty = \tau$ also leads to limiting $\omega^{\pm 1}$ response, here τ_0 and τ_∞ may differ greatly, and for the range $\tau_\infty^{-1} < \omega < \tau_0^{-1}$, non-Debye fractional exponent response may appear, and usually does so.

Note that CPE response fails the above test of physical realism at both frequency extremes; so does ZC response for $\psi < 1$. On the other hand, Davidson–Cole low-frequency-limiting response is realistic, but not its high-frequency-limiting response. One might reasonably ask, if all these models are not entirely physically realistic, why are they discussed and used for fitting? The reason is that it is rare for a single-response process, say that associated with an electrode reaction (R_R and C_R in the ideal nondistributed case), to be so isolated in its frequency range that one can follow its response alone to very high or low (relative) frequencies. Because of the usual presence of other processes yielding response near or even overlapping in frequency with that of the process of immediate concern, one cannot usually follow the response of the process in question very far into its wings where $\omega^{\pm 1}$ limiting response finally must appear. Further, one usually finds that experimental limitations preclude measuring far into the high-frequency wing of the lowest-frequency process present. In essence, what we can't measure doesn't matter—at least until we can measure it! Nevertheless, a theoretical model which does incorporate proper limiting behavior is clearly superior in that respect to one that doesn't.

Although dielectric response data often leads to temperature-independent ψ's (so that the time–temperature or frequency–temperature superposition law holds), this is by no means always the case (Jonscher [1983]). Further, conductive-system response, as in ionic hopping conductors, often leads to appreciable temperature dependence of ψ. Surprisingly, ψ_ε and ψ_Z temperature responses, when apparent, are usually found to be quite different, with ψ_ε increasing with increasing temperature and ψ_Z decreasing.

In recent years, several important variations of the Kohlrausch–Williams–Watts model have been produced and attention has been drawn to a nearly constant component of bulk dielectric response in low conductivity electrolytes. This work is discussed in Section 4.2.2.3.

Fitting Ambiguity and a New Semiempirical Model. Although all the models we have discussed in this section and in Section 2.2.2.2 are distinct and separate, and although they may be associated with different physical processes, it turns out that there is a high degree of practical fitting ambiguity between most of them. Response differences between several unsymmetric models with the same ψ value are demonstrated in Figure 2.2.11. Here Debye response is included for comparison and DAE_1 (which involves the parameter ϕ rather than ψ) refers to the semiempirical distribution-of-activation-energies model discussed below. But the situation is different when "data" derived from one model involving a given ψ, say ψ_a, are fitted

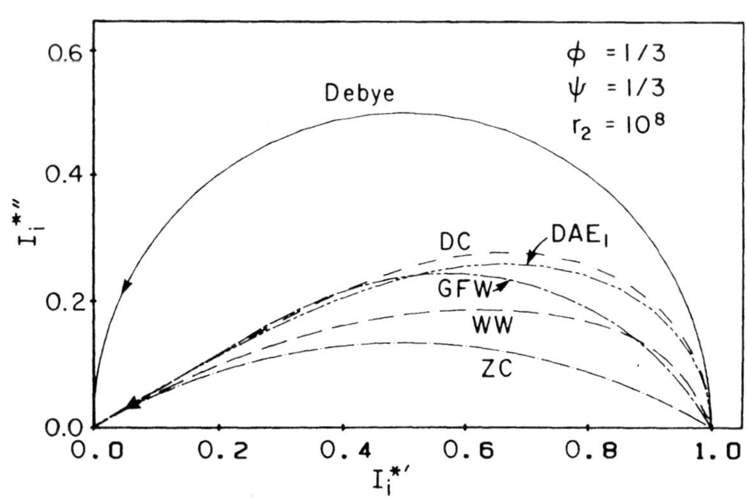

Figure 2.2.11. Complex plane response of the normalized I_i^* response of various distributed models.

by CNLS to another model, yielding a ψ estimate for this model, say ψ_b. It turns out, as we shall demonstrate below (see also Macdonald [1985d]), that when ψ_a and ψ_b are allowed to be different, one model can often fit another within 1% or so (usually better than most experimental data are known) over quite wide frequency and magnitude ranges. When such ambiguity is present, as it usually is for practical less-than-perfect data, it will often be easiest to fit with the simplest model, whether or not it is physically reasonable for the material–electrode system considered, and then relate the fitting results to a more appropriate, but more complex, model.

Such a more complex model is the DAE, involving an exponential density distribution of activation energies. Its rationale and results are described in detail in Macdonald [1963, 1985c,d]. Let us distinguish three forms of it. First is the DAE_1, which involves a single exponential density distribution and leads to unsymmetrical response (Macdonald [1985c]). Second is the DAE_2, which involves two joined, complementary exponential distributions and leads to symmetric behavior in the complex plane. Finally, the general DAE involves two joined, noncomplementary exponential distributions and spans the range of shapes from the DAE_1 to the DAE_2 (Macdonald [1985d]). Of course, the DAE is most generally given in normalized I_k form.

Although the frequency response of the DAE model can only be expressed in integral form (associated with a hypergeometric function) for arbitrary ϕ, relatively simple closed-form response has been given for many values of ϕ. Such closed-form, discrete-ϕ response is useless, however, for accurate CNLS fitting using this model. Therefore, the full-integral DAE model as well as nearly all of the other distributed element models discussed so far, have been built into the general CNLS fitting program, LEVM/LEVMW. Thus, any of these models can be used to fit experimental frequency response data or "data" derived from another model.

Some of the model-fitting ambiguity mentioned above is demonstrated in the next figures. Further discussion of DAE–Jonscher ambiguity appears in Macdonald [1985d]. First, it is worthwhile to categorize the models discussed by their complex plane symmetry as in Table 2.2.3. The symmetric and asymmetric curves give closed arcs in the complex I_k plane, but the CPE and GJ2 yield only open spurs in this plane (and, as mentioned earlier, cannot be normalized in the usual I_k way). The fitting ambiguity with which we are concerned here applies only within a given column of

Table 2.2.3. Summary of Main Models Discussed, Showing Their Symmetry Characteristics in the Complex Plane

Symmetric	Asymmetric	General
ZC	DC	CPE
GFKJ	GFW	GJ2
DAE_2	WW	DAE
	DAE_1	

Table 2.2.3. We cannot expect to get a good fit of asymmetric WW data, for example, with a symmetric model such as the ZC. Note that all models will generally show some region of frequency where CPE-like response appears. In this region, the CPE model is clearly sufficient. It is not this ambiguity with which we are concerned but rather with the holistic response, that which includes regions beyond and below that where CPE response alone dominates.

We have found that any symmetric model of Table 2.2.3 can be very well fitted by any other symmetric model. Figures 2.2.12 and 2.2.13 show two-dimensional

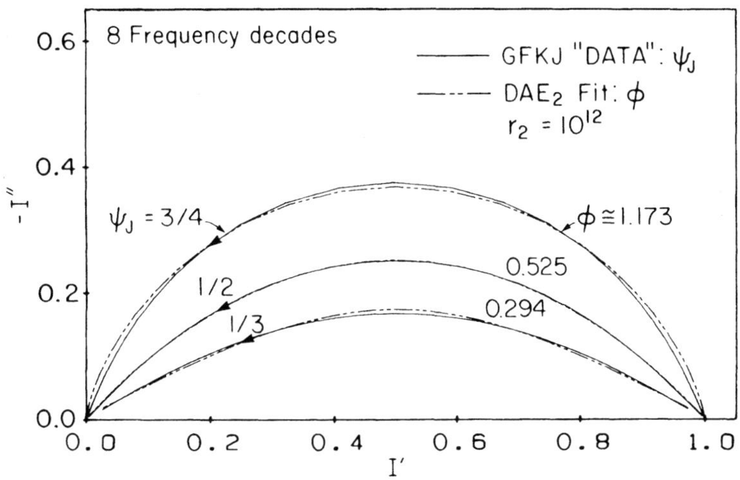

Figure 2.2.12. Complex plane comparisons of the response of the GFKJ and DAE_2 models when the DAE_2 is fitted to GFKJ response with CNLS.

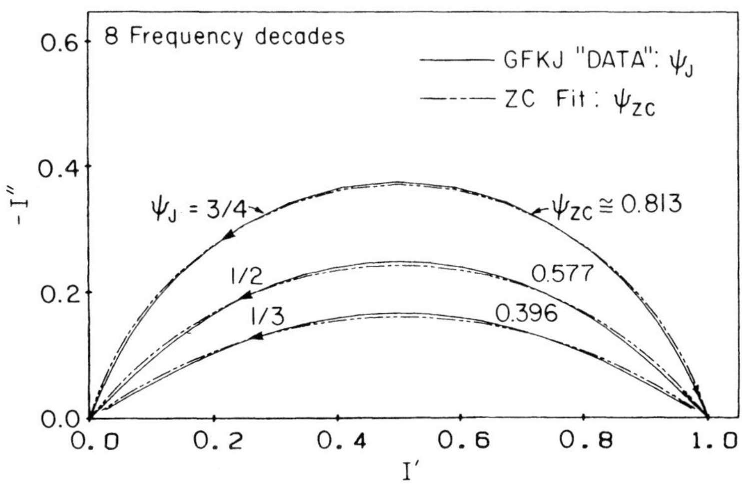

Figure 2.2.13. Complex plane comparisons of the response of the GFKJ and ZC models when the ZC is fitted to GFKJ response with CNLS.

plots obtained from full CNLS fitting of the DAE_2 and ZC models to GFKJ model "data" for several different ψ values. The fits are so good that only with extremely accurate experimental data (better than those usually available) could one decide unambiguously between any of the three symmetric models on the basis of CNLS fitting alone. Further, the DAE has been found to yield a very good fit of the GJ2 model as well (Macdonald [1985d]).

Figures 2.2.12 and 2.2.13 of course do not show frequency response explicitly. It is found to agree exceptionally well also at each point, at least until one moves far away from the peak frequency. Some results for DAE_2 fitting to GFKJ data for $\psi_J = \frac{1}{3}$ are shown in Figure 2.2.14. The unity weighting used in Figures 2.2.12 and 2.2.13 yields a better fit near the peak, and proportional weighting (see Section 3.3.2) leads to better agreement in the skirts of the curve. Here $s \equiv \omega\tau_0$ is a normalized frequency variable (Macdonald [1985c,d]).

The situation is somewhat less ambiguous for the special asymmetric models of Table 2.2.3. Although the DC and DAE_1 models can well fit each other, neither one can fit WW model response adequately over a wide ψ_i range. For $0.7 \leq \psi_{WW} \leq 1$, the DAE_1 can fit WW results quite well, but better fits are obtained for this region and below using the general DAE model to fit WW "data." Figure 2.2.15 shows the

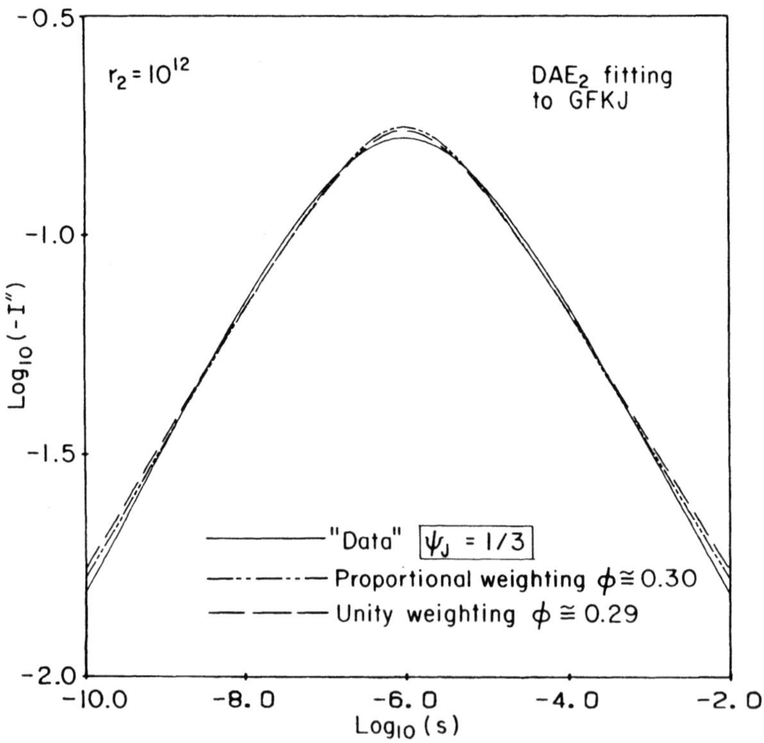

Figure 2.2.14. Frequency response curves comparing the results of fitting the DAE_2 model to GFKJ "data." The logarithm of $-I''$ is plotted vs. the logarithm of a normalized frequency variable s.

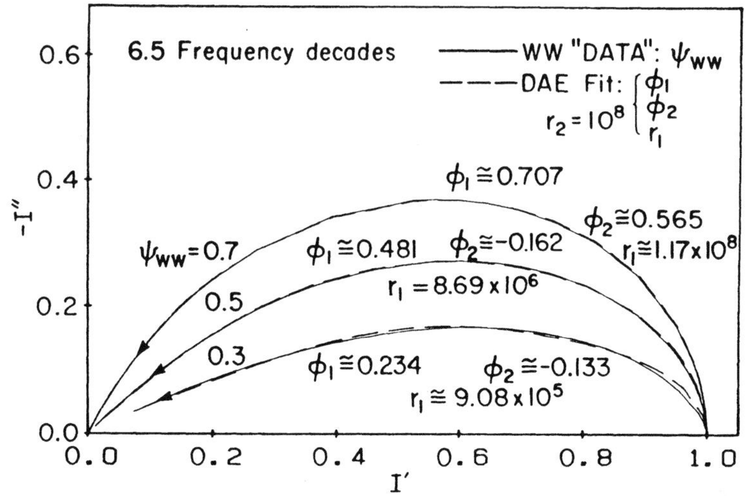

Figure 2.2.15. Complex plane comparisons of the response of the WW and DAE models when the DAE is fitted to WW response with CNLS.

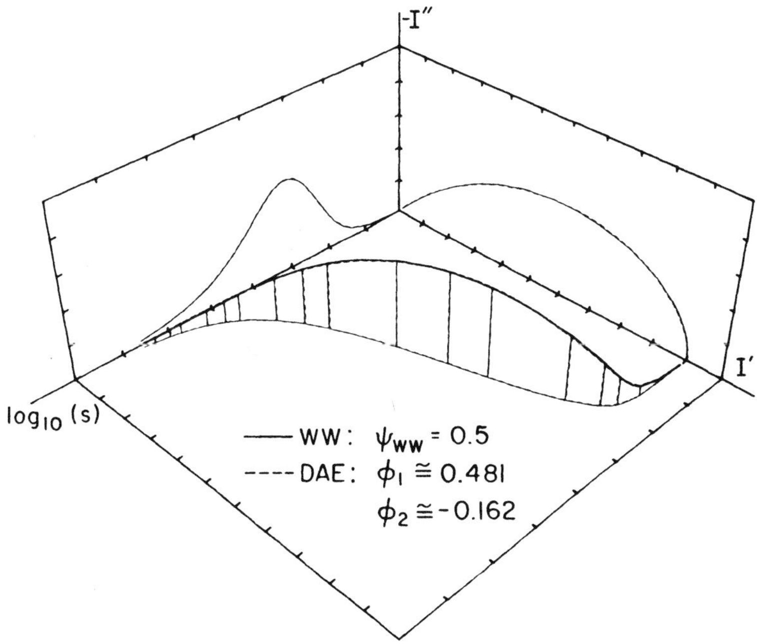

Figure 2.2.16. Three-dimensional plot with perspective showing the excellent agreement between WW "data" and the results of fitting these data with the DAE model.

results of such CNLS fitting plotted in the complex plane, and Figure 2.2.16 shows them with 3-D plotting (see Section 3.3.1). Again the fits are so close that only with superb data could one unambiguously discriminate between the two models. One reason to prefer the various DAE models to the others, however, is that the former yield explicit temperature dependences for the ϕ_i's which enter the model, while no such ψ_i temperature dependence is a part of the other models. When the actual fractional frequency response exponents observed in a set of experimental data are found to vary with temperature, as they often do for both conductive and dielectric systems, it is thus natural to try DAE fitting and see if the ϕ_i estimates found depend on temperature in one of the ways predicted by the theory (Macdonald [1985c,d]). If such agreement is established, much can be learned about the detailed response of the system. Finally, it should be mentioned that the exponential DAE model has recently been simplified and its predictions compared to those of a symmetric or asymmetric Gaussian DAE model (Macdonald [1987]). Again fitting ambiguity is sometimes found.

Chapter 3

Measuring Techniques and Data Analysis

3.1 IMPEDANCE MEASUREMENT TECHNIQUES

Michael C. H. McKubre
Digby D. Macdonald

3.1.1 Introduction

Until the advent of digital computers, all electrochemical studies involved the processing and analysis of analog signals in either the time domain or the frequency domain. Typical examples of analog signal analysis include the use of ac coupled bridges and of Lissajous figures for determining interfacial impedance. In both instances, the desired information (e.g. balance of a bridge) is obtained in purely analog format, and no need exists for converting signals into digital form.

When describing analog instrumental methods, it is convenient to classify techniques according to the type of excitation functions employed, particularly with respect to the independent variable. For example frequency domain impedance measurements are carried out using a small-amplitude sinusoidal excitation with frequency as the independent variable. Alternatively, the perturbation and response may be recorded in the time domain with time as the independent variable, and the impedance as a function of frequency can then be extracted by time-to-frequency conversion techniques such as Laplace or Fourier transformation. Time domain methods characteristically use digital-processing techniques; frequency domain methods have traditionally used analog techniques, although digital processing is becoming common in synthesis and analysis of sinusoidal signals.

The application of a sine wave excitation to a system under test often is the easiest method of determining the system transfer function. Here we are concerned with measuring or inferring a transfer function for an electrochemical cell as a first

Impedance Spectroscopy, Second Edition, edited by Evgenij Barsoukov and J. Ross Macdonald
ISBN 0-471-64749-7

step in determining reaction mechanistic and kinetic parameters (Macdonald [1977], Gabrielli [1981], Macdonald and McKubre [1981], Macdonald and McKubre [1982]).

By way of review, the transfer function of a system can be determined as the output divided by the input

$$G(j\omega) = X_{\text{out}}(j\omega)/X_{\text{in}}(j\omega) \tag{1}$$

For the special case where the output signal is the system voltage and the input (or excitation function) is the current, the transfer function is the system impedance

$$G(j\omega) = E(j\omega)/I(j\omega) = Z(j\omega) \tag{2}$$

Since the output may be changed in both amplitude and phase with respect to the input, we must express the impedance as a complex number

$$Z(j\omega) = Z' + jZ'' \tag{3}$$

where primed and double-primed variables refer to in-phase and quadrature components, respectively.

It is important to note that we are using the formalism of linear systems analysis; that is, Eq. (2) is considered to hold independently of the magnitude of the input perturbation. Electrochemical systems do not, in general, have linear current–voltage characteristics. However, since any continuous, differentiable function can be considered linear for limitingly small input perturbation amplitudes (Taylor expansion), this presents more of a practical problem than a theoretical one.

In the following section we present a number of standard methods of measuring a system impedance or a frequency domain transfer function. In applying any of the methods described, the perturbation must be of a sufficiently small magnitude that the response is linear. Although the condition of linearity may be decided from theoretical considerations (Bertocci [1979], McKubre [1981], McKubre [1983], McKubre and Syrett [1986]), the most practical method is to increase the input signal to the maximum value at which the response is independent of the excitation function amplitude.

3.1.2 Frequency Domain Methods

3.1.2.1 Audio Frequency Bridges

In the past, impedance measurements using reactively substituted Wheatstone bridges at audio frequencies have been the easiest to accomplish. Consequently, great emphasis has been placed historically on electrochemical processes having characteristic impedance spectra in the audio frequency range 20–20,000 Hz, namely, double-layer capacitive and moderately fast reaction kinetic effects at plane parallel electrodes.

The mathematics and methodology of such measurements are well understood (Hague [1957], Armstrong *et al.* [1968]). However, considerable use still may be made of passive audio frequency bridge measurements in this age of active circuitry, principally in high-precision applications. Following a brief review of bridge cir-

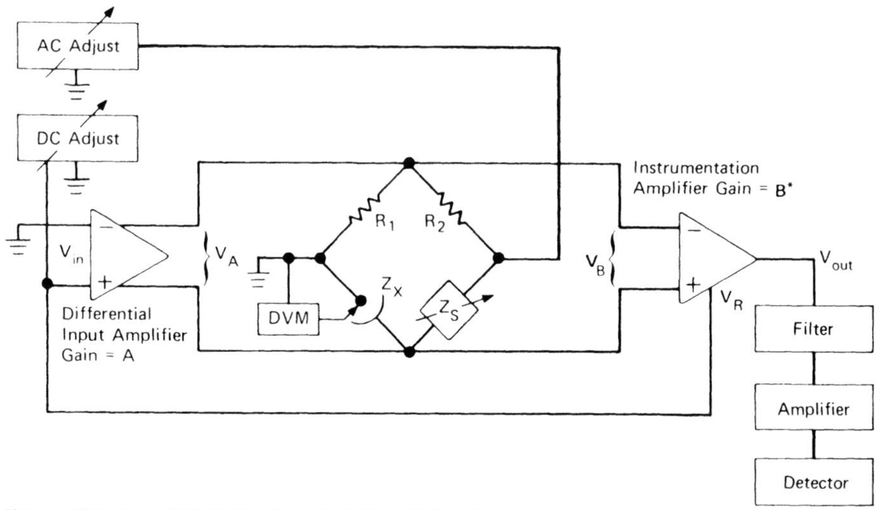

$^*V_{out} = BV_B - V_R = (AB-1)\,V_{in}$. For AB=1, V_{out} (DC) = 0.

Figure 3.1.1. Audio frequency bridge modified to include working electrode dc potential control.

cuits, we will restrict our discussion to the limitations imposed by the use of each type of bridge, since these will influence the point at which an experimentalist will select a more complex measuring device.

Figure 3.1.1 shows schematically the familiar representation of an audio frequency bridge adapted for use with an imposed dc potential. The condition of balance for the bridge shown is

$$Z_x = (R_1/R_2)Z_s \tag{4}$$

where subscripts x and s refer to unknown and standard impedances, respectively. A variety of RCL combinations are possible for Z_s; in the commonly used Wien bridge (Hague [1957]) Z_s takes the form of series variable resistance and capacitance standards, which are adjusted alternately until the real and imaginary components of the voltage at the null detector simultaneously are zero. For this null condition the real and imaginary components of the unknown impedance may be calculated as

$$Z'_x = (R_1/R_2)R_s \tag{5}$$

$$Z''_x = (R_1/R_2)\omega C_s \tag{6}$$

The form of Eqs (5) and (6) has led to the widespread and unfortunate practice of tabulating and plotting measured impedance data in terms of the complex pair (R_s, $j/\omega C_s$) even when a Wien bridge has not been used. The impedance notation (Z', jZ'') is significantly less ambiguous and will be used here.

High-Frequency Limitations. The upper operating limit is imposed primarily by reactivity and nonlinearity of available resistive standards (chiefly inductive effects) and the effects of stray capacitive shunts. By using a Wagner earth (Hague [1957], Armstrong *et al.* [1968]), the latter effect can often be reduced sufficiently to allow sensible measurements at frequencies up to 10^5 Hz. However, the importance of Wagner earthing varies greatly with the magnitude of the impedance being measured (Hague [1957]). In general, elimination of stray capacitance is most important at high frequencies when measuring small capacitances or large resistances (i.e. for small-area electrodes).

Low-Frequency Limitations. The null detection system traditionally used with an audio frequency bridge consists of an amplifier, filter, and ac voltmeter. This combination imposes three limitations at low frequencies:

- Null detection with a magnitude voltmeter or oscilloscope is most sensitive when the resistive and reactive components of the unknown impedance are of the same magnitude, since the total bridge out-of-balance signal contains terms proportional to each. For an impedance bridge used to measure the electrical properties of electrochemical cells, this fact imposes a limit on accuracy at low frequencies since the reactive terms, which are primarily capacitive, dominate the cell admittance with decreasing frequency. Increasing the gain to observe the resistive component more precisely results in saturation of the detection system with the reactive out-of-balance signal.
- A significant source of noise at the detector may result from harmonic distortion originating in the oscillator or caused by nonlinearity in the system under test or in subsequent amplifiers. In such cases, the signal at balance consists mainly of the second harmonic. At high frequencies, this signal can be removed effectively by appropriate signal conditioning with bandpass, low-pass, or notch filters (McKubre and Macdonald [1984]). At low frequencies, however, analog filters of bandwidth less than 10 Hz are less easy to construct and control.
- Another major source of noise at low frequencies is mains pickup. This may amount to hundreds of millivolts superimposed on the test signal unless major efforts are made at shielding and ground loop suppression. Usually, unless an adequate notch filter is used in addition, the experimentalist must be satisfied with reduced precision at frequencies below about 100 Hz.

These three effects can be reduced to a large extent by using a phase-sensitive detector (PSD) to measure separately the real and imaginary components of the bridge out-of-balance signal. By separate amplification of the in-phase and quadrature components, differential sensitivities in excess of 100 : 1 can be attained. The advantages and limitations conferred by the use of PSDs are described in Section 3.1.2.6.

In normal operation, a PSD is completely insensitive to the second harmonic, but most commercial instruments have the additional facility of being able to select a reference signal at twice the fundamental frequency. By this means the extent of second-harmonic distortion can be measured. This distortion often reflects not an error signal (i.e. noise) but an expected response induced by nonlinearity of the system under test (McKubre [1981, 1983], McKubre and Syrett [1984]).

In addition, and unlike traditional bandpass filters, a PSD has a bandpass characteristic with bandwidth that decreases with decreasing frequency and frequently can be used within ±5 Hz of 50- or 60-Hz mains pickup.

When phase-sensitive null detection is used, the practical low-frequency limit becomes a function of the particular form of bridge chosen. For the Wien bridge, this limit is imposed by the selection of suitably large adjustable capacitance standards at frequencies below about 20 Hz.

Limitations of Imposed Potential. A considerable limitation on the use of this form of bridge is that it necessitates the use of a two-terminal cell. Although it is often possible to construct a cell in which the working-electrode impedance greatly exceeds that of the counter electrode, potentiostatic conditions cannot be established adequately with this type of bridge. Closely associated with this limitation is the fact that in normal use, the cell current and voltage vary with the settings of the resistive and reactive standards.

In electrochemical applications, these combined limitations may be severe. Figure 3.1.1 shows one of a variety of possible methods by which an imposed working-electrode dc potential can be adjusted to the desired value without influencing the detector circuit. The method shown can be used at frequencies less than the normal operating frequency limit of ac coupled amplifiers.

3.1.2.2 Transformer Ratio Arm Bridges

The high-frequency limitation imposed on the operation of reactively substituted Wheatstone bridges by unavoidable stray capacitances prompted the development of the transformer ratio arm bridge (Calvert [1948]). By substituting a transformer for orthodox ratio arms, a bridge was produced for which the impedance ratio is proportional to the square of the number of turns and which was capable of accepting heavy capacitive loads with virtually no effect on the voltage ratio.

The operation of a transformer ratio arm bridge is shown schematically in Figure 3.1.2. Briefly, voltage 180° out of phase is fed from the secondary winding of the input "voltage" transformer to the cell or unknown impedance and to resistance and capacitance standards. The "arms" of the bridge consist of a series of ratio taps of the primary windings of an output "current" transformer. The standard and unknown impedances are connected to the output transformer in such a way that a detector null is achieved when the sum of the flux induced by the unknown and standard currents in the output transformer is zero. In this condition

$$\frac{r_1}{Z_x} = \frac{r_2}{R} + j\omega C r_3 \tag{7}$$

for all V_{in}, where r_1, r_2, and r_3 are ratios (usually decade), separately selected.

The advantages of using this type of bridge are as follows:

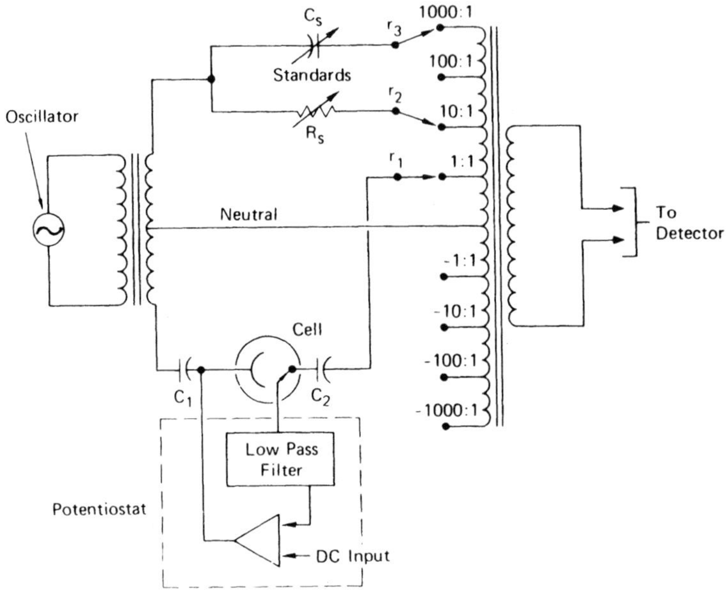

Figure 3.1.2. Transformer ratio arm bridge with dc potenticostatic control.

- Error resulting from the impurity of standard variables can be virtually eliminated. Because ratios are selectable over a wide range (usually 1000 : 1), standards can be small. Also, with decade-spaced transformer ratios, standards need be variable only over a range of about 11 : 1. Consequently, standards can be used that closely approximate ideality (e.g. air-gap capacitors and non-reactively wound metal resistance), and one standard can be used to measure a wide impedance range.
- By the use of precision transformers as ratio arms, one can obtain highly accurate ratio values that are essentially independent of frequency well into the megahertz range.
- The bridge is highly insensitive to the presence of stray capacitance. Figure 3.1.3 shows the reason: C_1, C_2, and C_3 can cause no measurement error – C_1 because it merely produces a reactive potential drop that is common to the unknown and standard circuits, and C_2 and C_3 because at balance no potential drop appears across them. Now C_u represents the capacitance across the unknown terminals and its effect is canceled by trimming capacitor C_t on the standard side. Here C_t is adjusted at each measurement frequency by disconnecting the standard and balancing the bridge. Similarly, effects of the stray capacitances to earth virtually disappear if the neutral terminal is grounded (Calvert [1948]; see also Figure 3.1.2).
- Impedances may be measured in all four quadrants by selecting positive or negative ratios. Of particular importance is the use of pure capacitive standards to measure unknowns with a positive (inductive) reactance.

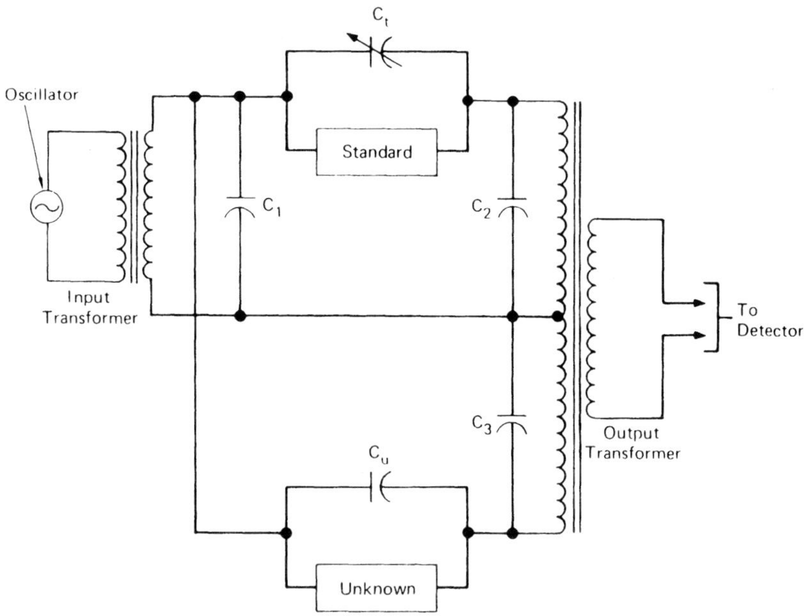

Figure 3.1.3. The effect of stray capacitances in the transformer ratio arm bridge.

High-Frequency Limitations. In normal use for electrochemical cells, the effective upper operating limit is imposed by effects external to the bridge. These effects, which have been described in detail by Armstrong *et al.* [1968], consist primarily of transmission line effects in connecting cables, the effect of residual series inductance in leads and the cell, and (normally desired) impedance dispersion effects of solid electrodes. In the latter group, edge effects (Sluyters [1970]) and transmission line effects due to surface roughness (de Levie [1963, 1965*b*, 1967]) become dominant with increasing frequency. In electrochemical systems for which the interfacial impedance is the desired parameter, measurement precision becomes limited by the dominance of the uncompensated electrolyte resistance in the total measured impedance. This effect has prompted the use of very small electrodes for which the ratio of uncompensated resistance to interfacial impedance is reduced (Zeuthen [1978]).

Series leakage inductances in the transformers within the bridge result in an impedance measurement error that is proportional to frequency. This effect has been examined by Calvert [1948], but is seldom likely to impose high-frequency limitations in electrochemical applications.

Low-Frequency Limitations. The use of input and output transformers results in cell current and voltage, and thus detector signals that decrease with decreasing frequency. This effect becomes apparent only at low audio frequencies and imposes a practical lower limit of the order of 100–200 Hz with commercial bridges.

Limitations of Potential Control. The limitations of potential control for a transformer ratio arm bridge are similar to those imposed in classical bridge measurement. That is, it is not possible to apply the ac potential via a reference electrode and potentiostat circuit only to the interface of interest. The measured impedance necessarily includes series terms associated with the lead and electrolyte resistance and the counter electrode impedance.

Dc potentials can be applied to the interface of interest by using a circuit of the form shown within the dashed lines in Figure 3.1.2, since at moderate frequencies the low-pass filter will not observe the ac component. However, direct current must be excluded from the bridge windings by the use of blocking capacitors C_1 and C_2. The impedance of these also will be included in the measured "cell" impedance.

3.1.2.3 Berberian–Cole Bridge

An active null admittance measuring instrument that incorporates many of the advantages of the transformer ratio arm technique, while obviating many of the disadvantage of passive bridges, has been reported by Berberian and Cole [1969]. Figure 3.1.4 shows a form of this bridge modified to measure impedance and to remove some of the limitations of the earlier instrument (McKubre [1976]).

The basic operation is as follows. The external variable decade standards are R_1 and C, while R' and R'' are internal and fixed. With reference to Figure 3.1.4, at all times,

$$i_1 + i_2 + i_3 = 0 \tag{8}$$

$$i_1 = AV_A/R_1 \quad (V_A = IZ) \tag{9}$$

$$i_2 = AV_A(j\omega C) \quad (V_A = IZ) \tag{10}$$

$$i_3 = BV_B/R' \quad (V_B = -IR') \tag{11}$$

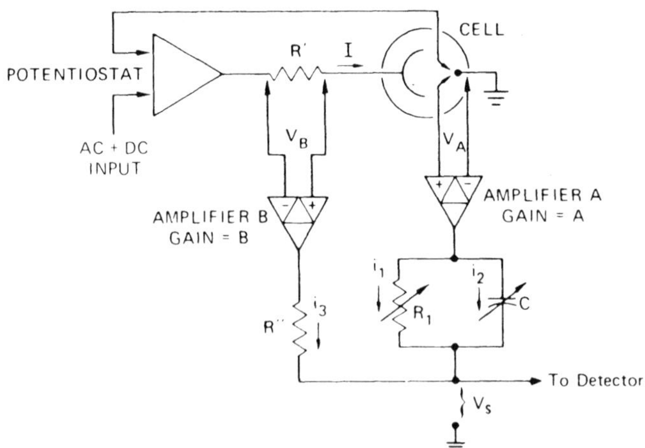

Figure 3.1.4. Modified Berberian–Cole bridge shown as a three-terminal interfacial-impedance-measuring system, with potentiostatic control of the working electrode.

where Z is the impedance between the working electrode and the reference electrode and I is the current flowing through the cells. Therefore, for the condition of balance at the summing point

$$BIR'/R'' = AIZ/R_1 + AIZ(j\omega C) \tag{12}$$

Removing I and solving for the unknown impedance yields

$$Z = \frac{BR'R_1}{AR''} \frac{(1 - j\omega R_1 C)}{1 + \omega^2 R_1^2 C^2} \tag{13}$$

The advantages of this method apply principally at low (audio and subaudio) frequencies. It is important to note that the device shown schematically in Figure 3.1.4 is a bridge only in the sense that external variables are adjusted to produce an output null.

The principal advantages of the Cole–Berberian bridge are as follows:

- Because of the use of buffer amplifiers, null adjustment does not vary the potential across (or current through) the unknown impedance, as is the case for classical and transformer ratio bridge measurement.
- Measurements can be made on two, three, or four terminal cells, allowing the isolation of the impedance component of interest from the total cell impedance. This is not possible with a passive bridge, and it is frequently infeasible to construct a cell for which the impedance of interest is much greater than all series terms. This is particularly difficult when measuring the impedance of an electrode of large area, when measuring impedance in a highly resistive electrolyte, or when the impedance of interest is that of a highly conductive electrolyte.
- Measurements can be made effectively down to 0 Hz. Because the bridge shown in Figure 3.1.4 is direct coupled, the low-frequency limits are those of the null detection system and the patience of the experimenter.
- Measurements can be made in the presence of a dc bias under potentiostatic control, without the use of blocking capacitors.
- Impedance can be measured over an extremely wide range, from below $10^{-3}\,\Omega$ to over $10^9\,\Omega$.
- Error resulting from the impurity of standards can be virtually eliminated because standards can be selected according to ideality, not magnitude of the components.
- By using differential gain for the real and reactive standards, a suitable range of measurement can be selected for each impedance component separately. This feature is incorporated in Figure 3.1.5.
- Impedances may be measured in all four quadrants (RC, $-RC$, RL, $-RL$) using resistance and capacitance standards alone.

Because the gains of amplifiers A and B perform the same function as the ratios in a transformer ratio arm bridge, the two techniques have many features in common.

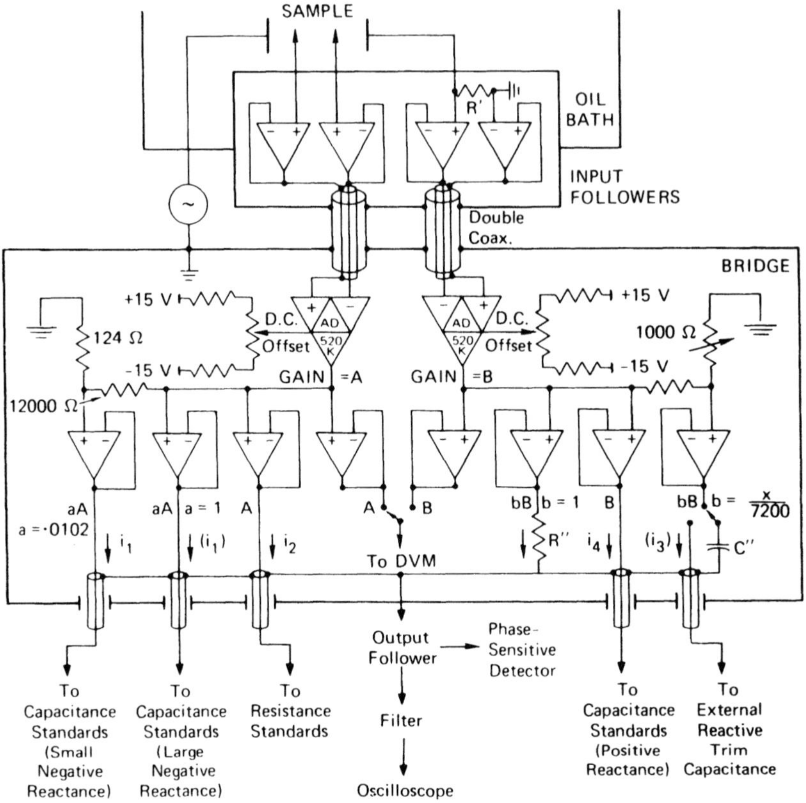

Figure 3.1.5. Schematic diagram of a working (modified) Berberian–Cole bridge shown as a four-terminal impedance-measuring system.

High-Frequency Limitations. Inaccuracies at high frequencies can occur because of errors in the gain functions A and B with decreasing amplifier open-loop gain (McKubre and Macdonald [1984]). Figure 3.1.5 shows, schematically, a practical bridge of the Berberian–Cole type. Gain errors in the voltage followers are negligible, and, since amplifiers A and B are identical devices and their gains appear as a ratio in Eq. (13), inaccuracies in this term are partially compensated. Nevertheless, the upper operating frequency limit for the bridge shown in Figure 3.1.5 is about 10 kHz, depending somewhat on the magnitude of the unknown impedance. This device is capable of 0.01% measurement accuracy for both impedance components between 1.0 Hz and 1 kHz, and 0.1% accuracy in the peripheral decades (0.1–1.0 Hz, 1–10 kHz).

Low-Frequency Limitations. As stated previously, the low-frequency operating limit is imposed by the detection system. At frequencies down to 0.5 Hz, a two-component PSD performs an ideal null detection function (McKubre [1976]). At frequencies below 0.1 Hz, a low-pass filter and oscilloscope or picoammeter can be used (Berberian and Cole [1969]).

Potential Control. Although it is possible to impose ac potentiostatic control at the interface of interest, the presence of a dc bias will result in a signal in the active bridge circuits. Dc offset must be adjusted to near zero to prevent overloading in subsequent gain stages. For a cell under dc potentiostatic control, this requirement may necessitate frequent offset adjustment of the current amplifier B.

3.1.2.4 Considerations of Potentiostatic Control

An essential element of electrode kinetics is the characteristic dependence of electrode reaction rate of the electrode potential. Thus, for many electrode studies, the use of the potentiostatic control is the most convenient method of obtaining relevant kinetic and mechanistic parameters. A limitation of passive bridge methods in general is their inflexibility with regard to potential control, so that in many cases the experimenter must forgo the advantages of simplicity and sensitivity associated with bridge measurement to impose ac and/or dc potentiostatic control at a single interface. The "direct" methods permit effective potential control while retaining the relative simplicity of operation of many of the bridge techniques.

If the cell current and voltage are measured with regard to their magnitude and phase relations, the impedance can be determined directly from Eq. (2). Figure 3.1.6 shows, in simplified form, a circuit that allows the direct measurement of impedance under potentiostatic control.

It is necessary at the outset to separate phase shifts associated with the cell impedance from those attributable to the potentiostat control loop. Commercial potentiostats normally are optimized for fast step response, and the potentiostatting function becomes substantially in error for sinusoidal inputs, with increasing frequency. Analyses have been performed of the frequency-dependent errors introduced by the potentiostatting function for a variety of potentiostats with varying loads (Brown *et al.* [1968], McKubre and Macdonald [1984]). However, the fidelity of the potentiostatting function with respect to an ac test signal superimposed on a dc control level is seldom of significance *provided that* the system under test is linear (the fundamental assumption of the use of ac methods in electrochemical kinetic studies) and that the ac voltage is measured directly as the potential difference between the working and a suitably placed reference electrode, and *not* at the input to the potentiostat.

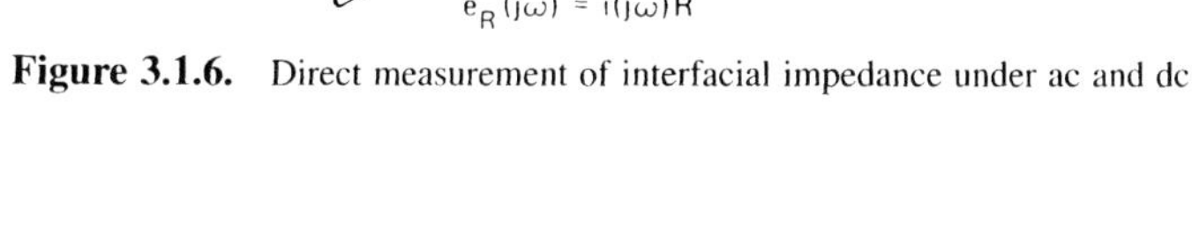

Figure 3.1.6. Direct measurement of interfacial impedance under ac and dc potentiostatic control.

In fact, so-called high-speed potentiostats are often undesirable for use in high-frequency impedance measurements at an electrode–aqueous-electrolyte interface. The reactive impedance at such an interface reduces to that of the double-layer capacitance at limiting high frequencies. Thus, one may have 3 A/cm^2 of out-of-phase current flowing at 10^6 Hz to an electrode with 50 μF/cm^2 of double-layer capacitance if the potentiostat is able to maintain a 10 mV ac perturbation at that frequency. Such high current densities may result in severe nonlinearities, and one often will prefer the reduced amplitude and phase shift of a narrow-bandwidth potentiostat when the voltage is measured at the point of interest (e, not e_i in Figure 3.1.6).

3.1.2.5 Oscilloscopic Methods for Direct Measurement

By recording $e(j\omega)$ and $i(j\omega)$ (as the voltage drop across a series resistance R_s; see Figure 3.1.6) with a twin-beam oscilloscope, the magnitude of the impedance can be calculated from the ratio of the two peak-to-peak voltages and the directly observed phase angle. Figure 3.1.7 shows the oscilloscope traces for $e(j\omega)$ and $e_R(j\omega)$ that result from the imposition of a sine wave between the working and reference electrodes.

The real and imaginary components can be calculated (with reference to Figure 3.1.7) as

$$|Z| = \frac{R_s|e(j\omega)|}{|e_R(j\omega)|} \tag{14}$$

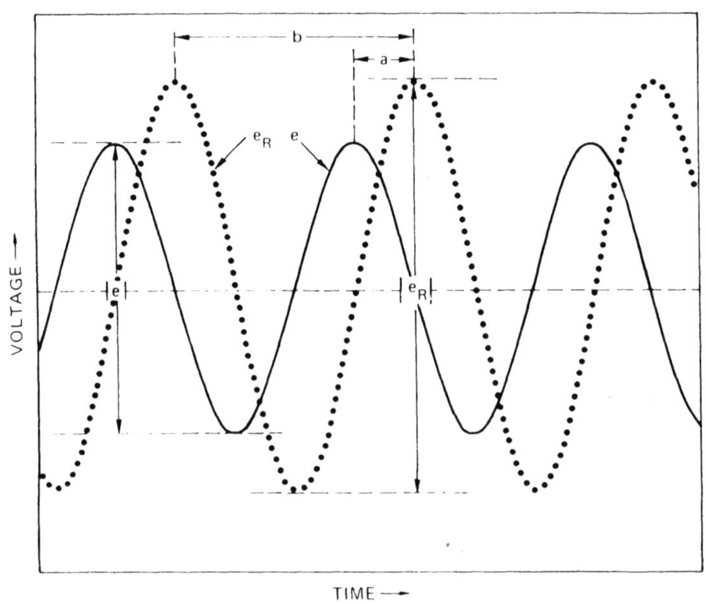

Figure 3.1.7. The direct measurement of impedance using a twin-beam oscilloscope.

$$Z' = |Z|\cos(\phi) \tag{15}$$

$$Z'' = |Z|\sin(\phi) \tag{16}$$

The time base of available storage oscilloscopes limits low-frequency measurements to about 10^{-2} Hz. High-frequency limitations are imposed by effects external to the oscilloscope, principally stray capacitance and transmission line effects in the leads and cell. Measurements can often be made at frequencies above 10^5 Hz.

The primary limitation of this technique is precision. Oscilloscope linearity is seldom better than 1%, and it is difficult to measure phase angles directly with a precision of better than 2°. Measurements usually can be accomplished with an uncertainty in Z' and Z'' of ±3% of $|Z|$.

A single-beam oscilloscope or an "*X–Y*" recorder also can be used to measure impedance parameters directly by the method of Lissajous figures.

Elimination of t between expressions for e and i of the form

$$e = |e|\sin(\omega t)$$

$$i = |i|\sin(\omega t + \phi)$$

leads to an equation of the form of an ellipse when e and i are plotted orthogonally (e applied to the "*X*" plates and i applied to the "*Y*" plates), and the components of the impedance can be calculated from the dimensions of the ellipse. With reference to Figure 3.1.8,

$$|Z| = \Delta e/\Delta i \tag{17}$$

$$\sin(\phi) = \Delta i'/\Delta i = \alpha\beta/(\Delta i \Delta e) \tag{18}$$

where Z' and Z'' can be calculated from Eqs (15) and (16).

Limitations of oscilloscopic recording are essentially those of precision as described above for two-channel measurement. However, since time is not an explicit variable, time base limitations do not apply when recording Lissajous

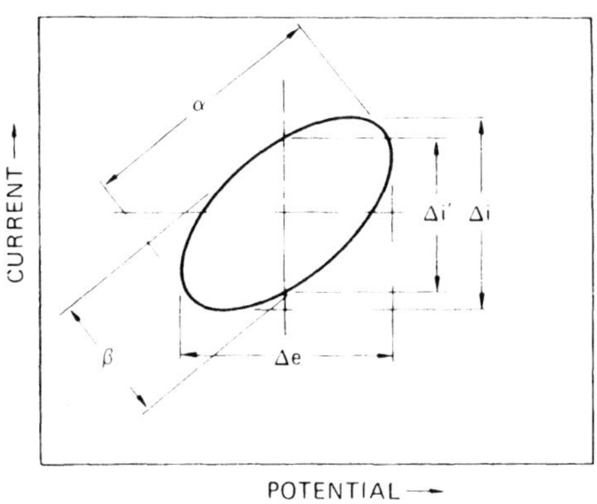

Figure 3.1.8. Lissajous figure for the evaluation of impedance.

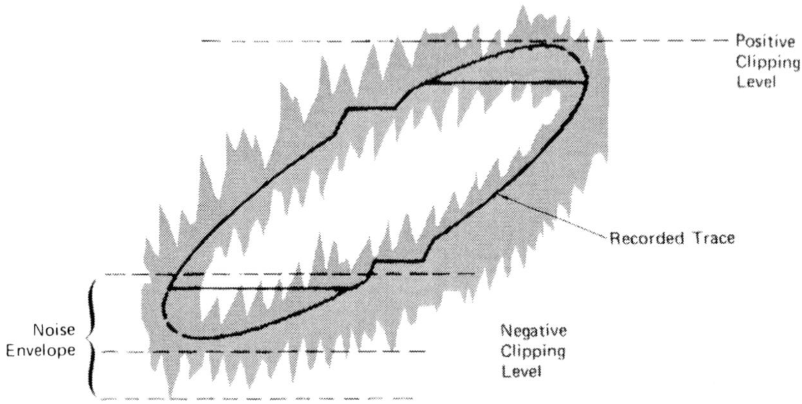

Figure 3.1.9. The errors in electromechanical Lissajous figure recording due to the presence of input noise.

figures. Low-frequency limitations are imposed by electrochemical instabilities in the system under test and electrical instabilities (particularly dc offset drift) in the attendant circuitry. Electromechanically "X–Y" recording can be used to achieve a precision better than 1% of $|Z|$ at frequencies from 1 Hz to below 10^{-3} Hz.

Considerable caution is necessary when applying this last method. Electrochemical systems are susceptible to external noise pickup. The use of high gain, without appropriate electrical filtering, to amplify low-level since-wave voltage and current perturbations may result in severe errors in the dimensions of the ellipse traced on an electromechanical "X–Y" plotter because the mechanical damping of the plotter may disguise the fact that the input amplifiers are overloaded by the "high"-frequency (>10 Hz) noise envelope. This effect is shown schematically in Figure 3.1.9 for 50- or 60-Hz mans pickup in the "Y" amplifier. Errors may, of course, occur in both channels. This phenomenon often is reflected as skewing or tracking of the recorded ellipse, but may result in a stable erroneous trace. To prevent errors in the calculated impedance values, appropriate electronic low-pass or notch (50- or 60-Hz mains pickup) filtering must be used at an early stage of amplification.

3.1.2.6 Phase-Sensitive Detection for Direct Measurement

The real and imaginary components of a voltage can be measured directly with respect to a reference signal with a phase-sensitive detector (PSD). Because of the requirements for linearity, small input signals must be used to measure electrochemical impedances, and noise problems often make it impractical to use either e or e_R (Figure 3.1.6) as a reference signal. Accordingly, e and e_R must be measured alternately in terms of a coherent reference signal of arbitrary phase and the impedance determined from the complex quotient

$$Z = \frac{e}{e_R} R = \frac{(e' + je'')}{(e'_R + je''_R)} R \tag{19}$$

To understand the advantages inherent in this method, it is appropriate to discuss briefly the detection technique.

Phase-sensitive detection may be accomplished by the sequential operation of multiplexing and time-averaging circuits. The multiplexer serves effectively to multiply the input sine wave e_i with a reference square wave e_{ref}. We can represent e_{ref} in terms of its Fourier components

$$e_{\text{ref}} = \frac{4}{\pi}\left[\sin(\omega_r t) + \frac{1}{3}\sin(3\omega_r t) + \frac{1}{5}\sin(5\omega_r t) + \cdots\right] \tag{20}$$

and the input sine wave can be written as $e_i = |A^0|\sin(\omega_i t + \phi)$, where $|A^0|$ is the input signal amplitude, ω is the angular frequency, and subscripts r and i refer to the reference and input signals.

The multiplexer output will be

$$\begin{aligned} e_{\text{mpx}} = e_{\text{ref}} e_i = \frac{2|A^0|}{\pi}\Big\{&\cos[(\omega_i - \omega_r)t + \phi] + \frac{1}{3}\cos[(\omega_i - 3\omega_r)t + \phi] \\ &- \frac{1}{3}\cos[(\omega_i + 3\omega_r)t + \phi] + \cdots\Big\} \end{aligned} \tag{21}$$

In normal practice, ω_r and ω_i are derived from a common source (i.e. $\omega_r = \omega_i$), and the multiplexer output is

$$\begin{aligned} e_{\text{mpx}} = \frac{2|A^0|}{\pi}\Big\{&\cos(\phi) - \cos(2\omega_r t + \phi) + \frac{1}{3}\cos(-2\omega_r t + \phi) \\ &- \frac{1}{3}\cos(4\omega_r t + \phi) + \cdots\Big\} \end{aligned} \tag{22}$$

Only the first term in Eq. (22) is time-independent and, when applied to the time-average circuit, will result in a nonzero output

$$e_{\text{out}} = \frac{2}{\pi}|A^0|\cos(\phi) \tag{23}$$

This is obviously a phase-sensitive dc output voltage, which is a maximum at $\phi = 0$.

The PSD output is frequency-selective since the time average of Eq. (21) for $\omega_r \neq \omega_i$ is zero. The important exception to this statement is for $\omega_i = 3\omega_r$, $5\omega_r$, $7\omega_r$, and so forth. That is, a PSD responds to odd-order harmonics of the input signal. This contribution diminishes with the order of the harmonic.

Time-averaging may be accomplished by analog or digital means. In the vast majority of commercial instruments, an analog low-pass smoothing circuit is used with a front-panel-adjustable time constant. This arrangement offers the advantage of simplicity and flexibility in high-frequency operation. The upper frequency limit is commonly 10^5 Hz. The low-frequency limit of analog time-averaging devices is imposed by the practical details of low-pass filter design (Sallen and Key [1955]) smoothing capacitor ideality, current leakage in buffer amplifiers, and external asynchronous (nonrandom) noise effects. The low-frequency limit of commercial instruments is commonly in the range 0.5–10 Hz. Impedance usually can be measured with 0.1% precision in both components over the specified frequency range.

By using digital integration methods, the low-frequency response can be extended to below 10^{-3} Hz. In this method, the average is taken digitally over an integral number of cycles (McKubre and Hills [1979]). At very low frequencies, information relating to e and i taken over a single cycle can be used to calculate the real and imaginary impedance components with a precision of 0.1%.

3.1.2.7 Automated Frequency Response Analysis

In general, direct methods can be used to acquire impedance data significantly more rapidly than bridge methods. This is particularly true for digitally demodulated, phase-sensitive detectors, for which only a single cycle is required. Nevertheless, in unstable systems, such as rapidly corroding specimens, acquisition rate is an important consideration, and a major criticism of PSD methods is that these must be performed frequency by frequency. Fortunately, this often is not a serious hindrance when such equipment is automated. In the past decade, a number of experimenters have used automated "frequency response analyzers" as digitally demodulated, stepped-frequency impedance meters. Typical of this class are the Solartron 1170 and 1250 series frequency response analyzers (FRAs).

FRAs determine the impedance by correlating the cell response $S(t)$ with two synchronous reference signals, one of which is in phase with the sine-wave perturbation and the other shifted 90° in phase (Gabrielli [1981], Gabrielli and Keddam [1974], Armstrong *et al.* [1968, 1977]). A typical FRA is shown schematically in Figure 3.1.10. The sine-wave perturbation function $P(t)$ applied to the cell may be represented as

$$P(t) = P^0 \sin(\omega t) \tag{24}$$

where P^0 is the amplitude and ω is the frequency. Likewise, the cell response may be written as

$$S(t) = P^0|Z(\omega)|\sin[\omega t + \phi(\omega)] + \sum_m A_m \sin(m\omega t - \phi_m) + N(t) \tag{25}$$

where $|Z(\omega)|e^{j\phi(\omega)}$ is the transfer function of the cell and the first term on the right side of Eq. (25) is the fundamental component. However, because of the nonlinear nature of electrochemical systems, the response will also contain harmonics. Also, electrochemical studies are normally carried out in environments electronically "cluttered" by signals due principally to pickup from main power sources. The harmonic and noise contents of the cell response are represented by the second and third terms, respectively, on the right side of Eq. (25).

The real and imaginary components of the impedance are given by the integrals

$$H'(\omega) = \frac{1}{T}\int_0^T S(t)\sin(\omega t)dt \tag{26}$$

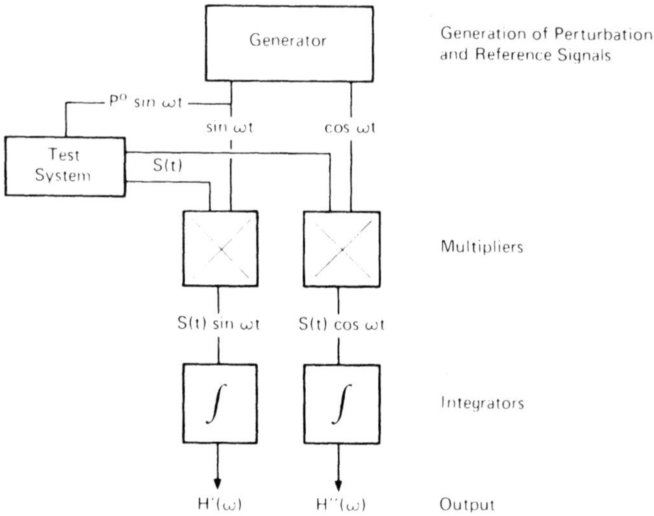

Figure 3.1.10. Schematic of transfer function analyzer.

$$H''(\omega) = \frac{1}{T}\int_0^T S(t)\cos(\omega t)dt \tag{27}$$

Substituting Eqs (26) and (27) into Eqs (24) and (25), we obtain

$$\begin{aligned} H'(\omega) &= P^0|Z(\omega)|\int_0^T \sin[\omega t + \phi(\omega)]\sin(\omega t)dt \\ &+ \frac{1}{T}\int_0^T \sum_m A_m \sin(m\omega t - \phi_m)\sin(\tau t)dt + \frac{1}{T}\int_0^T N(t)\sin(\omega t)dt \end{aligned} \tag{28}$$

$$\begin{aligned} H''(\omega) &= P^0|Z(\omega)|\int_0^T \sin[\omega t + \phi(\omega)]\cos(\omega t)dt \\ &+ \frac{1}{T}\int_0^T \sum_m A_m \sin(m\omega t - \phi_m)\cos(\tau t)dt + \frac{1}{T}\int_0^T N(t)\cos(\omega t)dt \end{aligned} \tag{29}$$

If the noise is completely random (i.e. asynchronous), then the last integrals in Eqs (28) and (29) are equal to zero provided that they are carried out over infinite time. If the integration is carried out over N_f periods of the sinusoidal perturbation, the equivalent filter selectively is given by (Gabrielli [1981])

$$\Delta f / f_1 = 1/N_f \tag{30}$$

where f_1 is the center frequency in hertz and Δf is the bandwidth. For example if the integration is carried out over 10 periods, then at f_1 = 1000 Hz and 1 Hz, Δf is 100 Hz and 0.1 Hz, respectively. On the other hand, if the integration is carried out over 100 periods, the bandwidths are reduced to 10 Hz and 0.01 Hz. Clearly, the ability of a transfer function analyzer to reject asynchronous noise improves greatly as the number of periods over which the integration is performed is increased. However, the price is an excessively long data acquisition time, during which the

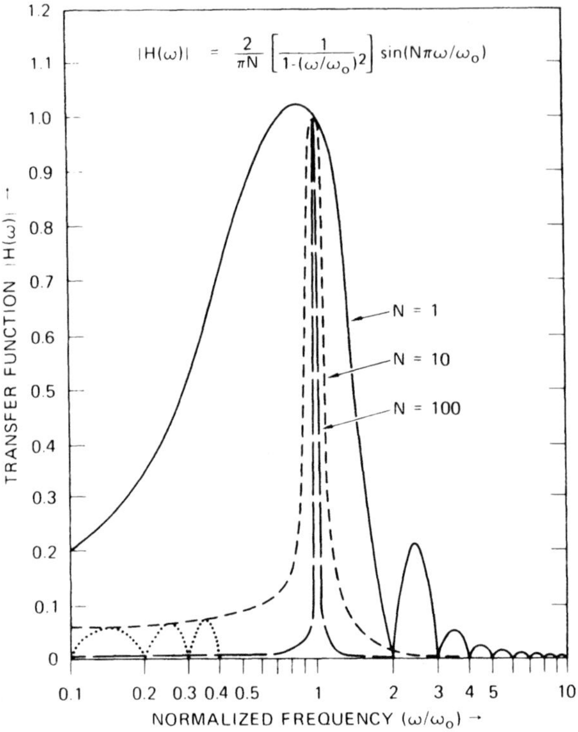

Figure 3.1.11. Frequency response analyzer transfer function vs. normalized frequency, as a function of number of integration cycles.

stability condition may be violated (see Section 3.1.2.9). Figure 3.1.11 shows the transfer function of an FRA as a function of the number of integration cycles performed.

As far as the harmonics are concerned, the integrals in Eqs (28) and (29) may be expanded to read

$$\int_0^T \sin(m\omega t - \phi_m)\sin(\omega t)dt = \cos(\phi_m)\int_0^T \sin(\omega t)\sin(m\omega t)\mathrm{dt} - \sin(\phi_m)\int_0^T \sin(\omega t)\cos(m\omega t)dt \tag{31}$$

$$\int_0^T \sin(m\omega t - \phi_m)\cos(\omega t)dt = \cos(\phi_m)\int_0^T \cos(\omega t)\sin(m\omega t)\mathrm{dt} - \sin(\phi_m)\int_0^T \cos(\omega t)\cos(m\omega t)dt \tag{32}$$

Also noting that

$$\int_0^{k\pi T} \sin(nx)\sin(mx)dx = \begin{cases} 0 \text{ if } m, n \text{ integers}, m \neq n \\ k\pi/2 \text{ is } m, n \text{ integers}, m = n \end{cases} \tag{33}$$

$$\int_0^{k\pi T} \sin(nx)\cos(mx)dx = \begin{cases} 0 \text{ if } m, n \text{ integers, } m+n \text{ even} \\ 2k\pi/(m^2 - m^2) \text{ if } m, n \text{ integers, } m+n \text{ odd} \end{cases} \quad (34)$$

Then the integrals involving the harmonics in Eqs (31) and (32) are identically equal to zero provided that the integrals are carried out over multiples of 2π. Accordingly, FRAs effectively reject the harmonics. Application of the above identities to the fundamental components in Eqs (28) and (29) therefore yields the real and imaginary outputs from the integrators as

$$H'(\omega) = P|Z(\omega)|\cos[\phi(\omega)] \quad (35)$$

$$H''(\omega) = P|Z(\omega)|\sin[\phi(\omega)] \quad (36)$$

which may be scaled to give directly the real and imaginary components of the cell impedance.

FRAs are also readily used to determine the harmonics contained within the output from the cell. This is done by multiplying the reference signal to the multipliers (but not to the cell) by the harmonic coefficient (2 for the second harmonic, 3 for the third, and so forth). The ability of FRAs to characterize the harmonics provides a powerful tool for investigating nonlinear systems; a topic that is now being actively developed (McKubre [1983]).

FRAs provide a very convenient, high-precision, wide-bandwidth method of measuring impedances in electrochemical systems. Commercial instruments are available which provide up to $4\frac{1}{2}$ digits of precision in the real and imaginary components, in frequency ranges covering 10^{-4} to 10^6 Hz. These are direct-measuring devices and therefore are not susceptible to limitations on imposed potentiostat control.

The primary limitation is one of cost. The basic FRA may cost in the order of \$20,000 (1986), and an additional investment for microcomputer and data storage facility is necessary to accommodate the higher rates of data collection made possible by the use of a FRA. A more subtle difficulty often occurs, as these devices are capable of operating with $4\frac{1}{2}$ digits of precision (the data dutifully recorded by the microcomputer) whether or not the instrument is connected correctly, or at all, to the electrochemical cell. Considerable familiarity with electrical systems is necessary in order to get accurate impedance data, particularly at higher frequencies. It is highly desirable that an oscilloscope be used in parallel with the two input channels of an FRA, in order to monitor continuously the form of the input and output signals.

3.1.2.8 Automated Impedance Analyzers

There are a number of automated and semiautomated "impedance analyzers" on the market. Although these are intended primarily for network and network component analysis, they have a limited applicability for measurements in electrochemical systems.

Generally, this class of ac analyzer operates with a so-called autobalance bridge. The desired signal (comprising both ac and dc components) is applied to the

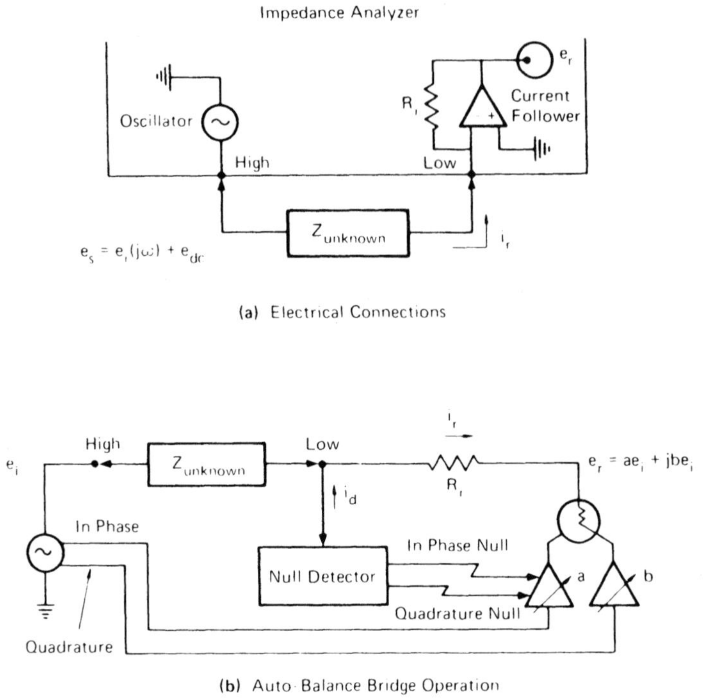

Figure 3.1.12. Direct measurement of impedance using an impedance analyzer.

unknown impedance, as is shown in Figure 3.1.12*a*. The current follower effectively constrains all the current flowing through the unknown impedance i_r to flow through the range resistor R_r, presenting a virtual ground at the terminal marked "low" (for a further description of the use of current followers, see McKubre and Macdonald [1984]). For this condition the impedance can be measured as

$$Z_{\text{unknown}} = R_r \frac{e_i}{e_i} \tag{37}$$

The complex ratio e_i/e_r is measured in a manner very similar to that described in Sections 3.1.2.2 and 3.1.2.3 for transformer ratio and Berberian–Cole bridge circuits, in which in-phase and quadrature factions of the input signal are summed with the unknown output signal (current) until the result is zero. One method of accomplishing this is shown in Figure 3.1.12*b*. The oscillator that produces the input perturbation signal e_i also outputs in-phase and quadrature (90° out-of-phase) reference signals that are proportional in amplitude to e_i. These are fed to a summing circuit and summed with the unknown current until the current to the detector, i_d, is zero. At this condition the low-potential terminal is at ground voltage, and

$$Z_i = \frac{R_r}{a = jb} \tag{38}$$

Thus, the unknown impedance can be determined directly from the value of the range resistor, R_r, and the attenuation factors a and b imposed by the null detector to achieve the null condition.

The advantages of this method are that relatively high speed and high precision are attainable. Being a null method, the effects of stray capacitances are somewhat reduced, although, unlike in a "true" bridge, currents do flow through the unknown impedance at the null condition. This method is usable up to very high frequencies (tens or hundreds of megahertz), well beyond the range of interest in aqueous electrochemistry.

The intrinsic disadvantage of this method is its two-terminal nature: the facts that a dc potential cannot be applied to the electrode of interest with respect to a suitable reference electrode and that the potential e_i across the specimen varies during the balance procedure. Since the in-phase and quadrature null signals usually are derived from a PSD, instruments of this type are limited at low frequencies to approximately 1 Hz due to the instability of analog filters with longer time constants.

3.1.2.9 The Use of Kramers–Kronig Transforms

The use of a frequency domain transformation first described by Kramers [1929] and Kronig [1926] offers a relatively simple method of obtaining complex impedance spectra using one or two ac multimeters. More important, retrospective use of Kramers–Kronig (KK) transforms allows a check to be made on the validity of an impedance data set obtained for linear system over a wide range of frequencies. Macdonald and Urquidi-Macdonald [1985] have applied this technique to electrochemical and corrosion impedance systems.

The KK transforms of interest in analyzing corrosion and electrochemical systems are

$$Z'(\omega) - Z'(\infty) - \left(\frac{2}{\pi}\right)\int_0^\infty \frac{xZ''(x) - \omega Z''(\omega)}{x^2 - \omega^2}dx \tag{39}$$

$$Z'(\omega) - Z'(0) = \left(\frac{2\omega}{\pi}\right)\int_0^\infty \left[\left(\frac{\infty}{x}\right)Z''(x) - Z''(\omega)\right]\frac{1}{x^2 - \omega^2}dx \tag{40}$$

$$Z''(\omega) = -\left(\frac{2\omega}{\pi}\right)\int_0^\infty \frac{Z'(x) - Z'(\omega)}{x^2 - \omega^2}dx \tag{41}$$

$$\phi(\omega) = \left(\frac{2\omega}{\pi}\right)\int_0^\infty \frac{\log|Z(x)|}{x^2 - \omega^2}dx \tag{42}$$

$$R_p = \left(\frac{2}{\pi}\right)\int_0^\infty \frac{Z''(x)}{x}dx \tag{43}$$

These equations show that the real component of the impedance can be calculated from the imaginary component and vice versa, the phase angle $\phi(\omega)$ can be computed from the magnitude of the impedance, and the polarization resistance (R_p) can be extracted from the imaginary component [Eq. (43)] in addition to being derived directly from the real component of the impedance

$$R_p = Z'(0) - Z'(\infty) \tag{44}$$

The use of these expressions to validate impedance data will not be discussed in detail here (see Section 3.1.3.4 and Macdonald and Urquidi-Macdonald [1985], Urquidi-Macdonald *et al.* [1985]). Instead, we note that if the magnitude of the impedance is measured over an effectively infinite bandwidth, then the real and imaginary components can be calculated. These data are then used to compute the polarization resistance directly from Eq. (44) and indirectly from the imaginary component according to Eq. (43). The application of this latter method to calculating R_p in concentrated potassium hydroxide solution at 25°C is show in Figures 3.1.13 and 3.1.14. In the first figure, the complex-plane diagram for this system is

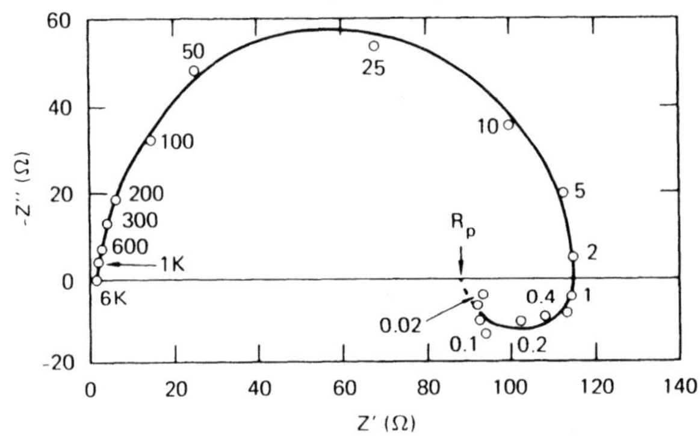

Figure 3.1.13. Impedance diagram for Al–0.1 P–0.1 In–0.2 Ga–0.01 Tl alloy in 4 M KOH at 25°C and at the open-circuit potential (−1.760 V vs. Hg/HgO). The parameter is frequency in Hz.

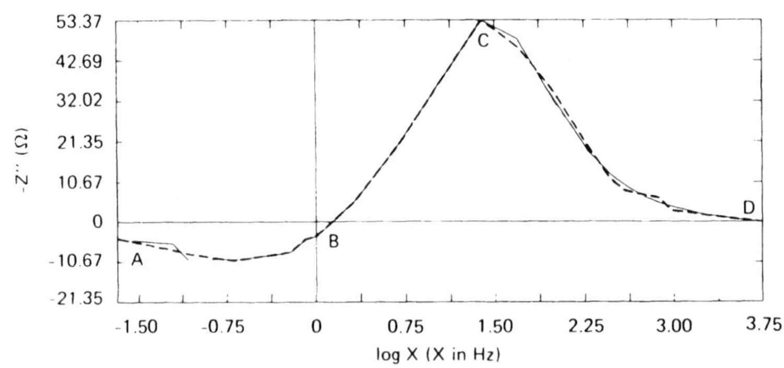

Figure 3.1.14. Plot of Z'' vs. log X(X in Hz) for Al–0.1 P–0.1 In–0.2 Ga–0.1 alloy in 4 M KOH at 25°C under open-circuit conditions (E = −1.76 V vs. Hg/HgO). Dashed line = polynomial fit [Eq. (45)], solid line = experimental data.

shown, illustrating the inductive behavior at low frequencies and the extrapolation to the real axis to determine $Z'(0)$ and $Z'(\infty)$. In the second figure, the imaginary component is plotted as a function of log x, together with a fitted polynomial

$$Z''(x) = \sum_{k=0}^{n} a_k x^k \tag{45}$$

which is then used to evaluate the integral in Eq. (40). This procedure yields a value for R_p of 82.4 Ω, compared with 90 ± 5 Ω determined from the real components. This difference is insignificant from a corrosion-monitoring viewpoint; it probably arises from changes in the interface during the period of data acquisition.

As indicated earlier, KK transforms not only can be used to check a data set for internal consistency, but also provide a simple method of obtaining impedance data. Briefly, the magnitude of an unknown impedance is often very easily measured as the scalar ratio of the magnitudes of the voltage across, and current passing through, the unknown element. It is considerably more difficult to measure the phase information, but this can be calculated from the impedance magnitude spectrum using the appropriate KK transformation.

A circuit to allow the measurement of an unknown impedance magnitude spectrum is shown schematically in Figure 3.1.15; although a two-terminal configuration is shown, there is no reason why this method cannot be applied with potentiostatic control. To obtain an impedance magnitude spectrum, the frequency of the ac oscillator is simply stepped or swept through the desired frequency range, the ac voltage and current recorded, and the ratio taken as a function of frequency. Since a machine transformation is necessary to obtain the phase spectrum (which can be used in conjunction with the magnitude information to yield the real and imaginary components in Cartesian coordinates), the most practical implementation of this method is to use a computer interfaced with, for example, an IEEE-488 con-

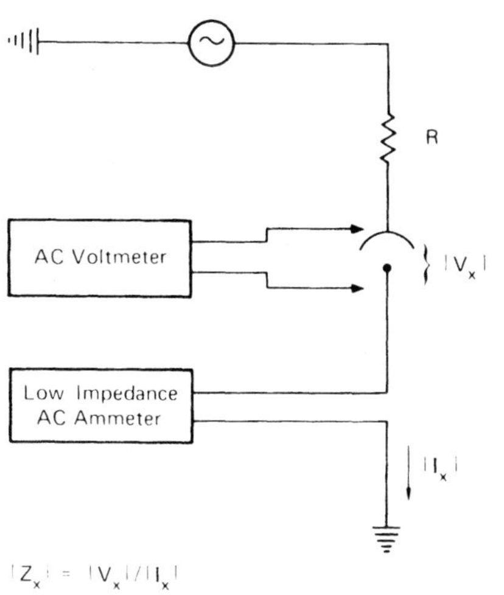

Figure 3.1.15. Circuit to obtain an impedance magnitude spectrum.

trolled oscillator and multimeters to output frequencies and record the current and voltage information directly (see Section 3.1.3.3 for a description of the IEEE-488 interface).

The major advantages of this method are simplicity and low cost; a practical system can be configured for under $5000 if a microcomputer is used to synthesize the sine wave. Since ac amp/voltmeters operate basically as dc devices (following a rectifying front end), they can operate to very high frequencies, and reliable measurements can be made well into the megahertz frequency range, the high-frequency limitation usually depending only on stray capacitances and transmission line effects external to the measurement circuit.

Practical limitations are imposed at low frequencies, however, where the rectification-smoothing function necessary to transduce the ac voltage magnitude to a dc level becomes inaccurate. Ac voltmeters typically become seriously in error at frequencies below 20 Hz. To obtain an accurate KK transform, it is necessary to extend the measurement frequency range significantly beyond the limits of frequency needed to elucidate the equivalent circuit under test. Thus, the method described here is not appropriate for aqueous electrochemical systems for which the diffusional impedance is prominent. This method can be useful for systems in which the lowest frequency of interest is greater than 50 Hz or so, as is usually the case for solid ionic conductors, oxide films, and semiconductor surfaces.

A more subtle limitation is imposed by the use of the method described here in that all of the four assumptions implied in the use of KK transforms are subsumed when the magnitude-to-phase transformation is made. That is, the unknown impedance is given the properties of linearity, invariance, and causality whether or not they apply, and there is no independent check of this assumption. In current practice, this limitation is not very severe since experimenters frequently report and draw conclusions from impedance data sets, normally derived, that have not been subjected to the scrutiny of the KK rules or other simple tests of experimental validity.

3.1.2.10 Spectrum Analyzers

Spectrum analyzers are instruments that are optimized to characterize signals in the frequency domain; the requirements of signal analysis are subtly different from those of linear network analysis, the former requiring low noise and low distortion over a wide range of frequencies (bandwidth), the latter being optimized to give accurate amplitude and phase measurements over a wide range of input-output voltages (dynamic range). Nevertheless, spectrum analyzers can be used to measure impedances rapidly at audio and higher frequencies, using a variety of input excitation functions. In this section we will describe the functioning of the three major classes of spectrum analyzer: parallel filter, swept filter, and dynamic.

The classical function of a spectrum analyzer is to measure the power (or amplitude) of a signal at a number of discrete points, or in discrete frequency bands, within a defined frequency range. Normally, the frequency bands are linearly or, more commonly, logarithmically spaced within the spectrum of interest. A very simple method to accomplish this goal is to apply the unknown signal to a parallel array of filters,

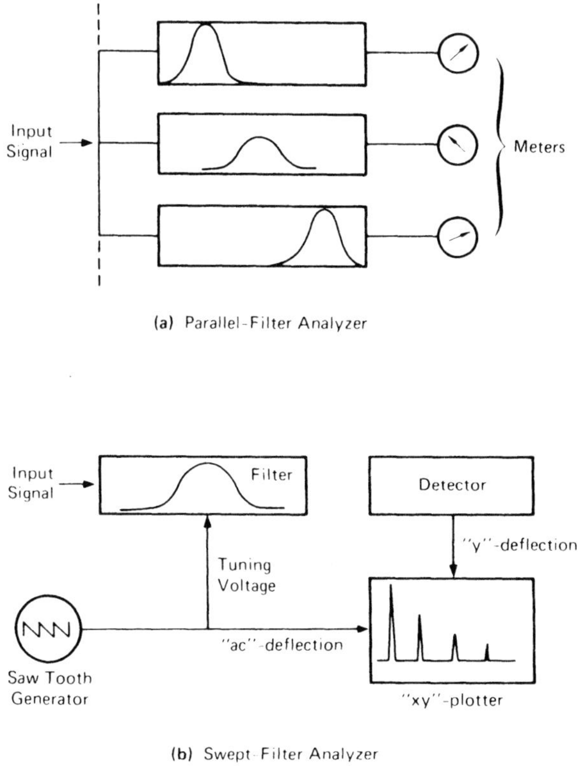

Figure 3.1.16. Spectrum analyzers.

each tuned to pass a defined (and narrow) frequency band. If these bandpass filters are arranged to be uniformly spaced with minimal overlap, as shown in Figure 3.1.16*a*, then the output of one or more voltmeters applied sequentially or simultaneously to the parallel array of "*N*" filters will be an "*N*" point analog of the input frequency spectrum. The advantages of this method are simplicity and speed. If, however, a highly accurate analog of the spectrum is needed, then a large number of closely spaced, narrow-bandwidth filters are required. As the bandwidth is reduced, such filters become expensive and unstable, and the cost of such an analyzer becomes greater as the resolution is increased.

One way to avoid the need for a large number of expensive filters is to use only one filter and to sweep it slowly through the frequency range of interest. If, as in Figure 3.1.16*b*, the output of the filter is plotted against the frequency to which it is tuned, then one obtains a record of the spectrum of the input signal. This swept-analysis technique is commonly used in radio frequency and microwave spectrum analysis. However, the filter has a finite response time, and the narrower the bandwidth of the filter, the longer it takes to respond. To avoid amplitude errors, one must sweep the filter slowly through the frequency range of interest, and the advantage of speed afforded by spectrum analyzers is compromised.

There is a basic trade-off between parallel- and swept-filter spectrum analyzers. The parallel-filter analyzer is fast but has limited resolution and is expensive. The swept-filter analyzer can be cheaper and have higher resolution, but the measurement takes longer (especially at high resolution). Furthermore, since the swept-filter analyzer does not observe all frequencies simultaneously, it cannot be used to analyze transient events.

A disadvantage common to both classes of spectrum analyzer discussed so far is that they do not measure absolute amplitudes accurately, and they do not measure phase at all. Although this last limitation can be circumvented by the use of KK transformations (see Section 3.1.2.9), these instruments generally are poor choices for linear circuit (ac impedance) analysis.

Another kind of analyzer has been developed which offers the best features of parallel- and swept-filter spectrum analyzers. So-called dynamic signal analyzers use analog-to-digital conversion followed by frequency-to-time-domain transformation, usually using hard-wired computational machines, to mimic the function of a parallel-filter analyzer with hundreds of filters, and yet are cost-competitive with swept-filter analyzers. In addition, dynamic spectrum analyzers are capable of measuring amplitude *and phase* accurately; these are basically time domain instruments, and their function will be discussed in Section 3.1.4.

3.1.3 Time Domain Methods

3.1.3.1 Introduction

With the advent of high-speed digital computers, a clear trend toward digital signal processing has become apparent. The advantage of digital over analog data processing is purely mathematical; a far wider range of mathematical computations can be performed in the digital mode than on analog signals. Digital signal processing using hard-wired devices has also expanded rapidly over the past decade and is likely to find even more application in the years to come.

Since the world of electrochemistry is an analog one, the use of digital computation methods must be preceded by analog-to-digital conversion. One of the most important experimental aspects of this process is the method by which the computer interacts with the analog experiment: the computer interface. Having achieved the digital state, the range of computational algorithms used to extract ac impedance information is very diverse. Although a thorough discussion of these topics is beyond the scope of this chapter, in this section we discuss briefly the techniques that are now in common use for analog-to-digital conversion, computer interfacing, and digital signal processing, with reference to the measurement of the ac impedance parameters.

From the definitions given in Section 4.4.2, it is apparent that the interfacial impedance can be calculated from the perturbation and response in the time domain, in which the excitation can be any arbitrary function of time. In principle, any one of several linear integral transforms can be used (Macdonald and McKubre [1981]) to convert from the time domain into the frequency domain, but the two most commonly used are the Laplace and Fourier transforms:

$$F(s) = \int_0^\infty F(t)e^{-st}\,dt \tag{46}$$

$$F(j\omega) = \frac{1}{\sqrt{2\pi}}\int_{-\infty}^{\infty} F(t)e^{-j\omega t}\,dt \tag{47}$$

where s is the Laplace frequency. Noting that $s = \sigma + j\omega$, Eq. (46) leads to

$$F(j\omega) = \int_0^\infty F(t)e^{-j\omega t}\,dt \tag{48}$$

which is referred to as a single-sided Fourier transform. By transforming the time domain voltage [$E(t)$] and current [$I(t)$] to yield the frequency domain quantities [$E(j\omega)$ and $I(j\omega)$], the impedance may be calculated as

$$Z(\omega) = \frac{E(j\omega)}{I(j\omega)} \cdot \frac{I^*(j\omega)}{I^*(j\omega)} \tag{49}$$

where $I^*(j\omega)$ is the complex conjugate of $I(j\omega)$.

As noted above, any arbitrary time domain excitation can be used to measure the system impedance provided that the excitation is applied and the response recorded over a sufficiently long time to complete the transforms over the desired frequency band. Thus, potential and current steps and various noise excitations have been extensively used (Sierra-Alcazar [1976], Pilla [1970, 1972, 1975], Doblhofer and Pilla [1971], Smith [1966, 1971, 1976], Creason and Smith [1972, 1973]), particularly in the field of ac polarography. More recently, these same methods have been applied in corrosion science (Pound and Macdonald [1985]; Smyrl ([1985a,b]; Smyrl and Stephenson [1985]) to obtain impedance spectra, but more importantly to estimate polarization resistance for rapidly corroding systems. In the work of Pound and Macdonald [1985], various time-to-frequency transformation techniques were evaluated, including the discrete Fourier transform, the fast Fourier transform (FFT), the Laplace transform, and an algorithm that duplicates the mathematical operation of an FRA (see Section 3.1.2.7). All these techniques involve the recording of the perturbation and response in digital form in the time domain before signal processing in either software or hardware. Regardless of the mode of processing, the accuracy of transformation depends critically on acquiring data records having the desired characteristics of length and sampling frequency.

3.1.3.2 Analog-to-Digital (A/D) Conversion

The conversion of analog signals into digital form, and ultimately into binary-word representation, is now a common practice in electrochemistry, particularly for interfacing analog instruments, such as potentiostats, with digital recording and processing equipment (e.g. computers). The essential operation desired is to convert the value of an analog signal into a binary word whose magnitude is proportional to the signal being sampled. This process involves two operations: sampling and quantization. The first involves momentarily "freezing" the analog signal in time to

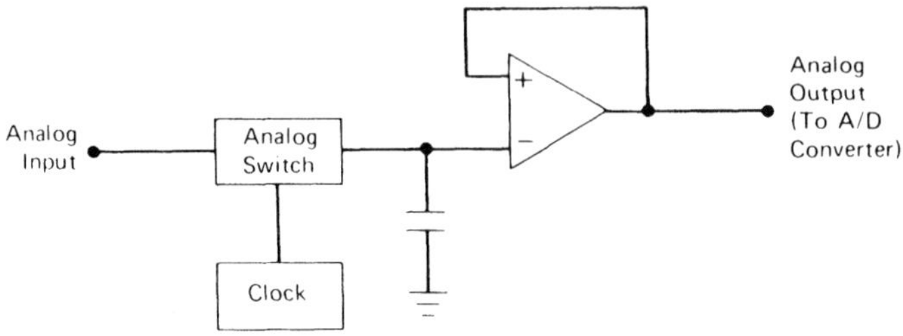

Figure 3.1.17. Schematic diagram of a sample-and-hold amplifier.

produce a discrete value. This value is then converted to its binary representation during the "quantization" step, after which the cycle is repeated.

Sampling is normally achieved by using "sample and hold" amplifiers of the type shown schematically in Figure 3.1.17. In this circuit, a signal to the analog switch (e.g. 4066 CMOS) connects the analog input to the amplifier. Provided that the capacitance to ground is sufficiently small, the capacitor will charge to the analog input voltage with good fidelity. Removal of the control signal effectively disconnects the input from the capacitor so that the analog output assumes the value of the input at the instant the switch was opened. The above cycle is then repeated, with the sample rate being determined by the control signal from the clock; it is necessary in the case of A/D conversion, however, that the hold time be sufficiently long for the quantization step to generate an accurate digital representation of the analog input.

A number of quantization techniques are available, and the selection of the optimum analog-to-digital converter (ADC) for a particular application is properly based on considerations of resolution (precision), accuracy (initial and drift with time and temperature), ease of interfacing, cost, and convenience (availability, size and power requirements). To select an ADC, it is useful to understand exactly what types are available and how they work. The listing in Table 3.1.1, although far from complete, does include the most popular ADCs, especially those currently produced in an integrated circuit form.

We will discuss here only successive-approximation and integration ADCs in any detail. The tracking A/D and voltage-to-frequency converter can be looked on as variations of the successive-approximation and the integration design techniques; in these types, the digital data is available on a virtually continuous basis.

The principle of the successive-approximation technique is shown in Figure 3.1.18. In this particular example we wish to convert the analog input voltage (10.3 V) into its floating-point, 8-bit binary form to the nearest 0.0625 V. The process involves eight successive steps, in which the field is divided into halves and a bit of 1 or 0 is assigned to each step, depending in which half of each field the analog value lies. For example in the first step 10.3 lies between 8 and 16, so that 1 is assigned as the most significant bit. However, in the second conversion step the analog input lies in the 8–12 field rather than in the upper half (12–16), so that 0 is assigned to the second most significant bit. This process is repeated until the desired

Table 3.1.1. Property of Common Analog-to-Digital Converters

Type	Advantages	Disadvantages	Typical Uses
Successive approximation	High speed Flexibility	Precision expensive Susceptible to noise	Multiplexing 100 Hz–1 MHz/ channel
Integrating	High accuracy Low cost Low sensitivity to noise	Low speed	DC–100 Hz Digital voltmeters
Tracking (counter-comparator)	High tracking speed	Susceptible to noise	DC–100 Hz
Multicomparator (flash)	High speed High resolution	Expensive	1 MHz and up
Voltage–frequency converter	Fast response Continuous output	Moderate precision	Telemetry

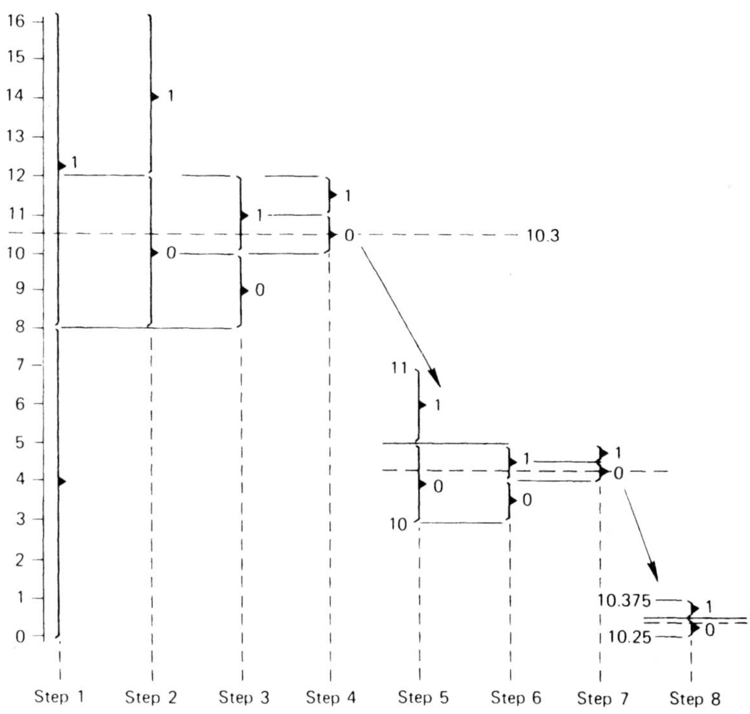

Figure 3.1.18. Successive-approximation conversion of an analog signal to its 8-bit binary representation. Analog input = 10.3, equivalent binary output = 1010.0100 ≡ 10.25.

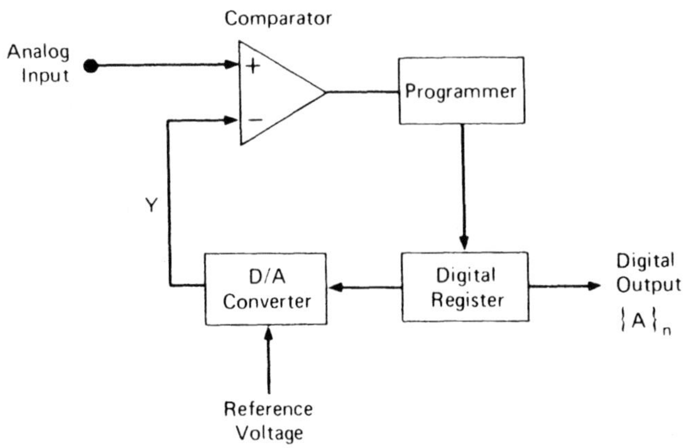

Figure 3.1.19. Schematic diagram of a successive-approximation A/D converter.

precision is achieved. Clearly, only n steps are required to quantize an unknown voltage into its $\{A\}_n$ binary form.

A block diagram of a successive-approximation ADC is shown in Figure 3.1.19. The circuit converts each successive approximation into an analog signal Y, which is equivalent to the center of each division shown in Figure 3.1.18. The analog input (A) is then compared with Y in the following cycle; if A is greater than Y, the comparator swings to positive saturation and a "1" is loaded into the register.

The principal advantage of the successive-approximation technique is high speed, and conversion rates in the megahertz range are possible. The principal disadvantages are limited accuracy and precision. Accuracy is limited because, as with all wide-bandwidth (high-speed) devices, the technique is very susceptible to external noise sources, and a noise spike coinciding with any of the more significant bit conversions can result in large errors; to some extent this difficulty can be removed by averaging multiple conversions, at the expense of speed. The precision is limited by the number of bits converted. Thus a 16-bit ADC (8, 10, 12, 14, and 16 bits are commonly used) will have a minimum uncertainty of 1 part in 2^{16} or 0.0015% and can achieve that level of precision only if the voltage being measured is close to the ADC maximum. However, even when using a 2-V device, the sensitivity of a 16-bit ADC is 0.03 mV, which is adequate for most electrochemical measurements.

A second A/D conversion technique offers greater immunity to noise and almost unlimited precision, as well as reducing the need for sample-and-hold circuitry at the signal input. The concept of the "dual-slope" or up/down integrating ADC is simple. A current proportional to the input signal charges a capacitor for a fixed length of time; the capacitor is then discharged by a current proportional to a reference voltage until the starting point is crossed. Figure 3.1.20 shows a schematic representation of the implementation of this technique. The input voltage e_i is applied to the input resistor R_i. With the switches in position 1, op amp OA1 forces a current $I_i = e_i/R_i$ to charge the plates of capacitor C. Thus, with the switches in position 1, and after time t_i, the output voltage e_o will be given by

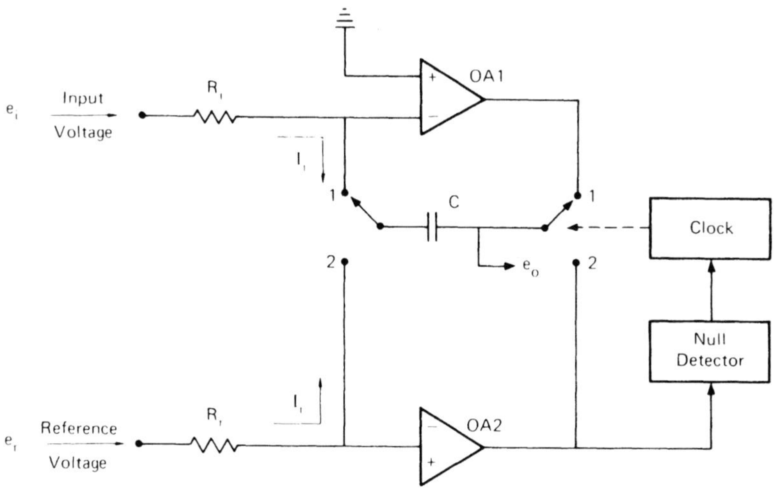

Figure 3.1.20. Schematic diagram of an up/down integrating D/D converter.

$$e_o = e_{\text{initial}} + \frac{1}{RC}\int_0^{T_i} e_i dt \tag{50}$$

where e_{initial} is the output at $t = 0$. For simplicity, if the initial voltage is zero and the integral is replaced by the average in the input time window, $\bar{e}_i$,

$$e_o = \frac{\bar{e}_i t_i}{R_i C} \tag{51}$$

After an accurately clocked interval t_i, the clock sets the switches to position 2, and the negative reference current is applied to discharge the capacitor. Thus,

$$e_o = \frac{\bar{e}_i t_i}{R_i C} - \frac{e_r t}{R_r C} \tag{52}$$

and at the condition of null, when $e_o = 0$,

$$e_i = \frac{R_i t}{R_r t_i} e_s \tag{53}$$

and the averaged input voltage can be calculated very precisely from the accurately known values of resistances and time.

Although good-quality components (especially the capacitor) must be used for reasonable accuracy, only the reference need be an expensive, high-quality component. Speed is an obvious limitation because of the long count time required. (For example one must count to 2000 and effectively do 1000 successive comparison tests at the null detector to achieve 3-digit or 10-bit resolution.)

Dual-slope integration has many advantages. Conversion accuracy is independent of both the capacitor value and the clock frequency, because they affect both the upslope and the downramp in the same ratio. The averaging mode and the fixed averaging period also grant excellent immunity to noise, and an integrating ADC has "infinite" normal-mode rejection at frequencies that are integral multiples of $1/t_i$. In practical terms, if the ADC is set to integrate over exactly n cycles of some extraneous and periodic noise source (e.g. main frequency) then the integral will be zero, as if the spurious signal had been completely filtered out at the input.

Throughput rate of dual-slope converters is limited to somewhat less than $\frac{1}{2}t_i$ conversions per second; the sample time t_i is determined by the fundamental frequency to be rejected. For example if one wishes to reject 60 Hz and its harmonics, the minimum integration time is 10.167 ms, and the maximum conversion rate is somewhat less than 30 Hz.

3.1.3.3 Computer Interfacing

The details of computer interfacing are so intimately connected to the details of programming itself that a discussion of arbitrary, low-level interfacing is best suited to a treatise on software than to one on hardware. Interfaces that operate at a high level, with their details of operation obscured to the user by a "driver" program, however, are of significant importance to the experimenter interested in the implementation of ac methods. Most common among the high-level interfacing systems is the general-purpose interface bus (GPIB), also known as the IEEE-488 (or IEC-625 or Hewlett-Packard Interface Bus (HPIB)). This interface standard is becoming capable of almost universally connecting computers with digital multimeters, transient recorders, Fourier analyzers—in short, with all those tools needed to implement the ac impedance method in the time domain.

IEEE standard 488–1978 interface represents a highly flexible, moderate-speed system that is well suited to general laboratory use. The IEEE-488 interface bus (IB) consists of 16 signal lines. Eight lines are used for data, five for bus management, and three lines are used to establish a temporary communication link, or "handshake," between two devices that are properly attached to the bus. Because there are eight data lines, an 8-bit byte can be communicated in each handshake cycle. Thus 16-bit, 24-bit, etc. words (either instructions or data) can be communicated with sequential handshake cycles. This method is often referred to as "bit parallel, byte serial" transmission (Colloms [1983]).

A very large number of devices can be connected simultaneously to the interface bus. Each device is given a unique address, which is used in establishing a handshake. Handshake is used to ensure that data is transferred from a source to one or more designated acceptors. Figure 3.1.21 shows a portion of the bus structure, and Figure 3.1.22 shows the handshake sequence in detail. The three signals used for handshake are: data valid (DAV), not ready for data (NRFD), and no data accepted (NDAC). The DAV line is driven by the sender, while the NFRD and NDAC lines are driven by the receiver. The handshake procedure ensures that all listeners are ready to receive data, that the data on the eight data lines is valid, and that the data has been accepted by all listeners. Data will be sent only as fast as it can be accepted by the slowest receiver.

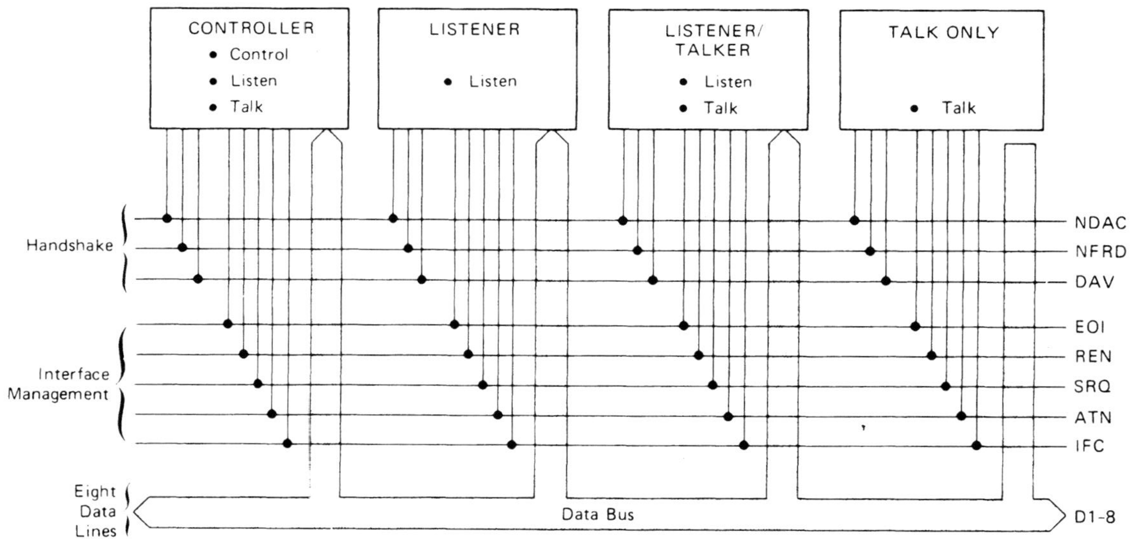

Figure 3.1.21. A section of an IEEE-488 interface bus, showing the major classes of user devices.

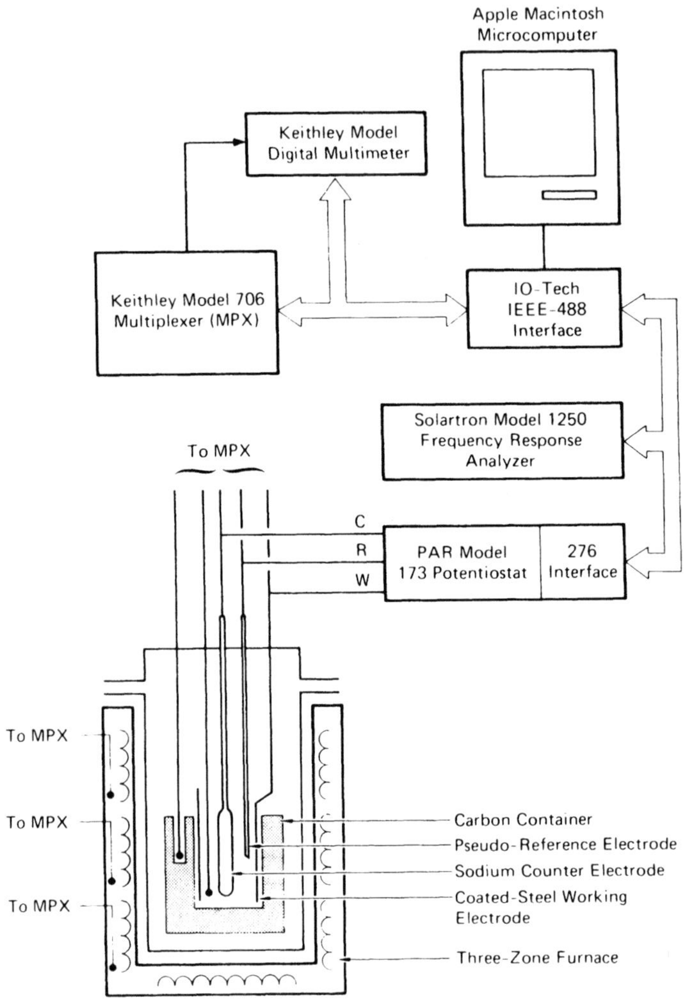

Figure 3.1.22. Use of the IEEE-488 interface for dc and ac measurement and control.

The IEEE-488 IB is designed to interface with the four major types of devices shown in Figure 3.1.21. A master or controller sends commands over the bus, using the bus control (uniline) and data lines (multiline). Normally, one controller is present (e.g. a computer or microprocessor), but if more are present, only one may exercise control at any time. A controller issues a system initiation command, interface clear (IFC), and designates which devices are talkers and which are listeners. The controller has complete control of the attention line (ATN). When ATN is "true," the controller is issuing messages or commands.

A listener receives data over the IB, following an acceptor handshake. Addressed listeners respond to controller commands. Listen-only devices are intended for use in a circuit with no controller. An example of a listener might be a digitally controlled analog potentiostat.

A talker is a device capable of sending data over the IB to a controller or listener. An unaddressed talk-only device, such as a digital voltmeter, may represent a problem in a circuit with a controller, since this device may continue talking when the controller requires attention (ATN "true"), resulting in a garbled message.

Most commercial devices intended for use with an IEEE-488 IB are combined talker-listeners, capable of receiving instructions, setting the data collection mode and experimental conditions, and returning data to the controller.

IEEE-488 systems of considerable complexity have been developed for electrochemical data acquisition and experimental control. One such system, shown in Figure 3.1.22, uses a microcomputer to monitor temperature and dc signals with an IEEE-488 multiplexer and multimeter, to control and IEEE-488 potentiostat, and to output and input data for an IEEE-488 frequency response analyzer, in order to measure impedances in a sodium/sodium-polysulfide cell at elevated temperatures (McKubre and Sierra-Alcazar [1985]).

3.1.3.4 Digital Signal Processing

In principle, any one of several integral transforms can be used to convert data collected in the time domain into the frequency domain for subsequent analysis. Because of the similarity of the various transformation techniques, it is convenient first to discuss briefly the interrelationships of the various transform functions.

The general linear integral transformation of a function $F(t)$ with respect to a kernel $K(t, q)$ is given as (Bohn [1963], Crain [1970])

$$\bar{F}(q) = \alpha \int_a^b K(t,q)F(t)dt \tag{54}$$

where a and b define the transform interval and q is the transformation variable. The kernels and the limits of integration frequently adopted for the transforms of interest are summarized in Table 3.1.2. It is clear that the methods are very closely related. In particular, the reader will note that the imaginary-axis Laplace transformation

$$\tilde{F}(j\omega) = \int_0^\infty F(t)e^{-j\omega t}dt \tag{55}$$

is in fact a single-sided Fourier transform. Also, since the form of the Laplace variable ($s = \sigma + j\omega$) dictates that both frequency domain and transient responses can

Table 3.1.2. Linear Integral Transforms

Transform	Kernel $K(t, q)$	α	a	b
Laplace	e^{-st}	1	0	∞
Fourier				
Infinite	$e^{-j\omega t}$	1	$-\infty$	$+\infty$
Infinite	$e^{-j\omega t}$	$\frac{1}{2\pi}$	$-\infty$	$+\infty$
Single-sided	$e^{-j\omega t}$	1	0	$+\infty$
Segment	$e^{-j\omega t}$	1	0	$+\infty$

be obtained, it is clear that the Fourier transform is a special case of the more general Laplace transform. The one remaining linear transform of interest, the "Z" transform, to our knowledge has not been used for the analysis of interfacial impedance. Accordingly, this transform will not be considered in this discussion.

In this chapter, we are concerned with machine implementation of transform techniques, either software or hardware, to obtain data in a form convenient for ac impedance analysis. In recent years, the advent of hard-wired (dedicated) Fourier transform units (Reticon [1977]) and the fast Fourier transform algorithm, or FFT (Cooley and Tukey [1965], Hartwell [1971]), have concentrated practical interest almost exclusively on the Fourier transform.

Again, in the formalism of linear systems analysis, the transfer function is the mathematical description of the relationship between any two signals. In the special case where the signals of interest are the input (current excitation) and output (voltage response) of a linear electrical system, the transfer function is equivalent to the system impedance.

Mathematically,

$$G(j\omega) = \frac{E(j\omega)}{I(j\omega)} = \frac{\tilde{F}[E(t)]}{\tilde{F}[I(t)]} = Z(j\omega) \tag{56}$$

where $G(j\omega)$ is the system transfer function, $\tilde{F}$ denotes the Fourier transform, E is the system voltage, and I is the current. The variable $j\omega$ indicates that this is a complex frequency domain parameter, and t indicates a time domain parameter. Equation (56) indicates that the ratio of the Fourier transforms of the measured time domain voltage and current is equal to the impedance.

Obviously, two stages of data manipulation are required to obtain $Z(j\omega)$ as a function of frequency from the response of the system to an arbitrary time domain perturbation: first, the input and response functions must be sampled and recorded in the time window of interest, then the transform of each must be computed and the complex ratio calculated. In hard-wired Fourier transform units, the acquisition subsystem is an integral part of the unit, and this function normally can be ignored. If the experimenter is using a computer or microcomputer to perform these functions, the concepts of analog-to-digital conversion and computer interfacing (described briefly in Sections 3.1.3.2 and 3.1.3.3) must be used.

With either hard-wired or software-programmed logic, the most common method of obtaining $\tilde{F}[E(t)]$ and $\tilde{F}[I(t)]$ uses the FFT algorithm first devised by Cooley and Tukey [1965] as a method for obtaining a discrete digital approximation of the infinite Fourier transformation from a finite data record. The digital nature of the transformation, however, and the finite length of the time record give rise to a number of properties of the FFT that must be recognized in order to minimize distortion of the derived impedance data (Smith [1976], Creason and Smith [1972, 1973]).

The FFT algorithms demand that the time record contain 2^n words, where n is an integer. This requirement is easily satisfied by simply adjusting the digitizing sampling rate and/or the length of the record. However, the sampling theorem states that the highest-frequency component that can be completely characterized in terms of amplitude and phase must have a frequency of less than half the sampling rate. On the other hand, the lowest frequency that is accessible is the reciprocal of the total sampling period. These limitations are readily illustrated by considering a standard FFT array of 1024 words. If this array is collected over 0.7 s, the lowest frequency is 1.43 Hz, whereas the highest frequency is $0.5 \times (1024/0.7) = 731.4$ Hz. Clearly, somewhat less than three decades of frequency are accessible from a single 1024-point FFT. This may be construed as a serious limitation of the FFT algorithm, but it is possible to apply the transformation to successive segments, thereby extending the total frequency range to many orders of magnitude.

The finite length of the data record may cause broadening of the Fourier spectrum relative to the actual spectrum. The phenomenon, which is frequently referred to as "leakage," may be minimized (but not eliminated) by increasing the length of the time record as much as possible or by modifying the way in which the time record is truncated. Also, the leakage error can be reduced to zero if the waveform is periodic within the time record since the components whose frequencies match those computed are not subjected to leakage error.

A third source of error is due to a phenomenon known as "aliasing," which arises because of the discrete nature of the data record. In this case, the error is induced by components whose frequencies are greater than the $\frac{1}{2}x$ sampling rate maximum imposed by the sampling theorem. These higher-frequency components are incorrectly included as lower-frequency components when executing the FFT. Aliasing is easily avoided by simply ensuring that the data-sampling frequency is greater than twice the highest frequency in the exciting waveform. This can be achieved by using a low-pass filter to remove the unwanted high-frequency components, thereby giving rising to the use of bandwidth-limited excitation.

A number of other operational problems exist when using the FFT algorithm. The most important of these, as far as electrochemistry is concerned, is due to the inherently nonlinear nature of the system. When Eq. (56) is used to measure the impedance with an arbitrary time domain input function (i.e. not a single-frequency sinusoidal perturbation), then the Fourier analysis will incorrectly ascribe the harmonic responses due to system nonlinearity, to input signal components which may or may not be present at higher frequencies. As a consequence, the "measured" impedance spectrum may be seriously in error.

Up to this point we have described methods in which impedance is measured in terms of a transfer function of the form given by Eq. (56). For frequency domain methods, the transfer function is determined as the ratio of frequency domain voltage and current, and for time domain methods as the ratio of the Fourier or Laplace transforms of the time-dependent variables. We will now describe methods by which the transfer function can be determined from the power spectra of the excitation and response.

In addition to Eq. (56), the transfer function $G(j\omega)$ can be calculated for the cross-power spectra of the input and the output, which in turn can be calculated from the linear spectra of the input and output. Thus,

$$G(j\omega) = \frac{P_{yx}(j\omega)}{P_{xx}(j\omega)} = \frac{S_y(j\omega)S_x^*(j\omega)}{S_x(j\omega)S_x^*(j\omega)} \tag{57}$$

where $P_{yx}(j\omega)$ is the average cross-power spectrum of the input and output, $P_{xx}(j\omega)$ is the average power spectrum of the input, $S_x(j\omega)$ and $S_y(j\omega)$ are the linear spectra of the input and output, respectively, and * denotes the complex conjugate.

By invoking the equivalence of Eqs (56) and (57), it is apparent that the information required to calculate the operational impedance is contained in the input and output linear magnitude spectra S_x and S_y. In practice these are cumbersome to compute. The power and cross-power spectra give the same basic information, are faster to compute, and can be applied to measurements to which linear magnitude spectra cannot (Roth [1970]).

Calculations of power spectra are most conveniently performed via the correlation functions. The auto- and cross-correlation functions for time domain input $[x(t)]$ and output $[y(t)]$ functions are

$$R_{xx} = \frac{1}{T}\int_0^T x(t)x(t+\tau)dt \tag{58}$$

$$R_{xy} = \frac{1}{T}\int_0^T x(t)y(t+\tau)dt \tag{59}$$

where T is the time interval over which the correlation is required and τ is a time displacement or delay. In essence, the correlation function yields a time-averaged quantity having greatly improved signal-to-noise characteristics. The value of self- and autocorrelation before transformation is therefore clear.

The significance of the correlation functions in transfer function analysis becomes apparent from the following equations:

$$P_{xx}(\omega) = \tilde{F}[R_{xx}(t)] \tag{60}$$

$$P_{yx}(\omega) = \tilde{F}[R_{yx}(t)] \tag{61}$$

Thus, it is possible to calculate the frequency domain power spectra [and hence $Z(\omega)$] from the Fourier-transformed auto- and cross-correlation functions. The application of correlation techniques for the determination of electrochemical impedance data has been used by Blanc *et al.* [1975], Barker [1969], and Bindra *et al.* [1973], using both random noise input functions and internally generated noise.

A number of significant advantages are inherent in this method:

- The correlation technique is an averaging method and thus affords the same type of insensitivity to asynchronous system noise as phase-sensitive detection.
- In common with other transform methods, $G(j\omega)$ is determined for all frequencies simultaneously and in the time required for the lowest frequency alone by conventional methods. Thus, impedance can be measured down to relatively low frequencies in time-varying systems, and impedance parameters can be measured as a function of time in, for example, a rapidly corroding environment.
- Correlation analysis can be performed on internally generated noise in the complete absence of an external excitation function. Because the ionic events

that produce this noise are not synchronized to an external trigger, the correlation function in this case contains no phase information, but may be considered analogous to the magnitude of the impedance. This technique is potentially an extremely powerful one, allowing equilibrium and steady state conditions to be approached very closely.

- The coherence function provides an internal check on the validity of the measurement. In this regard, it is important to note that methods which determine impedance as the ratio of the imposed input to the observed output do so without regard to the degree of causality between the two signals. Thus, for example, in a system exposed to mains noise or containing electrolyte pumped in an oscillatory or peristaltic fashion, a component of the output signal power results from frequencies characteristic of the environment or system but not of the applied input. Another frequent cause of error in a measured electrochemical impedance is nonlinearity of the interfacial reaction impedance at large perturbations. Thus excitation at frequency ω_0 results in harmonic distortion and a component of output power at frequencies $2\omega_0$, $3\omega_0$, and so on, which may invalidate the "impedance" measured at these frequencies.

The coherence function $\gamma_{xy}^2(\omega)$ can be calculated to determine the validity of a transfer function measurement if the extent of extraneous input and nonlinearity is not known. This function is defined as

$$\gamma_{xy}^2(\omega) = \frac{\overline{P_{xy}(\omega)}^2}{P_{xx}(\omega) P_{yy}(\omega)} \tag{62}$$

where bars denote average quantities and P_{yy} is the auto power spectrum of the output signal $y(t)$. Coherence function values range between 0 and 1. A coherence value of 1 means there is only one input and the system is linear.

The primary limitation of this method of impedance measurement is cost.

3.1.4 Conclusions

In an age of computerized instrumentation, ac impedance and other measurement results are often presented to the user with four or more digits of precision, with little reminder of the intrinsic limitations of the measurement or computational techniques used. Even when considerable care is given to the electrical connections of the system under test and to analysis of subsequently produced data, the operation of the instrument is often transparent to the user. Since the choice of analyzer may determine acquisition precision, time, and other important parameters of data collection, it is of some value to the careful experimenter to understand as fully as possible the method of operation of the impedance analyzer. It is hoped that this chapter is useful in contributing to this understanding.

As a final note of caution, even carefully performed experiments may be subject to systematic error; the system under test may be intrinsically nonlinear, or it may be subject to periodic oscillations, to drift with time, or to other extraneous effects. The results of such perturbations may not be obvious to the experimenter, even when the input and output waveforms are closely monitored with an oscilloscope. It is therefore desirable that impedance data be screened routinely for systematic error. Two screening methods, the Kramers–Kronig integrals and the coherence function, have been described in this chapter, and their use is recommended.

3.2 COMMERCIALLY AVAILABLE IMPEDANCE MEASUREMENT SYSTEMS

Brian Sayers

3.2.1 Electrochemical Impedance Measurement Systems

Electrochemical impedance tests usually investigate the interface between an electrode material and a solution (for example corrosion tests may investigate different coated metals in a salt solution, while battery/fuel cell tests may investigate different electrode materials in an electrolyte). Electrochemical impedance tests provide complementary information to that obtained from dc electrochemical techniques such as cyclic voltammetry, pulse voltammetry, ohmic drop analysis, and chronoamperometry.

3.2.1.1 System Configuration

Electrochemical impedance measurement systems used for the analysis of the ac properties of electrochemical cells typically consist of a potentiostat (sometimes called an electrochemical interface) together with a frequency response analyzer (FRA) or a spectrum analyzer, or even a combination of the two. The potentiostat provides buffered connections to the cell under investigation together with circuitry for applying a controlled voltage or current stimulus and for the measurement of the dc properties of the cell. The FRA is connected through the potentiostat to the cell and therefore the bandwidth of the potentiostat is a very important consideration for accurate high frequency analysis.

A PC running application software is typically used to provide control of the instrumentation, storage of data, display of graphical results and detailed analysis functions (including access to equivalent circuit fitting and data analysis programs such as LEVM—refer to Section 3.3 Data Analysis). The instruments are connected to the PC via a standard interface such as GPIB (IEEE-488), USB, Ethernet, serial port (RS232), parallel port or in some cases via the internal PCI Bus in the PC or by a proprietary interface. The USB and Ethernet approaches are gaining popularity due to the low cost of connection to the PC (these interfaces are available in most PCs as standard or can easily be added at very low cost). High data transfer rates are available via Ethernet and USB. Ethernet also provides the possibility of distributing instrumentation and control PCs throughout a factory or laboratory, allowing data to be easily analyzed in a more comfortable environment well away from corrosive chemicals and noisy equipment.

3.2.1.2 *Why Use a Potentiostat?*

The potentiostat is a very important part of an electrochemical impedance test system. The potentiostat consists of the following elements:

- a power amplifier that can supply or take power from a cell, for instance when charging or discharging a battery or fuel cell;
- one or more high input impedance voltage measurement reference inputs, usually connected to the cell via a standard reference electrode;
- a current measurement input connected to the working electrode on the cell;
- a control loop that maintains the required voltage across the cell (when operating in potentiostatic control mode) or current through the cell (when operating in galvanostatic mode);
- connections to a frequency response analyzer to allow impedance analysis of the cell.

One important capability of the potentiostat is to maintain the required dc conditions on the cell while the frequency response analyzer is performing the impedance (ac) analysis. The actual dc conditions required depend on the application. For tests on fuel cells it may be necessary to set up a particular dc steady state current to investigate the impedance of the cell under load conditions. In other cases it may be necessary to run the impedance test at the "open circuit" potential of the electrochemical cell, in which case the open circuit voltage is measured (Figure 3.2.1)

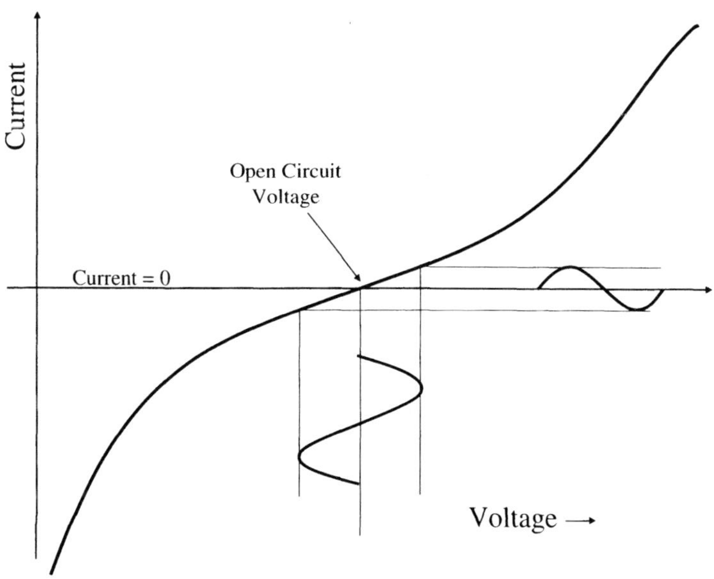

Figure 3.2.1. Setting up the dc conditions on the cell for an impedance test.

and then held on the cell while the impedance test progresses. In other cases, it may be required to impose a particular dc potential offset relative to the open circuit potential of the cell to investigate the impedance under conditions that encourage a particular form of corrosion to take place. Whenever running impedance tests it is very important to operate in a linear regime and under conditions of stability on the cell. If the cell changes significantly during the time taken to run the impedance test, the results taken at the start of the test (usually the high frequency part of the sweep) may not be consistent with the results at the end of the test (usually the low frequency data). This can cause strange impedance artifacts which, if the reason is not known, may easily lead to misinterpretation of the data. A simple test for stability is to repeat the impedance test to see if the results are consistent.

3.2.1.3 Measurements Using 2, 3 or 4-Terminal Techniques

Most modern potentiostats provide at least four connections to the cell. These connections typically consist of a counter electrode (CE) that provides current to the cell, a working electrode (WE) that provides measurement of the current through the cell, and at least two reference electrode inputs (RE) for voltage measurement. Potentiostats that have 4-terminal connections are capable of 2, 3 or 4-terminal testing of electrochemical cells (Figure 3.2.2). The 2-terminal test technique is used mainly for the measurement of high impedance materials where the impedance of cables is not significant (see Section 3.2.2.3).

The 3-terminal technique is often used for corrosion or general electrochemical tests using older equipment where 4-terminal connection to the cell was not provided. There is a counter electrode providing current to the cell, a working electrode that is used for measuring the current through the cell, and a connection for the use of a standard reference electrode within the cell. It is important to realize that since the measurement of the differential voltage across the cell also includes the voltage

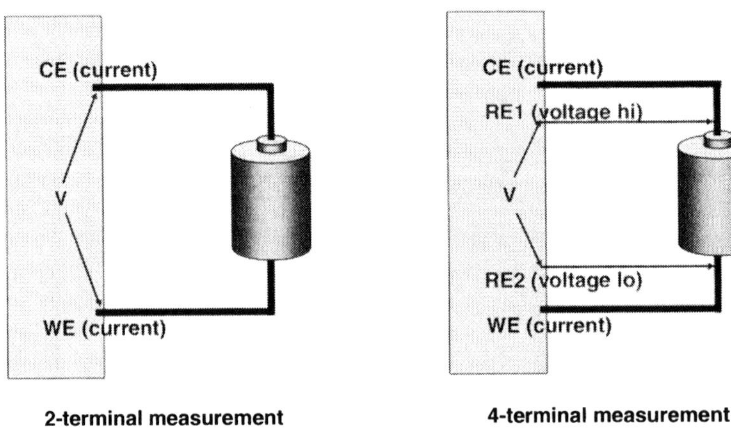

Figure 3.2.2. 2-terminal/4-terminal measurement techniques.

drop in the working electrode connection cable; this type of system can give measurement errors when measuring low impedance cells such as batteries and fuel cells. The 4-terminal measurement technique is therefore usually preferred.

The 4-terminal measurement capability is extremely important when measuring low impedance devices such as batteries, fuel cells and ultra-capacitors. When measuring very low impedance cells, the impedance of the connection cables may well be of the same order of magnitude as the impedance of the cell itself. If 2-terminal (or even 3-terminal) connections are used for this type of measurement, the instrumentation will be unable to differentiate between the impedance of the cables and of the cell and this can lead to large measurement errors. The 4-terminal technique ensures that the voltage drop measured by the instrument is measured directly across the cell and does not include the voltage drop in the CE and WE cables and therefore provides accurate measurement of the cell voltage. When the actual cell voltage is measured using the 4-terminal technique and is combined with accurate measurement of the current flowing through the cell, it becomes possible to characterize very low impedance test cells.

3.2.1.4 Measurement Resolution and Accuracy

Frequency resolution is an important parameter for some applications and this is a function of the design of the waveform synthesizer in the FRA or spectrum analyzer. Using state-of-the-art techniques it is possible to design very high resolution waveform synthesizers which are able to provide 26 bit, or even higher, frequency resolution (i.e. 1 in 67,000,000). Using these waveform generator techniques, very fine frequency sweeps with as little as 100 mHz steps can be achieved at 1 MHz (this is essential when testing highly resonant devices such as crystals, for example in quartz crystal microbalance (QCM) applications). Of course it is important that the analyzers are also able to measure at such fine increments of frequency and this is all part of the overall system design.

Frequency accuracy, as opposed to resolution, depends on the stability of the on-board frequency reference, which is usually a crystal oscillator, and these are usually stable to 10 ppm if the temperature is held stable. For more precise applications, some instruments allow a very accurate external frequency reference to be used.

Impedance measurement accuracy depends on the measurement of ac voltage and current through the cell. The 4-terminal technique is extremely important for measurement of low impedance devices as explained earlier. Other considerations are the voltage and current ranges that are available in the potentiostat and FRA, the bandwidth of the potentiostat (this is dependent on the range selection), and the measurement accuracy of the frequency response analyzer.

Usually, measurement of ac voltage is not so much of a problem and bandwidths of 1 MHz and beyond can fairly easily be achieved with careful design. Current measurement, however, usually presents much more of a problem. It is therefore extremely important to check the specified bandwidth of the potentiostat for the particular measurement that is required. The bandwidth of any potentiostat will tend to be widest in its mid-impedance range (probably when measuring cells of around 1 kiloohm impedance). For measurements on batteries and fuel cells, however, the impedance being measured is very much lower (typically much less than 1 ohm and even down to milliohms). In this case, since the cell impedance is low, the potentiostat is usually measuring on one of its high current ranges (measured current through the cell = applied voltage/cell impedance), which usually has lower frequency bandwidth. At the other end of the impedance scale, when measuring high impedance devices such as anticorrosion coatings, the potentiostat is usually operating on a low current range where again the bandwidth is limited compared to the mid-impedance range bandwidth. It is very important, therefore, to study and understand impedance measurement specifications that are provided by instrument manufacturers and select instrumentation that is appropriate for the measurements to be performed.

It is not sufficient to only consider the bandwidth of the electrometer (the voltage measurement circuits); the whole system, especially including the bandwidth of the current measurement system, should be considered for the specific cells that are likely to be tested. Wherever possible, the instruments should be tested on the actual cells to be analyzed to make sure that they are able to make the required measurements.

3.2.1.5 Single Sine and FFT Measurement Techniques

Single sine and FFT (fast Fourier transform) measurement techniques are widely used for the impedance analysis of electrochemical cells. In terms of accuracy the single sine frequency response analysis technique (sometimes referred to as single sine correlation) is unsurpassed. This technique involves applying a low amplitude pure sinusoidal voltage or current waveform to the cell at a particular frequency and then calculating the impedance of the cell at that frequency from the measured ac voltage across the cell and the ac current passing through the cell. This measurement process is repeated at a number of frequencies in order to fully characterize the cell impedance across the frequency range of interest. The FFT technique provides an alternative method for measuring the impedance of the cell by applying a stimulus waveform containing multiple frequency components (for example multisine, random noise, or step/pulse waveforms), and then calculating the impedance from the measured voltage and current time domain waveforms. This process involves performing FFT computations to transform the measured time domain results into the frequency domain. Both of these techniques are examined in more detail and the advantages and disadvantages of each technique are assessed later in this section.

For the single sine analysis technique, a low amplitude sinusoidal stimulus (usually around 10 mV) is used in order to operate within a linear region on the cell. This is very important since the nonlinear nature of electrochemical cells can lead to impedance measurement inaccuracies when using higher amplitude signals. Figure 3.2.3 shows what would happen if a sine wave stimulus of too high amplitude were used to make measurements on a cell which has a nonlinear voltage vs. current response curve. The resulting ac current waveform in this case is heavily

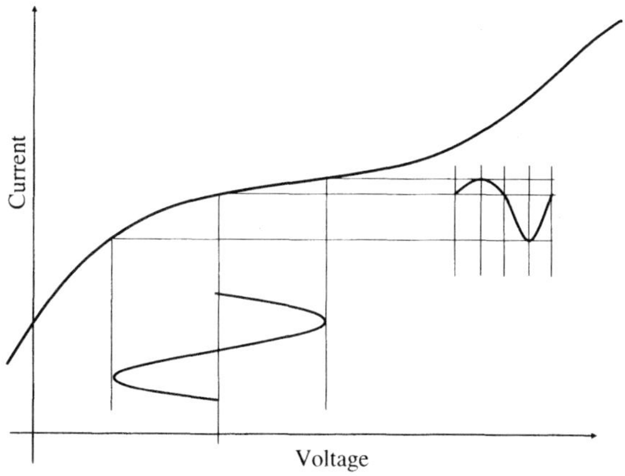

Figure 3.2.3. Non-linear electrochemical cells.

distorted (certainly not a pure sine wave), giving rise to additional frequency components (harmonic distortion). Since the single sine correlation technique applies a single frequency sine wave and measures only at the same frequency, the effect of harmonic distortion introduced by measurement of cells in a nonlinear regime is minimized. However, in order to obtain accurate impedance results it is *essential* to operate within a linear regime on the cell by using a low enough amplitude ac stimulus. As will be seen later in this section, the FFT (or harmonic analysis) approach can be used to provide useful information about whether the measurement results are obtained in a linear or nonlinear measurement regime.

Refer to section 3.1.2.7 for more information about the single sine correlation process. Even measurement of one cycle of the waveform to be analyzed provides rejection of all harmonic frequencies, as can be seen from the upper trace in Figure 3.2.4. Additional cycles can be averaged to provide rejection of noise and subharmonics from the waveform, as demonstrated by the other two traces which show the improved rejection of spurious signals for 10 and 100 cycles of integration.

For a complete impedance characterization of the cell, the frequency of the stimulus sine wave is swept across the frequency range of interest and measurements are taken at each frequency point. Frequency response analyzers give very accurate measurements but are limited in their measurement speed due to having to sweep the measurement frequency. This has limited their use for measurements on fast changing systems. However, with the increased use of digital signal processor (DSP) technology, this is becoming less of a limitation, with very fast frequency sweeps now being available, at least at the high frequency end of the spectrum. At low frequency the measurement time is limited by the necessity to measure at least one cycle of the waveform (so at 1 mHz, a single measurement takes 1000 s).

The use of FFT (fast Fourier transform) spectral analysis can be the perfect complement to the single sine analysis technique. The FFT can perform very fast measurements since it is not necessary to sweep the stimulus frequency in order to measure multiple frequencies. With the introduction of DSP technology, very fast

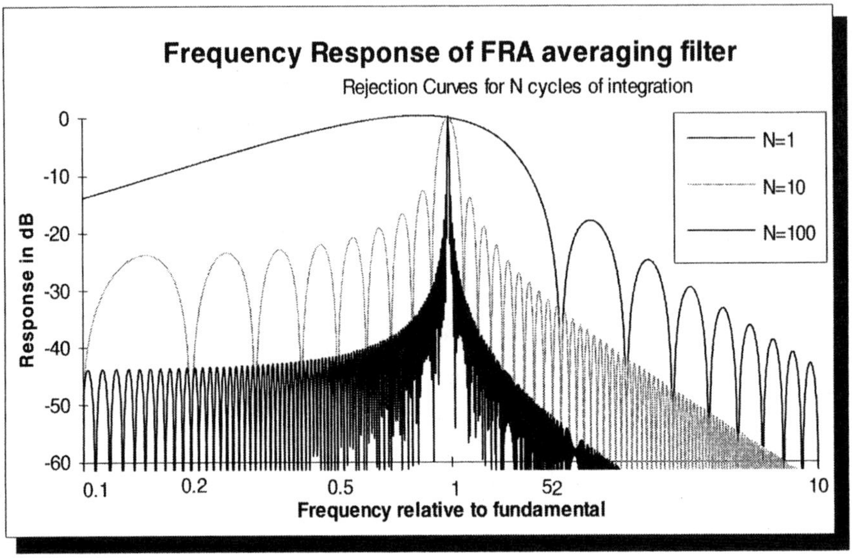

Figure 3.2.4. Single sine correlation (rejection of noise and harmonics).

measurements at multiple frequencies can be performed. Typically a multisine stimulus waveform is constructed that includes a number of discrete frequency components added in such a way that the crest factor (spikiness) of the final waveform is kept as low as possible. This is important since high signal levels corresponding to peaks in the multisine waveform may lead to harmonic distortion effects. The crest factor of the waveform is optimized by adjusting the phase of each sinusoidal frequency component relative to the others until the waveform is as smooth as possible and is free from large spikes. The resulting multisine waveform is applied to the cell and measurements of ac voltage and current are taken over a user-defined whole number of cycles of the lowest frequency present in the stimulus waveform. Since the stimulus waveform contains multiple frequencies, the measured waveforms will also contain the same frequencies. The FFT algorithm computes the results from all frequencies simultaneously, producing an impedance spectrum. The advantages for FFT analysis are: fast multifrequency analysis, fewer problems when measuring time variant systems (since measurements can be very fast—depending on the lowest frequency to be analyzed), and a consistent set of data since all frequencies are measured at the same time.

The choice of frequencies that are included in the multisine stimulus waveform is an interesting area for discussion. In the case where all frequencies are stimulated in the applied multisine waveform, the FFT analysis is not able to differentiate between signals that are the response of the cell to the same frequency in the stimulus waveform, or signals that are simply harmonics of the stimulus frequencies due to the nonlinearity of the cell, and this can lead to very poor results. Of course it is essential to operate in a linear regime on the cell and if that rule is observed the FFT impedance results can be much improved. However, if all frequencies are used in the stimulus waveform, in order to keep the overall signal level low to operate in a linear regime on the cell, the amplitude at each frequency will need to be very low, leading to possible additional problems with noisy results.

Much improved measurement can be obtained by reducing the number of stimulus frequencies in the multisine waveform. Some commercially available analyzers have either a preprogrammed list of frequencies or allow the multisine frequencies to be selected by the user. In this case, many frequencies in the FFT analysis band are deliberately not stimulated. The applied multisine frequencies are usually chosen so that each stimulus frequency is not coincident with the main harmonics of lower frequencies. In this case, the main components of harmonic distortion from each frequency in the stimulus waveform do not interfere with other stimulus frequencies (though of course it is not possible to avoid all significant harmonic frequencies). Without doubt, the impedance results from this approach are much improved compared to those obtained by the method where all frequencies are stimulated.

It is also possible (and this is a major benefit of the FFT or harmonic analysis approach), by measuring the results at the nonstimulated frequencies to investigate whether the cell is operating in a linear regime. If the nonstimulated frequencies are showing a significant response then it is necessary to adjust the experimental conditions, for instance by reducing the applied ac level until the nonstimulated responses are sufficiently low.

In addition, the fact that very much fewer frequencies are contributing to the multisine stimulus waveform allows a higher amplitude signal to be used at each frequency while keeping the overall waveform amplitude within the linear regime on the cell. Typically 10–15 frequencies may be selected (usually with an approximately logarithmic relationship) for each decade leading to perhaps 50 frequencies being stimulated across the entire measurement frequency band, while analyzing over 8000 frequencies using the FFT algorithm to check for nonlinearity and then focusing on the 50 original frequencies to provide the impedance response. This approach can lead to good quality impedance data while utilizing the advantage of the FFT measurement speed.

Another possible problem area of the FFT approach that should be discussed is that of dynamic range. Since all frequencies are measured at the same time, there can be problems with dynamic range, especially when measuring over several decades of frequency. This can be seen when using an FFT based instrument to measure the impedance of a capacitor over a wide range of frequency. Since the impedance of a capacitive sample changes by an order of magnitude for every decade of frequency, the current measurement input needs to be able to resolve the wide range of current levels that are present at the same time. This can be a problem for measurements on corrosion coatings for example, which may cover a very wide range of impedance when tested over several decades of frequency. In other applications the range of impedance being measured may not be as large and the dynamic range issue may not be a significant problem.

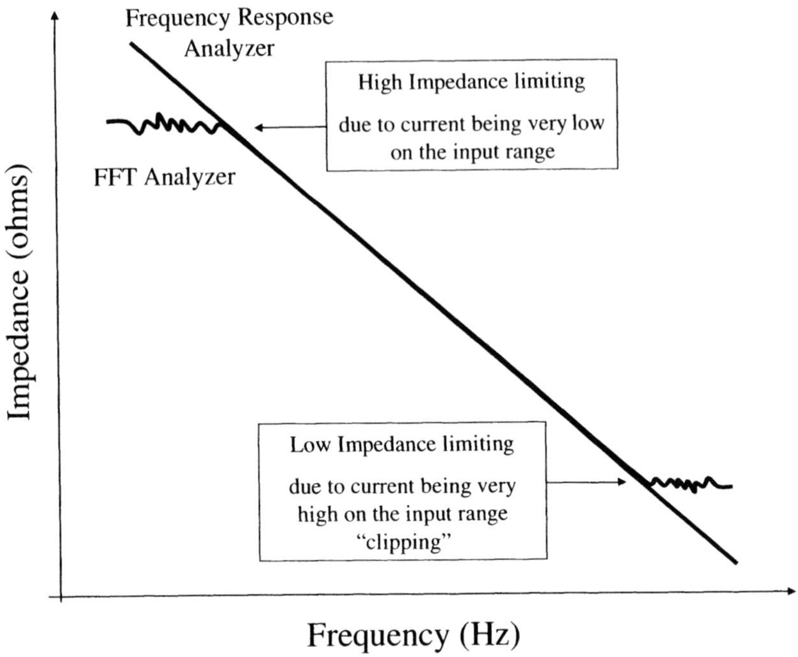

Figure 3.2.5. FFT analyzer vs. frequency response analyzer.

In the case of the capacitive sample, the problem is to measure the high current signal present at the high frequency end of the spectrum where the capacitor is low impedance at the same time as measuring the low current signal at the low frequency end of the spectrum where the capacitor is high impedance. If many decades of frequency are being analyzed at the same time, then the high frequency current component may be some orders of magnitude bigger than the low frequency current component. This may give rise to noisy or "limited" measurements at the low frequency end of the spectrum and "clipping" at the high frequency end of the spectrum (Figure 3.2.5). This problem can be avoided by running several FFTs in sequence each on different input ranges over different bands of frequency and then patching the results together, though this reduces the speed of measurement and the results are no longer simultaneous. There may also be steps at the edges of each frequency band since the integration time may be different in each case. The single sine technique does not suffer from this limitation since optimum voltage and current ranges can be selected at each frequency in the sweep.

Great care must also be exercised in the acquisition of data that is to be used for FFT analysis. In order to avoid problems with aliasing of frequencies (see section 3.1.3.4 for a discussion of this problem), it is essential to apply rigorous analog and digital filtering techniques, and the measurement sample rate must be sufficiently high (at least two times the maximum frequency of interest) otherwise out of band signals (aliases) may appear in the analysis frequency range. Subsampling (sometimes called undersampling) techniques are sometimes used where the sample rate is many times lower than the frequency that is to be analyzed and this can give errors due to unwanted interference frequencies contributing to the impedance results.

In summary, the single sine analysis technique is a pure technique that gives rise to very stable, repeatable measurements and is able to minimize the effects of noise and distortion from the cell. Speed improvements for this technique are available due to the introduction of DSP-based frequency response analyzers. The single sine technique can also give harmonic information that can be used to check for nonlinearity. The FFT technique is complementary to the single sine technique and gives very fast impedance results that are less affected by time variant cells and can incorporate information about cell nonlinearity. However, if not applied with care the FFT can sometimes be more susceptible to nonlinear effects, alias frequencies, noisy results, or by dynamic range limitations. Systems that offer both techniques throughout the full frequency range of the instrumentation are able to extract the maximum information from test cells.

3.2.1.6 Multielectrode Techniques

Some commercially available electrochemical impedance test systems have the ability to measure not only the overall impedance of a complete electrochemical cell, but also the impedance contribution of various component parts of the cell allowing, for example, direct comparison of different anode or cathode materials in a battery. Multielectrode measurement techniques may also be used to investigate individual cells in a battery or fuel cell stack (Figures 3.2.6 and 3.2.7).

It is important to realize that some systems may only provide the use of auxiliary electrodes for dc voltage measurement (it is not possible to use these particular systems for the measurement of impedance from the auxiliary electrodes). Careful selection of equipment is therefore important if it is necessary to measure the impedance of parts of cells in this way.

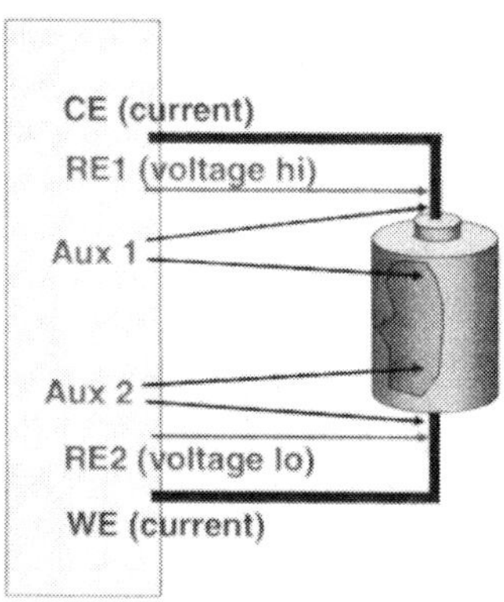

Figure 3.2.6. Auxiliary voltage measurements (measurement of electrodes in a cell).

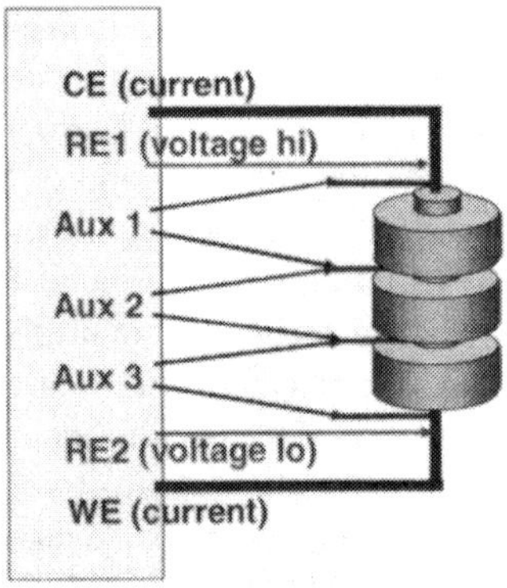

Figure 3.2.7. Auxiliary voltage measurements (measurement of individual cells in a battery).

For anode/cathode investigations on batteries or fuel cells it is necessary to have extra connections built into the cell itself to allow connection to the auxiliary voltage inputs on the instrumentation. This provides separate measurement of the ac voltage drop across the anode, cathode, and separators. With modern measurement equipment it is becoming possible to make simultaneous measurements of the overall battery impedance and of the individual electrodes within the cell using single sine or FFT techniques.

By combining the ac voltage measurements from the auxiliary voltage inputs together with the measurement of the ac current through the whole cell, it is possible to measure the impedance contribution of the anode, cathode, and the impedance of the whole cell. By overlaying the results it is easy to see the contribution to the overall cell impedance that is made by the anode and cathode. Measurements from the main and auxiliary channels can be made simultaneously to ensure that the results represent a consistent data set.

For measurements of individual cells in a battery or fuel cell stack, the main channel connections are placed across the whole stack with the auxiliary electrodes connected across each individual cell allowing the overall impedance and the impedance of each cell to be investigated. This method of obtaining individual cell impedance measurements from the auxiliary voltage input signals, in combination with the charge/discharge facilities provided by the potentiostat for the standard cycling of the cells, provides early detection of bad cells in a stack.

3.2.1.7 Effects of Connections and Input Impedance

Connections to electrochemical cells are usually made via a potentiostat/galvanostat. There are particularly critical points on the cell where great care needs to be taken in the method of connection. Usually the counter electrode (CE) is where the stimulus waveform is applied to the cell and this is not a particularly critical part of the system. However, the points where voltage and current are measured are vital to the accuracy of the results.

In the case of voltage measurement, it is critical that current that has passed through the cell does not "leak" into the voltage inputs since this can give rise to errors in both current and voltage measurement. For this reason, the reference inputs usually consist of very-high-input impedance, low capacitance buffer amplifiers. The method of connection of the voltage measurement circuits to the point of interest in the cell is also critical. Typically, potentiostats provide either "driven shield" voltage measurement connections to the cell, or they have external voltage-follower buffer amplifiers (sometimes referred to as electrometers) that are positioned close to the measurement points in the cell. Using the latter technique, it is necessary to position electronics close to the cell which can be a disadvantage when performing temperature tests on the cell since the electronics may also be subject to temperature variations which can lead to measurement errors or in extreme cases may damage the electronics. If, however, tests are only being performed at room temperature this technique gives good results.

Using the driven shield technique, all of the electronics is positioned inside the potentiostat avoiding any problems of cell temperature affecting the measurement electronics. The driven shield technique ensures that whatever voltage signal is seen at the measurement point within the cell is accurately reproduced on the shield of the connection cable, so that the voltage difference between the cable inner and shield is permanently maintained at zero volts (Figure 3.2.8). Therefore, there is zero current flow between cable inner and shield. Using this technique, a cable that has a real capacitance of around 100 pF/m actually exhibits very low capacitance of the order of a few pF/m, which allows accurate high-frequency ac voltage measurements to be made on the cell.

For very sensitive measurements on very low impedance (or very high impedance) cells it may be necessary to shield the cell from external interference using a Faraday cage which is connected to the instrument ground and it may also be necessary to take particular care regarding the placement of cables to avoid electrical interference and cross-talk between cables.

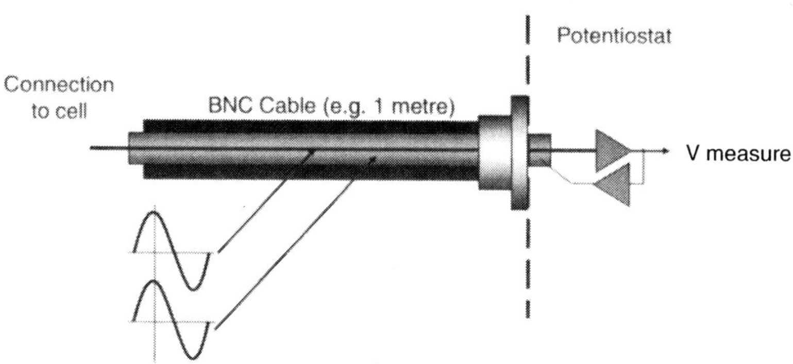

Figure 3.2.8. Driven shields (zero voltage drop between cable inner and shield).

3.2.1.8 Verification of Measurement Performance

Usually, equipment manufacturers supply a known test circuit to verify the performance of the equipment. The test circuit is typically a combination of resistors and capacitors that give a known impedance frequency response curve. Of course this only tests the particular voltage and current ranges that are appropriate for the circuit in the test box. It may well be the case that a particular system performs well for that level of impedance but has problems when measuring other cells. Batteries are difficult to simulate in a test box since they often have very low impedance and also have a dc cell voltage. The impedance response of the battery also changes dramatically at different stages of partial charge/discharge so it is difficult to investigate the performance of the equipment for these tests. However, tests can be performed on accurate low value resistors or large capacitors to simulate different aspects of the battery in a more controlled way and gain an insight into the measurement performance of the system. Of course the voltage/current rating of the components must be compatible with the tests that are to be performed otherwise accuracy of the result could be affected or the test component may fail or even explode.

Other tests that are of great value are to check the stability and linearity of the cell under the imposed test conditions. The stability may easily be checked by repeating the impedance tests on the same cell within a short period of time and in the same environmental conditions to check if the impedance results are repeatable and stable. If this is not done, time can be wasted in trying to interpret artifacts that are simply due to an unstable cell. In addition, it is very important to test for cell linearity by referring to the tests using harmonic or FFT analysis that were suggested in section 3.2.1.5.

3.2.1.9 Floating Measurement Techniques

Most modern potentiostats allow measurements on grounded devices such as pipelines, metal storage tanks, and even laboratory tests using grounded autoclaves or similar devices. For measurements to be properly made in these difficult conditions it is necessary for the instrumentation to have "floating" measurement capability. In this case, the internal circuitry in the potentiostat is referred to an internal floating ground rather than actual ground. For measurements on grounded structures (Figure 3.2.9) the instrument is set to "floating" measurement mode (usually by software control or by a manual switch on the unit) and the counter electrode (CE) is connected directly to the grounded structure while the working electrode (WE) that measures current flowing through the structure is allowed to "float" to the required voltage under the control of the potentiostat. In this case, the whole structure becomes a large counter electrode which supplies the required test current to the WE where it is accurately measured. The WE is controlled by the potentiostat and floats to an offset voltage relative to the grounded CE in order to supply the required current (if operating in current controlled mode) or offset potential (when operating in voltage controlled mode). In the case of a nonfloating measurement system, where

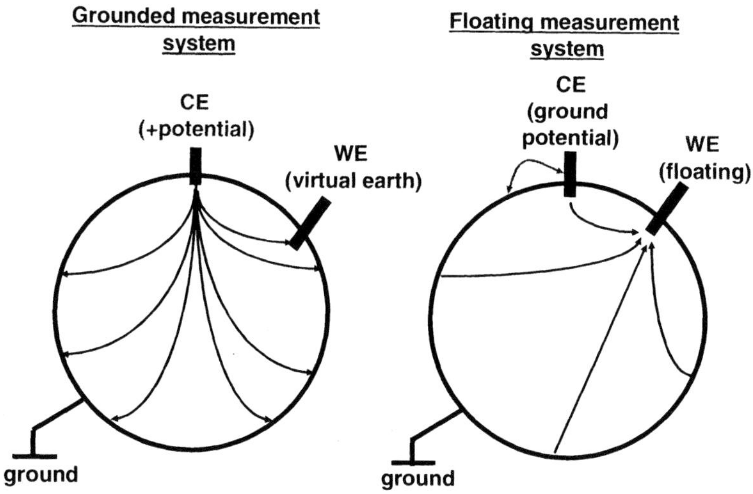

Figure 3.2.9. Grounded and floating measurement systems.

CE is instead set to a particular voltage relative to ground while the WE is held close to ground potential (nonfloating), much of the current supplied be the CE goes directly to ground via the structure under test, completely avoiding (short-circuiting) the WE current measurement input and leading to very high current levels and overloads in the system. The floating measurement capability is therefore extremely important for some applications.

Floating measurements are also used to avoid ground loops where equipment and test cell are grounded at different points leading to noise and interference in the measurements. If the cell is grounded but the equipment is floating then this problem can be avoided.

It has been known for earth connections to be removed from nonfloating systems to convert them into floating systems. This is a very dangerous practice and is not recommended from the safety point of view since the whole equipment could become "live" which could lead to injury or fatalities.

3.2.1.10 Multichannel Techniques

There are many applications where multichannel impedance measurement systems are particularly useful. The throughput of testing in a laboratory can be increased either by the use of multiplexed systems where multiple cells are connected and are automatically tested in sequence by, for example, a potentiostat and FRA, or by the use of true parallel measurement systems where each cell has access to its own potentiostat and FRA. The parallel system, of course, is the more efficient method since all cells can be tested simultaneously; however the equipment required for this is more expensive since there are separate potentiostats and FRAs on every channel. The introduction of multichannel systems, however, has seen a reduction in cost of these parallel measurement systems since the PC, software, power supplies, communications, and casework are shared between a number of channels in the system.

Multichannel systems also allow more advanced tests to be performed, for example testing a segmented fuel cell where one electrode has been partitioned to investigate different electrode materials or to do impedance mapping of the fuel cell electrode. Different materials can also be simultaneously screened in a common corrosive environment using a multichannel potentiostat system with several working electrodes and one counter electrode and a reference electrode.

Multichannel potentiostats can often be connected in parallel to provide tests on high power devices when needed while allowing the flexibility to be used as a multichannel system on lower power devices by reconfiguring the connections to the cells.

3.2.2 Materials Impedance Measurement Systems

Whereas electrochemical tests usually investigate the interface between electrode material and solution (for example a metal in salt solution for corrosion tests, or an electrode in an electrolyte for battery tests), materials impedance tests tend to focus on the properties of the material in isolation. For materials impedance tests, the electrodes are simply a method for providing electrical contact to the material. The material under investigation might be a solid (ferroelectric, piezoelectric, ceramic, polymer, etc.) or a liquid (liquid crystal materials, oils, pharmaceutical products, etc.). In some cases, the impedance properties of the material may be under investigation as a function of temperature (e.g. ceramic materials in high temperature fuel cells or satellite materials over a wide range of temperature). Materials that are used in insulator applications are also investigated using impedance techniques at close to their breakdown potential by the use of high voltage amplifiers.

3.2.2.1 System Configuration

Typical, commercially available, materials impedance test systems consist of either a self-contained impedance analyzer that is capable of measuring the ac voltage drop across a material and the ac current through the sample or, for very high impedance materials, a more specialized system made up of a very sensitive ultra-high-resolution current to voltage converter interface unit connected to a frequency response or impedance analyzer. The measurement principles are the same in both cases; an ac stimulus signal is applied to the sample under test, the ac voltage and current are both measured, and the impedance of the sample is obtained. A PC is used to control the instrumentation and collect impedance results for storage on disk and for display in graphical form. Equivalent circuit fitting routines such as LEVM are widely used for further analysis of the impedance results.

Self-contained impedance analyzers are usually connected to the sample using 2-terminal or 4-terminal techniques, depending on the impedance of the sample to be measured. Specialized sample holders are available from various manufacturers that can be used for measurements of liquid and solid materials samples. Self-contained impedance analyzers usually cover measurements of medium to low impedance samples (100 megaohms to 10 milliohms being a typical measurement range). Ultra-sensitive amplifiers are often added to the system for the investigation of very high impedance materials and these typically offer an impedance range of tens of ohms to over 10^{14} ohms. The frequency range of medium frequency impedance analyzers and frequency response analyzers is typically from tens of μHz to tens of MHz. Radio Frequency (RF) impedance analyzers are also available to extend the frequency range up to tens of GHz using waveguide sample measurement techniques.

3.2.2.2 Measurement of Low Impedance Materials

The same arguments apply to testing low impedance materials as those for testing low impedance electrochemical cells. It is very important to use 4-terminal test techniques to ensure that the impedance of the connection cables does not invalidate the measurement of the material itself (refer to section 3.2.1.3 for more details). Typically, a stand alone 4-terminal impedance analyzer is used for this type of measurement. A Faraday cage may also be required to screen very low impedance samples against interference since the voltage levels measured on low impedance samples is often very small (maybe in the mV region).

3.2.2.3 Measurement of High Impedance Materials

For measurement of high impedance materials, 2-terminal test connections to the sample are usually used. This is because the impedance of the sample is generally much greater than the impedance of the test cables; therefore, any errors introduced by the cables are unlikely to significantly affect the sample measurement results. This is true at low frequency in particular but at high frequency, specialist techniques are often needed to minimize errors due to cables and these will be discussed later in this section.

It is important in many applications to have very sensitive current measurement capability for measurements at low frequency (less than 1 Hz) where the impedance of a wide range of materials becomes very high. This is especially true for ceramic and other insulator materials.

For the measurement of high impedance (or ultra-high impedance insulator) materials, it is often necessary to add a specialized materials test interface into the system. The impedance interface typically comprises of a very sensitive virtual earth current to voltage converter capable of resolving current in the subpico amp region ($<10^{-12}$ amps). This type of measurement system is able to measure very high impedance (insulator) materials in the range up to 100 teraohms (10^{14} ohms) allowing characterization of ceramic, polymer and ferroelectric materials. The interface is connected to a frequency response or impedance analyzer that provides the analysis of the ac voltage across the material and the signal output of the current to voltage converter which is proportional to the current flowing through the sample. The impedance of the sample can then be calculated. For measurement of ultra high impedance samples, a Faraday cage is again recommended in order to screen the sample from interference. The Faraday cage can usually be connected to the screen of one of the sample connection cables since this is usually connected to earth inside the instrument.

3.2.2.4 Reference Techniques

Impedance tests on materials are often performed over a wider frequency range than those on electrochemical cells. In particular, measurements at the high frequency end of the spectrum (>1 MHz) are more difficult to perform with high accuracy, and measurement errors due to cables and instrumentation become more pronounced in this frequency range, requiring specialized reference and normalization techniques to reduce or eliminate these effects and obtain accurate and repeatable results.

Impedance interfaces often provide the facility for automatically switching between measurements of the sample to be measured and measurements of a low loss calibrated reference capacitor (Figure 3.2.10). An ideal reference capacitor would have a completely "flat" response (constant capacitance) across the entire frequency range. This "ideal" capacitor cannot be achieved in reality since there will always be some parallel resistance in the capacitor even though this can, in practice, be an extremely high value. However, the difference between an ideal and nonideal capacitor is sufficiently small for most purposes and the reference capacitor is a very useful tool that can be used to quantify errors due to cables and instrument measurement errors. The deviation of the capacitor from its ideal response due to cables

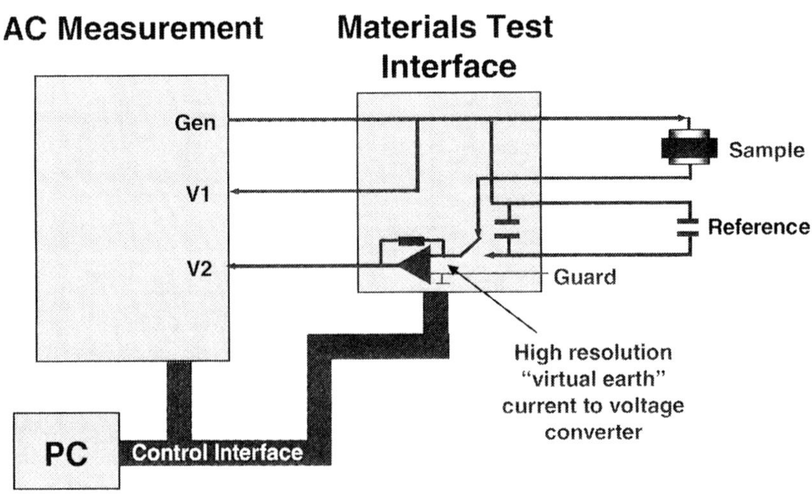

Figure 3.2.10. Connection of ac measurement unit and materials test interface.

and measurement errors is measured and recorded and this deviation is used to apply a correction to the measured results from the material under test.

Materials test interfaces often have a selection of reference capacitors available inside the unit that can be automatically switched to match a wide range of different materials samples. Measurement is usually taken of the sample and then the closest available reference capacitor (of similar capacitance value) is chosen for the reference measurement. In this case the measurements can be taken on the same voltage and current ranges and any errors due to instrumentation or connection cables can be virtually eliminated. The internal reference capacitors can even be changed automatically as the sweep progresses to provide the best reference capacitance matching when measuring dispersive materials where the measured capacitance of the material changes significantly during the sweep.

Usually external reference capacitors can be connected to the interface allowing an ultra-stable reference to be used if one is available. This also provides best matching of cable errors since the same type of connection cables can be used on the sample and on the reference. The reference technique is especially useful when using external amplifiers for high voltage tests, though the reference capacitor must be chosen with care to withstand the applied voltage level.

3.2.2.5 Normalization Techniques

An alternative technique is to use normalization to improve measurement accuracy. In this case, the capacitance of the sample is measured across the range of frequency of interest. A capacitance value is then chosen to most closely match the measured capacitance of the sample. The sample is removed from the sample holder, converting it into an empty cell (air) capacitor and the spacing of the sample holder electrodes is adjusted until the same capacitance value is measured. The same frequency sweep is then performed on the "empty cell" recording results at the same list of frequencies as the original sweep on the sample and also recording the dimensions of the empty cell. The "empty cell" results may then be used to normalize the measured results from the sample.

The "empty cell" measurement may only need to be taken once if a series of samples of similar capacitance are required to be measured, saving valuable test time.

3.2.2.6 High Voltage Measurement Techniques

High voltage amplifiers are sometimes added to the system to allow impedance testing at close to the dielectric breakdown of the material under test. Amplifiers can provide ac or dc stimulus signals at beyond 1 kV. Care must be taken to attenuate the signals back down to within the measurement voltage range of the equipment in order to avoid damage to the instrumentation. Usually, current-limited power supplies are required to ensure that high current does not enter the equipment, especially in the case when the sample breaks down.

3.2.2.7 *Temperature Control*

Some systems allow temperature control to be added to the materials impedance test system to characterize the way that materials change with temperature. There are a wide range of temperature controllers available from various manufacturers and these can often be coupled with furnaces for very high temperature tests to >1,000°C or with nitrogen or helium cryostats for low temperature testing of the sample (Figure 3.2.11). Cryostats and furnaces provide closed loop control of temperature using temperature sensors positioned close to the sample. The temperature controller compares the temperature being reported by the sensor with the required temperature and automatically adjusts the power supplied to the heater coil in order to achieve the desired sample temperature.

In the case of cryostats, liquid nitrogen or helium is used to cool the sample and a heater coil is used to raise the temperature of the sample relative to the temperature that would otherwise be set by the nitrogen or helium. It is usually preferable that the cryogen does not come into contact with the sample to ensure that there is no contamination or condensation on the sample which may, in some cases, affect the results. It is also preferable for the sample chamber not to be filled with air to avoid possible oxidation problems. For this reason it is often the case that the sample is placed in a separate chamber in the cryostat in an inert gas environment.

There are often two temperature sensors in the system, one is positioned close to the heater coil and provides very stable control of temperature since there is immediate feedback of any temperature changes in the coil. The other is positioned

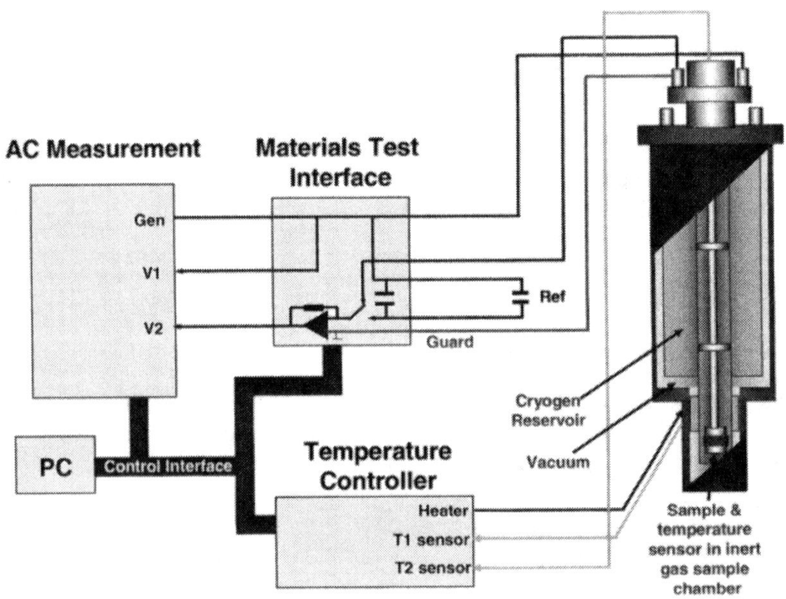

Figure 3.2.11. Cryostat measurement set-up.

close to the sample in order to provide an accurate record of the actual sample temperature (this may be slightly different from the heater coil temperature).

3.2.2.8 Sample Holder Considerations

The sample holder is also a very important part of a materials impedance measurement system. Fringing and stray capacitance effects can easily give measurement errors if inappropriate sample holders are used. Tests on dielectric (or high impedance) materials are often performed using 2-terminal connections, and for the most accurate results, sample holders that have guard electrodes are used in order to minimize fringing effects. Ideally, field lines should be parallel throughout the sample for accurate measurements but Figure 3.2.12 shows how the field lines at the edge of the sample become distorted, leading to measurement errors. For this reason, sample holders have been specifically developed which provide an earthed guard ring surrounding the actual measurement electrode. The "Lo" electrode is connected to the instrument current measurement input while the guard ring is connected to the shield of the current measurement input which is typically connected to earth inside the instrument. The current measurement circuit uses a "virtual earth" current-to-voltage converter which maintains the "Lo" electrode on the sample holder close to earth potential, while the guard electrode is connected to actual earth. Since there is no potential difference between the guard and the measurement electrode, the field lines are kept parallel and evenly distributed throughout the part of the sample that is to be measured. The current that passes through the edge of the sample, which experiences fringing and stray capacitance effects, simply goes direct to earth and is not measured, giving very accurate measurement of current through the central part of the sample. A general rule is that the width of the guard ring should be at least two times the thickness of the material in order to avoid fringing problems. The same techniques can be applied to liquid sample holders and also to sample holders that are used in high or low temperature tests as part of a furnace/cryostat system.

This section was written by Brian Sayers.

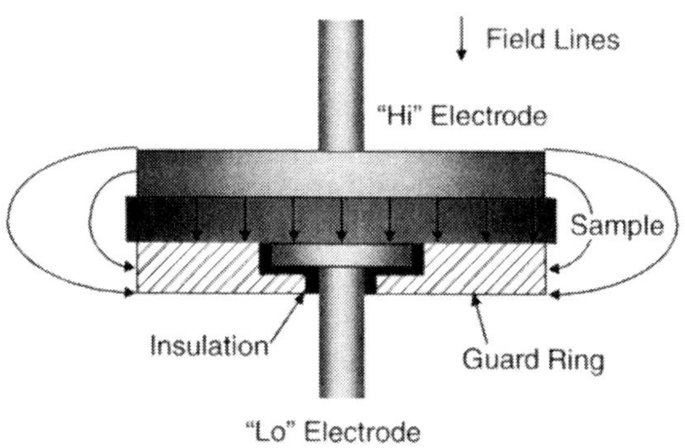

Figure 3.2.12. Sample holder with guard ring.

3.3 DATA ANALYSIS

J. Ross Macdonald

3.3.1 Data Presentation and Adjustment

3.3.1.1 Previous Approaches

In this section we shall first summarize a number of previous methods of data presentation and then illustrate preferred methods. A common method of showing data has been to plot the imaginary parts (or sometimes their logarithms when they show considerable variation) of such quantities as Z, Y, M, or ε vs. ν or $\log(\nu)$. More rarely, real parts have been plotted vs. ν. Such plotting of the individual parts of Z or M data has itself been termed *impedance or modulus spectroscopy* (e.g. Hodge *et al.* [1976], Almond and West [1983*b*]). As mentioned earlier, however, we believe that this approach represents only a part of the umbrella term *impedance spectroscopy* and that complex plane and 3-D plots can much better show full-function frequency dependence and interrelationships of real and imaginary parts.

Let us consider some of the above plots qualitatively for the simplest possible cases: a resistor R in parallel or in series with a capacitor C. Let the single time constant $\tau \equiv RC$ and define the Debye function $D(\omega\tau) = D' - jD'' = [1 + j\omega\tau]^{-1}$, which leads to a semicircle in the complex plane. The $D'' = (\omega\tau)/[1 + (\omega\tau)^2]$. A plot of this functions vs. ω yields a peak at $\omega_m = \tau^{-1}$ and an eventual dropoff proportional to ω at $\omega << \omega_m$ and to ω^{-1} at $\omega >> \omega_m$. Thus the final slopes of $\log(D'')$ vs. $\log(\omega\tau)$ are $+1$ and -1. It is easy to show that for the parallel connection; $Z = RD(\omega\tau)$, $-Z'' = RD''(\omega\tau)$, $M = (C_c/C)\,(j\omega\tau)\,D(\omega\tau)$, and $M'' = (C_c/C)\,D''(\omega\tau)$. For the series connection, one finds $Y = (j\omega C)\,D(\omega\tau)$, $Y'' = GD''(\omega\tau)$, $\varepsilon = (C/C_c)\,D(\omega\tau)$, and $\varepsilon'' = (C/C_c)D''(\omega\tau)$. These results demonstrate that under different conditions $-Z''$, M'', Y'', and ε'' all exhibit $D''(\omega\tau)$ response. Further, in some sense, the pairs Z and $M = (j\omega C_c)Z$, and ε and $Y = (j\omega C_c)\varepsilon$ are closely related. Real materials often do not lead to IS results of the simple $D(\omega\tau)$ semicircle form, however, but frequently involve a distorted or depressed semicircle in the complex plane which may arise from not one but several relaxation times or from a continuous distribution of relaxation times. Under such conditions, curves of $-Z''$ vs. ν, for example, are often appreciably broader than that following from $D(\omega\tau)$.

It has been customary in much past work to plot $-Z''$ or M'' vs. ν and either not give estimates of parameters, such as R and C, leading to the response or to estimate them roughly by graphical means. It is often found that the frequency at the peak, ω_m, is at least approximately thermally activated. But this frequency involves both R and C, quantities which may be separately and differently thermally activated. Thus instead of using the composite quantity ω_m, which may be hard to interpret properly, we believe it to be far preferable to find estimates of all the parameters entering into an equivalent circuit for the situation. As we shall see, such estimates can best be obtained from CNLS fitting.

One other type of plotting has been very common in the dielectric field in earlier years, namely plotting of $\tan(\delta)$ vs. ν, where $\delta = -\theta = \tan^{-1}(\varepsilon''/\varepsilon')$. Thus $\tan(\delta) = \varepsilon''/\varepsilon'$. Since energy loss is proportional to ε'' and energy storage to ε', δ is often termed the *loss angle*. When ε is given by Eq. (20) in Section 2.2 with $\psi_\varepsilon = 1$, so there is only a single time constant and $D(\omega\tau)$ response is involved, the simplest case of interest, one readily finds that

$$\tan(\delta) = \frac{(\varepsilon_s - \varepsilon_\infty)(\omega\tau)}{\varepsilon_s + \varepsilon_\infty(\omega\tau)^2} \tag{1}$$

which again gives a Debye-type peaked curve, with maximum value $(\varepsilon_s - \varepsilon_\infty)/[2\sqrt{\varepsilon_s\varepsilon_\infty}]$ occurring at $\omega\tau = \sqrt{\varepsilon_s\varepsilon_\infty}$. But here again there is no longer any good reason to plot $\tan(\delta)$ vs. ν when much more about the total response can be learned from a 3-D plot of ε or Y. Incidentally, the use of $\tan(\delta)$ plots is entirely absent from a recent compendium of dielectric theory and behavior (Böttcher and Bordewijk [1978]).

3.3.1.2 Three-Dimensional Perspective Plotting

It should be clear from the above discussion that we strongly believe that all IS data should be examined and presented using three-dimensional perspective plotting. One then automatically obtains the plots of real and imaginary parts as projections in the coordinate planes, a normal complex plane plot in the real–imaginary plane and an overall 3-D curve showing the response in proper 3-D perspective. Sometimes when the variation of the quantity plotted is extreme, it is desirable to replace the real and imaginary axes by log(real) and log(imaginary) axes.

A typical simple circuit is shown at the top of Figure 3.3.1, and 3-D plots for its impedance response appear in the middle and bottom of the figure. The two 3-D graphs are for different viewing angles. Note that a log(frequency) axis has been added at right angles to the ordinary $-\mathrm{Im}(Z)$, $\mathrm{Re}(Z)$ complex plane plot, allowing frequency response to appear explicitly. We shall sometimes use f and sometimes ν to indicate frequency in the succeeding graphs. The heavy curve in Figure 3.3.1 is the 3-D response line, its vertical projections to the $\log(f)$–$\mathrm{Re}(Z)$ plane are shown dotted, and the curve in the complex plane plot is actually a semicircle here.

There is an alternative to the usual 3-D plotting of Z in rectangular form which we have been discussing. Instead, we may express a quantity such as Z or M in polar form, involving, say, $|M|$ and θ. If one defines the three axes in a 3-D plot as $(X, Y, Z) = (\theta, |M|, \log(\nu))$, one will obtain a different looking 3-D plot and different projection curves. If the $|M|$ axis is replaced by $\log|M|$, one even has a 3-D Bode plot. Note that $\ln(M) = \ln[|M|\exp(j\theta)] = \ln|M| + j\theta$, a complex number, so the $\ln|M|$, θ plane is a consistent complex plane, whereas that involving M and θ is not, since M and θ do not form the parts of a complex number. Thus, there is some reason to prefer the second of these approaches to the first. The use of the projection curves $\ln|Z|$ vs. ν and θ vs. ν has been proposed and encouraged by Cahan and Chen [1982]. Although there may be instances where both the standard rectangular and polar 3-D representations are worthwhile plotting and examining, we believe that the standard rectangular one with either real and imaginary axes or log

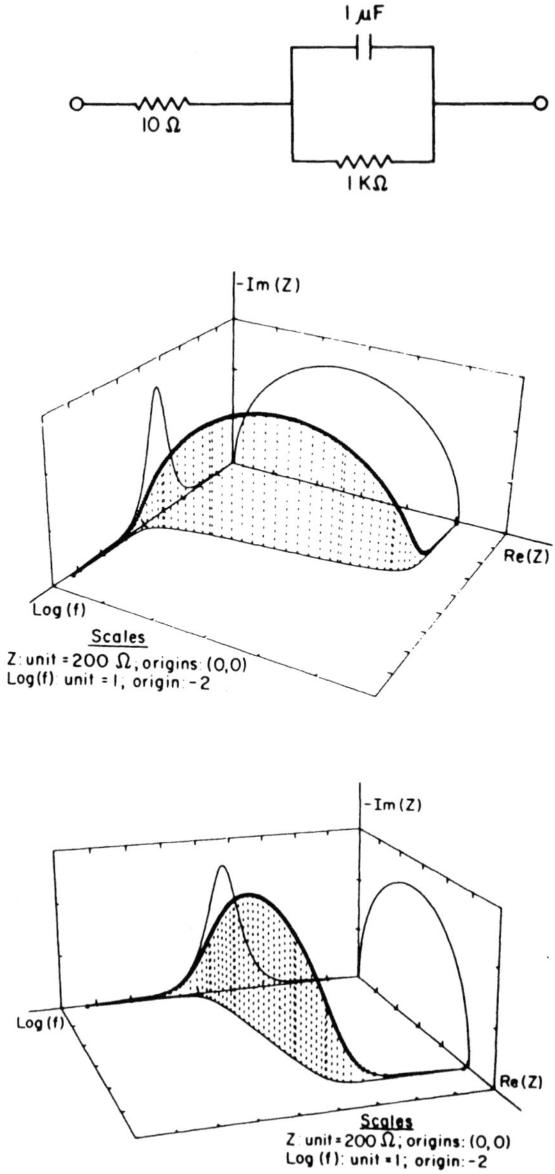

Figure 3.3.1. A simple circuit and 3-D plots of its impedance response. The 3-D plots are for different viewpoints. (Reprinted from J. R. Macdonald, J. Schoonman, and A. P. Lehnen, *Solid State Ionics* **5**, 137–140, 1981.)

axes is usually quite sufficient and shows the entire frequency response of the function considered in a way that yields projection curves of the kind that have been widely used in the past and about which there exists a larger body of interpretative knowledge.

There still remains the problem of which ones of the four functions Z, M, Y, and ε to plot in 3-D. Certainly, one should always plot the one of these that was directly measured. If, in addition, all four are plotted and examined, as a good general rule, emphasis should be put on the one (or ones) which show the most detail and cover the entire frequency range with best resolution.

We shall illustrate the above by showing results for IS impedance data for a single-crystal Na β-alumina with blocking gold electrodes, measured at T = 83K. These data (Almond and West [1981], Bruce *et al.* [1982]) were kindly provided by Dr. West [1983]. This set, and data for eight higher temperatures, have been re-analyzed with CNLS fitting (Macdonald and Cook [1984, 1985]).

Figure 3.3.2 shows a 3-D impedance plot of the data, using straight lines between data points. Solid dots are employed for the data points themselves and open circles for their projections in the three planes. The $\log(\nu)$ scale starts at 1 here and in the succeeding plots, and its scale interval (between tick marks) is always also 1. The zero points of $-Z''$ and Z' occur at the origin. Three important conclusions may be drawn from these results. First, it is clear that much of the higher-frequency data are too small to be adequately resolved in this plot. Second, it seems likely that the lowest-frequency point is in error. Third, although this anomaly shows up clearly in the 3-D curve and in the complex-plane curve, it is entirely absent from the $-Z''$ vs. $\log(\nu)$ and the Z' vs. $\log(\nu)$ projection curves. Were these the only curves

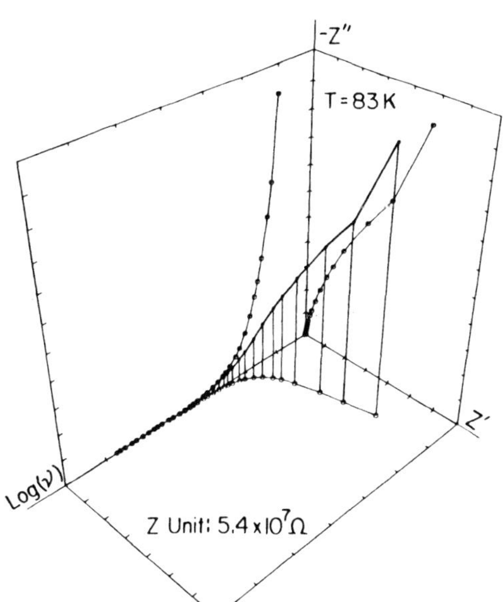

Figure 3.3.2. A 3-D plot of the impedance–$\log(\nu)$ response of single-crystal Na β-alumina at T = 83K. (Reprinted from J. R. Macdonald and G. B. Cook, *J. Electronanal. Chem.* **193**, 57–74, 1985.)

plotted, as would often be the case in the absence of 3-D plotting, the anomaly would not be discovered. More 3-D plots are presented in Section 3.3.2.

3.3.1.3 Treatment of Anomalies

What should one do about anomalies of this kind, points which do not seem to lie close to a smooth curve? If the experiment can be repeated, that should be done, and averaged or best data used. In the present instance, measurement at a few more frequencies between the present lowest and next lowest point would yield intermediate points which would help clarify whether the last point is badly off or not. If the experiment cannot be repeated, then outliers of appreciable magnitude, such as the lowest-frequency point in the present plot, should be omitted (or weighted very low) in subsequent CNLS fitting.

Figure 3.3.3 presents a log-impedance 3-D plot of the same data. It has the virtue of allowing all the data to appear with the same relative resolution, but it clearly reduces anomalies. Although we could show the higher-frequency data alone on a separate regular 3-D plot to achieve better resolution, it turns out for the present data that a 3-D *M* plot (Figure 3.3.4) covers the full data range with adequate resolution. Here, we see two more important anomalous regions not apparent on the earlier curves or in the papers of Almond and West. First, the one in the middle arises (West [1983]) from overlapping data taken with two different measuring devices which evidently gave inconsistent results in their regions of overlap. This anomaly only

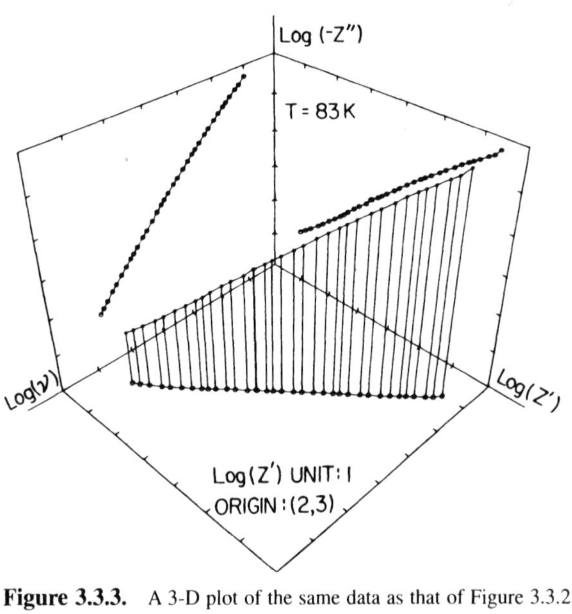

Figure 3.3.3. A 3-D plot of the same data as that of Figure 3.3.2, but with logarithmic transformation of the impedance. (Reprinted from J. R. Macdonald and G. B. Cook, *J. Electronanal. Chem.* **168**, 335–354, 1984.)

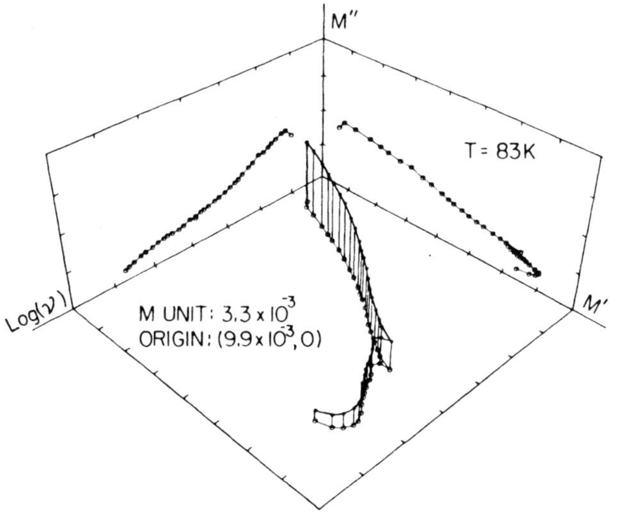

Figure 3.3.4. A 3-D modulus function plot of the Na β-alumina, $T = 83\mathrm{K}$ data. Here $M = j\omega C_c Z$. (Reprinted from J. R. Macdonald and G. B. Cook, *J. Electroanal. Chem.* **193**, 57–74, 1985.)

shows up clearly in the 3-D curve and in the M' vs. $\log(\nu)$ curve. It was apparently not recognized until the present 3-D M plotting was carried out (Macdonald and Cook [1984]). In this instance, the plot shows that it would certainly have been desirable to recalibrate the measuring instruments and repeat the experiment. Second, there is a probable anomaly for the highest-frequency three points. Although M plots emphasize high frequencies as compared to Z plots, there is no physical reason to believe that M' should begin to decrease at high frequencies (Macdonald and Cook [1984]), and it would be difficult to justify putting in elements in a fitting equivalent circuit which would cause it to do so. Therefore, in subsequent CNLS fitting of these data, at least the highest-frequency three points, should be omitted.

Thus far we have recommended that when appreciable outliers or other anomalies appear in IS data, the experiment be repeated one or more times or, when this is impractical, particularly anomalous points be omitted. But there is one further procedure which has often been found useful in other areas and could be applied here as well. This procedure is smoothing; it should, however, only be carried out after the averaging and pruning mentioned above. All IS frequency data curves should, ideally, follow smooth curves with no abrupt changes in slope, except at possibly sharp cusps where two arcs join each other in the complex plane (e.g. Figure 1.3.1*e*). Even here no sharp change is present unless the processes represented by the two arcs occur in widely separated frequency regions, and even then it is not present in the 3-D curve itself.

When large anomalies have been eliminated, it is still often valuable to reduce the remaining smaller deviations (noise, experimental error) from a smooth curve by smoothing. The resulting smoothed data will then generally lead to more accurate estimates of the parameters involved in a CNLS fitting. There are many methods available for smoothing. We suggest that the holistic approach of smoothing using B-splines calculated by a least squares approximation would be particularly appropriate for the type of complex data obtained in IS (de Boor [1978]). Unfortunately, it does not yet seem practical to apply this procedure to both real and imaginary parts simultaneously, as would certainly be most appropriate; instead, one would smooth, say, Z' vs. ν [or possible $\log(\nu)$] and Z'' vs. ν results separately and then recombine the results to obtain $Z(\omega)$. This procedure would have another virtue. Besides producing an analytic approximation to the Z data, it would yield $\partial Z/\partial \nu$ results with no further calculation. These results, now expressed in analytic form, in turn would simplify CNLS fitting of the smoothed $Z(\omega)$ data to a specific model or equivalent circuit. Much work still remains to be done to develop this approach, which could, in fact, be incorporated as an optional initial part of a CNLS fitting program. It may sometimes prove useful, when major anomalies (which cannot be directly remedied) appear more evident in M or Y 3-D plots than in the Z plot, to carry out the actual smoothing on the transformed data, then convert the smoothed Z for use in CNLS fitting.

3.3.2 Data Analysis Methods

3.3.2.1 Simple Methods

When non-overlapping semicircular arcs appear in, say, the impedance plane, one can directly estimate the associated R and C values from the left and right intercepts of the arc with the real axis and the value of ω at the peak of the arc, $\omega_m = (RC)^{-1}$. This procedure is quite adequate for initial estimates, but it yields no uncertainty measures for the parameters and does not check that the frequency response along the arc is consonant with that for an R and C in parallel. Further, experimental arcs rarely approximate exact semicircles well. There have been many graphical methods proposed for the analysis of impedance and dielectric data (e.g. Cole and Cole [1941], Vetter [1967], Macdonald [1974*a*,*b*], Böttcher and Bordewijk [1978]). These methods often consist of plotting some function of the data vs. some function of frequency with the expectation of obtaining a straight line whose parameters may be related to the desired parameters of the equivalent circuit. Although these parameters may be estimated from the straight line by ordinary least squares fitting, this is not usually necessary if the estimates are to be used as initial values in subsequent CNLS fitting. Often subtraction of the effects of some estimated parameters is used to help in the estimation of further parameters. But subtractive methods are notoriously inaccurate. Again, all quantities are usually estimated without using all the available data simultaneously and without yielding uncertainties.

An improved geometrical, iterated–subtractive curve fitting method for resolving two or more overlapping arcs (which may be depressed) has been described by Kleitz and Kennedy [1979] and has been further developed and automated by Tsai and Whitmore [1982]. An algebraic method of estimating parameters for systems which exhibit pseudoinductance (i.e. negative capacitance and resistance) and lead to an arc in the first quadrant of the impedance plane followed by an arc in the fourth quadrant has been proposed by D. D. Macdonald [1978*c*] (see also Macdonald and Franceschetti [1979*b*]. Finally, a sophisticated least squares method for finding the best circle through a set of data points in the plane appears in Macdonald, Schoonman, and Lehnen [1982]. For obtaining initial parameter estimates, the simple methods described in Section 1.3 are usually sufficient.

3.3.2.2 Complex Nonlinear Least Squares

None of the above methods uses all the data simultaneously, and they are often restricted to the analysis of limited situations (e.g. two possibly overlapping arcs). Nevertheless, when applicable, these methods are useful for initial exploration of the data and for initial parameter estimates for use in CNLS fitting. The much more general and powerful method of complex nonlinear least squares was first applied to IS in the work of Macdonald and Garber [1977], and further discussion of the method and demonstrations of its high resolving power and accuracy appears in Macdonald, Schoonman, and Lehnen [1982] and Macdonald [1991]. Later, a somewhat similar CNLS program was developed by Tsai and Whitmore [1982], but it does not include the important feature of allowing arbitrary or analytical weighting of the data and has no built-in procedure for avoiding or recognizing local minima in the sum of squares to be minimized. The current version of the very flexible and general computer program, LEVM, for weighted CNLS fitting is available for free downloading and use (Macdonald and Potter [1987], Macdonald and Solartron [2003]).

Why should one be interested in using complex nonlinear least squares fitting? After all, as already mentioned, the Kronig–Kramers (KK) relations (Macdonald and Brachman [1956]) ensure that if one knows the real or imaginary part of a minimum-phase function over all frequencies, one can calculate the value of the other part at any frequency. This may suggest to the reader that ordinary nonlinear least squares fitting of data for either the real or imaginary part vs. ν should be sufficient. But we never have data over all frequencies, and all experimental data are contaminated with error. Thus, application of the KK relations to the real or imaginary part of actual experimental data often does not allow the measured values of the other part (which contain errors) to be calculated very closely. It therefore appears reasonable to attempt to fit all of the data simultaneously to a given model, a model which involves a set **P** of unknown parameters (e.g. circuit elements) which may enter nonlinearly in the formula for the measured function of frequency, impedance, admittance, and so on. Although the real and imaginary parts of this physically realizable function are connected in a holistic, averaged way with each other through the KK relations, it is usually a good approximation to assume that the random (nonsystematic) errors in each of these quantities are uncorrelated with each other. Since practical fitting models nearly always are minimum-phase and thus satisfy the KK relations, the achievement of a good CNLS fit of data to the model automatically ensures that the KK relations hold, and it is thus unnecessary to carry out the lengthy integrations necessary to check KK satisfaction directly. When no appropriate mathematical model or equivalent circuit is available, however, CNLS fitting is not useful.

In such cases, KK analysis turns out to be particularly helpful, as demonstrated in the discussion of Section 4.4.5.

Complex nonlinear least squares avoids most of the weaknesses of earlier methods since it fits all the data simultaneously and thus yields parameter estimates associated with all, rather than half, the data. In addition, it provides uncertainty estimates for all estimated parameters, showing which ones are important and which unimportant in the model or equivalent circuit used for fitting; and finally, it allows one to fit a very complex model, one having 5, 10, or even more unknown (free) parameters. None of the other methods can do this adequately, especially when several of the time constants of the model are close together.

Here, we shall briefly describe a slightly more general fitting method than ordinary weighted CNLS, but for simplicity we shall still refer to it as CNLS. Consider a theoretical model expression $f_t(\omega; \mathbf{P})$ which is a function of both angular frequency ω and a set of model parameters $\mathbf{P}$. Consider $i = 1, 2, \ldots, k$ data points associated with ω_i. We need not specify the number of parameters here, but we do assume that they enter f_t nonlinearly in general. Now suppose that f_t is separated into two parts, $f_t^a(\omega; \mathbf{P})$ and $f_t^b(\omega; \mathbf{P})$, which both depend on the same set of parameters. For $f_t(\omega; \mathbf{P}) = Z_t(\omega; \mathbf{P})$, for example, the two parts might be $f_t^a = Z_t'$ and $f_t^b = Z_t''$, or $f_t^a = |Z|$ and $f_t^b = \theta$, where θ is the phase angle of Z_t. The least squares procedure involves minimizing the sum of squares function

$$S = \sum_{i=1}^{k} \left\{ w_i^a \left[f_{ei}^a - f_t^a(\omega_i; \mathbf{P}) \right]^2 + w_i^b \left[f_{ei}^b - f_t^b(\omega_i; \mathbf{P}) \right]^2 \right\} \tag{2}$$

where w_i^a and w_i^b are the weights associated with the ith data point and f_{ei}^a and f_{ei}^b are experimental data values. When, say, $f_t^b(\omega; \mathbf{P}) \equiv 0$, the procedure reduces to ordinary nonlinear least squares.

Since the above generalization of the ordinary nonlinear least squares method is so minor, it is a simple matter to modify a standard nonlinear least squares program to implement Eq. (2). In any nonlinear least squares procedure, however, there is a problem of ensuring that the minimum found in parameter space is the absolute minimum. This problem, which may become serious when the number of free parameters is large, has been attacked in the work of Macdonald, Schoonman, and Lehnen [1982] by using two programs in series. The first does not require inversion of the system matrix and thus nearly always converges. The second program, which uses the results of the first as input, does not usually converge, however, unless it can find a set of parameter values which lead to an absolute minimum in S or at least a good local minimum. The first program often helps it to do so. In spite of the power of this serial method, the achievement of an absolute minimum, giving the least squares estimates of the parameters as well as their uncertainty estimates (standard deviation estimates), is always simplified and facilitated by using the best available estimates of the parameter values as input to the first program. Some of the simpler analysis methods mentioned above are often useful in providing such initial estimates.

The procedure described above is not really a CNLS approach unless f^a and f^b are the real and imaginary parts of a complex variable. But as we have seen, $|Z|$ and θ are not, although $\ln|Z|$ and θ and Z' and Z'' are. Since we sometimes are interested in fitting data in the $|Z|$, θ form rather than $\ln|Z|$, θ, this distinction is worth making, although it makes no formal difference in the minimization of Eq. (2).

3.3.2.3 Weighting

The problem of what weights to use is not always an entirely well-defined one. The simplest choice, termed *unweighted* or *unity-weighted*, is to set all w_i^a and w_i^b values equal to unity. But if values of f_e^a and f_e^b vary over several orders of magnitude, as is often the case, only the larger values will contribute appreciably to the sum S, resulting in poor parameter estimates. A reasonable procedure, when it can be done, is to replicate the experiment 5 or 10 times and determine the w_i's from the standard deviations from the mean for each point. The general relations to use are $w_i^a = (\sigma_i^a)^{-2}$ and $w_i^b = (\sigma_i^b)^{-2}$, where the σ's are the experimental standard deviations.

When replication is impractical and there is no direct information on the best individual w_i's to employ, it has usually proved most satisfactory to assume that the relative errors of the measured quantities are constant. This approach has been termed proportional or P weighting. It is equivalent to setting

$$\sigma_i^a = gf_{exi}^a \quad \text{and} \quad \sigma_i^b = gf_{exi}^b \tag{3}$$

where g is a proportionality constant which is usually taken to be unity and whose value makes no difference in the parameter estimates. Here P weighting is particularly needed when the data exhibit large variation. In the Z case, S would become

$$S = S_Z = \sum_{i=1}^{k} \left\{ \left[\frac{Z'_{ei} - Z'_t(\omega_t;\mathbf{P})}{Z'_{ei}} \right]^2 + \left[\frac{Z''_{ei} - Z''_t(\omega_t;\mathbf{P})}{Z''_{ei}} \right]^2 \right\} \tag{4}$$

An alternative choice, which yields nearly the same parameter estimates if the relative errors are small (i.e., the squared terms above), is to take $\sigma^a = gf_t^a(\omega_i;\mathbf{P})$ and $\sigma_i^b = gf_i^b(\omega;\mathbf{P})$.

3.3.2.4 Which Impedance-Related Function to Fit?

The next problem is which function Z, Y, M, or ε to fit. The answer is that it is most sensible from a statistical point of view to fit the data in measured rather than transformed form. Suppose that Z is measured in rectangular form. When both Z and the associated $Y = Z^{-1}$ data are separately fitted with P weighting, it is found that there are often significant differences between the parameter estimates obtained from the two fits. This is not unexpected; the operation of taking an inverse (complex or not) on data with errors generally introduces a bias in the fitted results; it is for this reason that the directly measured results should be fitted directly.

In most automated measurements the rectangular components are measured directly, but sometimes the modulus and phase angle are directly obtained. These are the two quantities, $f_e^a = |f_e|$ and $f_e^b = \theta = \sphericalangle f_e$, which would then appear in Eq. (2) for S. Again in the absence of measured uncertainties, P weighting would usually be most appropriate. In this case $\sigma_i^a = g|f_{ei}|$ and $\sigma_i^b = g\theta_{ei}$. It is worth mentioning that it has sometimes been suggested that with data in rectangular form, a modified

P weighting be used in which $\sigma_i^a = \sigma_i^b = g|f_{ei}|$. Such weighting leads to parameter estimates from Z and Y fitting which are generally much closer together than those obtained with ordinary P weighting. Thus with this weighting, it makes no significant difference whether the data are fitted to the model in Z or Y form. Since, however, this weighting seems physically unrealistic and blurs a distinction which we in fact expect, it seems to have little to recommend it. Further, when ordinary P weighting is used for fitting of data in either rectangular or polar form, it is easy to show that the ω factors occurring in M or ε cancel out of the S function, and thus fitting of Z and M then yield exactly the same set of parameter estimates and relative fitting residuals, and fitting of Y and ε also yield the same set of parameter estimates and relative residuals. Of course, the Z and Y sets will be different unless the data are exact for the model considered. Thus when P weighting is employed, one should fit to Z or Y depending on which was measured directly and should fit in rectangular or polar form, again depending on which form is directly measured.

3.3.2.5 The Question of "What to Fit" Revisited

The remaining problem is what equivalent circuit or equation to fit by CNLS (see Sections 2.2.2.3 and 2.2.3.4). If it is expected that the data arise from an experiment described by a known analytic model, then of course fitting to the $Z(\omega)$ or $Y(\omega)$ predictions of this model would be appropriate. In the more usual case where a complete, appropriate model is unknown, the first step is to examine 3-D plots of the data and attempt to identify the effects of specific processes appearing in different frequency regions. An equivalent circuit may then be put together to try to describe these processes and their interactions (see Sections 2.2.3.3 and 2.2.3.5). When CNLS fitting with this circuit is then carried out, one would hope to find little evidence of systematic error (leading to large, serially correlated residuals), small relative standard errors for all free parameters, and small relative residuals for the data points, so the overall relative standard deviation of the fit, S_F, is small. If relative standard deviations of the parameters are of the order of 30% or more, the associated parameters are not well determined by the data and should be removed from the equivalent circuit. Generally, one would keep modifying the equivalent circuit and doing CNLS fitting until the above criteria are as well satisfied as possible, under the general criterion of using as simple an equivalent circuit with as few individual elements as practical.

3.3.2.6 Deconvolution Approaches

There is an alternative to some of the above trial-and-error procedure that is sometimes worth trying. Suppose that the impedance-plane plot shows a wide arc which is not exactly a displaced semicircle. It might possibly be best described by several discrete time constants not too far apart (e.g. Armstrong *et al.* [1974], Badwal and de Bruin [1978]) or by a continuous distribution of time constant. In the first case, the equivalent circuit would involve several individual parallel RC's in series, and in the second it might involve one or more ZARC functions (CPE and R in parallel) in series. Although the best of these choices could be discovered by carrying out several CNLS fits, a more direct method would first be to use deconvolution of the

$Z''(\omega)$ data to find an estimate of the distribution function of time constants implicit in the data (Franklin and de Bruin [1983], Colonomos and Gordon [1979]). Such a distribution, if sufficiently accurate, will separate out the various time constants present, even if they are completely invisible in 3-D plots, and by the width of the individual relaxations apparent in the distribution suggest whether they may be best described by discrete circuit elements or by continuous distributions in the frequency domain. From the values of relaxation time τ where relaxation peaks occur, τ_p, one may also calculate the approximate frequency region $\omega_p = \tau_p^{-1}$ where the relaxation produces its maximum effect. These results may then be used to construct an appropriate equivalent circuit and estimate initial values of the parameters for subsequent CNLS fitting.

The basic equations for obtaining the distribution of relaxation times, $g_z(\tau)$, at, say, the impedance level, start with the defining relation (Macdonald and Brachman [1956])

$$Z(\omega) = R_0 \int_0^{\infty} \frac{g_z(\tau)dt}{1 + j\omega\tau} \tag{5}$$

where R_0 is the $\omega \to 0$ value of $Z(\omega)$. This relation can be put in convolution form by several transformations. Let us use normalized quantities and pick some frequency ω_0 which is approximately the central value of all frequencies measured. Let $\omega_0 \equiv 2\pi\nu_0$, $\tau_0 \equiv \omega_0^{-1}$, $\omega\tau_0 \equiv \exp(-z)$, $\tau \equiv \tau_0\exp(s)$, and $G_z(s) \equiv \tau g_z(\tau)$. We have here introduced the new logarithmic variables s and z. Then Eq. (5) becomes

$$Z(z) = R_0 \int_{-\infty}^{\infty} \frac{G_z(s)ds}{1 + j\exp[-(z - s)]} \tag{6}$$

This equation may now be separated into real and imaginary parts, each giving an independent expression involving $G_z(s)$, the desired quantity. Although both may be used, the imaginary part of Z generally shows more structure than the real part, and it is customary to calculate $G_z(s)$ from Z'' data rather than from Z' data. The expression for Z'' following from Eq. (6) is

$$Z''(z) = -(R_0/2)\int_{-\infty}^{\infty} G_z(s)\operatorname{sech}(z - s)ds \tag{7}$$

now in standard convolution form. The process of deconvolution to find $G_z(s)$ and thus $g_z(\tau)$ is generally a complicated one but can be carried out by computer when needed. Although two different methods are described by Franklin and de Bruin [1983] and Colonomos and Gordon [1979], current superior procedures using LEVM are discussed and illustrated in Macdonald [2000a, 2000c], see also Lasia [1999].

3.3.2.7 Examples of CNLS Fitting

As a first example of CNLS fitting, the circuit shown in Figure 3.3.5 was constructed with lumped elements whose values were measured on an impedance bridge (top figures) (Macdonald, Schoonman, and Lehnen [1982]). This circuit leads to very little structure in either the Z or Y 3-D plots shown in Figures 3.3.6 and 3.3.7. The

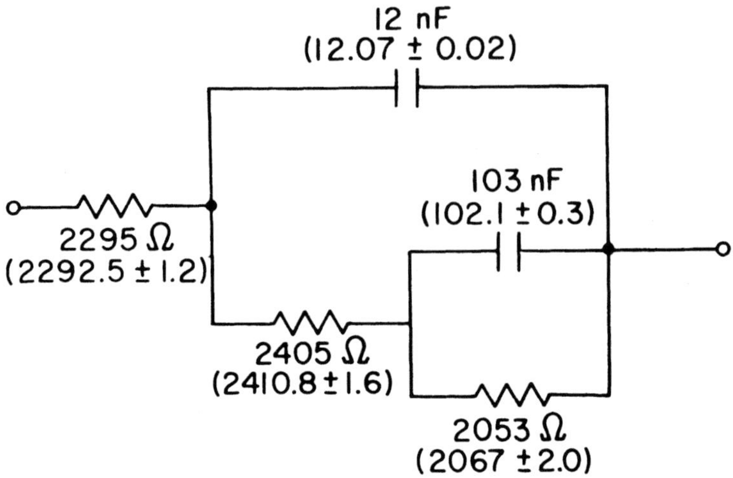

Figure 3.3.5. Test circuit involving lumped circuit elements. Nominal values are the numbers on top, while those in parentheses are CNLS estimates. (Reprinted from J. R. Macdonald, J. Schoonman, and A. P. Lehnen, *Solid State Ionics* **5**, 137–140, 1981.)

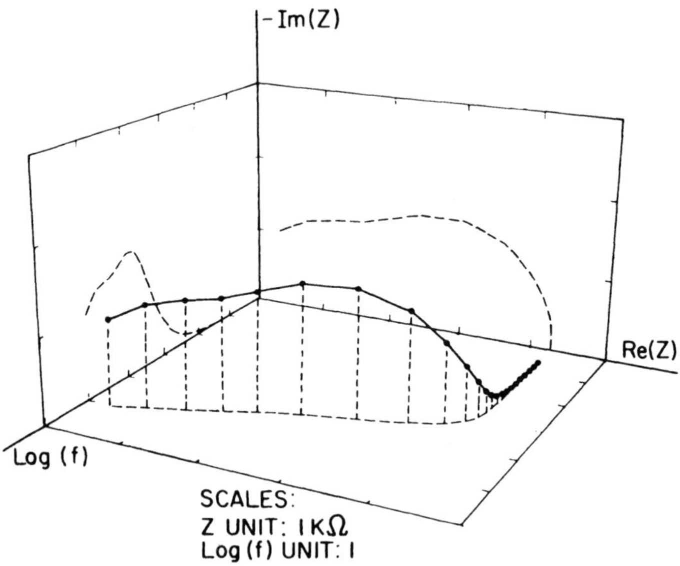

Figure 3.3.6. Perspective 3-D plot of the *Z* response of the circuit of Figure 3.3.5. (Reprinted from J. R. Macdonald, J. Schoonman, and A. P. Lehnen, *Solid State Ionics* **5**, 137–140, 1981.)

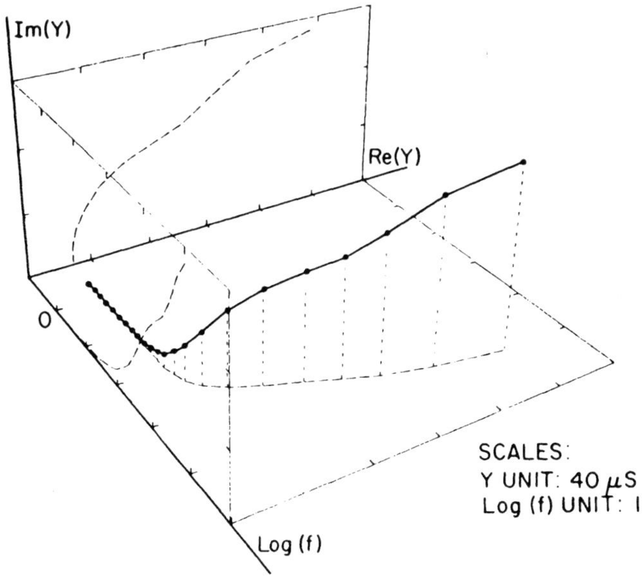

Figure 3.3.7. Perspective 3-D plot of the Y response of the circuit of Figure 3.3.5. (Reprinted from J. R. Macdonald, J. Schoonman, and A. P. Lehnen, *Solid State Ionics* **5**, 137–140, 1981.)

bottom numbers shown for each element in Figure 3.3.5 are the CNLS Y-fitting values estimated from all the data; they are probably more accurate than the nominal values. Clearly the resolution and accuracy are very good here in spite of the appearance of little structure in the 3-D plots. But this was a situation where the proper circuit was initially known.

Figure 3.3.8 shows the results of fitting impedance data for β-PbF_2 at 474K (Macdonald, Schoonman, and Lehnen [1982]) with CNLS to the circuit at the top of the figure. Initially the form of the circuit which would best fit the data was unknown, so several different ones were tried. The use of a CPE in the circuit, as shown, allowed quite a good fit to the data to be obtained and led to well-determined parameter estimates. The deviations between the dotted and dashed projection lines at the lowest frequencies show that the fit is not perfect in this region, however.

Finally, Figure 3.3.9 shows a 3-D log impedance plot for the same data shown in Figure 3.3.3. Here the fit (dotted curves) appears to agree very well indeed with the data, but an M plot (as in Figure 3.3.4) showing both data and fit predictions shows regions of appreciable disagreement (Macdonald and Cook [1984, 1985]). Since it has become customary to plot curves such as $\log(-Z'')$ vs. ν or $\log(\nu)$ and $\log(Z'')$ vs. ν or $\log(\nu)$ when data variation is large and to compare data and fitted results in this form, we wish to warn that this is often insufficient; discrepancies of some importance may be obscured by the logarithmic transformation. Even though several 3-D plots with different scales may be required to show data of wide variability plotted linearly, it is generally a good idea to examine them all when data and predicted values are to be compared. Regions of discrepancy then yield

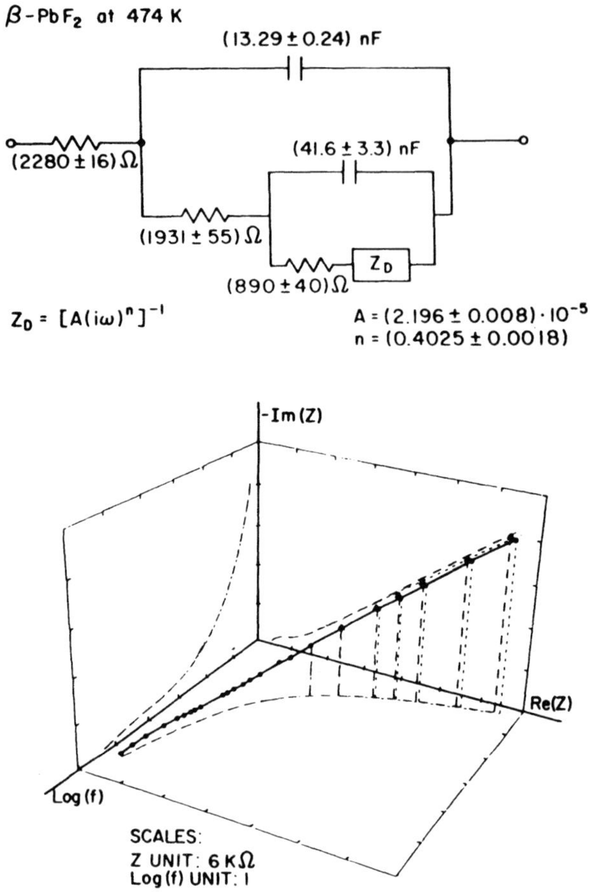

Figure 3.3.8. At the top is a circuit used to fit β-Pbf$_2$ data at 474K. Parameter values and their standard deviations estimated from CNLS fitting are shown. The bottom part shows a 3-D perspective plot of the Z data (solid line and short dashes) and predicted values and curves (long dashes). (Reprinted from J. R. Macdonald, J. Schoonman, and A. P. Lehnen, *Solid State Ionics* **5**, 137–140, 1981.)

immediate information about needed modifications in the fitting circuit. Finally, although the idea hasn't been much exploited as yet, it is a simple matter to plot relative residuals themselves in 3-D. For Z data one would plot $(-Z_e'' + Z_t'')/Z(_e'')$ on the vertical axis and $(Z_e' - Z_t')/Z_e'$ on the horizontal real axis and employ the usual $\log(\nu)$ axis. For P weighting, one would expect the resulting 3-D curve to be well bounded (magnitudes of most residuals comparable), but the plot should clearly show any regions of correlated residuals, indicating the presence of systematic errors and the need for improving the fitting circuit.

3.3.2.8 Summary and Simple Characterization Example

In the IS field, workers ordinarily spend much time and effort in preparing materials and measuring them under closely controlled conditions. But frequently their

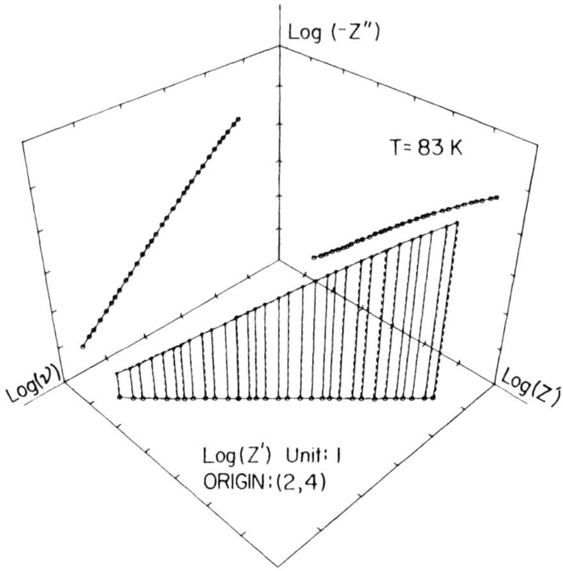

Figure 3.3.9. Perspective 3-D Z plot showing both data (solid lines) and CNLS fit (dotted lines) for Na β-alumina data at 83K. Compare Figure 3.3.3 for the data alone. (Reprinted from J. R. Macdonald and G. B. Cook, *J. Electroanal. Chem.* **168**, 335–354, 1984.)

subsequent attention to data analysis is quite inadequate and does not do justice to the work done. Although some kind of a data presentation is usually included, it is rare when sufficient plots are incorporated to resolve all the data well and show its shapes for different kinds of plots. We have tried herein to illustrate some of the virtues of 3-D perspective plotting and strongly urge its increased use in the IS field.

Even in the rare cases where the data are presented adequately, one often finds either no discussion of an appropriate model or equivalent circuit or just the statement that the data fit a given equivalent circuit without either a comparison of the original data and the circuit predictions or a listing of estimates of the values of the parameters in the equivalent circuit. In most cases where such a listing is included, the estimates have been obtained by approximate methods and the reader is given no measure of their accuracy and applicability. Finally, even when an equivalent circuit is presented and used, one rarely finds any discussion of why it is the most appropriate circuit to use or any comparison with other possible fitting circuits.

In summary, we first advocate 3-D plotting of data in various forms, followed by data adjustment and smoothing when warranted. Then crude approximate analysis methods may be employed to estimate initial values of the parameters which enter into an equivalent circuit thought appropriate. Next, CNLS fitting with weighting should be carried out using this and any other possibly likely allied circuits in order to find the simplest and best-fitting one. Then the data and the fitting results should be compared in 3-D and the final parameter estimates and their estimated standard deviations presented.

Even when all the above procedures have been completely carried out, there is a final stage of analysis which should always be included when possible. This stage is the essence of characterization: the passage from good equivalent-circuit macroscopic element estimates to estimates of microscopic parameter values. It is, of course, unnecessary if one is fitting data directly to an impedance function involving microscopic parameters. But in the more usual case of fitting to an equivalent circuit, this stage is the heart of the whole enterprise. A general approach to such macroscopic–microscopic transformation for unsupported systems has been outlined in detail in Macdonald and Franceschetti [1978], and the method is illustrated in, for example, Macdonald, Hooper, and Lehnen [1982] and Macdonald and Cook [1984, 1985].

Here we shall illustrate the method for a simple unsupported situation, that where the equivalent circuit of Figure 2.2.8b applies with only C_∞, R_∞, C_R, and R_R present. Assume that univalent charge of only a single sign, say, positive, is mobile and the partially blocking electrodes are identical with known spacing l. The equations presented in Section 2.2.3 become (per unit area)

$$C_\infty = \varepsilon\varepsilon_0/l \tag{8}$$

$$R_\infty = l/F\mu_p c_p^0 \tag{9}$$

$$C_R = (\varepsilon\varepsilon_0 F^2 c_p^0/RT)^{1/2} \tag{10}$$

and

$$R_R = RT/(F^2 c_p^0 k_p^0) \tag{11}$$

Assume that either ε, the bulk dielectric constant, is known or that C_∞ has been determined from CNLS fitting of the data. In either case, $\varepsilon\varepsilon_0$ can be obtained and used in Eq. (10) to obtain from the C_R estimated value an estimate of c_p^0, the bulk concentration of the mobile positive charge carriers. Then this value of c_p^0 may be used in Eq. (9) along with the estimated value of R_∞ to obtain an estimate of μ_p, the mobility of the positive charges. Finally, the c_p^0 estimate may be used in Eq. (11) in conjunction with the R_R estimate to calculate k_p^0, the effective reaction rate for the reacting positive charges. Thus from the four macroscopic estimates C_∞, R_∞, C_R, and R_R, one obtains estimates of the four microscopic quantities ε, c_p^0, μ_p, and k_p^0. These values and their dependences on controllable variables such as ambient temperature should then finally lead to valuable insight into the electrical behavior of the material–electrode system.

Chapter 4

Applications of Impedance Spectroscopy

4.1 CHARACTERIZATION OF MATERIALS

N. Bonanos
B. C. H. Steele
E. P. Butler

4.1.1 Microstructural Models for Impedance Spectra of Materials

4.1.1.1 Introduction

In polycrystalline solids, transport properties are strongly affected by microstructure, and impedance spectra usually contain features that can be directly related to microstructure. This is clearly illustrated in ceramic electrolytes such as zirconia and β-alumina. Much work on correlating microstructure and electrical properties can be directly traced to the pioneering study of Bauerle [1969]. The main purpose of this research effort has often been to optimize electrical conductivity for use in solid state electrochemical devices, such as solid oxide fuel cells (SOFC) and sensors. In parallel, impedance spectroscopy (IS) has been developed as a technique for the study of materials not intended for electrical applications, where transport properties are incidental to the main application, for example structural ceramics. In this section we discuss microstructural models describing grains and grain boundaries of differing phase composition, suspensions of one phase within another, and porosity, while in Section 4.1.3, we give examples of the combined use of IS and electron microscopy. With the two techniques combined, it is possible to derive information of microstructural relevance that is not accessible using one of the techniques alone.

Impedance Spectroscopy, Second Edition, edited by Evgenij Barsoukov and J. Ross Macdonald
ISBN 0-471-64749-7

The electrical properties of heterogeneous media have been modeled for over 100 years. Meredith and Tobias [1962], Mitoff [1968], and McLachlan *et al.* [1990] have given clear accounts of their scope and validity. However, since these articles cover the case where the conductivity or permittivity are real, which normally means dc conductivity or permittivity of loss-free dielectrics, they are not directly applicable to IS. AC properties have been discussed by Wimmer, Graham, and Tallan [1974] with special reference to ceramics. The dielectric literature has been reviewed by van Beek [1965], while Dukhin and Shilov [1974] have described models that include the effects of the interfacial double layer. Sihvola [1999] has produced a comprehensive survey of the properties of mixed phase systems with coverage of the historical and theoretical background.

As this section emphasizes materials properties, the results are presented in terms of bulk intensive quantities, such as the complex conductivity ψ, the complex resistivity ρ, the permittivity ε, and the modulus M. A distinction is made between two kinds of properties, both bulk intensive: (i) macroscopic properties, such as the resistivity of a sample due to grain boundaries (after correction for length/area), which are directly measurable, and (ii) microscopic properties, such as the resistivity of the grain boundary phase, which can only be estimated with the help of a microstructural model. The macroscopic quantities are represented by Latin symbols and the microscopic ones by Greek symbols.

The quantity used for expressing models of this type is the complex conductivity, defined as

$$\psi_i = \sigma_i + j\omega\varepsilon_i \tag{1}$$

where σ_i is the dc conductivity and ε_i the permittivity of phase i. The role of the model is to provide a hypothetical microstructure, for which the total complex conductivity, ψ_t, may be calculated. Having obtained $\psi_t = \psi_t(\omega)$, this may be converted to another representation, such as the complex resistivity, $\rho \equiv \psi^{-1}$. In many cases, it is possible to find an equivalent circuit that represents the model. This has the advantage that the circuit parameters can be estimated from the impedance spectrum. Equation (1) is equivalent to assigning a Voigt element to each phase. As will be shown below, some of the models for two-phase microstructures are equivalent to a circuit consisting of two elements in series (Figure 4.1.1), while others predict a more complex behavior. The models fall into two types: layer models and effective medium models.

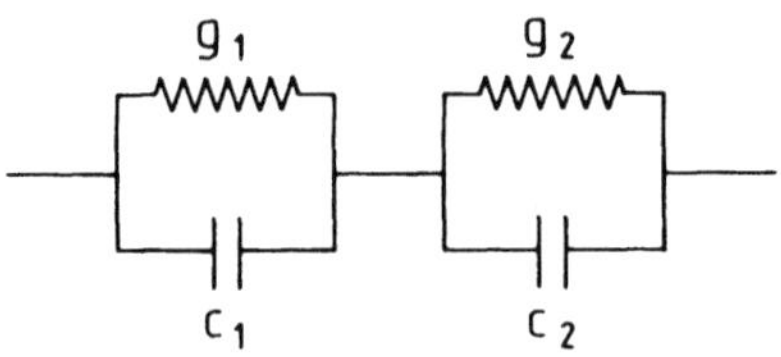

Figure 4.1.1. Circuit equivalent of a two-phase microstructure.

4.1.1.2 Layer Models

The earliest model used to describe the electrical properties of a two-phase mixture is the *series layer model* (Maxwell [1881]) shown microstructurally in Figure 4.1.2*a*. The two phases are assumed to be stacked in layers parallel to the measurement electrodes, with total thicknesses of each phase proportional to the volume fractions x_1 and x_2. The series layer model is described by the equation

$$\psi_t^{-1} = x_1\psi_1^{-1} + x_2\psi_2^{-1} \tag{2}$$

or, in terms of the complex resistivities

$$\rho_t = x_1\rho_1 + x_2\rho_2 \tag{2a}$$

which expresses a linear mixing rule. The series layer model is equivalent to the circuit of Figure 4.1.1, in which

$$\begin{aligned} g_1 &= \sigma_1/x_1, g_2 = \sigma_2/x_2 \\ c_1 &= \varepsilon_1/x_1, c_2 = \varepsilon_2/x_2 \end{aligned} \tag{3}$$

The impedance and modulus spectra for this model in general consist of two semicircles. The diameters of these are g_1^{-1}, g_2^{-1} in the impedance and c_1^{-1}, c_2^{-1} in the modulus spectra. The time constants of the two phases are defined as $\tau_1 = c_1/g_1$ and $\tau_2 = c_2/g_2$. If these time constants differ as a result of differences in c, then the arcs will be well resolved in the impedance spectrum. If they differ as a result of g, they will be resolved in the modulus spectrum (Hodge *et al.* [1976]). In practice, good resolution is not obtained in both Z and M spectra.

It is interesting to compare the series model (Figure 4.1.2*a*) with the corresponding parallel model (Figure 4.1.2*b*), in which the layers are stacked across the electrodes. For the *parallel layer model*, the complex conductivity follows a linear mixing rule

$$\psi_t = x_1\psi_1 + x_2\psi_2 \tag{4}$$

The equivalent circuit would be that of Figure 4.1.3*a* with values

$$\begin{aligned} g_1 &= x_1\sigma_1, g_2 = x_2\sigma_2 \\ c_1 &= x_1\varepsilon_1, c_2 = x_2\varepsilon_2 \end{aligned} \tag{5}$$

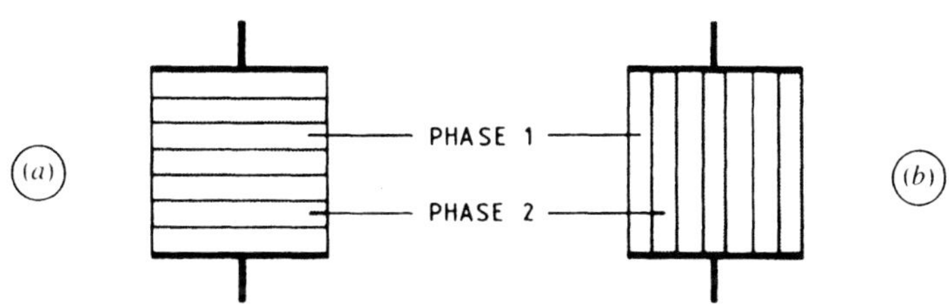

Figure 4.1.2. Hypothetical two-phase microstructures described by simple layer models: (*a*) series layer model; (*b*) parallel layer model.

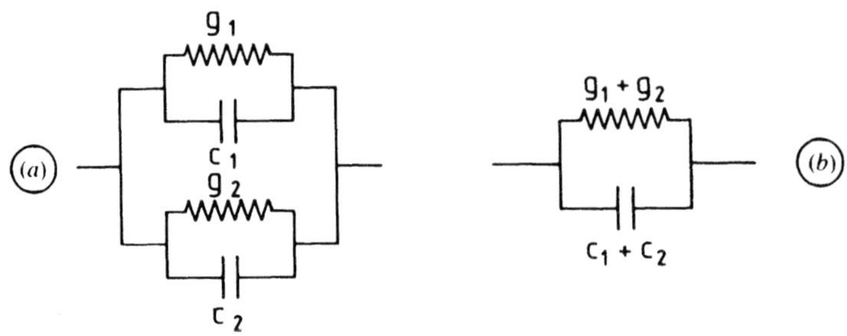

Figure 4.1.3. Circuit equivalent of the parallel layer model. Two parallel connected Voigt elements (*a*) transform to a single Voigt element (*b*).

The behavior of this circuit differs qualitatively from the previous one, because conductances g_1, g_2 and capacitances c_1, c_2 are in parallel. Thus the circuit is equivalent to that of Figure 4.1.3*b*, which shows only one relaxation. For the microstructure of Figure 4.1.2*b* the individual relaxations cannot be resolved by any method, graphical, CNLS, or other. Although at first glance this model would appear to be as plausible as the series layer model, it fails to describe the behavior of grain boundaries in ceramics.

A third model, originally suggested by Beekmans and Heyne [1976], has been termed the *brick layer model* (van Dijk and Burggraaf [1981], Verkerk, *et al.* [1982]). It is more realistic, treating the microstructure as an array of cubic grains, separated by flat grain boundaries, as shown in Figure 4.1.4*a*. The grains have a side D and the grain boundaries a thickness d, where $d << D$, and so the volume fraction of the grain boundary phase is $3d/D$. The current flow is assumed to be one-dimensional, and curvature of the current paths at the corners of the grains is neglected. In this case the two paths available to the current are either through grains *and* across grain boundaries or along grain boundaries, as depicted in the exploded diagram shown in Figure 4.1.4*b*. Depending on the relative magnitudes of σ_{gi} and σ_{gb}, one of the two paths will normally dominate.

Case (i): $\sigma_{gi} >> \sigma_{gb}$. Conduction along the grain boundaries is negligible, and conduction through the grains and across grain boundaries dominates. The behavior can be described in terms of complex resistivity:

$$\rho_t = \rho_{gi} + \frac{1}{3} x_{gb} \rho_{gb} \tag{6a}$$

or, for consistency with the rest of this section, in terms of the complex conductivity:

$$\psi_t = \left(\psi_{gi}^{-1} + \frac{1}{3} x_{gb} \psi_{gb}^{-1} \right)^{-1} \tag{6b}$$

Case (ii): $\sigma_{gi} << \sigma_{gb}$. Conduction along the grain boundaries is dominant:

$$\psi_t = \psi_{gi} + \frac{2}{3} x_{gb} \psi_{gb} \tag{7}$$

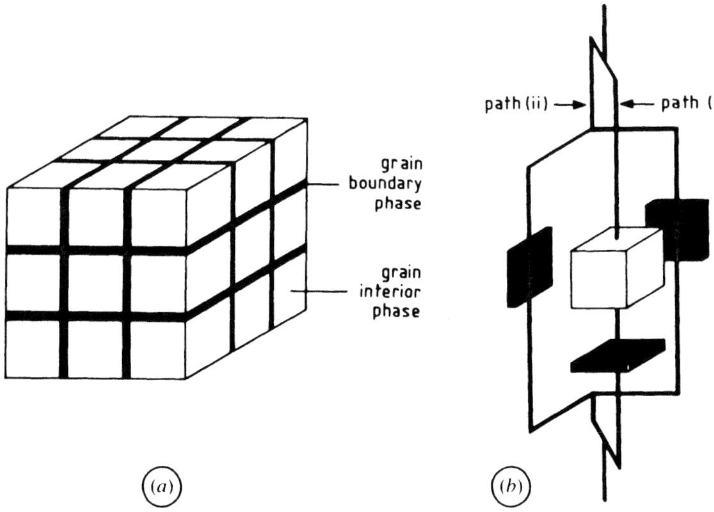

Figure 4.1.4. Brick layer model for a two-phase ceramic: (*a*) Overall view, showing array of cubic grains, separated by flat grain boundaries. (*b*) Exploded view of a single cell, showing parallel electrical paths: (i) through grains and grain boundaries, and (ii) along grain boundaries.

Case (iii): general. According to Näfe [1984], the two paths (i) and (ii) may be combined into a network describing the polycrystalline properties for all ratios of σ_{gi}/σ_{gb}. Using our own notation this means:

$$\psi_t = \left(\psi_{gi}^{-1} + \frac{x_{gb}}{3}\psi_{gb}^{-1}\right)^{-1} + \frac{2}{3}x_{gb}\psi_{gb} \tag{8}$$

We examine each case separately:

Case (i). For $\sigma_{gi} >> \sigma_{gb}$, the brick layer model is equivalent to the series layer model but with a one-third weighting of the grain boundary resistance. This reflects the fact that grain boundaries in only one of the three orientations (i.e. normal to the current) have a blocking effect. The circuit equivalent of the brick layer model is, therefore, that of Figure 4.1.1 with parameters

$$\begin{aligned} g_{gi} &= \sigma_{gi}, c_{gi} = \varepsilon_{gi} \\ g_{gb} &= 3\sigma_{gb}/x_{gb}, c_{gb} = 3\varepsilon_{gb}/x_{gb} \end{aligned} \tag{9}$$

A useful expression can be derived from these equations

$$x_{gb} = 3\frac{c_{gi}}{c_{gb}}\frac{\varepsilon_{gb}}{\varepsilon_{gi}} \tag{10}$$

or, expressed in terms of grain size D and the grain boundary thickness d,

$$\frac{d}{D} = \frac{c_{gi}}{c_{gb}}\frac{\varepsilon_{gb}}{\varepsilon_{gi}} \tag{11}$$

By making the assumption $\varepsilon_{gb} = \varepsilon_{gi}$, the thickness or volume fraction of the grain boundary phase can be estimated, if the grain size is known. This applies only when the boundary capacitance arises from the presence of a second phase which, as explained later, is not always the case. Often one refers loosely to g_{gi} and g_{gb} as grain interior and grain boundary conductivities. A more appropriate term would be *macroscopic* conductivities, as they are corrected for the macroscopic shape of the sample (length/area). For the brick layer model, and for $d << D$, g_{gi} is indeed the microscopic conductivity of the grain interior. By contrast, g_{gb} is usually 100 to 1000 times higher than σ_{gb} because x_{gb} is small. Therefore, although we can have $g_{gb} > g_{gi}$, this should not be taken to imply $\sigma_{gb} > \sigma_{gi}$.

Case (ii). When the current is mainly carried along grain boundaries, the brick layer model is equivalent to the parallel layer model (Figure 4.1.2*a*), but with a two-thirds weighting on the grain boundary conductance term. A ceramic sample will then show a single arc in the impedance or modulus spectrum revealing little or nothing about its microstructure.

Case (iii). Näfe's [1984] model combines the two extremes in one expression, and is valid at high or low conductivity ratios, as it reduces to the expressions given for cases (i) and (ii). We have reservations about using this model over the entire range of σ_{gi}/σ_{gb}, since the assumption of current flowing via separate pathways is not tenable when $\sigma_{gb} \cong \sigma_{gi}$.

Thus it is possible, on the basis of the impedance spectrum alone, to differentiate between two situations: conducting grains with blocking grain boundaries or poorly conducting grains with highly conducting grain boundaries.

Blocking of Ions—Easy Paths. The brick layer model assumed the grain boundary region separating the grains to be continuous. However, it is often observed that the activation energies for the two conductivities, g_{gb} and g_{gi}, are equal or very similar. This led Bauerle [1969] to suggest, in connection with zirconia ceramics, that there were regions of the grain boundary where good intergranular contact was established; these are called *easy paths* (Figure 4.1.5*a*). The circuit chosen by Bauerle (Figure 4.1.5*b*) expresses the idea that mobile species (oxide ions) are sequentially blocked at grain interiors and grain boundaries. g_{ep} is the conduction through easy

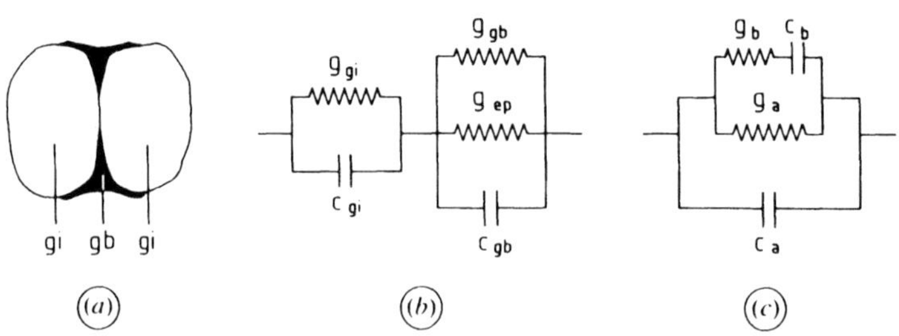

Figure 4.1.5. Easy path model for a two-phase ceramic: (*a*) Schematic representation of grains separated by a discontinuous grain boundary phase. (*b*) Series circuit equivalent according to Bauerle [1969]. (*c*) Parallel circuit equivalent according to Schouler [1979].

paths. Since $\varepsilon_{gb} \cong \varepsilon_{gi}$, the small intergranular contact areas do not affect c_{gb}. A slightly different model, proposed by Schouler [1979] (Figure 4.1.5*c*), divides the ionic current into two paths, one of which (g_b, c_b) is blocked capacitatively, while the other (g_a) is not. The proportion of blocked ionic current is then given by

$$\beta = g_b / (g_a + g_b) \tag{12}$$

The same ratio β could be expressed in terms of a series model as:

$$\beta = r_{gb} / (r_{gi} + r_{gb}) \tag{13}$$

Both models explain the identical activation energies of r_{gi} and r_{gb} and, as they are equivalent (see Chapter 1), there is no *a priori* reason to prefer one to the other. One might prefer to visualize constriction resistances according to Bauerle [1969], when the grain boundary coverage is known to be high, or partial blocking according to Schouler [1979], when the grain boundary coverage is low, in the form of isolated islands.

Effect of Space Charge. As mentioned earlier, grain boundary effects arise not only from the presence of second phases. The electrical behavior of interfacial regions often differs from that of the grain interior due space charge effects, studied by Maier and co-workers. Maier [1986] has shown that the conductance of a polycrystal contains three components: that of the grain interior (gi), the grain core (gc), and a region next to the grain boundary, dominated by space charge (sc). The space charge extends over a length, λ, from the core, known as the Debye length. For the simplest case, of a pair of defects with the same charge, the Debye length is given by the expression:

$$\lambda = \sqrt{\frac{\kappa \varepsilon_0 RT}{2 c_\infty z^2 F^2}} \tag{14}$$

where κ is the dielectric constant of the medium
z is the number of charges on the defect
c_∞ is the molar concentration of defect far from the interface
and ε_0, R, T and F have their usual meanings.

In the Debye model, the defect concentration is an analytic function of the distance from the boundary and, therefore, so is the conductivity, assuming that the mobility is constant. The defect concentration depends on the difference in local free energy between a defect at the boundary and a defect in the bulk and is parameterized by a quantity in the range −1 to 1, describing the sign and strength of the effect. Figure 4.1.6*a* shows a defect concentration profile, in which the concentration is normalized with respect to the bulk concentration and the distance from the interface is normalized with respect to the Debye length.

For current perpendicular to a grain boundary, the following expression holds for the impedance[1]:

[1] The notation used in this expression has been altered for consistency with the rest of the section.

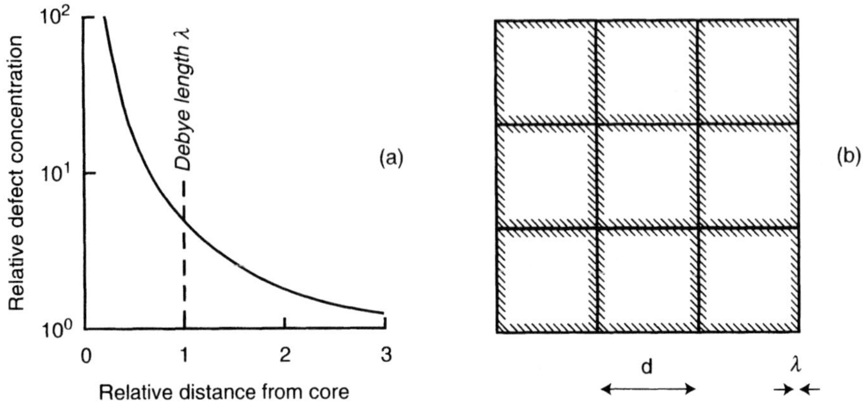

Figure 4.1.6. (*a*) Defect concentration profile near an interface, according to the Debye model. (*b*) Model microstructure for a polycrystalline solid, after Maier [1987].

$$Z^{\perp} = Z_{gi}^{\perp} + Z_{gc}^{\perp} + \Delta Z_{sc}^{\perp} \tag{15}$$

while for current parallel to a grain boundary, a related admittance expression is derived:

$$Y^{\parallel} = Y_{gi}^{\parallel} + Y_{gc}^{\parallel} + \Delta Y_{sc}^{\parallel} \tag{16}$$

Both terms prefixed by Δ in these equations can be obtained by integrating the relevant concentration profiles over the distance from the grain boundary core. By considering a system composed of square grains (Figure 4.1.6*b*) and summing the contributions of the perpendicular and parallel paths (while neglecting the squared term in x_{gb}), the following expression is derived for the complex conductivity:

$$\psi_t = \frac{\psi_{gi}\psi_{gb}^{\perp} + \frac{2}{3}x_{gb}\psi_{gb}^{\parallel}\psi_{gb}^{\perp}}{\psi_{gb}^{\perp} + \frac{1}{3}x_{gb}\psi_{gi}} \tag{17}$$

where x_{gb} refers to the combined volume fraction of the core and space charge region, i.e. $x_{gb} = x_{gc} + x_{sc}$. This is equivalent to Eq. (8), re-written so as to distinguish between parallel and perpendicular paths and with the squared term in x_{gb} removed, namely:

$$\psi_t = \left(\psi_{gi}^{-1} + \frac{1}{3}x_{gb}\psi_{gb}^{\perp -1}\right)^{-1} + \frac{2}{3}x_{gb}\psi_{gb}^{\parallel} \tag{18}$$

A salient point in this work is that, by specifying a variation in conductivity with distance from the grain boundary core, conductivity profiles in both orientations can be integrated and a phenomenological description obtained that is consistent with the brick-layer model. On one hand, this is encouraging, because it shows that the simple brick-layer model is able to reproduce some of the more subtle effects in ceramics. On the other hand, it suggests that conventional studies of ceramics, especially those of pure ceramics with low grain boundary phase content, may completely miss the effect of the space charge layer. A recommendation arising from the above is that several ceramic samples should be prepared with the same composition and with the widest possible range of grain sizes. However, it must be recognized that this recommendation cannot always be turned into practice.

The above treatment assumes that the defects are mobile and distributed in response to the thermodynamic driving force. It should be mentioned that this is not always the case: ceramics often contain defects frozen in from higher temperature treatment (e.g. sintering), in which case some of the defect concentrations will be fixed. The two cases, respectively known as the Debye and Schottky cases, produce different interfacial properties, including different characteristic lengths. For a concise discussion of the two cases see Maier [1996]. Further relevant information can be found in Kim *et al.* [2003] and Maier [2004].

Effects of Grain Shape and the Presence of Distributed Easy Paths. Fleig and Maier [1998] have approached the question of grain boundary impedance by finite element modeling. They have modeled various periodic grain structures with grains of different shapes and aspect ratios. They have compared the results to the predictions of the brick-layer model. The modeling results published so far cover only two-dimensional systems. Figure 4.1.7 shows two model structures comprising grains of high symmetry and the resulting impedance spectra, together with the spectrum for the brick-layer case. For these examples, the discrepancy from the brick layer model in terms of r_{gb} was of the order of 5%. For another case, namely of a serrated grain structure, where grain boundaries intersected the current lines at an angle, the macroscopic grain boundary conductivity varied with the slope angle, α, as follows:[2]

$$g_{gb} = \frac{d}{D} \sigma_{gb} \sin \alpha \tag{19}$$

This differs from the brick-layer model, for which the sine term is absent. This should be expected, since for this model, $\alpha = 90°$. Note that the above relation holds even when the serrations are on a much smaller scale than that of the grains. Fleig and Maier [1998] also explored the effect of a specific serrated microstructure on the temperature dependence of the macroscopic grain boundary conductivity. They assumed activation energies of 1.0 and 0.5 eV for σ_{gi} and σ_{gb} respectively, fed these values into the model and observed that $\Delta H_{g_{gb}}$ was slightly different from $\Delta H_{\sigma gb}$. In the temperature region near the point where $g_{gi} = g_{gb}$, $\Delta H_{g_{gb}}$ increased to 0.6 eV; outside this region, it returned to the expected value of 0.5 eV. Fleig [2000] examined the spatial distribution of easy paths in grain boundaries and found that con-

[2] This has been converted to our preferred notation.

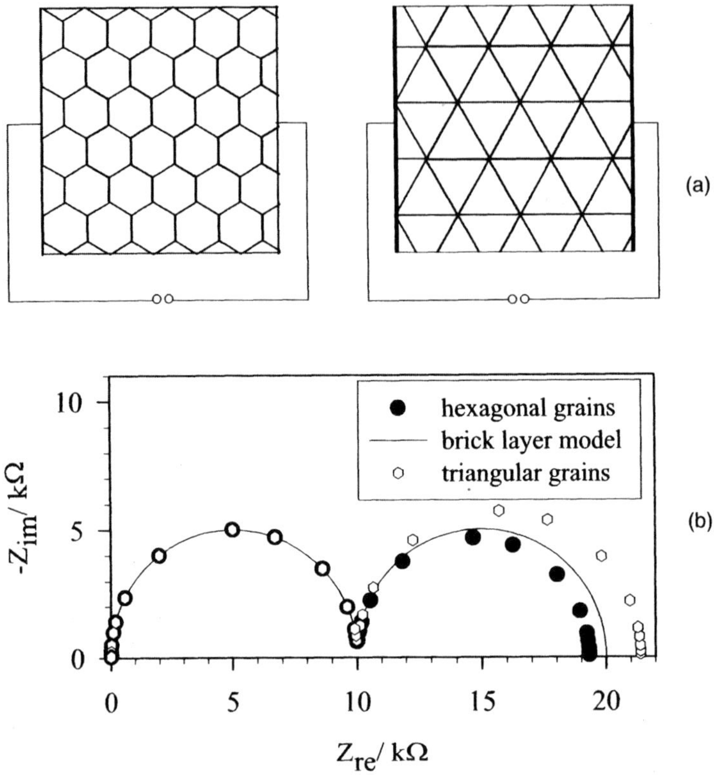

Figure 4.1.7. (*a*) Model grain boundary structures and the impedance spectra resulting from finite element modeling of these structures. The spectrum for the square, brick-layer case is shown for comparison (*b*). After Fleig and Maier [1998]. (Courtesy of the Electrochemical Society).

duction across the boundary was significantly greater with many small paths than with a single large one *when these paths occupied the same total grain boundary area*. Fleig [2002] examined a large number of computer-generated realistic microstructures and confirmed that deviations from the brick layer model were usually less that 10% and noted that the grain boundary relaxation frequency for the model reflected the correct input value.

The work demonstrates that the layer model is acceptable, provided the grain size distribution is narrow and the grain shape is roughly isotropic. In the case of grain boundary blocking with easy paths, the grain boundary conduction is not directly proportional to the area of the easy path contacts, but depends also on their spatial distribution. Regarding the effect of space charges, we conclude that, in general, simple studies will miss these effects, although some help could be provided by studying ceramics with a range of grain sizes.

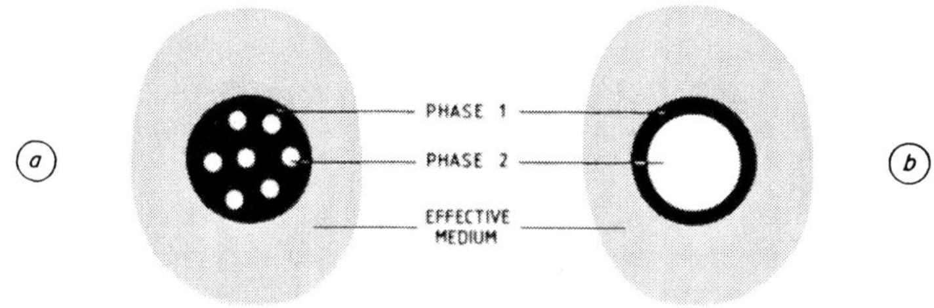

Figure 4.1.8. Hypothetical microstructures that may be described by the effective medium model: (*a*) Continuous matrix of phase 1 containing a dilute dispersion of spheres of phase 2. (*b*) A grain boundary shell of phase 1 surrounding a spherical grain of phase 2.

4.1.1.3 Effective Medium Models

The Maxwell–Wagner Model. The layer models presented so far have the advantage of clarity, but have been derived under unrealistic assumptions concerning the current distribution. Alternative models have been developed that take into account the real current distributions, based on the *effective medium technique*. This may be briefly described as follows. One imagines a continuous medium (the effective medium) of conductivity ψ_{eff}. A portion of this is removed and replaced by an equal portion of the heterogeneous system consisting of two (or more) phases. By applying the constraint that the current distribution in the effective medium is not altered by this operation, an expression is derived for ψ_{eff}. The total conductivity of the two phase system ψ_t is then equal to ψ_{eff}. It should be noted that this approach makes certain implicit assumptions about the microstructure, especially regarding the continuity of the constituent phases. Figure 4.1.8 shows two possible microstructures with a continuous phase (1) and a discontinuous phase (2).

The first exact expression of this type was derived by Maxwell [1881] for the dc conductivity of a dispersion of spheres in a continuous medium. Maxwell Garnett [1904] derived a similar expression for dielectric and optical properties. Wagner [1914] extended Maxwell's model to the complex domain and this model has thereafter been known as the Maxwell–Wagner model.[3] It gives the following expression for complex conductivity:

$$\psi_t = \psi_1 \frac{2\psi_1 + \psi_2 - 2x_2(\psi_1 - \psi_2)}{2\psi_1 + \psi_2 + x_2(\psi_1 - \psi_2)} \tag{20}$$

Despite its complexity compared to expressions derived from layer models, this expression has proven to be versatile in describing the impedance spectra of

[3] There is confusion regarding the name that should be attributed to this expression. Garnett's given names were James Clerk Maxwell, but he appears to have incorporated the last one into his surname. Furthermore, while K.W. Wagner extended Maxwell's formula to the complex variable, the term Maxwell–Wagner is also associated with a slightly different phenomenon of interfacial polarization. We have used the term to describe Eq. (20), following the practice of Fricke [1953].

heterogeneous media. It is assumed that distortions to the electric field caused by the particle are local, i.e. that neighboring particles experience a uniform field. This has traditionally been taken to mean that the suspension must be dilute in phase 2.[4] An embodiment of this model is shown in Figure 4.1.8*a*, in which the system consists of spheres suspended in a continuous medium. Simulated complex resistivity and modulus spectra given in Figure 4.1.9*a*,*b* for $x_2 = 0.25$ and for phases having conductivities $\sigma_2 >> \sigma_1$ and identical permittivities. In this case, the relaxations of the two phases are best resolved in the modulus plot.

Brailsford and Hohnke [1983] have applied the Maxwell–Wagner model to grain boundaries in two-phase systems. Their microstructural model, shown in Figure 4.1.8*b*, consists of a spherical grain of radius r_2 surrounded by a shell of outer radius r_1, which represents the grain boundary and has a volume fraction $x_1 = 1 - (r_2/r_1)^3$. The authors observe that for $x_1 \rightarrow 0$ and $\psi_2 << \psi_1$, the effective medium model becomes identical to case (i) of the brick layer model, namely Eq. (6). Further, we have found that for $x \rightarrow 0$ and $\psi_1 >> \psi_2$, it reduces to case (ii) of the brick layer model, namely Eq. (7).[5]

The assumed conditions of a low grain boundary phase volume fraction and large low grain boundary conductivity are realistic, as far as ceramic electrolytes are concerned. It is therefore encouraging to see that models derived from two different grain geometries give the same results under these conditions. This suggests that the estimation of grain boundary properties from electrical circuit parameters should be subject to a relatively small error as a result of assumptions about the shape of the grains, if they are isotropic.

Simulated complex resistivity and modulus spectra based on Eq. (20) for the blocking grain boundary situation (Figure 4.1.8*b*) are shown in Figure 4.1.9*c* and *d*. It is evident that for the parameters chosen ($\sigma_2 >> \sigma_2$) *either* the impedance *or* the modulus spectra resolve the microstructural components. For an appreciable volume fraction of high conductivity suspended phase (Figure 4.1.8*a*), the modulus spectrum is preferable. For the situation with a low volume fraction of continuous grain boundary phase (Figure 4.1.8*b*) the impedance or complex resistivity spectra give good resolution.

Brailsford and Hohnke [1983] make two further observations regarding porosity for the case $\sigma_{gi} << \sigma_{gb}$. If the volume fraction of pores is x_p, then for intragranular pores,

$$g_{gi} = \frac{\sigma_{gi}}{1 + \frac{3}{2} x_p} \tag{21}$$

[4] The maximum volume fraction x2 for which Eq. (20) is valid has not been accurately specified, with values as low as 0.1 and as high as 0.5 having been proposed. A sensible upper bound might be 0.30 since above this the system approaches the percolation limit for spherical particles. On the other hand, in the following paragraph, we mention a case where the model is valid for $x_2 \cong 1$.

[5] For consistency with the rest of the section, we have used different subscripts than have Brailsford and Hohnke.

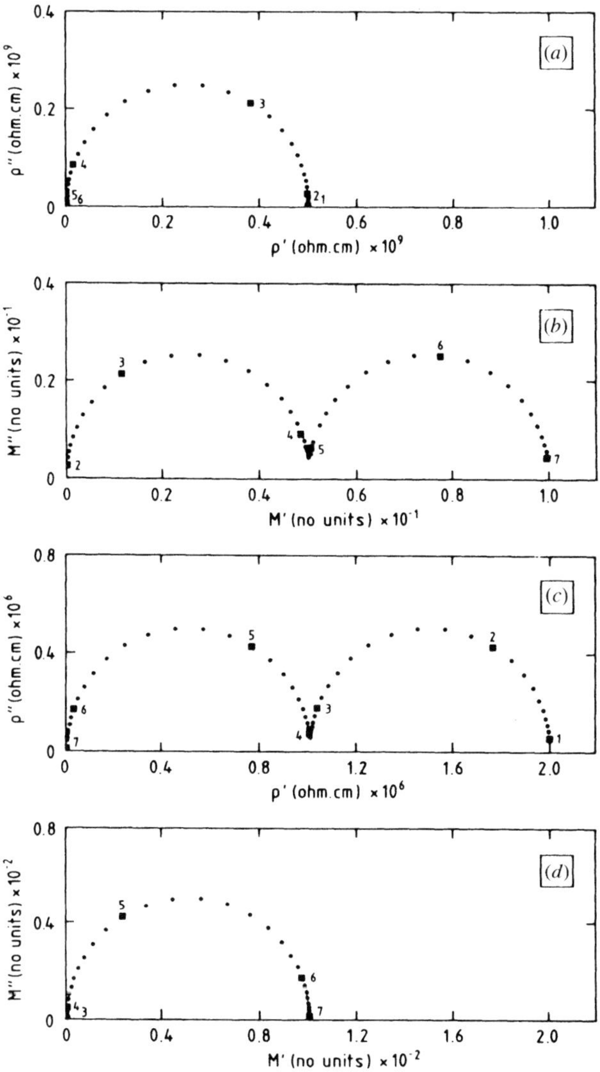

Figure 4.1.9. Simulated impedance and modulus spectra for a two-phase microstructure, based on the effective medium model. Values of the input parameters are given in Table 4.1.1. (*a*, *b*) Spectra for a matrix of phase 1 containing 25% by volume of spheres of phase 2. Resolution is achieved in the modulus spectrum (*b*) but not the impedance spectrum (*a*). (*c*, *d*) Spectra for a spherical grain of phase 2 surrounded by a grain boundary shell of phase 1. The ratio of shell thickness to sphere radius is 10^{-3}. Resolution is achieved in the impedance spectrum (*c*) but not the modulus spectrum (*d*).

The same amount of porosity distributed intergranularly has a larger effect on the grain boundary conductivity, highlighting the importance of the spatial distribution of pores:

$$g_{gb} = \frac{3\sigma_{gb}}{x_{gb} + \frac{3}{2}x_p} \tag{22}$$

Comparing the last two equations to Eq. (9) shows that the presence of pores modifies the impedance spectra (complex conductivity spectra) by altering the diameters of the grain interior or grain boundary arc. Nevertheless, in neither of these cases do pores introduce a new arc or other feature in the impedance spectrum. Thus, stating that the effect of pores can be seen on the impedance spectrum, does not imply that from an empirical impedance spectrum the degree or type of porosity can be established. This deduction could only be made if the electrical properties σ, ε were known beforehand for both the grain and for grain boundary phases, a condition that is unlikely to be fulfilled.

Inspection of Figure 4.1.9 suggests that Eq. (20) generates spectra that are similar to those of simple RC circuits. This is indeed the case. In fact, Bonanos and Lilley [1981] showed that the Maxwell–Wagner model is formally identical to the two-element circuit of Figure 4.1.1, but with values of g_1, g_2, c_1, and c_2 that can be expressed as rather complicated functions of σ_1, σ_2, ε_1, ε_2 and x_2 (Table 4.1.1).

The Fricke Model for Two-Phase Dispersions. Expressions similar to those of Maxwell have been derived for ellipsoidal particles of random orientation (Fricke [1932]) and for aligned ellipsoidal particles (Fricke [1953]). The expressions contain form factors which depend on the axial ratio of the ellipsoids and their orientation with respect to the electric field. The case of random orientation is the most interesting, as it describes a realistic ceramic microstructure and results in the following equation:

$$\psi_t = \psi_2 + \frac{(\psi_1 - \psi_2)(1 - x_2)}{1 + (x_2/3)\sum_{n=1}^{3}(\psi_1 - \psi_2)/(\phi_n\psi_1 + \psi_2)} \tag{23}$$

where

ψ_t = complex conductivity of the dispersion
ψ_1, ψ_2 = complex conductivities of the matrix and the dispersed phase, respectively

Table 4.1.1. Input parameters used to calculate the spectra shown in Figure 4.1.9

	σ_1	σ_2	ε_1	ε_2	
Volume Fraction x_2	S/cm		F/cm		Figure
0.25	10^{-9}	10^{-6}	8.85×10^{-13}	8.85×10^{-13}	(a), (b)
0.997	10^{-9}	10^{-6}	8.85×10^{-13}	8.85×10^{-13}	(c), (d)

Note: The permittivities ε_1, ε_2 correspond to a dielectric constant of 10.

x_2 = volume fraction of the dispersed phase

$\phi_{1,2,3}$ = form factors which depend on the axial ratios of the ellipsoidal particles, defined by the semiaxes a, b, c, where $a \geq b \geq c$.

Selected values of ϕ are given in Table 4.1.2. A simulated modulus spectrum for Eq. (23) is shown in Figure 4.1.10. The parameters used to generate this are listed in Table 4.1.3. A plot of complex resistivity (not shown) displays only the low-frequency arc, with a very small distortion at high frequency.

For the parameters chosen, there are two readily resolvable arcs in the modulus spectrum Figure 4.1.10. The low-frequency arc corresponds to the low-conductivity continuous phase and is apparently a perfect semicircle with its center on the real axis. The high-frequency arc corresponds to the discontinuous phase and is composed of three relaxations corresponding to the three possible orientations of the ellipsoids. In Figure 4.1.10 these are not well resolved, but cause the arc to be nonideal. The two-phase dispersion can be represented by either of the two circuits in Figure 4.1.11; the values shown in Table 4.1.4 are the results of NLLS fits to the spectrum.

The Bruggeman Effective Medium Models. Bruggeman [1935] is credited with two very significant contributions to the modeling of two-phase systems, generally known as the Bruggeman asymmetric and symmetric models.[6] The asymmetric model is an extension of the Maxwell–Wagner model to high volume fractions of discontinuous phase. This model treats the discontinuous phase as embedded in a medium which is itself modified by the presence of the discontinuous phase. Using the same notation as before, the Bruggeman asymmetric model is described by Eq. (24).

Table 4.1.2. Selection of Form Factors According to Fricke [1953] for Use in Eq. (23)

a/b	b/c	ϕ_1	ϕ_2	ϕ_3	Particle Shape
1	1	2.00	2.00	2.00	Spherical
1	2	4.79	1.42	1.42	
1	4	12.2	1.16	1.16	Spheroidal
1	6	21.6	1.09	1.09	
2	2	7.9	2.50	0.66	Ellipsoidal
2	6	20.7	6.9	0.209	

Note: These factors have been calculated from the axial ratios of the ellipsoidal particles a, b, c, where $a \geq b \geq c$.

[6] As is common in this area, these expressions have been credited to several sources. For example the symmetrical Bruggeman formula is also known as the Polder–van Santen formula. It has also been attributed to Landauer [1974], who pointed out the percolation aspect. For a more complete discussion of the equations and their nomenclature, the reader is referred to Sihvola [1999].

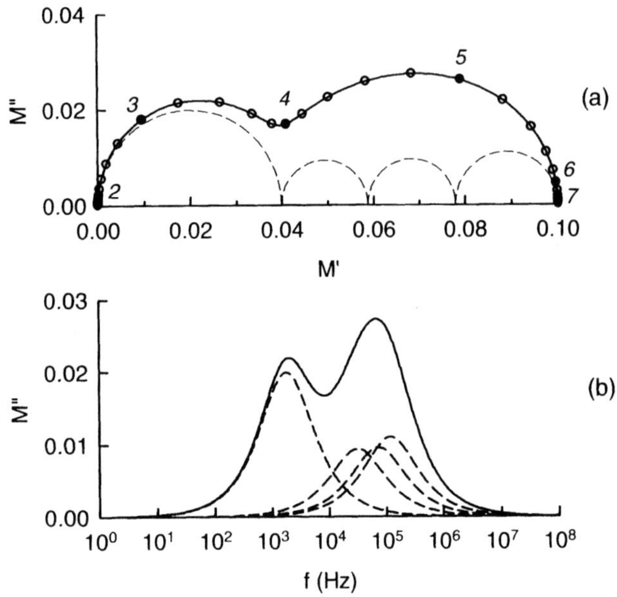

Figure 4.1.10. Simulated impedance and modulus spectra for a two-phase microstructure comprising a matrix of phase 1 with 25% by volume of randomly oriented ellipsoids of phase 2. (*a*) Impedance spectrum showing only one arc. (*b*) Modulus spectrum resolving two arcs, one being nonideal due to anisotropy of the ellipsoids. Values of input parameters are given in Table 4.1.3.

Table 4.1.3. Input Parameters Used for the Simulation Shown in Fig. 4.1.10, According to the Fricke [1953] Model for Ellipsoids

Conductivities: $\sigma_1 = 10^{-8}$ S/cm, $\sigma_2 = 10^{-6}$ S/cm
Permittivities: $\varepsilon_1 = \varepsilon_2 = 8.85 \times 10^{-13}$ F/cm
Ratios of ellipsoid semiaxes: $a/b = 2$, $b/c = 2$
Form factors: $\phi_1 = 7.90$, $\phi_2 = 2.50$, $\phi_3 = 0.66$
Volume fraction of suspended phase: $x_2 = 0.25$

$$\frac{\psi_1}{\psi_t}\left(\frac{\psi_t - \psi_2}{\psi_1 - \psi_2}\right)^3 = (1 - x_2)^3 \tag{24}$$

Since this expression has an excellent pedigree, and is explicit in the volume fraction, x_2, it has often been used to estimate the phase volume fractions of composite conductors or dielectrics where the conductivity and permittivity, respectively, are real. The expression has not been often used with complex variables, no doubt because it is implicit in ψ_t and, therefore, difficult to evaluate. Tuncer *et al.* [2001, 2002] have used a numerical solution, while Sihvola [1999] has given a series expansion that can be used for complex variables. An extended version of the series is given below (Sihvola [2003] private communication).

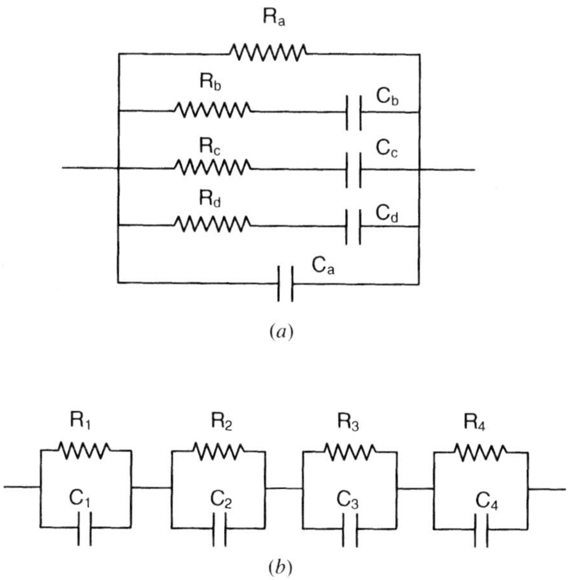

Figure 4.1.11. Two equivalent circuits, for a matrix of phase 1 containing randomly oriented ellipsoids of phase 2 according to model proposed by Fricke [1953]: (*a*) parallel circuit; (*b*) series circuit.

Table 4.1.4. Parameter Values for the Circuits Shown in Figure 4.1.11, Obtained by NLLS Fitting to the Spectrum of Figure 4.1.10; the Products $f_a{\cdot}f_b{\cdot}f_c{\cdot}f_d$ and $f_1{\cdot}f_2{\cdot}f_3{\cdot}f_4$ are Equal

Element	r_i (Ωcm)	c_i (F/cm)	f_i (Hz)	Element	r_i (Ωcm)	c_i (F/cm)	f_i (Hz)
a	4.2×10^7	8.9×10^{-13}	4.3×10^3	1	4.0×10^7	2.2×10^{-12}	1.8×10^3
b	9.8×10^6	8.3×10^{-13}	2.0×10^4	2	1.1×10^6	4.7×10^{-12}	3.1×10^4
c	2.0×10^7	7.8×10^{-14}	1.0×10^5	3	4.8×10^5	4.6×10^{-12}	7.2×10^4
d	1.2×10^7	2.4×10^{-13}	5.5×10^4	4	3.4×10^5	4.0×10^{-12}	1.2×10^5

$$\psi_t = \psi_1 + 3\psi_1 r x_2 + 3\psi_1 r^2 \frac{\psi_1 + 2\psi_2}{\psi_2 + 2\psi_1} x_2^2 + 2\psi_1 r^4 \frac{\psi_1 + 5\psi_2}{\psi_2 + 2\psi_1} x_2^3 + 3\psi_1 r^4 \frac{5\psi_2^3 - 21\psi_1\psi_2^2 + 6\psi_2\psi_1^2 + \psi_1^3}{(\psi_2 + 2\psi_1)^3} x_2^4 \cdots \qquad (25a)$$

Where

$$r = \frac{\psi_2 - \psi_1}{\psi_2 + 2\psi_1} \qquad (25b)$$

Evaluation of this expression generates a spectrum rather similar to that for the Maxwell–Wagner model, but with a different weighting of the two phases. Figure 4.1.12 shows a modulus spectrum for the same input parameters as those that were used to produce the spectrum in Figure 4.1.9*b*.

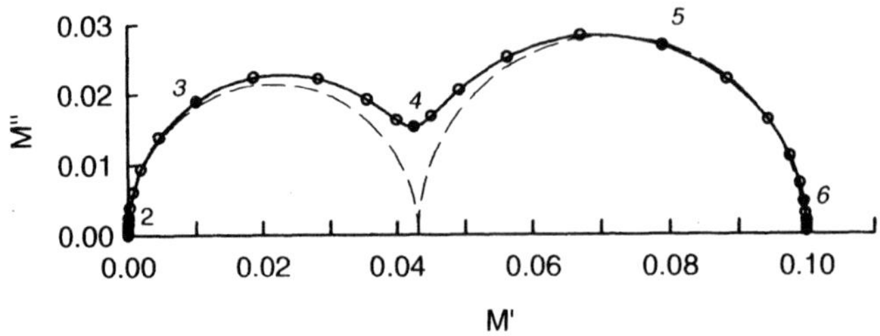

Figure 4.1.12. Modulus spectrum for the Bruggeman asymmetric dispersed phase model with $\sigma_1 = 1 \times 10^{-8}\,\mathrm{S\,cm^{-1}}$, $\sigma_1 = 1 \times 10^{-6}\,\mathrm{S\,cm^{-1}}$ and $x_2 = 0.3$. The calculation was made using the series expansion Eq. (25) due to A. Sihvola Labels indicate log(f).

The symmetric Bruggeman model does not make a distinction between continuous and discontinuous phases and, therefore, the relevant expression is symmetrical in ψ_1, ψ_2. It seems to have been first applied to the complex variable by Landauer [1978]. Coverdale *et al.* [1995] give an in-depth discussion of its relevance to impedance spectra of composite materials. The symmetric Bruggeman model is described by Eq. (26).

$$\psi_m = \frac{1}{4}\left\{q + \left(q^2 + 8\psi_1\psi_2\right)^{\frac{1}{2}}\right\} \tag{26a}$$

where

$$q = (3x_1 - 1)\psi_1 + (3x_2 - 1)\psi_2 \tag{26b}$$

and

$$x_1 + x_2 = 1 \tag{26c}$$

While the model was originally based on effective medium considerations, it is instructive that Wu and Liu [1997] were able to derive an identical equation using a resistor network.

Figure 4.1.13 shows a modulus spectrum generated using (Eq. 26) using the same input parameters as were used for the previous figure. Unlike the case of the Maxwell–Wagner model, the spectrum cannot be reproduced by a simple RC circuit (the dashed arcs shown in the figure give only an approximate fit). Eq. (26) has the important property that, as the volume fraction of the high conductivity phase approaches $\frac{1}{3}$, its effect becomes dominant, as can be easily verified by substituting two unequal values of (real) dc conductivity. Thus the model provides an elementary and mathematically economical description of phase percolation.

Finite Element Modeling of Mixed Phase Systems. Another approach to the problem is to generate lattices populated with phases of different dielectric constant/conductivity and to compute their properties by the method of finite elements. Tuncer *et al.* [2001, 2002] have pursued this for a variety of pseudorandom and peri-

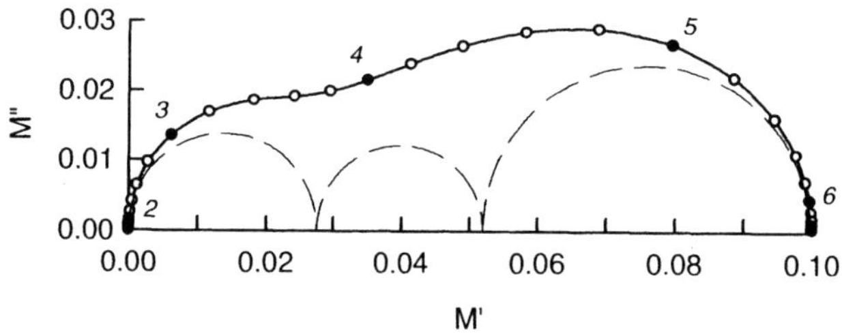

Figure 4.1.13. Modulus spectrum for the Bruggeman symmetric model (Eq. 20) for dispersed phases using the same input parameters as were used for Figure 4.1.12. Labels indicate log(f).

odic lattices and compared the results obtained with the predictions of mathematically precise models, including the Bruggeman asymmetric model and others presented in this section. The articles are a good introduction to the literature on mixed phases as well as covering computational aspects of the subject.[7]

Nonideal Behavior. The above examples illustrate the way in which microstructural features affect the impedance spectra. This should not, however, be taken to mean that nonideal behavior (i.e. departure from a single arc centered on the real axis of Z or M) necessarily implies phase heterogeneity. The assumption that σ and κ are frequency-independent is not justified for real materials, see for example Jonscher [1975c, 1983]. Frequency-dependent conductivity can be caused by factors other than of microstructure, even for ceramics such as zirconia (Abélard and Baumard [1982]). However, the resulting nonidealities are, in general, less prominent than the microstructural effects discussed here. The examples also illustrate the importance of choosing the right representation. In cases where grain boundary effects are dominant, the impedance spectrum is the best starting point, while for mixtures of phases with different conductivity, the modulus is more useful. In the author's opinion, the value of the modulus representation has been often underrated, and investigators are urged to use it more extensively, especially where mixed phases are concerned.

4.1.1.4 Modeling of Composite Electrodes

In the models discussed above, the constituent phases have different values of conductivity, but the conductivities are of the same type, i.e. they are both ionic or both electronic; therefore, no special conditions apply at the boundaries between the phases. Examples of systems where this situation holds are polyphase zirconia ceramics (oxide ion conductors), discussed in Section 4.1.3, and PTCR materials (electronic conductors).

[7] Tuncer *et al.* [2001, 2002] displayed the results in the spectrum of complex dielectric susceptibility, χ, defined by the relation $\chi \equiv \kappa - 1$, rather than the modulus ($M \equiv \kappa^{-1}$), adopted in this section.

There are, however, a number of important systems where this situation does not hold, for example ceramic–metal composites, and ceramic composites of electronic and ionic conductors, used as electrodes in solid oxide fuel cells. Composite electrodes are important in a solid oxide fuel cell (SOFC), as they provide the contact area necessary for the electrode processes to occur.[8] This is usually visualized as the three-phase boundary (TPB), the boundary line where electronic conductor, ionic conductor, and pores meet. A composite cathode is shown schematically in Figure 4.1.14, after Costamagna *et al.* [1998]. The processes occurring in a composite electrode are briefly as follows:

- electron transport from the electronic current collector to the TPB;
- electrochemical reaction at the TPB and gas diffusion in the pores;
- ion transport from TPB to the electrolyte, via the ionic conductor.

These reactions generate electrochemical impedances due to charge transfer, gas or solid state diffusion, etc. Since these impedances appear specifically at the boundaries between dissimilar phases, the composites cannot be fully described by simple effective medium models, even if these impedances are approximated by linear resistive elements. As pointed out by several authors, in the mixture of electronic and ionic phases there are clusters connected to (i) both current collector and electrolyte, (ii) only to the electrode, and (iii) isolated clusters. Clusters of all three types are visible in Figure 4.1.14.

Kenjo *et al.* [1991] have described composite electrodes using the finite length pore model shown in Figure 4.1.15. The electronic and ionic conductor paths run in parallel, with pores providing access to the gas phase via the electrolyte, which is unrealistic, but may not be of any great consequence. Essentially, the composite

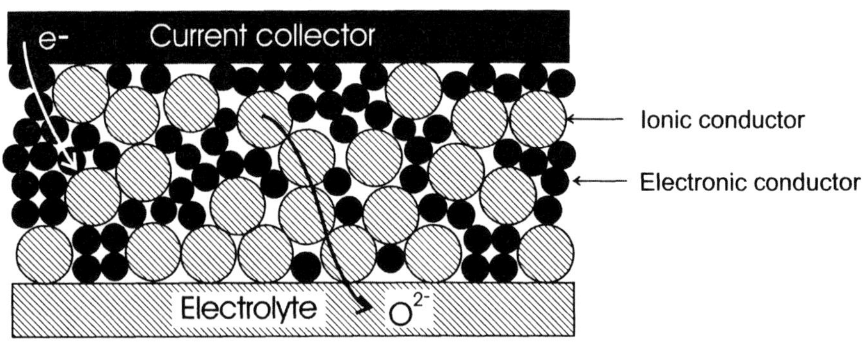

Figure 4.1.14. Schematic diagram of composite cathode, after Costamagna *et al.* [1998].

[8] The electronic conductors commonly used for SOFCs are lanthanum strontium manganite (LSM) for cathodes and nickel metal for anodes. The anodes are prepared *in situ* by reduction of a nickel oxide/electrolyte composite.

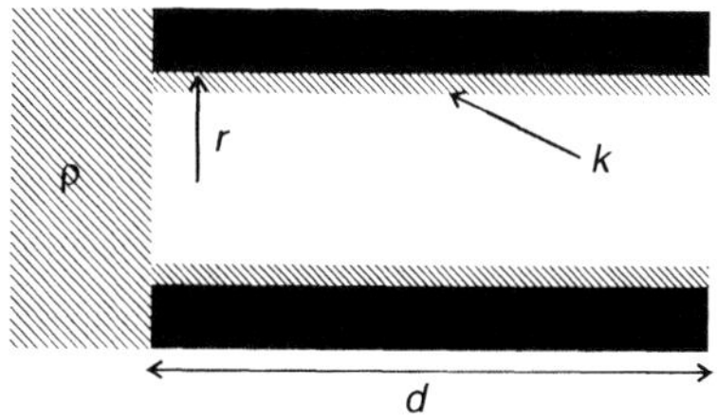

Figure 4.1.15. Visualisation of a composite electrode as a finite length pore, with electronic conductor paths (black) and ionic conductor paths (hatched) running in parallel. After Kenjo *et al.* [1991].

is regarded as consisting of type (i) clusters only. For this model, the polarization resistance is given by Eq. (27)

$$R_p = \sqrt{\rho k r/2}\,\coth\sqrt{2d^2\rho/(kr)} \tag{27}$$

where ρ is the electrolyte resistivity, k is the contact resistance, d is the pore length, and r is the pore radius. The resistivity of the electrode material is taken as zero.[9] R_p increases with the pore radius and decreases with the pore length, to an asymptotic value of $\sqrt{\rho k r/2}$. The pore radius and pore length in Figure 4.1.15 roughly correspond to the electrode thickness and particle size in Figure 4.1.14. It is not clear how the volume fractions of the electrode and electrolyte materials could be built into this model. Kenjo *et al.* [1991] treated only the dc case but, if an interfacial capacitance were to be included, the model would become equivalent to a finite transmission line terminated in a resistance. The impedance spectrum of such a circuit is shown in Figure 2.1.13.

A refinement of the above approach is to treat the composite as a quasihomogeneous system with specific electronic/ionic conductivities and gas diffusion properties, with the electrochemical reaction distributed uniformly over its volume. The problem has been solved in one dimension by Costamagna *et al.* [1998], but their solution applies to the steady state and is not relevant to the present discussion, which concentrates on the frequency domain. Furthermore, intuition would suggest that a one-dimensional model would not completely describe a system as complex as that depicted in Figure 4.1.14.

There seems to be no alternative to the generation of stochastic composite structures and their solution by computer methods. In a series of papers, Sunde has treated the bulk resistivity and polarization resistance for the steady-state (Sunde [1996a, 1996b]) and the electrode impedance in the frequency domain (Sunde [1997, 2000]). Although this work focused on nickel/zirconia anodes, the methodology is equally valid for other composite electrodes. The first step was to use the Monte Carlo

[9] In the above expression, ρ is in Ωcm and k in Ωcm^2. Kenjo *et al.* [1991] have mainly used an alternative definition for k, which contains the factor $r/2$ and, therefore, has units of Ωcm^3.

method to generate pseudorandom packings of spherical electrode and electrolyte particles and identify neighbors. Contacts between dissimilar phases were assigned electrochemical impedances describing charge transfer, double layer capacitance, reactant adsorption, etc., while contacts between similar phases were taken to be purely ohmic.[10]

The simplest case is illustrated here, where the contact impedances contained terms for charge transfer and double layer capacitance. Figure 4.1.16 shows impedance spectra for such an electrode for two volume fractions, ϕ, of the electrode phase. In the first case ($\phi = 0$), which describes a single phase electrode, the spectrum corresponds to a summation of many RC elements in parallel. In the second case ($\phi = 0.3$), which describes a composite electrode, a new arc appears at the high frequency end of the spectrum. For the simulation in hand, the high frequency arc can be fitted to a ZARC ($n \approx 0.9$), while the low frequency arc corresponds closely to a parallel RC element ($n = 1$). Note that in the case of $\phi = 0.3$, the polarization resistance ($Z_{f\to 0} - Z_{f\to\infty}$) decreases by a factor of about four, which elegantly illustrates the technological advantage of composite electrodes in solid oxide fuel cells.

The presence of a new feature may be qualitatively explained in terms of the three types of clusters mentioned above. At low frequency only clusters of type (i) and (ii) are active, while, at high frequency, the double layer capacitance reduces the magnitude of the contact impedance and type (iii) clusters progressively "switch in", producing a second arc. The degree of overlap of the arcs depends on the volume fraction of the electrode phase, an effect for which Sunde provides a tentative explanation: at low volume fractions, where the overall impedance is dominated by the electrolyte, type (iii) clusters have a significant effect and generate a well-

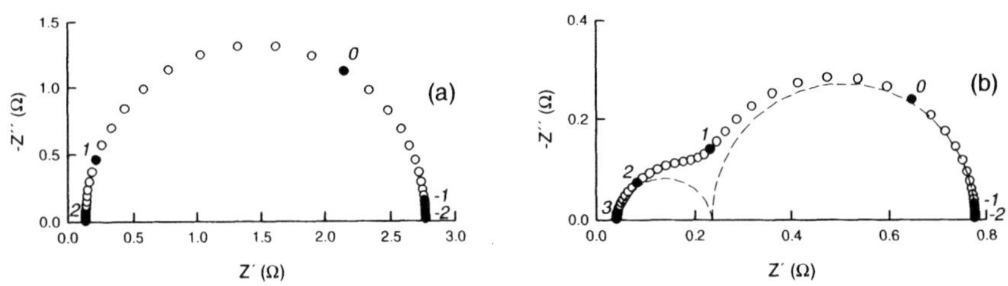

Figure 4.1.16. Impedance spectra of an electrode with two volume fractions, ϕ, of the electrode phase, for the case where the contact impedances contain only charge transfer resistance and double layer capacitance: (*a*) $\phi = 0$ (single phase electrode) and (*b*) $\phi = 0.3$ (composite electrode). For details of the simulation see Sunde [2000]. The labels show the logarithm of the normalized frequency.

[10] It is important to distinguish between the bulk resistivity of the composite, which arises when it is contacted via to electronic conductors, and the electrode impedance, which arises when it is placed between an ionic and an electronic conductor, as in an SOFC.

resolved arc, while at higher volume fractions, where the electrode phase is near or above percolation, the effect amounts only to a distortion of the original electrode arc.

A valuable property of such modeling experiments is that all currents and potentials are stored and available for further analysis. A possible way of visualizing the function of the composite electrode would be to sum the partial electronic and ionic currents at various depths within the composite electrode and plot them as a function of position. The partial currents could be used to identify the region of maximum charge transfer in the composite.

4.1.2 Experimental Techniques

4.1.2.1 Introduction

A wide range of materials can be usefully characterized by IS, namely electrical and structural ceramics, magnetic ferrites, semiconductors, membranes, polymeric materials, and protective paint films. The measurement techniques used to characterize materials are generally simpler than those used for electrode processes. Impedance spectra are usually independent of applied potential (both ac amplitude and dc bias) up to potentials of 1 V or more. Consequently, it is unnecessary to fix the potential of electrodes, as is the case with potentiostatic experiments, and two-electrode symmetrical cells are commonly used.

Considerations of Frequency and Impedance Range. As discussed in Section 4.1.3, the frequency range to be chosen depends on the relaxation frequencies f_n of phases present in the sample under study and also on the microstructure. The highest relaxation frequency is normally that of the grain interior relaxation and is given by $f_{gi} = \sigma_{gi}/(2\pi\varepsilon_{gi})$, corresponding to the apex of the grain interior impedance arc. Conductivities vary enormously from one material to another, and with temperature for any one material. By contrast, the dielectric constants of most compounds, apart from ferroelectrics, lie within the range of 5 to 100. Taking a dielectric constant 30 as typical (this is close to the value for cubic zirconia) the relationship between conductivity and relaxation frequency is illustrated in Figure 4.1.17, with the shaded regions showing commonly accessible ranges of conductivity and frequency.

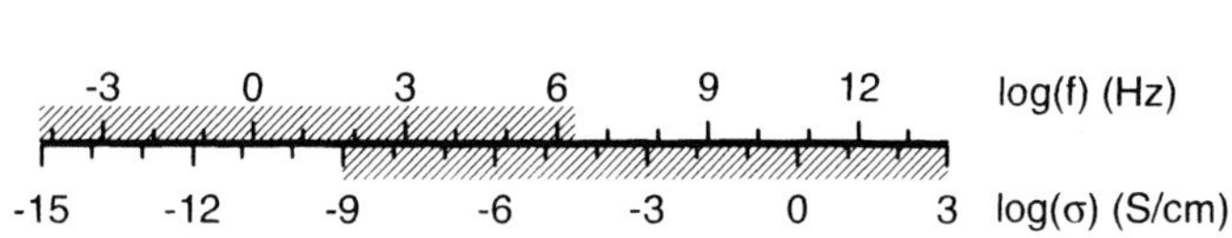

Figure 4.1.17. Relationship between relaxation frequency and conductivity and for a solid having a dielectric constant of 30.

4.1.2.2 Measurement Systems

Most of the measurement equipment required for impedance studies can be bought off the shelf. The equipment falls into three categories:

(i) frequency response analyzers (FRAs);

(ii) component analyzers, impedance analyzers, and ac bridges;

(iii) potentiostats, galvanostats, and general electrochemical interfaces.

Frequency Response Analyzers. These instruments are designed measure a voltage ratio, rather than impedance, but can be adapted to measure impedance by the addition of auxiliary components—in the simplest case, a standard resistor. A detailed account of the circuitry, which was included in the first edition of this volume, would be out of place today; here it will be dealt with in brief, to illustrate the principles involved.

The FRA generates a sinusoidal voltage of variable frequency and amplitude and measures the in-phase and out-of-phase components of the voltage at two inputs, "X" and "Y." The voltages are expressed as complex quantities $V_X = V'_X + jV''_X$ and $V_Y = V'_Y + jV''_Y$. The simplest circuit for measurement of impedance uses the standard resistor in a voltage divider configuration, shown in Figure 4.1.18. In this case, input "X" is used to measure the generator output and input "Y" to measure the midpoint of the divider. The unknown impedance is then given by:

$$Z_u = \frac{V_y}{V_x - V_y} Z_s \tag{28}$$

The main disadvantage of this configuration is that input "Y" is connected in parallel to the sample. The inputs of the FRA have a finite impedance, Z_{in}, typically 1 MΩ in parallel with 30 to 50 pF and, unless $Z_s << Z_{in}$, this leads to an error in Z_u. The problem can be reduced by placing fast unity-gain amplifiers (buffers) at the FRA inputs. The situation can also be helped by interchanging the standard and the unknown, so that the input impedance of "Y" appears across the standard. If the standard resistance is made small, the effect of Z_{in} is minimized. This, however, adds complexity and results in a low voltage signal at "Y".

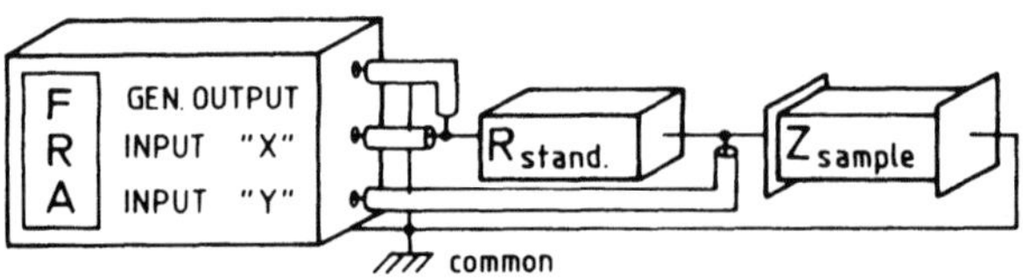

Figure 4.1.18. Impedance adaptor for frequency response analyzer (FRA) based on voltage divider network.

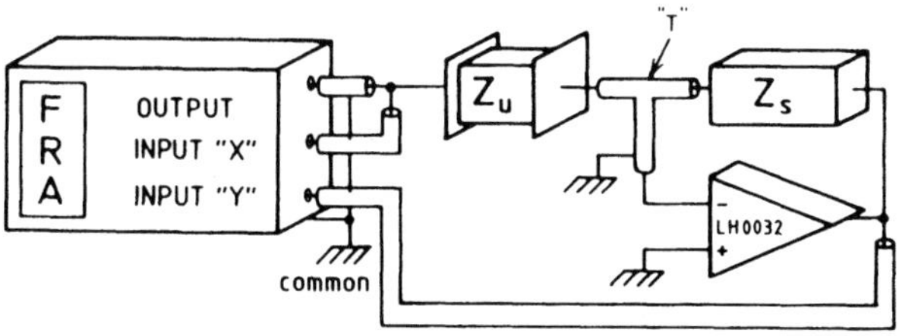

Figure 4.1.19. Impedance adaptor for frequency response analyzer (FRA) incorporating a current-to-voltage converter.

An alternative approach that eliminates the above problems uses the active current-to-voltage converter shown in Figure 4.1.19, with the standard resistor placed in the feedback loop of a fast operational amplifier. The amplifier delivers an output, such that the junction "T" in the diagram is maintained at ground potential (*virtual ground*). Consequently, any stray conductance/capacitance from "T" to ground is excluded from the measurement. Since the effect of capacitances has been eliminated, the leads to the sample can be screened, reducing the effect of electrical noise. In this configuration, the unknown impedance is given by:

$$Z_u = -\frac{V_x}{V_y} Z_s \tag{29}$$

The complex quotient V_x/V_y is computed by the impedance analyzer. Normally, Z_s is a pure resistor, in which case, Z_u has a simple relationship to this quotient. In principle, any circuit of well-defined impedance may be used (e.g. parallel RC combination), bearing in mind, of course, that the multiplication of V_x/V_y and Z_s must be performed in the complex domain. It is good practice to select a standard resistor of magnitude comparable to the real part of the unknown. Obviously, a trial measurement is needed in order to select the most suitable standard resistor.

The frequency range of an active current-to-voltage converter is limited mainly by the gain and phase performance of the operational amplifier, which falls off at high frequencies. An analysis of this effect (T. Goldrick [1982], unpublished) shows that, for certain simplifying assumptions, the apparent impedance Z_m is related to the standard and unknown as follows:

$$Z_m = Z_u + j\frac{f}{f_o}(Z_s + Z_u) \tag{30}$$

where f is the measurement frequency and f_o the frequency at which the open-loop gain of the amplifier falls to unity. This formula can be re-written as:

$$Z_m = Z_u + j2\pi f L_s \tag{31}$$

where

$$L_s = \frac{Z_s + Z_u}{2\pi f_0} \tag{32}$$

For the case where the standard and unknown resistances are both real, the introduced error is equivalent to that of an inductor in series with the unknown. The pseudoinductance will, of course, appear in addition to any physical inductance introduced by the leads.

Impedance Analyzers. Component analyzers, impedance analyzers, and ac bridges have built-in current measuring circuits, which are designed to minimize the above problems. These instruments are usually configured for four-wire connection, which minimizes stray series resistance and inductance, the dominant source of error for low impedance samples.

Impedance analyzers do not have the capability of controlling the potential with respect to a reference electrode—this requires the use of electrochemical interfaces, as described in Section 3.2. They can, nevertheless, be used to measure electrochemical under current load. In the configuration of Figure 4.1.20, the ac current generated by an impedance analyzer in galvanostatic mode is added to the (higher) dc current generated by an external galvanostat (S. Højgaard Jensen and J. Poulsen, unpublished work [2003]). This configuration takes advantage of the superior high frequency performance of the impedance analyzer, as compared to an electrochemical interface, which is essential for separating the series electrolyte resistance and electrode polarization effects.

While commercial measurement systems generally perform better than the improvised impedance adaptors described in Figures 4.1.18 and 4.1.19, they are not

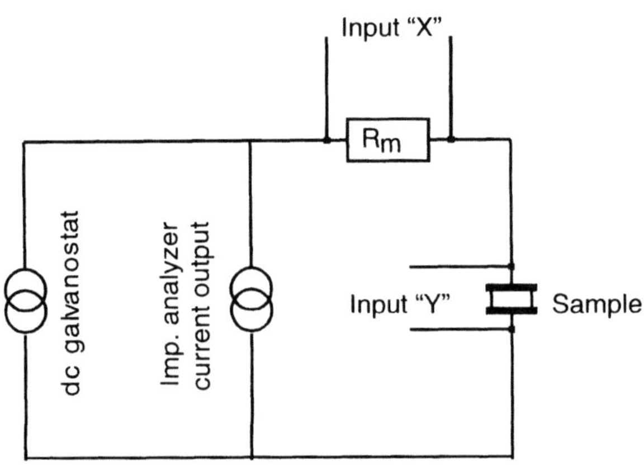

Figure 4.1.20. Measurement on an electrochemical cell under load. The dc current generated by the external galvanostat is added to the ac current generated by the impedance analyzer. The measurement resistor R_m, is a low inductance, metal strip resistor, rated to the necessary dc current. (S. Højgaard Jensen and J. Poulsen, unpublished work [2003]).

immune to errors, especially at the extremes of their specified range. The burden of obtaining error-free data cannot be carried wholly by the manufacturers, who may have designed the instrument for a different application than the one for which it is used.

Measurement Rigs. Important considerations in the design of a rig for IS of materials are:

(i) provision of a controlled temperature and chemical environment for the sample;

(ii) provision of the best possible electrical path to the sample.

The importance of temperature control is illustrated by considering a sample with an activation energy of 1 eV for the conductivity: at 800°C a change of 1°C would produce a change of 1% in conductivity. At 300°C, the same error would produce a change of 3.5%. A temperature stability of 1°C or better can be easily achieved with present-day three-term controllers, but this alone is not sufficient to ensure good temperature control: temperature gradients in the rig must be kept to a minimum and the temperature sensor must be in thermal contact with the sample.

The inner assembly of a general-purpose conductivity rig is shown in Figure 4.1.21. It is based on a ceramic "shoe" that slides over a ceramic supporting tube. A spring-loaded harness of Nichrome or Kanthal wire provides a load on the contacts. Contact is made with the sample via a platinum foil on either side. For situations where the electrode response is to be measured, the sample can be contacted

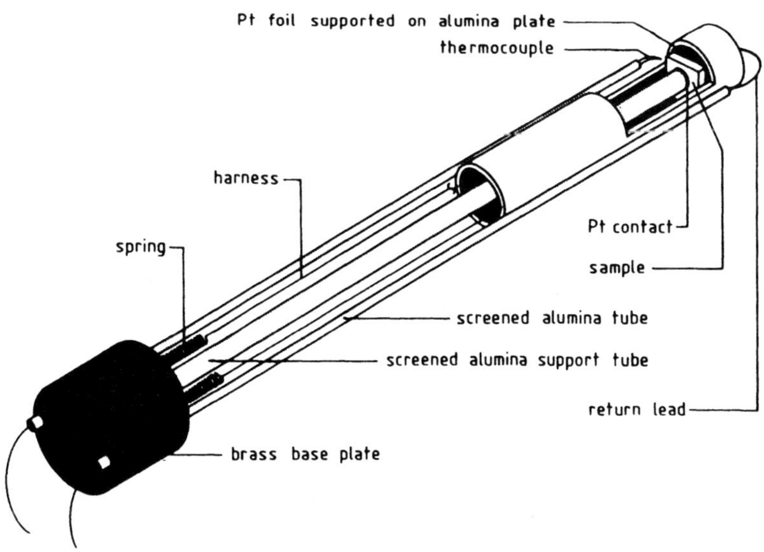

Figure 4.1.21. Rig used for the characterization of ceramic samples.

by means of "castellated" alumina pellets (Figure 4.1.22) wrapped in platinum mesh. This allows good contact while providing the gas with access to the electrodes. A four-wire configuration is recommended for this cell, as it reduces the magnitude of the series stray resistance and stray inductance.

Measurement Errors and their Correction. One of the major sources of error is uncompensated or stray components. The components can be effectively connected in parallel or in series with the sample. The most important parallel component is stray capacitance, C_{sp}. In the first place, this can be minimized by screening leads to within a short distance from the sample and grounding metal parts of the measurement cell. C_{sp} cannot be estimated by a measurement on the empty cell; without a sample, the capacitance of the airgap remains. Instead, the unknown is replaced with a sample of known dielectric constant and shape and the calculated capacitance is subtracted from the measured one to give C_{sp}. Correction for the stray capacitance can then made by subtracting $2\pi f C_{sp}$ from the imaginary part of the measured admittance. In a well-designed cell the stray parallel conductance G_{sp} should be negligible. In cases where G_{sp}, is significant, it is not straightforward to correct for it after the measurement, as G_{sp} varies with the measurement conditions more strongly than C_{sp} does.

The series stray components are the inductance L_{ss} and resistance R_{ss}. Both of these can be kept low by using a four-wire connection. While the resistance can usually be eliminated, the inductance cannot; it depends on the length of leads and their positioning. Figure 4.1.23 shows two wiring configurations that could be used

Figure 4.1.22. Castellated pellet used to support platinum mesh in impedance measurement rig.

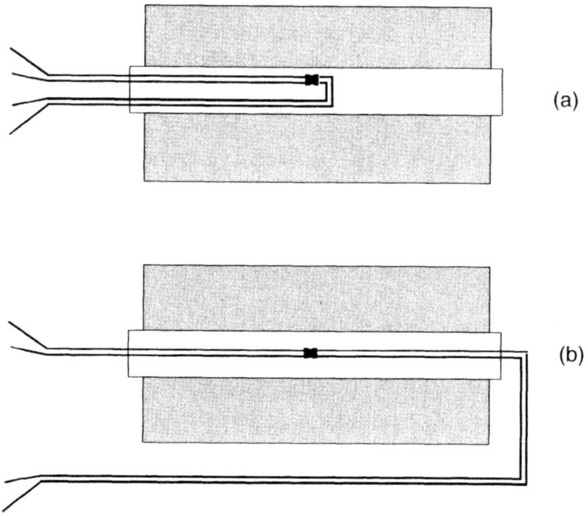

Figure 4.1.23. Two possible wiring configurations for a tube furnace. Configuration (*b*) produces an inductance many times greater than that of (*a*) and should be avoided.

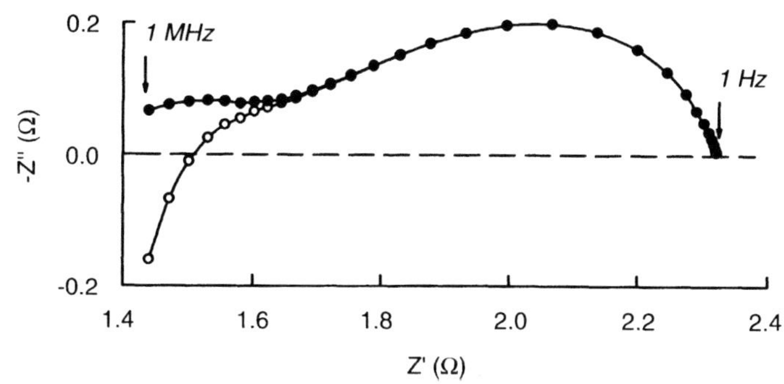

Figure 4.1.24. Effect of inductance error on the impedance spectrum of a symmetrical cell. Open symbols: raw data; closed symbols: data corrected for stray inductance of 36×10^{-9} H. The sample is a YSZ tape of area 0.21 cm^2 with an LSM composite cathode on both sides, measured in air, 850°C.

with a tube furnace. Configuration (b) produces an inductance that is 10 to 20 times greater than with (a), and should be avoided at all cost (Primdahl and Hendriksen [1996]). The components R_{ss} and L_{ss} are measured with the cell shorted at the sample position. Correction for the series stray components is made by subtracting R_{ss} from the real part of the impedance and $2\pi f\, L_{ss}$ from the imaginary part. The effect of correction for inductance is shown in Figure 4.1.24 for a symmetrical cathode sample described in Section 4.1.3.

4.1.2.3 Sample Preparation—Electrodes

The impedance of a sample is the product of the complex resistivity and the length over area ratio, l/A; therefore samples should be prepared with parallel faces and a well-defined cross section. For measurement of grain interior properties, surface finish is not critical, as this only affects the electrode impedance (Armstrong *et al.* [1973]). If the grain boundary and electrode arcs overlap, their resolution can be improved by polishing before applying the electrodes. Electrodes are applied by painting, vacuum evaporation, or sputtering. The types of electrodes commonly used are described below.

Precious metal paints, also known as pastes or dags, are good general-purpose electrodes for IS studies. The most commonly used precious metal paints are platinum, palladium, gold, and silver, all of which are commercially available. Nickel and other transition metals are not generally available as paints. It the case of platinum, certain commercial products contain inorganic fluxes or glass to assist sintering and these are best avoided, as they give rise to high electrode impedances. Flux-free platinum paints are preferable, as long as the samples can tolerate the curing temperatures of over 1000°C that these products require. If high temperature, oxidizing, conditions are not part of the study, carbon paints can be used.

Silver and gold, which have relatively high vapor pressures, can be easily applied by vacuum evaporation, while metals of lower vapor pressure, such as platinum, palladium, and stainless steel, can be deposited by radio frequency sputtering; in this case, however, the film thicknesses are limited to about 100 nm. Before electrode deposition, the samples must be cleaned and heat treated. Failure to remove surface impurities will result in loss of adhesion when samples are subsequently heated during the measurement.

The choice of electrodes also depends on the temperature of the investigation and the type of conductivity that is being investigated. Platinum and gold electrodes perform well at high temperatures, but at temperatures below 600°C, they are relatively blocking to oxygen. If the measurement temperature is lower than this, silver electrodes are preferable, as they generally have lower electrode impedances. Figure 4.1.25 shows impedance spectra for yttria-stabilized zirconia (an oxide ion conductor) with platinum and silver electrodes, at 300°C. The electrode impedances are considerably lower for the silver electrodes.

The preparation of electrodes is only the first step towards a successful experiment. The measurement system must also be kept free of elements such as silicon or phosphorus, which, under high temperature reducing conditions, form volatile compounds that attack the electrodes, for example platinum, to produce brittle and poorly-conducting compounds.

4.1.2.4 Problems Associated With the Measurement of Electrode Properties

In the work discussed so far, the main objective was to characterize the bulk properties; the electrodes were there only to make this measurement possible. In other

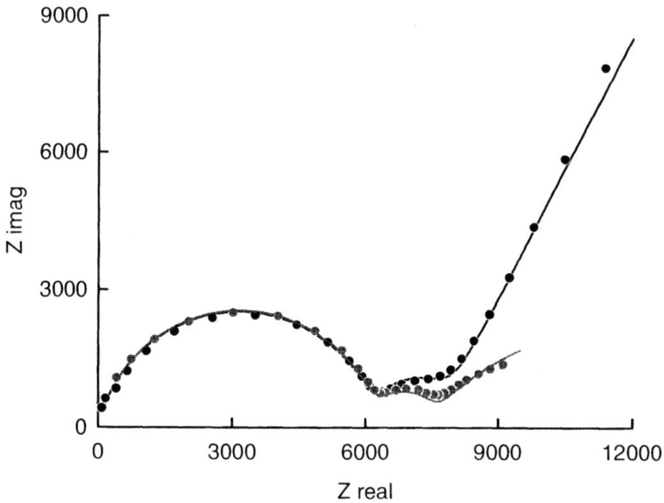

Figure 4.1.25. Impedance spectra for yttria-stabilized zirconia oxide ion conductor with platinum (black) and silver (grey) electrodes, at 300°C. The electrode impedances are lower for the silver.

situations, the bulk properties are already known and the objective is to characterize the electrodes: this applies, for example, in the development of electrodes for SOFCs. In SOFCs, the current-potential characteristics of the electrodes are often found to be linear, and the steady state properties of the electrodes can be adequately represented by a linear polarization resistance, R_p. The characterization of electrodes in these systems can be performed in one of two ways:

1. On symmetrical cells equipped with two nominally identical electrodes, using a small ac signal, at zero dc polarization. In this case, the electrode polarization obtained from the impedance plot must be divided by two, since it is caused by the two electrodes in series. The measurement is simple and can be performed with an impedance analyzer.
2. On three-electrode cells, i.e. cells equipped with working electrode, counterelectrode, and reference electrode (WE, CE, and RE) under potentiostatic or galvanostatic control. In this case, the small ac signal is superimposed on a dc polarization. The measurement requires an electrochemical interface (potentiostat or galvanostat) coupled to an FRA (see Section 3.2).

While the latter case offers greater flexibility, it is also susceptible to rather subtle errors, which arise from the positioning of the RE in relation to the WE. These errors need to be considered if reliable electrode performance is to be obtained on solid electrolyte cells. With liquid electrolytes, the RE can be made as small as necessary and placed as close to the WE as required. With solids, the electrode must be located on the electrolyte surface, or, if the electrolyte thickness permits, in a suitably fashioned recess (see Figure 4.1.26*b*). Here, the discussion is confined to the planar geometry shown in Figure 4.1.26*a*.

The problem arises because of non-uniform current distribution at the WE, or more precisely, different nonuniform current distributions in the case of $f \to \infty$ and $f \to 0$. If the cell could be constructed so that the current density was uniform over the WE, the measurement would yield the correct value of R_p. However, any misalignment of the WE and CE produces a distortion in the equipotential lines, of the type shown in Figure 4.1.27. Accordingly, a RE placed at the side of the WE will not be intersected by the equipotential passing through the midpoint of the electrolyte, but will read a potential either too high, or too low. At $f \to 0$ the electrode impedance is equal to R_p, while, at $f \to \infty$, the electrode impedance is negligible, due to the displacement current flowing through the double layer capacitance. The different current distributions obtained in the two cases will cause a shift of the equipotential lines. This will interfere with the correct estimation of R_p when this is obtained using the expression $R_p = |Z_{f\to 0}| - |Z_{f\to\infty}|$ (See also Section 4.3.3).

The error has been analysed Winkler *et al.* [1998] using finite-element modelling in the real domain (see also Primdahl and Hendriksen [1996]). The total current and RE potential were calculated for the low and high frequency limits and used to estimate the apparent value of R_p. The resulting error is plotted in Figure 4.1.28 as a function of the displacement between WE and CE, based on realistic values of electrolyte conductivity and polarisation resistance of the electrodes, and a value of 9×10^{-3} for the ratio of electrolyte thickness/electrode width. The figure shows that a displacement of one electrolyte thickness can result in an error of over 50%, for input parameters typical of solid oxide fuel cells at operating temperature. Since electrolytes have thicknesses of 150 μm or less, it is clear that the construction of a fully reliable three electrode cell of planar geometry is a considerable technical challenge. It was further shown that, even with perfectly aligned electrodes, errors still arise, if the polarisation resistances of the WE and CE do not match. Since the above simulation covered only the two extrema of frequency, the shape of the electrode arc cannot be reconstructed from this analysis. Distortions in the three-electrode spectra due to a size mismatch between WE and CE were described by Reinhardt and Göpel [1998]. Errors in three electrode and four electrode cells are covered in a series of papers by Hsieh *et al.* [1996a, 1996b, 1997].

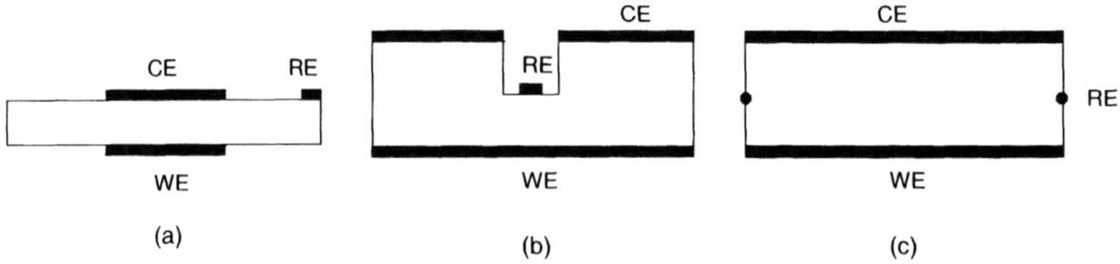

Figure 4.1.26. Commonly used geometries of three-electrode solid electrolyte cells.

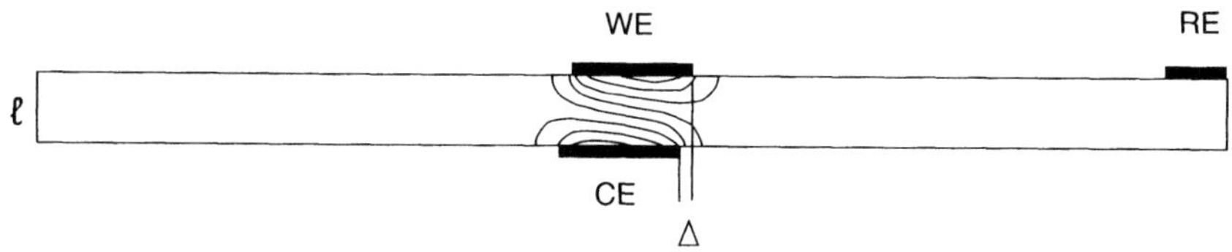

Figure 4.1.27. Distortion in the equipotential lines due to misalignment of the CE and WE in a thin electrolyte cell.

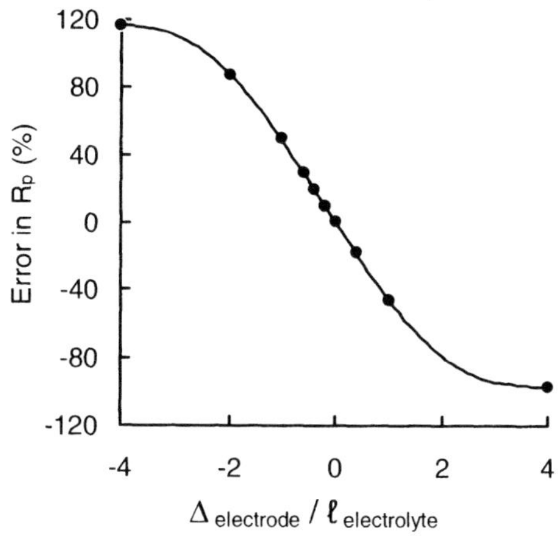

Figure 4.1.28. Error in the apparent polarization resistance due to the relative displacement of WE and CE in a planar cell. The error is calculated for an electrolyte resistivity of $10\,\Omega\text{cm}$ and a polarization resistance $0.16\,\Omega\text{cm}^2$ for both the WE and CE. A positive displacement is defined as a displacement of the WE in the direction of the RE, as in Figure 4.1.27 (for further details see Winkler *et al.* 1998).

An authoritative analysis of these errors has been provided by Adler [2002], who has described the thin electrolyte case using finite element analysis in the complex domain. This revealed distortions of the electrode arc, both as a result of electrode displacement, and of mismatch in the polarisation resistances between the WE and CE. The magnitude and sign of the predicted errors in R_p are in agreement with those reported by Winkler *et al.* [1998]. Figure 4.1.29 shows Adler's simulations, for the case when theoretical polarisation resistances of the WE and CE are in the ratio 1/3. The figure shows the theoretical impedance of the WE, and the calculated impedance spectra, for WE displacements towards the RE, and away from the RE. As expected, the displacements affect the magnitudes of both R_s and R_p. Furthermore, Adler showed that the distortions due to mismatch in R_p arise *in the case of perfectly aligned electrodes*, an effect that could be loosely described as "interference" of the WE from the CE.

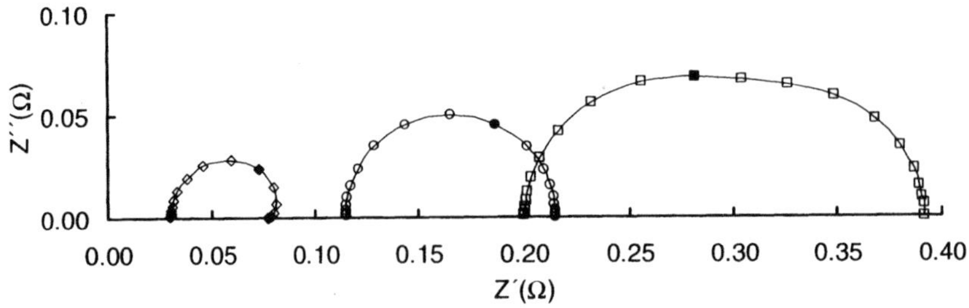

Figure 4.1.29. Distortion of the electrode arc in a thin electrolyte cell due to displacement of the WE. (○) No displacement (□) Displacement *from* the RE by one electrolyte thickness, (◊) Displacement *towards* the RE by one electrolyte ethickness. For details see Adler *et al.* [2002]. Solid markers show the data for a frequency 1 Hz. (Courtesy of the Electrochemical Society).

From the above, and from our practical experience with thin electrolyte cells, we conclude that the problems associated with three-electrode measurements are so complex and severe that the technique cannot be used routinely for the evaluation of electrode properties, and recommend the two remaining alternatives:

i) to prepare the electrodes on a thick electrolyte, for example using the geometry of Figure 4.1.26*c*.
ii) to use symmetrical cells and perform impedance measurements at open circuit conditions.

4.1.3 Interpretation of the Impedance Spectra of Ionic Conductors and Interfaces

4.1.3.1 Introduction

This section covers the interpretation of experimental impedance spectra of two-phase ceramics, solid electrolytes, and their electrode interfaces. Examples are given from the authors' work and from the literature, with preference to cases where the original data were available. Many of the examples refer to zirconia ceramics, since these materials give well-resolved spectra and exhibit interesting polymorphic phenomena, as well as the ionic conductivity that underlies their use in solid oxide fuel cells and other electrochemical devices. In many of these examples, spectra are presented for temperatures well below normal operating temperatures of the materials, in order to bring the spectral features into the instrumental frequency range (see Figure 4.1.17).

Microstructural Aspects of Zirconia Ceramics. ZrO_2 is polymorphic and can have one of three crystal structures: monoclinic (*m*) at ambient temperatures, tetragonal (*t*) at 1170 to 2370°C, and cubic (*c*) between 2370°C and its melting point of 2700°C. On cooling from the temperatures of 1300 to 1800°C, normally used for sintering ceramics, ZrO_2 adopts the monoclinic structure. The transformation from *t* to *m* is accompanied by a significant volume change; as a result, pure ZrO_2 is unsuitable for the preparation of ceramics. The effect can be suppressed by doping with oxides such as CaO, MgO, and Y_2O_3 (Subbarao [1981], Subbarao and Maiti [1984]), with which it forms extensive solid solutions. At dopant concentrations of ~10 mol % the solid solutions have a cubic structure and are called fully stabilized zirconia (FSZ). At 2–6 mole % Y_2O_3 their structure remains tetragonal, provided the grain size is around 500 nm or less (Gupta *et al.* [1977]) and the materials are sometimes referred to as tetragonal zirconia polycrystals (TZP). These systems can be sintered at lower temperatures than cubic zirconia and have a very high mechanical strength and toughness. Intermediate level of stabilizer result in a *c* matrix, with a fine dispersion of *t* and/or *m* precipitates (Kobayashi *et al.* [1981]); which is termed partially stabilized zirconia (PSZ). The *t* precipitates have the property of transforming to *m* under the influence of a stress field and in doing so retard the propagation of cracks; the effect is known as transformation toughening (Garvie *et al.* [1975]).

When elements of valence 2 or 3 are introduced into zirconia, they occupy Zr^{4+} lattice sites, generating vacancies at oxygen sites. These provide zirconia with its well-known ionic conductivity, leading to its use in high temperature electrochemical cells (Subbarao and Maiti [1984], Steele [1976], Steele *et al.* [1981]). FSZ is mainly used for electrochemical application, PSZ mainly for structural applications, while TZP is used both as a solid electrolyte and a structural ceramic.

Bauerle's Circuit Equivalent. The use of IS for the characterization of ceramic solid electrolytes became routine after Bauerle [1969] showed that, for stabilized zirconia with platinum electrodes, the polarizations of the grain interiors, grain boundaries, and electrode–electrolyte interface could be resolved in the admittance plane. He presented a circuit equivalent for this arrangement which has since been applied to many other systems (Figure 4.1.30) with RC elements corresponding to grain interiors (gi), grain boundaries (gb), and electrode (e) connected in series.[11] The estimation of the circuit parameters was not straightforward due to Bauerle's choice of the admittance plane. Subsequent workers have generally used the impedance plane, where a direct relationship exists between the spectrum and the circuit (Armstrong *et al.* [1974], Schouler *et al.* [1981]). Examples of admittance and impedance spectra for a sample of ZrO_2: 6 mole % Y_2O_3 at 240°C are given in Figure 4.1.31*a,b*. A simulated impedance plot is shown in Figure 4.1.31*c* using the parameters given in Table 4.1.5. The agreement between experiment and simulation is satisfactory for the grain interior and grain boundary arcs both in shape and distribution

[11] The circuit shown in Figure 4.1.30 contains an extra capacitor for the grain interior element, not included by Bauerle, because the grain interior relaxation lay above the frequency range covered.

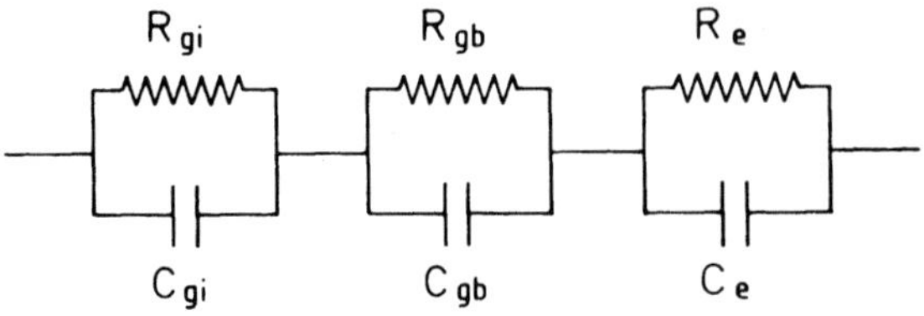

Figure 4.1.30. Circuit equivalent for a ceramic electrolyte according to Bauerle [1969] and modeling the impedance of the grain interiors (gi), grain boundaries (gb), and electrode (e).

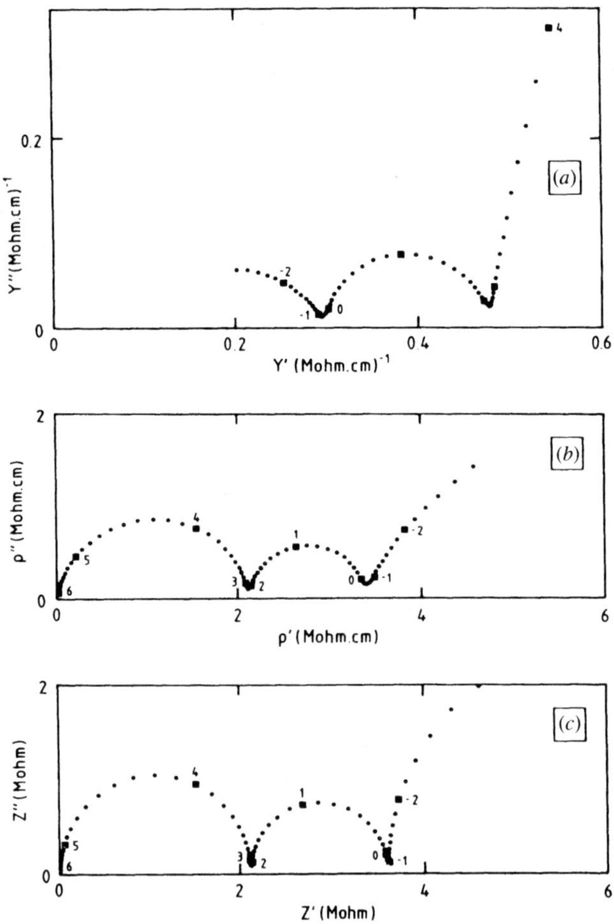

Figure 4.1.31. Comparison of admittance and impedance spectra for a zirconia solid electrolyte (ZrO_2: 6 mole % Y_2O_3) at 240°C: (*a*) Experimental admittance spectrum. (*b*) Experimental impedance spectrum. (*c*) Simulated impedance spectrum, using the circuit of Figure 4.1.30 and parameter values given in Table 4.1.5.

Table 4.1.5. Circuit Parameters Used to Generate the Impedance Spectrum of Figure 4.1.31*c*

$C_{gi} = 4.8\,\text{pF}$	$C_{gb} = 1.7\,\text{nF}$	$C_e = 2.0\,\mu\text{F}$
$R_{gi} = 2.1\,\text{M}\Omega$	$R_{gb} = 1.5\,\text{M}\Omega$	$R_e = 5.0\,\text{M}\Omega$

of frequencies, indicating that the circuit is adequate for representing the salient features of the frequency response.

4.1.3.2 Characterization of Grain Boundaries by IS

Understanding the origin of polarizations in ceramics is important, not only in order to optimize the electrical conductivity, but also to achieve a better understanding of the microstructure, that in turn influences other properties such as the mechanical strength. The fabrication of ceramics may involve adding small amounts of impurities that form liquids at the sintering temperature. In zirconia ceramics, trace amounts of silica and alumina fulfill this function, even if not added deliberately. On cooling, aluminosilicate phases, both amorphous and crystalline, form at the grain boundaries with adverse effects on the ionic conductivity, as shown by Bauerle [1969], Schouler *et al.* [1973], Schouler [1979], Kleitz *et al.* [1981], and Bernard [1981]. Noting that small additions of Al_2O_3 could reduce the grain boundary resistance, Drennan and Butler [1973] suggested that the Al_2O_3 particles scavenge SiO_2 to form discrete particles of $Al_6Si_2O_{13}$ (mullite). Subsequent grain boundary migration during grain growth leaves these particles in the interior of the grains, where their influence on the ionic conductivity is smaller. This effect is in accordance with Eqs (21) and (22), describing the effect of high-resistivity inclusions on the resistivity of ceramics.

Microstructure–Grain Boundary Property Correlations. The way in which the grain boundary structure affects the impedance spectrum is illustrated by comparing 3 mole % Y_2O_3 tetragonal and 6 mole % mainly cubic ceramics. Figure 4.1.32*a* shows a transmission electron micrograph of the 3 mole % specimen. The grain boundary phases were continuous with a thickness of 10 nm or less; from images obtained at lower magnifications, the grain size was estimated as 0.5 to 1 μm. Selected area diffraction patterns (SADP) showed the grain boundary films to be amorphous and energy-dispersive x-ray microanalysis detected SiO_2, Al_2O_3, and Y_2O_3 at levels of 65, 20, and 70 wt %, respectively (Butler and Bonanos [1985]). Figure 4.1.32*b* shows a transmission electron micrograph of the 6 mole % specimen with a grain size of ~10 μm. Pockets of second phase are seen along the grain boundaries—their lenticular shape indicates that they originate from nonwetting liquid phases present at high temperatures. Selected area electron diffraction analysis shows both crystalline and amorphous grain boundary phases; both are discrete, allowing partial intergranular contact.

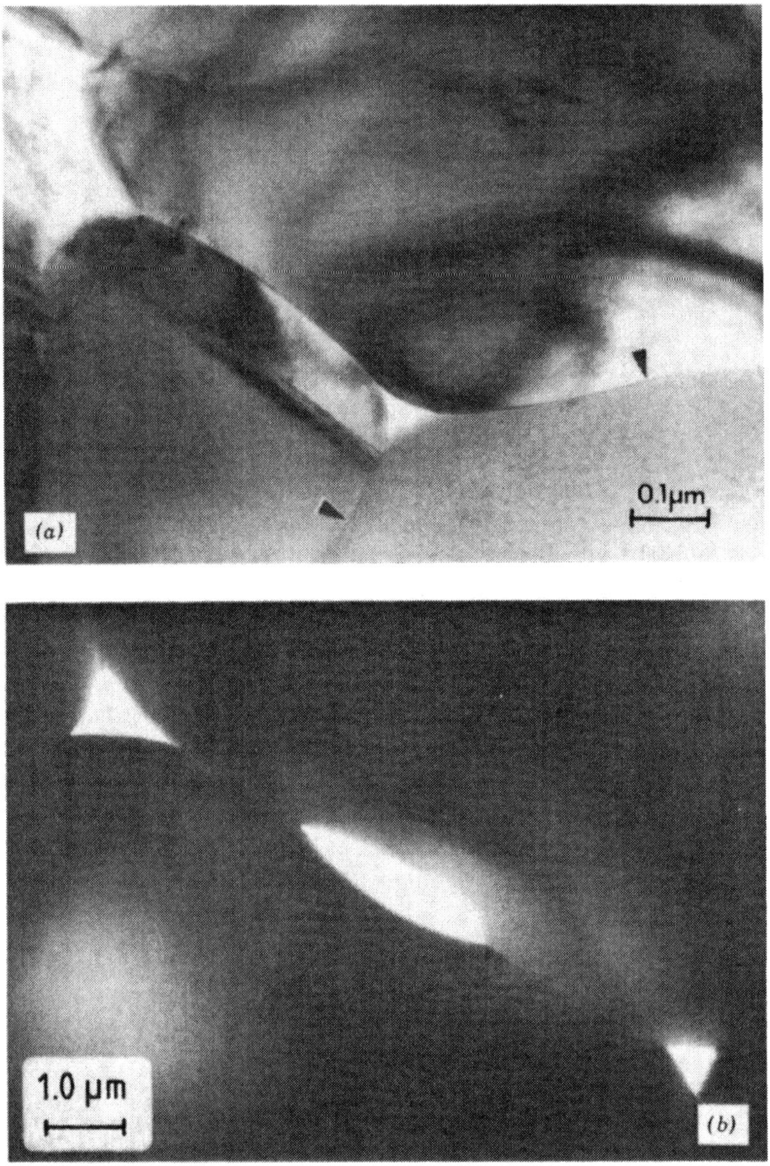

Figure 4.1.32. Transmission electron micrographs of two zirconia ceramics: (*a*) Tetragonal zirconia ceramic (ZrO_2: 3 mole % Y_2O_3) showing thin, continuous grain boundary phase (arrowed). (*b*) Partially stabilized ceramic (ZrO_2: 6 mole % Y_2O_3) showing discrete, lenticular grain boundary phase.

While high-resolution transmission electron microscopy (TEM) is a powerful technique, it samples only a tiny fraction of the grain boundary area, potentially allowing some grain boundary defects to be missed. A definitive statement as to the absence of grain boundary phases cannot be made—the phases could be too thin or too sparse to detect. For electrically conducting materials, IS averages over the entire sample and thus provides a valuable complement to TEM.

Impedance spectra of the two zirconia ceramics are shown in Figure 4.1.33 for 300°C. Both systems exhibit a Bauerle-type behavior. The spectrum of the 3 mole % material (Figure 4.1.33*a*) is dominated by a grain boundary arc, due to the presence of a continuous grain boundary phase. In the 6 mole % material (Figure 4.1.33*b*), the grain boundary arc is relatively small. It seems that this material has been optimized for electrical conductivity and that the grain boundary resistance has been reduced by making the grain boundary phases discrete.

Further indirect information about the topology of the grain boundary phases can be obtained from the temperature dependence of the quantities r_{gi} and r_{gb}. In crystalline ionic conductors in the extrinsic region, the conductivity is thermally activated and described by:

$$\sigma = (\sigma_0/T)\exp(-\Delta H_m/kT) \tag{33}$$

where ΔH_m is the activation enthalpy for migration. The Arrhenius plot linearizes this equation and ΔH_m is obtained from its slope. Arrhenius plots of r_{gi} and r_{gb} for the TZP and FSZ ceramics (Figure 4.1.34) are linear over the temperature range

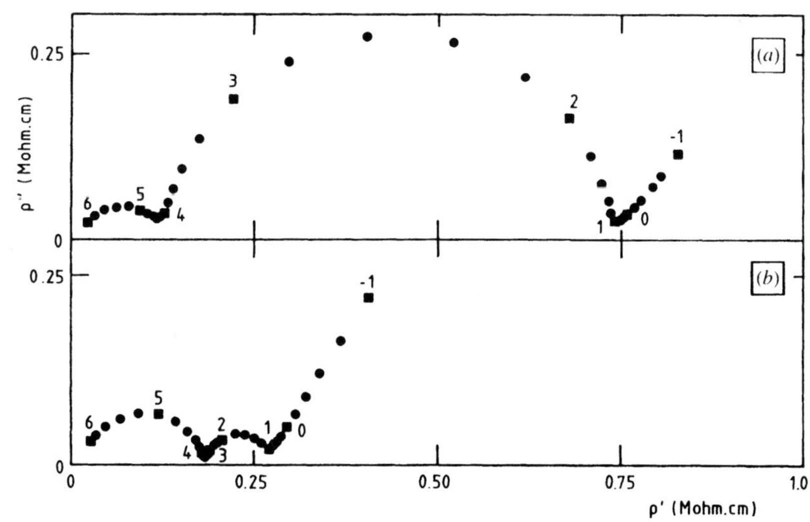

Figure 4.1.33. Impedance spectra for two zirconia ceramics, obtained at 300°C using sputtered platinum electrodes: (*a*) Tetragonal zirconia ceramic (ZrO_2: 3 mole % Y_2O_3) with large grain boundary arc due to continuous grain boundary phase. (*b*) Partially stabilized ceramic (ZrO_2: 6 mole % Y_2O_3) with small grain boundary arc due to discrete grain boundary phase. (Courtesy of Silicates Industriels.)

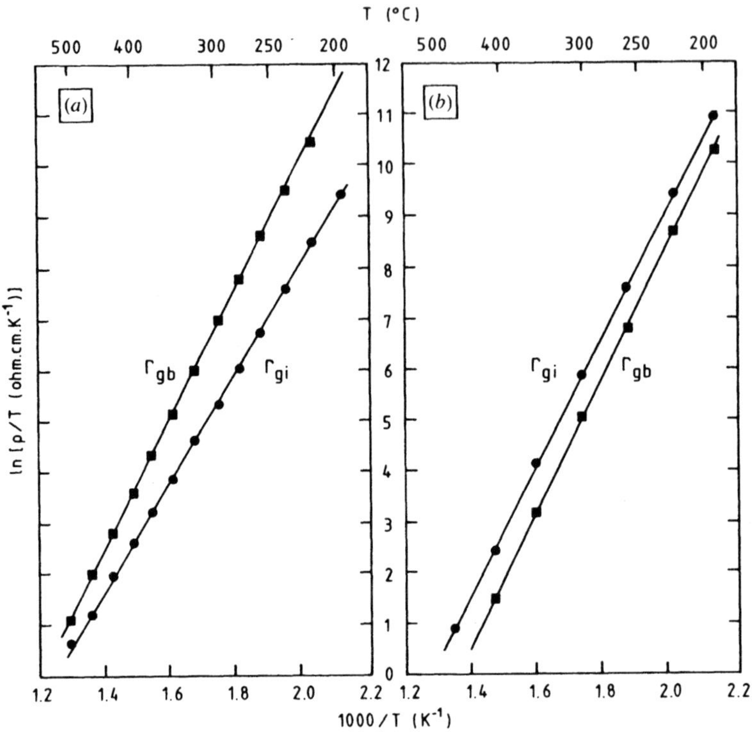

Figure 4.1.34. Arrhenius plots of the grain interior and grain boundary resistivities for two zirconia ceramics: (*a*) Tetragonal zirconia ceramic (ZrO_2: 3 mole % Y_2O_3) in which the lines have different slopes, as expected from the brick layer model. (*b*) Partially stabilized ceramic (ZrO_2: 6 mole % Y_2O_3) in which the slopes of the lines are similar, as expected for discrete grain boundary phase. (Courtesy of Silicates Industriels.)

examined (200–500°C). In the tetragonal ceramic (Figure 4.1.34*a*) the slopes are different (higher for r_{gb}), as expected on the basis of the brick layer model without easy paths. For the mainly cubic ceramic (Figure 4.1.22*b*) the slopes are quite similar, suggesting that a partial blocking model is more appropriate to this material. The conclusions based on analysis of the impedance spectra support TEM studies in assigning specific microstructural models to each ceramic. Table 4.1.6 gives the activation enthalpies for the two materials, and for a third PSZ ceramic, intended for electrical applications.

The treatment of the capacitative elements of the circuit equivalent differs according to which model applies. For the bricklayer tetragonal ceramic, the grain boundary capacitance can be used to calculate the grain boundary thickness using Eq. (11). Not having any information on the dielectric constants of the grain boundary phase, we assume $\varepsilon_{gi} = \varepsilon_{gb}$,[12] and calculate a grain boundary thickness of 4 nm, compatible with the 10 nm estimated from the electron micrograph.

[12] The error introduced by this assumption is smaller than the spread in the distribution of grain size, that is, about a factor of two.

Table 4.1.6. Activation Energies for Conduction for Three $ZrO_2:Y_2O_3$ Ceramics of Different Composition and Structure

Sample	Composition	Structure	ΔH_{gi} (eV)	ΔH_g (eV)
Y-TZP	3.0 m% Y_2O_3	Tetragonal	0.92	1.09
Y-PSZ	4.7 m% Y_2O_3	Cubic + tetragonal	1.07	1.15
Y-PSZ	6.0 m% Y_2O_3	Cubic + tetragonal	1.07	1.12

For the mainly cubic 6 mole % material this type of calculation is inappropriate, since the grain boundary phases are discontinuous. Instead, one can use Eq. (13) (partially blocked transport) to calculate a blocking coefficient β. A value of 0.3 is found, implying grain boundary coverage of 30% in second phase. This seems to be in accord with the appearance of the boundaries in Figure 4.1.32 and in other micrographs examined. Thus, for the two ceramics, TEM and IS give compatible information regarding the dimensions and continuity of the grain boundary phases.

Materials with no Grain Boundary Impedance Arc. Single crystals are an obvious example of such materials. Single crystals of stabilized zirconia are grown commercially for artificial gems by a process known as *skull melting*. Bonanos and Butler [1985] characterized crystals of ZrO_2: Y_2O_3 (2.2 and 3.4 mole %) by x-ray diffraction, TEM, and IS. Transmission electron micrographs of the 2.2 mole % crystal (Figure 4.1.35) showed striated areas characterized by a reorientation of

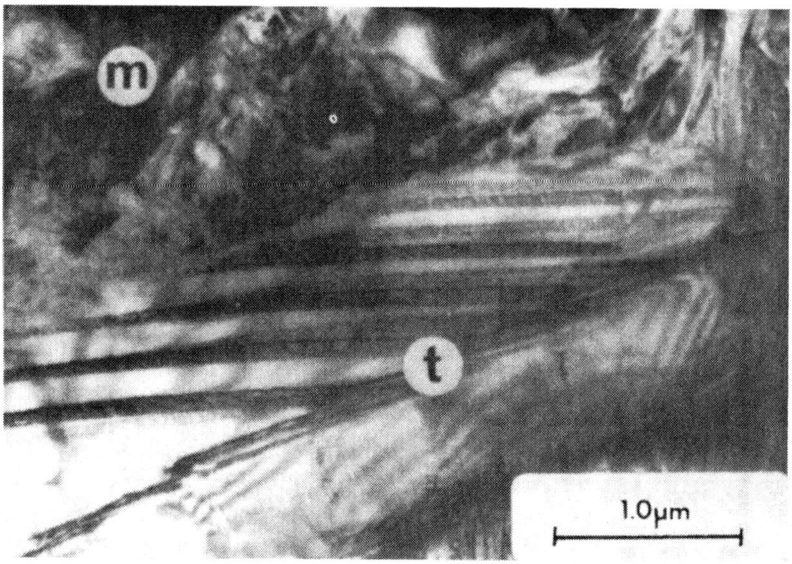

Figure 4.1.35. Transmission electron micrograph of a single crystal (ZrO_2: 2.2 mole % Y_2O_3), showing separate regions of monoclinic (m) and tetragonal (t) phases. The micrograph was obtained at 1 MV. (Courtesy of Silicates Industriels.)

the crystals known as *twinning*. SADP of these areas revealed the presence of both *m* and *t* phases. The *t* phase was similar in morphology to a nontransformable phase *t* observed by Lanteri *et al.* [1983] in crystals of similar composition. In the 3.4 mole % crystal only the *t* phase was observed.

The impedance spectra of the two crystals are shown in Figure 4.1.36. For the 2.2 mole % crystal (Figure 4.1.36*a*) the spectrum shows a large, distorted grain interior arc and an electrode arc. The depressed arc is due to the combination of tetragonal and monoclinic phases; indeed, it appears to be composed of two poorly resolved arcs. The 3.4 mole % crystal displays one high-frequency arc due to the *t* phase. The resistivity of 0.10 MΩcm is close to the value of 0.13 MΩcm measured for 3 mole % Y-TZP. The two spectra illustrate the sensitivity of IS towards phase composition and lend confidence to its use in investigating relatively complex microstructures.

Although the three-arc response is observed in most crystalline solid electrolytes, cases are known where the grain boundary arc is absent or at least too small to be seen. Figure 4.1.37 shows the impedance spectrum of ZrO_2:(Y_2O_3 + MgO) measured by Slotwinski, Bonanos, and Butler [1985]. The large arc with ρ''_{max} at about 10 kHz can be identified as a bulk property from the fact that it passes through the origin, and from the associated capacitance of about 4 pF/cm. The abnormally small grain boundary arc was explained by reference to a transmission electron micrograph (Figure 4.1.38). Unlike other zirconia-based systems, in ZrO_2:(Y_2O_3 + MgO) there are finely twinned crystallites along substantial lengths of grain boundary. Selective area electron diffraction patterns of this phase showed a tetragonal symmetry, resembling the patterns of the *t* phase, which has a relatively high ionic conductivity. It is suggested that *t*-phase dispersed along the grain boundary area effectively short circuits the grain boundary impedance.

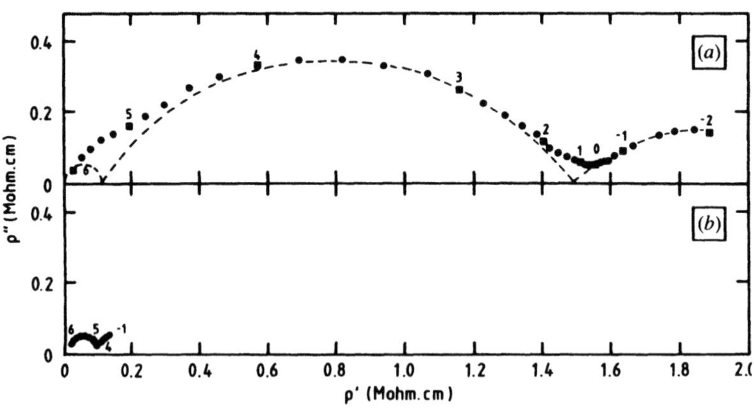

Figure 4.1.36. Comparison of the impedance spectra obtained at 300°C for two zirconia–yttria single crystals. (*a*) Tetragonal + monoclinic single crystal (ZrO_2: 2.2 mole % Y_2O_3) showing large bulk arc. (*b*) Fully tetragonal single crystal (ZrO_2: 3.4 mole % Y_2O_3) showing small bulk arc.

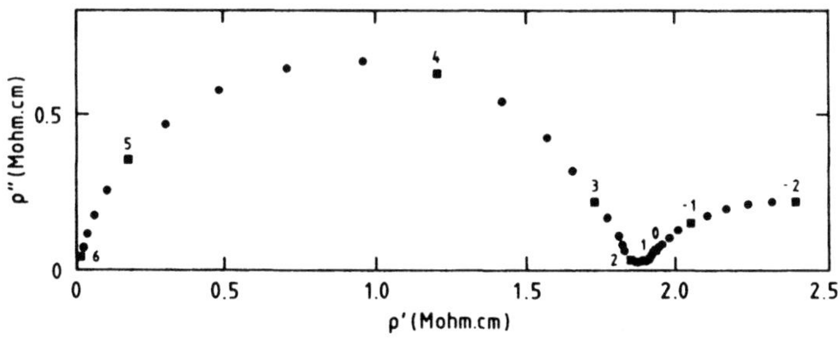

Figure 4.1.37. Impedance spectrum for a partially stabilized zirconia ceramic of composition ZrO_2: (7.0 mole % MgO + 1.5 mole % Y_2O_3), obtained at 300°C, not resolving a grain boundary arc. (Reprinted from R. K. Slotwinski, N. Bonanos, and E. P. Butler, *J. Mat. Sci. Lett.* **4**, 641–644, 1985, courtesy of Chapman & Hall.)

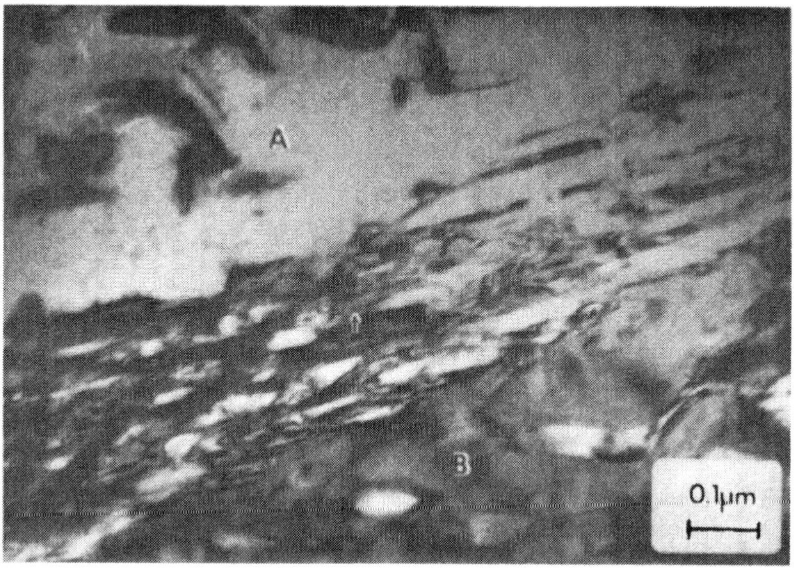

Figure 4.1.38. Transmission electron micrograph of partially stabilized zirconia ceramic ZrO_2: (7.0 mole % MgO + 1.5 mole % Y_2O_3) showing two grains (labeled A, B) and the intervening tetragonal grain boundary phase (arrowed). (Reprinted from R. K. Slotwinski, N. Bonanos, and E. P. Butler, *J. Mat. Sci. Lett.* **4**, 641–644, 1985, courtesy of Chapman & Hall.)

Another example of a ceramic lacking a grain boundary response is found in dense Bi_2O_3:Er_2O_3 ceramics (C. P. Tavares and N. Bonanos, 1984, unpublished work). Figure 4.1.39 shows the impedance spectrum of a pellet of Bi_2O_3 (17 mole % Er_2O_3). Here, the absence of grain boundary impedance can be tentatively explained by the absence of grain boundary phases altogether. Unlike zirconia, bismuth oxide has a considerable solid solubility for silica and alumina, common

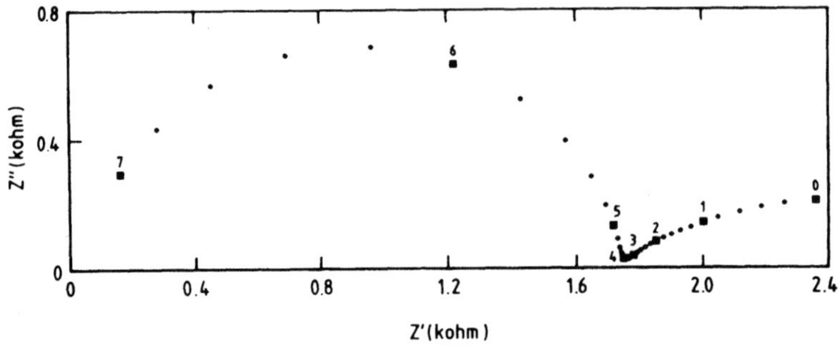

Figure 4.1.39. Impedance spectrum obtained at 300°C for a pellet of Bi_2O_3: 17 mole % Er_2O_3 showing no grain boundary arc. (The pellet was supplied by Dr C. P. Tavares of Basic Volume Ltd., London, England).

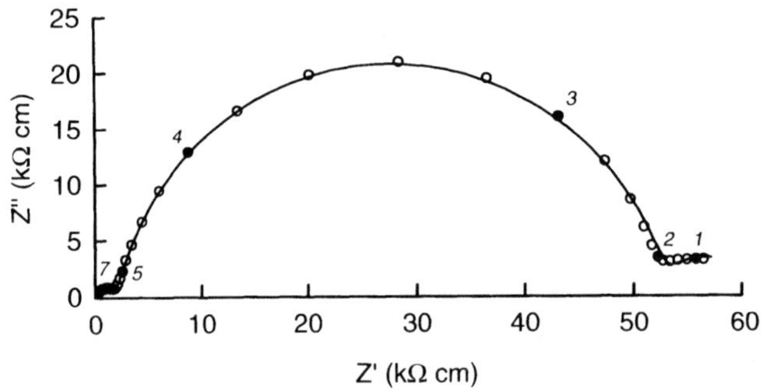

Figure 4.1.40. Complex resistivity spectrum for Bi_2UO_6 at 200°C (air, Ag electrodes).

impurities in ceramics and, therefore, they would not be expected to segregate during sintering and subsequent cooling.

Effect of Anisotropy in Grain Shape and Conductivity. In the ceramics discussed so far, the shape and conductivity of the grains were isotropic. This, is not, however, a rule for ceramics. Two examples are given of materials whose conduction or grain structure are anisotropic.

The first example concerns the polymorphic compound Bi_2UO_6 which, at ambient temperature, has a monoclinic structure, which transforms to triclinic at ~730°C. The monoclinic structure comprises layers of U–O and double Bi–O layers, and may be regarded as a superstructure of fluorite. Bonanos [1989] identified Bi_2UO_6 as an oxide ion conductor and, based on impedance spectroscopy, reported grain interior conductivities of 5×10^{-4} at 200°C and $10^{-1}\,\mathrm{S\,cm^{-1}}$ at 400°C. An impedance spectrum is shown in Figure 4.1.40 for 200°C. The large grain boundary arc (an order of magnitude larger than the grain interior arc) might suggest the presence

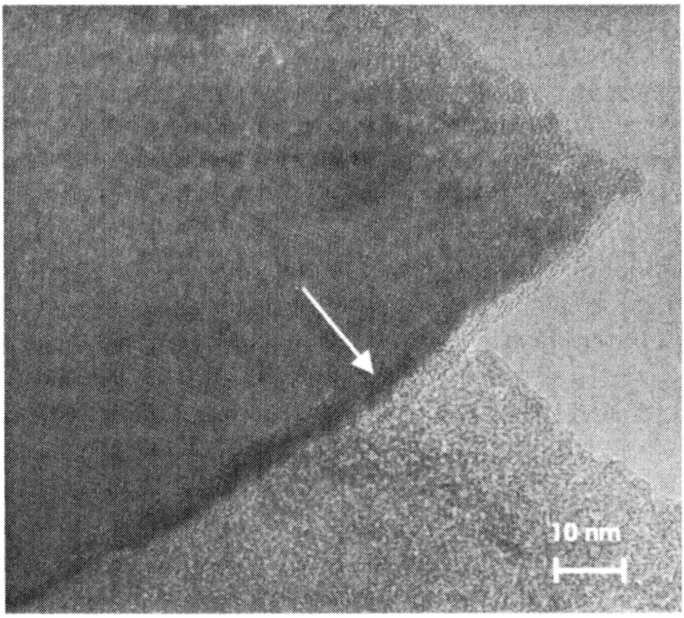

Figure 4.1.41. High resolution transmission electron micrograph of a grain boundary in Bi_2UO_6. Courtesy of D. White, BP Research Centre, Sunbury and J. L. Hutchinson, University of Oxford, UK. For further details see White, Ramdas, Hutchinson, and Billyard [1989].

of second phase at grain boundaries. However, high resolution TEM (Figure 4.1.41) showed the grain boundaries to be free of second phases to a scale of a few nm. To reconcile these observations, Bonanos [1989] suggested that the grain boundary impedance arose because of the random orientation of grains having two-dimensional conductivity. In a subsequent structural/electrical study of this system, Vannier *et al.* [1999] reported the total conductivity as $2 \times 10^{-2}\,S\,cm^{-1}$ at 400°C. However, since the grain interior arc was not resolved, the exceptionally high value of σ_{gi} claimed by Bonanos was neither confirmed nor refuted.

Sodium-conducting β-alumina ceramics are a second example of anisotropic materials: they consist of elongated grains in which the Na^+ ions migrate along specific crystallographic planes. Early studies by Whittingham and Huggins [1971] and by Powers and Mitoff [1975] established activation energies of 0.2 for the grain interior and 0.4 eV for the grain boundary conductivities. Lilley and Strutt [1979] performed IS over the temperature range −135 to 400°C. Figure 4.1.42*a* reveals a large grain boundary resistance at −135°C,[13] which was interpreted as due to two-

[13] The experimental conditions were chosen to allow resolution of the grain interior arc over the widest possible temperature range. The conductivity of $5 \times 10^{-3}\,S\,cm^{-1}$ at ambient temperature corresponds to a relaxation frequency of the order of 10^8 Hz.

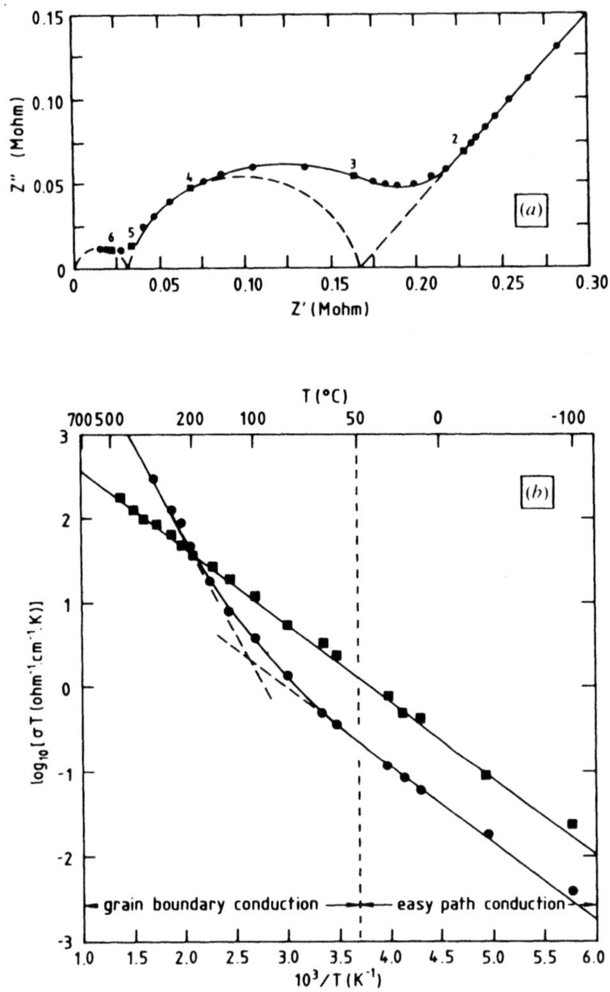

Figure 4.1.42. Application of impedance spectroscopy to a study of sodium β-alumina by Lilley and Strutt [1979]: (*a*) Impedance spectrum obtained at −135°C using evaporated gold blocking electrodes. (*b*) Arrhenius plot of σ_{gi} and σ_{gb} showing transition between grain boundary and easy path conduction. (Reprinted from E. Lilley and J. E.Strutt, *Phys. Stat. Sol.* (*a*) **54**, 639–650 [1979], courtesy of Akademie-Verlag.)

dimensional conductivity of grains in random orientation. Arrhenius plots of the conductivities (Figure 4.1.42*b*) fell into two regions: a low temperature region, where the activation energies were identical at 0.18 eV, and a high temperature region, where the grain boundary activation energy was 0.45 eV. According to Lilley and Strutt, at low temperatures, the grain boundary conduction was dominated by easy paths (Figure 4.1.43*a*) while, at high temperatures, transport through the grain boundary overtook the easy path mechanism. If this interpretation is correct, a transition between easy-path and true grain boundary conduction in the same system would be a clear indication that IS can reveal deep information, when performed over the necessary range of frequency and temperature. Table 4.1.7 gives the capacitance values reported by these authors.

In a contemporaneous TEM/IS study of the same system, De Jonghe [1979] concluded that a simple RC network was unsuitable for describing the varying cross

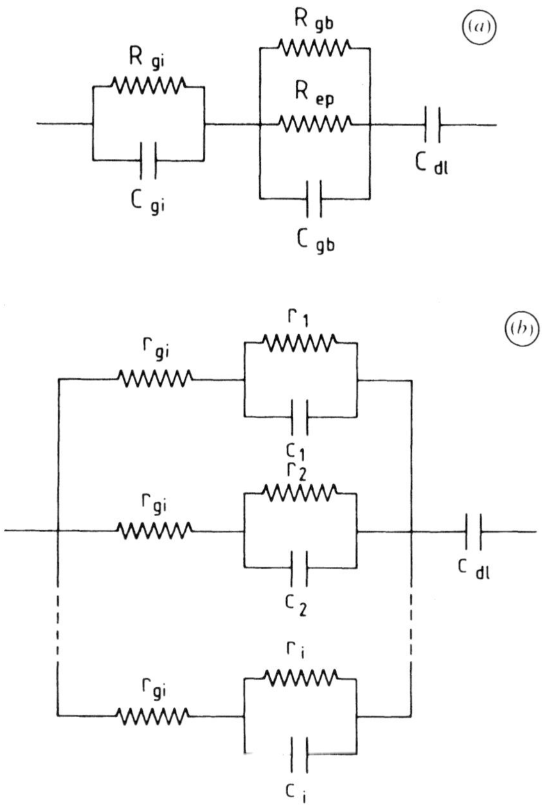

Figure 4.1.43. Circuits proposed for modeling the impedance spectrum of polycrystalline sodium β-alumina: (*a*) Easy path model according to Lilley and Strutt [1979]; (*b*) Multielement model according to De Jonghe [1979].

Table 4.1.7 Capacitance Values for β-Alumina at Various Temperatures, as Determined by Lilley and Strutt [1979]

Temperature (K)	C_{gi} (F)	C_{gb} (F)	C_{el} (F)
138	4×10^{-12}	7×10^{-10}	1.5×10^{-7}
149	—	8×10^{-10}	1.5×10^{-7}
371	—	8×10^{-10}	1.0×10^{-7}
652	—	—	0.8×10^{-7}

section and orientation of the grains in ceramics. As an alternative, he proposed a circuit having a series-parallel configuration (Figure 4.1.43*b*). De Jonghe simulated the effect of grain boundary blocking by varying the grain boundary parameters r_i, c_i along the branches, while holding r_{gi} constant. By assigning activation energies 0.17 eV and 0.35 eV to the microscopic parameters ΔH_{rgi} and ΔH_{rgb}, he simulated a range of temperatures and compared the macroscopic values of r_{gb}, c_{gb} with those obtained by a parallel summation of the r_i, c_i branches. The macroscopic values deviated from the microscopic values at all but the lowest temperatures, moreover the apparent grain boundary capacitance increased with temperature. De Jonghe concluded that, for systems with grain shape significant anisotropy, the information obtained by IS was of limited fundamental value. He also stressed the difficulty in interpreting c_{gb} in microstructures with discontinuous grain boundary phases.

While De Jonghe's cautionary advice is worth heeding, the actual trend in C_{gb} predicted by his model is not mirrored in the results of Strutt and Lilley [1979] (Table 4.1.7) where the capacitances were found to be constant over the temperature range studied. Accordingly, the De Jonghe model may be regarded as a worst-case analysis of a complex microstructure with grain anisotropy.

While the first edition of this monograph contained a defense of IS against the above and other criticisms, such a discussion would be less relevant today, in view of the advancement in finite element simulations (Coverdale *et al.* [1995], Fleig [2000, 2002], Tuncer *et al.* [2001, 2002]).

4.1.3.3 Characterization of Two-Phase Dispersions by IS

In the systems discussed so far, the continuous phase was concentrated in grain boundary regions, such that its blocking properties were noticeable, even at low volume fraction. This sub-section covers two-phase dispersions, where the volume fractions of continuous and discontinuous phase are closer, for example 0.70 and 0.30. Two examples are discussed, one where the time constants of the phases are well resolved, and another where they are not.

Characterization of a Two-phase Alkali Halide System. Alkali halide crystals have been studied as model systems for defect aggregation and solid state precipitation. In the $NaCl/CdCl_2$ system, crystals grown from the melt contain a phase of composition $CdNa_6Cl_8$, *Suzuki phase*, in which 12.5% of the cation sites are vacant. This phase has a higher Na^+ conductivity than the matrix in which it forms. A transmission electron micrograph of a crystal with Suzuki phase precipitates is shown in Figure 4.1.44. The volume fraction of the dispersed phase in this system is about 0.28.

Impedance spectra of such crystals, reported by Bonanos and Lilley [1981], displayed only bulk and electrode arcs. The same data plotted in the modulus plane (Figure 4.1.45) revealed two overlapping arcs: a low frequency arc ascribed to the matrix and a high frequency one ascribed to the dispersed phase. Using the Maxwell–Wagner effective medium relation (Eq. 20), the modulus spectra were modeled and the microscopic conductivities of the two phases were evaluated for

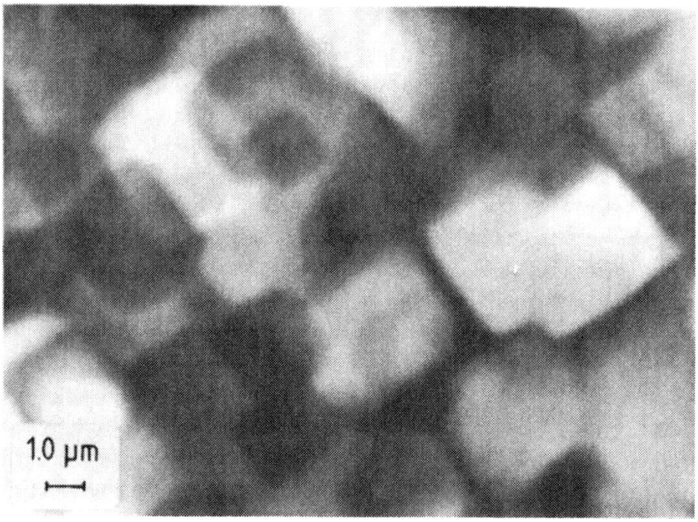

Figure 4.1.44. Transmission electron micrograph of a NaCl: $CdCl_2$ single crystal with precipitates of the Suzuki phase $CdNa_6Cl_8$. The micrograph was taken by A. L. Guererro and E. P. Butler at a voltage of 1 MV, at liquid helium temperature. (Reprinted with permission from N. Bonanos and E. Lilley, Conductivity Relaxation in Single Crystals of Sodium Chloride Containing Suzuki Phase Precipitates, *J. Phys. Chem. Solids*, **42**, 943–952. Copyright 1981 Pergamon Journals Ltd.).

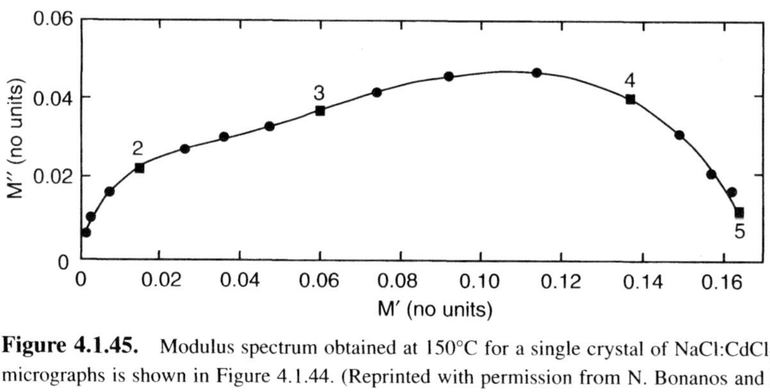

Figure 4.1.45. Modulus spectrum obtained at 150°C for a single crystal of NaCl:$CdCl_2$, of which a micrographs is shown in Figure 4.1.44. (Reprinted with permission from N. Bonanos and E. Lilley, Conductivity Relaxation in Single Crystals of Sodium Chloride Containing Suzuki Phase Precipitates, *J. Phys. Chem. Solids*, **42**, 943–952. Copyright 1981 Pergamon Journals Ltd.).

several volume fractions of dispersed phase. The ionic conductivity of the dispersed phase was similar to that determined on polycrystalline samples.

In the above system, the analysis was possible because the relaxation frequencies of the matrix and dispersed phases differed by a factor of about 30. In other cases, it is impossible to identify the relaxations of individual phases due to proximity of the time constants. Even in these cases, the impedance spectroscopy in conjunction with microstructural characterization can be used to glean information regarding the conductivities of individual phases. This is illustrated below.

Resistivity Analysis in Polyphase Ceramics. As mentioned earlier, the grains in PSZ consist of a matrix of cubic zirconia with a fine dispersion of tetragonal particles. Figure 4.1.46 shows the microstructure of ZrO_2 (Y_2O_3 + MgO) PSZ. The appearance of the elongated particles, distributed equally along three normal axes, is reminiscent of the Fricke model for a dispersion of ellipsoids (Section 4.1.1). The simulated spectrum for this model (Figure 4.1.10) indicates that (depending on the symmetry of the precipitates) up to four different relaxations could arise within the grain interior, in addition to that of the grain boundary.[14] Prolonged heat treatment (aging) of PSZ ceramics at high temperature is known to cause transformation of the precipitates from *t* to *m*, with a small increase in size, but size, no significant change in shape or chemical composition. Since the *c*, *t*, and *m* phases have different conductivities, these microstructural changes should be reflected in the electrical properties.

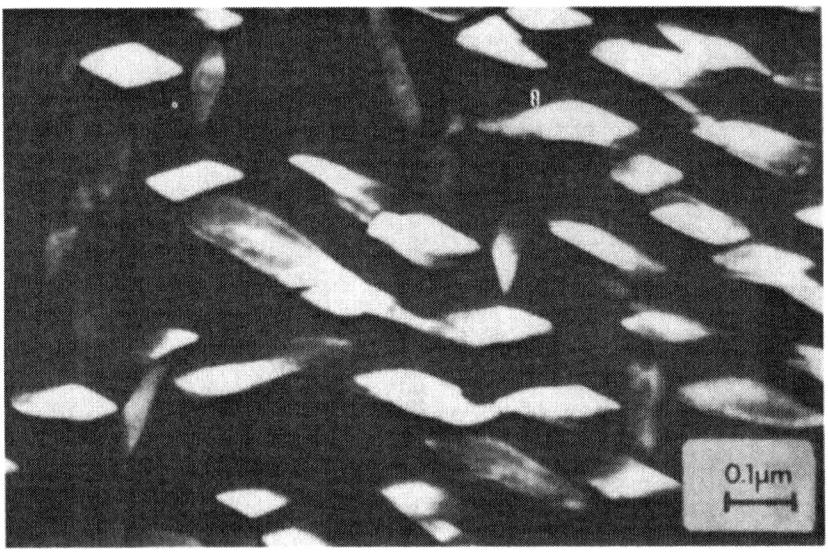

Figure 4.1.46. Transmission electron micrograph of grain interior of a partially stabilized zirconia ceramic ZrO_2: (7.0 mole % MgO + 1.5 mole % Y_2O_3), aged for 40 h at 1400°C. Bright areas correspond to the tetragonal phase.

[14] The number would depend on the symmetry of the precipitates.

This approach has been used by Bonanos *et al.* [1984a] to study PSZ of composition ZrO_2 with 8 mole % CaO (Ca-PSZ). XRD and TEM showed the $t \rightarrow m$ transformation to be complete after 20 h of heat treatment at 1400°C, while quantitative x-ray diffraction analysis gave a value of 0.3 for the volume fraction of the t component. Electron microscopy showed the t particles to be ellipsoidal, with axial ratios a/b and b/c of ~2, for which Table 4.1.2 gives the corresponding form factors, $\phi_1 = 7.9$, $\phi_2 = 2.5$, $\phi_3 = 0.66$. In short the parameters ϕ_i and x_2 in Eq. (23) are given and the only remaining variables are the complex conductivities ψ_1, ψ_2, and ψ_t.

Figure 4.1.47 shows impedance spectra for Ca-PSZ after aging at 1400°C for different times. Distinct relaxations of the matrix and dispersed phases are not resolved in these spectra, which are of a simple Bauerle type, nor are they resolved in the corresponding modulus spectra (not shown), presumably due to the proximity of the relaxation frequencies. Thus, multielement circuits (Figure 4.1.9) cannot be used to represent these data and the grain interior response must be approximated by a single RC element. Nevertheless, Eq. (23) can be used to interpret the dc conductivity of the grain interiors g_{gi} (for convenience, the inverse, r_{gi}, was used).

From a microstructural analysis and from the evolution of r_{gi}, aging was found to occur in three stages.

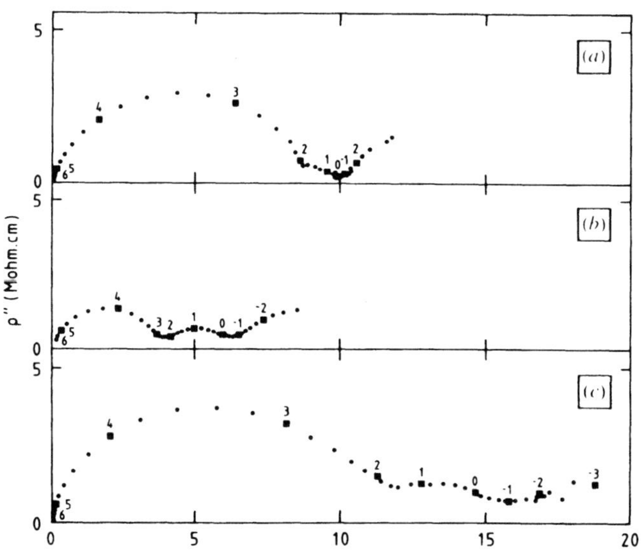

Figure 4.1.47. Impedance spectra obtained at 300°C for partially stabilized zirconia ceramics (ZrO_2: 8 mole % CaO): (*a*) as fired; (*b*) aged for 15 h at 1400°C; (*c*) aged for 30 h at 1400°C. (Reprinted from N. Bonanos, R. K. Slotwinski, B. C. H. Steele, and E. P. Butler, *J. Mat. Sci.*, **19**, 785–793, courtesy of Chapman & Hall.)

Stage I: an increase in volume fraction of *t* phase, resulting in a sharp fall in r_{gi} since $\sigma_t > \sigma_c$ (Figure 4.1.47*a* to 4.1.47).

Stage II: a period of precipitate growth at constant volume fraction with no change in r_{gi}.

Stage III: a transformation of the precipitates from *t* to *m* without change in the chemical composition of the matrix or the precipitates (Figure 4.1.47*b* to 4.1.47*c*), with no change in σ_c.

A numerical analysis of the conductivity changes during these transitions showed that physically consistent solutions to Eq. (23) could be obtained only for $\sigma_t > \sigma_c > \sigma_m$ and allowed an estimation of σ_t and σ_c to within an order of magnitude. Since CaO-stabilized TZP was not available at the time of that paper (and to the author's knowledge have not been prepared since), the analysis of a polyphase material was the only way of evaluating the conductivity of this system.

The Siugnificance of Grain Boundary Phases Placed in Context. While grain boundary phases have a dominant effect on the impedance spectrum, it should be remembered that, even in the absence of second phases, the electrical properties of interfacial regions differ from those of the interior, as illustrated by the space charge effects discussed in Section 4.1.1. Dopants segregated to interfaces during high temperature processing and frozen in during later cooling also modify the electrical resistance of the interfaces. Finally, unlike bulk properties, which are invariably linear, the grain boundaries in oxide semiconductors often have nonohmic characteristics, which form the basis for positive temperature coefficient resistors (PTCR) (Levinson [1981]). Interfacial segregation and space charges in ionic conductors have been invoked by Heyne [1983], Burggraaf *et al.* [1985], and Steele and Butler [1985] to explain the appearance of grain-boundary-like arcs in nominally pure ceramics. In many situations, second phases, segregation, and space charges all together contribute to a rich variety of electrical behavior.

4.1.3.4 Impedance Spectra of Unusual Two-phase Systems

In a volume such as this, most examples are unavoidably drawn from systems that are fairly well understood. As a counter-balance, two examples are mentioned that are, in some way, unusual.

The first example concerns a composite of (Sc_2O_3 + Y_2O_3) stabilized ZrO_2, a pure ionic conductor, and $La_{0.85}Sr_{0.15}Mn_{1.10}O_\xi$, a pure electronic conductor, at a volume fraction 0.30 for the latter. Figure 4.1.48 shows an impedance spectrum obtained at ambient conditions. It resembles that of a classical solid electrolyte, with a "bulk" resistivity of $5 \times 10^6\,\Omega$cm and a "bulk" capacitance of $4 \times 10^{-12}\,\text{F}\,\text{cm}^{-1}$.

The "grain boundary" and "electrode" features are associated with Warburg components of $3 \times 10^{-8}\,\text{S}\,\text{cm}^{-1}\,\text{s}^{1/2}$ and $1.3 \times 10^{-5}\,\text{S}\,\text{cm}^{-2}\,\text{s}^{1/2}$ respectively. The known properties of the zirconia phase do not, however, fit such an interpretation: the resistivity, extrapolated from higher temperature is at least $10^{12}\,\Omega$cm. This would suggest

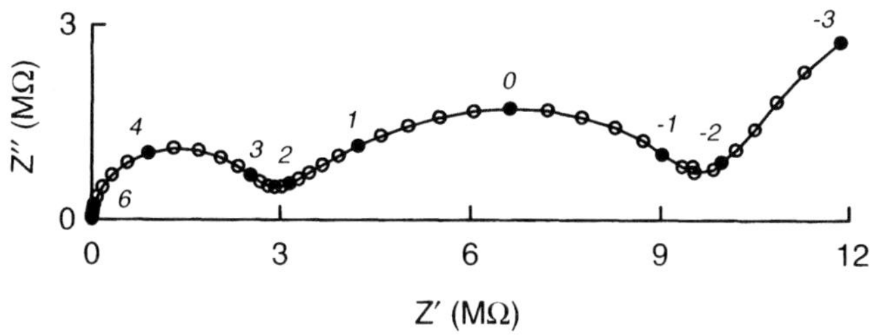

Figure 4.1.48. Impedance spectrum at ambient conditions for a composite of (Sc_2O_3 + Y_2O_3) stabilized ZrO_2, and $La_{0.85}Sr_{0.15}Mn_{1.10}O_\xi$ at a volume fraction 0.30. The spectrum resembles that of a classical solid electrolyte, but the "bulk conductivity" exceeds the expected value by a factor of 10^6.

that the enhanced bulk conductivity is due to the presence of a high conductivity phase close to the percolation threshold. Unfortunately, the effective medium models presented in Section 4.1.1 cannot be used to describe this composite, as they apply to phases of the same type of conductivity, i.e. they do not include any charge transfer impedances at the interfaces between ionic and electronic conductor.

The second example is a concentrated aqueous dispersion of dihydrogenated tallow dimethylammonium chloride (DHTDMAC) a lamellar liquid crystal commonly used in fabric conditioners. Figure 4.1.49 shows an impedance spectrum obtained at ambient temperature in the frequency range 10^2 to 10^9 Hz. It shows two overlapping arcs with capacitance/CPE elements of $8 \times 10^{-12}\,\mathrm{F\,cm^{-1}}$ and $8 \times 10^{-9}\,\mathrm{S\,cm^{-1}s^{1/2}}$. Despite the superficial similarity of the spectra with the grain/grain boundary spectra of solids, the substance is actually a viscous liquid consisting of spherical aggregates, called *liposomes*. These are made up of thick regions of

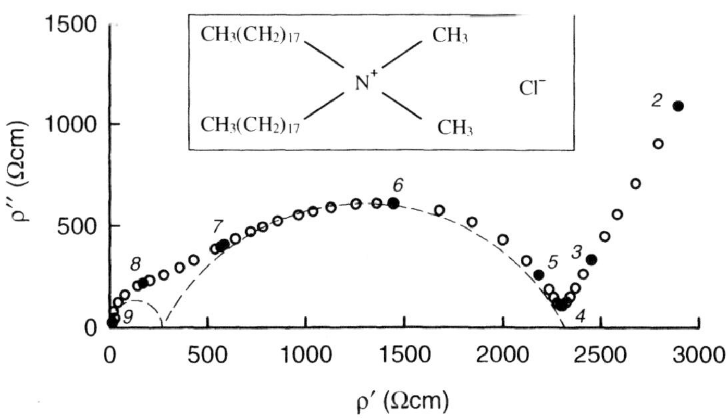

Figure 4.1.49. Impedance spectrum of dihydrogenated tallow dimethylammonium chloride (see inset), obtained at ambient temperature in a coaxial cell with gold electrodes. Labels indicate log(f).

aqueous phase separated by thinner bilayers of DHTDMAC (Clint [1992]). One may imagine that the bulk arc is due to the conductive aqueous regions and the grain-boundary-like arc is due to the resistive bilayers. Thus, while it is often said that impedance spectroscopy resolves the grain boundary properties in ceramics, it would be more correct to say that this spectrum is sensitive to structure, be that the structure of a solid or a liquid.

4.1.3.5 Impedance Spectra of Composite Electrodes

This section presents two examples of composite cathodes for SOFC. For these systems, the most commonly used cathode material and solid electrolyte are LSM and YSZ, respectively. Electrode impedances appear over a small space scale, characteristic of atomic dimensions, and as a result their spectra differ from those of bulk materials in several ways listed below.

1. Their properties are dependent on both the electrolyte and the electrode material and on the microstructure (grain size, porosity, etc).
2. The associated capacitances are of the order of $10^{-6}\,\mathrm{F\,cm^{-2}}$, or more when they arise from processes other than charge storage at the double layer.
3. Their impedances are in general nonlinear, i.e. dependent on the amplitude of the applied ac signal and the dc bias. Measurements are performed at small signal amplitudes in order to stay within the linear regime.
4. Electrode reactions usually involve a gas phase and, therefore, electrochemical impedances will depend on the gas composition and transport within this phase.
5. As stated in Section 4.1.3, measurements on cells with reference electrodes are subject to substantial errors, in the case of thin electrolyte cells.

Figure 4.1.50 shows the structure of a composite electrode. The composite is made of a mixture of electrolyte and electronically conducting phases and has a thickness of 5 to 50 μm. Since this layer usually has insufficient in-plane electronic conductivity for current collection, it is covered with a current collecting layer of porous electronic conductor. This can be made of the same substance as the electronic component of the composite, or another substance of high electronic conductivity. For laboratory testing, precious metal pastes are convenient for this purpose. The thickness of the current collecting layer is typically in the region of 50 μm.

The impedance spectra of composite cathodes reveal several processes, not all of which are fully resolved in the impedance spectrum. Some of the processes are observed also in point contact electrodes, while some are specific to the composite electrode[15]. Jørgensen and Mogensen [2001] have surveyed the dominant impedance

[15] Point contact electrodes are made by contacting a point of electrode material on a dense electrolyte. They are simpler than composite cathodes due to (a) their well defined geometry, allowing the estimation of the three-phase boundary length and (b) the absence of complex diffusional processes.

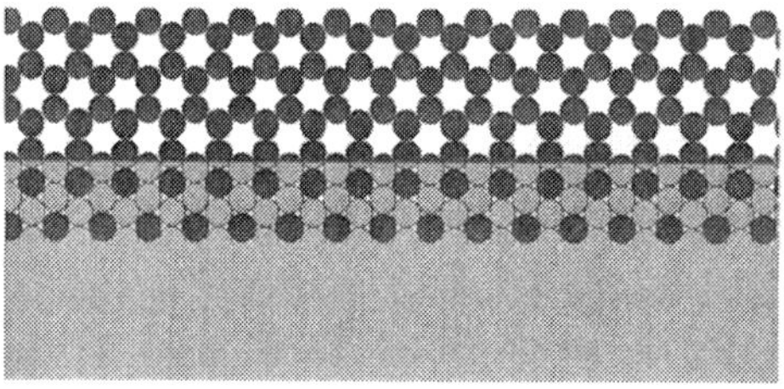

Figure 4.1.50. Composite electrode, consisting of a mixture of ionically and electronically conducting phases, with porous current collecting layer.

arcs of composite cathodes (not including the electrolyte series resistance) and have classified them as follows, in descending order of their relaxation frequency:

A and B. Conductivity relaxations, related to the microstructure of the composite. Their activation energies are about 1 eV, similar to those of the grain or grain boundary conductivities and are independent of the oxygen partial pressure (P_{O_2}). One or both of these arcs may be absent in some composite cathodes, but are absent with point contact electrodes.

C. This arc is ascribed to the main cathodic process, namely adsorption and dissociation of oxygen molecules on the surface of the electrode material. Activation energies are in the range 1.8 to 2.0 eV and it is dependant on P_{O_2} with a power law $(P_{O_2})^q$, with q in the range −0.14 to −0.50. The associated capacitances are in the region 10^{-4} to $10^{-1}\,\mathrm{F\,cm^{-2}}$. This feature is observed in composite cathodes and point contact electrodes alike.

D. Gas phase diffusional impedance. This arc is observed mainly in high performance cathodes, and increases with cathodic polarization (negative applied potential). The observed impedance is in accordance with the presence of a stagnant gas layer close to the electrode surface.

E. Inductive loop. It is not clear whether this arc represents a new feature or is a manifestation of an activation process within arc C. The feature is observed with point contact electrodes. It is ascribed to adsorbed oxygen intermediates, or segregation of chemical species at the TPB.

The above classification implicitly assumes a circuit equivalent composed of ZARC elements. There are, however, alternatives, namely the Gerischer[16] element, as used by Holtappels *et al.* [1998]. This is illustrated in Figure 4.1.51 for the spectrum of a composite cathode at 850°C. In Figure 4.1.51*a* the electrode response is

[16] For the definition of this element see Section 2.2.2.2.

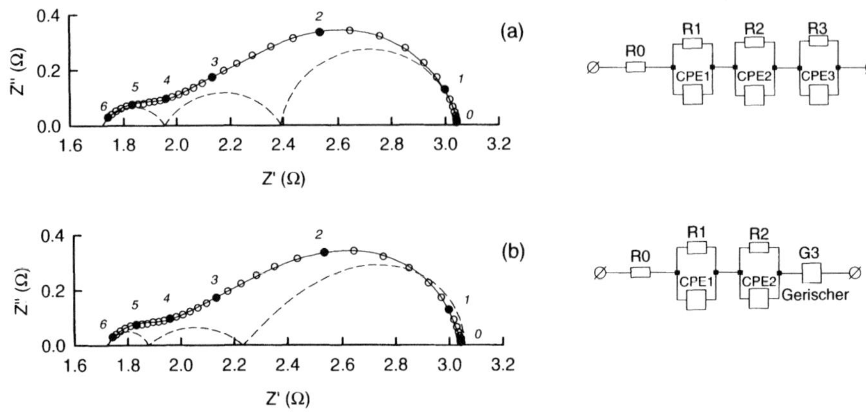

Figure 4.1.51. Impedance spectrum of a symmetrical cell with composite cathode on both sides, fitted to two different circuits. The solid lines show the total fit, while the dashed lines show the impedances of the individual elements. The composite layer consists of ZrO_2 (8 mol% Y_2O_3) and $La_{0.85}Sr_{0.15}Mn_{1.10}O_\xi$ in equal proportions and the current collector is porous platinum paste. The measurement was made in air, at 850°C. The data were corrected for the series inductance of the measurement rig, determined in short circuit.

simulated using a circuit with three ZARC elements, while in 4.1.51*b* one of the ZARCs is replaced by a Gerischer element. There is no appreciable difference in the goodness of fit, as judged by the solid lines, the χ^2 values in both cases were about 6×10^{-6}, and the estimation of the polarization resistance differs by 2%.[17]

The ability of two different circuits to model the same data set has both positive and negative implications. On one hand, it facilitates the process of electrode development, since the values of the polarization resistance estimated by different investigators can be compared, even if they were not based on exactly the same circuit. On the other hand, it complicates the interpretation of the spectral features in terms of mechanisms, because arcs cannot be paired to processes on a one-to-one basis. It is, therefore, essential that a fundamental investigation covers the widest possible range of experimental conditions (namely temperature, gas composition), including conditions that are far from those of the planned application, so that all the electrode processes "get a chance to appear" in the impedance spectrum.

The second example concerns functionally graded cathodes, consisting of five layers of YSZ/LSM with increasing content of LSM (20–100%) (Holtappels and Bagger [2002]). The cathodes were prepared on both sides of a YSZ tape (thickness ~150 μm), characterized by IS and found to have a very high performance, with

[17] The circuit containing the Gerischer element is not able fully to reproduce the spectrum at the low frequency end, although in the data of Holtappels *et al.* [1998], this problem did not arise.

polarization resistances of $100\,m\Omega\,cm^2$ or less at 850°C (see also Figure 4.1.24). Close examination of the results revealed that the electrolyte resistance associated with the cathodes was higher, by about $50\,m\Omega\,cm^2$, than that expected from the known conductivity of the electrolyte; this was attributed to the resistance of the YSZ-rich layer adjacent to the electrolyte. To investigate this further, impedance studies were performed at temperatures below the operating temperature (Bonanos *et al.* [2002]).

Figure 4.1.52 shows impedance spectra at two temperatures, with simulated data for a circuit based on four ZARCs (for Figure 4.1.52*a* one of the ZARCs is replaced by a resistor). By starting at low temperatures and using the parameters estimated at one temperature as initial guesses for the next one, convergence was obtained in all cases.

The spectra illustrate the difficulty in estimating R_0. At 700°C (Figure 4.1.52*a*) the arc R_1Q_1 is very flat. The simulation suggests that this arc should intersect the real axis at 10^8 Hz, which is outside the frequency range used. The spectrum at 400°C, however, falsifies this expectation: rather than approaching the real axis, the spectrum develops into a new, high frequency arc, R_0Q_0. This development could

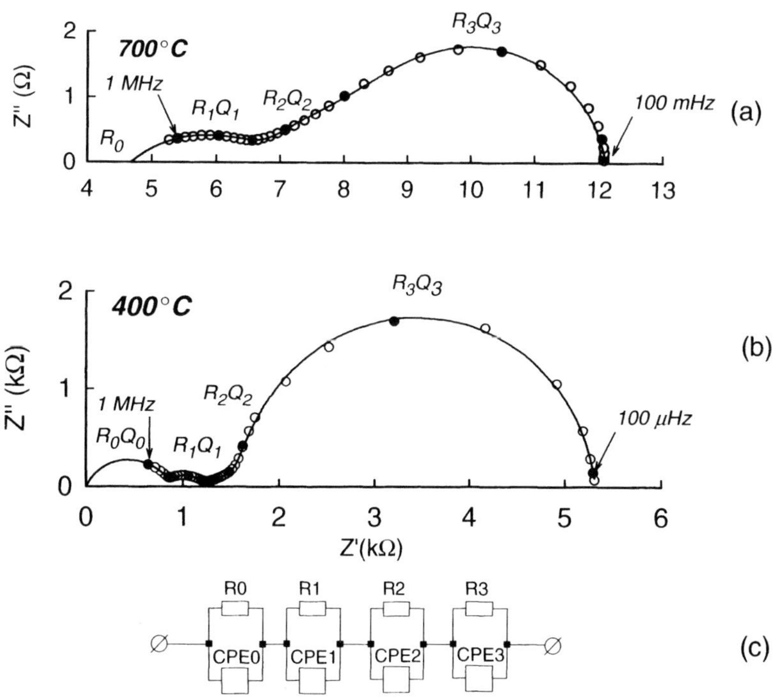

Figure 4.1.52. (*a*), (*b*) Impedance spectra of cathode consisting of five layers of YSZ/LSM, with graded thickness and composition. (*c*) Circuit equivalent. The cathodes were prepared on both sides of a YSZ tape. The measurements were performed in air (Bonanos *et al.* [2002]).

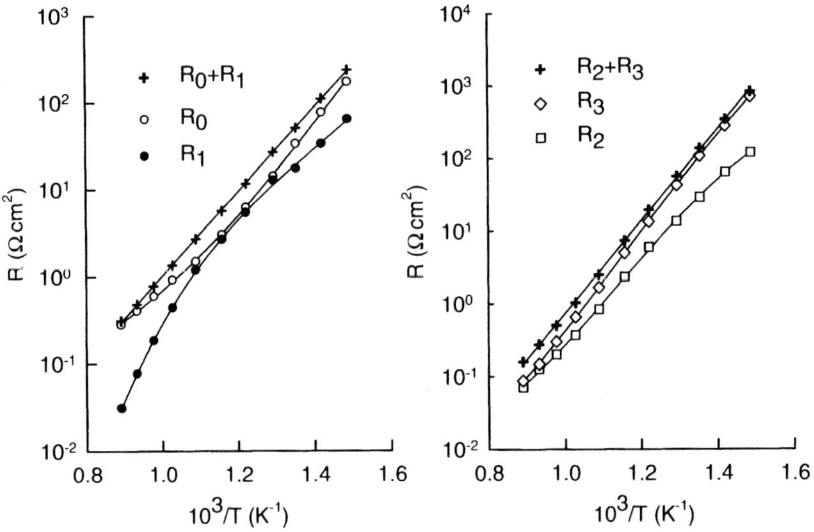

Figure 4.1.53. Temperature dependence of the resistive elements obtained by fitting the circuit of Figure 4.1.52(*c*) to the data of a multilayer composite cathode. (*a*) R_0 and R_1; (*b*) R_2 and R_3. The plots for R_0 and R_1 curve in opposite directions, indicating a systematic error of resolution, while their sum gives a straight line. This may be due to a "smearing out" of the electrode/electrolyte interface in the graded structure. The corresponding plots for R_2 and R_3 also show a (slightly) greater curvature than their sum. The activation energies for ($R_0 + R_1$) and ($R_2 + R_3$) are 0.96 and 1.25 eV, corresponding to electrolyte and electrode processes respectively.

not have been observed in the spectrum obtained at high temperature, graphically, or by CNLS fitting. The frequency range is simply too limited to reveal this arc.

The temperature dependence of the resistors R_0 to R_3 is shown in Figure 4.1.53; for convenience, R_0, R_1 and R_2, R_3 are shown separately. The plots for R_0 and R_1 (Figure 4.1.53*a*) have significant curvature in opposite directions. This is certainly due to a systematic error in the resolution, because their sum, $R_0 + R_1$, gives a good straight line with an activation energy of 0.96 eV, the value expected for YSZ.[18] This leads to the conclusion that R_0 and R_1 are both manifestations of the electrolyte resistance but that they cannot be resolved completely in the impedance spectra, even at low temperatures. This may be qualitatively described as a "smearing out" of the electrode/electrolyte interface due to the graded nature of this system. The corresponding plots for R_2 and R_3 are shown in Figure 4.1.53*b*. The Arrhenius plots are curved, as observed for two-layer composite cathodes (Juhl *et al.* [1996]). As with the case of R_0 and R_1, the sum $R_2 + R_3$ gives a better straight line (activation energy of 1.25 eV), than either R_2 or R_3, suggesting that these processes too may be not be fully separable.

[18] Estimated from $\ln[(R_0 + R_1)/T]$ versus T^{-1}.

The problems of resolution described above are explained by the proximity of characteristic time constants, rather than experimental error. The problem is fundamental to the material and could not be solved using another technique, such as current interrupt. The understanding of a system with four relaxation frequencies but only two temperature dependences seems to need a more advanced model than that afforded by an electrical circuit.

4.1.3.6 Closing Remarks

The work presented here has covered three main aspects of IS, namely modeling, experimental techniques, and the application of the technique to systems with complex microstructures. Since it is not possible to summarize this work in a paragraph, we will draw attention to three themes that recur in this section.

- Choice of appropriate model: IS is not a technique that can or should be applied without prior knowledge of the system. Impedance spectra must be interpreted in the context of a model, be this a simple brick-layer model for a ceramic, or an advanced one based on electrode kinetics. When used in conjunction with electron microscopy, IS provides information about structure, and especially grain boundary structure. The microstructural information and the models derived from this are what make the conclusions of IS unequivocal.
- Choice of equipment and experimental conditions: the instrumental frequency range needs to cover the dominant relaxations of the system. It is as important to choose the temperature of the system as it is to choose the right instrument. It is also important to make the measurements over a range of conditions considerably wider than those expected in operation.
- Choice of formalism: the formalism (Z, M, etc.) should be chosen to make the correspondence between measured spectrum and model as simple as possible. In cases where grain boundaries or electrodes are blocking, the preferred choice will normally be the impedance formalism—in other cases, others will be better. It is advisable to examine data in more than one formalism.

Acknowledgments

The following are warmly thanked: E. Lilley, R. K. Slotwinski, S. Sunde, J. Fleig, and S. B. Adler for providing original data from experiments and simulations; A. Sihvola and T. Goldrick for mathematical analysis; E. Ringgaard and J. Maier for critical comments; and C.P. Tavares for supplying a sample of Bi_2O_3 : Er. Work on measurement systems is credited to C. D. Waters and J. Poulsen. M. Mogensen is thanked for contributions to the section on composite cathodes. Countless discussions with members of the Wolfson Unit for Solid State Ionics, Imperial College,

London (1979–1984) and the Fuel Cells Group, Materials Research Department, Risø National Laboratory, Denmark have indirectly contributed to this section.

This section was written by N. Bonanos, B. C. H. Steele, and E. P. Butler and revised by N. Bonanos. The revised section is dedicated to the memory of B. C. H. Steele.

4.2 CHARACTERIZATION OF THE ELECTRICAL RESPONSE OF HIGH RESISTIVITY IONIC AND DIELECTRIC SOLID MATERIALS BY IMMITTANCE SPECTROSCOPY

J. Ross Macdonald

4.2.1 Introduction

For at least several decades, the effects of charged-particle motion in doped semiconductors, amorphous materials, polycrystals, single crystals, inorganic glasses, and polymers have been of much interest to both experimentalists and theorists. In fact, J. C. Phillips [1994] has characterized the problem of relaxation in complex disordered systems as the most important unsolved problem in physics! For ionically conducting materials such as solid electrolytes, the dynamics of the mobile ions have usually been investigated by analyzing the frequency response of the material over a wide range of frequencies, sometimes as wide or wider than 10^{-5} Hz to 10^{12} Hz. Such investigations thus usually involve immittance spectroscopy measurements and techniques. In this section, the main emphasis is on ionic conductors because of their technological importance in such areas as batteries, fuel cells, electrochromic displays, energy storage in capacitors, sensors, and even bionics. Because the electrical response of ionic conductors is rarely of simple Debye-relaxation character except in limiting cases, one must be concerned with its generalization: dispersive response.

Conductive-system dispersive response may be associated with a distribution of relaxation times (DRT) at the complex resistivity level, as in the work of Moynihan, Boesch, and Laberge [1973] based on the assumption of stretched-exponential response in the time domain (Eq. (118), Section 2.1.2.7), work that led to the widely used original modulus formalism (OMF) for data fitting and analysis. In contrast, dielectric dispersive response may be characterized by a distribution of dielectric relaxation times defined at the complex dielectric constant or permittivity level (Macdonald [1995]). Its history, summarized in the monograph of Böttcher and Bordewijk [1978], began more than a hundred years ago. Until relatively recently, however, these two types of dispersive response were not usually distinguished, and conductive-system dispersive response was often analyzed as if it were of dielectric character, even when this was not the case. In this section, material parameters will be expressed in specific form appropriate to the level concerned.

Conductive-system dispersion (CSD) usually involves thermally activated conduction extending to zero frequency plus an always-present bulk dielectric constant, $\varepsilon_{D\infty}$, usually taken to be frequency-independent in the experimental range. Dielectric-system dispersion (DSD) often involves dielectric-level response with only weak temperature dependence, and it may or may not involve a non-negligible frequency-independent leakage resistivity, $\rho_{C\infty} = \rho_{dc} \equiv \rho_0 \equiv 1/\sigma_0$. There may be cases where separate processes lead to the simultaneous presence within an experimental frequency range of both types of dispersion, but this is rare for most solid electrolytes. Further complications are present when conduction involves both mobile ionic and electronic charges, neither of whose effects are negligible (Jamnik [2003]). Here only ionic, dipolar, and vibronic effects will be further considered, with the main emphasis on conductive rather than on dielectric dispersion.

Since conductive-system dispersive response may be transformed and shown graphically at the complex dielectric level, and dielectric dispersion may be presented at the complex resistivity level, frequency-response data alone may be insufficient to allow positive identification of which type of process is present, since there may be great similarity between the peaked dispersion curves that appear in plots of $\rho''(\omega)$ and of $\varepsilon''(\omega)$ or of $\varepsilon_s''(\omega) \equiv \varepsilon''(\omega) - (\sigma_0/\omega\varepsilon_V)$. Here, ε_V is the permittivity of vacuum. This quantity has usually been designated as ε_0, as in other parts of this book. Its designation here as ε_V avoids ambiguity and allows clear distinction between it and $\varepsilon(0) = \varepsilon'(0) \equiv \varepsilon_0$, the usage in the present section.

Even CNLS data fitting at a specific temperature may not always allow unambiguous discrimination between CSD and DSD responses. But if data are available over a range of temperatures, discrimination is straightforward. Then, one generally finds that ρ_0/T (or ρ_0) and τ_o, the characteristic relaxation time of a model exhibiting thermally activated CSD, involve the same activation enthalpy (usually termed the activation energy) (e.g. Macdonald [2002a]). This is an effective quantity when the process considered involves a distribution of activation energies. Dielectric dispersion response may not be thermally activated but when it is, τ_o certainly does not have the same activation energy as that of an independent leakage resistivity ρ_0. A detailed study of discrimination between the two types of dispersion appears in Macdonald [1999a].

4.2.2 Types of Dispersive Response Models: Strengths and Weaknesses

4.2.2.1 Overview

Conductive-system dispersive response involving mobile charge may be conceptually associated with the effects of three processes:

1. electrode effects, which are particularly important at low frequencies (see Section 2.2.3.1) but may not be negligible at very high ones (Macdonald [2002a,b]);

2. ionic hopping effects, usually significant at mid-range frequencies (Macdonald [2002a,b]);
3. nearly constant loss effects primarily evident at sufficiently low temperatures over the usual frequency range or at high frequencies for higher temperatures (Ngai [1999], Ngai and León [2002], Macdonald [2002c, 2003b]).

Three different kinds of models have been proposed for describing these responses. A summary of some of the pertinent history of attempts to characterize the situation appears in Roling *et al.* [2001]. We shall consider here only models for the above behaviors that may be associated with mobile charge effects. The first and most desirable would be a fully microscopic model that accounted for all the above processes, since they are all directly or indirectly associated with mobile charge in conductive-system materials. Unfortunately, this many-body problem involving all interactions is currently insoluble.

A second approach involves approximate microscopic models whose log–log $\sigma'(\omega)$ slope continuously increases toward a value of unity until a high-frequency plateau is reached. No account of electrode effects is included in these approaches. In most other models, their high-frequency slope is related to a model parameter and quickly increases to a constant value less than unity as the frequency increases and before a final plateau begins to appear (Macdonald [1997b, 2002d]).

The third approach involves a composite model involving separate parts: one accounting for ionic hopping; a parallel contribution representing the effect of the endemic bulk dielectric constant, $\varepsilon_{D\infty}$; possibly a part describing nearly constant loss; and finally a series response model to account for electrode effects. For fitting most limited-range data, only two or three of these parts are usually required and excellent data fits are generally found using appropriate models. We shall therefore consider some composite models in detail.

It is noteworthy that most comparisons and fits of models to experimental data deal only with $\sigma'(\omega)$ response. An advantage of this procedure is that $\sigma'(\omega)$ and $\varepsilon''(\omega) \equiv \sigma'(\omega)/\omega\varepsilon_v$ are the only ones of the eight real and imaginary parts of the four immittance levels that are independent of the presence of $\varepsilon_{D\infty}$: $\sigma'(\omega)$ fitting is thus simpler than fitting with any of the four complex immittance-level responses or with the six other real and imaginary parts. Such an approach does not allow estimation of $\varepsilon_{D\infty}$, however, and it not only forfeits the error-averaging inherent in CNLS fitting but also the latter's test for the applicability of the Kronig–Kramers transformations.

The following discussion does not include consideration of all reasonable models that have been proposed and used for conductive-system fitting, but only some widely used ones and ones of particular theoretical importance.

4.2.2.2 Variable-slope Models

The Mismatch and Relaxation Model. Although some apparent theoretical defects inherent in the mismatch-and-relaxation model of Funke [1998], have been pointed out (Macdonald [1999b]), they have neither been explicitly recognized nor directly resolved. A recent empirical modification of this approach (Funke *et al.* [2002]) seems, however, to avoid some of the problems of the earlier work. Further, new work of Funke and Banhatti [2004] corrects further weakness in the model, although it still contains some empirical elements and thus cannot be considered a full microscopic response model.

The Symmetric Hopping Model. This model (Dyre and Schrøder [2000]) ignores Coulomb interactions, claims to be of universal character in the extreme disorder limit, and yields response rather similar to that of the mismatch-and-relaxation model. Of the several approximate but specific microscopic hopping realizations of the microscopic model considered by Dyre and Schrøder, the diffusion-cluster-approximation one led to best results, although it involves low-frequency-limiting response in disagreement with the physically realistic dependencies of the real and imaginary parts of the ac conductivity on ω^2 and ω, respectively (Odagaki and Lax [1980], Macdonald [1996, 1997b, 2001a]. The mathematical complexities of both the mismatch-and-relaxation model and the diffusion-cluster-approximation one makes data fitting and the estimation of values of model parameters difficult, and thus no CNLS fitting of data to estimate such parameters seems to have been published so far.

Comparisons of the variable-slope models with real-part conductivity data have rarely involved responses with a variation of $\sigma'(\omega)/\sigma_0$ greater than three decades starting from a low-frequency experimental value of this ratio of nearly unity, and even for such a limited range they usually show increasing disagreement with experiment toward the high end of this ratio where the relative frequency is large. In contrast, the results of a PK1-model (defined in the next section) fit of accurate synthetic data calculated for the microscopic diffusion-cluster hopping model and involving a range of $\sigma'(\omega)/\sigma_0$ greater than seven decades yielded a value of S_F, the relative standard deviation of the fit, of less than 0.01 and showed no deviation between $\sigma'(\omega)/\sigma_0$ data and fit points on a log–log plot, as well as no apparent slope variation (Macdonald [2001b]).

It is therefore clear that since the variable-slope models have not been compared with data that would allow discrimination between their predictions and those of simpler composite models, the variable-slope approaches, while of theoretical interest, are currently less appropriate for data fitting and analysis than are simpler and well-fitting composite models.

4.2.2.3 Composite Models

The ZC Power-law Model. Although we discuss some single dispersive-response models here, in practice they must always take account of $\varepsilon_{D\infty}$ and of possibly some other effects as well and so the overall model is always composite. A frequently used fitting model is the ZARC one of Eq. (22), Section 2.2. It is now more often designated as the ZC and, when written at the complex conductivity level, it may be expressed as $\sigma(\omega) = \sigma_0[1 + (i\omega\tau^{ZC})\gamma_{ZC}]$, where $0 < \gamma_{ZC} \leq 1$. The exponent γ_{ZC} has often been written as n and is the high-frequency-limiting log–log slope of the model. It has usually been found to have a value in the range $0.6 \leq \gamma_{ZC} \leq 0.7$.

The real part of the ZC model has been termed Jonscher or universal dynamic response, but the word "universal" is inappropriate since CNLS fits with the ZC or with its $\sigma'_{ZC}(\omega)$ part have been shown to be much poorer than those with other composite models (Macdonald [2000b, 2003a]). Finally, the identification of τ_{ZC}, or its real-part-fitting counterpart, as the inverse of the hopping radial frequency of the charge carriers has also been shown to be unsuitable (Macdonald [2003a]), and a more appropriate choice, the CK1 model, is discussed below.

OMF and CMF Kohlrausch Response Models. Consider now the general definition of the I_k normalized frequency response quantity of Eq. (3), Section 2.2, with $k = D$, 0, and 1. For $k = D$, U_k in that equation is the complex dielectric constant, $\varepsilon(\omega)$, and for the other two values, U_k is the complex resistivity, $\rho(\omega)$. Now I_k may be calculated from either a distribution of relaxations times or from a temporal correlation function: see Macdonald [1996, 2002d] and Section 2.1.2.3. Although the $\rho_{0\infty}$ and $\rho_{1\infty}$ quantities entering into the definition of U_0 and U_1 are usually either zero or negligibly small, they may be large enough to affect the frequency response of the model at very high frequencies (Macdonald [2002d]). They will be taken zero for most of the present work. Then it follows that we may write $\rho_0(\omega) = \rho_0 I_0(\omega)$ and $\rho_1(\omega) = \rho_0 I_1(\omega)$, where we ignore the distinction between ρ_{00} and ρ_{01}.

The stretched-exponential temporal response of Eq. (63), Section 2.1, a versatile and theoretically plausible correlation function, is one whose corresponding frequency behavior is now called Kohlrausch–Williams–Watts or just Kohlrausch [1854] model response, denoted here by Kk. It is also now customary to replace the α of the stretched-exponential equation by β or β_k, with $k = D$ or 0. The $k = D$ choice may be related to KD-model dispersive frequency response involving a distribution of dielectric relaxation (properly "retardation") times, and the $k = 0$ and 1 choices to two different distributions of resistivity relaxation times and thus to K0 and K1-model responses, respectively. Note that the β_1 parameter of the important K1 model is not directly related to stretched exponential temporal response, as are the other Kohlrausch models, but the DRTs of the K0 and K1 models are closely related (Macdonald [1997a]). Further, although the KD and K0 models are identical in form, they apply at different immittance levels and so represent distinct response behaviors.

No closed form expressions are available for the frequency responses of the Kk models for arbitrary β_k values but algorithms for calculating such responses and for fitting data with them are included in the free LEVM CNLS fitting program (Macdonald and Potter [1987], Macdonald [2000a]) and are very accurate for $0.3 \leq \beta_k \leq 0.7$ and somewhat less accurate outside this range. Further, LEVM also includes closed-form exact-response expressions for the choices $\beta_k = 1/3$ and $1/2$.

Although defects in the 1973 OMF K1-model approach of Moynihan and associates [1973] have been pointed out for the last 10 years, papers continue to be published that use the OMF and ignore criticisms of it. It is therefore worthwhile to discuss it and its corrected version, the corrected modulus formalism (CMF), in order to make the issues involved clear to the reader, who can then make an informed choice between the two approaches. Although they both use the K1 response model, the OMF and CMF approaches are nevertheless crucially different.

Since the OMF response model was originally derived at the modulus level, let us begin by writing for the K0 model, $M_0(\omega) = i\omega\varepsilon_V\rho_0 I_0(\omega)$. In contrast, the OMF analysis (Moynihan *et al.* [1973]) led to the following result for the $M_1(\omega)$ response of the K_1-model in terms of $I_0(\omega)$,

$$M_1(\omega) = i\omega\varepsilon_V\rho_0 I_1(\omega) = [1 - I_{01}(\omega)]/\varepsilon_Z \tag{1}$$

where ε_Z was defined as $\varepsilon_{D\infty}$, now written by supporters of the OMF as ε_∞. The subscript 01 is used here to indicate that $I_{01}(\omega)$ is just $I_0(\omega)$ in form but involves β_1 rather than β_0.

The OMF K1 model of Eq. (1), derived from a purely conductive-system correlation function, improperly mixes together conductive-system and dielectric-system responses through its identification of ε_Z as $\varepsilon_{D\infty}$. This identification leads to a world of problems (e.g. Macdonald [1996, 2002a, 2004]) vitiating this approach and implying that the OMF should be replaced by the CMF or by a superior model.

The CMF correction is simple: ε_Z in Eq. (1) is defined as the limiting dielectric constant $\varepsilon_{C1\infty} \equiv \varepsilon_{C1}(\infty)$, a purely conductive-system non-zero quantity associated only with charge-carrier motion and defined below. Except for the explicit introduction of ε_Z, the essence of the 1973 OMF derivation of Eq. (1) appeared in the earlier work of Macdonald and Barlow [1963]. Incidentally, for the K0 model, $\varepsilon_{CO\infty} \equiv \varepsilon_{CO}(\infty) = 0$. For both the K0 and CMF K1 models, one therefore needs to account for the endemic presence of $\varepsilon_{D\infty}$ by including a free dielectric-constant fitting parameter, ε_x, in the composite fitting model, now designated the CK0 model for K0 response and the CK1 for the CMF K1 situation. Then for the K0 model $\varepsilon_\infty = \varepsilon_x$, and for the CK1 $\varepsilon_\infty = \varepsilon_{C1\infty} + \varepsilon_{D\infty}$. The separate existence of $\varepsilon_{C1\infty}$ is not recognized by users of the OMF. Note that CK0 and CK1 fits of the same data lead to nearly the same estimates of ε_∞.

It has sometimes been found useful to replace the ideal capacitance represented by $\varepsilon_x = \varepsilon_{D\infty}$ by a parallel constant-phase element, the PCPE, $\varepsilon_{PC}(\omega) \equiv A_{PC}(i\omega)^{-\gamma_{PC}}$, with $0 \le \gamma_{PC} < 1$, reducing to a nearly ideal capacitance when $\gamma_{PC} << 1$ so that $A_{PC} \cong \varepsilon_{D\infty}$. The resulting composite model has been designated the PK1. A series CPE, the SCPE, $\sigma_{SC}(\omega) \equiv \varepsilon_V A_{SC}(i\omega)^{\gamma_{SC}}$ with $0 \le \gamma_{SC} \le 1$, has often been found satisfactory for modeling electrode effects, and it represents the effect at the complex resistivity level of a completely blocking series capacitance when $\gamma_{SC} = 1$. When SCPE response is combined with that of the CK1, the result is written as the CK1S model. For the data fitting described in the next section, it turns out that a more complicated model is needed to represent electrode effects more exactly.

The OMF K1 was derived by considering electric field decay at constant dielectric displacement and is thus a macroscopic response model. It has been shown, however, that the CMF K1, with $\varepsilon_Z = \varepsilon_{C1\infty}$, is completely isomorphic in form with the famous stochastic-transport microscopic analysis of Scher and Lax [1973a], a continuous-time, random-walk hopping model. The extended version of this model (Macdonald [2002d]) leads to response of exactly the form shown in Eq. (1), involving a quantity equivalent to $I_{01}(\omega)$ derived by Fourier transform from an initially unspecified correlation function associated with a waiting time distribution for hopping. It is the specific stretched-exponential choice for this function that leads to explicit K1 response. These considerations show that the K1 may be derived by considering either macroscopic or microscopic processes, and such generality possibly accounts for the ability of the CK1 to fit a variety of conductive-system frequency-response data exceptionally well (e.g. Macdonald [2000b, 2002a, 2003a]).

The OMF expression for $\varepsilon_Z = \varepsilon_\infty$ may be written (Macdonald [1996, 2001c, 2002d])

$$\varepsilon_\infty = \sigma_0 \langle \tau \rangle_{01} / \varepsilon_V = \varepsilon_{Ma} \langle x \rangle_{01} = \varepsilon_{Ma} \beta_{1O}^{-1} \Gamma(\beta_{1O}^{-1}) \tag{2}$$

where the averages are over the resistivity DRT for the K1 model, and the OMF β_1 is designated as β_{1O} to distinguish it from that of the CMF, β_{1C}. Here the Maxwell quantity ε_{Ma} is

$$\varepsilon_{Ma} \equiv \sigma_0 \tau_o / \varepsilon_V \tag{3}$$

$x \equiv \tau/\tau_o$; and τ_o denotes the characteristic relaxation time of the K1 model, and it will be used for other models as well. The part of Eq. (2) involving the gamma function is only appropriate in the absence of cutoff of the K1 distribution of relaxation times (Macdonald [1996, 2001c]).

In contrast, for the CMF K1 dispersion model, $\varepsilon_Z = \varepsilon_{C1\infty}$, where

$$\begin{aligned} \varepsilon_{C1\infty} &= \varepsilon_{Ma} / \langle x^{-1} \rangle_1 = \varepsilon_{Ma} \langle x \rangle_{01} = \varepsilon_{Ma} \beta_{1C}^{-1} \Gamma(\beta_{1C}^{-1}) \\ &= \left[\gamma N (qd)^2 / (6 k_B \varepsilon_V) \right] / T = A/T \end{aligned} \tag{4}$$

and N is the maximum mobile charge number density; γ is the fraction of charge carriers of charge q that are mobile; and d is the rms single-hop distance for the hopping entity. The high-frequency-limiting effective dielectric constant, $\varepsilon_{C1\infty}$, associated entirely with mobile-charge effects, is likely to arise from the short-range vibrational and librational motion of caged ions.

Comparison of CMF equations with those of the Scher–Lax hopping model (Macdonald [2002d]) shows that the K1 mean relaxation time, $\langle \tau \rangle_{01} \equiv \tau_o \langle x \rangle_{01}$, is identical with the mean hopping time of the microscopic model, also defined as the mean waiting time for a hop. The term involving N in Eq. (4), not included in the OMF, is fully consistent with the Scher–Lax model predictions. In practice, fits of the same data with the OMF K1 and with the CK1 of the CMF approach lead to very different estimates of τ_o and of β_{1O} and β_{1C}.

We expect that the quantities in the square brackets of Eq. (4) are usually temperature independent, so the fitting parameter A is then itself independent of temperature. It follows that in the usual case where τ_o is thermally activated, $T\sigma_0$ is activated with the same activation energy (Macdonald [2002a]). The presence of the N term of Eq. (4) shows that as the ionic concentration approaches zero, $\varepsilon_{C1\infty} \to 0$ and so $\varepsilon_\infty \to \varepsilon_{D\infty}$, requiring that $\varepsilon_{Ma} \to 0$ as well, in accordance with CMF fit results. The situation is different for the OMF expression of Eq. (2), however. In this case, OMF fits show that both ε_∞ and ε_{Ma} approach the same constant value, that of $\varepsilon_{D\infty}$. There is then no dispersion, and the response reduces to that of single-time-constant Debye behavior.

Fits of frequency-response data for a variety of materials, temperatures, and concentrations lead to β_{1C} estimates all very close to 1/3. But OMF fits, particularly of data in $M''(\omega)$ form, the usual OMF approach, invariably yield appreciably larger values of β_{1O}, ones that approach unity as the ionic concentration decreases or as the temperature increases. Such dependence led most users of the OMF to conclude that the correlation between charge carriers decreased as β_{1O} increased. But constancy of β_{1C} and the lack of Coulomb interactions in the well-fitting CMF microscopic model fail to support this supposition. For most data, it has been found that CK1 fits are superior to CK0 fits of the same data, but even in situations where these fits are comparable, CK1 ones are preferable to CK0 ones because $\beta_{1C} \cong 1/3$ estimates are virtually independent of temperature and ionic concentration, while CK0 β_0 estimates depend strongly on these variables (Macdonald [2002a, 2003a]).

Note that OMF data fitting with LEVM leads to estimates of the free parameters ρ_0, τ_o, and β_{1O}, and ε_∞ may then be calculated using Eq. (2). When β_{1C} is taken constant at the value of 1/3, CMF fits yield estimates of ρ_0, τ_o, and $\varepsilon_x \cong \varepsilon_{D\infty}$, and $\varepsilon_{C1\infty}$ may then be calculated using Eq. (4), with $\varepsilon_{C1\infty} = 6\varepsilon_{Ma}$ for this value of β_{1C}. Although ρ_0 estimates are usually nearly the same for the two types of fits of the same data, as are also calculated values of ε_∞, β_{1O} is always appreciably larger than 1/3, and CMF τ_o estimates are generally at least an order of magnitude smaller than those from OMF fits.

When the OMF approach is used to fit experimental data, a fatal flaw appears, one that invalidates any conclusions based on such fitting results. For good data, all CMF fits yield closely the same estimates of τ_o and β_{1C}, independent of the immittance level for the data. This is not the case, however, for OMF fits. They lead to inconsistent results such that fits of the data in $M(\omega)$ or $M''(\omega)$ form yield characteristically large values of β_{1O}, usually falling in the range $0.45 \le \beta_{1O} \le 0.55$ for mid-range temperatures and concentrations, while fits of the same data in $\sigma'(\omega)$ form yield values close to 1/3. As mentioned earlier, since $\varepsilon_{D\infty}$ has no effect on $\sigma'(\omega)$ response, K1 and CK1 fits at this level must yield the same estimates, and OMF and CMF fits are then equivalent. A table of such comparisons and further discussion of OMF problems appear in Macdonald [2004] and make it evident that the OMF treatment of $\varepsilon_{D\infty}$ as an intrinsic part of the K1 dispersive conductive-system model is incorrect.

Coupling and Cutoff Models. The Ngai coupling model (Ngai [1979, 1998]), discussed in Macdonald [1998, 2005a], has been used in many conductive-system data analyses by Ngai and his associates. It assumes that for times longer than t_c (a temperature-insensitive cross-over time of the order of 1 ps) the temporal response of the system is of stretched-exponential character, and for shorter times it is of ordinary exponential character. In its applications to frequency response behavior, the coupling model has made use of OMF estimates of β_{1O}, although the frequency-response model directly corresponding to stretched-exponential behavior is the K0, not the K1, and generally $\beta_{1O} \neq \beta_0$.

A superior alternative, the cutoff model, avoids this inconsistency, makes no use of the OMF, and is based on a cutoff of the K1 distribution of relaxation times at $\tau = t_c$. It does not involve the OMF assumption that the correlation between charge carriers decreases as β_{1O} increases for response at frequencies below $\omega_c = 1/t_c$, and it properly undergoes a transition to simple Debye response for frequencies greater than ω_c. Further, as shown in Macdonald [2005a], it leads not only to a smoother frequency-response transition around $\omega = \omega_c$ but also to satisfaction of the physical requirement that the K1 $\tau_o(T)$ never decreases below t_c as the measurement temperature becomes high. This requirement is not met by the $\tau_o(T)$ of the coupling model approach, suggesting that it should be superseded by the cutoff model. Both the coupling model and the cutoff one lead to non-Arrhenius behavior of $\sigma_0(T)$, with a transition from a low-temperature Arrhenius activation energy to a smaller apparent energy at high temperatures (Macdonald [1998, 2005a], León *et al.* [1998]).

Rationalization of the Barton, Nakajima, and Namikawa Relation. The Barton [1966], Nakajima [1972], and Namikawa [1975] empirical relation, usually designated by BNN, has played a useful role for some time in the analysis of dispersed frequency response data (e.g. Dyre [1988], Macdonald [1996], Dyre and Schrøder [2000], Porto *et al.* [2000]). It involves a loosely defined parameter, p, expected to be of order 1, and Nakajima and Namikawa believed that it arose from correlation between electrical conduction and dielectric polarization, apparently because it involved both measured dc conductivity and a dielectric strength quantity $\Delta\varepsilon$.

But as we have seen, for a conductive system both σ_0 and $\Delta\varepsilon = \varepsilon'(0) - \varepsilon'(\infty) = \varepsilon_0 - \varepsilon_\infty$ may arise entirely from mobile charge effects and not involve bulk dielectric effects at all. Then $\Delta\varepsilon = \Delta\varepsilon_{C1} \equiv \varepsilon_{C10} - \varepsilon_{C1\infty}$ for the CK1 model, and $\Delta\varepsilon = \Delta\varepsilon_{C0} \equiv \varepsilon_{C00}$ for the CK0 one. It was indeed pointed out by Macdonald [1996] that the K1 conducting-system model could lead to a quantitative value for p, one that depended on the value of β_{1C}.

Here it is shown that the BNN expression is most reasonably interpreted as arising entirely from charge motion, and if the K1 fit value of $\beta_{1C} = 1/3$ is a universal value, then the value of p is fully defined and the BNN equation is just a natural consequence of the apparent universal applicability or quasiuniversality of the conductive-system CK1 model with $\beta_{1C} = 1/3$. For ion-conducting homogeneous glasses and single crystals with charge motion allowed in all three dimensions it has been shown theoretically, in two independent ways that 1/3 is the only possible value of β_{1C} and that the resulting high-frequency-limiting-response power-law exponent is 2/3 (Macdonald [2005b], Macdonald and Phillips [2005]). Consistent with these results, it follows that CK0 model fits of such response lead to $\beta_0 = 2/3$ when the data extend to sufficiently high frequencies.

The BNN equation may be expressed as

$$\Delta\varepsilon = p^{-1}(\sigma_0/\varepsilon_V\omega_p) = p^{-1}(\tau_p/\tau_0)\varepsilon_{Ma} = p^{-1}(\nu_o/\nu_p)\varepsilon_{Ma} \tag{5}$$

Here as usual, τ_o is the characteristic response time of a fitting model such as the CK1. Further, $\omega_p \equiv 2\pi\nu_p = 1/\tau_p$, where ν_p is the frequency at the peak of the dielectric loss curve, $\varepsilon_s''(\nu)$, and $\nu_o \geq \nu_p$. For $\beta_{1C} = 1/3$, the K1 model leads to $\varepsilon_{C1\infty} = 6\varepsilon_{Ma}$ and to $\varepsilon_{C10} = 60\varepsilon_{Ma}$ (Macdonald [2001c, 2005b]). Therefore, $\Delta\varepsilon = 54\varepsilon_{Ma}$ and one may write for this situation $p = (\nu_o/\nu_p)/54$.

Sidebottom [1999] noted the similarity between the BNN equation and a scaling factor he proposed. This similarity arises because his result, appropriate for situations where the frequency response shape of the model is temperature independent, the situation for the K1 model with a constant $\beta_{1C} = 1/3$ value, is a simplification of scaling factors associated with K0 and K1 models with variable β_k, as discussed in Macdonald [2001c]. Of course with accurate CNLS fitting, scaling is unnecessary. The success of the Sidebottom scaling approach is further indirect evidence of the widespread applicability of the CMF CK1 model with fixed $\beta_{1C} = 1/3$.

From nearly exact calculations of K1 model $\varepsilon_s''(\omega)$ synthetic data derived from the parameter estimates of experimental data fits of the next section, with the electrode contributions present or removed, one finds that the ν_o/ν_p ratio is about 95 and 89, respectively, leading to p estimates of about 1.77 and 1.65. The 1.65 value is the appropriate one for K1-alone response and is universal to the degree that Eqs (1) and (4) are applicable and $\beta_{1C} = 1/3$. Although many data fits suggest that this value of β_{1C}, is a constant for CK1 fits, one would expect that as $\beta_{1C} \to 1$, p should also approach unity in the limit, and, for example, when $\beta_{1C} = 0.5$, one obtains $p \cong 1.27$.

Over the years since the introduction of the BNN equation, published p values have mostly fallen in the range of 0.5 to 10 but are often close to unity. Accurate estimation of p directly from experimental data is uncertain when electrode effects are significant and/or when the data range is too small to lead to good estimates of ε_0 and ε_∞. It is therefore appropriate to calculate p values from parameter values estimated from data fitting.

Although Hunt [1992] concluded that p cannot have a universal value, the present 1.65 value is consistent with most of the many BNN-related p estimates for experimental data presented by Dyre and Schrøder [2000] in their Figure 3, ones mostly slightly larger than unity. Such agreement is further evidence of the appropriateness of the CK1 model for many different materials. Earlier, Dyre [1988] quoted an estimate of p for a CTRW model different from the present Scher–Lax K1 one of only 0.42, while for their microscopic symmetric hopping model Dyre and Schrøder [2000] listed a value of 1.5 ± 0.4. The present results show that if CMF fitting is used, there is no need for the BNN since it is an automatic consequence of the applicability of such fitting. When CK1 CNLS fit parameters are available, however, the BNN equation with $p = 1.65$ may be used to obtain an accurate estimate of ν_p for the conducting-system part of the data alone.

Finally, Porto *et al.* [2000] have recently suggested that the BNN relation cannot apply for an appreciable range of concentrations because data fits show that $\Delta\varepsilon$ does not scale as N/T. But Eq. (4) shows that for the CMF K1 model $\varepsilon_{C1\infty}$ does indeed scale in this fashion and involves d^2 as well. Further, at constant β_{1C}, ε_{C10} and thus $\Delta\varepsilon$ also do so (Macdonald [2001c, 2002a]). Therefore, this criticism does not seem appropriate. To test the matter, estimates of p were calculated from CK1 CNLS fits of $x_cK_2O{\cdot}(1 - x_c)GeO_2$ germanate glasses with the relative ionic concentration, x_c, equal to 0.2 and 0.02, data kindly provided by Drs. Jain and Krishnaswami [1998]. The p estimates were 1.64 and 1.65, respectively, thus well verifying the appropriateness of the BNN equation over a considerable concentration variation.

Nearly Constant Loss Models. Nearly constant loss (NCL) is evidenced by a power-law dependence of $\sigma'(\omega)$ on frequency with an exponent very close to unity, leading to $\varepsilon_s''(\omega)$ loss response that varies only slightly over a substantial frequency range. It may appear directly at low temperatures or may contribute significantly to $\sigma(\omega)$ response at the high end of the measured frequency range. In the first case, NCL is dominant and thermally activated hopping response is completely negligible (Macdonald [2001a, 2003b]. In the second case, hopping is dominant over most of the frequency range.

Excellent reviews and discussions of NCL behavior in ionically conducting glasses appear in Ngai [1999] and Roling *et al.* [2001]. Although most authors believe that NCL arises from the restricted motion of caged ions or groups of atomic species, very few quantitative NCL models have been proposed. An important early composite one may be written as $\sigma'(\omega) = \sigma_0[1 + (\omega\tau_o)^n] + A\omega^s$, with $0 < n \leq 1$ and $s \approx 1$ (Lee *et al.* [1991], Nowick *et al.* [1998]). The first term represents universal dynamic response, as discussed earlier, and constant loss occurs when $s = 1$, not a viable situation for a finite frequency range.

Although this composite model implies the additivity of hopping and NCL effects, the appropriateness of such additivity has been challenged by León *et al.* [2001] and Rivera *et al.* [2002]. They suggested an alternate serial (not series) picture in which NCL ceases to exist when hopping begins and ions begin to exit their cages. This is not a quantitative model, and their work dealt primarily with $\sigma'(\omega)$ response. Fitting of both synthetic and experimental complex data provides strong evidence, however, that additivity should not be rejected, and analysis using a quantitative complex model such as the PK1 suggests that hopping and NCL effects can exist simultaneously in a crossover region of finite length (Macdonald [2001a–c, 2002a]). Here, the parts of the model are in parallel electrically and additivity is ensured. The PCPE part of the expression models NCL behavior and can extend over an unrestricted frequency range.

Although a PCPE may be used to model NCL data with equal slopes for both $\sigma'(\omega)$ and $\varepsilon_s''(\omega)$ since they both involve the same $\gamma_{PC} << 1$ exponent, some data may be better represented by such power-law response for $\sigma'(\omega)$ but by a function that yields a very close approximation to constant loss for the $\varepsilon_s''(\omega)$ part of the response (Nowick *et al.* [1998]). In the absence of hopping, just the series combination of an ideal capacitor and a CPE can yield such behavior with very nearly constant loss over several decades of frequency (Macdonald [2001a]).

It was first shown in 2002 that the CPK1 composite model, where both C and a PCPE are in parallel with K1, could be used to represent frequency-independent undispersed $\varepsilon_{D\infty}$ behavior, hopping behavior, and NCL (Macdonald [2002a]). This work, in turn, suggested that the model could be made more physically plausible by an effective medium approach, one that might be able to represent both kinds of NCL behavior, as well as possibly non-negligible hopping effects. The resulting effective medium model, the EMK1, indeed met this objective well (Macdonald [2003b]). It is based on the assumption of a background involving a constant $\varepsilon_{D\infty}$ term and a volume fraction, η, of "inclusions" associated with ions vibrating over a limited region and represented by a PCPE. Thus, even in the absence of the K1 part, $\varepsilon_{EM}(\omega)$ is complex.

The EMK1 model, unlike the CPK1 one, leads to physically plausible low-frequency-limiting slopes for $\varepsilon(\omega)$ and other immittance functions, as well as equal or superior fits to those of the CPK1. As a first approximation, η is set equal to the relative ionic concentration, x_c. Synthetic data for $\varepsilon''_{EM}(\omega)$ extending over many decades of frequency and for a wide range of η values showed that although there is no finite range of exact constant-loss behavior, such response is well approximated for η near 0.25. In addition, when the response is approximated by a power-law model, the resulting very small exponent may be either positive for $\eta << 1$ or negative for $\eta \geq 0.25$ over the higher-frequency region of the response.

There are two important questions arising from the present model discussions. First, a microscopic model needs to be developed that leads to $\beta_{1C} \cong 1/3$ and is less approximate than the Scher-Lax one and second, a microscopic model is also needed that yields response like the present effective medium model and takes explicit account of the detailed interactions, electromagnetic and otherwise, between vibrating ions and bulk dipoles.

4.2.3 Illustration of Typical Data Fitting Results for an Ionic Conductor

CNLS fitting has been little used by most workers who have analyzed frequency-response data for solid ionic conductors. The majority of published work deals primarily with either $\sigma'(\nu)$ or $M''(\nu)$ response, but not usually with both or with simultaneous fitting of real and imaginary parts of an immittance data set. An apparent advantage of the fitting and analysis of $\sigma'(\nu)$ data alone is that it and its direct transform, $\varepsilon''(\omega) \equiv \sigma'(\omega)/\omega\varepsilon_V$, are the only immittance-level parts that include no effects from $\varepsilon_{D\infty}$, as already mentioned in Section 4.2.2.1. But much more can generally be learned by considering full complex response at other immittance levels. Here we will only deal with data that do not extend to high enough frequencies or low enough temperatures to require a nearly constant loss contribution.

It is therefore worthwhile to illustrate, for a typical data set, the usefulness of CNLS fitting and of various plots of the results. For generality, the data set selected is one for which both bulk dispersion and electrode effects are non-negligible. It was kindly provided by Dr Carlos León and involves the fast ionic conductor $Li_{0.5}La_{0.5}TiO_3$, measured at T = 225 K (León *et al.* [1998]). This set is designated hereafter as LLTO.5. Fitting was carried out using the O circuit of the LEVM program. To allow independent work with this data set, its full LEVM input file, 225Z36EL, has been included in the LEVM FITTESTS folder of test files for the O circuit.

Figure 4.2.1 shows the full O circuit. For LEVM, only those circuit elements that are given non-zero values are used in fitting. Here "DE" designates a distributed circuit element, one that can be selected from a large number of different elements available in LEVM. DED involves a dielectric distributed element, such as the DSD, a dielectric-system dispersive element. Similarly, DEC designates a conductive distributed element, such as the CSD. In LEVM, the series inductance shown in the figure may be replaced by a short circuit, a resistor, or a capacitance, C_S.

Since it was initially established that the LLTO.5 data sets for different temperatures involved thermally activated response and were therefore of CSD charac-

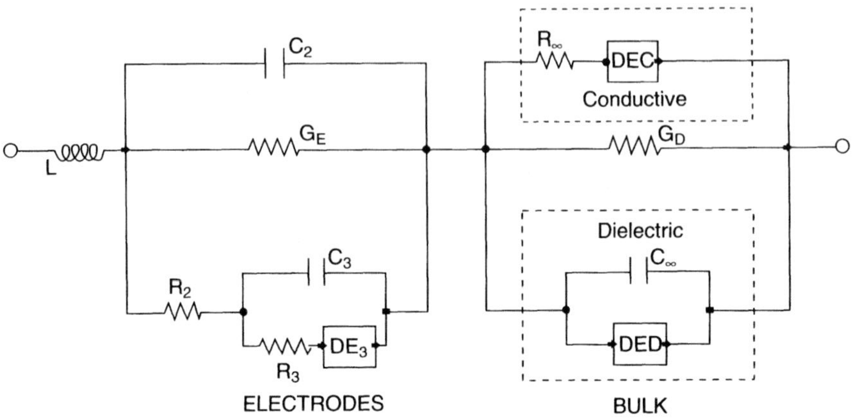

Figure 4.2.1. The LEVM fitting circuit O. It may be used as shown for fitting immittance data in raw or specific form. The DE blocks may each be selected as any one of the many available distributed-circuit-element response models.

ter, we begin by carrying out CMF fits of the T = 225 K data. The bulk response was thus represented by the K1 model in the DEC part of the circuit. It involves the parameters ρ_0, τ_o, and β_{1C}, but, as usual, a β_{1C} value of 1/3 yielded best results. Therefore, this value was taken as fixed for all the present fits. When the CMF CK1 model was employed, $\varepsilon_{D\infty}$ was represented by the C_∞ element of the circuit. As usual, R_∞ was found to contribute nothing to the fits and was thus not used thereafter.

For blocking electrodes the simplest element to represent their effect is a series capacitance, C_S, but electrode processes are generally too complicated for adequate representation by a single capacitance. The next level of complexity, often found adequate, is to use a series constant-phase element, the SCPE, in the DE_3 position of the circuit. A recent analysis of the use of a CPE for modeling electrode behavior appears in Bisquert *et al.* [1998]. For the present data, for which electrode effects are far from negligible, it was found that they were best represented by a SCPE in parallel with the C_3 capacitance of the circuit, all in series with C_S, involving a total of four free fitting parameters. The full CK1 model including these free electrode-related parameters is termed the CK1EL and involves a total of seven free parameters.

The CK1EL CNLS fit of the data at the complex resistivity level using LEVM with proportional weighting led to the estimates $\varepsilon_{D\infty} \cong 83.08$, $\rho_0 \cong 1.784 \times 10^5$ ohm-cm, and $\tau_o \cong 4.488 \times 10^{-8}$ s. In addition, the estimate for the γ_{SC} parameter of the SCPE was about 0.641. The fit also led to the estimates $\varepsilon_0 \cong 254$, $\varepsilon_{C10} \cong 171$, $\varepsilon_{C1\infty} \cong 17.1$, $\varepsilon_\infty \cong 100$, and $\Delta\varepsilon = \Delta\varepsilon_{C1} \cong 154$. Exactly the same parameter values were obtained for proportional-weighting fitting at the complex modulus level. The relative standard deviation of the overall fit, S_F, was 0.0072, indicating an excellent result.

When electrode effects were represented only by a SCPE, the CK1S model, S_F increased appreciably to 0.015. This fit led to a larger γ_{SC} estimate of about 0.897 and to the slightly different estimates for $\varepsilon_{D\infty}$, ε_0, ε_{C10}, $\varepsilon_{C1\infty}$, ε_∞, and $\Delta\varepsilon$ of about 79,

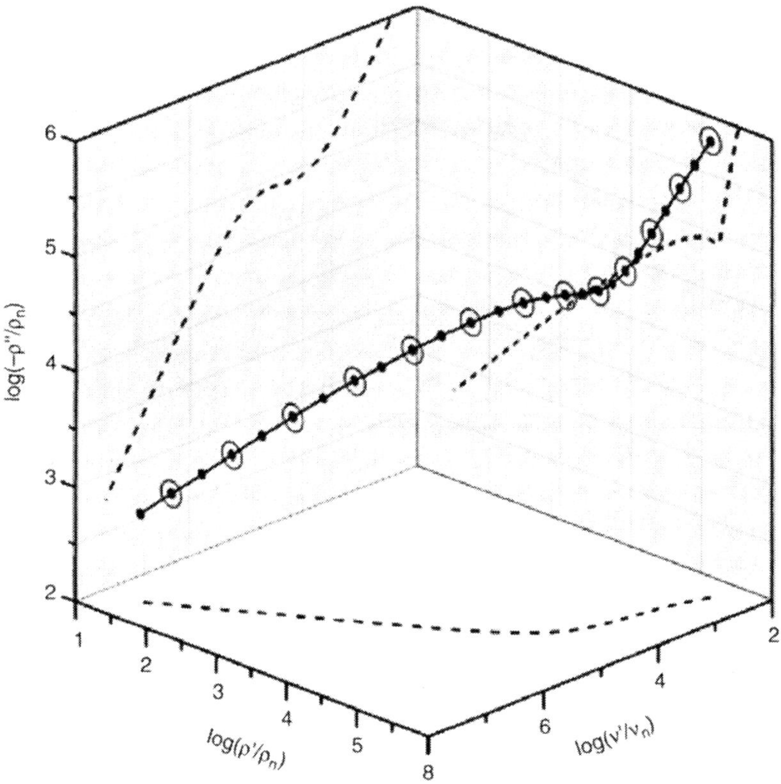

Figure 4.2.2. Three-dimensional log–log–log plot, with planar projections. The solid lines and solid circles show the data, and open circles identify points from CK1EL-model CNLS fitting of the $Li_{0.5}La_{0.5}TiO_3$ data (denoted LLTO.5 hereafter) at the complex resistivity level. The quantities with a subscript "n" in the axes names of this and subsequent figures are of unity magnitude and are included to make the arguments of the logarithms dimensionless as they should be.

254, 175, 17.5, 97, and 157, respectively. In the limit of low frequencies, the four-parameter model for electrode behavior is dominated by the blocking capacitor, C_S. Its value, expressed in dielectric-constant form was more than 30 times larger than the CK1EL estimate of ε_0.

Figure 4.2.2 presents a 3-D log–log–log plot of complex-resistivity data as well as fit points for the CK1EL model fit. The projections in the three planes involve only the data. The 3-D line shows every other one of the data points and every fourth fit point. Since the open-circle fit points enclose their corresponding data points symmetrically, no deviations are evident. The projection lines at the two back planes clearly show the transition to electrode-related power-law behavior toward the low end of the frequency scale. Space restrictions preclude presentation here of the three other 3-D immittance plots.

Figure 4.2.3 shows the behavior of the real and imaginary parts of the complex modulus. In addition to the CK1EL-fit lines, those for the CK1 and K1 parts of the

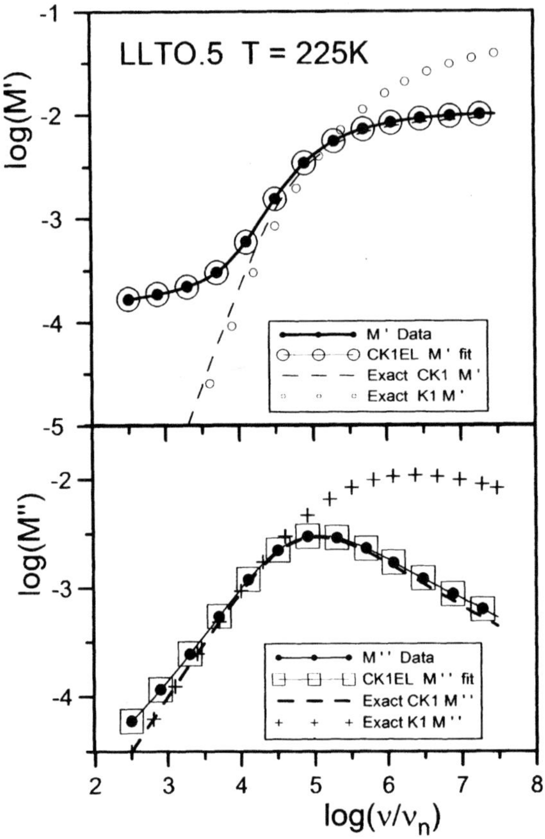

Figure 4.2.3. Log–log complex modulus data and fit results for $M'(\omega)$ and $M''(\omega)$ obtained from fitting the $M(\omega)$ complex data with the CK1EL model. In addition, predictions for the CK1 and K1 parts of the full model are shown.

model are also shown. They were calculated using, in LEVM, the appropriate parameter values found from the full CK1EL fit, and thus they are virtually exact representations of the model behavior for these values. The present M' results show that electrode effects are dominant at low frequencies and have only a minor effect at the high-frequency end of the data range. As one would expect, the difference between the CK1 and K1 results, associated entirely with $\varepsilon_{D\infty}$, becomes great at the high-frequency end.

It is often been stated that a virtue of plotting and analyzing data in M'' form is the resulting suppression of electrode effects. The present results demonstrate such suppression near the M'' peak, but it clearly diminishes as the frequency departs from the peak value. Further, since the same parameter estimates are obtained for both $\rho(\omega)$ and $M(\omega)$ fits when proportional weighting is employed, the suppression is graphical but not significant for least-squares fitting. Finally, it is evident that the peak of the K1 M'' curve appears at much higher frequencies than that of the data

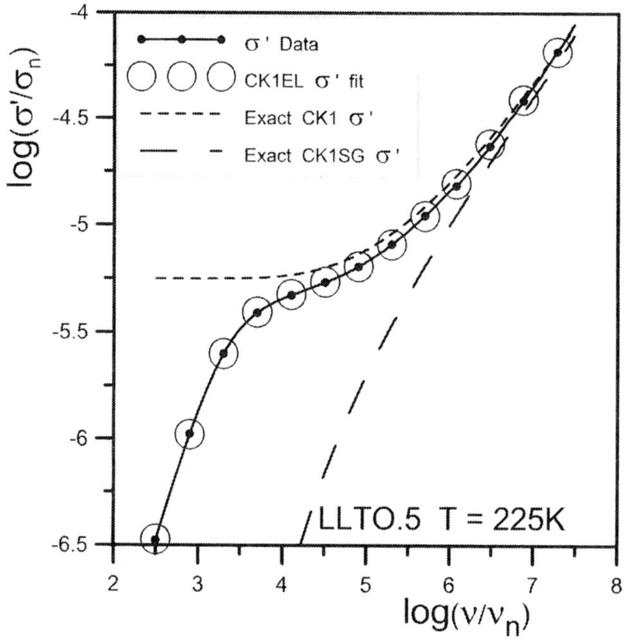

Figure 4.2.4. Log–log $\sigma'(\omega)$ data and fit results from CNLS fitting as in Figure 4.2.3. The CK1SG results eliminate both electrode effects and those of σ_0.

and the CMF CK1 one and the breadth of the K1 curve at half height is also much larger. This difference is associated with the CK1 value of β_1 of 1/3 and a value greater than 0.5 found for OMF fitting of the $M(\omega)$ data, as expected from the discussion in Section 4.2.2.

Figure 4.2.4 compares $\sigma'(\omega)$ data and CK1EL fit values, as well as individual contributions to the full model. For this immittance level, there is no effect from $\varepsilon_{D\infty}$, so here CK1EL and K1EL fit results are equivalent. The CK1SG results were obtained by first setting the G_D parameter of the Figure 4.2.1 circuit to $-\sigma_0$. This, together with the K1 parameter estimates obtained from the CK1EL fit, were then used in LEVM to calculate the resulting exact response of the combination and thus to eliminate the effect of σ_0. It is evident that, as expected, at the high frequency end of the range the $\sigma'(\omega)$ response is nearly entirely associated with the ac part of the K1 model, with only a small contribution from electrode effects apparent. Further, the data curve shows that no accurate value of σ_0 could be directly estimated from it, making it essential that all fits should account for electrode effects.

Rather than present $\sigma''(\omega)$ fitting results, it is appropriate to show those for the corresponding $\varepsilon'(\omega)$, related to $\sigma''(\omega)$ by a factor of $1/\varepsilon_V\omega$. Such results are presented in the top part of Figure 4.2.5. It is clear that the $\varepsilon'(\omega)$ data curve alone does not allow one to obtain a reasonable estimate of ε_0 from it. Removal of the electrode effects obtained from the full CK1EL fit leads to the low- and high-frequency plateau values ε_0 and ε_∞, respectively, while subsequent removal of $\varepsilon_{D\infty}$ leads to the limiting conductive-system K1-model quantities ε_{C10} and $\varepsilon_{C1\infty}$. It is again evident that elec-

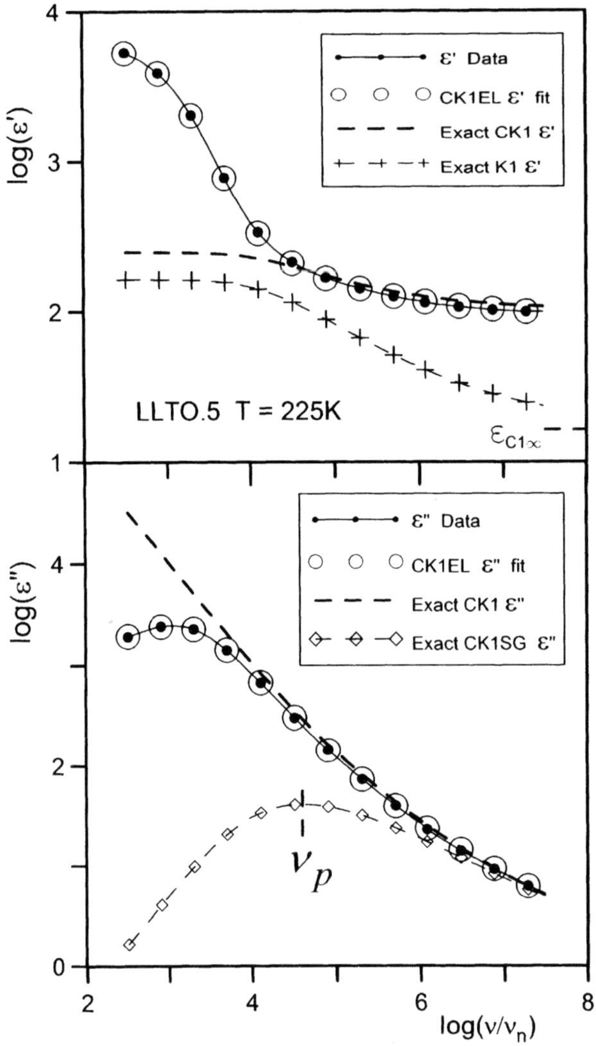

Figure 4.2.5. Log–log $\varepsilon(\omega)$ data and fit results from CNLS fitting as in Figure 4.2.3. The exact CK1 response is that without electrode effects, and the K1 response eliminates the effect of $\varepsilon_{D\infty}$ as well and shows the approach of the data toward the limiting $\varepsilon_{C1\infty}$ value. The peak of the CK1SG $\varepsilon_s''(\omega)$ curve is denoted by ν_p.

trode effects play a minor but not completely negligible role at high frequencies. The bottom part of the present figure shows $\varepsilon''(\omega)$ results, where again $\varepsilon_{D\infty}$ plays no role. The frequency at the peak of the CK1SG curve, that for $\varepsilon_S''(\omega)$, is shown by ν_p and is needed for the calculation of the BNN quantity p.

Finally, Figure 4.2.6 is a linear-scale complex-resistivity-plane plot. Here, to allow greater resolution, lower-frequency points than those shown have been

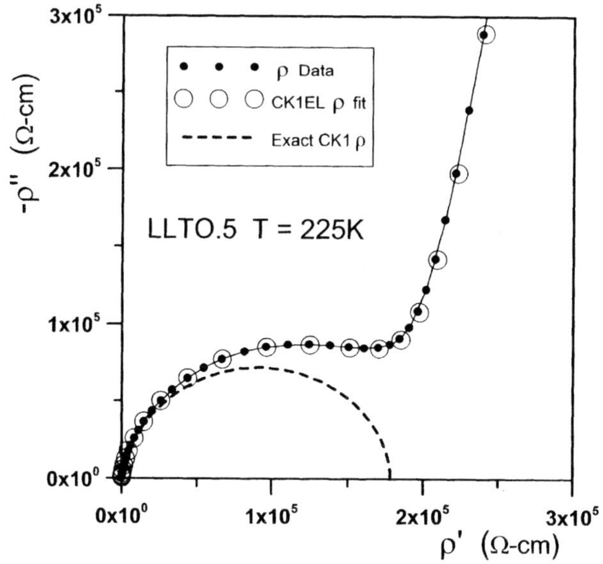

Figure 4.2.6. $\rho(\omega)$ data and fit results from CNLS fitting as in Figure 4.2.2. These complex-plane results demonstrate clearly the dominance of electrode effects up to quite high frequencies.

omitted. The results indicate that electrode effects remain important over much of the frequency range, even toward the higher frequencies. Comparison of corresponding data and fit points shows some very minor discrepancies for the spur part of the response, ones that are too small to be evident in log–log plots. The low-frequency end of the CK1-only part of the response approaches the axis at 90° as it should and defines the value of ρ_0. It is important to note that extrapolation of the electrode spur line down to the ρ' axis leads directly to an excellent estimate of ρ_0. This can be useful when the temperature is so high that little or none of the bulk arc is included in the measurement range and CNLS fitting may not have been carried out. However, such extrapolation fails for mixed ionic and electronic conduction situations.

In a full data analysis, one would first determine the most appropriate model and then use it to carry out fits for each different temperature available. Here, only partial results for fits of the present T = 225 K data with a few other models will be discussed. First, S_F values for CNLS proportional weighting fits with the CK0EL, OMF K1EL, and "DSD" EDAEEL models were all close to 0.007; excellent fits. Here, the EDAE model involves an exponential distribution of activation energies fitted at the complex dielectric level and assuming dielectric-system dispersion. Since the fits were all comparable, selection of a best model must depend on other criteria.

The CK0EL model led to CNLS estimates of ε_0, $\varepsilon_{C10} = \Delta\varepsilon$, and ε_∞ of about 255, 145, and 97, respectively. Since this model always involves $\varepsilon_{C0\infty} = 0$, it does not yield a separate estimate of $\varepsilon_{D\infty}$, but its β_0 and γ_{SC} estimates were about 0.487 and 0.635, respectively. Note that with $\beta_{1C} = 1/3$, $\beta_0 + \beta_{1C} \neq 1$ here. Even when the CK1EL and CK0EL models yield comparable fits and nearly the same estimates for some common parameters, the former, with fixed $\beta_1 = 1/3$, should be preferred because it yields not only a comparable fit with fewer free parameters but because it also leads to separate estimates of both $\varepsilon_{C1\infty}$ and $\varepsilon_{D\infty}$.

For the OMF K1EL model, $\varepsilon_0 = \varepsilon_{C10}$, $\Delta\varepsilon$, and $\varepsilon_\infty = \varepsilon_{C1\infty}$ values were all calculated from the CNLS fit parameters, leading to estimates of about 233, 118, and 115, respectively. They thus agree less well with the CK1EL and CK0EL fit results. The above $M(\omega)$ fit results used proportional weighting, but $M''(\omega)$ NLS fits with either proportional or unity weighting led to closely similar estimates. The β_{1O} and γ_{SC} values estimated for these fits were about 0.604 and 0.607, respectively. Finally, an OMF K1EL fit of the $\sigma'(\omega)$ part of the data, with electrode parameters fixed at their K1EL $M(\omega)$-fit values, led to ε_0, $\Delta\varepsilon$, and ε_∞ estimates of about 175, 156, and 18.5, respectively. The last value is clearly an estimate of the CK1 $\varepsilon_{C1\infty}$ quantity here. Further, the β_{1O} estimate was 0.338, very close to the fixed value of 1/3 for the CK1EL fitting. The stark inconsistency between the OMF $M(\omega)$ and $\sigma'(\omega)$ β_{1O} estimates, also observed in all other such published comparisons, is a clear indication of the failure of the OMF to take proper account of $\varepsilon_{D\infty}$. Therefore, it is a particularly inappropriate fitting model and should not be used.

Although the present data involve CSD rather than DSD behavior, it is of interest to fit at the dielectric level with a DSD model, one that involves a ρ_0 parameter separate from the dispersion model. The asymmetric EDAE model, available in LEVM, is appropriate for this situation and involves the bulk parameters $\Delta\varepsilon$, ε_∞, τ_o, and γ_E, where γ_E falls in the range $0 < \gamma_E \leq 1$. CNLS fitting using the EDAEEL model with proportional weighting led to estimates of the above quantities of 131, 109, 9.11×10^{-6} s, and 0.473, respectively. The prediction for ε_0 is therefore 240 and the estimate for ρ_0 was 1.73×10^5 ohm-cm.

The standard deviations of parameters common to both the CK1EL and EDAEEL fits were appreciably larger for the latter than for the former even though their overall S_F values were nearly the same. Not only does the EDAEEL model involve two more free parameters than does the CK1EL one, but its separate treatment of ρ_0 is inappropriate for a CSD situation. The present results clearly indicate that for the LLT0.5 CSD data, and probably for most such data, the CK1 model with $\beta_1 = 1/3$ is the most appropriate bulk fitting and analysis model. Its BNN p value was found to be 1.65, as was that for the EDAE fit, and that for the K0 was about 1.33.

4.3 SOLID STATE DEVICES

William B. Johnson
Wayne L. Worrell

In this section examples of several different applications of impedance spectroscopy (IS) will be presented. Four different devices have been chosen: solid electrolyte chemical sensors, secondary (rechargeable) batteries, photoelectrochemical devices, and semiconductor–insulator–electrolyte sensors. In each subsection one selected application of IS will be briefly summarized to indicate the utility of this technique in determining the parameters important to that device. These sections are not intended to provide an extensive review of the area, but rather to show the power of the technique to solid state researchers. Thus this section is not a description of every device to which IS has been applied. Devices not discussed here which have been studied using IS include, among others, ion selective membranes (Sandifer and Buck [1974], Buck [1980, 1982]) and high-temperature steam electrolyzers (Schouler *et al.* [1981]).

Before beginning a detailed discussion it will be helpful to indicate the advantages and limitations that IS has in general application. Particularly desirable features of IS inherent to each specific applications will be discussed in the subsections below. The most important advantage is the ability to determine all of the time constants associated with a given interface in one experiment. That means that it is possible to determine diffusive, electrochemical, and chemical rate constants for a process from a single impedance spectrum. Further, the impedance is measured with a small ac signal, and a dc bias voltage can be superimposed with the ac signal so that the impedance and the rate information can be determined under various conditions. Such potential control is particularly important for electrochemical systems because the applied potential influences the rate of electron transfer at the interface. By measuring the impedance in such systems as a function of applied potential (i.e. dc bias), it is possible to determine the importance of the electrochemical reaction step to the overall rate of the reaction.

Although the equipment necessary to measure impedance spectra is readily available from many different suppliers, it remains expensive ($20,000–40,000 in 2005 dollars) even before the purchase of the nearly mandatory computer required for control and data analysis. Another disadvantage of IS is that very careful cell design is required to minimize stray capacitances and inductances. In addition to requiring three-electrode cell arrangements, as with all electrochemical systems, lead effects, including length, shielding, and the nature of all electrical contacts leading to and from the sample, must be considered. Ideally, the impedance of the cell should be measured under the actual experimental conditions but in the absence of the sample. These results can be used to verify or correct the experimental results. However, it is usually sufficient to minimize the stray impedances so that their values are negligible in comparison to the sample impedances.

A final drawback of the technique is the cumbersome data analysis which is required to obtain the desired physical quantities from the impedance spectra. A model electrical circuit which approximates the physical process being examined must be formulated. The model parameters are then obtained by determining parameter values which give the best fit to the impedance data. Finally, the model must be correlated with the physical system to establish the reliability of the model and to establish that the model values determined from the fit are physically reasonable. If not, the model may have to be modified and the entire analysis process repeated. In Chapter 3 there is a detailed discussion of this entire procedure. Clearly the required analysis is not always straightforward and is usually quite involved.

4.3.1 Electrolyte–Insulator–Semiconductor (EIS) Sensors

Electrolyte–insulator–semiconductor (EIS) sensors are one of a larger class of chemically sensitive electronic devices (Zemel in Janata and Huber [1985]) which meld integrated circuit technology with traditional chemical sensor technology. The EIS device is composed of a doped semiconductor, normally Si, acting as a substrate for a thin insulating layer, normally an oxide or nitride, which can be immersed in an electrolyte containing a fixed concentration of an ionic species to be measured. General reviews of the construction (Huber in Janata and Huber [1985]), thermodynamics (Janata in Janata and Huber [1985]), and operation (Abe *et al.* [1979], Lauks and Zemel [1979], Bergveld and De Rooij [1981]) of such devices are available. Only a brief overview of the area will be given here so that those unfamiliar with these devices will be able to appreciate the application of IS described below.

A schematic of a typical EIS sensor is shown in Figure 4.3.1*a*. The operation of the sensor can easily be understood by considering the solid state analog of it, the metal oxide semiconductor (MOS) capacitor shown in Figure 4.3.1*b*. In the MOS device the capacitance is controlled by applying an external voltage between the gate and substrate. When there is a negative voltage relative to a *p*-type substrate the capacitance will be large because the holes in the substrate will be attracted to the insulator–semiconductor interface, giving rise to a wider region of dielectric material through which charge is separated. If the gate voltage is increased toward zero, the space charge layer in the capacitor will become narrower so the capacitance will decrease. As the voltage goes positive, the space charge layer in the semiconductor becomes narrower until it eventually disappears, and electrons build up at the surface of the semiconductor, forming what is normally called an *inversion layer*. This process is shown schematically in Figure 4.3.1*c*. For an *n*-type substrate, the capacitance curve is inverted (Figure 4.3.1*d*), as when $V_G < 0$, an inversion layer forms (holes at surface of *n*-type Si), and when $V_G > 0$, there is a wider space charge layer.

The EIS functions in exactly the same fashion except the gate is formed by a reference electrode in solution rather than a metal contact. The ability of the EIS to respond to ions in solution results from a modification of the electric charge distribution at the insulator–liquid and/or the insulator–semiconductor interface. Thus, at a given value of the reference potential the capacitance of the device will change depending on the ionic concentration in the solution. For example for the simple device shown in Figure 4.3.1*a*, it has been shown that the capacitance will respond to a change in pH in the solution (Bergveld [1970], Siu and Cobbold [1979], Leroy *et al.* [1982], Bousse [1982], Bousse and Bergveld [1983], among others).

One real advantage of these sensors lies in the fact that an ion-selective membrane can act as a gate directly on a field effect transistor (FET) (Janata and Huber [1985]). These ion-selective field effect transistors (ISFET, shown schematically in Figure 4.3.2*a*) again are the analog of a solid state device, the metal oxide semiconductor field effect transistor (MOSFET, Figure 2.2.2*b*).

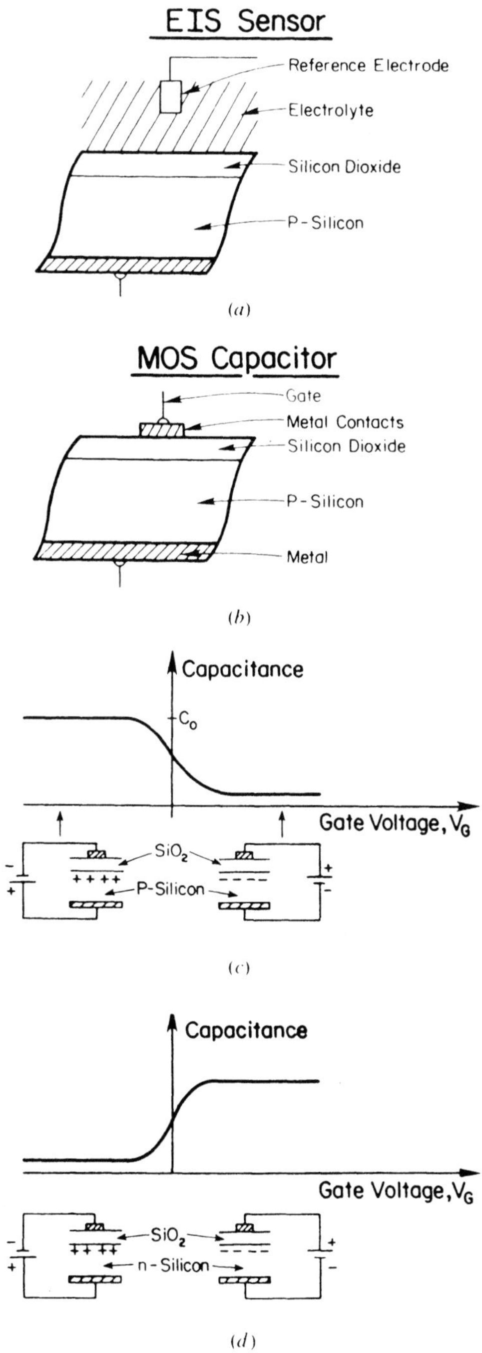

Figure 4.3.1. A schematic of a typical (*a*) EIS sensor and (*b*) an analogous MOS capacitor. (*c*) The capacitance of a MOS capacitor as a function of gate voltage for a *p*-type substrate, and (*d*) the equivalent response for an *n*-type substrate.

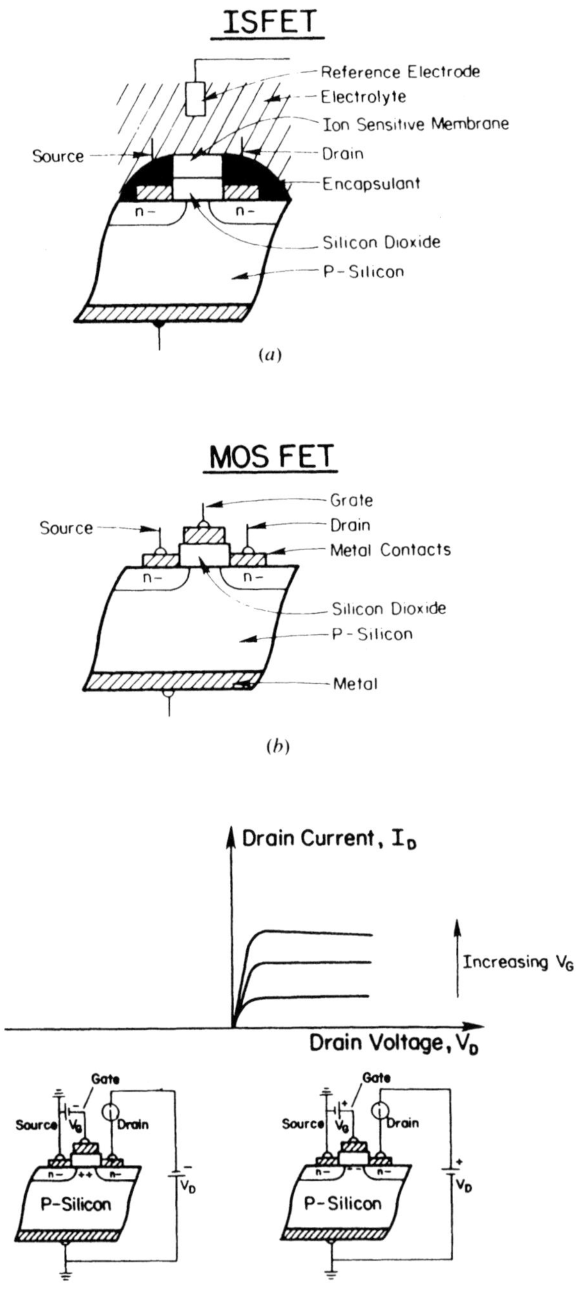

Figure 4.3.2. A schematic of a typical (*a*) ISFET device and (*b*) the analogous MOSFET device. (*c*) The current in the drain circuit as a function of the drain voltage at various different gate voltages. When the gate voltage is negative (left portion of Figure) the *n*–*p*–*n* junction will not conduct and there will be negligible drain current.

A MOSFET operates by controlling the concentration of charge carriers in *p*-type substrate between two *n*-type regions, called the source and the drain. When the gate is negatively biased with respect to the substrate, the region between the two *n*-type regions below the gate has no free electrons and the conductivity between the source and drain is very low. But as the voltage is increased until it becomes positive, at some point an inversion layer will form so there will be electrons available to form a channel. The conductivity between the source and drain will then increase. If there is a drain voltage supplied in this case, then a drain current can exist. At a given gate voltage, as the drain voltage is increased the drain current will saturate because the inversion layer is no longer of uniform thickness and becomes pinched off at the drain end. The transfer curve shown in Figure 4.3.2*c* results. As before, the ISFET operates in the same fashion except that an ion-selective membrane and reference electrode operate as a gate.

Although the impedance characteristics of the MOS devices are reasonably well understood (Nicollian and Goetzberger [1967], Nicollian and Brew [1982]), IS has not been applied nearly as widely to the EIS or ISFET devices. In this section the IS results of one of the simplest EIS devices, the Si–SiO_2–electrolyte pH sensor (Barabash and Cobbold [1982], Bousse and Bergveld [1983], Diot *et al.* [1985]) will be used to illustrate the relative advantages of the technique.

An equivalent circuit for an EIS device has been derived (Bousse and Bergveld [1983], adapted by Diot *et al.* [1985]). From left to right in Figure 4.3.3*a*, it con-

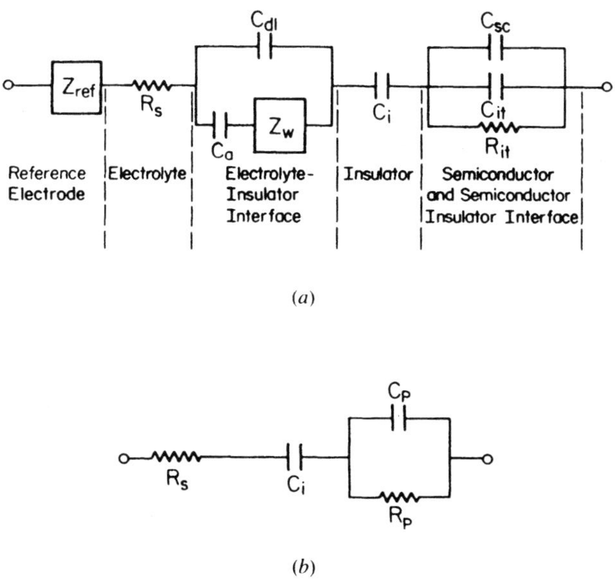

Figure 4.3.3. The equivalent circuit of EIS sensor shown in Figure 4.3.1*a* (after Bousse and Bergveld [1982]). (*b*) The reduced equivalent circuit for a MIS where the reference electrode–electrolyte interface impedances are small. Here R_p and C_p are the equivalent R and C elements associated with the combined impedance of the semiconductor and semiconductor-insulator interface. (After Diot *et al.* [1985].)

sists of the reference electrode impedance Z_{ref}; the electrolyte solution resistance R_s; the electrolyte–insulator interface impedance, which is composed of the double-layer capacitance C_{dl}, a diffusion impedance associated with the ionic species in solution (hydrogen ions for pH sensor), Z_w, and the SiO_2–electrolyte interface capacitance C_a; the insulator capacitance C_i; and the semiconductor–insulator interface impedance, which is composed of the space charge capacitance in the semiconductor, C_{sc}, as well as a capacitance and resistance C_{it} and R_{it}, respectively, associated with the interface states at this interface. At high frequency the electrolyte–insulator impedance is small with respect to C_i, as is the impedance of the reference electrode, so the equivalent circuit reduces to that shown in Figure 4.3.3*b*, where C_p represents the combined response of the C_{sc} and C_{it}.

In the EIS structure the value of C_p depends upon the biasing of the device as described in Figure 4.3.1. In forward bias, C_p will be large with respect to C_i because an accumulation layer exists in the semiconductor (Figure 4.3.1*d*, $V_G < 0$). In this case, the measured capacitance at high frequency, which is a series combination of C_{sc} and C_i, reduces to $C_i(1/C_i + 1/C_{sc} \cong 1/C_i)$, which is independent of applied potential. By evaluating the relative voltage in the space change layer and the oxide (Diot *et al.* [1985], Sze [1985]), it can be shown that

$$E - E_{fb} = eN_D\varepsilon_0\varepsilon_i\left[(C_i/C)^2 - 1\right]\big/2C_i^2 \tag{1}$$

where E is the applied potential, E_{fb} is the flat-band potential, e is the charge on an electron, N_D is the doping level, ε_0 and ε_i are the vacuum permittivity and relative dielectric constant, respectively, C_i is the insulator layer capacitance, and C is the total measured capacitance. The value of C varies with applied potential because the semiconductor capacitance, C_{SC}, depends on E through the surface voltage ψ_s (see below). Thus, the values of N_D and E_{fb} can be determined from the slope and intercept, respectively, of the linear portion of a plot of E vs. $(C_i/C)^2 - 1$. An example of such a result is shown in Figure 4.3.4.

Additional information about the semiconductor can be obtained from the interface capacitance C_{it}, which arises because each interface state stores a charge. A surface potential C_{it} can be defined as the potential at the semiconductor–insulator interface which causes the center of the band gap of the semiconductor [the Fermi level of the intrinsic material, $(E_f)_i$] to shift to a new value (Figure 4.3.5). This surface potential arises whenever the applied potential causes charge to build up at the interface. For example, for an *n*-type material when $E = E_f - (E_f)_{ref}$ is very much less than zero, $(E_f)_i$ will cross E_f as shown in Figure 4.3.5, leading to an accumulation of holes at the interface, that is, inversion as described above for the MOS devices. Now ψ_s can be calculated from the capacitance data described above by means of (Nicollian and Goetzberger [1967])

$$\psi_s = \int_{E_{fb}}^{E} (1 - CC_i^{-1})dE \tag{2}$$

Thus, the important electrical characteristics of the semiconductor can be determined.

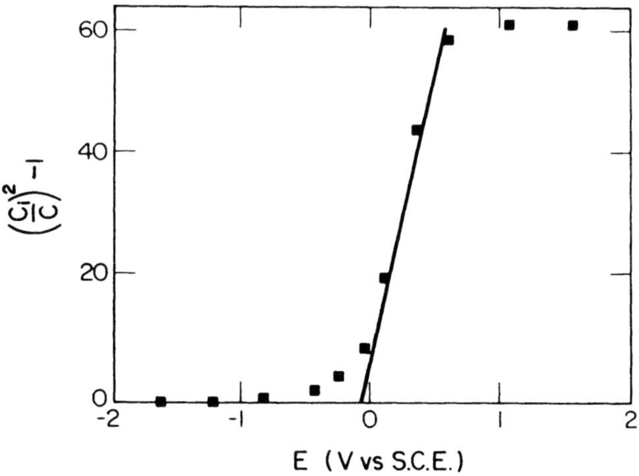

Figure 4.3.4. A plot of $(C_i/C)^2 - 1$ for n-type Si/SO$_2$ EIS, where C_i is the insulator capacitance derived from high-frequency data and C is the total capacitance of the system. The oxide thickness is 94 nm and pH = 2.5. The values of the doping density N_d and the flat-band potential E_{fb} calculated from the linear portion of the curve are $2.2 \times 10^{20}/\text{m}^3$ and −0.06 V, respectively (Diot *et al.* [1985]).

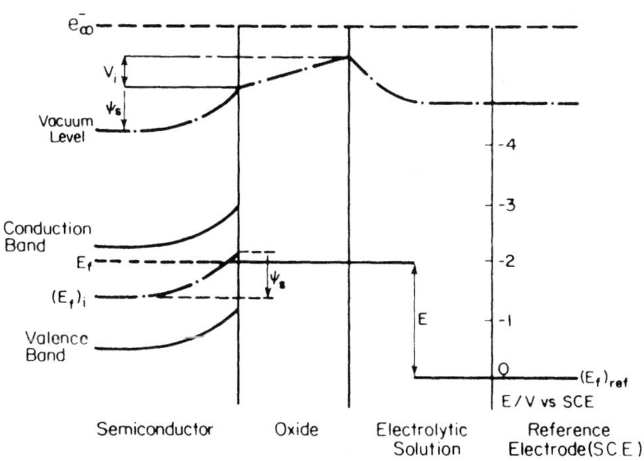

Figure 4.3.5. A simplified energy band diagram of an n-type EIS device which has an applied voltage E such that the semiconductor oxide interface is in the inversion regime [shown by crossing of E_f and $(E_f)_i$] leading to a buildup of holes at the interface. Two energy scales are shown, one referenced to an electron at infinity (e_∞^-) and the other referenced to the saturated calomel electrode (SCE). The surface potential ψ_s, the voltage drop across the insulator, V_i, and the Fermi levels of the reference electrode, the semiconductor under an applied voltage E, and the intrinsic semiconductor [$(E_f)_{\text{ref}}$, E_f, and $(E_f)_i$ respectively] are all shown. (After Diot *et al.* [1986].)

Further information about the interface states can also be extracted from the impedance data. From measured values of the total conductance G and capacitance C, the interface conductance can be calculated at any given potential after correcting for the solution resistance according to (Diot *et al.* [1985])

$$\frac{G_p}{\omega} = \frac{C_i^2 G}{\omega\left[(G/\omega)^2 + (C_i - C)^2\right]} \tag{3}$$

The insulator capacitance can be determined as described above. Thus, at any given reference potential the surface conductance G_p can be determined as a function of frequency. An example for typical results (Diot *et al.* [1985]) of a Si–SiO_2–electrolyte EIS is shown in Figure 4.3.6. Alternatively, G_p/ω can be calculated as the reference potential is swept and then converted to the representation in Figure 4.3.6.

Using these data, it is possible to calculate the number of interface states N_{it} at the semiconductor–insulator interface. In addition, a characteristic time constant τ_{it} associated with these states can be calculated. It is the time constant associated with the exponential decay of the interface states back to equilibrium after a perturbation. Qualitatively, when a small ac potential applied to the system swings in one direction, the electrons will be promoted from the interface states into unoccupied states in the silicon band and then demoted back to them as it swings in the other sense. The characteristic time for the electrons (in an n-type material) to decay back to the equilibrium configuration is τ_{it}.

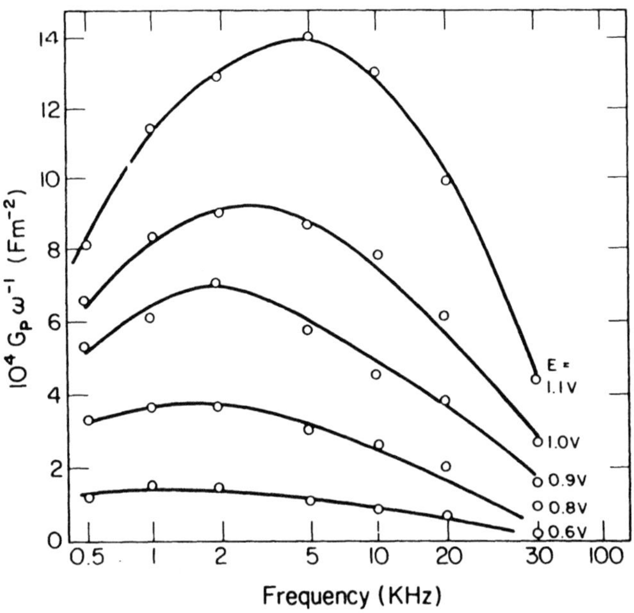

Figure 4.3.6. Plot of G_p/ω vs. frequency for a p-type Si/SiO_2 EIS at pH = 4.0 with doping density $N_A = 5.0 \times 10^{21}/m^3$ and oxide thickness of 60 mm. Here G_p is the total equivalent conductance associated with the semiconductor-semiconductor interface as described in Figure 4.3.3b. (Diot *et al.* [1985].)

The interface conductance can be understood, then, simply as the energy loss associated with this RC circuit such that $R_{it}C_{it} = \tau_{it}$ or $G_{it} = C_{it}/\tau_{it}$. The loss will be a maximum when the applied frequency reaches resonance with the characteristic time. In that case, G_p/ω will reach a maximum value. For the case of weak inversion in the semiconductor, the maximum will occur when $\omega\tau_{it} = 1$ and $(G_p/\omega)_{\max}$ will equal $C_{it}/2$. The number of interface states N_{it} is given by C_{it}/e.

Using the techniques above, the effect of pH in the electrolyte has been exampled on the Si–SiO_2–electrolyte EIS to ascertain the nature of the interaction between the hydrogen ions and the device (Bousse and Bergveld [1983], Diot *et al.* [1985]). One immediate advantage of IS appears in the ability to measure very-low-interface state concentrations, lower than $10^{15}\,m^{-2}(eV)^{-1}$ (Diot *et al.* [1985]). Plots of the N_{it} (calculated as described above) vs. the surface potential [calculated from Eq. (2)] have also been made as a function of pH and show that the effect of pH is very small (Figure 4.3.7). This implies that the SiO_2–electrolyte interface is the one responsi-

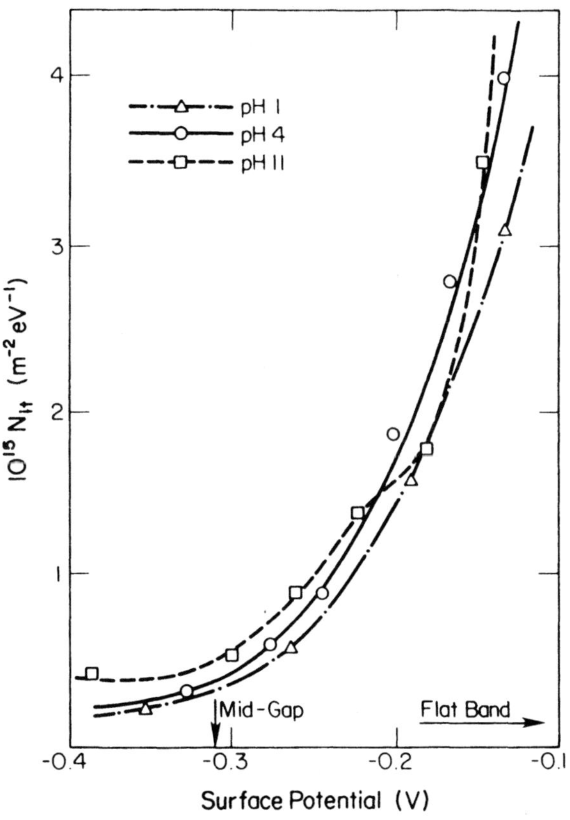

Figure 4.3.7. The number of interface states N_{it} as a function of surface potential ψ_s at three different pH values for a *p*-type Si/SiO_2 EIS (doping density $N_A = 4 \times 10^{-21}/m^3$ and oxide thickness of 92.5 nm). (Diot *et al.* [1985].)

ble for the change in potential of the device, not the SiO_2–Si interface (Diot *et al.* [1985]). Based upon a similar conclusion, a theoretical model which depends only on the sensitivity of the SiO_2 to the ionic concentration in the electrolyte has been used to successfully model the Si–SiO_2–electrolyte capacitance as a function of pH (Bousse and Bergveld [1983]). In addition, the general shape of Figure 4.3.7 is identical to that observed in MOS devices, thus reinforcing the contention that MOS and EIS devices function in exactly the same manner. Measurements of τ_{it} (Diot *et al.* [1985]) are also consistent with this supposition.

To reiterate, IS can be used to determine the important semiconductor electrical characteristics [E_{fb} and N_d (or N_a)], the insulator characteristics (C_i), and the nature of the semiconductor–insulator interface states (N_{it} and ψ_s). The technique is quite sensitive, allowing interface state concentration measurements below 10^{15} $m^{-2}(eV)^{-1}$. Results from several different studies (Barabash and Cobbold [1982], Diot *et al.* [1985]) verify that the EIS device behavior is identical to that of the MOS device except that the metal gate is replaced by an electrolyte containing a reference electrode. Finally, and most importantly to device operation, in the Si–SiO_2–electrolyte device the electrolyte–insulator interface is shown (Diot *et al.* [1985], Bousse and Bergveld [1983]) to be the one that responds to changes in pH rather that the Si–SiO_2 interface. A major advantage of IS is its ability to gather such detailed interfacial information which is not easily accessible with other measurement techniques.

4.3.2 Solid Electrolyte Chemical Sensors

The role of IS in the development and characterization of solid electrolyte chemical sensors (SECSs) is rapidly expanding. SECSs are electrochemical cells designed to measure the concentration or pressure of chemical species in gases or fluids. IS is emerging as an extremely useful technique to investigate the critical parameters which determine the electrolyte and electrode performances in these sensors.

The most successful SECSs are those which use zirconia-based electrolytes to measure oxygen concentrations. The three most common applications of these electrolytes are to measure oxygen concentrations of steel melts and in combustion gas environments and to control the air–fuel ratio in automobile engines. In the latter two applications, there is increasing interest in lowering the sensor temperature below 600°C, the current minimum temperature of operation because of low ionic conductivity and slow charge-transfer reactions at electrode–electrolyte interfaces.

An excellent example of the advantages and limitations of IS is the use of this technique to examine the effect of various electrode materials on the properties of zirconia-based oxygen sensors at temperatures below 600°C (Matsui [1981], Badwal [1983], Mizusaki *et al.* [1983], Badwal *et al.* [1984]). The most common electrode material is platinum. However, the charge-transfer reaction (I) at the electrode–electrolyte interface is restricted to regions at or near lines of three-phase (gas–electrode–electrolyte) contact:

$$\frac{1}{2}O_2(\text{gas}) + 2e^-(\text{electrode}) = O^{-2}(\text{electrolyte}) \tag{I}$$

Because of this, a finely dispersed, porous electrode structure is formed on the electrolyte surface to maximize the regions of three-phase contact. However, an optimum pore structure is very difficult to maintain due to electrode sintering upon exposure to elevated temperatures.

Several authors (e.g. Matsui [1981], Badwal [1983], Mizusaki *et al.* [1983], Badwal *et al.* [1984]) have used IS to investigate the effects of different electrode materials and their pretreatment temperatures upon oxygen sensor performance at low temperatures. It is particularly interesting to compare the properties of gold and silver electrodes with the commonly used platinum electrodes. A typical impedance spectra of a zirconia-based oxygen sensor at 500°C is characterized by two semicircles, as shown in Figure 4.3.8. (Matsui [1981]). The semicircle in the low-frequency range shows a characteristic distortion depending largely on the electrode material and preparation. The intersection of the low-frequency semicircle at the extreme right side of the abscissa (3000 Ω) is determined by the resistance arising from the oxygen electrode reaction (I) and is represented in the equivalent circuit by R_3. The values of R_1 and R_2 in Figure 4.3.8 represent the bulk and grain boundary resistance, respectively. As shown in Figure 4.3.8, the centers of the semicircle are usually below the real axis. In general this may result from two factors as described in Section 1.3, a constant-phase element such as that arising either from diffusion (the Warburg impedance) or from a distribution of time constants around an ideal value.

Oxygen sensor electrodes can experience temperatures as high as 900°C during cell preparation because of the necessity to remove organic impurities in the

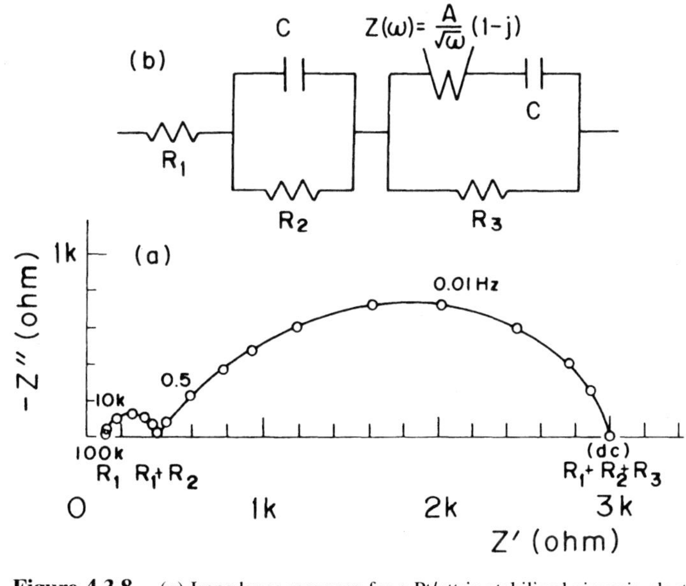

Figure 4.3.8. (*a*) Impedance response for a Pt/yttria-stabilized zirconia electrolyte with additives/Pt cell at 500°C and (*b*) the corresponding equivalent circuit. (Matsui [1981].)

platinum paste electrodes and/or to ensure adherent platinum films on the zirconia electrolyte. The exposure time and temperature can affect and significantly increase the electrode impedance due to a reduction in the three-phase contact regions for reaction (I), which is caused by sintering of the finely dispersed, porous electrodes at high temperatures. Thus, oxygen cells with similar platinum paste electrodes but having different exposure times and temperatures will exhibit different complex-impedance spectra and electrode resistances. The IS data can therefore be used to optimize the sintering times and temperature to provide an electrode with better properties.

The electrode preparation technique is another important factor in determining the electrode resistance. For example the difference between porous platinum electrodes prepared from a platinum paste (A) and a sputtering technique (B) is shown in Figure 4.3.9 (Mizusaki *et al.* [1983]), which shows only the low-frequency part of the complex-impedance spectrum. Although both cells were held at 900°C for 50h in 1 atm oxygen, the resistance of the sputtered platinum electrode is less than that of the one prepared from platinum paste. However, the resistance of the oxide ($U_{0.5}Sc_{0.5}O_{2+x}$) electrode (C) is about an order of magnitude less than that of the platinum electrodes. These oxide electrodes significantly decrease the electrode resistance by increasing the interfacial area for charge-transfer reaction (I). Both oxygen ions and electrons are mobile in these electrodes (Badwal *et al.* [1984]), and reaction (I) can occur over the entire gas–electrode interfacial area. Scanning electron micrographs of the three electrodes shown in Figure 4.3.9, taken before and after heating at 900°C, clearly indicate substantial sintering of the platinum electrodes, while only small morphological changes are observed with the oxide electrodes (Badwal [1983]).

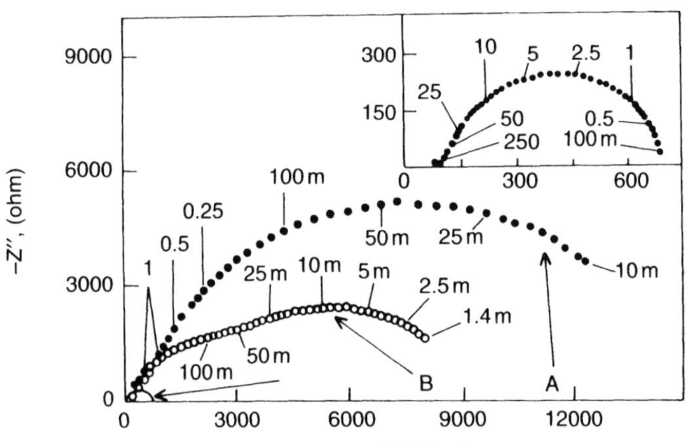

Figure 4.3.9. Complex impedance response for (*a*) 6082 Pt paste electrodes, (*b*) sputtered Pt electrodes 0.9 μm thick, and (*c*) $U_{0.5}Sc_{0.5}O_{2+x}$ electrodes (also inset on enlarged scale) at 600°C in 100% oxygen. All cells were given a prior heat treatment at 900°C for 50h. Numbers on the arcs are frequencies in hertz. (Badwal [1983].)

The impedance spectra (only the low-frequency region) for three noble metals (Ag, Au, Pt) electrodes are shown in Figure 4.3.10 (Badwal *et al.* [1984]). The results clearly indicate a significant difference between the silver and the gold electrode resistance in an oxygen sensor cell at 600°C. Although the resistance of the silver electrode is only slightly smaller than that of the platinum electrode, the resistance of the latter electrode significantly increases upon exposure to high temperatures. These sintering effects are not as severe for the silver electrodes because the appreciable solubility of oxygen in silver enables reaction (I) to occur over the entire electrode–electrolyte interfacial area rather than only at or near the three-phase contact region, as in the case with the platinum and gold electrodes.

Figure 4.3.10 clearly indicates that silver is a better electrode material in low-temperature oxygen sensors. However, significant volatility and microstructural changes of silver can occur at elevated temperatures, particularly at or above 900°C (Badwal *et al.* [1984]). Silver has been combined with platinum to form a Pt–Ag electrode, which possibly could exhibit the advantages of both metals. As shown in Figure 4.3.11 (Matsui [1981]), a Pt–Ag electrode (circles with centers) does have a significantly lower resistance than that of the platinum one (open circles). The

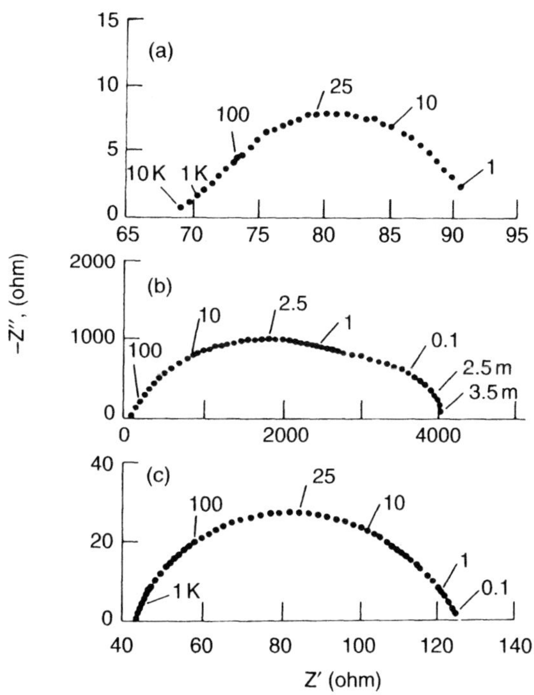

Figure 4.3.10. Complex impedance response at 600°C in 1 atm of oxygen after heating at 600°C for 50 h for the cells (*a*) Ag/yttria-stabilized zirconia/Au, (*b*) Au/yttria-stabilized zirconia/Au, and (*c*) Pt/yttria-stabilized zirconia/Pt. The numbers on the arcs are frequencies in hertz. (Badwal *et al.* [1984].)

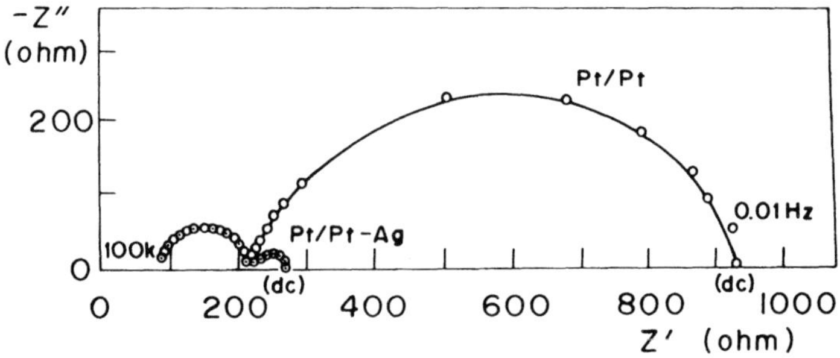

Figure 4.3.11. The impedance response for two kinds of electrodes using a tube of yttria-stabilized zirconia as the electrolyte. The arcs in the high-frequency range coincide, but the low-frequency arcs show a significantly lower resistance for the Pt/Ag electrode. (Matsui [1981].)

impedance spectra shown in Figure 4.3.11 and zirconia-cell results at 300°C indicate that the Pt–Ag electrode could be a very useful electrode in a low-temperature oxygen sensor (Matsui [1981]).

The use of impedance spectra to determine the optimum electrode materials and preparation procedures for low-temperature oxygen sensors is only one example of the application of this technique in solid electrolyte sensors. For example impedance spectra have already been used to examine the properties of zirconia stabilizers such as yttria and calcia in low-temperature zirconia electrolyte oxygen sensors (Badwal [1983]). The use of this technique in the development and characterization of other solid state sensors should increase significantly in the next few years.

4.3.3 Photoelectrochemical Solar Cells

Photoelectrochemical solar cells (PESCs) are devices which harness light energy and convert it into electrical or chemical energy by means of an electrochemical reaction at an interface. A general review of the electrochemistry of these devices can be found in most electrochemistry texts (e.g. Bard and Faulkner [1980]), but a cursory description will be given here for those unfamiliar with these devices. Most PESCs are composed of a semiconductor–electrolyte interface with an appropriate redox couple in solution. For an *n*-type semiconductor, when light with energy greater than the band gap strikes the interface, photons are absorbed and electron-hole pairs are created in the semiconductor. Some of these electron-hole pairs will simply recombine in the bulk, dissipating their energy thermally by the creation of phonons, by photon emission and so on. However, some proportion of the holes created at the interface will be available to oxidize the reduced species in solution, liberating an electron in the semiconductor which can flow in the external circuit. This photocurrent is absent in the dark where the concentration of holes is very low, so no reaction with the species in the solution is possible. The behavior of *p*-type semiconductors under irradiation is analogous; however, in this case electrons assist a reduction process in the solution and a current is produced by holes in the semiconductor.

In practice, the electrochemical behavior of semiconductor–electrolyte interfaces is far more complex than that described above (for a good review, see Boddy [1965]). One of the complications arises because the semiconductor surface at the electrolyte–semiconductor interface is not equivalent to that in the bulk. In particular, the energy states localized at the surface for holes and/or electrons are different than those present in the bulk. These surface states may arise in several ways—for example through pretreatment (etching, polishing, etc.) of the semiconductor surface before immersion in the electrolyte. The surface states can be detrimental to the PESC efficiency if they increase the recombination of the electron-hole pairs in the semiconductor, thus reducing the number of holes (electrons for *p*-type material) available for chemical reaction with the redox species in solution.

Another complication arises because the semiconductor may chemically or electrochemically react with the electrolyte after immersion, leaving a layer on the surface of the semiconductor which has different electrical or electrochemical characteristics (e.g. an insulating layer) from the semiconductor. Because the photocurrent under illumination is very sensitive to the semiconductor–electrolyte interface, these surface perturbations not only change the electrochemical behavior but they can, in extreme cases, completely inhibit the photoresponse.

Impedance spectroscopy offers an excellent tool to examine the existence of surface states or other modifications of the ideal semiconductor–electrolyte interface. The general response of such interfaces was reviewed as early as 1965 (Boddy [1965]). Dutoit *et al.* [1975] found that the capacitance of these interfaces at a given dc potential was dependent on the measuring frequency for CdSe, CdS, and TiO_2 in several different aqueous and nonaqueous electrolytes. Tomkiewicz [1979] and McCann and Badwal [1982] have made more thorough investigations of the impedance response of several different technologically important semiconductor–electrolyte interfaces. The capacitance of a semiconductor–electrolyte junction has also been measured as a function of incident wavelength and used to characterize energy levels in semiconductors (Haak and Tench [1984], Haak *et al.* [1982]). One particular study (Shen *et al.* [1986]) will be examined in more detail here to illustrate the kinds of effects that can be resolved using IS.

The impedance response of *n*-$CuInSe_2$ in polyiodide solutions has recently been used to understand the behavior of this material in a PESC (Shen *et al.* [1986]). Typical current potential response curves for *n*-$CuInSe_2$–polyiodide solutions are given in Figure 4.3.12 (Shen *et al.* [1986]) in which the effect of various pretreatments is shown. Polishing + etching or polishing + etching + oxidation significantly improve the photoresponse over simple polishing. For example, from Figure 4.3.12, at −0.2 V vs. Pt the photocurrent increases by approximately a factor of two after each additional pretreatment. IS was used in combination with electroreflectance (Shen *et al.* [1986]) to understand this behavior.

The impedance response obtained for polished + etched crystals and polished + etched + oxidized crystals (Figure 4.3.13*a* and *b*, respectively) show qualitatively different behavior, the principal one being the addition of at least one more time constant in the oxidized case as manifested by (at least) one additional peak in the imaginary part in Figure 4.3.13*b*. Such behavior is reasonable since there is an additional interface between the oxide and semiconductor.

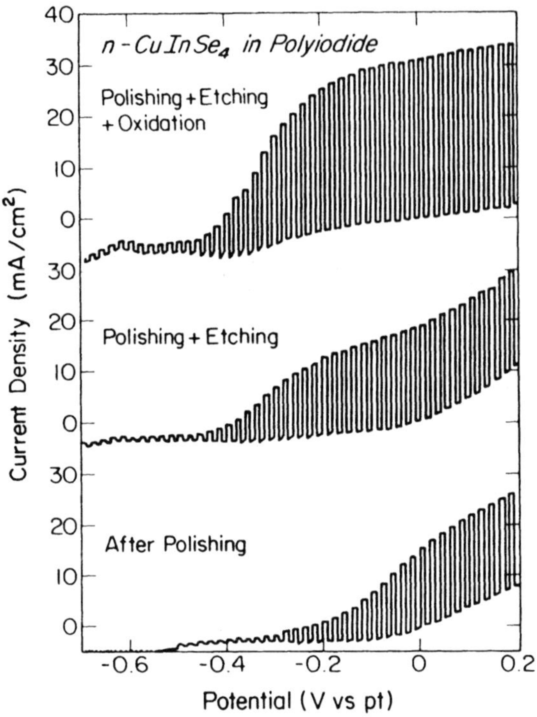

Figure 4.3.12. The effect of surface preparation of the current–potential response curves of n-CuInSc$_2$ in a solution of 6 M KI + 0.1 M InI$_3$ + 0.0125 M I$_2$ at pH 6.0. The square-wave response results from using a chopped white light source of intensity 100 mW/cm^2. Etching was in a 2% Br$_2$–methanol solution for 60 s; oxidation was for 2 h at 150°C. (Shen *et al.* [1986].) Reprinted by permission of the publisher, The Electrochemical Society, Inc.

Using the equivalent circuits shown in Figure 4.3.13*a* and *b*, the high-frequency data were analyzed to determine the capacitance associated with the two fastest time constants in the polished + etched material and the fastest time constant in polished + etched + oxidized material. The low-frequency data were not analyzed because their physical interpretation was not clear. For the data shown in Figure 4.3.13*a* and *b*, the fastest time constant has been associated with the space charge layer in the semiconductor (C_{SC} in Figure 4.3.13*a* and C_1 in Figure 4.3.13*b*). In the polished + etched material, the next fastest is that associated with surface states on the semiconductor-electrolyte interface. It was assumed that the surface states are characterized by one time constant which does not significantly overlap with the time constants of any other states. It has been pointed out (McCann and Badwal [1982]) that should there be overlap of states with time constants close to one another, the time constants become essentially continuous, and a frequency-dependent resistance and capacitance must be used to model the interface. Here, though, the essential features of the interface appear to be adequately described without resorting to such elements.

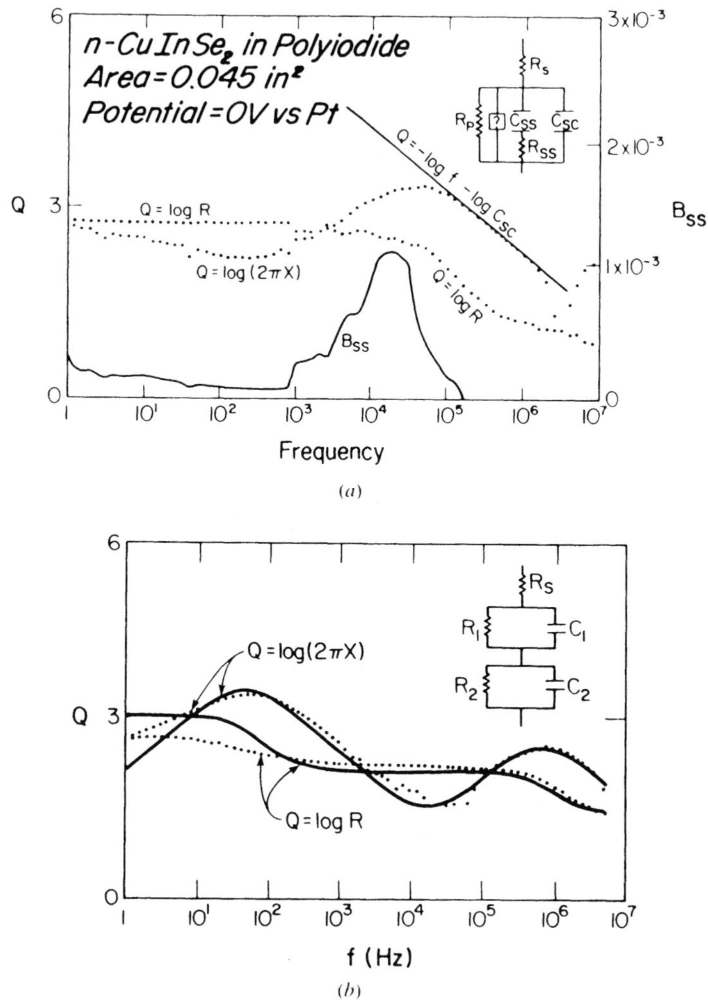

Figure 4.3.13. The impedance response and equivalent circuit of n-$CuInSe_2$ in the same solution as described in Figure 4.2.12 for: (*a*) Polished + etched sample. B_{ss} is the imaginary portion of the measured admittance less ωC_{sc}. The value of C_{sc} is calculated from the linear part of the high-frequency portion of the imaginary response of the impedance. The low-frequency response was not analyzed. (*b*) Polished + etched + oxidized sample. The solid line is a theoretical fit assuming the equivalent circuit shown. (Shen *et al.* [1986].) Reprinted by permission of the publisher, The Electrochemical Society, Inc.

In the case of the polished + etched + oxidized sample, C_1 was associated with the space charge layer capacitance. No further data were used. Thus, C_1 is representative of the change in capacitance of the space charge layer from the presence of the oxide layer.

Analyzing the effect of applied potential on the capacitance arising from the surface states, C_{ss}, in the polished + etched material (Figure 4.3.14) led to the

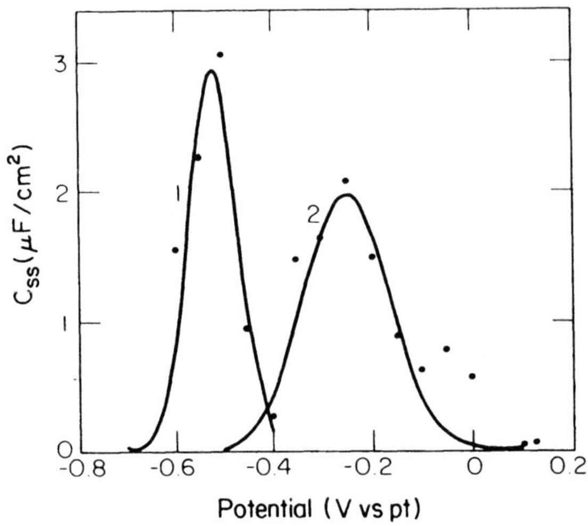

Figure 4.3.14. The variation of the capacitance associated with the surface states C_{ss} as a function of potential. The solid lines are a theoretical fit to two Gaussian line shapes as described in text. (Shen *et al.* [1986].) Reprinted by permission of the publisher, The Electrochemical Society, Inc.

conclusion that there were two surface states, one centered 0.17 eV below the conduction band [–0.69 – (–0.72) in Figure 4.3.14] and the other at 0.45 eV below the conduction band. Assuming a Gaussian distribution of surface states (Tomkiewicz [1979]), the area density of both states was calculated to be less than 1% of a monolayer. Thus, it was concluded that one major effect of etching was to remove most of the surface states. This conclusion is consistent with electroreflectance results (Shen *et al.* [1986]) on the same system which show that in unetched samples the surface states pin the Fermi level, while after etching the surface states are nearly completely removed.

To determine the effect of oxidation, a Mott–Schottky plot of the space charge capacitance before and after oxidation was compared. In these plots, which were originally derived for a metal–semiconductor interface (Schottky [1939, 1942], Mott [1939]) but hold equally well for the metal–electrolyte interface, a linear relationship is predicted between the applied potential and one over the square of the capacitance arising from the space charge layer in the semiconductor. The slope is inversely proportional to the effective donor or acceptor concentration in the semiconductor. For the semiconductor–electrolyte interface (Bard and Faulkner [1980]),

$$\frac{1}{C_{sc}^2} = \frac{2}{e\varepsilon_0\varepsilon N_d}\left(-\Delta\phi - \frac{kT}{e}\right) \tag{4}$$

where $\Delta\phi$ is the difference between the applied potential and the flat-band potential $E - E_{fb}$, C_{sc} is the space charge capacitance, ε is the dielectric constant, ε_0 is the permittivity of free space, k is Boltzmann's constant, T is absolute temperature, and N_d is the concentration of donors or acceptors. For the n-$CuInSe_2$ electrode the

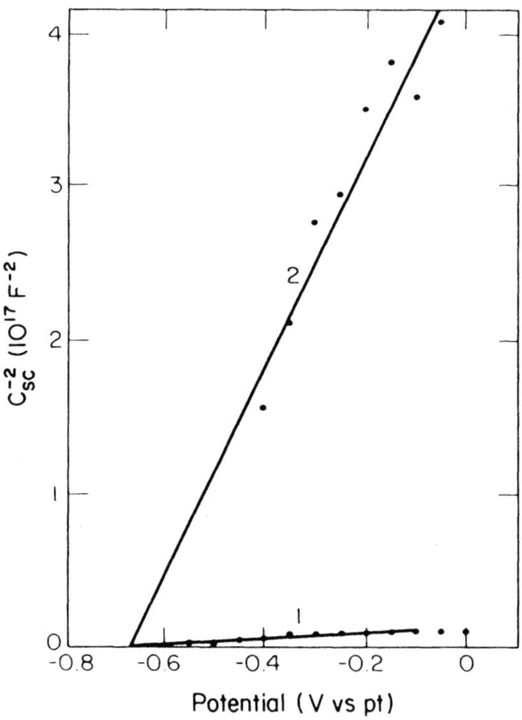

Figure 4.3.15. Mott–Schottky plots of the space charge capacitance C_{sc} (curve 1) as derived from data like those shown in Figure 4.3.9*a* and the capacitance associated with the high-frequency response, C_l (curve 2) derived from data like those shown in Figure 4.3.9*b*. The flat-band potential is the same in both cases (0.69 V), but the doping level, as calculated from the slope of the lines, is an order of magnitude lower for curve 2 (polished + etched + oxidized sample) than for curve 1 (polished + etched sample). (Shen *et al.* [1986]). Reprinted by permission of the publisher, The Electrochemical Society, Inc.

Mott–Schottky plot (Figure 4.3.15) shows that the polish + etch + oxidation procedure does not change the flat-band potential, but the effective doping level decreases by nearly one order of magnitude from that observed in the polished + etched material.

Several conclusions can be drawn from these data. First, the oxidation produced a layer that does not alter the electrical characteristics of the semiconductor since the flat-band potential did not change. Second, the oxide layer decreases the doping level, thus increasing the width of the space charge layer. This wider layer in turn leads to higher photocurrent because most of the light is absorbed within the space charge layer so that recombination of charge carriers in the bulk is reduced. Finally, by applying a simple model of photoresponse, Tomkiewicz [1979] determined that surface recombination arises from the surface state at 0.17 eV below the conduction band. Electroreflectance measurements (Shen *et al.* [1986]) are consistent with this conclusion. Thus, the improved response caused by etching can be explained by the decrease in density of these surface states observed in the impedance results after etching.

Several key features of this study should be emphasized. IS clearly can be used to successfully model a semiconductor–electrolyte interface in a PESC. The ability to probe the physics of this interface using IS while controlling the applied potential can allow significant insight into the important parameters of the device. In particular, the surface states at the semiconductor–electrolyte interface may be determined, as can their relative importance after several different pretreatments or in different cell configurations. The electrical characteristics of the interface, for example the flat-band potential and the space charge capacitance, can also be determined.

The work described above also shows that it is not always necessary to analyze the entire frequency spectrum (i.e. determine the complete equivalent circuit) of a cell in order to obtain significant insights into its operation if it is possible to associate a particular region of the spectrum with a meaningful physical quantity. In the case of the semiconductor–electrolyte interface described above, a strong theoretical background describing the expected behavior along with other experimental findings (electroreflectance, current–potential curves) on the system permitted such a limited but meaningful analysis. Further, a more detailed analysis of the results would probably have led to a more complete description of the operation of the PESC. This fact was recognized by the authors themselves (Shen *et al.* [1986]). Although a limited analysis may allow significant insights, it should be made with extreme caution, as the results could lead to erroneous conclusions. A complete detailed analysis of the entire frequency spectrum is far preferable and leads to a more complete understanding of the device operation.

4.3.4 Impedance Response of Electrochromic Materials and Devices

Gunnar A. Niklasson
Anna Karin Johsson
Maria Strømme

4.3.4.1 Introduction

Electrochromic materials have the ability to change their optical properties when an electrical voltage is applied across them. The optical properties and hence the color can be altered gradually and reversibly, i.e. the original state is recovered if the polarity of the voltage is reversed. For practical use, the electrochromic material must be incorporated in a system, or a device, which allows for the application of a voltage over it. Depending on the application at least one of the contacts must exhibit a high optical transmittance. Research on electrochromic materials became widespread around 1970, due to a widely cited paper by Deb [1969]. The aim of the early work was to develop devices for display technology. Today, electrochromic devices are of interest for a wide range of applications and commercial activities are being started up (Lampert [2003]). Much of the emphasis of the applied work has shifted to energy-efficient windows with variable solar and luminous transmittance; they are able to reduce the influx of solar energy into a building. This will lead to large reductions in the need for cooling and air conditioning in warm and temperate climates (Azens and Granqvist [2003]). Overheating and reduced comfort is prone to occur because windows and glass façades allow too much solar energy to penetrate into buildings. Works on electrochromic materials and devices are numerous and have been reviewed several times. The comprehensive reviews by Granqvist ([1995, 2000]) sum up most of the work prior to 2000.

In this section we concentrate on inorganic electrochromic materials and devices that use this class of materials. The electrochromic effect is caused by ion and electron insertion into the materials. Figure 4.3.16 shows a schematic picture of an electrochromic device. It consists of five layers interfaced between two substrates, or alternatively deposited on a single substrate. We consider first the configuration with two substrates. They are coated by a transparent conducting layer and then by an electrochromic layer. The two sides of the device are laminated by an ion conductor, preferably a polymer. The electrochromic layers are chosen to be complementary. One of them is cathodic (i.e. it colors under cation/electron insertion) and the other is anodic (coloring under cation/electron extraction). This means that both will color upon the application of a voltage between the transparent contacts, and both will bleach when the polarity is reversed. One of the electrochromic layers may be replaced by an ion storage layer that can accommodate inserted ions and electrons without changing its optical properties. The alternative design, the so-called all-thin-film device consists of five layers consecutively deposited on one substrate. The material requirements are similar for this kind of device, except for the ion conductor, which in this case has to be an inorganic thin film.

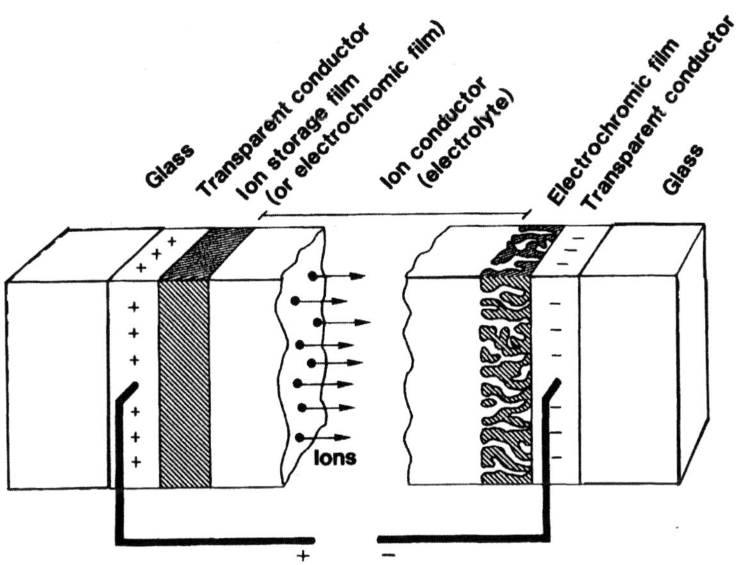

Figure 4.3.16. Schematic illustration of an electrochromic device.

Optical properties as well as ion and electron transport are the physical properties of electrochromic materials that are of most importance for the performance of a device. The requirements on the optical behavior can be summarized as follows. One of the electrochromic materials should change color upon ion-electron intercalation while the other electrochromic material should show complementary behavior, i.e. change color upon ion-electron extraction, or remain optically invariant during the whole intercalation–deintercalation cycle. The optical properties of the ion conductor material are not allowed to change while ions are transported through it. Inorganic electrochromic materials mostly belong to the class of transition metal oxides. Coloration occurs, during ion/electron insertion or extraction, as the Fermi level moves from a position in the band gap into the conduction or valence band. In most cases, the absorption process is due to electronic transitions between localized states close to the band edges. The electron–phonon interaction is of major importance in this process.

The requirements for the electrical properties are somewhat more complex. A good electrochromic material should be a relatively fast ionic and electronic conductor. Since the optical modulation is an effect of the changing number of electrons in the material, and because of the requirement of charge neutrality, charges of opposite signs must be able to move. Hence, the electrochromic material must be a mixed ionic and electronic conductor. The ion conductor, on the other hand, should ideally not conduct electrons at all, in order to prevent the device from short-circuiting. When investigating the ability of a material to act as one of the layers in an electrochromic device it is, thus, important to extract information, not only about the dc conductivity, but also about the mobility, revealing the speed by which the ions move (Strømme Mattsson and Niklasson [1999]). Further, the energy barriers felt by the ions when moving between the ion conductor and the electrochromic layers should be as low as possible.

Impedance spectroscopy is widely recognized as a versatile technique to obtain detailed information on ion and electron transport. The first works on the electrochemical impedance of electrochromic coatings in contact with a liquid electrolyte appeared around 1980. Measurements on Li intercalated tungsten oxide films were carried out by Ho *et al.* [1980]. They developed a basic model of the impedance response and showed that it can be represented by the Randles equivalent circuit. A similar circuit was used by Glarum and Marshall [1980] in an early work on iridium oxide films. The determination of the circuit elements from experimental data yielded information on the ion diffusion in the coating, as well as the properties of the electrolyte–film interface. The work of Ho *et al.* [1980] was generalized by Franceschetti and Macdonald [1982], who introduced an additional adsorption process. It was assumed that an ion combines with an electron from the conduction band of the coating to form an adsorbed intermediate species before insertion into the electrochromic film. This process leads, under certain conditions, to the occurrence of two semicircles at high frequencies in complex impedance (Z) plots. Unfortunately, this theory has been rarely used in the field of electrochromic materials (Pyun *et al.* [1996]), and the occurrence of two semicircles in the Z plots has sometimes been ascribed to electron injection at the back contact of the film (Yoshiike *et al.* [1984]). A systematic approach for deriving generalized equivalent circuits for mixed conductors from the basic transport equations has been established by Jamnik and Maier [2001]. The case of a mixed conductor with one electrode blocking to ions and the other blocking to electrons (Jamnik [2003]) is of obvious relevance for electrochromic materials. Below, we will use the basic Randles circuit, with some additional features, to illustrate how basic information on the electrochemical response of electrochromic materials can be extracted from impedance spectra.

4.3.4.2 Materials

Cathodic Electrochromic Materials. Amorphous tungsten oxide is a widely studied cathodic electrochromic material. The history of induced color changes in tungsten oxides dates back to the days of Berzelius in the early 19th century, while electrochemical studies were initiated around 1930, as discussed by Granqvist [1995]. The optical properties of WO_3 switch from a transparent state to a colored blue state upon insertion of small cations like H^+, Li^+, and Na^+. The coloration is actually due to simultaneous insertion of charge-balancing electrons into the conduction band through an outer circuit. Electronic transitions between localized states give rise to a polaron absorption. On the other hand, polycrystalline tungsten oxide switches from a transparent state to a near-infrared absorbing one and then to a reflecting state upon ion/electron insertion. The reflecting state is due to free electrons in the conduction band. Published research on tungsten oxide coatings is very extensive; see for example the reviews of Granqvist [1995, 2000] and the review on charge transport by Monk [1999]. Most electrochromic devices employ tungsten oxide as a cathodic material and no serious contenders have appeared despite long-standing efforts.

Titanium dioxide exhibits optical properties very similar to those of tungsten oxide. Electrons in the conduction band become localized by the electron–phonon interaction and give rise to polaron absorption. Coatings of titanium oxide are less stable in an electrochromic device than films of tungsten oxide, and have therefore not been used so much.

Anodic Electrochromic Materials. The most commonly used anodic electrochromic materials are nickel oxide (Svensson and Granqvist [1986]) and iridium oxide (Gottesfeld *et al.* [1978]). They switch from a transparent state to a colored one upon extraction of protons. Charge-balancing electrons are simultaneously extracted from the valence band. The films are probably a mixture of oxide and hydroxide components in the bleached state, since there needs to exist a reservoir of protons in the films. Due to the high cost of iridium, the use of nickel oxide is favored for large scale applications. Recently, a class of mixed nickel oxides with enhanced modulation between the transparent and the colored state have been discovered (Avendano *et al.* [2003]). Intercalation of Li into nickel oxide films has been attempted, but the optical properties are not modulated very much (Decker *et al.* [1992]). The mechanism of optical absorption is not known in detail. However, in the case of nickel-based oxide films, it is known that the oxygen-rich compounds NiOOH and Ni_2O_3 are both strongly absorbing for visible light.

Transition metal oxides that do not change their transparency, or color very little, under ion/electron insertion and extraction can also be used as a counter electrode in electrochromic devices employing tungsten oxide as a cathodic material. There has been particular interest in oxides based on vanadium pentoxide and cerium oxide. Pure V_2O_5 as well as a mixture of vanadium and titanium oxide are of interest. Cerium-based mixed oxides, in particular cerium–zirconium oxide (Veszelei *et al.* [1999]), exhibit less optical absorption, but the stability is not sufficient for many applications.

Ionic Conductors. As mentioned above, the ion conductor used in electrochromic device can be of two types, that is a polymer electrolyte or an inorganic thin film. A large number of polymer electrolytes, protonic as well as lithium conducting, have been tested in electrochromic devices. A review of recent results for devices, using a polymer electrolyte, was given in Granqvist [2000]; see also references therein. In Section 4.3.4.5 below we report data on a laminated devices using polyethylene glycol mixed with the salt lithium trifluoromethanesulfonate, as the ion conductor. In good polymer electrolytes the dc conductivity usually dominates the impedance response for frequencies below the MHz range. At higher frequencies relaxations can appear. The dielectric properties of polymer electrolytes is a large field of research, and a treatment is outside the scope of this work. Reviews have been given by Bruce [1987] and Greenbaum *et al.* [1990].

Inorganic ion conductors are found among the transition metal oxides, as well as lithiated metal oxides and fluorides. We refer to Granqvist [2000], and references given therein, for data on recent devices that have used inorganic ion conductors. Among the transition metal oxides, tantalum pentoxide has been used in window devices close to the commercial stage (O'Brien *et al.* [1999]). In Section 4.3.4.5 below, we report on an all-thin-film device with zirconium oxide as ion conductor. Both these oxides exhibit reasonable proton and lithium ion conductivities (Jonsson *et al.* [1999], Frenning *et al.* [2001]).

4.3.4.3. Experimental Techniques

In this section, we show how impedance spectroscopy can be used to extract quantities necessary for the evaluation of the viability of a material as a component in an electrochromic device. We also show how transient current measurements can be used for the extraction of electrical conduction parameters that cannot be obtained easily from impedance spectroscopy.

In the first subsection we discuss how the dc conductivity, σ_{dc}, can be accurately extracted from ordinary two-electrode impedance spectroscopy, i.e. when electronically conducting electrodes are attached to the top and bottom surface of the film under investigation.

In the second subsection we will show how data obtained from electrochemical three-electrode impedance spectroscopy can be used to obtain information about the chemical diffusion coefficient, $\tilde{D}$, as well as about energy barriers at the interfaces. In electrochemical three-electrode impedance spectroscopy the film under study is backed by an electronic conductor, which is connected to a counter electrode and a reference electrode via an outer circuit. The film is immersed in an electrolyte containing ions that will be forced into the film. The ac potential is superimposed on an equilibrium potential, U_{Eq}, applied between the film, which acts as the working electrode, and the reference electrode, while the current is measured between the film and the counter electrode. The equilibrium potential decreases as more cations are inserted into the film. The extraction of cations is accompanied by an increasing equilibrium potential. Electrochromic materials display quite high U_{Eq}, for example they are in the region of 3 V vs. the Li electrode for tungsten and nickel oxide. It should not be possible for ion conductor materials to intercalate at the operating potentials of the electrochromic materials. For example, significant intercalation starts at about 2 V vs. Li in tantalum and zirconium oxide films.

In the third subsection, the determination of ion density and mobility will be in focus, both by using impedance spectroscopy and transient current measurements.

Extracting the dc Conductivity. There are several ways to extract a value of the dc conductivity from impedance spectroscopy data. A method that is built on the amply documented power-law response at frequencies above the dc regime (Jonscher [1983]) is presented here. If we assume that σ' has an approximate power-law behavior within some frequency range, it can be expressed as

$$\sigma' = \sigma_{dc} + B\omega^{m-1} \tag{5}$$

where B and m are constants. By rewriting the derivative of σ', we obtain

$$\sigma_{dc} = \sigma' - \frac{1}{m-1}\frac{d\sigma'}{d\ln\omega} \tag{6}$$

and σ_{dc} can be extracted by plotting σ' vs. $d\sigma'/d\ln\omega$. In order to determine for which frequency region this plot should be made, one may first plot σ' and $d\sigma'/d\ln\omega$ vs. ω separately. The region for which Eq. (6) should be applied is the one where $d\sigma'/d\ln\omega$ has a good power law behavior and is proportional to ω^{m-1} and where σ' starts to level off towards low frequencies as a consequence of σ_{dc} becoming the dominant contribution (Strømme Mattsson *et al.* [1999]). The procedure also yields good estimates of σ_{dc} when the frequency response is an approximate power law (Strømme Mattsson *et al.* [1999]).

Determination of Parameters from Randles Circuit. Electrochemical three-electrode impedance spectra taken on electrochromic materials can very often be fitted to the Randles equivalent circuit (Randles [1947]) displayed in Figure 4.3.17. In this circuit R_{hf} denotes the high frequency resistance of the electrolyte, R_{ct} is the charge-transfer resistance associated with the ion injection from the electrolyte into the electrochromic film and $Z_{\tilde{D}}$ is a Warburg diffusion impedance of either semi-infinite, or finite-length type (Ho *et al.* [1980]). The CPE_{dl} is a constant phase element describing the distributed capacitance of the electrochemical double layer between the electrolyte and the film having an impedance that can be expressed as

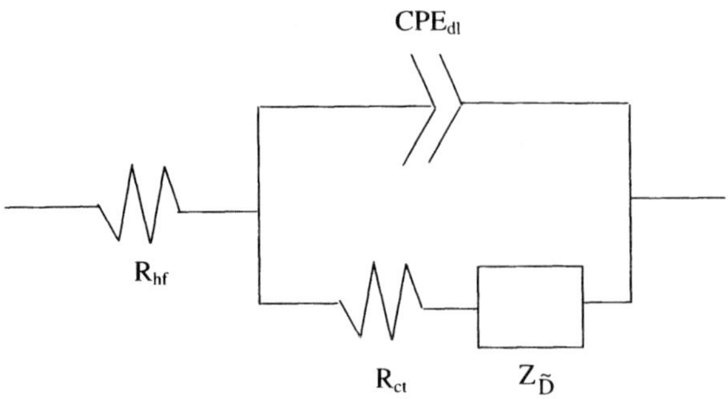

Figure 4.3.17. The Randles equivalent circuit, with resistance R_{ct} of a charge-transfer process and the diffusional impedance $Z_{\tilde{D}}$. R_{hf} and CPE_{dl} are the high frequency resistance and the double layer distributed capacitance, respectively.

$$Z_{CPE_{dl}} = \frac{1}{\varsigma(i\omega)^k} \tag{7}$$

where ς is a parameter with the dimension Fs^{k-1}, and k is a positive number between zero and unity. Further, it has been shown that the number k in Eq. (7) can be directly associated with the fractal dimension of the electrochromic film surface (Strømme Mattsson *et al.* [1996a]).

Fitting an impedance spectrum to the Randles circuit is a convenient way of obtaining quantitative values of the height of the barrier the ions have to pass when entering the electrochromic film (R_{ct}) and the chemical diffusion coefficient, which gives the speed with which the intercalated ion-electron couples move ($\tilde{D}$ extracted from $Z_{\tilde{D}}$). By recording impedance spectra at different temperatures, the activation energies of the charge transfer and the diffusion process can be obtained, as will be demonstrated in Section 4.3.4.4 below.

Determining the Ion Density and Mobility. In this section we present methods to obtain the ion density and mobility from two-electrode measurements with ion blocking contacts. We review methods based on impedance spectroscopy and transient currents. It is often of interest to investigate not only the conductivity of the ions inside the different layers of an electrochromic device, but also the *density*, n, of moving ions and their *mobility*, μ. Provided that electrode polarizations (space-charge effects) dominate the impedance response of the layer under investigation, this can be done by analyzing the impedance spectra, or rather the permittivity spectra, using the method of Schütt and Gerdes [1992a, b]. In their model, developed for singly charged positive ions moving under the influence of relatively low applied voltages (≤≈0.03 V), the real part of the effective dielectric constant is given by (Schütt and Gerdes [1992b])

$$\varepsilon' = \varepsilon'_s\left(1 + \frac{\delta}{1 + (\omega\tau\delta)^2}\right). \tag{8}$$

Here ε_s' is the real part of the relative permittivity in the low-frequency bulk relaxation regime, $\delta = d/\sqrt{D\tau}$ where d is the film thickness, D the diffusion coefficient of the moving ions, and $\tau = \varepsilon_0\varepsilon_s'/\sigma_{dc}$ is the electrical relaxation time. The density and mobility of the moving ions can be obtained from the high-frequency part of ε' immediately below the frequency range of bulk polarization (ε_s') by picking a frequency ω_X, where $\varepsilon'(\omega_X) = X\varepsilon_s'$. Provided that the approximation $1 + (\omega_X\tau\delta)^2 \approx (\omega_X\tau\delta)^2$ holds, the ion density is obtained by using the Nernst–Einstein relation ($D = \mu kT/e$, where kT is the thermal energy and e the unit charge) as

$$n = \left(\frac{\sigma_{dc}}{\sqrt{(X-1)}\varepsilon_0\varepsilon_s'\omega_X}\right)^4 \frac{\varepsilon_0\varepsilon_s' kT}{e^2 d^2} \tag{9}$$

and μ is then straight forwardly obtained from

$$\sigma_{dc} = ne\mu\,. \tag{10}$$

When both migration, diffusion, as well as space-charge effects, significantly contribute to the ionic motion inside a film, impedance spectroscopy can still be used to extract values of n and μ (Frenning *et al.* [2003]). In this case the equations become significantly more complex, and profound computations are needed to solve them. The approach presented by Frenning *et al.* [2003] may, however, be used not only for single layer ion conductors but also to model the impedance response of a complete three-layered device. For low ion concentrations space-charge effects can be neglected and instead diffusion and migration govern the current response (Frenning and Strømme [2001]). In this case one may use transient current measurements instead of impedance spectroscopy to extract n and μ as we shall see below.

In the mid eighties, Greeuw and Hoenders [1984] presented an analytical solution for the transient ionic displacement current in a one-dimensional dielectric slab, neglecting space-charge effects but taking into account both diffusion and migration. In their model the slab is placed between two blocking electrodes. The slab, which is exposed to a static electric field, contains no ions to begin with, but a source-function emits identical ions, with a release frequency υ, from interface traps at one end of the slab into the bulk of it. The measured current as a function of time in this model is given by (Greeuw and Hoenders [1984])

$$I(t) = \frac{eD}{d}[c(0,t) - c(d,t)] + \frac{\nu Ne}{d}[1 - \exp(-\nu t)]. \tag{11}$$

Here c(0, t) is the time-dependent ion concentration at the electrode from which the ions are released, and c(d, t) is the same quantity at the electrode they move toward. The total number of released ions is N. c(0, t) and c(d, t) are given explicitly by Greeuw and Hoenders [1984].

This model was initially used to estimate the density and mobility of alkali ions in ion implanted SiO_2 layers (Greeuw and Verway [1984]). It should be noted that the same model may be used to extract a value of the mobility of Li ions intercalated from an electrolyte into an electrochromic film under the application of a potential step (Strømme Mattsson *et al.* [1999b], Strømme Mattsson [2000]). The mobility obtained in such an electrochemical transient current experiment represents an average over the compositions present in the film, during the time span the potential step is applied and is a very relevant measure for device applications (Strømme Mattsson [2000]).

If we instead consider the case when the potential across the sample is changed from zero to a specified value taking the sample from a uniform charge distribution to a polarized state, the equation describing the transient current becomes much simpler. Neglecting space-charge effects and assuming that the initial current decay is dominated by migration of ions toward an ion-blocking contact, both n and μ can be obtained by fitting the measured current response to (Watanabe *et al.* [1985], Strømme Mattsson and Niklasson [1999])

$$\mathrm{I(t)} = \frac{A\sigma_{\mathrm{dc}}U}{\mathrm{d}}\exp\left(-\frac{\mu U}{\mathrm{d}^2}\mathrm{t}\right). \tag{12}$$

Here A is the electrode area and U the applied potential. n is extracted from σ_{dc} using Eq. (10). The above equation holds when there are only identical ions present. The equation is, however, easily adapted to the case when more than one type of ion reside in the sample simultaneously by just adding one additional term—identical to the one already at the right hand side of the equation—per ion type (Watanabe *et al.* [1985]). This procedure is especially useful when studying Li ion movements in device layers where one suspects that protons are also present and contributing to the current response (Strømme Mattsson and Niklasson [1999]). Also, in the case of potential step experiments, a model for the ionic current response has been developed for the situation when migration, diffusion, and space-charge effects all together contribute to the ionic motion (Frenning and Strømme [2001]). And, just as for impedance spectroscopy measurements, the model can be used to analyze the transient current response of a full three-layered device (Frenning *et al.* [2003]).

4.3.4.4 Experimental Results on Single Materials

In this section we present some examples of impedance spectra, in order to illustrate the concepts introduced above. We show how the methods in Section 4.3.4.3 may be used to characterize electrochromic materials and inorganic ion conductors. Data on titanium dioxide and tantalum oxide are emphasized, because they illustrate especially well several of the concepts treated above. Of course, we also review the application of impedance spectroscopy to the widely used electrochromic materials tungsten oxide and nickel oxide.

Cathodic Electrochromic Materials—Fluorinated Ti Oxide. Figure 4.3.18 shows two electrochemical three-electrode impedance spectra taken at different temperatures on a heavily intercalated Li containing fluorine doped Ti oxide film (Strømme Mattsson *et al.* [1997]). The impedance response corresponds to that of the Randles circuit with a $Z_{\bar{D}}$ of finite-length type. Details about the film preparation and the measurement conditions can be obtained from Strømme Mattsson *et al.* [1996c, 1997]. The high frequency semicircle clearly has a center below the real

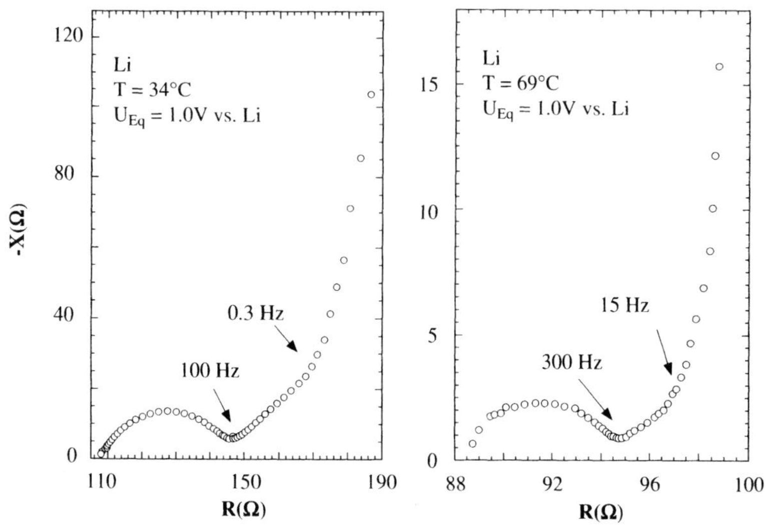

Figure 4.3.18. Electrochemical three-electrode impedance complex-plane plots for the real (R) and the imaginary (X) parts of the impedance of fluorine doped Ti oxide films recorded at the displayed temperatures at an equilibrium potential of 1.0 V vs. a Li reference electrode (corresponding to ~0.85 Li/Ti unit) (Strømme Mattsson *et al.* [1997]). Explicit frequency readings are shown at a few selected data points.

axis, hence it is appropriate to use a constant phase element for C_{dl}. The low frequency response is a model case of a Warburg diffusion element of finite-length type. The impedance spectra taken at different temperatures and different Li compositions (equilibrium potentials) can be fitted to the Randles circuit with excellent agreement. By this procedure, the results for the circuit elements, displayed in Figure 4.3.19 were obtained (Strømme Mattsson *et al.* [1996c]).

Figure 4.3.19 a shows that both the charge transfer resistance, as well as its activation energy, decreases with decreasing equilibrium potential (increasing ion content in the film). These quantities are associated with the barrier that the Li ions have to transverse to enter the film. The more Li ions in the film, the easier it gets for additional ions to enter. From the way the activation energy changes, one can also draw conclusions about the symmetry properties of the energy barrier (Strømme Mattsson *et al.* [1996c]). From Figure 4.3.19*b* we observe that the power-law parameter k is rather independent of temperature and of Li ion composition (equilibrium potential). This is in accordance with the fact that the distributed capacitance exponent described by this parameter only depends on the surface geometry (Strømme Mattsson *et al.* [1996a]). Finally, Figure 4.3.19*c* shows that the chemical diffusion process is thermally activated with the activation energy virtually independent of Li ion content (equilibrium potential). The value of ~0.5 eV is comparable to that of Li ion diffusion in amorphous WO_3 films (Nagai *et al.* [1986]), but somewhat higher than the activation energy for proton diffusion (Randin and Viennet [1982]). The activation energy being independent of the Li content in the film indicates that the

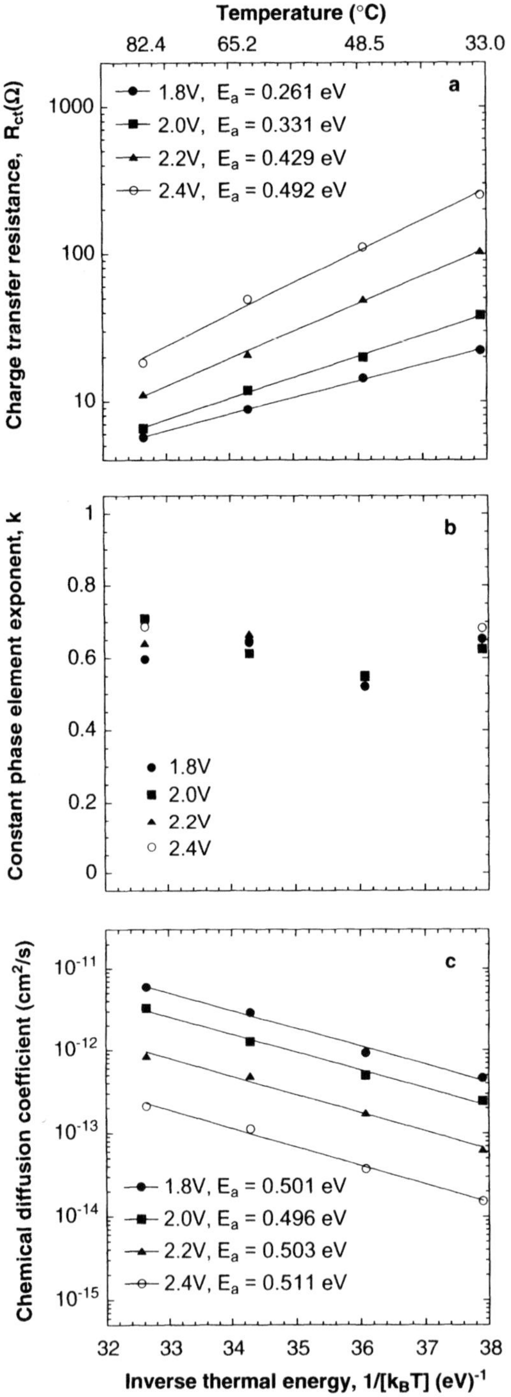

Figure 4.3.19. Magnitude of the quantities R_{ct} (*a*), *k* (*b*), and $\tilde{D}$ (*c*) for Li ions in fluorine doped Ti oxide vs. inverse thermal energy (Strømme Mattsson *et al.* [1996c]). Data were obtained from fits of impedance spectra of the type shown in Figure 4.3.18 to the circuit in Figure 4.3.17. The displayed activation energies E_a in part (*a*) and (*c*) were obtained from exponential fits (lines) to the individual data points.

ions move unaffected by each other so that they are only influenced by their surroundings. In fact, it was shown (Strømme Mattsson *et al.* [1996c]), that the obtained activation energy for diffusion gave a Li ion "jump length" corresponding to the distance between neighboring cells in the anatase structure of the nanocrystallites (Strømme Mattsson *et al.* [1996b]) in the films under study.

Cathodic Electrochromic Materials—Tungsten Trioxide. Figure 4.3.20 shows electrochemical impedance spectra on both amorphous and crystalline Li containing WO_3 films together with fits to the Randles circuit (Strømme Mattsson [2000]). For the amorphous film the high frequency semicircle overlaps with the diffusion response. In the case of the crystalline film, only a part of the semicircle due to C_{dl} and R_{ct}, can be observed. As is obvious from the displayed spectra, the charge transfer resistance is much larger for the crystalline sample than for the disordered one at an equilibrium potential of 2.9 V vs. the Li reference electrode used in the experiment. Impedance spectra were taken at several equilibrium potentials, and in all cases the impedance response corresponded to that of the Randles circuit with a $Z_{\tilde{D}}$ of semi-infinite type.

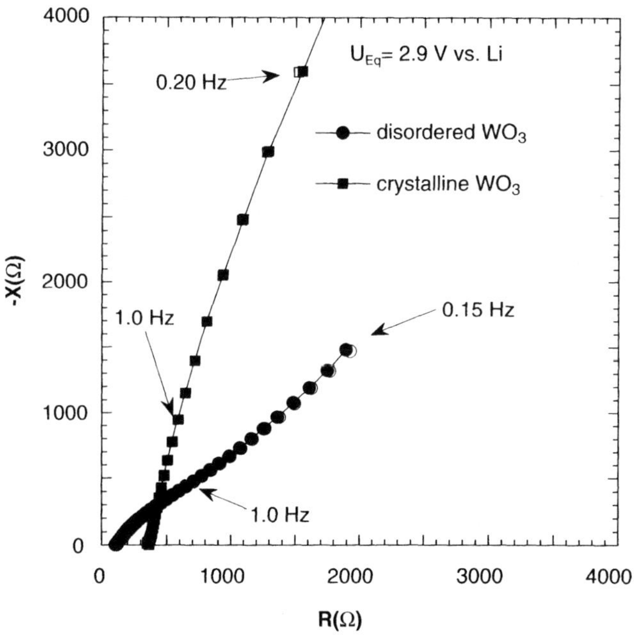

Figure 4.3.20. Electrochemical three-electrode impedance complex-plane plots for the real (R) and the imaginary (X) parts of the impedance of crystalline (filled squares) and disordered (filled circles) WO_3 recorded at 2.9 V vs. Li (Strømme Mattsson [2000]). This equilibrium potential corresponds to ~0.004 Li/W unit for the crystalline film and 0.03 Li/W for the amorphous one (Strømme Mattsson [2000]). Included in the figure are also fits (open squares for crystalline and open circles for disordered WO_3) to the Randles circuit in Figure 4.3.17. Explicit frequency readings are shown at a few selected data points.

In Figure 4.3.21 electrochemical transient current measurements on the same types of films are displayed together with fits to Eq. (11). Prior to the transient current measurement the WE potential was set to 4.5 V vs. the Li reference electrode, thus making sure that no Li ions resided in the films prior to the potential step (Strømme Mattsson [2000]). From the fits, values for the chemical diffusion coefficients of the tungsten oxide films were extracted. The chemical diffusion coefficient from impedance spectroscopy and the average diffusion coefficient from the transient current measurements are presented in Figure 4.3.22, while Figure 4.3.23 shows the charge transfer resistance, as obtained from impedance spectroscopy, associated with the barrier the Li ions have to pass to enter the WO_3 films. The results are discussed in further detail in Strømme Mattsson [2000], where information about the film preparation and the measurement conditions are given. As is obvious from the figures presented above, however, there is a marked difference between the parameters for the amorphous and crystalline samples. There is also a significant difference in the information carried by the impedance spectroscopy and the transient current diffusion coefficients.

Tungsten trioxide is a very widely studied electrochromic material and the impedance response has been measured several times. The results are consistent in that they can invariably be fitted to a Randles circuit, but details in the spectra vary.

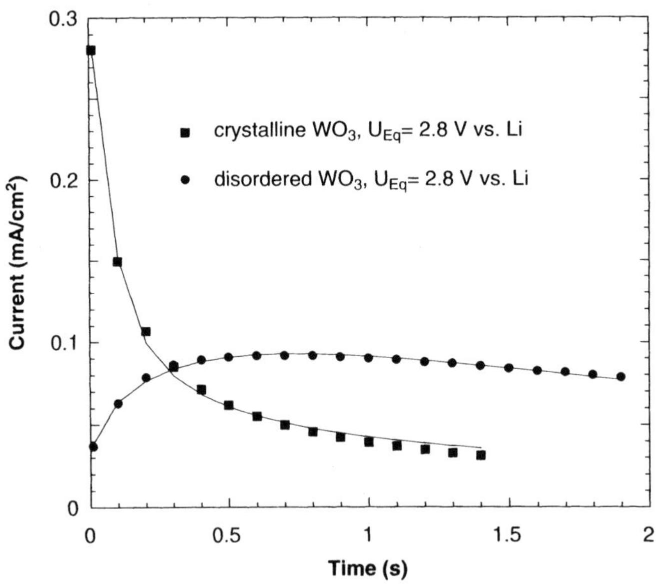

Figure 4.3.21. Current response of Li intercalation into crystalline (squares) and disordered (circles) WO_3 after a potential of 2.8 V vs. Li was applied (Strømme Mattsson [2000]). Included in the figure are also fits (full lines) to Eq. (11). The applied potential corresponds to an equilibrium concentration of ~0.009 Li/W unit for the crystalline film and 0.06 Li/W for the amorphous one (Strømme Mattsson [2000]).

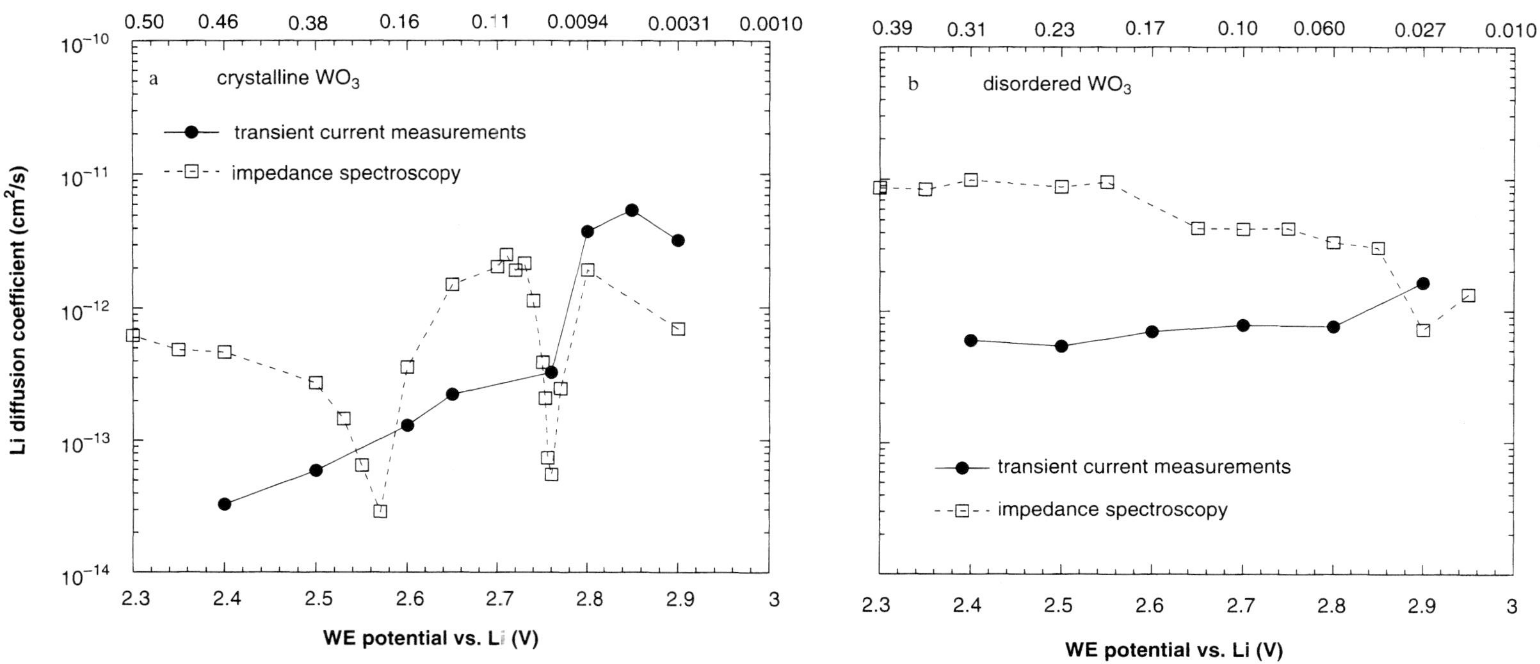

Figure 4.3.22. Diffusion coefficients for Li in crystalline (*a*) and disordered (*b*) WO_3 as obtained from electrochemical transient current measurements and from electrochemical impedance spectroscopy (Strømme Mattsson [2000]). The diffusion coefficient obtained by impedance spectroscopy is the chemical diffusion coefficient relevant for a specific composition (specified on the upper × axis), while that obtained from the current measurements is an average diffusion coefficient for all compositions lower than the composition value specified on the upper × axis.

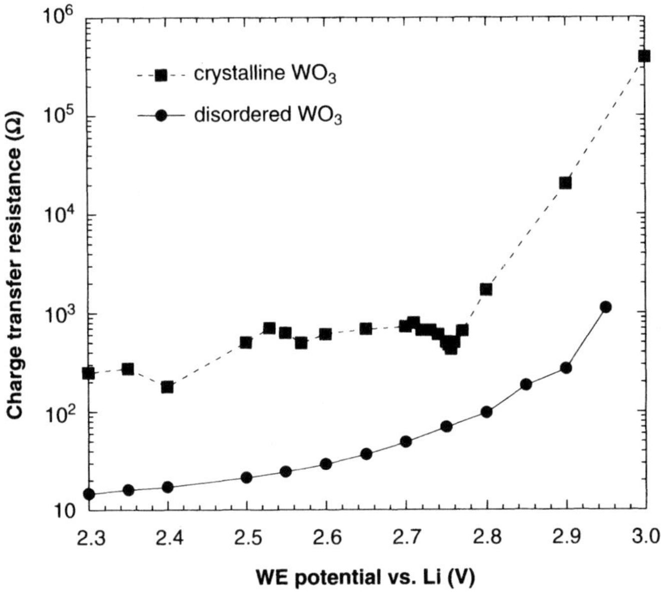

Figure 4.3.23. Charge transfer resistance for Li ion intercalation into WO_3. Data are obtained by electrochemical impedance spectroscopy from fits of the type shown in Figure 4.3.20.

In particular, values of the ion diffusion coefficient and the other circuit parameters depend not only on the crystallinity, but also on the deposition conditions of the films. Since tungsten oxide easily becomes non-stoichiometric when deposited as a thin film, the stoichiometry probably also plays a role. Data have been presented for films produced by thermal evaporation (Ho *et al.* [1980], Bohnke and Bohnke [1990], Lee *et al.* [2000, 2001]), electron-beam evaporation (Yoshiike *et al.* [1984]), sputtering (Pyun and Bae [1996]), sol-gel (Wang and Bell [1999], Sharma *et al.* [2002]), and various other techniques (Bohnke and Bohnke [1988]). Our data presented here are in general agreement with other studies of sputtered films, as well as with the recent detailed study of Lee *et al.* [2000, 2001].

There are a few questions, however, that deserve a further study. Sometimes the Warburg element is not of the finite-length type, but exhibits a power-law response instead. The reason for this behavior is not clear; it has been attributed to the WO_3/substrate interface (Bohnke *et al.* [1993]), or to an additional kinetic step in the bulk of the film (Fabregat-Santiago *et al.* [2001]).

Counter Electrode Materials. The anodic electrochromic materials and ion storage materials have not been so widely studied as the cathodic electrochromic materials discussed above. In general, the main features of the impedance spectra are similar to those shown above. The impedance response of nickel oxide films with protons (Gorenstein *et al.* [1990]) as well as with lithium ions (Decker *et al.* [1992], Artuso *et al.* [2002]) as the intercalated species, has been studied. The spectra can be fitted by a Randles circuit, with a finite length Warburg element (Gorenstein *et al.* [1990]). The charge transfer resistance decreases and the low frequency limiting capacitance as well as the diffusion impedance both increase with increasing equilibrium potential, i.e. during proton extraction. The ion diffusion is thus slowed down as the hydrated nickel oxide colors. Impedance data for iridium oxide films were also modeled within the framework of a Randles circuit (Glarum and Marshall [1980]). In a detailed study of lithium intercalation into vanadium oxide films (Bae and Pyun [1995]), the impedance spectra were interpreted with the extended Randles circuit of Franceschetti and Macdonald [1982].

Inorganic Thin Film Ion Conductors—Tantalum Oxide. We give here a detailed account of the impedance response of thin tantalum pentoxide films. Firstly, the data illustrates several of the methods treated in Section 4.3.4.3. Secondly, Ta_2O_5 is a very interesting ion conductor for applications, in particular because it is possible to make the electronic leakage current very low. We also make some comments on ZrO_2.

Figure 4.3.24 shows a series of two-electrode impedance spectroscopy measurements on a proton containing β-Ta_2O_5 sample (Strømme Mattsson *et al.* [1999a]).

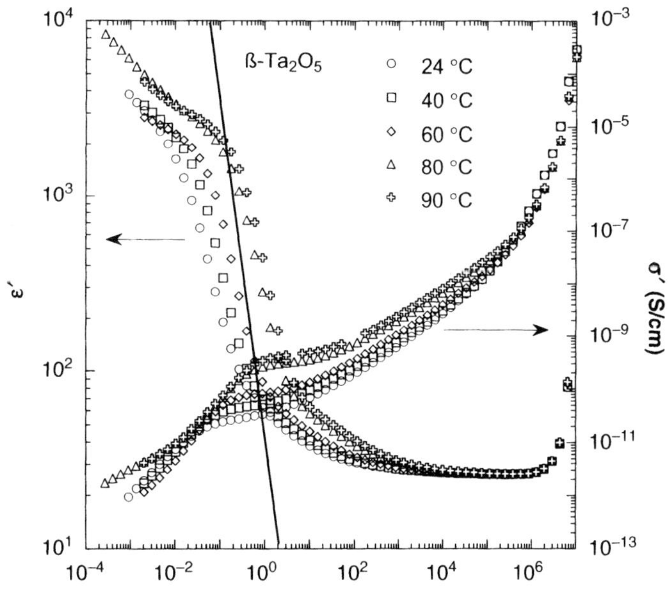

Figure 4.3.24. Real parts of the dielectric permittivity (ε') and the ac conductivity (σ') as a function of frequency for a β-Ta_2O_5 sample (Strømme Mattsson *et al.* [1999a]). The measurements were performed at the displayed temperatures. Also included in the figure is a straight line with a slope of ω^{-2}.

The data are presented in terms of the real parts of the dielectric permittivity (ε') and the ac conductivity (σ'). The measurements were performed at temperatures between 24 and 90°C. By applying Eq. (6) to these data, one obtains that the dc conduction process has an activation energy of 0.42 eV with the room temperature value of σ_{dc} being 25.3 pS/cm (Strømme Mattsson *et al.* [1999a]).

The behavior of both ε' and σ' in the frequency region between 10^{-2} and 10^2 Hz clearly shows a space-charge limited response (Schütt and Gerdes [1992]). A straight line with a slope of ω^{-2} is included in the figure to show that the Schütt and Gerdes model may be used to extract approximate values of n and μ. We observe that ε', when its magnitude is slightly below 10^3, is approximately parallel to the line. The somewhat smaller slope of ε' as compared to the line can partly be attributed to surface roughness of the Ta_2O_5 samples (Schütt and Gerdes [1992]) and partly to the fact that the factor $(\omega_X\tau\sigma)^2$ in Eq. (8) is only ~5 (as can be deduced from the calculations below). Applying Eq. (9) to the room-temperature ε' and using an X value of $941/\varepsilon_s' = 36.4$ ($\varepsilon_s' = \varepsilon_\infty = 25.8$), which corresponds to a ω_X value of $0.0235\,\text{Hz} * 2\pi = 0.148\,\text{s}^{-1}$, gives the proton density in the sample as n $\approx 3.5 * 10^{17}$ protons/cm^3. Further, Eq. (10) gives the proton mobility as $\mu \approx 4.5 * 10^{-10}\,\text{cm}^2/\text{Vs}$. These numbers imply that $(\omega_X\tau\delta)^2 \approx 5$ and that the relation $1 + (\omega_X\tau\delta)^2 \approx (\omega_X\tau\delta)^2$ required for the validity of Eq. (9) is not completely satisfied. The obtained proton density and mobility must therefore be considered as rough order of magnitude estimates.

An independent determination of n and μ can be obtained from transient current measurements. Figure 4.3.25 shows the current response after a potential of 3V is applied across the proton-containing Ta_2O_5 sample, taking the sample from a uniform charge distribution to a polarized state (Strømme Mattsson *et al.* [1999a]). Assuming that the initial current decay is dominated by migration of ions toward an ion-blocking contact, both n and μ can be obtained by fitting the measured current response to Eq. (12). This procedure gives $n \approx 1.3 * 10^{17}$ protons/cm^3 and $\mu \approx 5.5 * 10^{-10}\,\text{cm}^2/\text{Vs}$, which corresponds very well to the parameters obtained from the space-charge limited impedance response.

Ta_2O_5 has also been investigated for its Li ion conduction properties (Strømme Mattsson and Niklasson [1999]). Figure 4.3.26 shows a transient current experiment on the above-discussed β-Ta_2O_5 sample after Li ions have been intercalated into the film. The current response is clearly made up of two mechanisms involving two more or less overlapping peaks. The presence of peaks here indicates that the charge carriers are initially trapped (Strømme Mattsson and Niklasson [1999]). Both protons and Li ions were uniformly distributed in the sample prior to the measurement, thus Eq. (12) is applied to analyze the current decay after the peaks. An exponential function is fitted to the measured current in a time window immediately after the first peak, most likely pertaining to protons. In this time range, contributions from the other charge carrier is negligible. Similarly, the current decay following the second peak, most likely caused by Li ions, is fitted to another exponential function. From this procedure we find that the proton mobility μ_p equals $3.24 * 10^{-10}\,\text{cm}^2/\text{Vs}$ and the Li ion mobility $\mu_{Li} = 1.17 * 10^{-11}\,\text{cm}^2/\text{Vs}$. These numbers were verified by using the position of the peaks to estimate the speed with which the ions traversed

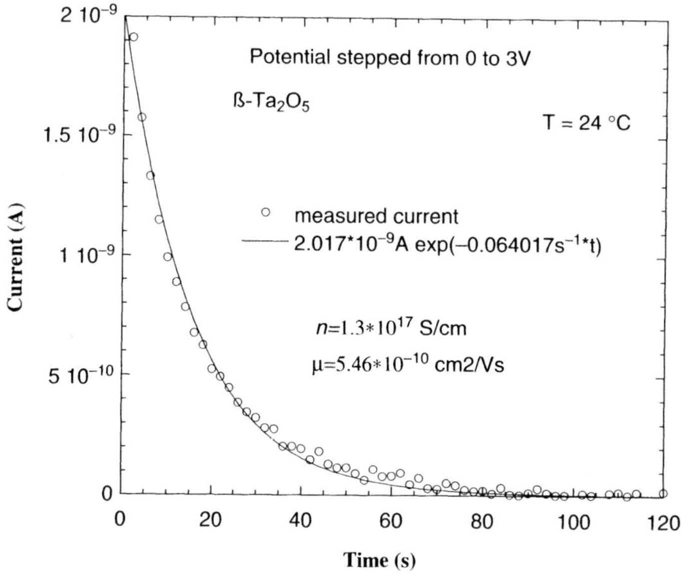

Figure 4.3.25. The transient current (circles) in β-Ta_2O_5 after a potential step. The displayed equation is the exponential fit (full line) to the data. From this fit the shown proton density n and mobility μ could be extracted using Eqs (12) and (10).

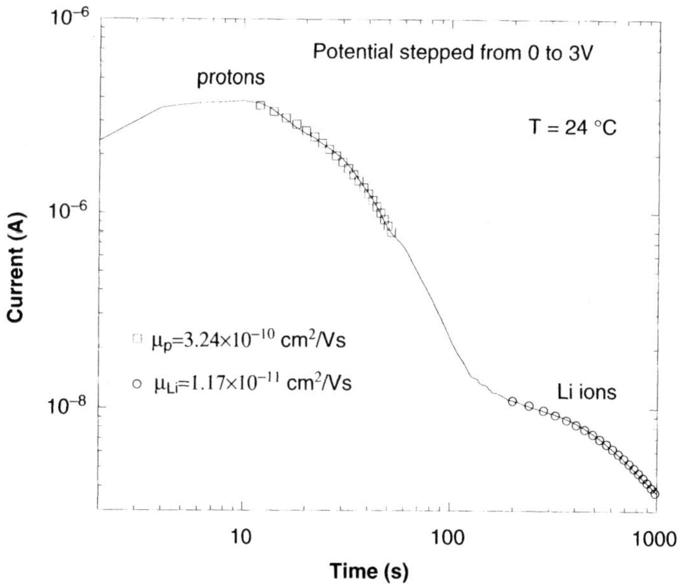

Figure 4.3.26. The transient current (full line) in lithiated β-Ta_2O_5 after a potential step. The proton and Li ion mobility μ_p and μ_{Li}, respectively, are obtained from a fit to Eq. (12) (squares for μ_p and circles for μ_{Li}).

the sample (Strømme Mattsson and Niklasson [1999]). The proton mobility has obviously decreased after Li ions were introduced, which can be interpreted as a mixed alkali effect (Bunde *et al.* [1991]). The Li ion mobility is almost one order of magnitude lower. We have hence shown that the two methods for determining ion density and mobility, outlined in Section 4.3.4.3, are mutually consistent.

It has been shown that Li ions and their charge balancing electrons enter the β-Ta_2O_5 at a higher energy than that for intercalation into, for example, WO_3 (Strømme Mattsson and Niklasson [1999]). This finding, together with the fact that Li ions in β-Ta_2O_5 have a mobility of the same order as in WO_3 (cf. Figure 4.3.22), makes the Ta oxide potentially very interesting as a Li ion conductor in a WO_3 based electrochromic device.

Inorganic Thin Film Ion Conductors—Zirconium Oxide. Electrochemical three-electrode impedance spectroscopy was performed for proton intercalation on sputtered ZrO_2 by Jonsson *et al.* [1999]. The data can be fitted very well to the Randles equivalent circuit. When the charge transfer resistance R_{ct} is extracted, it is found to be very high at low intercalation levels and then drops several orders of magnitude at medium intercalation levels. The reason for the high R_{ct} could be a layer grown on the surface of the film that slows down the transport of ions from the electrolyte into the film. For some reason it becomes easier for the protons to enter the film at higher ion content, either the surface film disappears or the protons find an easy way through it. The diffusion coefficient was determined to be in the range of 10^{-13}–10^{-12} cm^2/s, which is only about one order of magnitude lower than in Ta_2O_5.

4.3.4.5 Experimental Results on Electrochromic Devices

Impedance Spectroscopy. Impedance spectroscopy has been carried out on devices with WO_3 as the cathodic electrochromic layer, counter electrodes of iridium oxide, polyaniline or Prussian blue, and polymers as electrolytes (Katsube *et al.* [1986], Friestad *et al.* [1997]). The equivalent circuit for a whole device becomes very complicated. In the works quoted above simplified, Randles-type circuits were used for the two electrochromic layers, while the ion conductor was modeled by a pure resistance, or neglected. Extraction of device parameters from the data fitting was reported. However, it is clear that in many cases it will be difficult to distinguish the contributions from the different layers in a device, in particular if the migration impedances, ion diffusion impedances, etc. are of the same order of magnitude. When it comes to characterizing electrochromic devices, impedance spectroscopy is a very time-consuming process, since a spectrum down to low frequencies should be taken at a number of equilibrium potentials. Thus we believe that transient current measurements in many cases offer a faster alternative that sometimes allows a simple determination of diffusion coefficients.

Transient Ion Currents. Transient current measurements, according to Section 4.3.4.3, have been performed on laminated devices (Jonsson *et al.* [2004]) as well as on all-thin-film (ATF) devices (Frenning *et al.* [2003]). The transient current response for a typical configuration of a laminated electrochromic device is shown in Figure 4.3.27. The device consisted of cathodic tungsten oxide and anodic nickel oxide films that were laminated with a 50 μm thick layer of polymer electrolyte prepared by mixing polyethylene glycol and lithium trifluoromethanesulfonate. Both Li ions and protons may act as charge carriers in this device. As can be seen in the figure very different responses are obtained depending on the direction of the electric field, i.e. in which direction the charge carriers are moving. The solid curve shows the response when the device is not polarized, and a potential of −1.5 V is applied. The current decreases rapidly during the first 0.3 s, after which it levels out during approximately 20 s before a rapid decrease sets in again. The dotted curve shows the current after the potential has been switched to 1.5 V and ions move from the WO_3 layer to the nickel oxide layer. The initial current is higher and it decreases slower than for the unpolarized case but after 1 s a more rapid decrease sets in. The potential was switched once more, from 1.5 V to −1.5 V, and the ions moved from the nickel oxide layer back to the WO_3 layer. Here, two areas of low slope are found, the first after 0.02 s and the second after 1 s. After 100 s all three cases show a steady decrease and a current density less than 10^{-5} A/cm^2. The charge carriers move with

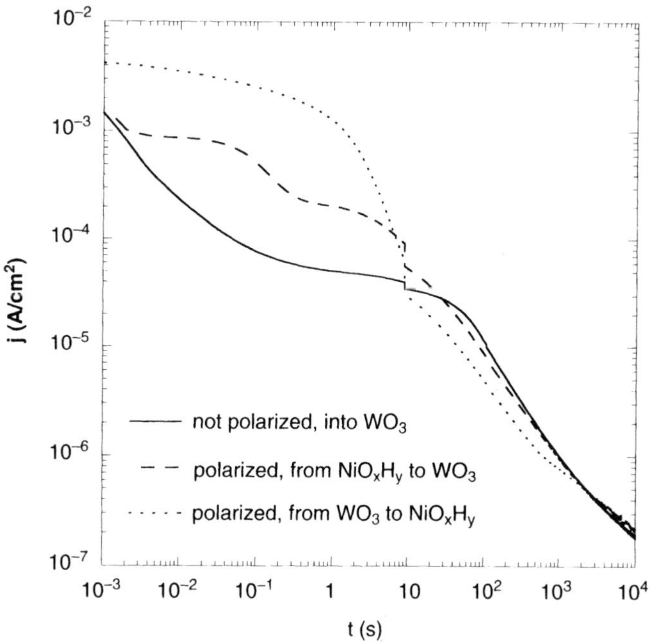

Figure 4.3.27. Transient current versus time response of laminated tungsten oxide–polymer electrolyte–nickel oxide electrochromic device for different directions of the ion movement. The break in the curve at 9 s is due to measurement artifacts.

different velocities in the different layers of the device (Granqvist [1995], Wu *et al.* [2000], Avendano [2002]), which influences the current and may give rise to peaks or shoulders. When ions move through the device they encounter two interfaces that can act as potential barriers with certain activation energies (Atanasov *et al.* [1999]). This may be depicted as if the ions get trapped at the interface and then are released at a certain release rate. In a laminated device, the complementary ion in the salt, in this case trifluoromethanesulfonate, is also present. This ion will stay in the electrolyte however, and thus not pass any of the interfaces.

The ATF devices consisted of nanocrystalline thin films of hydrated NiO and ZrO_2 and X-ray amorphous WO_3. A potential step of 1.5 V was applied over the devices and the resulting current was measured as a function of time. The main charge carriers through this device are protons, introduced into the nickel oxide during deposition. In Figure 4.3.28 the transient current versus time response is shown for coloration and bleaching, respectively. For the coloration measurement the device was polarized to assure that all protons were in the NiO prior to the measurement and similarly in the bleaching process all protons resided in the WO_3 layer prior to the application of the potential step. The initial features, up to 10–100 s, show similarities to those of the laminated device. In both cases the current is higher

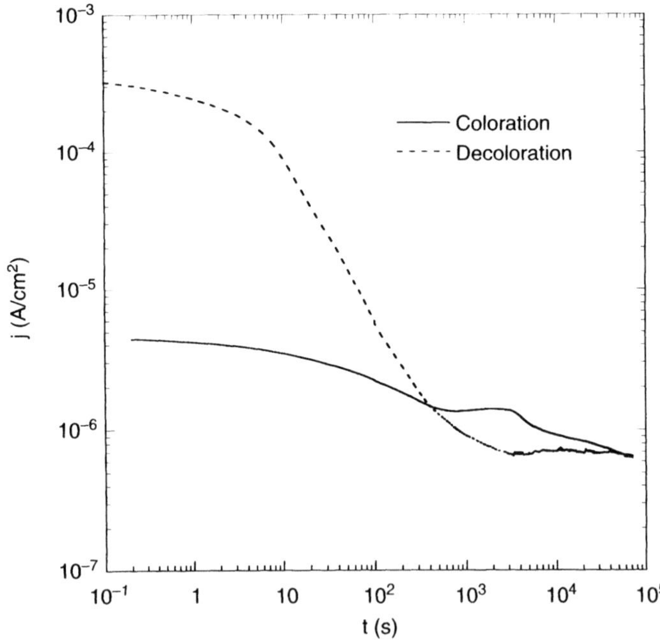

Figure 4.3.28. Transient current versus time response for an all-solid state device, using tungsten oxide and nickel oxide as electrochromic layers and zirconium oxide as an ion conductor. The transient current is shown for both the coloration and decoloration process.

for ion movement from the WO_3 to the nickel oxide. This suggests that the proton conductivity is higher in WO_3 than in NiO. For longer times the response of the ATF device shows a well-defined peak. When protons are transported from the NiO layer into the WO_3 layer in the coloration process, the peak is situated at about 3000 s. When the protons are transported in the opposite direction, a much smeared out peak is observed at about 10,000 s. Subsequently, the current decreases, because the ions have arrived at a blocking electrode. The peak has been modeled by Frenning *et al.* [2003], and its position corresponds to an ion mobility of the same order of magnitude as in a ZrO_2 film.

This behavior can be understood within two scenarios. In the first scenario, the proton conductivity is lower in the ion conductor than in the WO_3 and NiO layers. When protons are transported through the device they are slowed down in the ion conductor and the current thus decreases. After the protons enter the electrochromic oxides their speed increases and the current goes up. At the back contact the protons are blocked and the current decreases again, thus a peak is seen. Since the proton conductivity is higher in WO_3 than in NiO the peak is less pronounced in the bleaching process compared to the coloration. In the other scenario, the proton conductivity does not differ by orders of magnitude between the three oxide layers. Instead there are barriers at the interfaces between the ion conductor and the electrochromic oxides acting as ion traps. When the ions are released the current increases and the protons pass through to the back contact. In the laminated device no peak at long times is observed, probably because the ion conductor effectively serves as a short for the ions in this case. In the ATF device, after approximately 100,000 s the current takes the same constant value for ion movements in both situations. This value most likely represents a leakage electron current through the ZrO_2 ion conductor.

The transient current versus time response for all-solid-state devices and laminated devices show some interesting common features. The initial response is similar in the two devices although the response is slower in the all-solid-state device. In both types of devices the ion transport is markedly quicker from WO_3 into NiO than in the opposite direction. Also, pronounced peaks and shoulders are only present when ions are transported from the NiO layer towards the WO_3 layer. This asymmetric behavior indicates that barriers of different magnitude are present at the interfaces between the oxide layers and that these barriers strongly affect the transient current response.

The investigation of transient ionic currents in electrochromic devices is still in the initial stage. The theory of Frenning *et al.* [2003] constitutes a good starting point for detailed comparisons with experiments, but further developments are probably needed in order to reach a thorough understanding.

4.3.4.6 Conclusions and Outlook

Today, there exists a good understanding of the impedance response of electrochromic materials. Experiments on oxide electrodes deposited onto an electron-conducting substrate and immersed in an electrolyte, invariably show the characteristic behavior of the Randles circuit. By fitting the data to this circuit, detailed information on ion chemical diffusion coefficients in the oxide film, as well as processes occurring at the interfaces can be obtained. Most investigations have been carried out on tungsten oxide. In the future, more work will probably be directed towards other materials.

Some details of the Randles circuit require further study. First, the double layer capacitance is in many cases more properly modeled with a constant phase element. This gives information on the mesoscopic structure of the oxide–electrolyte interface. Also, in some cases, the diffusion impedance contains a power-law behavior. The reason for this is controversial: is it due to the back contact or rather an indicator of an unknown kinetic process in WO_3? It should also be mentioned that the adsorption process proposed by Franceschetti and Macdonald [1982] has not been studied in detail, and the systematic equivalent circuit approach of Jamnik [2003] has only been rarely used.

Much information about the kinetics of electrochromic devices can be obtained from impedance spectroscopy or transient current measurements. In particular, the transient current technique is a promising method for device testing. It is based on the well-known transient ion current technique used in studies of ion conduction and it is less time consuming than impedance spectroscopy. More efforts should be devoted to development of theory as well as to detailed comparison with experimental data.

A rather neglected area is studies of the so-called color impedance. To investigate this property, an ac signal is applied to the electrochromic film and the resulting transmittance modulation is measured. This method should give information not only on the ion and electron transport, but also on the kinetics of the coloration process. Initial work has been reported on WO_3 (Gabrielli *et al.* [1994], Kim *et al.* [1997]) and NiO (Cordoba-Torresi *et al.* [1990]) films. It appears that the kinetics of coloration is not determined solely by the ion diffusion (Gabrielli *et al.* [1994]). The technique may open up interesting possibilities for further studies of the electrochromic coloration process.

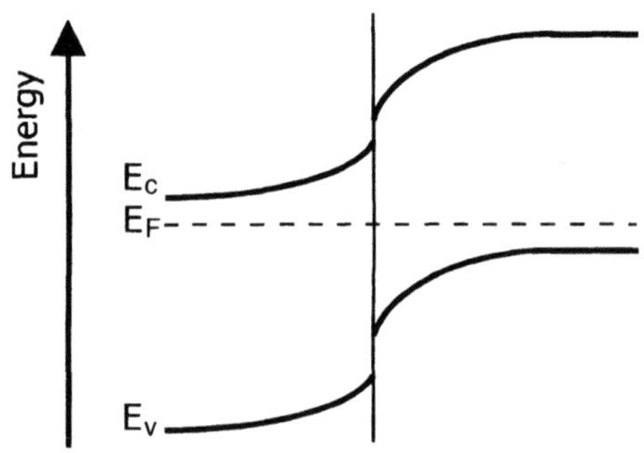

Figure 4.3.29. Energy band diagram of a junction between an n-type and a p-type semiconductor.

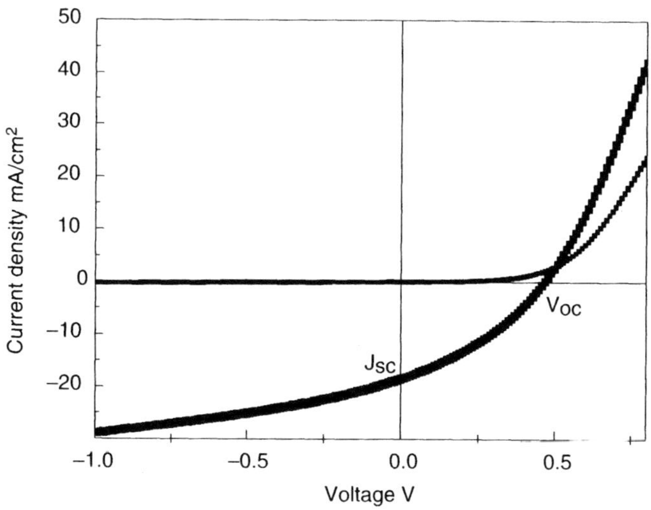

Figure 4.3.30. Typical steady-state photocurrent voltage curve. The thin line is the current in the dark, which shows diode behavior. The thick line is the current under standard AM1.5 solar irradiation.

4.3.5 Time-Resolved Photocurrent Generation

Albert Goossens

4.3.5.1 Introduction—Semiconductors

A semiconductor is a material in which the electrical conduction is in between that of an insulator and that of a metal. Usually, the electrical resistivity of a semiconductor ranges between 10^{-5} and $10^{5}\,\Omega$cm. A good semiconductor shows no ionic conduction whatever. To understand the nature of semiconductors, one needs to study the quantum-physical properties of the electrons. When atoms are brought together to form a lattice, the valence electron orbitals overlap. This orbital overlap implies that electrons are no longer located on individual atoms, but delocalize over all constituent atoms. This electronic delocalization is stronger when the orbital overlap is larger. In a typical semiconductor, the overlap is so large that the valence orbitals merge together completely and form an energy band. This is called the valence band. Likewise, the lowest empty atomic orbitals also merge together and form the conduction band. The lowest energy state of the conduction band has an energy E_c, while the highest energy state in the valence band has energy E_v. There is an energy gap in between these bands called the bandgap energy, E_g, and $E_g = E_c - E_v$.

For typical semiconductors the bandgap is between 1 and 3 eV. The electronic population of the energy bands is determined by the Fermi–Dirac distribution function, i.e.

$$f_{FD}(E) = \frac{1}{1 + \exp[(E - E_F)/kT]} \tag{13}$$

In this expression E_F is the Fermi-energy level, which determines the population statistics. In a non-degenerate semiconductor $E_c - E_F$ and $E_F - E_v$ are much larger than kT and the Fermi–Dirac distribution can be approximated by the Boltzmann distribution functions for the conduction and valence bands, i.e.

$$\begin{aligned} f_B(E_c) &= \exp[-(E_c - E_F)/kT] \\ f_B(E_v) &= \exp[-(E_F - E_v)/kT] \end{aligned} \tag{14}$$

The effective density of states of the conduction band and valence band are N_C and N_V, respectively. The concentrations of conduction band electrons, n, and valence band holes, p, are given by

$$\begin{aligned} n &= N_C \exp[-(E_c - E_F)/kT] \\ p &= N_V \exp[-(E_F - E_v)/kT] \end{aligned} \tag{15}$$

Thermal excitation of valence band electrons yields conduction band electrons, which can be written in the Kröger–Vink notation, i.e.

$$0 = e' + h^{\bullet} \tag{16}$$

In this expression e' is a conduction-band electron and $h^{\bullet}$ is a valence-band hole. If this reaction is in thermodynamic equilibrium, the following condition holds

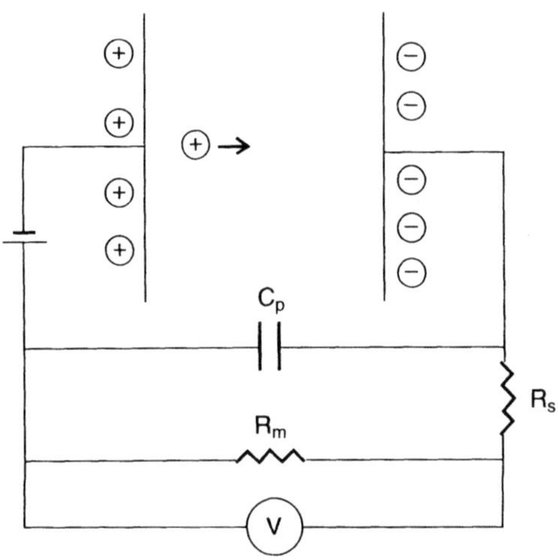

Figure 4.3.31. Typical set-up for time-of-flight experiments. The capacitances of photoelectrode and the shunt add up. R_s is the series resistance and R_m the resistance over which the voltage is measured.

$$K_{eh} = [e'][h^{\bullet}] = np \tag{17}$$

Here $n \equiv [e']$ and $p \equiv [h^{\bullet}]$. Upon substitution of Eq.15 in Eq.17 the equilibrium constant K_{eh} is found

$$K_{eh} = np = N_C N_V \exp[-E_g/kT] \tag{18}$$

Intrinsic semiconductors are undoped and the concentration of conduction band electrons equals that of valence band holes, i.e. $n = p$. In this case, the Fermi energy level is located near the centre of the bandgap.

To get a feeling for quantities, it is useful to take a closer look at silicon, which is the most important semiconductor. Silicon has a bandgap of 1.1 eV and effective density of states $N_C = 2.8 \times 10^{19}\,\mathrm{cm}^{-3}$ and $N_V = 1.04 \times 10^{19}\,\mathrm{cm}^{-3}$ (Sze [1981]). For intrinsic silicon at room temperature the concentrations of conduction band electrons and valence band holes are about $10^{10}\,\mathrm{cm}^{-3}$, which is very low when considering that the concentration of atoms is twelve orders of magnitude larger. Therefore, intrinsic semiconductors at room temperature exhibit a very small electronic conductivity. The conductivity can be increased considerably by applying a suitable doping element. When a few parts per million of phosphorous is dissolved into silicon, the concentration of conduction band electrons increases six orders of magnitude; the concentration of holes reduces by the same amount. The opposite is true when boron is used as doping element. In that case p increases many orders of magnitude and n reduces accordingly.

A small concentration of a well-chosen doping element affects the concentrations of conduction-band electrons and valence-band holes significantly. Since Eq.15 is also valid for doped semiconductors, the position of the Fermi energy can be shifted from the midgap position towards either of the bands. If an electron donor is applied as doping element $n > p$ holds, and the Fermi energy shifts towards the conduction band; these semiconductors are called n-type. If an electron acceptor is applied as doping element $p > n$ holds, and the Fermi energy shifts towards the valence band; the semiconductor is called p-type.

The Fermi-energy level is equal to the electrochemical potential of the electrons, i.e.

$$E_F = \tilde{\mu}_e = \overline{\mu}_E + kT \ln[e'] - q\phi \tag{19}$$

$\overline{\mu}_e$ is the chemical potential of the electrons in either the conduction band or valence band deep inside the bulk of a material; q is the elementary charge, and ϕ is the galvanic potential.

In general, the electrochemical potential of the electrons of two different materials is not the same. When a junction is made between n- and p-type semiconductors, or between a semiconductor and a metal, the system will reach thermodynamic equilibrium by equalizing the electrochemical potentials (the Fermi-energy levels) of the two materials.

Electrons flow from the material with a high Fermi energy to the one with a low Fermi energy until the gradient of the Fermi energy is zero. This redistribution of electrons leads to the formation of a space-charge region, which is the region in the proximity of the interface in which local charge neutrality is no longer observed. In the n-type material, the concentration of conduction band electrons is depleted and an effective positive background space charge, Q_{SC}, remains. In the p-type material the valance-band holes are depleted and a negative background charge is present. The positive and negative space charges are equal to ensure overall charge neutrality.

In a junction between an n-type and p-type semiconductor the positive-charged and negative-charged space-charge regions are connected together. This charged interface exhibits a space charge capacitance, C_{SC}, which is defined as dQ_{SC}/dV and can be found by solving the Poisson equation with adequate boundary conditions (Rhoderick and Williams [1988]). In the Mott–Schottky approach, C_{SC}^{-2} is considered as a function of the applied dc voltage, V. The Mott–Schottky equation gives the relation between the reciprocal of the square of the space charge capacitance and the applied dc voltage, i.e.

$$C_{SC}^{-2}(V) = \frac{2}{\varepsilon_0 q A^2}\left(\frac{1}{\varepsilon_0 N_D} + \frac{1}{\varepsilon_p N_A}\right)\left(V - V_{fb} - \frac{kT}{q}\right) \tag{20}$$

in which ε_0 is the permittivity of free space; ε_n and ε_p the dielectric constants of the n-type and p-type materials; N_D and N_A the concentrations of the electron donors and acceptors; V is the externally applied potential; and V_{fb} is the built-in potential, which is associated with the flat-band potential. The presence of space-charge region is associated with a built-in electrical field, pointing from the n-type side to the p-type side of the junction. This electric field is also associated with the built-in potential.

To investigate the electrical properties of pn junctions, impedance spectroscopy is a powerful tool. First of all, the formation of a space-charge region is associated with a junction capacitance, which can be measured. A good approach is to measure the impedance spectrum at a range of applied voltages. By fitting the spectra to a well-chosen equivalent circuit the space-charge capacitance is found as a function of the applied dc. voltage. Following Eq. (44), a Mott–Schottky plot can be constructed of C_{SC}^{-2} versus V. For ideal junctions a Mott–Schottky plot is linear. From the slope of a Mott–Schottky plot, $\{(\varepsilon_n N_D)^{-1} + (\varepsilon_p N_A)^{-1}\}^{-1}$ can be determined. In case of a metal-semiconductor Schottky contact, the same equation can be used, except for the small kT/q term, with the notion that the metal resembles a very-highly-doped semiconductor. In this case, the doping density of the semiconductor can be determined. The intercept of a Mott–Schottky plot with the voltage axis equals the built-in voltage within a factor of kT/q. The intercept voltage is called the flat-band potential, V_{fb}. The interested reader can find more on basic properties of semiconductors in classical text books Sze [1981], Rhoderick *et al.* [1988], Bube [1992], Tanner [1995], Solymar and Walsh [1999]).

4.3.5.2 Steady-State Photocurrents

If a junction between n-type and p-type semiconductors is irradiated with light a photocurrent can be generated. If photons with sufficient energy are absorbed in a semiconductor, electrons are excited from the valence band into the conduction band. The concentrations of conduction-band electrons and valence-band holes increase by this optical pumping process. Now the system is no longer in thermodynamic equilibrium, and the Eqs (13) to (20) no longer hold. The optical-generated electrons and holes are mobile and lead to a photocurrent. If generation of electron-hole pairs takes place in the space charge region, the built-in electric field decouples the electrons from the holes and accelerates them in opposite directions, giving rise to a photocurrent. If electron-hole pair generation occurs outside the space charge region, it is possible that the minority carrier, i.e. holes in n-type material and electrons in p-type material, reach the space-charge region by diffusion. Upon arrival, the built-in electric field accelerates the charge carrier away and a photocurrent is generated. The first type of current is based on drift of free charge carriers, the second one on diffusion.

In general, both the drift and diffusion components contribute additively to the photocurrent. The steady-state efficiency of the photogeneration process can be found by solving the drift and diffusion equations and adding the currents. When assuming that the built-in field is located primarily in either the n-type or p-type side of the junction, the Gärtner equation for monochromatic photocurrent generation applies, which reads

$$J_{ph} = -q\Phi\left[1 - \frac{e^{-aW}}{1+aL}\right] \tag{21}$$

in which J_{ph} is the photocurrent density, Φ the photon flux (number of impinging photons per second per cm^2), W is the width of the space charge region, a the absorption coefficient, and L the minority-carrier diffusion length (Gärtner [1959]).

W depends on the applied voltage V according to

$$W(V) = \left(\frac{2\varepsilon_0\varepsilon_n}{qN_D}\right)^{1/2}\left(V - V_{fb} - \frac{kT}{q}\right) \tag{22}$$

which implies that the generated photocurrent is voltage dependent.

Here, it is assumed that the applied field fully drops over the n-type side of the junction. Eq. (21) is only valid if electron-hole pairs do not recombine in the space-charge region, or at the surface. For practical devices this assumption usually does not hold, but to avoid complications we shall not consider these types of electron-hole pair recombination here.

In a steady-state photocurrent experiment, a monochromatic light source is used to excite the semiconductor and the photocurrent and photovoltage are measured (Green [1992]). If the load resistance is zero, the short-circuit photocurrent is measured and the photovoltage is zero. If the load is infinite, the photocurrent is zero and the open-circuit photovoltage is measured.

Steady-state techniques are convenient to obtain a general idea about the generation and collection of charge carriers. But in steady-state experiments only overall processes can be observed; details of the charge carrier dynamics are out of sight. To study these details one needs to apply time-resolved techniques. There are two approaches: time-of-flight techniques, and intensity-modulated techniques. In the first method an irradiation pulse or step is applied, usually making use of a pulsed laser. In the second method, the intensity of the light source is harmonically modulated. Below both techniques shall be discussed in more detail.

4.3.5.3 Time-Of-Flight

In a typical time-of-flight (TOF) experiment the semiconductor is kept under bias control. One of the contacts is transparent, being the injecting contact. The other contact is the collecting contact. By applying a bias voltage, V, the contacts are charged and the system behaves as a parallel plate capacitor (Fig. 4.3.32). Before the laser pulse is applied, the charge on the injecting contact is Q_i and that on the collection contact is Q_c. Charge neutrality requires that $Q_i = -Q_c$. The capacitance is

$$C = \frac{Q}{V} = \frac{\varepsilon\varepsilon_0 A}{L} \tag{23}$$

At the injecting contact, minority charge carriers are generated instantaneously by applying a laser pulse. The spatial distribution of these charge carriers must be kept small, which can be achieved by choosing a wavelength for which the absorption coefficient is large. The initial condition, therefore, is a collection of minority charge carriers with a total charge, ΔQ, confined in space and time.

Let us suppose that we deal with an n-type semiconductor, for which valence band holes are the minority carriers. A positive voltage is applied on the injecting contact. Immediately after the laser pulse, the holes reside near the injecting contact. They induce an image charge, which changes the charge on the injecting contact from Q_i to $Q_i - \Delta Q$. To ensure that the internal electric field is not affected, the intensity of the laser pulse must be reduced until $\Delta Q << Q_i$ holds. Forced by the electric field, the holes drift away from the injection contact towards the collecting contact. Since the electric field, E, is constant the holes drift with a constant velocity, v, given by

$$v = \mu_h E = \mu_h \frac{(V - V_{fb})}{L} \tag{24}$$

In this expression the presence of a built-in field of V_{fb}/L has been taken into account.

While moving from the injecting towards the collecting contact, the image charges on these contacts change in time. At $t = 0$, the charge on the injecting contact is $Q_i - \Delta Q$ and that on the collecting contact $I = -I$. After a period τ, the holes arrive at the collecting contact. At $t = \tau$ the charge on the injecting contact is Q_i, and that on the collecting contact is $Q_c + \Delta Q$. Accordingly, the charge on the collecting

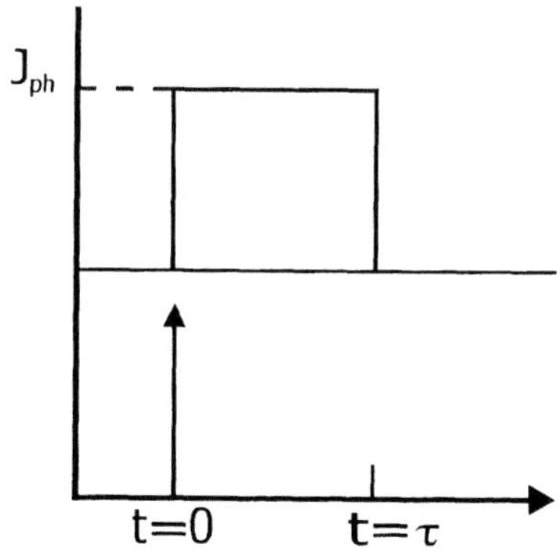

Figure 4.3.32. Internal photocurrent transient in an ideal TOF experiment.

contact changes linearly from Q_c to $Q_c + \Delta Q$ over the time interval τ. The current that runs through the external circuit is $J_{ph} = \Delta Q/\tau$. This photocurrent sets on at $t = 0$ and turns off at $t = \tau$, as is shown in Figure 4.3.32.

In the ideal case, as described above, the photocurrent transient drops abruptly to zero at $t = \tau$. The following equation holds

$$\tau = \frac{L}{v} = \frac{L}{\mu_h E} = \frac{L^2}{\mu_h (V - V_{fb})} \tag{25}$$

The best way to determine the hole drift mobility μ_h is to construct a plot of $1/\tau$ versus V. This plot will be linear with a slope of $\frac{\mu_h}{L^2}$. Even more reliable results are obtained if one constructs a graph of the obtained slopes versus L^{-2}. Further information of the principles of the time-of-flight technique can be found in Scher and Montroll [1975], Bässler [1993], and Schroder [1998].

In reality, the interpretation of a time-of-flight experiment is more complex than outlined above. There are several aspects that have been neglected so far. Some important aspects are discussed below.

It has already been mentioned that the condition $\Delta Q << Q_i$ must hold to ensure that the electric field strength is not affected by the optically-generated charge carriers. If this condition is not fulfilled, the electric field strength is not constant and can become a function of space and time. This will complicate the interpretation of TOF signals.

Diffusion. In the above the diffusion contribution to charge migration has been neglected. This is only valid if the transit time, τ, is much smaller than the effective diffusion transit time: $L^2/2D$, where D is the diffusion coefficient.

$$\tau = \frac{L}{v} = \frac{L}{\mu_h E} << \frac{L^2}{2D} \tag{26}$$

Using the Einstein relation between the mobility and the diffusivity we write

$$\tau = \frac{L}{v} = \frac{L}{\mu_h E} << \frac{qL^2}{2kT\mu_h} \Rightarrow E >> \frac{2kT}{qL} \Rightarrow (V - V_{fb}) >> \frac{2kT}{q} \tag{27}$$

Accordingly, the diffusion contribution can be neglected only if the above condition holds. If this condition is not fulfilled, the spatial distribution of charge carriers broadens during the transit time. In this case, the current transient no longer drops abruptly to zero at $t = \tau$ but is smoothed somewhat.

Trapping. In our simple model, all generated charge carriers have the same mobility. This is only true for ideal, defect free semiconductors. In defect semiconductors, electrons and holes can be trapped in sub-bandgap electronic states. If the energy of these states is within a few kT of the conduction or valence band, the charges can be emitted to one of the bands and they can be transported towards the collecting electrode by a hopping process. This will slow down the carrier transport considerably. It can be modeled by introducing an effective mobility, which is much smaller than the mobility of free charge carriers. If there is only one dominant trap state, a time of flight experiment shows two waves, the first one coming from the free charges and the second one from the trapped charges. In most cases trapping of charge carriers leads to a broadening of the current drop at $t = \tau$, showing a current tail beyond the transit-time threshold. This tail is denoted the post-transit tail and can be used to determine the presence of sub-bandgap electronic states.

Recombination. Minority carriers can recombine with majority carriers during the transit time, leading to a loss of charge ΔQ in time. In a simple approach, recombination is expressed as

$$\frac{d\Delta Q(t)}{dt} = \frac{-\Delta Q(t)}{\tau_R} \tag{28}$$

which yields

$$\Delta Q(t) = \Delta Q_0 \exp[-t/\tau_R] \tag{29}$$

in which ΔQ_0 is the charge generated at $t = 0$.

Since $J_{ph} = \Delta Q/\tau$, we write

$$J_{ph}(t) = \frac{\Delta Q(t)}{\tau} = \frac{\Delta Q_0}{\tau} \exp[-t/\tau_R] \tag{30}$$

In the time interval between $t = 0$ and $t = \tau$, the current is no longer constant but decays exponentially in time. The effects of diffusion, trapping, and recombination of charge carriers is illustrated in Figure 4.3.33.

Convolution. A final complication must be addressed. Since the photocurrent is driven into a capacitive component, the shape of the measured transient is affected. The measured photocurrent response function is convoluted with the response function of the measuring system. If the response function of the measuring system is known, deconvolution of the two contributions to the current response can be accomplished using Laplace transform algorithms. The internal photocurrent, J_{ph}, as shown in Figure 4.3.33, has a Laplace transform given by

$$J_{ph}(t) \leftrightarrow J_{ph}(s) = \frac{\Delta Q}{\tau}\left\{\frac{1-\exp(-s\tau)}{s}\right\} \tag{31}$$

The voltage is measured over the network presented in Figure 4.3.31, which has an impedance given by

$$Z(\omega) = \frac{R_m}{1 + k\omega(R_s + R_m)C} \tag{32}$$

In this expression R_m is the resistance over which the voltage is measured, R_s is the series resistance of the device, and C is the sum of the capacitances of the photo electrode and the shunt capacitor. We write $R_m + R_s = R$ and RC being the characteristic response time of the system.

Taking the Laplace transform gives for the measured voltage

$$V_{meas}(s) = Z(s)J_{ph}(s) \tag{33}$$

For simplicity we consider only the behavior immediately after the onset of the photocurrent, i.e. for the limit $s\tau >> 1$. Then

$$V_{meas}(s) = Z(s)J_{ph}(s) = \frac{R_m}{1+sRC}\frac{\Delta Q}{\tau s} = \frac{\Delta Q}{\tau}\frac{R_m}{s(1+sRC)} \tag{34}$$

Taking the inverse Laplace transform gives

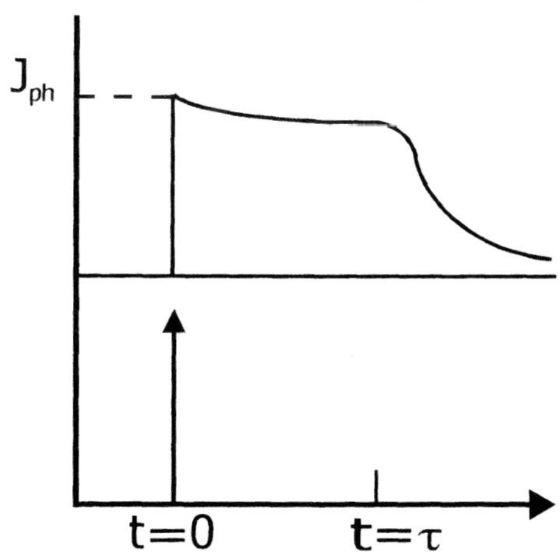

Figure 4.3.33. Effects of diffusion, trapping, and recombination of charge carriers on the TOF response.

$$V_{meas}(t) = \frac{\Delta Q}{\tau} R_m (1 - \exp[-t/RC]) \tag{35}$$

There are two limiting cases, $\tau >> RC$ and $\tau << RC$. For $\tau >> RC$, the measuring system is fast and can follow the internal photocurrent. In this limit $t/RC >> 1$ and

$$V_{meas}(t) = \frac{\Delta Q}{\tau} R_m \tag{36}$$

For $\tau << RC$, the measuring system is slow and cannot follow the internal photocurrent. In this limit $t/RC << 1$ and the measured voltage reads

$$V_{meas}(t) = \frac{\Delta Q}{\tau} R_m \frac{t}{RC} \tag{37}$$

In this case the *RC* circuit acts as a current integrator and the voltage rises linearly in time. When neither of these limiting cases hold one should use the Laplace transform method. Or one can also use the approach elaborated by Donovan and Kreouzis [2000].

The time-of-flight technique has been applied specifically to semiconductors with high defect densities, because the charge carrier mobilities in these materials are low. In particular, amorphous silicon is studied extensively with TOF and, recently, also organic semiconductors.

4.3.5.4 *Intensity-Modulated Photocurrent Spectroscopy*

Another way to investigate the dynamics of electrons and holes in semiconductors is to apply intensity-modulated photocurrent spectroscopy (IMPS). This technique is based on the application of a small-signal harmonic perturbation of the irradiation intensity. As in the case of time-of-flight, two electrical contacts are applied: the injecting contact, which is optically transparent, and the collecting contact. Irradiation is applied through the injecting contact. In the vicinity of this contact, charge carriers are generated. Since the light intensity is harmonically modulated, the generation rate of the electron-hole pairs is a sinusoidal function of time. The generated charge carriers migrate to the collecting contact, leading to photocurrent generation, which is also a sinusoidal function of time. The phase of the current response is shifted with respect to the phase of the impinging irradiation. This phase shift, and the amplitude of the photocurrent, are functions of the applied modulation frequency. If one measures the phase shift and the amplitude of the photocurrent over a broad frequency range the optoelectrical response function of the system can be determined.

To define the above in more rigid mathematical expressions, we write for the modulated irradiation intensity

$$\Phi(\omega) = \overline{\Phi} + \overline{\Phi} \sin(\omega t) \tag{38}$$

in which $\bar{\Phi}$ is the continues background irradiation, in photons cm^{-2}s^{-1}, and $\tilde{\Phi}$ is the modulation amplitude. The angular frequency ω equals 2π times the modulation frequency.

When $\tilde{\Phi}$ is sufficiently small the photocurrent is a sinusoidal function of time and reads

$$J_{ph}(\omega) = \bar{J}_{ph} + \tilde{J}_{ph} \sin(\omega t + \varphi) \tag{39}$$

In this expression $\tilde{J}_{ph}$ and φ are the amplitude and phase of the modulated photocurrent density. Both are frequency dependent. In a frequency response analyzer $\tilde{J}_{ph}$ and φ are measured as a function of the modulation frequency.

For further analysis of the response, one can make use of the familiar concepts of impedance spectroscopy. The only difference is that while in impedance spectroscopy (IS) the input signal is a voltage, in IMPS the input signal is a modulated light beam. But from a theoretical point of view this is unimportant and linear-response theory can readily be applied in both cases. As a matter of fact, the differences between IMPS and IS are even less than one might expect. In IS a modulated voltage, superimposed on a stationary bias voltage, is applied across the contacts. This time-varying voltage modulates the difference of the electrochemical potential of the electrons between the two electrodes. As a result, a modulated electrical current, superimposed on a stationary current, flows between the contacts. In IMPS one irradiates the sample through the injecting contact. Electron-hole pairs are generated, which leads to a change of the electrochemical potential of the electrons at the contacts. As a result, a modulated photocurrent, superimposed on a stationary current, will flow through the system.

It is possible to use the conventional complex impedance, which is defined for impedance spectroscopy as

$$Z(j\omega) = \frac{\tilde{V}(j\omega)}{\tilde{I}(j\omega)} \tag{40}$$

in which $j = \sqrt{-1}$. The admittance is $Y(j\omega) = 1/Z(j\omega)$. In the case of IMPS we write

$$Z_{ph}(j\omega) = \frac{\tilde{\Phi}(j\omega)}{e\tilde{J}_{ph}(j\omega)} \quad \text{and} \quad Y(j\omega) = \frac{1}{Z_{ph}(j\omega)} \tag{41}$$

It is good to realize that we have normalized the irradiation intensity and the photocurrent to particles per second per unit area. In this way the IMPS impedance and admittance are dimensionless.

The concept of impedance is only applicable if the system under study complies with the conditions of stability, causality, finity, and linearity. It is good practice to verify whether these conditions hold before giving further interpretation to the data. A stable system has a $Z(j\omega)$ independent of time. Causal systems do not show any response if the input signal is zero (in the dark for IMPS). All practical systems are finite. Linear systems show a response linearly related to the input; the current density is linearly related to the irradiation intensity. A convenient technique is the use of Kramers–Kronig relationships. If the data set complies with these relationships, linear response theory can be applied. If this is not the case, one should question whether the concept of impedance is justified. From here on we shall deal with systems that comply to the above requirements, allowing the use of linear-response theory. An excellent introduction to the IMPS technique is given by Peter [1990]. One of the first reports on the use of IMPS to study semiconductor properties dates back to 1981 (Oheda [1981]). A few years later, IMPS was applied to study the reaction kinetics in photoelectrochemical systems, electron-hole recombination, anodic thin films, and electron transport in disordered semiconductors.

Electron Drift in a Constant Electric Field. As an example, let us consider the system discussed in the time-of-flight section. In this system, charge carriers are generated close to the injecting contact and drift to the collecting contact under the force of a constant electric field. As discussed above, the current response on a laser pulse has a constant value of $J_{ph} = \Delta Q/\tau$ for $0 < t < \tau$, and drops instantly to zero at $t = \tau$. The input signal is a delta function and the output response is a step function. Linear-response theory shows that the system function $H(s)$ is the Laplace transform of the impulse response function $h(t)$. In our example:

$$h(t) = \frac{\Delta Q}{\tau} F(t-\tau) \tag{42}$$

in which F is the function that is 1 at $0 < t < \tau$ and zero for $t < 0$ and $t > \tau$.

The Laplace transform of $h(t)$ is the system function $H(s)$, which reads

$$h(t) = \frac{\Delta Q}{\tau} F(t-\tau) \quad \rightarrow \quad H(s) = \frac{\Delta Q}{\tau}\left\{\frac{1-\exp(-s\tau)}{s}\right\} \tag{43}$$

The impedance $Z(j\omega)$ is identical to the system function with $s = j\omega$.

$$Z(j\omega) = \frac{\Delta Q}{\tau}\left\{\frac{1-\exp(j\omega\tau)}{j\omega}\right\} \tag{44}$$

For low frequencies $\omega\tau << 1$

$$Z(j\omega) = \Delta Q \tag{45}$$

For high frequencies $\omega\tau >> 1$

$$Z(j\omega) = \frac{\Delta Q}{j\omega\tau} \tag{46}$$

It is possible to interpret the optoelectrical impedance in terms of an electrical equivalent circuit. When doing so, the system resembles a resistor $R = \Delta Q$ at low frequencies and a capacitor $C = \tau/\Delta Q$ at high frequencies. At intermediate frequencies the system shows a more complex behavior. One should be careful to have a mental picture of a resistor and a capacitor when using equivalent circuit approach for interpreting IMPS data. Maybe it is better to look upon it in a more abstract mathematical way.

The impedance as defined by Eq. (44) can be plotted in the usual ways. In Figure 4.3.34 a Nyquist plot is presented.

Electron Diffusion. It is also possible that there is no internal electric field and that electrons migrate exclusively by diffusion. This is the situation in nanoporous TiO_2 electrodes as used in dye-sensitized solar cells (O'Regan and Grätzel [1991], Hagfeldt and Grätzel [1995], Bach *et al.* [1998], Grätzel [2000]). In recent years, these systems have been studied extensively and we shall use the results of these studies to demonstrate the strength of the IMPS technique. Without going into detail, when light is absorbed in a dye-sensitized solar cell, electrons are injected into the conduction band of a nanoporous TiO_2 electrode (Figure 4.3.35). These electrodes are composed of nanometer sized TiO_2 crystals with a size of about 32 nm. The nanocrystals are sintered together and form a film about 22 μm thick. On the surface, dye molecules are grafted. The molecules have the function of absorbing sunlight and injecting an electron from the first excited state of the molecule into the conduction band of TiO_2. The nanoporous electrode is immersed in a liquid non-aqueous redox electrolyte, which takes care of fast regeneration (reduction) of the dye cations. At the counter electrode the electrolyte picks up electrons. The injected electron in the conduction band of TiO_2 migrates through the nanoporous network towards the front contact, which is a transparent conducting oxide (TCO). Since the nanoporous TiO_2 is immersed in a conducting medium, its internal surface has the

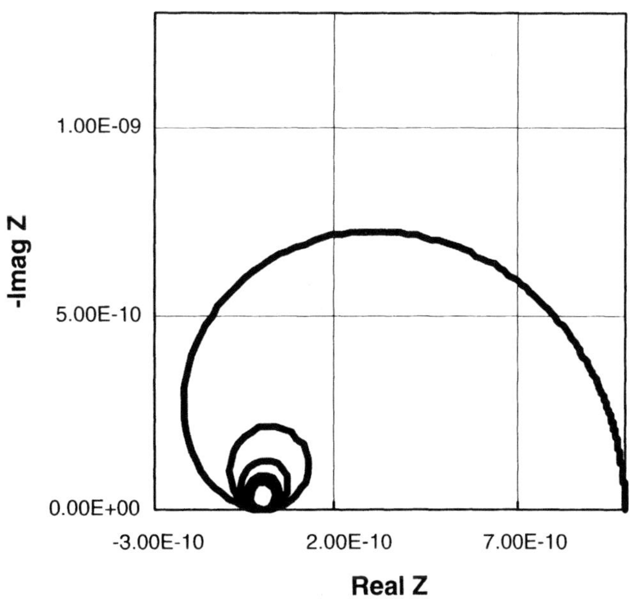

Figure 4.3.34. Nyquist plot of the IMPS impedance. On the vertical axis minus the imaginary component is plotted and on the horizontal axis the real component. ΔQ is taken as 10^{-9} C and $\tau = 5\,\mu$s.

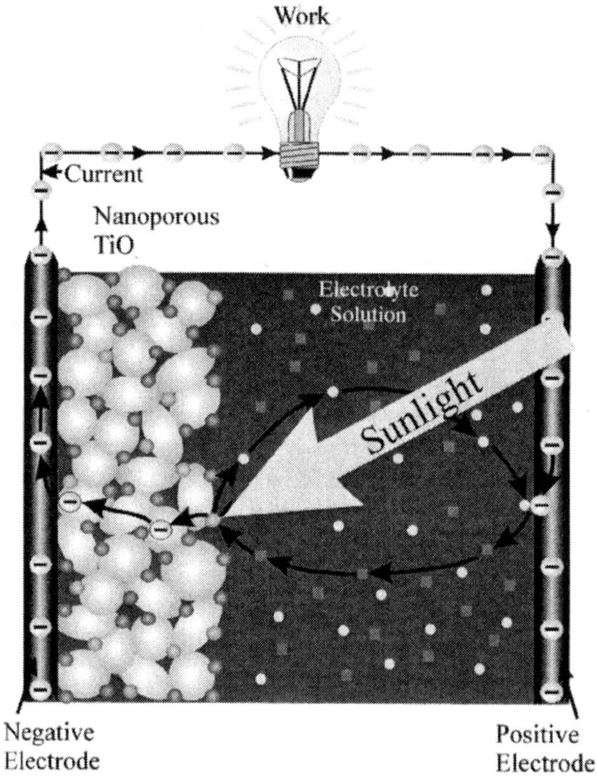

Figure 4.3.35. Schematic drawing of a dye-sensitized solar cell. The photoactive electrode is a $10\,\mu m$ thick film of sintered 20 nm anatase TiO_2 nanoparticles. On the surface light-sensitive dye molecules are applied. The electrode is immersed in a non-aqueous redox electrolyte.

same electrical potential everywhere. Furthermore, the size of the nanoparticles is much smaller than the Debye length, i.e. the screening depth of the electric field. Therefore, there is no internal electric field to drive the electrons; the only contribution to electron transport is diffusion. This is a slow process and it can take up to a second for an electron to pass $10\,\mu m$.

Electron transport and reaction kinetics in nanostructured photoelectrodes have been studied successfully with the IMPS technique. It appears that the diffusion coefficient is a strong function of the background light intensity. Without background irradiation, the diffusion coefficient, D, is about $10^{-8}\,cm^2\,s^{-1}$. With background light D can be four orders of magnitude larger. When more background light is applied, the electron injection rate rises, which speeds up the electron transport considerably. A striking observation is that an increase of the injection rate is almost exactly counterbalanced by the increase of the diffusion coefficient and the concentration of migrating charge carriers is practically independent on the presence of background light. Under a wide range of conditions, the electron concentration amounts to about one electron per nanoparticle.

In dye-sensitized solar cells the following equation holds

$$\frac{\partial\,\Delta n(x,t)}{\partial t} = a\Phi(x,t) + D\frac{\partial^2\,\Delta n(x,t)}{\partial x^2} - \frac{\Delta n(x,t)}{\tau} \tag{47}$$

Δn is the excess electron density due to irradiation: $\Delta n = n_{light} - n_{dark}$, a is the effective absorption coefficient, D the diffusion coefficient, and τ is the lifetime of the injected electrons. The first term on the right-hand side of Eq. (47) is the generation rate, the second term the diffusional electron flux, and the third term the recombination rate. This differential equation can be solved under the following boundary conditions. The electron consumption at the TCO|TiO_2 injecting contact is assumed to be infinitely fast, which yield a boundary condition

$$\Delta n(x=0,t) = 0 \tag{48}$$

At the other side of the nanoporous TiO_2 electrode the arriving electrons cannot escape into the electrolyte because this back reaction is kinetically hindered. They are bounced back into the porous network, which implies that the gradient of the electron density on the outside is zero. The following boundary condition holds

$$\left(\frac{\partial\,\Delta n}{\partial x}\right)_{x=\ell} = 0 \tag{49}$$

in which ℓ is the film thickness. The photocurrent is given by

$$J_{ph}(t) = -D\left(\frac{\partial\,\Delta n}{\partial x}\right)_{x=0} \tag{50}$$

Applying the boundary conditions Eqs (48) and (49), the electron transport Eq. (47), can be solved in the frequency domain and the harmonic photocurrent, driven by the harmonic oscillation of the irradiation intensity, reads

$$J_{ph}(\omega) = \frac{ae\Phi(\omega)}{a^2 - \lambda^{-2}}\left(\frac{\lambda^{-1}\,Sinh(\ell/\lambda) + ae^{-a\ell} - aCosh(\ell/\lambda)}{Cosh(\ell/\lambda)}\right) \tag{51}$$

in which $\frac{1}{\lambda} = \sqrt{\frac{1}{L^2} + \frac{j\omega}{D}}$ and $L = \sqrt{D\tau}$

L is the diffusion length and can be considered as the distance over which electrons can migrate within their lifetime τ.

If recombination can be neglected, $\tau \to \infty$ and $L \to \infty$, which implies that $\frac{1}{\lambda} = \sqrt{\frac{j\omega}{D}}$ and

$$J_{ph}(\omega) = \frac{ae\Phi(\omega)}{(a^2 - j\omega/D)}\left(\frac{ae^{-a\ell}}{Cosh(d\sqrt{j\omega/D})} + \sqrt{j\omega/D}\;Tanh\,(d\sqrt{j\omega/D}) - a\right) \tag{52}$$

The nanostructured electrodes can store a large number of electrons, which implies that the photocurrent is driven into a capacitive element. This introduces an additional time constant, the *RC* time, in the photocurrent response. To deal with it, Eqs (51) and (52) must be multiplied by the transfer function of the measuring system, presented in Figure 4.3.31. The transfer function of the measuring system and the measured signals in the frequency domain are given by Eqs (32) and (33), respectively.

The photocurrent measured externally, $J_{ext}(\omega)$, is the internally generated photocurrent multiplied by the transfer function of the measuring system, i.e.

$$V_{meas}(\omega) = J_{ph}(\omega)\frac{R_m}{1+j\omega(R_s+R_m)C} \tag{53}$$

in which R_m is the measuring resistance of the current meter, i.e. 50 Ω in our system, R_s the series resistance between the current generating device and the measuring unit, and C the cell capacitance. The quantity $(R_m + R_s)C$ is the *RC* time of the measuring system.

Figure 4.3.36*a* shows a Nyquist plot Y″ vs. Y′; a Bode plot is shown in Figure 4.3.36*b*, in which the frequency dependence is more clear. In Figure 4.3.36*b* a measured data set is presented along with the non-linear least squares fit of this dataset to Eqs (51) and (53). Also the Kramers–Kronig transfer is shown. Since all the curves fall on top of each other the following conclusions are justified: (i) linear-response theory can be applied; and (ii) the model Eqs (47) to (53) accurately describes the frequency response of the system.

With the aid of non-linear least squares fitting, the effective absorption coefficient a, the diffusion coefficient D, and the lifetime of the electrons τ, are found. These values can be used to determine the steady-state electron density, $\Delta n(x)$, which reads

$$\Delta n(x) = \frac{a\Phi/D}{a^2 - L^{-2}}\left(\frac{(e^{-\ell/L} - aLe^{-a\ell})}{2Cosh(\ell/L)}e^{x/L} + \frac{(e^{\ell/L} + aLe^{-a\ell})}{2Cosh(\ell/L)}e^{-x/L} - e^{-ax}\right) \tag{54}$$

If the recombination is slow, $\tau \to \infty$, the steady-state electron density is given by

$$\Delta n(x) = \frac{\Phi}{aD}(1 - ae^{-a\ell}x - e^{-ax}) \tag{55}$$

For two different types of TiO_2 nanoparticles, and for two different background light intensities the steady-state electron profiles have been calculated and are shown in Figure 4.3.37. If the background light intensity is increased by a factor 500, from 0.7 to 330 W m^{-2}, the diffusion coefficient increases from 14×10^{-12} to 1.2×10^{-9} m^2 s^{-1} for PK5585 anatase TiO_2 nanoparticles and from 19×10^{-12} to 1.5×10^{-9} m^2 s^{-1} for AK1 anatase TiO_2. The steady-state electron concentration is almost unaffected by these huge variations in light intensity. The diffusivity of the electrons is adjusted such that the concentration of conduction band electrons in TiO_2 nanostructured electrodes saturates to about one electron per nanoparticle. To date, this remarkable observation has not been fully explained.

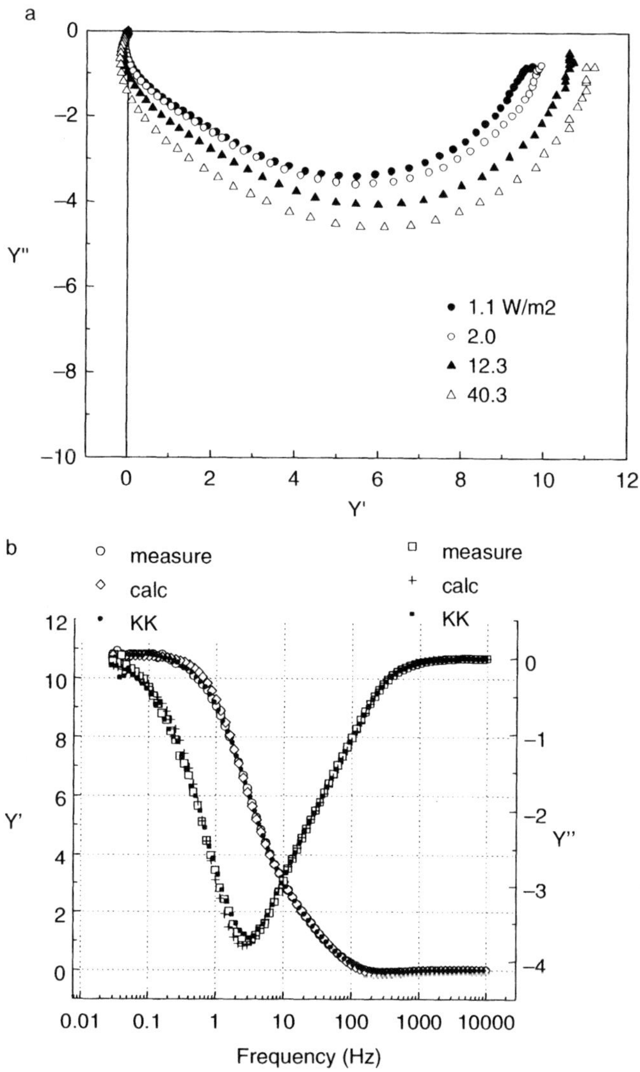

Figure 4.3.36. (*a*) Nyquist and (*b*) Bode plots of an IPMS spectrum of a dye-sensitized solar cell.

4.3.5.5 Final Remarks

The generation, separation, and recombination of electron-hole pairs, along with electronic transport can be studied in great detail with time-resolved photocurrent generation techniques. The examples given above are by no means limiting and do not cover all possible fields of applications. Yet, they demonstrate the enormous potential that these techniques offer. In the study of dye-sensitized solar cells the use of IMPS have proven invaluable and has lead to fundamental knowledge about

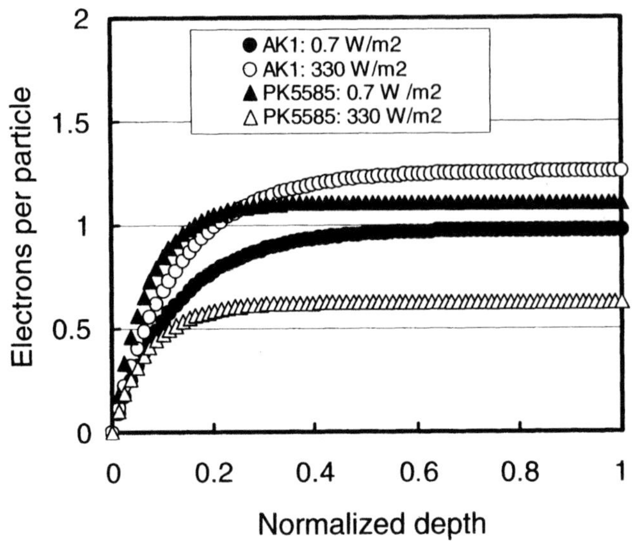

Figure 4.3.37. Steady-state electron concentration profiles in dye-sensitized solar cells made from different TiO_2 nanoparticles, i.e. AK1 and PK5585.

the transport of electrons in nanoporous electrodes. At the moment, solid-state physicists usually prefer the time-of-flight method above the intensity-modulated photocurrent method. In my opinion this is not always justified. Frequency-domain techniques offer a superior resolution when dispersive processes are involved. Data acquisition in the frequency domain allows for advanced digital processing and usually yields a much better signal-to-noise ratio. Furthermore, frequency-domain techniques make use of a small-signal perturbation, while time-domain experiments usually involve intense laser pulses. This may lead to non-linear response of the system, which is usually difficult to model. It is beyond doubt that the frequency domain photocurrent generation will continue to contribute to the elucidation of the physics and chemistry of photocurrent generation.

Acknowledgments

The author acknowledges Prof. dr. J. Schoonman for critically reading the manuscript.

4.4 CORROSION OF MATERIALS

Digby D. Macdonald
Michael C. H. McKubre

4.4.1 Introduction

Corrosion is defined as the spontaneous degradation of a reactive material by an aggressive environment and, at least in the case of metals in condensed media, it occurs by the simultaneous occurrence of at least one anodic (metal oxidation) and one cathodic (e.g. reduction of dissolved oxygen) reaction. Because these partial reactions are charge-transfer processes, corrosion phenomena are essentially electrochemical in nature. Accordingly, it is not surprising that electrochemical techniques have been used extensively in the study of corrosion phenomena, both to determine the corrosion rate and to define degradation mechanisms.

Of all of the electrochemical techniques that are available, impedance spectroscopy promises to be the most valuable because of its ability, in a single experiment, to detect interfacial relaxations covering a wide range of relaxation times. The application of this technique in corrosion science became possible, in the practical sense, only within the past decade, with the advent of techniques for measuring transfer functions at subhertz frequencies (see Section 3). Instruments and techniques are now available for measuring interfacial impedances at frequencies down to the 10^{-3}–110^{-4} Hz region, where relaxations involving adsorbed intermediates and diffusing species appear.

Over this same period, considerable development has taken place in the theoretical treatment of the impedance properties of corroding interfaces (Macdonald and McKubre [1981]). These theoretical developments have been especially important, since they serve to enhance the quantitative nature of the technique. Indeed, impedance spectroscopy has emerged as probably the most powerful technique currently available for identifying corrosion reaction mechanisms, and methods are now being developed to extract kinetic parameters (rate constants, transfer coefficients) for multistep reaction schemes.

In this section, we review the application of impedance spectroscopy to the study of corrosion phenomena. Emphasis is placed on illustrating how the method is applied to identify the different processes that occur at a corroding interface. We also review the use of impedance measurements for measuring corrosion rate, since this was the initial application of the technique in corrosion science and engineering. The use of impedance spectroscopy to analyze other cause and effect phenomena of interest in corrosion science, including electrochemical–hydrodynamic, fracture, and electrochemical–mechanical processes, is also discussed.

4.4.2 Fundamentals

The response of any physical system to a perturbation of arbitrary form may be described by a transfer function

$$H(s) = \overline{V}(s)/\overline{I}(s) \tag{1}$$

where s is the Laplace frequency and $\overline{\Phi}(s)$ and $\overline{I}(s)$ are the Laplace transforms of the time-dependent voltage and current, respectively (Goldman [1950]). In terms of the steady state sinusoidal frequency domain, the transfer function becomes

$$H(j\omega) = \frac{F[V(t)]}{F[I(t)]} = \frac{V(j\omega)}{I(j\omega)} \tag{2}$$

where F signifies the Fourier transform and $V(j\omega)$ and $I(j\omega)$ are the sinusoidal voltage and current, respectively. Provided that the system is linear, that causality is obeyed, and that the interface is stable over the time of sampling (see later), the transfer function may be identified as an impedance $Z(j\omega)$. Because they are vector quantities, $H(j\omega)$ and $Z(j\omega)$ are complex numbers containing both magnitude and phase information. From a theoretical viewpoint, the impedance (or, more generally, the transfer function) is one of the most important quantities that can be measured in electrochemistry and corrosion science. This is because, if it is sampled over an infinite bandwidth, it contains all the information that can be obtained from the system by purely electrical means.

An important requirement for a valid impedance function is that the system be linear. Theoretically, this implies that the real and imaginary components transform correctly according to the Kramers–Kronig relationships (discussed later in this section). Practically, linearity is indicated by the impedance being independent of the magnitude of the perturbation, a condition that is easily (although seldom) tested experimentally.

4.4.3 Measurement of Corrosion Rate

The Stern–Geary equation provides a direct relationship between the steady state corrosion current and the "dc" resistance across the interface (Stern and Geary [1957])

$$i_{\text{corr}} = \left[\frac{\beta_a\beta_c}{2.303(\beta_a + \beta_c)}\right]\left(\frac{1}{R_p}\right) \tag{3}$$

where β_a and β_c are the Tafel constants for the anodic and cathodic partial reactions, respectively, and R_p is the polarization resistance (Mansfeld [1976]). Because corroding interfaces are inherently reactive by nature, owing to the presence of capacitive, psuedoinductive, and diffusional impedance terms, it is evident that the polarization resistance is given only by the difference of the measured impedance at sufficiently low and high frequencies:

$$R_p = |Z(j\omega)|_{\omega\to 0} - |Z(j\omega)|_{\omega\to\infty} \tag{4}$$

Measurement of the series resistance at the high-frequency limit normally presents few problems, because $Z(j\omega)$ becomes nonreactive at frequencies as low as 10 kHz, in most cases. On the other hand, in the low-frequency region, reactance is commonly observed at frequencies in the vicinity of 10^{-3} Hz, so that special precautions must be adopted to obtain reliable data (Syrett and Macdonald [1979]). The need for these precautions is independent of the form of the perturbation applied to the interface. Accordingly, they apply equally well to the use of potential or current steps and triangular and sinusoidal voltage perturbations in the measurement of the polarization resistance, as well as to the determination of "steady state" current–voltage curves. Practically, therefore, it is necessary to use a sufficiently low-frequency (sinusoidal perturbation), low-voltage scan rate (small-amplitude cyclic voltametry) or to wait a long enough time (potential or current step perturbation) before acquiring the response data (e.g. current) for calculating the polarization resistance.

The problem of acquiring impedances at sufficiently low frequencies is amply demonstrated by the data (Syrett and Macdonald [1979]) for 90 : 10 Cu : Ni alloy corroding in flowing seawater (Figure 4.4.1). Thus, for an exposure time of 22 h, the impedance function can be defined over the entire bandwidth, and an accurate value for R_p may be obtained by probing the interface at frequencies above 0.01 Hz. On the other hand, at much longer exposure times, frequencies as low as 0.0005 Hz are not sufficient to completely define the interfacial impedance, and considerable extrapolation is required to acquire a value for R_p.

It is important to emphasize again that, because time domain functions can be synthesized as linear combinations of sinusoidal (sine and cosine) components (Fourier synthesis), this problem is not limited to impedance spectroscopy. Thus, failure to use a sufficiently low sweep rate in the case of small-amplitude cyclic voltametry (SACV) will also introduce significant error (Figure 4.4.2), depending upon which resistance is considered as being the parameter of interest (R_d or R_{app}, Figure 4.4.3; Macdonald [1978*a*]). Interestingly, our experience in using a variety of electrochemical monitoring techniques indicates that SACV is superior, in many respects, to impedance spectroscopy for determining the polarization resistance. Thus, quite reliable values for R_p for systems as reactive as that shown in Figure 4.4.1 generally can be obtained using a single voltage sweep rate of 0.1 mV/s, which is quite accessible using standard electrochemical instrumentation.

Because most impedance measurements are made sequentially at discrete frequencies, the total data acquisition time can be expressed as

$$T = \sum_i \frac{n_i}{f_i} \tag{5}$$

where n_i is the number of cycles at frequency f_i. The minimum acquisition time is obtained by setting $n_i = 1$. Therefore,

$$T_{\min} = \sum_i \frac{1}{f_i} \tag{6}$$

and it is apparent that the minimum acquisition time is dominated by the low-frequency components. For example the impedance data shown in Figure 4.4.1, for an exposure period of 164 h, required an acquisition time of more than 1 h. This contrasts with an acquisition time of 100 s required to obtain a reliable value for R_p using SACV with a voltage sweep rate of 0.1 mV/s and a peak-to-peak amplitude for the triangular voltage excitation of 5 mV. Because SACV does not readily

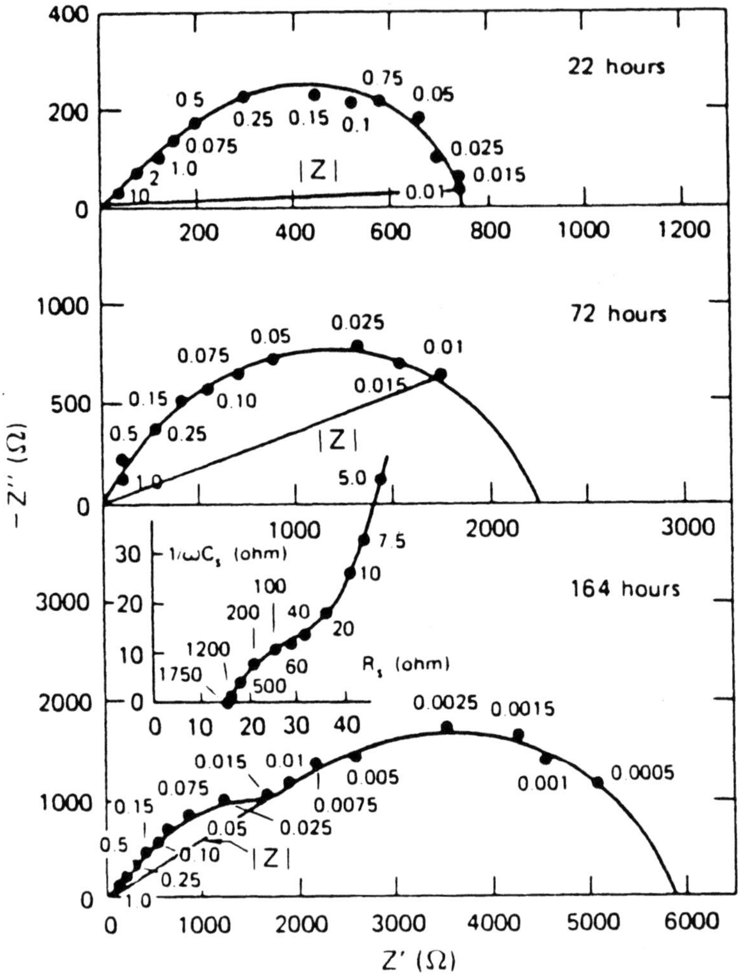

Figure 4.4.1. Complex plane impedance diagrams for 90 : 10 Cu : Ni alloy in flowing seawater as a function of exposure time. Flow velocity = 1.62 m/sec, $[O_2] = 0.045$ mg/l, specimen area = 11.05 cm^2, T = 26°C; exposure time = 50 h. (From B. C. Syrett and D. D. Macdonald, The Validity of Electrochmical Methods for Measuring Corrosion Rates of Copper-Nickel Alloys in Seawater. Reprinted with permission from *Corrosion*, **35**, 11, [1979], NACE, Houston, TX.) Numbers next to each point to frequency in hertz.

yield the mechanistic information afforded by impedance spectroscopy, the two methods are best regarded as being complementary in nature.

According to Fourier's theorem, all small-amplitude techniques must yield identical results (i.e. the same interfacial impedance), regardless of the form of the excitation. This is clearly the case for the system discussed above, as shown in Figure 4.4.4. In this figure, polarization resistance data, obtained using the impedance

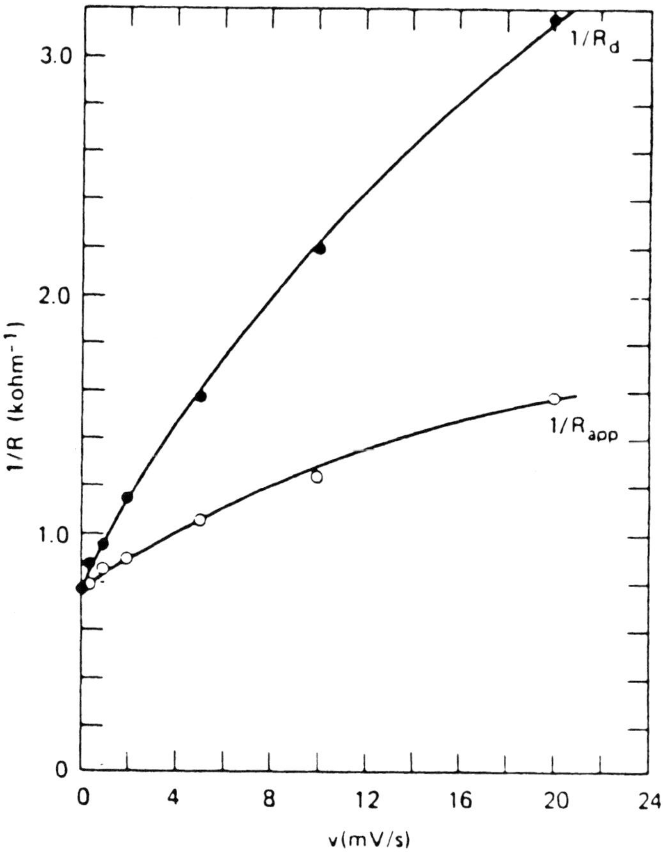

Figure 4.4.2. Plots of $1/R_d$ and $1/R_{app}$ as measured using SACV for 90 : 10 Cu : Ni in flowing seawater. Flow velocity = 1.62 m/s, $[O_2]$ = 0.045 mg/l, T = 26°C, exposure time = 50 h. (From D. D. Macdonald, An Impedance Interpretation of Small Amplitude Cyclic Voltammetry: I. Theoretical Analysis for a Resistive–Capacitive System, *J. Electrochem. Soc.*, **125**, 1443–1449 [1978]. Reprinted by permission of the publisher, The Electrochemical Society, Inc.)

spectroscopic, potential step, and SACV techniques, are plotted as a function of time for two copper–nickel alloys exposed to flowing seawater (Syrett and Macdonald [1979]). The fact that the polarization resistance data are independent of the technique used for their measurement implies that the experimenter has the freedom to tailor a perturbation for the measurement of interfacial impedance in order to achieve some desired experimental goal. One implementation of this concept is the application of a large number of sine-wave voltage signals simultaneously, so that the total data acquisition time is determined only by the lowest frequency, and not by the summation embodied in Eq. (5). These "structured noise" techniques are now being actively developed for corrosion-monitoring purposes.

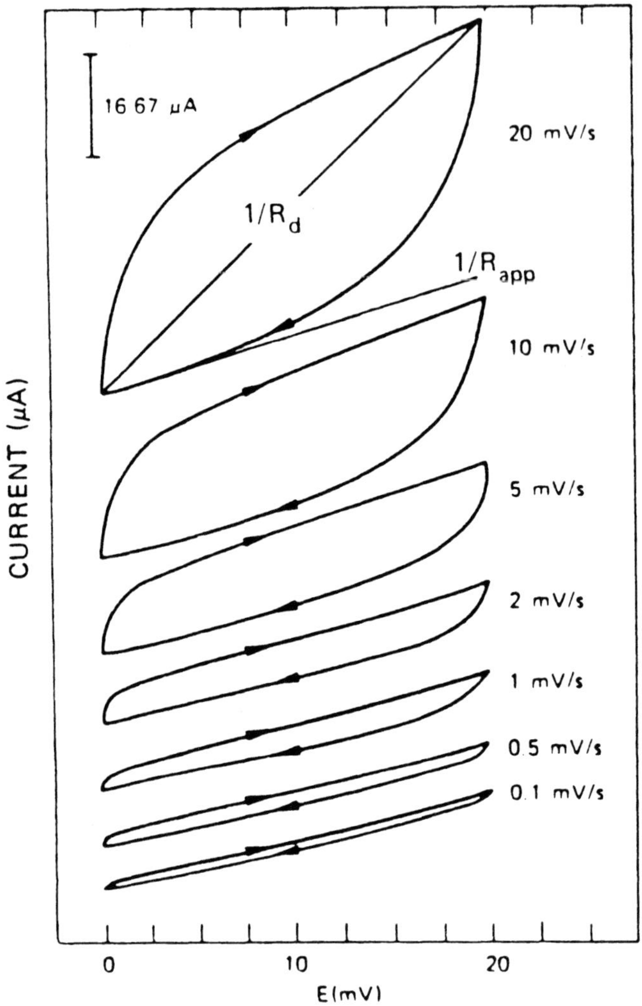

Figure 4.4.3. Small-amplitude cyclic voltamograms for 90 : 10 Cu : Ni alloy in flowing seawater. Experimental conditions are as listed in Figure 4.4.1. (From D. D. Macdonald, An Impedance Interpretation of Small Amplitude Cyclic Voltammetry: I. Theoretical Analysis for a Resistive–Capacitive System, *J. Electrochem. Soc.*, **125**, 1443–1449, [1978]. Reprinted by permission of the publisher, The Electrochemical Society, Inc.)

The structured noise method stems from the elegant work of Smith and coworkers (Smith [1966]), who developed a multifrequency technique for ac polarography. Subsequently, structured noise techniques have been used in corrosion studies by Smyrl and coworkers (Smyrl [1985*a*,*b*], Smyrl and Stephenson [1985]) and by Pound and Macdonald [1985]. In all cases, the perturbation applied to the system is of the form

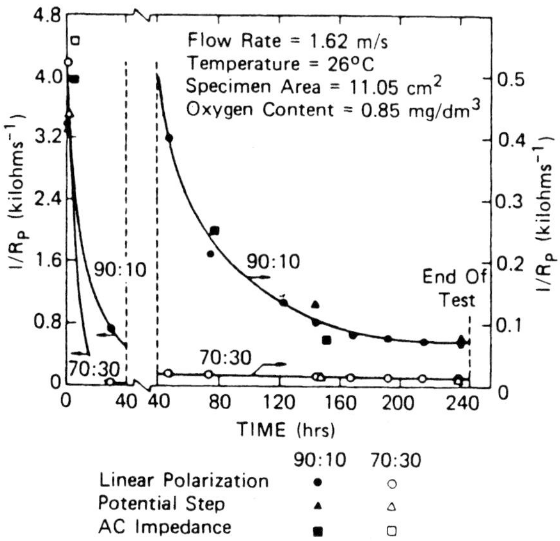

Figure 4.4.4. Corrosion rate (as I/R_p) vs. time for 90:10 Cu:Ni and 70:30 Cu:Ni in flowing seawater having an oxygen content of 0.85 mg/dm^3. (From B. C. Syrett and D. D. Macdonald, The Validity of Electrochmical Methods for Measuring Corrosion Rates of Copper–Nickel Alloys in Seawater. Reprinted with permission from *Corrosion*, **35**, 11, [1979], NACE, Houston, TX.) Numbers next to each point to frequency in hertz.

$$V(t) = \sum_i a_i \sin(\omega_i t + \phi_i) \tag{7}$$

where a_i is the amplitude, ω_i is the frequency, and ϕ_i is the phase. If these parameters are selected in a completely random fashion, the signal is referred to as *white noise*. However, because corroding interfaces are inherently nonlinear, considerable advantages exist in choosing values for a_i, ω_i, and ϕ_i such that certain experimental problems are avoided. For example nonlinearity produces harmonics of $2\omega_i$, $3\omega_i$, $\ldots n\omega_i$ in response to a perturbation at the fundamental frequency ω_i. Because the amplitude of a harmonic decreases rapidly with increasing n, harmonic intrusion may be avoided by ensuring that $\omega_j \neq n\omega_i$ ($n = 2, 3, \ldots$) or may at least be minimized by requiring that $n > 3$. Also, the power applied to the interface, which is proportional to the square of the amplitude of each component, may be tailored by choosing appropriate values for a_i. Regardless of the exact form of the perturbation employed, the impedance data are extracted from the perturbation and the response by Fourier or Laplace transformation (Pound and Macdonald [1985], Smyrl [1985a,b]). As an example of this technique, we show the data of Pound and Macdonald [1985] for carbon steel in acidified brine (Figure 4.4.5). The structured noise data are compared with those obtained using a frequency-by-frequency correlation technique (FRA). Clearly, the structured noise data are considerably more

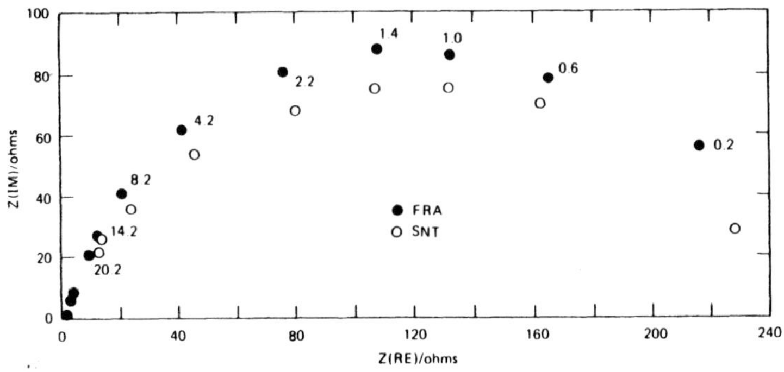

Figure 4.4.5. Nyquist plot of impedance data for 1018 steel in HCl-acidified 3% NaCl (pH = 3). Exposure time = 28 h, $E_{corr} = -0.682$ V (SCE).

scattered than are those obtained by the correlation method, but that is compensated for by the reduction in the data acquisition time.

4.4.4 Harmonic Analysis

The derivation leading to the Stern–Geary relationship [Eq. (3)] assumes that the corroding electrode responds linearly to the imposed electrical perturbation; that is, doubling the perturbing voltage amplitude results in a double current response (but an unchanged impedance). Since physical variables in all physically realizable systems must have a finite first derivative, it is always possible to achieve linear conditions by applying a perturbation of limitingly small amplitude.

The nonlinearity of the current–voltage relationship in corroding systems provides an opportunity to determine corrosion rates without the need to measure independently the Tafel constants. The reason is that the electrical perturbation, which is imposed on the system at a frequency of f, in a nonlinear system results in a response at $2f$, $3f$, $4f$, and so on, in addition to a dc component (McKubre [1983], Morring and Kies [1977], McKubre and Macdonald [1984], Bertocci [1979], Bertocci and Mullen [1981], Kruger [1903]). Neither the fundamental response (f_0) nor the total power response ($\Sigma_{h=0}^{\infty} hf$) can be analyzed to determine uniquely the corrosion rate (as opposed to the polarization resistance). Nevertheless, an analysis of the harmonic responses can be used to determine the unknown parameters in Eq. (3) and thus to measure corrosion rates in systems for which the Tafel coefficients are not known or at potentials removed from the free corrosion potential V_{fc}, as, for example, under conditions of an applied cathodic protection potential.

The origin of the harmonic response is shown schematically in Figure 4.4.6*a* for an input voltage sine wave at frequency f, superimposed on a current–voltage curve of the form

$$I = I_{fc}\{\exp[\beta_a(V - V_{fc})] - \exp[-\beta_c(V - V_{fc})]\} \tag{8}$$

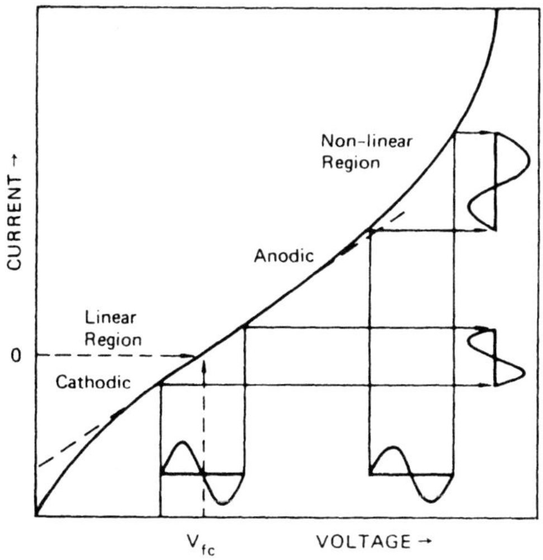

FIGURE 4.4.6.*a*. The source of a harmonic response: reflection of an input sine wave on a nonlinear current–voltage curve.

where β_a and β_c are the forward anodic and reverse cathodic Tafel coefficients, respectively, and I_{fc} is the free corrosion current flux, defined at the free corrosion potential (V_{fc}) as

$$I_{fc} = I_a = -I_c \quad (\text{at } V_{fc}) \tag{9}$$

Figure 4.4.6*a* shows a perturbing voltage sinusoid reflected about the dc current voltage response curve at V_{fc} and at some positive bias. In the linear region, this reflection results in an undistorted current response, with *V*/*I* being a constant (equal to the dc corrosion resistance).

The expected response in the time domain is shown schematically in Figure 4.4.6*b*. The output is generally shifted in phase with respect to the input due to reactive terms associated with diffusional and capacitive processes. The extent of the distortion in the nonlinear region can be quantified by performing a Fourier series analysis. As indicated in Figure 4.4.6*c*, when a sinusoidal perturbation of moderate amplitude is applied to a corroding electrode, the response will consist of a components at the same frequency (generally shifted in phase with respect to the input), as well as terms at integral multiples of the input frequency (the harmonics). Unless the input excitation is symmetric about V_{fc}, then the output also will show a dc offset that we term the *zero'th harmonic*. An offset is shown even for a symmetric perturbation if the *I*/*V* curve is not symmetrical—this is the basis of the faradic rectification effect. These harmonic response terms contain information sufficient to completely specify the current–voltage curve, in principle, at any dc voltage and thus to monitor the instantaneous corrosion rate even in the presence of an applied cathodic polarization.

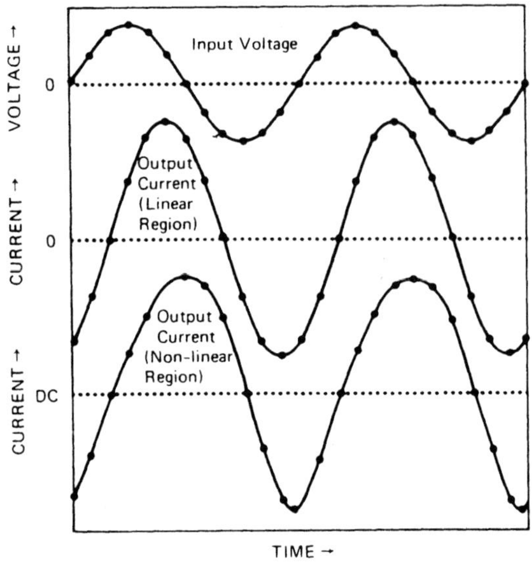

FIGURE 4.4.6.*b*. The source of a harmonic response: time domain representation of input and output waveforms.

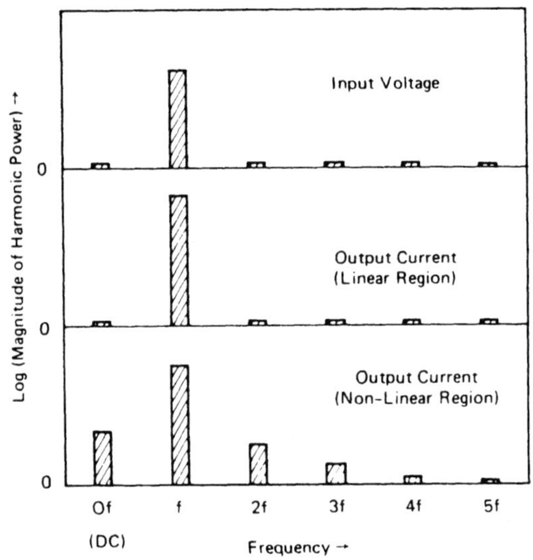

FIGURE 4.4.6.*c*. The source of a harmonic response: frequency domain representation of input and output waveforms.

The analysis of the harmonic response of a system of a sinusoidal current or voltage perturbation has received periodic attention in the electrochemical literature since the pioneering work of Warburg [1899] and Kruger [1903]. This effect has been studied as faradic rectification (Oldham [1957]), Barker *et al.* [1958], Barker [1958], Delahay *et al.* [1960], Iami and Delahay [1962], Bauer [1964]) and faradic distortion (Delahay [1954], Breyer and Bauer [1964], Smith [1966]), and the results of this form of analysis have been applied to the development of ac polarography (Morring and Kies [1977]). More recently, Bertocci [1979], Bertocci and Mullen [1981], and others (Chin and Venkatesh [1979] have investigated the effect of large-amplitude perturbations in increasing the corrosion rates of electrical conduit materials (so-called ac corrosion).

The theoretical treatments referenced above all suffer from a major deficiency. The nonlinear term of interest in corrosion (the electron transfer process) is contained within a circuit comprising other linear (electrolyte resistance) and nonlinear (double-layer capacitance and diffusional impedance) terms. Since the voltage dropped across nonlinear circuit elements cannot be considered to linearly superimpose, we cannot use the equivalent circuit method to isolate the impedance terms of interest. Properly, one must solve for the system as a whole, including diffusional and double-layer terms, and identify the harmonic components associated with the faradic process of interest.

The simplified theoretical treatment presented here is similar in form to that described previously (McKubre [1983], Bertocci [1979], Bertocci and Mullen [1981], Devay and Meszaros [1980], Devay [1982], Gill *et al.* [1983], Hladky *et al.* [1980], Rangarajan [1975], Ramamurthy and Rangarajan [1977], Rao and Mishra [1977], Callow *et al.* [1976], Devay and Meszaros [1980]).

We are interested in the current response of an electrode to a voltage perturbation of the form

$$V = V_0 + v\sin(\omega t) \tag{10}$$

Substituting Eq. (10) into Eq. (8) yields

$$I = I_{fc}\langle \exp\{\beta_a[\eta + v\sin(\omega t)]\} - \exp\{-\beta_c[\eta + v\sin(\omega t)]\}\rangle \tag{11}$$

where

$$\eta = V_0 - V_{fc} \tag{12}$$

One can make the substitution (Abramowitz and Stegun [1965], Bauer [1964])

$$\begin{aligned}\exp[z\sin(x)] = J_0(z) + 2\sum_{k=0}^{\infty}(-1)^k J_{2k+1}(z)\sin[(2k+1)x] \\ + 2\sum_{k=1}^{\infty}(-1)^k J_{2k}(z)\cos(kx)\end{aligned} \tag{13}$$

where $J_n(z)$ is a modified Bessel function of order n. The value of $J_n(z)$ can be calculated by means of the expression,

$$J_n(z) = (z/2)^n \sum_{k=0}^{\infty}\left[(z/2)^{2k}/k!(n+k)!\right] \tag{14}$$

The first term in Eq. (13) represents the expected dc response (zero'th harmonic or faradic rectification component) attributable to an ac perturbation. The second term gives the odd-order harmonic response, and the last term gives the even harmonics.

In the limit as $v \to 0$, all response functions except the fundamental disappear, and for $\eta = 0$ we obtain the familiar expression for the Stern–Geary (1957) relationship

$$\frac{dI_{fc}}{dV} = \frac{1}{R_p} = \frac{2.303 I_{fc}(\beta_a + \beta_c)}{\beta_a \beta_c} \tag{15}$$

Under all other conditions, the faradic current must be represented by a Fourier series of harmonic responses, as indicated by Eqs (12)–(14).

Substituting Eqs (13) and (14) into Eq. (11), we obtain an equation of the form

$$\begin{aligned} I/I_{fc} &= [\exp(\beta_a \eta)]({}_0C^+ + {}_1C^+ + {}_2C^+ + {}_3C^+ + \cdots) \\ &\quad - [\exp(-\beta_c \eta)]({}_0C^- + {}_1C^- + {}_2C^- + {}_3C^- + \cdots) \end{aligned} \tag{16}$$

where the harmonic series of constants $C^{\pm}$ are $\exp[\pm v \sin(\omega t)]$ evaluated according to Eq. (13) and presubscripts are used to denote the harmonic number.

Figures 4.4.7 and 4.4.8 demonstrate the influence of various corrosion parameters on the expected harmonic response, calculated from Eq. (16). Figure 4.4.7 shows the effect of the ac amplitude (v) on the magnitude of the response at large overvoltages; the responses are normalized by that at the first harmonic. Clearly, the application of perturbation levels as low as 50 mV can result in significant powers of the harmonic response at $0f$, $2f$, and $3f$.

The effect of dc potential and Tafel coefficient on the harmonic response is shown in Figure 4.4.8. The power of the current response normalized by the free corrosion current (I_{fc}), shown on a log–linear scale, parallels that of the dc current response with constant ratio between the harmonics. For the symmetric case, shown

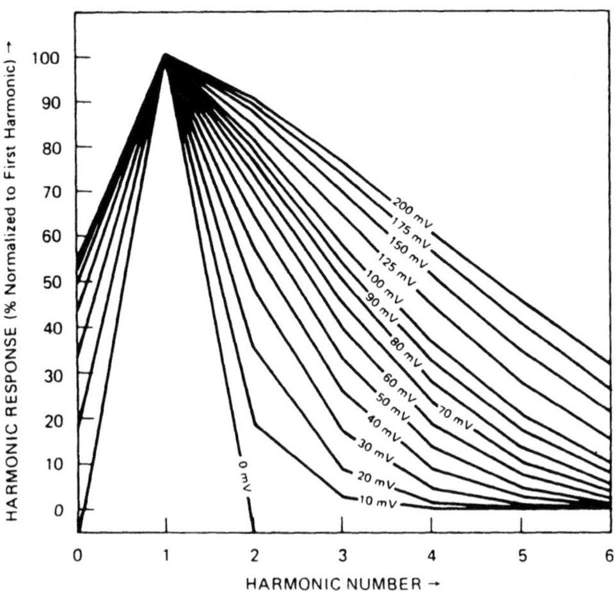

Figure 4.4.7. The effect of ac amplitude on the magnitude of the harmonic response for a two-electron process.

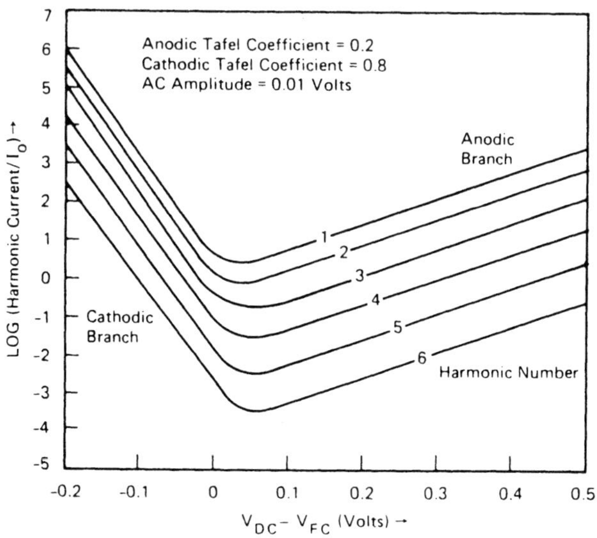

Figure 4.4.8. The effect of dc applied potential on the harmonic response for an asymmetric corrosion process.

in Figure 4.4.6*a*, the minimum in harmonic response occurs at the free corrosion potential. However, as the ratio of the reverse cathodic to forward anodic Tafel coefficients is increased, the potential at which the response is a minimum for each harmonic is increased. This phenomenon is shown in Figure 4.4.8 and has been suggested as the basis for a corrosion monitor (Gill *et al.* [1983]).

Precise measurements of the current responses at each of a number of harmonics can be used, in conjunction with Eq. (16), to evaluate the unknown terms in Eq. (11). A complete description of the faradic current–voltage response is sufficient to define the anodic partial current (the corrosion rate) at any potential; the parameters needed are the forward and reverse Tafel coefficients, the free corrosion current, and the free corrosion potential.

In applying the harmonic method to corrosion rate monitoring, three major sources of interferences must be overcome. A major source of error is due to the presence of an uncompensated series electrolyte resistance. The harmonic current responses due to the corrosion process are transformed by such a resistance into a voltage that appears as an input perturbation at the harmonic frequency, leading to an erroneous harmonic current response. A second source of error appears at large values of anodic or cathodic polarization, where the measured dc (and thus low-frequency impedance) response may be largely dominated by diffusional processes. Since the genesis of the harmonic response is considered to be in the faradic processes, it is necessary to deconvolve the diffusional from the charge-transfer impedance terms. A practical, although approximate, solution to the problem of uncompensated resistance and diffusional impedance is to completely determine the equivalent circuit for the corroding electrode by performing the impedance (fundamental or harmonic) study over a wide range of frequencies and mathematically correcting the data set retrospectively (McKubre and Syrett [1984]).

A more insidious problem is the limitation on precision imposed by the vanishingly small magnitude of the anodic (corrosion) component compared with the cathodic partial current at large values of cathodic bias, due to the exponential form of the current–voltage relationships for the anodic and cathodic half-reactions.

Methods by which these limitations can be minimized and the parameters of interest calculated are described by McKubre and Syrett [1984]. The ratio of the harmonic admittance to the fundamental admittance (both corrected as described above) is used to evaluate the desired corrosion parameters. Equation (15) can be expressed in the more appropriate form by noting that

$$\frac{{}_nY}{{}_mY} = \frac{{}_nI}{{}_1V} \cdot \frac{{}_1V}{{}_mI} = \frac{{}_nI}{I_{fc}} \cdot \frac{I_{fc}}{{}_mI} \tag{17}$$

$$\frac{{}_0Y}{{}_1Y} = \frac{[\exp(\beta_a\eta)]({}_0C^+) - [\exp(-\beta_c\eta)]({}_0C^-)}{[\exp(\beta_a\eta)]({}_1C^+) - [\exp(-\beta_c\eta)]({}_1C^-)} \tag{18}$$

$$\frac{{}_2Y}{{}_1Y} = \frac{[\exp(\beta_a\eta)]({}_2C^+) - [\exp(-\beta_c\eta)]({}_2C^-)}{[\exp(\beta_a\eta)]({}_1C^+) - [\exp(-\beta_c\eta)]({}_1C^-)} \tag{19}$$

$$\frac{{}_3Y}{{}_1Y} = \frac{[\exp(\beta_a\eta)]({}_3C^+) - [\exp(-\beta_c\eta)]({}_3C^-)}{[\exp(\beta_a\eta)]({}_1C^+) - [\exp(-\beta_c\eta)]({}_1C^-)} \tag{20}$$

The constants ${}_hC^{\pm}$ can be evaluated using Eqs (17) and (18); the unknown parameters η, β_a, β_c are calculated from the best fit of the measured admittance ratios to this system of equations.

4.4.5 Kramers–Kronig Transforms

At this point it is fitting to ask the question: "How do I know that my impedance data are correct?" This question is particularly pertinent in view of the rapid expansion in the use of impedance spectroscopy over the past decade and because more complex electrochemical and corroding systems are being probed. These give rise to a variety of impedance spectra in the complex plane, including those that exhibit pseudoinductance and intersecting loops in the Nyquist domain. By merely inspecting the experimental data, it is not possible to ascertain whether or not the data are valid or have been distorted by some experimental artifact. However, this problem can be addressed by using the Kramers–Kronig (KK) transforms (Kramers [1929], Kronig [1926], Tyagai and Kolbasov [1972], Van Meirhaeghe *et al.* [1976], Bode [1945], Macdonald and Brachman [1956]), as described by Macdonald and Urquidi-Macdonald [1985] and Urquidi-Macdonald, Real, and Macdonald [1986].

The derivation of the KK transforms (Bode [1945]) is based on the fulfillment of four general conditions of the system:

1. *Causality*. The response of the system is due only to the perturbation applied and does not contain significant components from spurious sources.

2. *Linearity*. The perturbation/response of the system is described by a set of linear differential laws. Practically, this condition requires that the impedance be independent of the magnitude of the perturbation.
3. *Stability*. The system must be stable in the sense that it returns to its original state after the perturbation is removed.
4. The impedance must be finite-valued at $\omega \to 0$ and $\omega \to \infty$ and must be a continuous and finite-valued function at all intermediate frequencies.

If the above conditions are satisfied, the KK transforms are purely a mathematical result and do not reflect any other physical property or condition of the system.

These transforms have been used extensively in the analysis of electrical circuits (Bode [1945]), but only rarely in the case of electrochemical systems (Tyagai and Kolbasov [1972], Van Meirhaeghe *et al.* [1976]).

The KK transforms may be stated as follows:

$$Z'(\omega) - Z'(\infty) = \left(\frac{2}{\pi}\right)\int_0^\infty \frac{xZ''(x) - \omega Z''(\omega)}{x^2 - \omega^2}dx \tag{21}$$

$$Z'(\omega) - Z'(0) = \left(\frac{2\omega}{\pi}\right)\int_0^\infty \left[\left(\frac{\omega}{x}\right)Z''(x) - Z''(\omega)\right]\cdot\frac{1}{x^2 - \omega^2}dx \tag{22}$$

$$Z''(\omega) = -\left(\frac{2\omega}{\pi}\right)\int_0^\infty \frac{Z'(x) - Z'(\omega)}{x^2 - \omega^2}dx \tag{23}$$

$$\phi(\omega) = \left(\frac{2\omega}{\pi}\right)\int_0^\infty \frac{\ln|Z(x)|}{x^2 - \omega^2}dx \tag{24}$$

where $\phi(\omega)$ is the phase angle, Z' and Z'' are the real and imaginary components of the impedance, respectively, and ω and χ are frequencies. Therefore, according to Eq. (4), the polarization resistance simply becomes

$$R_p = \left(\frac{2}{\pi}\right)\int_0^\infty \left[\frac{Z''(x)}{x}\right]dx \approx \left(\frac{2}{\pi}\right)\int_{x\,\text{min}}^{x\,\text{max}} \left[\frac{Z''(x)}{x}\right]dx \tag{25}$$

where χ_{max} and χ_{min} are the maximum and minimum frequencies selected such that the error introduced by evaluating the integral over a finite bandwidth, rather than over an infinite bandwidth, is negligible.

To illustrate the application of the KK transformation method for validating polarization resistance measurements in particular and for verifying impedance data in general, we consider the case of TiO_2-coated carbon steel corroding in HCl–KCl solution (pH = 2) at 25°C (McKubre [1985]). The complex plane diagram for this case is shown in Figure 4.4.9, illustrating that at high frequencies the locus of points is linear, but that at low frequencies the locus curls over to intersect the real axis. Application of Eq. (25) predicts a polarization resistance of 158.2 Ω compared with a value of 157.1 Ω calculated from the high- and low-frequency intercepts on the real axis (Macdonald and Urquidi-Macdonald [1985]).

By using the full set of transforms, as expressed by Eqs (21)–(24), it is possible to transform the real component into the imaginary component and vice versa (Macdonald and Urquidi-Macdonald [1985], Urquidi-Macdonald, Real, and

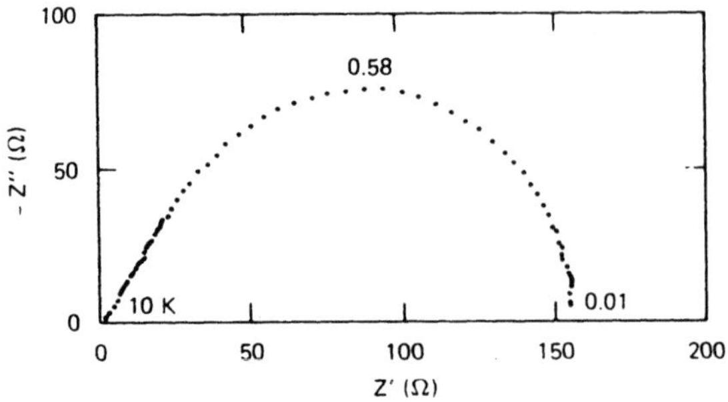

Figure 4.4.9. Complex plane impedance plot for TiO_2-coated carbon steel in HCl–KCl solution (pH = 2) at 25°C. The parameter is frequency is hertz. (From D. D. Macdonald and M. Urquidi-Macdonald, Application of Kramers–Kronig Transforms to the Analysis of Electrochemical Impedance Data: I. Polarization Resistance, *J. Electrochem. Soc.*, **132**, 2316–2319, [1985]. Reprinted by permission of the publisher, The Electrochemical Society, Inc.)

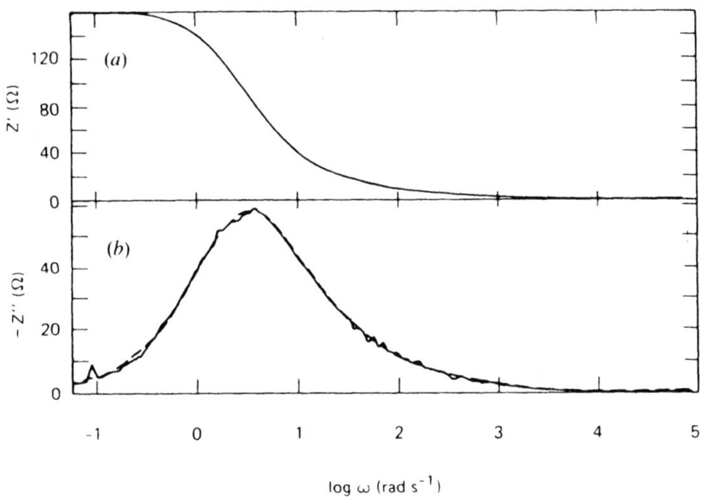

Figure 4.4.10. Kramers–Kronig transforms of impedance data for TiO_2-coated carbon steel in HCl/KCl solution (pH = 2) at 25°C: (*a*) Real impedance component vs. log ω. (*b*) Comparison of the experimental imaginary impedance component (– – –) with $Z''(\omega)$ data (—) obtained by KK transformation of the real component. (From D. D. Macdonald and M. Urquidi-Macdonald, Application of Kramers–Kronig Transforms to the Analysis of Electrochemical Impedance Data: I. Polarization Resistance, *J. Electrochem. Soc.*, **132**, 2316–2319, [1985]. Reprinted by permission of the publisher, The Electrochemical Society, Inc.)

Macdonald [1986]). These transforms therefore represent powerful criteria for assessing the validity of experimental impedance data. The application of these transforms to the case of TiO_2-coated carbon steel is shown in Figures 4.4.10 and 4.4.11. The accuracy of the transform was assessed by first analyzing synthetic impedance data calculated from an equivalent electrical circuit. An average error between the "experimental" and "transformed" data of less than 1% was obtained.

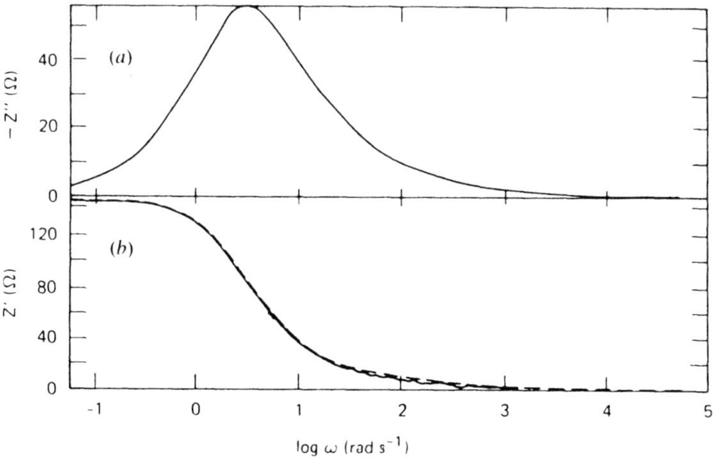

Figure 4.4.11. Kramers–Kronig transforms of impedance data for TiO_2-coated carbon steel in HCl/KCl solution (pH = 2) at 25°C: (*a*) Imaginary impedance component vs. log ω. (*b*) Comparison of the experimental real impedance component (– – –) with $Z''(\omega)$ data (—) obtained by KK transformation of the imaginary component. (From D. D. Macdonald and M. Urquidi-Macdonald, Application of Kramers–Kronig Transforms to the Analysis of Electrochemical Impedance Data: I. Polarization Resistance, *J. Electrochem. Soc.*, **132**, 2316–2319, [1985]. Reprinted by permission of the publisher, The Electrochemical Society, Inc.)

In this case, the residual error may be attributed to the algorithm used for evaluating the integrals in Eqs (21)–(23). A similar level of precision was observed on transforming McKubre's (1985) extensive data set for TiO_2-coated carbon steel in HCl/KCl (Macdonald and Urquidi-Macdonald [1985], Urquidi-Macdonald, Real, and Macdonald [1985]). Not all impedance data are found to transform as well as those for the equivalent electrical circuit and the TiO_2-coated carbon steel system referred to above. For example, Urquidi-Macdonald, Real, and Macdonald [1986] recently applied the KK transforms (21)–(23) to the case of an aluminum alloy corroding in 4 M KOH at temperatures between 25 and 60°C and found that significant errors occurred in the transforms that could be attributed to interfacial instability as reflected in the high corrosion rate.

4.4.6 Corrosion Mechanisms

4.4.6.1 Active Dissolution

Impedance spectroscopy has been applied extensively in the analysis of the mechanism of corrosion of iron and other metals in aqueous solutions. Typical work of this kind is that reported by Keddam *et al.* [1981], who sought to distinguish between various mechanisms that had been proposed for the electrodissolution of iron in acidified sodium sulfate solutions. Since this particular study provides an excellent review of how impedance spectroscopy is used to discern reaction mechanism, the essential features of the analysis are described below.

As the result of analyzing a large number of possible mechanisms for the dissolution of iron, Keddam *et al.* [1981] concluded that the most viable mechanism for this reaction involves three intermediate species

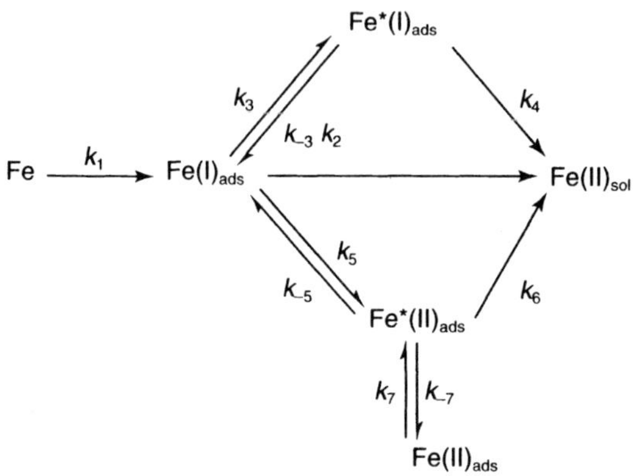

(26)

in which steps 4 and 6 are written in complete form as

$$\mathrm{Fe}*(\mathrm{I})_{\mathrm{ads}} + \mathrm{Fe} \xrightarrow{k_4} \mathrm{Fe}*(\mathrm{I})_{\mathrm{ads}} + \mathrm{Fe(II)}_{\mathrm{sol}} + 2e^- \tag{27}$$

and

$$\mathrm{Fe}*(\mathrm{II})_{\mathrm{ads}} + \mathrm{Fe} \xrightarrow{k_6} \mathrm{Fe}*(\mathrm{II})_{\mathrm{ads}} + \mathrm{Fe(II)}_{\mathrm{sol}} + 2e^- \tag{28}$$

In setting up the reaction model for this case, Keddam *et al.* assumed that the elementary steps obey Tafel's law, that the transfer coefficient (α) has a value between 0 and 1 and is independent of potential, and that the coverage by adsorbed species obeys the Langmuir isotherm. Designating the fractional coverages by the species $\mathrm{Fe(I)_{ads}}$, $\mathrm{Fe^*(I)_{ads}}$, and $\mathrm{Fe^*(II)_{ads}}$ to be θ_1, θ_2, and θ_3, respectively, and that of the passivating species $\mathrm{Fe(II)_{ads}}$ as θ_4, and assuming no overlap, then the current flowing across the interface may be expressed as

$$I = F[k_1\Sigma + (k_2 + k_5)\theta_1 + 2k_4\theta_2 + (2k_6 - k_{-5})\theta_3] \tag{29}$$

where

$$\Sigma = 1 - \theta_1 - \theta_2 - \theta_3 - \theta_4 \tag{30}$$

and k_i is the rate constant for the ith step defined by

$$k_i = k_{o,i} \exp\left(\frac{z\alpha F}{RT} \cdot E\right) \tag{31}$$

Mass balance relationships involving the adsorbed species results in the following expressions for the time dependencies of θ_1, θ_2, θ_3, and θ_4:

$$\beta_1 \frac{d\theta_1}{dt} = k_1\Sigma - (k_2 + k_3 + k_5)\theta_1 + k_{-3}\theta_2 + k_{-5}\theta_3 \tag{32}$$

$$\beta_2 \frac{d\theta_2}{dt} = k_3\theta_1 = k_{-3}\theta_2 \tag{33}$$

$$\beta_3 \frac{d\theta_3}{dt} = k_5\theta_1 - (k_{-5} + k_7)\theta_3 + k_{-7}\theta_4 \tag{34}$$

$$\beta_4 \frac{d\theta_4}{dt} = k_7\theta_3 - k_{-7}\theta_4 \tag{35}$$

where β is a constant that links the surface fractions to surface concentrations (mole/cm^{-2}). The value for β is ~10^{-8} mole/cm^2, which corresponds to about one monolayer. The steady state is characterized by $d\theta_i/dt = 0$, in which case

$$\bar{\theta}_1 = \frac{k_1 k_{-3} k_{-5} k_{-7}}{D} \tag{36}$$

$$\bar{\theta}_2 = \frac{k_1 k_{-3} k_{-5} k_{-7}}{D} \tag{37}$$

$$\bar{\theta}_3 = \frac{k_1 k_{-3} k_{-5} k_{-7}}{D} \tag{38}$$

$$\bar{\theta}_4 = \frac{k_1 k_{-3} k_5 k_7}{D} \tag{39}$$

where

$$D = k_1 k_{-3} k_5 k_7 + \{k_1[k_3 k_{-5} + k_{-3}(k_5 + k_{-5})] + k_2 k_{-3} k_{-5}\}k_{-7} \tag{40}$$

and hence the steady state current becomes

$$\bar{I} = 2F(k_2\bar{\theta}_1 + k_4\bar{\theta}_2 + k_6\bar{\theta}_3) \tag{41}$$

In order to derive the faradic impedance (Z_F) we note that for sinusoidal variations in the potential and in the surface coverages of reaction intermediates we may write

$$\delta E = |\delta E| e^{j\omega t} \tag{42}$$

$$\delta\theta_i = |\delta\theta_i| e^{j\omega t} \tag{43}$$

$$j = \sqrt{-1} \tag{44}$$

Thus, from Eq. (41) and defining Z_F as

$$z_F = \frac{\delta E}{\delta I} \tag{45}$$

we obtain the following expression for the faradic impedance:

$$\begin{aligned}\frac{1}{Z_F} = \frac{1}{R_T} - F[k_1 - k_2 - k_5]\frac{d\theta_1}{dE} + (k_1 - 2k_4)\frac{d\theta_2}{dE} \\ + (k_1 + k_{-5} - 2k_6)\frac{d\theta_3}{dE} + k_1\frac{d\theta_4}{dE}\end{aligned} \tag{46}$$

where

$$\frac{1}{R_t} = F[(b_1 + b_2)k_2\bar{\theta}_1 + 2b_4k_4\bar{\theta}_2 + [(b_5 + b_{-5})k_{-5} + 2b_6k_6]\bar{\theta}_3] \tag{47}$$

$$b_i = \alpha_i F/RT \tag{48}$$

The faradic impedance is readily obtained by first deriving expressions for $d\theta_i/dt$. This is done by taking the total differentials of Eqs (32)–(35). For example in the case of Eq. (33) we write

$$\delta\left(\beta_2 \frac{d\theta_2}{dt}\right) = k_3\delta\theta_1 + \theta_1\delta k_3 - k_{-3}\delta\theta_2 - \theta_2\delta k_{-3} \tag{49}$$

Since

$$\delta\left(\beta_2 \frac{d\theta_2}{dt}\right) = \beta_2 \frac{d(\delta\theta_2)}{dt} = \beta_2 j\omega\delta\theta_2 \tag{50}$$

and

$$\delta k_i = b_i k_i \delta E \tag{51}$$

we obtain

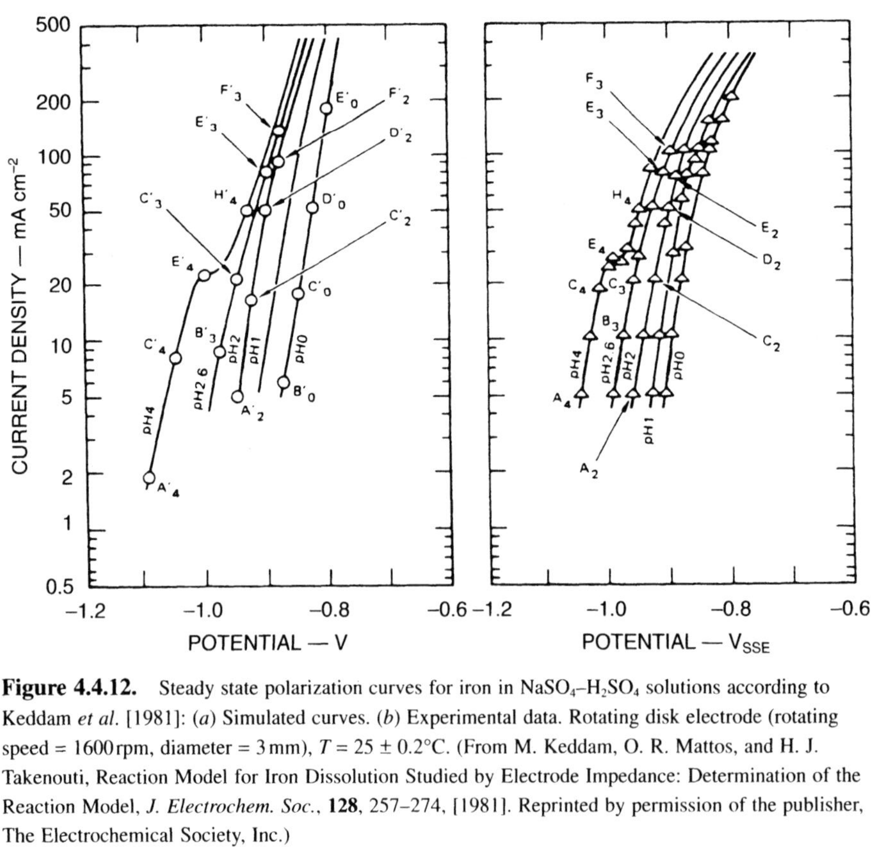

Figure 4.4.12. Steady state polarization curves for iron in $NaSO_4$–H_2SO_4 solutions according to Keddam *et al.* [1981]: (*a*) Simulated curves. (*b*) Experimental data. Rotating disk electrode (rotating speed = 1600 rpm, diameter = 3 mm), $T = 25 \pm 0.2°C$. (From M. Keddam, O. R. Mattos, and H. J. Takenouti, Reaction Model for Iron Dissolution Studied by Electrode Impedance: Determination of the Reaction Model, *J. Electrochem. Soc.*, **128**, 257–274, [1981]. Reprinted by permission of the publisher, The Electrochemical Society, Inc.)

$$(k_{-3} + j\omega\beta_2)\frac{\delta\theta_{-2}}{\delta E} - k_3\frac{\delta\theta_1}{\delta E} - \theta_1 b_3 k_3 + \theta_2 b_{-3} k_{-3} = 0 \tag{52}$$

Additional linear simultaneous equations may be generated from Eqs (32), (34), and (35), and the set may be solved for $\delta\theta_1/\delta E$, $\delta\theta_2/\delta E$, $\delta\theta_3/\delta E$, and $\delta\theta_4/\delta E$. These values are then substituted into Eq. (46) to calculate the faradaic impedance, which in turn yields the interfacial impedance as

$$Z_T = Z_F/(1 + j\omega C_{dl} Z_F) \tag{53}$$

where C_{dl} is the double layer capacitance.

Experimental and simulated [Eq. (29)] steady state current-voltage curves for iron in Na_2SO_4–H_2SO_4 solutions as a function of pH are shown in Figure 4.4.12, and simulated and experimental complex plane impedance diagrams at various points on these curves are shown in Figures 4.4.13 and 4.4.14, respectively (Keddam *et al.* [1981]). The close agreement between the steady state polarization curves is immediately apparent, even to the extent that the inflection in the curve for pH 4 is accurately reproduced. Examination of the impedance diagrams in Figures 4.4.13 and 4.4.14 show that the mechanism selected by Keddam *et al.* [1981] is capable of reproducing the essential features of the diagrams, including the number and type of relaxations, but not the details. However, the latter depend strongly upon the values selected for the rate constants, and are also probably affected by the isotherm selected for describing the adsorption of intermediate species onto the surface.

4.4.6.2 Active–Passive Transition

The sudden transition of a metal–solution interface from a state of active dissolution to the passive state is a phenomenon of great scientific and technological interest. This transition has been attributed to the formation of either a monolayer (or less) of adsorbed oxygen on the surface or to the coverage of the surface by a three-dimensional corrosion product film. In either case, the reactive metal is shielded from the aqueous environment, and the current drops sharply to a low value that is determined by the movement of ions or vacancies across the film.

The changes that typically occur in the complex-plane impedance diagram on increasing the potential through an active-to-passive transition are shown in Figure 4.4.15 (Keddam *et al.* [1984]). At point A, the high-frequency arm of the impedance is typical of a resistive-capacitive system, but the impedance locus terminates in a negative resistance as $\omega \to 0$. This, of course, is consistent with the negative slope of the steady state polarization curve. At higher potentials, the high-frequency locus is again dominated by an apparent resistive–capacitance response (see Section 4.4.6.3), but the low-frequency arm is not observed to terminate at the real axis in this case because of the very high value for the polarization resistance (horizontal I vs. E curve). The origin of the negative resistance can be accounted for theoretically (Keddam *et al.* [1984]) in terms of an increasing coverage of the surface by an adsorbed intermediate as the potential is increased. Thus, the low-frequency loop exhibited in Figure 4.4.15*a* is due to relaxations involving these surface species.

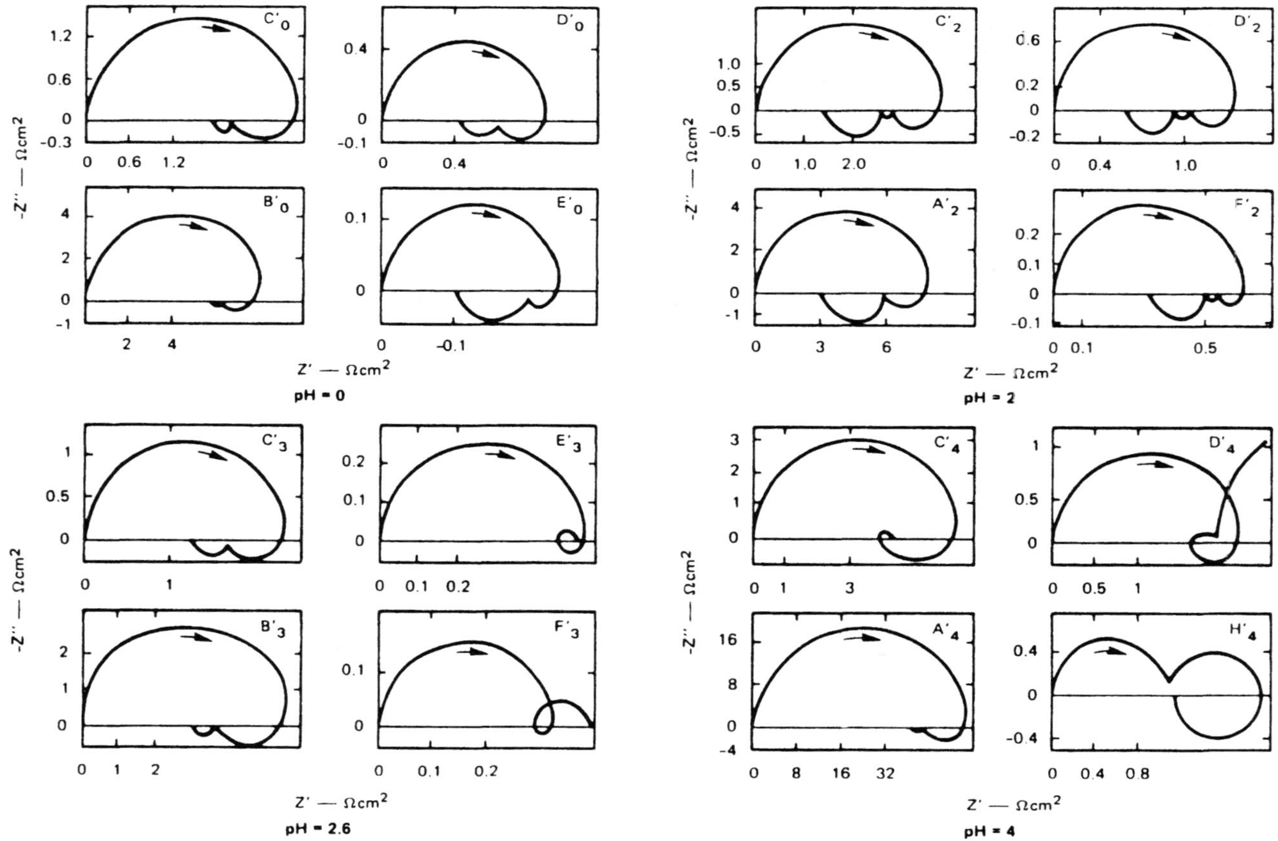
C'0
D'0
B'0
E'0
pH = 0
C'2
D'2
A'2
F'2
pH = 2
C'3
E'3
B'3
F'3
pH = 2.6
C'4
D'4
A'4
H'4
pH = 4
-Z'' — Ωcm2
Z' — Ωcm2

Epelboin and coworkers [1972] and Diard and LeGorrec [1979] have recognized a serious shortcoming of classical potentiostatic methods for investigating the active-to-passive transition. The problem arises because a potentiostat has a load line of negative slope in the I vs. E plane and hence is incapable of effectively defining the current–voltage characteristics of a metal–solution interface in the active-to-passive transition region. To overcome this limitation of potentiostatic control, Epelboin *et al.* [1972, 1975] and Diard and LeGorrec [1979] devised potential control instruments having negative output impedances, which are characterized by load lines having positive (and controllable) slopes. These negative impedance converters (NICs) have allowed "Z-shaped" active-to-passive transitions to be studied and the impedance characteristics to be determined, as shown by the data plotted in Figure 4.4.16. In contrast to the case shown in Figure 4.4.15*a*, the active-to-passive transition shown in Figure 4.4.16, as determined using a NIC, exhibits a change in the sign of dI/dE from negative to positive to negative as the current decreases from the active to the passive state. This change in sign, as reflected in the shape of the Z-shaped polarization curves, has been explained by Epelboin *et al.* [1975] in terms of coupling between mass transfer and surface reactions, although other explanations have also been advanced (Law and Newman [1979]).

4.4.6.3 The Passive State

The phenomenon of passivity is enormously important in corrosion science and engineering, since it is responsible for the relatively low corrosion rates that are observed for most engineering metals and alloys. It is not surprising, therefore, that passivity has been studied extensively using a wide variety of techniques, including IS. A brief account of these impedance studies is given below.

In discussing this subject it is convenient to delineate the processes that occur at the film–solution interface and those that take place within a passive film (Figure 4.4.17). In the first case, the processes are essentially ion exchange phenomena with the possibility of solution phase transport, whereas the second processes involve only transport. The movement of charged species within the film (anion vacancies $V_o^{\cdot\cdot}$ and cation vacancies $V_M^{x'}$) occurs, however, under the influence of both concentration and electrical potential gradients, with the electrical effects probably dominating, at least in the case of thin films. Accordingly, any analysis of the impedance characteristic of passive films must consider electromigration as well as diffusional transport.

The total impedance of the system of interphases shown in Figure 4.4.17 may be written as

Figure 4.4.13. Simulated complex plane impedance diagrams for the electrodissolution of iron in sulfate media as a function of pH according to Keddam *et al.* [1981]. The potentials for which the diagrams are calculated are shown in Figure 4.4.12. The arrows indicate the direction of decreasing frequency. (From M. Keddam, O. R. Mattos, and H. J. Takenouti, Reaction Model for Iron Dissolution Studied by Electrode Impedance: Determination of the Reaction Model, *J. Electrochem. Soc.*, **128**, 257–274, [1981]. Reprinted by permission of the publisher, The Electrochemical Society, Inc.)

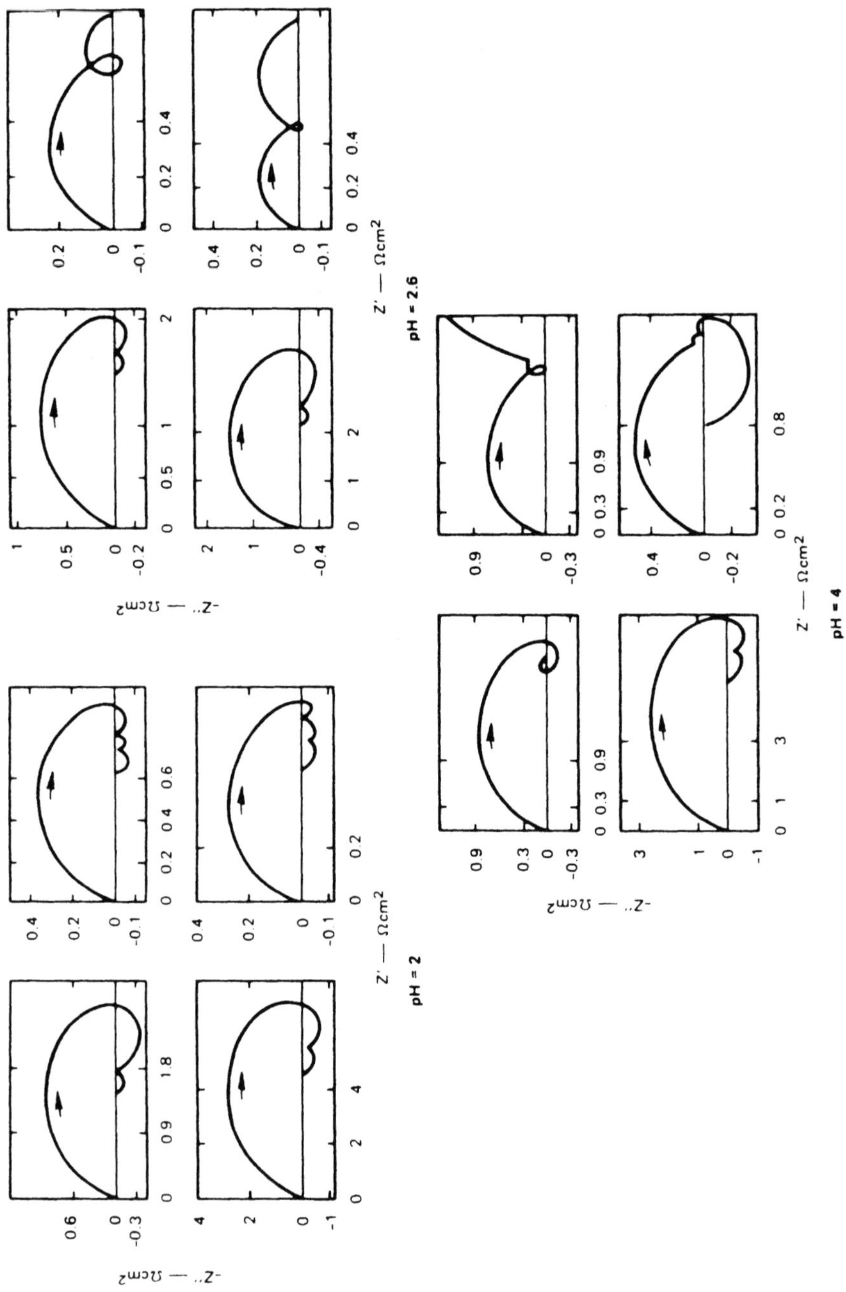
-Z'' — Ωcm²
Z' — Ωcm²
pH = 2
pH = 2.6
pH = 4

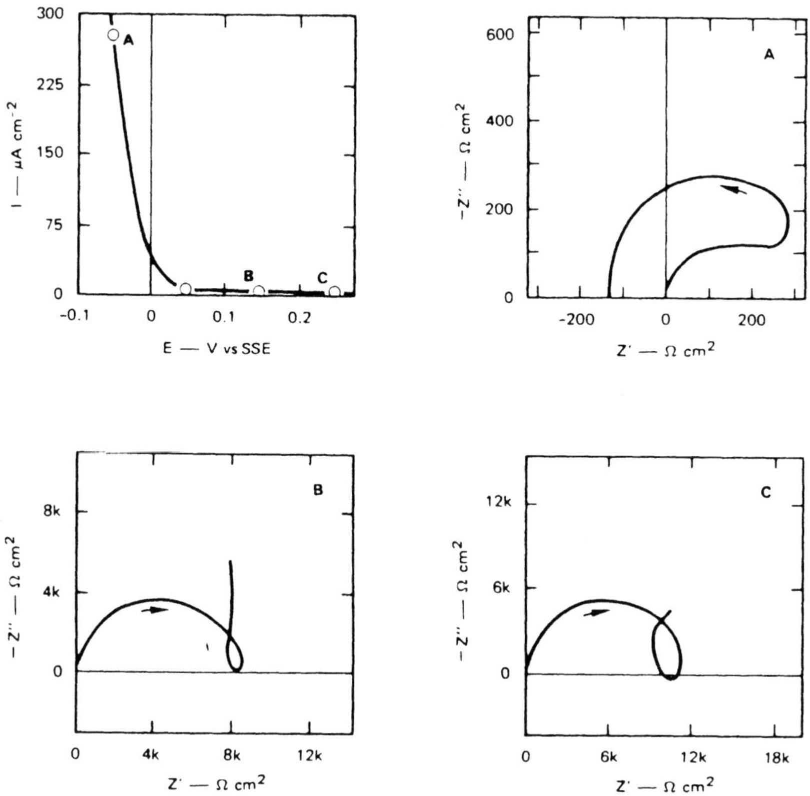

A: E = -0.050 V vs Hg/Hg_2SO_4, K_2SO_4 (sat)

B: E = 0.150 V vs Hg/Hg_2SO_4, K_2SO_4 (sat)

C: E = 0.250 V vs Hg/Hg_2SO_4, K_2SO_4 (sat)

Figure 4.4.15. Steady-state polarization curve and complex plane impedance diagrams at selected potentials through the active-to-passive transition for iron in 1 M H_2SO_4 as reported by Keddam, Lizee, Pallotta, and Takenouti [1984]. The arrows indicate the direction of decreasing frequency. (From M. Keddam, O. R. Mattos, and H. J. Takenouti, Reaction Model for Iron Dissolution Studied by Electrode Impedance: Determination of the Reaction Model, *J. Electrochem, Soc.*, **128**, 257–274, [1981]. Reprinted by permission of the publisher, The Electrochemical Society, Inc.)

Figure 4.4.14. Experimental complex plane impedance diagrams for iron in sulfate media as a function of pH according to Keddam *et al.* [1981]. The potentials at which the diagrams were measured are shown in Figure 4.4.12. The arrows indicate the direction of decreasing frequency. (From M. Keddam, O. R. Mattos, and H. J. Takenouti, Reaction Model for Iron Dissolution Studied by Electrode Impedance: Determination of the Reaction Model, *J. Electrochem. Soc.*, **128**, 257–274, [1981]. Reprinted by permission of the publisher, The Electrochemical Society, Inc.)

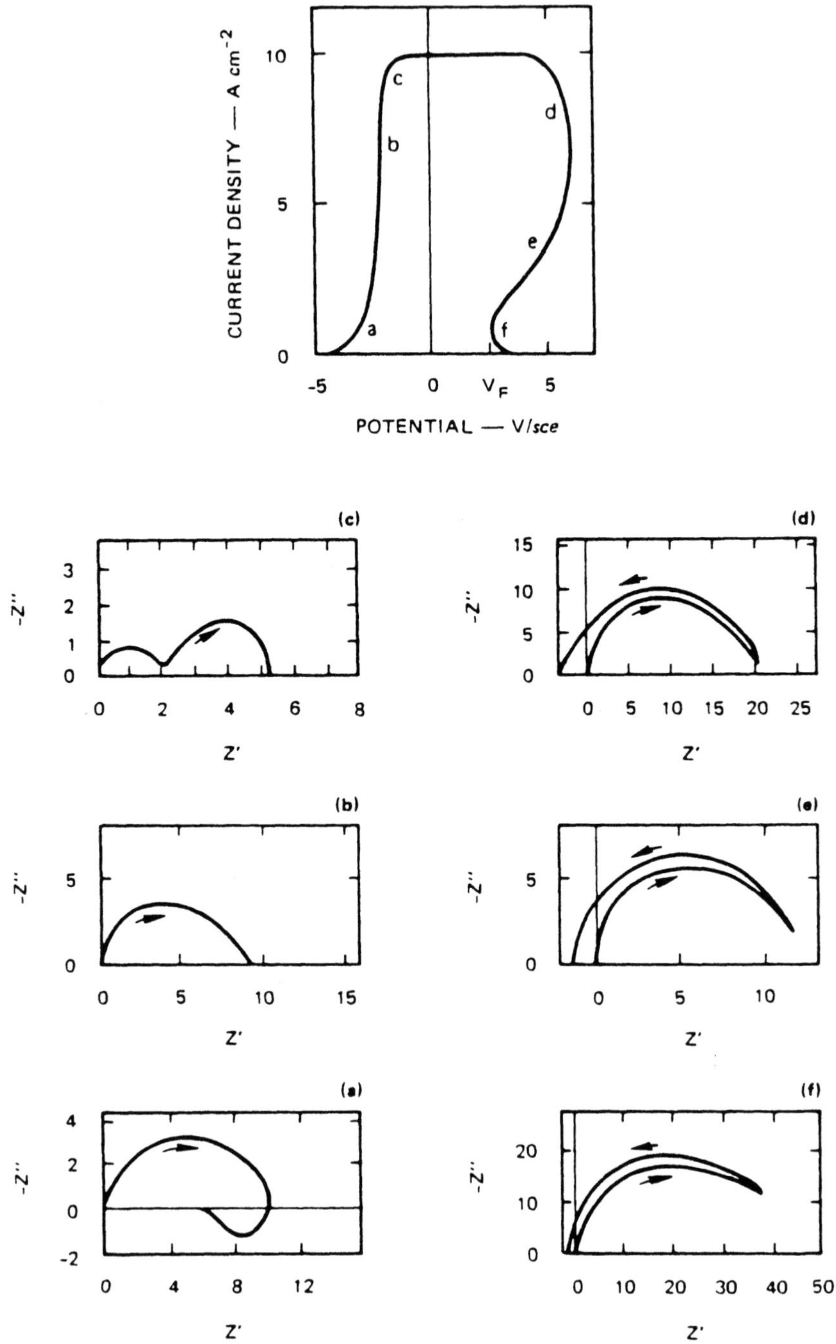

Figure 4.4.16. Impedance spectra for iron in 1 M H_2SO_4 at various potentials within the active dissolution and active-to-passive transition regions as determined using a negative impedance converter (NIC). Impedance values are given in ohms (electrode diameter = 0.5 cm), and the arrows indicate the direction of decreasing frequency. (After Epelboin *et al.* [1975]).

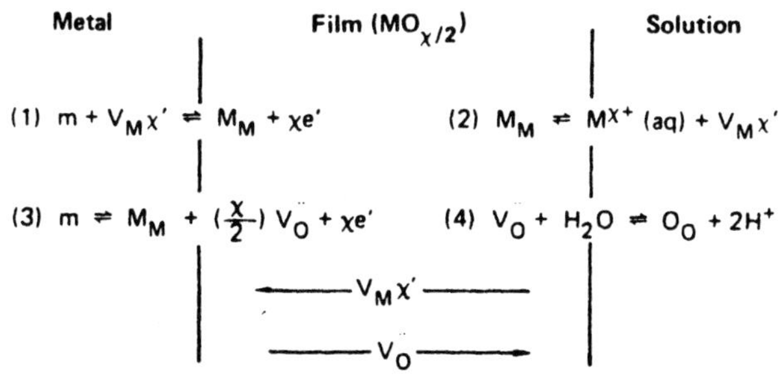

Figure 4.4.17. Schematic of physicochemical processes that occur within a passive film according to the point defect model. Here m = metal atom, M_M = metal cation in cation site, O_0 = oxygen ion in anion site, $V_M^{x'}$ = cation vacancy, $V_{\ddot{O}}$ = anion vacancy. During film growth, cation vacancies are produced at the film–solution interface but are consumed at the metal–film interface. Likewise, anion vacancies are formed at the metal–film interface but are consumed at the film–solution interface. Consequently, the fluxes of cation vacancies and anion vacancies are in the directions indicated.

$$Z_T = Z_{m/f} + Z_f + Z_{f/s} \tag{54}$$

where $Z_{m/f}$, Z_f, and $Z_{f/s}$ are the impedances associated with the metal–film interface, the film, and the film–solution interface, respectively. Because the elements are connected in series, the largest impedance will dominate the total impedance of the system. However, the impedance elements are frequency-dependent, so that each of the elements may dominate over different frequency ranges.

Metal–Film Interface. According to Armstrong and Edmondson (1973), the impedance of the metal–film interface can be described in terms of a capacitance (C'_∞) in parallel with two charge transfer resistances, one for the transfer of electrons (R_e) and the other for the transfer of cations (R_c) from the metal to the film. Accordingly,

$$Z_{m/f} = \frac{R_e R_c (R_e + R_c)}{(R_e + R_c)^2 + \omega^2 C^2 R_e^2 R_c^2} - j\frac{\omega^2 C R_e^2 R_c^2}{(R_e + R_c)^2 + \omega^2 C^2 R_e^2 R_c^2} \tag{55}$$

For most systems, particularly for diffuse metal–oxide junctions, we assume that the resistance to the movement of electrons across the interface is small compared with the resistance to the movement of cations, so that

$$R_e << R_c \tag{56}$$

In this case, Eq. (55) reduces to

$$Z_{m/f} = \frac{R_e}{1 + \omega^2 C^2 R_c^2} - j\frac{\omega C R_e^2}{1 + \omega^2 C^2 R_e^2} \tag{57}$$

Furthermore, the capacitance associated with this interface is probably that due to the space charge layer within the oxide. Therefore, over the frequency range of most interest to corrosion scientists (10^{-4}–10^4 Hz), $1/CR_e >> \omega$, so that

$$Zm/f \sim R_e \tag{58}$$

Accordingly, under these conditions, the impedance of the metal–film interface is likely to appear as a (small) frequency-independent resistance due to the transfer of electrons between the two phases.

The Film. A quantitative analysis of the impedance of a passive film has been reported by Chao, Lin, and Macdonald [1982], and the essential features of this treatment are reproduced here. The treatment is based upon their previously proposed (Chao *et al.* [1981], Lin *et al.* [1981]) point defect model for the growth and breakdown of passive films; the essential features of which are depicted in Figure 4.4.17.

In this model, it is assumed that the total current that is detected in an external circuit upon application of a voltage is the sum of four components: (1) electronic current due to the transport of electrons (e'); (2) electronic current due to the flow of electron holes ($\dot{h}$); (3) ionic current due to the transport of anion vacancies ($V_{\ddot{o}}$); and (4) ionic current due to the movement of cation vacancies ($V_H^{x'}$)

$$I = I_{e'} + I_{\dot{h}} + I_{V_o} + I_{V_M} \tag{59}$$

Therefore,

$$I/Z_f = 1/Z_e + 1/Z_h + 1/Z_0 + 1/Z_M \tag{60}$$

The total impedance of the film is therefore described in terms of the transport of vacancies in parallel with the electron and hole resistances, provided that electron or hole exchange processes do not occur at the film–solution interface. This situation exists in the absence of any redox couples in the solution.

The movement of anion and cation vacancies within the film, under the influence of concentration (C) and electrical potential (ϕ) gradients, is determined by Fick's second law:

$$\frac{\partial C}{\partial t} = D\frac{\partial^2 C}{\partial x^2} - DqK\frac{\partial C}{\partial x} \tag{61}$$

where

$$K = \varepsilon F/RT \tag{62}$$

$$\varepsilon = -d\phi/dx \tag{63}$$

Here q is the charge on the moving species ($-\chi$ for cation vacancies and +2 for oxygen vacancies for an oxide film of stoichiometry $MO_{\chi/2}$), D is the diffusivity, and F, R, and T have their usual meanings. The current observed in an external conductor due to the movement of the vacancies is given by Fick's first law, as applied to the metal–film interface:

$$I = qFJ = qF\left(-D\frac{\partial C}{\partial x} + DqKC\right)_{m/f} \tag{64}$$

According to the point defect model, and under conditions where the various equations can be linearized with respect to the applied ac voltage (V_{ac}), the concentration of vacancies at the metal–film and film–solution interfaces may be expressed as (Chao *et al.* [1982])

$$C_{V_0}(m/f) = [C_{V_0}(m/f)]_{dc} \cdot \frac{2F(1-\alpha)}{RT} \cdot V_{ac} \tag{65}$$

$$C_{V_M}(m/f) = [C_{V_M}(m/f)]_{dc} \cdot \frac{\chi F(\alpha-1)}{RT} \cdot V_{ac} \tag{66}$$

$$C_{V_0}(f/s) = [C_{V_0}(f/s)]_{dc} \cdot \frac{2F\alpha}{RT} \cdot V_{ac} \tag{67}$$

$$C_{V_M}(f/s) = [C_{V_M}(f/s)]_{dc} \cdot \frac{\chi F\alpha}{RT} \cdot V_{ac} \tag{68}$$

where the quantities in square brackets are constants, which are related to the thermodynamic parameters for the interfacial reactions shown in Figure 4.4.17.

The set of Eqs (61)–(68) are readily solved by Laplace transformation to yield the impedance

$$Z_f(j\omega) = \bar{V}_{ac}/\bar{I}_{ac},\ s = j\omega,\ j = \sqrt{-1} \tag{69}$$

where s is the Laplace frequency. The impedance so calculated is found to have the from

$$Z_T = \left(\frac{\sigma_M \sigma_0}{\sigma_M + \sigma_0}\right)\omega^{-1/2}(1-j) \tag{70}$$

where σ_0 and σ_M are given by

$$\sigma_0 = RT/F^2(32D)^{1/2}\{\lfloor C_{V_0}(m/f)\rfloor_{dc}(1-\alpha)\} \tag{71}$$

$$\sigma_M = RT/F^2(2x^4D)^{1/2}\{(\alpha-1)\lfloor C_{V_M}(m/f)_{dc}\rfloor\} \tag{72}$$

The reader will recognize Eq. (70) as being of the form of a Warburg impedance for two parallel moving species. Two limiting cases may be defined: (1) movement of cations vacancies alone ($\sigma_0 >> \sigma_M$, $Z_T \rightarrow Z_M$); and (2) movement of anion vacancies alone ($\sigma_M >> \sigma_0$, $Z_T \rightarrow Z_0$). Accordingly,

$$Z_T = \sigma_M \omega^{-1/2}(1-j) \quad \text{(cation vacancies)} \tag{73}$$

$$Z_T = \sigma_0 \omega^{-1/2}(1-j) \quad \text{(anion vacancies)} \tag{74}$$

Substitution of the appropriate expression for $[C_{V_0}(m/f)]_{dc}$ into Eq. (71) yields the Warburg coefficients for the movement of oxygen ion vacancies as

$$\sigma_0 = \frac{1}{I_{dc}}\left(\frac{D_0}{2}\right)^{1/2} \cdot \frac{\varepsilon}{1-\alpha} \tag{75}$$

The form of the equation for σ_0 is particularly interesting, because it suggests that if the electric field strength (ε) and α are constants (as assumed in the point defect model), then the product $\sigma_0 I_{dc}$ should be independent of the applied voltage across the system and the thickness of the film.

A plot of $-Z''$ vs. Z' for selected values of the various parameters contained in

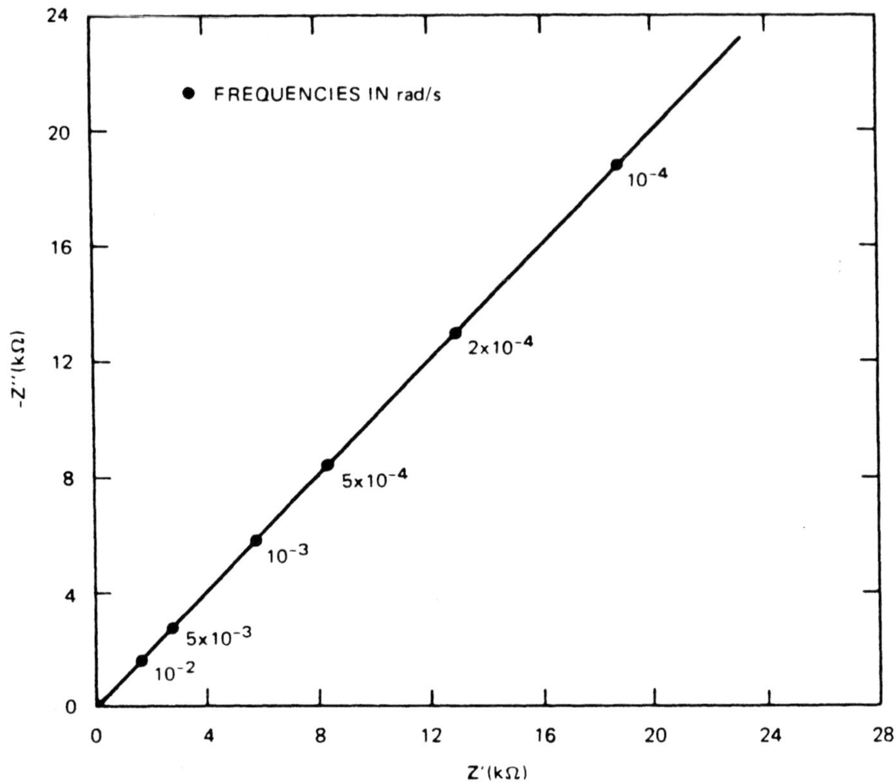

Figure 4.4.18. Complex impedance plane predicted by Eq. (74). $D_0 = 10^{-21}\,\text{cm}^2/\text{s}$, $I_{dc} = 1\,\mu\text{A/cm}^2$, $\varepsilon = 10^6\,\text{V/cm}$, $\alpha = 0.88$, area = $1\,\text{cm}^2$.

Eq. (74) is shown in Figure 4.4.18. As expected, the impedance locus is a straight line when

$$\varepsilon > D(F^2\varepsilon^2/R^2T^2) \tag{76}$$

However, for sufficiently low frequencies (Chao *et al.* [1982]), the impedance is predicted to intercept the real axis at a value of

$$R_{\omega=0} = \frac{RT}{4F^2DK\{2[C_{V_0}(m/f)]_{dc}(1-\alpha) - A'\}} \tag{77}$$

where

$$A' = A(RT/F\bar{V}_{ac}) \tag{78}$$

with

$$A = -\left(\frac{F\bar{V}_{ac}}{RT}\right) \cdot \frac{[C_{V_0}(f/s)]_{dc}a + [C_{V_0}(m/f)]_{dc}(1-\alpha)\exp\left[KL - (K^2 + s/DL)^{1/2}\right]}{\exp(KL)\sinh[(K^2 + s/DL^{1/2})]} \tag{79}$$

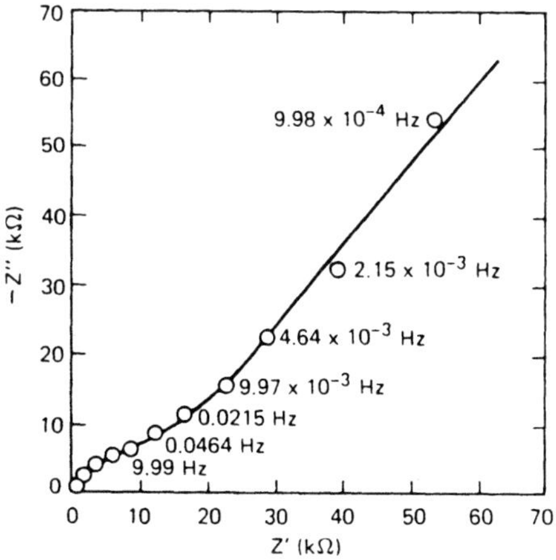

Figure 4.4.19. Complex plane impedance plot for Ni(111) in phosphate buffer ($[PO_4] = 0.1\,M$) at 25°C. $E = 0.1$ V (SCE), pH = 10. The frequency at which each point was measured is indicated. Electrode area = $0.998\,cm^2$.

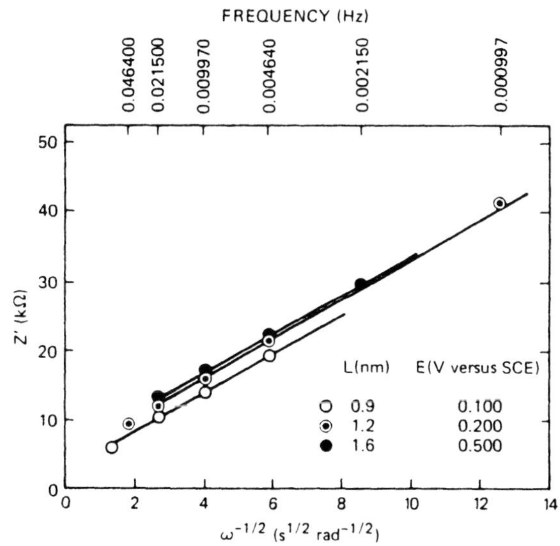

Figure 4.4.20. Randles plot of Z' vs. $\omega^{-1/2}$ for Ni(111) passivated in 0.1 M phosphate solution (pH = 9). L = film thickness measured ellipsometrically. Electrode area = $0.998\,cm^2$.

These equations show that the most critical parameters in determining the value of $R_{\omega=0}$ is the diffusivity of oxygen vacancies ($D \equiv D_0$) and the film thickness L; $R_{\omega=0}$ increases roughly exponentially with L and with $1/D$. Similar arguments can be made in the case of the transport of cation vacancies across a passive film.

The diagnostic features of this analysis have been used by Chao *et al.* [1982] in their investigation of the growth of passive films on nickel and Type 304 stainless steel in borate and phosphate buffer solutions. Typical complex plane and

Randles's plane plots for nickel in 0.1 N Na_2HPO_4 (pH = 9.1) and in 0.15 N H_3BO_3/0.15 N $Na_2B_4O_7$ (pH = 8.7) solutions are shown in Figures 4.4.19 and 4.4.20. The data shown in Figure 4.4.19 reveal a linear impedance locus at low frequencies, with a partially resolved semicircle at high frequencies. This latter characteristic is attributed to relaxations occurring at the film–solution interface, as discussed later. The Randles plots shown in Figure 4.4.20 provide further evidence for the low-frequency Warburg response predicted by Eqs (73) and (74). These plots also show that the experimentally measured Warburg coefficient $\sigma = dZ'/d\omega^{-1/2}$ is independent of the film thickness (as measured ellipsometrically) and of the applied voltage. Furthermore, the values of σ obtained for the phosphate (23.1 kΩ/s$^{1/2}$) and borate (8.53 kΩ/s$^{1/2}$) environments differ by nearly a factor of 3, as do the passive currents. This is predicted by Eqs (75) and (76), assuming that negligible differences exist between the passive films formed in these two solutions.

A more extensive impedance analysis of passive films formed on nickel in borate buffer solution has been reported by Liang, Pound, and Macdonald [1984]. In this study, impedance data were obtained for passive nickel over a wide range of applied potential and pH (Figure 4.4.21), and these data serve as a good test of the constancy of σI_{dc}, as predicted by Eqs (75) and (76). The data (Figure 4.4.21) show that this product is indeed constant, within experimental error, thereby supporting the original hypothesis of the point defect model that the electric field strength is independent of film thickness and applied voltage.

An important application of the data shown in Figure 4.4.21 and of the equation for the Warburg coefficient [Eq. (75)] is in the calculation of the diffusivity for anion vacancies within the film. In the case of passive polycrystalline nickel in borate and phosphate buffer solutions, Chao *et al.* (1982) computed a value of 1.3 $\times$ 10^{-21} cm^2/s for the diffusivity of oxygen ion vacancies. In a later study by Liang *et al.* [1984], a somewhat higher (and possibly more reliable) value of 1.5 $\times$

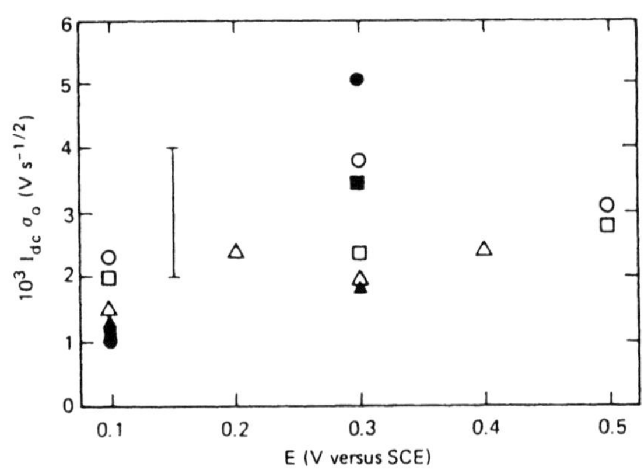

Figure 4.4.21. Plot of $I_{dc}\sigma_0$ vs. potential for Ni(111) in phosphate buffers ([PO_4] = 0.1 M) as a function of pH. Open circles = pH 7, closed circles = pH 8, open squares = pH 9, closed squares = pH 10, open triangles = pH 11, closed triangles = pH 12. T = 25°C, I = estimated error for each point, electrode area = 0.998 cm^2.

10^{-19} cm^2/s for this same quantity was calculated from the data shown in Figure 4.4.21 for passive films formed on single-crystal nickel (100) in borate buffer solution. The principal problem with these calculations lies in the accurate measurement of the passive current I_{dc}. Experience shows that this quantity can vary over several orders of magnitude depending upon how the surface is prepared, the method by which it is measured (potentiostatic vs. potentiodynamic techniques), and possibly the means by which the passive film is formed. Nevertheless, the values for D_0 given above are consistent with data extrapolated from high temperatures for a variety of oxides, and they appear to be eminently reasonable from a physicochemical viewpoint (Chao *et al.* [1982]).

The data shown in Figure 4.4.22 for Type 304 stainless steel appear to contradict the findings reported above for nickel. However, as noted by Chao *et al.* [1982], the potential-dependent and film-thickness-dependent Warburg coefficient can be accounted for by the fact that the composition of the passive film also changes with potential. A study of passive films on Fe-25 Ni-*X*Cr alloys by Silverman, Cragnolino, and Macdonald [1982] indicated that the quantity $\varepsilon/(1-\alpha)$ changes very little as the Fe/Cr ratio is varied over a wide range, but diffusivity data for oxygen ion vacancies in iron and chromium oxides at elevated temperatures very by many orders of magnitude, and in a manner that provides a qualitative explanation of the Randles plots shown in Figure 4.4.22.

Film–Solution Interface. One of the most comprehensive treatments of the impedance characteristics of the film—solution interface of a passive film is that reported by Armstrong and Edmondson [1973]. Their treatment essentially considers the ion exchange properties of an interface (Figure 4.4.23) by addressing the movement of anions and cations between the film surface and the solution as the applied potential is modulated over a wide frequency range.

Armstrong and Edmondson [1973] begin their analysis by noting that the time dependence of the excess of cations over anions in the surface layer (Figure 4.4.23) is given as

$$\frac{d\Gamma}{dt} = V_1 + V_2 - V_3 - V_4 \tag{80}$$

in which rates V_1, V_2, V_3, and V_4 can be expanded linearly as Taylor series in applied potential and excess cation concentration. Therefore,

$$V_i = V_{io} + \left(\frac{\partial V_i}{\partial E}\right)_\Gamma \Delta E e^{j\omega t} + \left(\frac{\partial V_i}{\partial \Gamma}\right)_E \Delta\Gamma e^{j\omega t}, \quad i = 1-4 \tag{81}$$

Equations (80) and (81) may be combined to yield

$$\Delta\Gamma = \frac{\left\{\left(\frac{\partial V_i}{\partial E}\right)_\Gamma + \left(\frac{\partial V_2}{\partial E}\right)_\Gamma - \left(\frac{\partial V_3}{\partial E}\right)_\Gamma - \left(\frac{\partial V_3}{\partial E}\right)_\Gamma\right\}\Delta E}{j\omega - \left(\frac{\partial V_1}{\partial \Gamma}\right)_E - \left(\frac{\partial V_2}{\partial \Gamma}\right)_E + \left(\frac{\partial V_3}{\partial \Gamma}\right)_E + \left(\frac{\partial V_4}{\partial \Gamma}\right)_E} \tag{82}$$

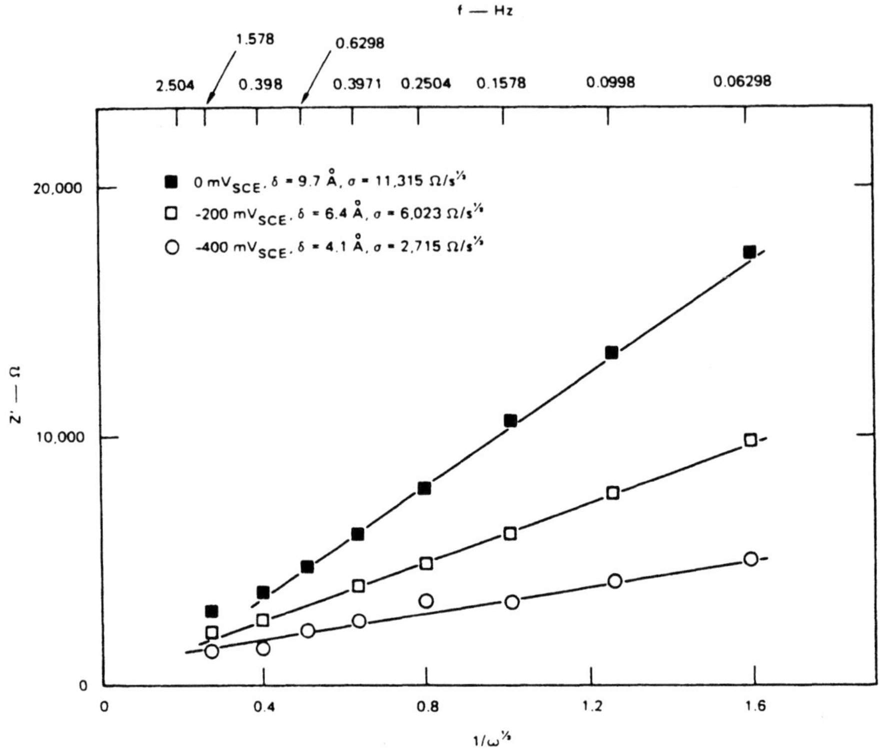

Figure 4.4.22. Dependence of Z' on $\omega^{-1/2}$ for Type 304SS passivated in 0.1 N Na_2HPO_4 solution (pH 9.1). (From C.-Y. Chao, L. F. Lin, and D. D. Macdonald, A Point Defect Model for Anodic Passive Films; III. Impedance Response, *J. Electrochem Soc.* **129**, 1874–1879, [1982]. Reprinted by permission of the publisher, The Electrochemical Society, Inc.)

If the number of electrons that flow through an external circuit upon the transfer of one species in reaction i (Figure 4.4.23) is n_i, then the faradic admittance of the film–solution interface is

$$Y_{f/s} = \frac{n_1FV_1 + n_2FV_2 + n_3FV_3 + n_4FV_4}{\Delta E \exp(j\omega t)} \tag{83}$$

which upon substitution of Eq. (81) becomes

$$Y_{f/s} = \sum_{i=1}^{4} n_i F\left(\frac{\partial V_i}{\partial E}\right)_\Gamma + \Delta\Gamma \sum_{i=1}^{4} n_i F\left(\frac{\partial V_i}{\partial \Gamma}\right)_E \tag{84}$$

where $\Delta\Gamma$ is given by Eq. (82). According to Armstrong and Edmondson [1973], it is convenient to define infinite-frequency charge-transfer resistances as

$$1/R_{\infty 1} = n_1 F\left(\frac{\partial V_1}{\partial E}\right)_\Gamma + n_2 F\left(\frac{\partial V_2}{\partial E}\right)_\Gamma \tag{85}$$

$$1/R_{\infty 2} = n_3 F\left(\frac{\partial V_3}{\partial E}\right)_\Gamma + n_4 F\left(\frac{\partial V_3}{\partial E}\right)_\Gamma \tag{86}$$

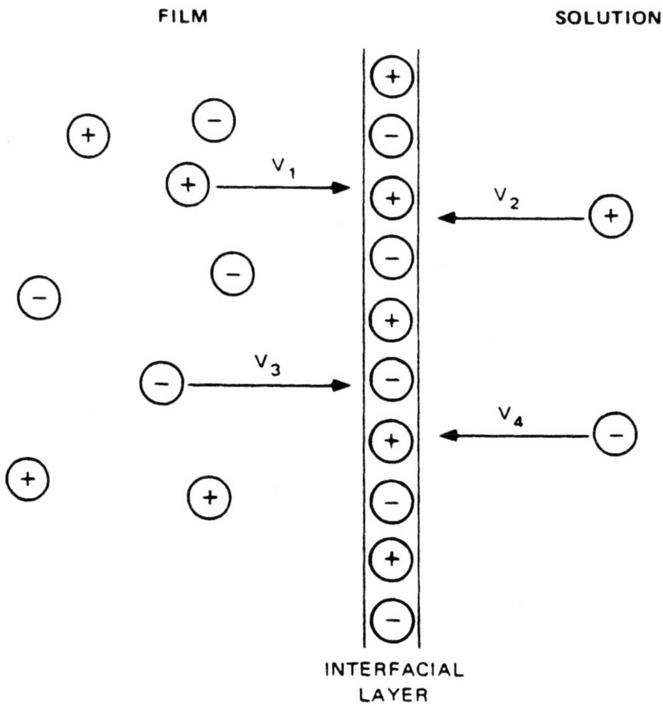

Figure 4.4.23. Model of film–solution interface according to Armstrong and Edmondson [1973]. (Reprinted with permission from R. D. Armstrong and K. Edmondson, The Impedance of Metals in the Passive and Transpassive Regions, *Electrochim. Acta* **18**, 937–943, [1973]. Copyright 1973 Pergamon Journals Ltd.)

and resistances at zero frequency as

$$1/R_{01} = \left[\left(\frac{\partial V_1}{\partial E}\right)_\Gamma + \left(\frac{\partial V_2}{\partial E}\right)_\Gamma - \left(\frac{\partial V_3}{\partial E}\right)_\Gamma - \left(\frac{\partial V_4}{\partial E}\right)_\Gamma\right] \cdot n_1 F\left(\frac{\partial V_1}{\partial \Gamma}\right)_E + n_2 F\left(\frac{\partial V_2}{\partial \Gamma}\right)_E \Big/ k \tag{87}$$

$$1/R_{02} = \left[\left(\frac{\partial V_1}{\partial E}\right)_\Gamma + \left(\frac{\partial V_2}{\partial E}\right)_\Gamma - \left(\frac{\partial V_3}{\partial E}\right)_\Gamma - \left(\frac{\partial V_4}{\partial E}\right)_\Gamma\right] \cdot n_3 F\left(\frac{\partial V_3}{\partial \Gamma}\right)_E + n_4 F\left(\frac{\partial V_4}{\partial \Gamma}\right)_E \Big/ k \tag{88}$$

where k defines the relaxation time τ as

$$\tau = 1/k \tag{89}$$

with

$$k = \left(\frac{\partial V_3}{\partial \Gamma}\right)_E + \left(\frac{\partial V_4}{\partial \Gamma}\right)_E - \left(\frac{\partial V_1}{\partial \Gamma}\right)_E - \left(\frac{\partial V_2}{\partial \Gamma}\right)_E \tag{90}$$

Accordingly, the faradic admittance becomes

$$Y_{f/s} = \frac{1}{R_{\infty 1}} + \frac{1}{R_{\infty 2}} + \frac{k}{k + j\omega} \cdot \left(\frac{1}{R_{01}} + \frac{1}{R_{02}} \right) \tag{91}$$

However, Armstrong and Edmondson [1973] claim that, in most cases, oxygen will be in equilibrium between the interface and the solution, a condition that can only be satisfied if $R_{\infty 2} = -R_{02}$. Thus, Eq. (91) becomes

$$Y_{f/s} = \frac{1}{R_{\infty 1}} + \frac{k}{k + j\omega} \cdot \frac{1}{R_{01}} + \frac{1}{R_{\infty 2}} \cdot \frac{j\omega}{k + j\omega} \tag{92}$$

and the total interfacial impedance is written as

$$Z_{f/s} = \frac{1}{Y_{f/s} + j\omega C_{\infty}} \tag{93}$$

where C_{∞} is the double-layer capacitance. The complex-plane impedance loci that can be generated by Eq. (93) according to the relative values of k and ω have been explored by Armstrong and Edmondson [1973], and their results are summarized below.

1. Large k ($k >> \omega$). In this case, Eq. (92) reduces to

$$Y_{f/s} = \frac{1}{R_{\infty 1}} + \frac{1}{R_{01}} + \frac{j\omega}{kR_{\infty 2}} \tag{94}$$

   and the impedance locus takes the form of a single semicircle in the complex plane (Figure 4.4.24) resulting from a resistance $R_{\infty 1}R_{01}/(R_{\infty 1} + R_{01})$ in parallel with the capacitance $C_{\infty} + kR_{\infty 2}$.

2. Small k ($k << \omega$), in which case Eq. (92) becomes

$$Y_{f/s} = \frac{1}{R_{\infty 1}} + \frac{1}{R_{\infty 1}} - \frac{jk}{\omega R_{01}} \tag{95}$$

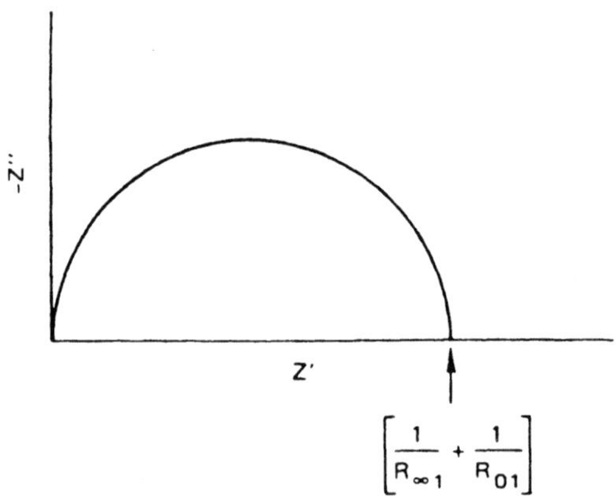

Figure 4.4.24. Film-solution interface according to Eq. (89) for large k. (Reprinted with permission from R. D. Armstrong and K. Edmondson, The Impedance of Metals in the Passive and Transpassive Regions, *Electrochim. Acta* **18**, 937–943, [1973]. Copyright 1973 Pergamon Journals Ltd.)

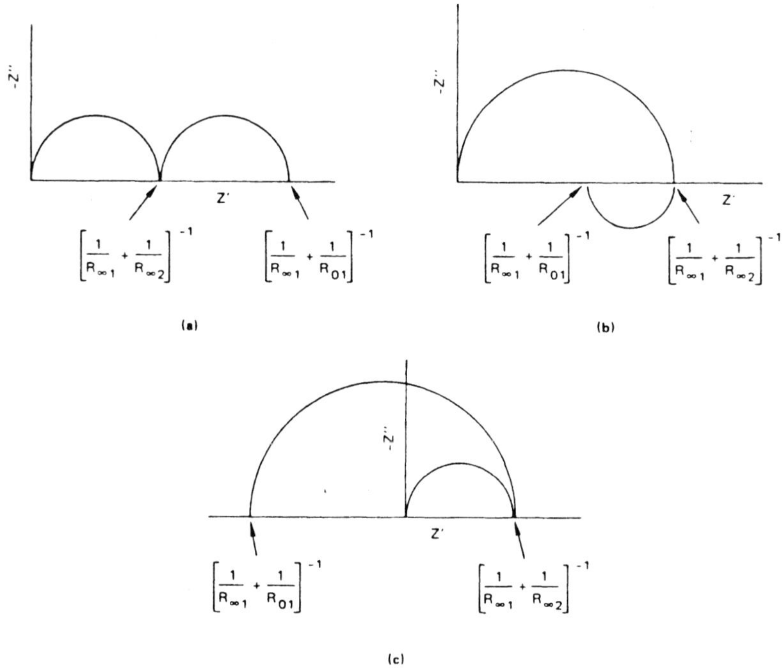

Figure 4.4.25. Complex plane impedance diagrams for film–solution interface according to Eq. (93) for small k. (After Armstrong and Edmondson [1973])

indicating that two semicircles will appear in the complex plane according to the relative values of A and B:

$$A = \frac{R_{\infty 1} + R_{01}}{R_{\infty 1} + R_{01}} \tag{96}$$

$$B = \frac{R_{\infty 1} + R_{\infty 2}}{R_{\infty 1} + R_{\infty 2}} \tag{97}$$

Thus, for $A > B > 0$ and $B > A > 0$, the complex plane impedance loci shown in Figures 4.4.25a and b are obtained, whereas for $B > 0 > A$ that shown in Figure 4.4.25c results. The latter case occurs because R_{01} may be positive or negative, depending upon the relative values of the differentials contained in Eq. (87). Provided that R_{01} is negative, but that $|R_{\infty 1}| > |R_{01}|$, second-quadrant behavior at low frequencies is predicted, terminating in a negative resistance at $\omega \to 0$. The low-frequency inductive response predicted in the second case ($B > A > 0$) is also of considerable practical and theoretical interest, because fourth-quadrant behavior is frequently observed experimentally.

At this point it is worthwhile pausing to consider the properties of the total impedance of the interphase system consisting of the metal–film interface, the film, and the film-solution interface. According to Eq. (54) and subsequent expressions for $Z_{m/f}$, Z_f, and $Z_{f/s}$, the total impedance becomes

$$Z_T = R_e + W + \frac{X}{X^2+Y^2} - j\left(W + \frac{Y}{X^2+Y^2}\right) \tag{98}$$

$$W = \frac{\sigma_M \sigma_0 \omega^{-1/2}}{\sigma_M + \sigma_0} \tag{99}$$

$$X = \frac{1}{R_{\infty 1}} + \frac{1}{k^2+\omega^2}\left(\frac{\omega^2}{R_{\infty 2}} + \frac{k^2}{R_{01}}\right) \tag{100}$$

$$Y = \frac{\omega k}{k^2+\omega^2}\left(\frac{1}{R_{\infty 2}} - \frac{1}{R_{01}}\right) + \omega C_\infty \tag{101}$$

These rather complicated expressions predict that a variety of impedance loci in the complex plane might be observed, depending upon the frequency range employed and the relative values of the parameters contained in Eqs (98)–(101).

By way of illustration, we calculate complex impedance diagrams for the case of a passive film in which only anion vacancies are mobile and for which k [Eq. (90)] is large. Thus, for $k >> \omega$, $\sigma_M >> \sigma_0$, and assuming that no redox reactions occur at the film—solution interface, then the total impedance becomes

$$Z_t = \left(\frac{a}{a^2+\omega^2 b^2} + \sigma_0 \omega^{-1/2}\right) - j\left(\frac{\omega b}{a^2+\omega^2 b^2} + \sigma_0 \omega^{-1/2}\right) \tag{102}$$

where

$$a = \frac{1}{R_{\infty 1}} + \frac{1}{R_{01}} \tag{103}$$

$$b = C_\infty + \frac{1}{kR_{\infty 2}} \tag{104}$$

Impedance diagrams for a passive film, computed using Eq. (102), are displayed in Figure 4.4.26. These diagrams show a partially resolved semicircle at high frequencies and a low-frequency Warburg response. These same general features are exhibited by the impedance data for passive Ni(III) in phosphate buffer solution, as shown in Figure 4.4.19. The calculated impedance spectra (Figure 4.4.26) show that as the kinetics of the interfacial ion exchange processes become slower (increasing R_{01}), the impedance becomes increasingly dominated by the nondiffusional component. In the limit of sufficiently slow interfacial reactions but fast transport of vacancies across the film, the impedance locus takes the form of a semicircle, which is similar to that expected for a purely capacitive (dielectric) response.

A major shortcoming of the theory developed above is that the various processes (e.g. ion exchange and vacancy transport) are assumed to be uncoupled. More realistically, the kinetic expression given by Eq. (80), for example, should be employed as a boundary condition for the solution of Eq. (61). In this way, coupling, which is expected to become most apparent at intermediate frequencies where the vacancy transport and ion exchange processes are of comparable importance, may be a significant factor in determining the impedance spectrum. Such coupling is included in the approach described in Section 2.2.3.3.

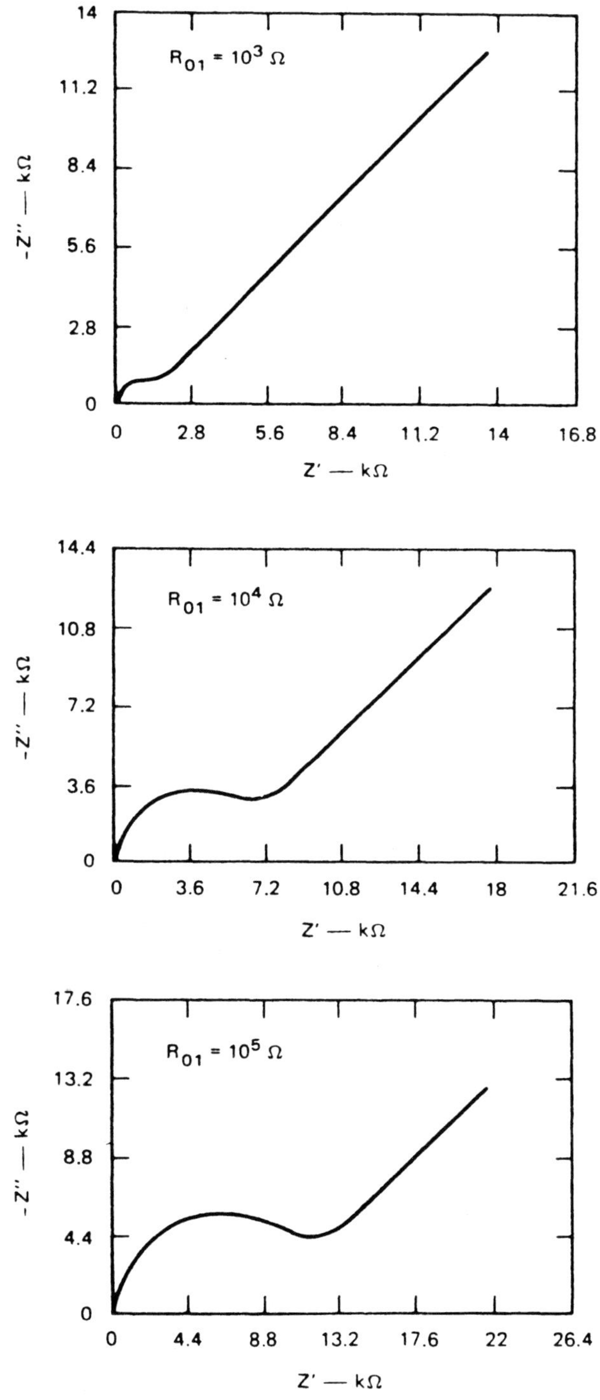

Figure 4.4.26. Theoretical complex plane impedance diagrams for a passive film according to Eq. (102). $C_\infty = 5 \times 10^{-5}$ F, $R_{\infty 1} = 10^4\ \Omega$, $R_{\infty 2} = 1\ \Omega$, $\sigma_0 = 10^3\ \Omega/s^{1/2}$.

4.4.7 Point Defect Model of the Passive State

Digby D. Macdonald

4.4.7.1 Introduction

Passivity is the phenomenon whereby a thermodynamically reactive metal (i.e. one whose Gibbs' energy of reaction with components of the environment, such as oxygen and/or water, is negative) attains kinetic stability because of the formation of a reaction product phase on the surface. The phase may be two-dimensional or three-dimensional, or may even be submono-layer if the adsorbed species block reactive sites. From the time that passivity was first described by Schonbien and Faraday in the 1830s (D. Macdonald [1992], Uhlig [1978]) this phenomenon has been of great theoretical and practical interest, the latter because passivity is the enabler of our metals-based civilization (D. Macdonald [1999]).

In order to illustrate the reactivity of the common metals in contact with terrestrial environments, Table 4.4.1 lists the Gibbs' energy changes for the reaction of various metals with oxygen and water vapor under prototypical terrestrial conditions [T = 25°C, $p_{O_2} = 0.21$ atm, $p_{H_2O} = 0.02532$ atm (RH = 80%), $p_{H_2} = 6.156 \times 10^{-42}$ atm], where the partial pressure of hydrogen has been calculated from the equilibrium $H_2O(l) = H_2(g) + \frac{1}{2}O_2(g)$ for the prevailing conditions. The reader will note that all of these metals are used in our current, metals-based civilization, either in pure form or as components of alloys.

Table 4.4.1. Gibbs Energies of Reaction of Metals with Oxygen and Water Vapor under Prototypical Terrestrial Conditions [T = 25°C, $p_{O_2} = 0.21$ atm, $p_{H_2O} = 0.02532$ atm (RH = 80%), $p_{H_2} = 6.156 \times 10^{-42}$ atm]

Metal	Reaction	ΔG, M/O_2 (kJ/mol)	ΔG, M/H_2O (kJ/mol)
Fe	$3Fe + 2O_2 = Fe_3O_4$	−1007.5	
	$3Fe + 4H_2O = Fe_3O_4 + 4H_2$		−1005.3
Cr	$2Cr + 1.5O_2 = Cr_2O_3$	−1047.3	
	$2Cr + 3H_2O = Cr_2O_3 + 3H_2$		−1045.7
Ni	$Ni + 0.5O_2 = NiO$	−209.7	
	$Ni + H_2O = NiO + H_2$		−209.1
Cu	$2Cu + 0.5O_2 = Cu_2O$	−145.9	
	$2Cu + H_2O = Cu_2O + H_2$		−145.4
Al	$2Al + 1.5O_2 = Al_2O_3$	−1576.5	
	$2Al + 3H_2O = Al_2O_3 + 3H_2$		−1574.9
Ti	$Ti + O_2 = TiO_2$	−885.6	
	$Ti + 4H_2O = TiO_2 + 4H_2$		−884.5
Zr	$Zr + O_2 = ZrO_2$	−1038.6	
	$Zr + 4H_2O = ZrO_2 + 4H_2$		−1037.5
Pt	$Pt + 0.5O_2 = PtO$	−43.0	
	$Pt + H_2O = PtO + H_2$		−42.5
Au	$2Au + 1.5O_2 = Au_2O_3$	83.7	
	$2Au + 3H_2O = Au_2O_3 + 3H_2$		85.3

Examination of the data summarized in Table 4.4.1 shows that the thermodynamic driving force for the formation of the lowest common oxide under the prevailing conditions is essentially the same, whether the metal reacts with gaseous oxygen or gaseous water (water vapor). (This is a remarkable finding, as it shows that corrosion would occur with roughly the same thermodynamic driving force in an anoxic world, provided that water is present). Furthermore, all of the common reactive metals have large negative values for the change in Gibbs' energy for the reactions under prototypical terrestrial conditions, demonstrating that the formation of the oxides from the metals are thermodynamically spontaneous processes. Even platinum, which is commonly regarded as being "inert", spontaneously forms the oxide PtO under these conditions. On the other hand, the change in Gibbs' energy for the formation of Au_2O_3 on gold is positive, indicating that the oxide will not form spontaneously on the metal. However, in this case, we cannot exclude the formation of a two-dimensional "AuO" phase on the surface (i.e. chemisorbed oxygen), as phases of this type generally form at a lower oxygen fugacity (and hence at a lower potential) than does the lowest bulk (three-dimensional) oxide.

Passivity is normally manifest as a sharp drop in the anodic current density to a constant value at a critical potential that is commonly referred to as the Flade potential (Figure 4.4.27). For many metals and alloys, the current density drops by three or more orders in magnitude leading to a corrosion rate in the passive state

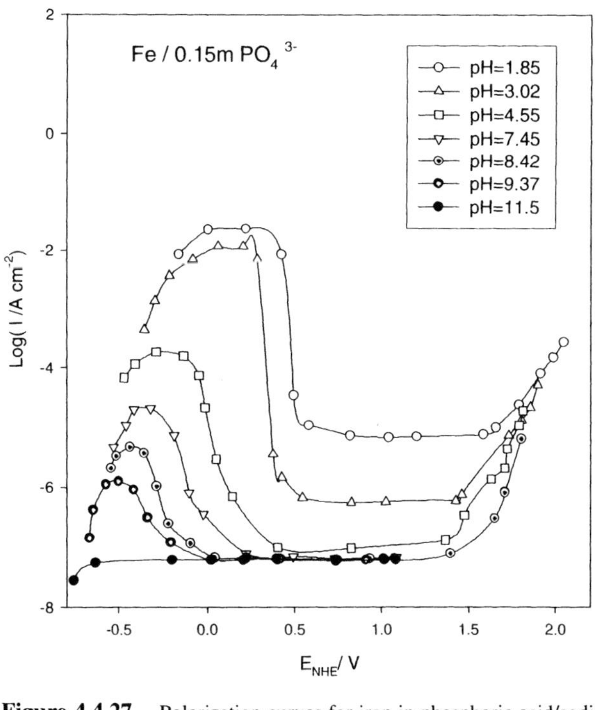

Figure 4.4.27. Polarization curves for iron in phosphoric acid/sodium hydroxide buffer solutions at 25°C as a function of pH (data taken from Sato and Cohen [1964]).

that is lower than that in the active state (at potentials more negative than the Flade potential) by the corresponding factor. Thus, a generally acceptable corrosion rate for components in industrial systems is 10 μm/year (0.01 mm/year); if that were increased by a factor of 1000 to 1 cm/year (10,000 μm/year), the use of metals in our metals-based civilization would be impractical. It is for this simple reason that passivity has been termed the "enabler of our metals based civilization" (D. Macdonald [1999]). Of course, in most practical situations, the overall corrosion reaction becomes cathodically controlled by mass transport of the cathodic depolarizer to the metal surface well before the higher rate is attained. Nevertheless, passivity is the single most important factor in enabling the use of reactive metals in terrestrial environments.

At higher potentials, passivity is observed to break down and the dissolution rate of the metal increases dramatically. This process is commonly due to the oxidative ejection of cations from the barrier layer; for example,

$$Cr_{Cr} + 4H_2O \rightarrow CrO_4^{2-} + V_{Cr}^{3'} + 8H^+ + 3e^- \tag{105}$$

and dissolution of the film

$$Cr_2O_3 + 5H_2O \rightarrow 2CrO_4^{2-} + 10H^+ + 6e^- \tag{106}$$

(D. Macdonald [1999]). Note that the Kroger–Vink notation is used in Eq. (105) to describe the various species in the system, with Cr_{Cr} and $V_{Cr}^{3'}$ indicating a Cr(III) cation in a normal site on the cation sub-lattice and a vacancy on the same sub-lattice, respectively. Reaction (105) leads to enhanced cation transmission through the barrier layer, and hence to a higher current density, while Reaction (106) results in a thinner barrier layer and also to a higher current (D. Macdonald [1999]). Of great significance is the fact that, in the case of both reactions, the oxidation sate of the chromium cation in the film increases by three, so that the rates of the reactions are highly potential dependent, as observed. Transpassive dissolution is observed in a great number of metal and metal alloy systems (Bojinov *et al.* [2002], Betova *et al.* [2002], Betova *et al.* [2004]); the critical requirement appears to be that an increase in the oxidation state of the principal cation within the barrier layer of the passive film (e.g. Cr_{Cr} in the Cr_2O_3 barrier layer on chromium-containing alloys) occurs as the potential is increased in the positive direction (D. Macdonald [1999]).

At still higher potentials, water decomposes oxidatively to produce oxygen. The kinetics of the oxygen evolution reaction have been extensively studied (Hoare [1968], Adzic *et al.* [1996]), and a discussion of this subject is well beyond the scope of this article. It is sufficient to state that oxygen evolution marks the upper extreme of the polarization range in aqueous systems, but it is also important to note that, in the case of valve metals (Ti, Nb, Zr, Al, etc), having rectifying passive films, oxygen evolution may not become a significant factor until very high voltages are applied (Young [1961]).

Examination of the findings of a great number of studies of the phenomenon of passivity leads to the following generalizations (D. Macdonald [1999] and references therein):

1. Passive films form as bilayer (Figure 4.4.28), or even multilayer structures, consisting of a defective oxide adjacent to the metal (the barrier layer) that forms by growth into the substrate and an outer layer that forms from the reaction of metal cations that are ejected from the barrier layer with species in the solution (including the solvent) or by the hydrolytic restructuring of the barrier layer at the barrier layer/outer layer (bl/ol) interface. Solution phase species are often incorporated into the outer layer, but not into the inner layer, whereas alloying elements from the substrate alloy may be incorporated into both layers (D. Macdonald [1999]).
2. For systems where the outer layer does not form, or where the outer layer presents little impediment to the transport of species to or from the barrier layer/outer layer (bl/ol) interface, the specific impedance in the absence of redox couples is very high (~10^6–$10^7\,\Omega\,\text{cm}^2$ for NiO on Ni, for example) but in the presence of redox couple [e.g. $Fe(CN)_6^{3-/4-}$] the impedance is often low. This demonstrates that the barrier layers may be good electronic conductors but, generally, are poor ionic conductors.
3. Given sufficient time, steady states are observed in the barrier layer thickness and in the current. While this point is still somewhat controversial, steady states have been demonstrated unequivocally for Ni, Zn, and W, amongst other metals.
4. The steady-state thickness of the barrier layer (and also possibly of the outer layer) varies linearly with the applied voltage. The logarithm of the current is also found to vary linearly with voltage, with a positive, finite slope being predicted, and observed, if the barrier layer is a cation conductor (e.g. NiO/Ni)). On the other hand, $ln(I_{SS})$ vs. V has a zero slope if the barrier layer is an interstitial cation conductor (ZnO/Zn) or an anion conductor (WO_3/W), but only if no change in oxidation state occurs upon the ejection of a cation from, or dissolution of, the barrier layer.
5. Metals possessing low oxidation states in the barrier layer tend to form cation-conducting (e.g. NiO/Ni) or cation interstitial-conducting (ZnO/Zn) barrier layers, while metals possessing high oxidation states form anion-conducting barrier layers (e.g. WO_3/W and ZrO_2/Zr). This generalization correlates with the relative energies of formation of the defects within the barrier layer.

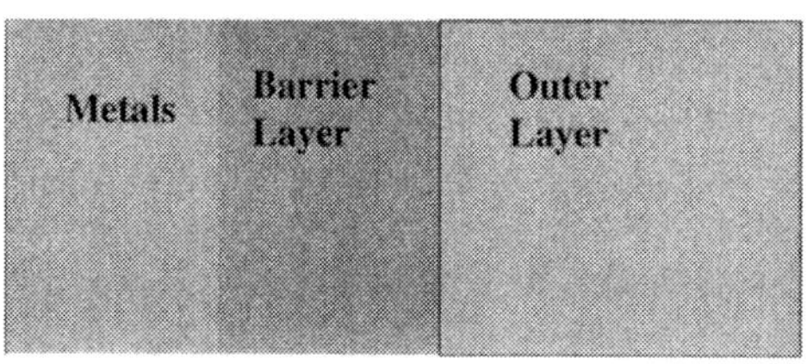

Figure 4.4.28. Schematic of the structure of a passive film on a metal (gray), comprising a defective barrier layer (orange) and a porous outer layer (green).

6. Cation vacancy conducting barrier layers tend to be p-type in their electronic character [e.g. NiO/Ni], while those that are cation interstitial conductors (e.g. ZnO/Zn) or anion conductors (W/WO_3) are n-type. This correlation corresponds to the well-known acceptor and donor properties of these defects.

While these six generalizations are not all encompassing, in that exceptions may exist, they are sufficient to differentiate between various theories that have been proposed for the growth of barrier oxide layers on metals and alloys. A number of models that have been developed to describe the growth of anodic oxide films on metals are listed in Table 4.4.2, together with some of their important features and predictions. Of the models listed, which were chosen because they make analytical predictions that can be tested and because they introduced new concepts into the theory of passivity, only the point defect model (PDM) in its latest form (D. Macdonald [1999], Pensado-Rodriguez *et al.* [1999a,b]) accounts for all of the observations summarized above.

4.4.7.2 Point Defect Model

The point defect model (PDM) was originally developed by Chao, Lin, and Macdonald (Chao *et al.* [1981], D. Macdonald [1999] and citations therein see Section 4.4.6.3), to provide an atomic scale description of the growth and breakdown of passive films on metal and alloy surfaces. Since its initial inception in the early 1980s, the PDM has undergone extensive modification, primarily through recognizing the importance of barrier layer dissolution and whether the defect generation and annihilation reactions at the interfaces, and the barrier layer dissolution reaction, are lattice conservative or nonconservative processes. The latest version of the model is based upon the following postulates.

The barrier layer is a highly defective oxide (or hydride, in the case of hydride forming metals—see below), in which point defects (cation and anion vacancies and cation interstitials) are generated and annihilated at the metal/barrier layer (m/bl) and the barrier layer/outer layer (bl/ol) interfaces, such that cation vacancies flow from the bl/ol interface to the m/bl interface, while metal interstitials and anion vacancies (oxygen or hydride) flow in the reverse direction (Chao *et al.* [1981]). On the other hand, the outer layer forms via the hydrolysis of cations ejected from the barrier layer (interstitials and/or cations from the cation sublattice) and subsequent precipitation of more-or-less stoichiometric phase (e.g. oxide, hydroxide, oxyhydroxides, carbonates, chlorides, and so forth) or by hydrolytic restructuring of the barrier layer itself (D. Macdonald [1999]). This picture is consistent with the general observations that minor alloying elements in the substrate metal are detected in both the barrier layer and the outer layer, but that species in the solution are generally detected only in the outer layer. This observation also suggests that solution-phase species do not penetrate the barrier layer.

The electric field strength within the barrier layer is buffered by internal (Esaki) band-to-band tunneling, such that the field strength is independent, or at least is only weakly dependent, upon the applied voltage (Chao *et al.* [1981], D. Macdonald [1999]). For mathematical convenience, it is assumed that the potential varies

Table 4.4.2. Comparison of Distinct Passive Film Growth Models

Model	Applicability	Field Strength	Rate Control	Film Dissolution	Interfacial Pot. Diff.	Electronic Structure
Verwey [1935]	Cation conductors	Unconstrained	Cation motion in oxide	No	No	Not addressed
Cabrera-Mott [1948]	Cation conductors	Unconstrained	Cation injection from metal into film	No	No	Not addressed
Vetter and Gorn [1973]	Iron	Unconstrained	Cation ejection	Partly	Yes	Not addressed
Sato and Cohen [1964]	Cation/anion conductors	Unconstrained	(i) Cooperative cation/ anion exchange [ ] (ii) Injection of O^{2-} into the film	No	No	Not addressed
Kirchheim [1987]	Iron	Unconstrained	Unclear	Yes	Yes	Not addressed
Chao *et al*. [1981]	Cation/anion conductors	Constrained by Esaki tunneling	Cation injection from metal or anion vacancy generation at the m/bl interface	No	Yes	Not addressed
D. Macdonald [1999]	Cation/interstitial/ anion conductors	Constrained by Esaki tunneling	Cation injection from metal or anion vacancy generation at the m/bl interface	Yes	Yes	Highly doped (degenerate), defect semiconductor
Pensado-Rodriguez, *et al*. [1999a,b]	Cation/ interstitial/anion conductors; hydride barrier layer on lithium	Constrained by Esaki tunneling	Cation injection from metal or anion vacancy generation at the m/bl interface	Yes	Yes	Not addressed

Metal | Barrier Oxide Layer | Solution

$$(1)\ m + V_M^{\chi'} \xleftrightarrow{k_1, k_{-1}} M_M + v_m + \chi e' \qquad (4)\ M_M \xleftrightarrow{k_4, k_{-4}} M^{\Gamma+} + V_M^{\chi'} + (\Gamma - \chi)e^-$$

$$(2)\ m \xleftrightarrow{k_2, k_{-2}} M_i^{\chi+} + v_m + \chi e^- \qquad (5)\ M_i^{\chi+} \xleftrightarrow{k_5, k_{-5}} M^{\Gamma+} + (\Gamma - \chi)e^-$$

$$(3)\ m \xleftrightarrow{k_3, k_{-3}} M_M + \frac{\chi}{2} V_{\ddot{O}} + \chi e^- \qquad (6)\ V_{\ddot{O}} + H_2O \xleftrightarrow{k_6, k_{-6}} O_O + 2H^+$$

$$(7)\ MO_{\chi/2} + \chi H^+ \xrightarrow{k_7} M^{\Gamma+} + \frac{\chi}{2} H_2O + (\Gamma - \chi)e^-$$

$(x = L_{ss})$ $\qquad$ $(x = 0)$

Figure 4.4.29. Interfacial defect generation/annihilation reactions that occur in the growth of anodic barrier oxide films according to the Point Defect Model (D. Macdonald [1999]). m = metal atom, $V_M^{\chi'}$ = cation vacancy on the metal sublattice of the barrier layer, $M_i^{\chi+}$ = interstitial cation, M_M = metal cation on the metal sublattice of the barrier layer, $V_{\ddot{O}}$ = oxygen vacancy on the oxygen sublattice of the barrier layer, O_O = oxygen anion on the oxygen sublattice of the barrier layer, $M^{\Gamma+}$ = metal cation in solution.

linearly with distance through the barrier layer. The potential drop across the bl/ol interface depends linearly on the applied voltage and pH (Chao *et al.* [1981]).

The defect generation and annihilation reactions that are envisioned to occur at the m/bl and bl/ol interfaces are depicted in Figure 4.4.29. Briefly, the barrier layer is viewed as being a highly defective, defect semiconductor, in which the defects (cation and anion vacancies and cation interstitials) act as the electronic dopants (D. Macdonald [1999]). The vacancies are generated and annihilated at the interfaces, which are normally separated by no more than a few nanometers. Under anodic polarization conditions, there exists a net flow of oxygen vacancies and cation interstitials from the m/bl interface (at $x = L$) to the bl/ol interface (at $x = 0$) and a net flow of cation vacancies in the reverse direction. Of the utmost importance in developing an understanding of the growth of the barrier layer, is the differentiation of the interfacial reactions as to whether they are lattice conservative or nonconservative processes (D. Macdonald [1999]). Thus, recognizing that the minimal unit of the barrier layer is $M_M(O_O)_{\chi/2}$ (i.e. $MO_{\chi/2}$), $M_M(V_O)_{\chi/2}$, $V_M(O_O)_{\chi/2}$, or even $V_M(V_O)_{\chi/2}$, Reactions (1), (2), (4), (5), and (6) in Figure 4.4.29 are lattice conservative, because their occurrence does not result in the movement of the boundary with respect to the laboratory frame of reference (i.e. does not result in the formation or destruction of the minimal unit of the barrier layer). These variants in representing the film are possible, because vacancies can be regarded as being real species. Note that the last of these variants is simply a vacancy "complex" (e.g. a vacancy pair), which is a commonly-invoked entity in describing the defective solid sate. Returning now to Figure 4.4.29, we see that Reactions (3) and (7) generate and destroy the barrier layer, respectively, so that these processes are lattice nonconservative. The steady state exists when the rates of these two nonconservative processes are equal, leading to the following expressions for the steady-state barrier layer thickness (L_{SS}) and current density (I_{SS}) (D. Macdonald [1999]) (see below).

4.4.7.3 Electrochemical Impedance Spectroscopy

Traditionally, EIS data have been interpreted in terms of minimum phase "electrical equivalent circuits" (EECs), comprising linear elements of resistance, capacitance, and inductance. However, it must be recognized that EECs are not models (the computational arm of a physical theory), but are simply analogs that may (or may not) faithfully mimic the system response (impedance). EECs add little to our understanding of the *physicoelectrochemistry* of interfacial processes and hence they should be used with great caution if the object of a study is to obtain mechanistic information on the atomic scale. In the present case, we use a highly modified version of the original point defect model that was initially developed to describe the growth and breakdown of passive films on metal surfaces and to interpret impedance data from passive systems (Chao *et al.* [1981, 1982], D. Macdonald and Smedley [1990]) (see Section 4.4.6.3) via reaction mechanism analysis (RMA), the goal of which is to derive values for the kinetic parameters (transfer coefficients and standard rate constants) for the reactions that occur in the system, and for other model parameters, via optimization. The latest version of the PDM is shown in Figure 4.4.29 (D. Macdonald [1999], D. Macdonald *et al.* [2004]) and, as noted above, differs substantially from the original model of Chao *et al.* [1981] in that kinetic effects at the interfaces are taken into account, the interfacial defect generation/annihilation reactions are classified according to whether they are lattice conservative or non conservative, dissolution of the barrier layer is recognized, and the formation of an outer, precipitated layer is incorporated in some of the analyses that have been performed (e.g. Pensado-Rodriguez *et al.* [1999a,b, 2001]). A major accomplishment of the later version of the PDM is that it accounts for steady states in the barrier layer thickness and the passive current (D. Macdonald [1999]) and for potentiostatic transients (D. Macdonald, Al-Rafaie, and Engelhardt [2001] D. D. Macdonald, M. Al-Rafie, and G. R. Engelhardt, "New Rate Law for the Growth and Reduction of Passive Films", J. Electrochem, Soc., *148*(9), BZ43 [2001]). Where comparisons have been made, the PDM is found to account quantitatively for those properties.

Variants of the PDM have been devised, most notably the "mixed conductor model" (MCM) that has been extensively developed in recent years by Bojinov and coworkers (Bojinov *et al.* [2000], Bojinov *et al.* [2002], Betova [2004], Bojinov *et al.* [2003], Betova *et al.* [2002]). The essence of the MCM is that conduction in a passive film (barrier layer) is due to both the electronic defects (electrons and electron holes) and crystallographic defects. The reader will note, however, that mixed conduction was part of the PDM from its inception in the early 1980s (Chao *et al.* [1981, 1982]). The MCM employs essentially the same reaction set as the PDM and uses the Kroger–Vink notation to represent the defects. However, the MCM employs the generalized transport equations of Fromhold and Cook [1967] in describing the motion of defects across the passive film, rather than the Nernst–Planck equations used in the PDM, but it has yet to be demonstrated (to the author's knowledge) that the Fromhold and Cook equations provides a significantly better description of the physics of the system. It should be noted that the transport of vacancies, at least, can be described in terms of coupled fluxes using the methods of irreversible thermodynamics (Zhang and D. Macdonald [1998]), but, again, it is not evident that this treatment offers any significant advantage over the more elementary Nernst–Planck equations. Another variant of the PDM (Castro [1994]) employs interfacial reactions that are not elementary in nature (i.e. the reactions can be decomposed into more elementary forms) and hence must be rejected theoretically upon the basis that reaction mechanisms should comprise elementary processes. Below is described, in some detail, an EIS analysis of the passive state on Alloy-22 in terms of the PDM in an environment that is of relevance to the disposal of high level nuclear waste (HLNW) in the Yucca Mountain, Nevada repository. However, before doing so, it is convenient to first summarize the progression of the PDM over the past two decades in its application to the analysis of impedance data (Table 4.4.3).

The reader will note that there has been a steady increase in the level of sophistication of the PDM, with the introduction of barrier layer dissolution, and the relaxation therein, interstitial cations, hydride barrier layers, and bilayer structures. Current work in the author's laboratory is focused on extending the PDM to consider vacancies in the substrate metal and their annihilation on grain boundaries, precipitates, and dislocations, and the incorporation of the porous outer layer in oxide/oxide bilayer structures.

With reference to Figure 4.4.29, the change in the film thickness with time is described in terms of the two lattice nonconservative reactions, as follows (D. Macdonald [1999],

$$\frac{dL}{dt} = -\frac{2\Omega}{\chi} J_o - \Omega k_7 C^n_{H^+} \tag{107}$$

in which Ω is the mole volume per cation of the barrier layer ($MO_{\chi/2}$), χ is the oxidation state of the cation in the barrier layer, Γ is the oxidation state of the cation in the outer layer/solution, J_o is the flux for oxygen vacancies within the barrier layer from the m/bl interface, where they are generated, to the bl/ol interface where they are annihilated, C_{H^+} is the hydrogen ion concentration in the solution at the film/solution interface, n is the kinetic order of the film dissolution reaction with respect to C_{H^+}, and k_7 is the rate constant for Reaction (7), Figure 4.4.29. Tables 4.4.4 and 4.4.5 display the rate constants for the seven interfacial reactions; in these expressions, α is the polarizability of the bl/ol interface (i.e. dependence of the potential drop across the interface on the applied potential), β is the dependence of the potential drop at the bl/ol interface on pH, $\gamma = F/RT$, and α_j and k_j^0 are the transfer coefficient and the standard rate constant, respectively, for the jth reaction shown in Figure 4.4.29 (D. Macdonald [1999]).

All interfacial reactions that result in a change in oxidation state contribute to the total current density, which is written as

$$\begin{aligned} I = F\big[&\chi J_c^{m/f} + (\Gamma - \chi)J_c^{f/s} - \chi J_i^{m/f} - (\Gamma - \chi)J_i^{f/s} - 2J_o^{m/f} \\ &+ (\Gamma - \chi)k_7 C^n_{H^+}\big] \end{aligned} \tag{108}$$

Table 4.4.3. Progression in the Development of the PDM in the Analysis of the Impedance of the Passive State

Reference	Defects Considered	Character of Interfacial Reactions	Relaxations	Comments
Chao *et al.* [1982]	Stainless steel. Cation vacancies + anion vacancies	Equilibrium, no dissolution of the barrier layer	Concentrations of cation and anion vacancies	Successfully accounted for phase angle > $\pi/4$.
D. Macdonald and Smedley [1990]	Nickel. Cation vacancies + anion vacancies	Reversible reactions with kinetic effects. No dissolution of the barrier layer	Concentrations of cation and anion vacanciess	Demonstrated the importance of kinetic effects
Pensado *et al.* [2001]	Lithium. Cation and anion vacancies, LiH hydride barrier layer, LiOH outer layer	Irreversible reactions with kinetic effects, barrier layer and outer layer dissolution	Concentrations of cation and anion vacancies and barrier layer thickness	First impedance analysis of bilayer structure and of hydride barrier layer. Cathodic reaction included in the model
D. Macdonald *et al.* [2004]	Alloy-22. Cation vacancies, cation interstitials, and anion vacancies	Irreversible reactions with kinetic effects, barrier layer and outer layer dissolution	Concentrations of cation and anion vacancies, cation interstitials, and barrier layer thickness	Inclusion of interstitials in the reaction mechanism

Table 4.4.4. Rate Constants $k_i = k_i^0 e^{a_i V} e^{-b_i L}$ for the interfacial Defect Generation and Annihilation Reactions Employed in the Point Defect Model

Reaction	a_i (V^{-1})	b_i (cm^{-1})	Units of k_i^0
(1) $m + V_M^{\chi'} \xrightarrow{k_1} M_M + v_m + \chi e'$	$\alpha_1(1-\alpha)\chi\gamma$	$\alpha_1\chi K$	$\frac{1}{s}$
(2) $m \xrightarrow{k_2} M_i^{\chi+} + v_m + \chi e'$	$\alpha_2(1-\alpha)\chi\gamma$	$\alpha_2\chi K$	$\frac{mol}{cm^2 s}$
(3) $m \xrightarrow{k_3} M_M + \frac{\chi}{2} V_{\ddot{O}} + \chi e'$	$\alpha_3(1-\alpha)\chi\gamma$	$\alpha_3\chi K$	$\frac{mol}{cm^2 s}$
(4) $M_M \xrightarrow{k_4} M^{\Gamma+} + (\Gamma - \chi)e'$	$\alpha_4\alpha\Gamma\gamma$		$\frac{mol}{cm^2 s}$
(5) $M_i^{\chi+} \xrightarrow{k_5} M^{\Gamma+} + (\Gamma - \chi)e'$	$\alpha_5\alpha\Gamma\gamma$		$\frac{cm}{s}$
(6) $V_{\ddot{O}} + H_2O \xrightarrow{k_6} O_O + 2H^+$	$2\alpha_6\alpha\gamma$		$\frac{cm}{s}$
(7) $MO_{\chi/2} + \chi H^+ \xrightarrow{k_7} M^{\Gamma+} + \frac{\chi}{2} H_2O + (\Gamma - \chi)e'$	$\alpha_7\alpha(\Gamma - \chi)\gamma$		$\frac{mol^{0.4}}{cm^{0.2} s}$

Table 4.4.5. Definition of the Standard Rate Constants for the Interfacial Defect Generation and Annihilation Reactions Employed in the Point Defect Model

Reaction	k_i^0
(1) $m + V_M^{\chi'} \xrightarrow{k_1} M_M + v_m + \chi e'$	$k_1^{00} e^{-\alpha_1\beta\chi\gamma pH} e^{-\alpha_1\chi\gamma\phi_{f/s}^0}$
(2) $m \xrightarrow{k_2} M_i^{\chi+} + v_m + \chi e'$	$k_2^{00} e^{-\alpha_2\beta\chi\gamma pH} e^{-\alpha_2\chi\gamma\phi_{f/s}^0}$
(3) $m \xrightarrow{k_3} M_M + \frac{\chi}{2} V_{\ddot{O}} + \chi e'$	$k_3^{00} e^{-\alpha_3\beta\chi\gamma pH} e^{-\alpha_3\chi\gamma\phi_{f/s}^0}$
(4) $M_M \xrightarrow{k_4} M^{\Gamma+} + (\Gamma - \chi)e'$	$k_4^{00} e^{\alpha_4\beta\Gamma\gamma pH} e^{\alpha_4\Gamma\gamma\phi_{f/s}^0}$
(5) $M_i^{\chi+} \xrightarrow{k_5} M^{\Gamma+} + (\Gamma - \chi)e'$	$k_5^{00} e^{\alpha_5\beta\Gamma\gamma pH} e^{\alpha_5\Gamma\gamma\phi_{f/s}^0}$
(6) $V_{\ddot{O}} + H_2O \xrightarrow{k_6} O_O + 2H^+$	$k_6^{00} e^{\alpha_6\beta\Gamma\gamma pH} e^{\alpha_6\Gamma\gamma\phi_{f/s}^0}$
(7) $MO_{\chi/2} + \chi H^+ \xrightarrow{k_7} M^{\Gamma+} + \frac{\chi}{2} H_2O + (\Gamma - \chi)e'$	$k_7^{00} e^{\alpha_7(\Gamma-\chi)\beta\gamma pH} e^{\alpha_7(\Gamma-\chi)\gamma\phi_{f/s}^0}$

in which $F = 96487\,C/mol$ is Faraday's constant, subscripts c, i, and o designate cation vacancies, cation interstitials, and oxygen vacancies, respectively, and superscripts *m/f* and *f/s* signify the metal/barrier layer (film) and barrier layer/outer layer (solution) interfaces. Note that in the coordinate system adopted in Figure 4.4.29, the fluxes of metal interstitials and oxygen vacancies are negative, while the flux of cation vacancies is positive.

Under steady state conditions, $dL/dt = 0$ and $J_c^{m/f} = J_c^{f/s} = J_c$, $J_i^{m/f} = J_i^{f/s} = J_i$, and $J_o^{m/f} = J_o^{f/s} = J_o = (\chi/2)k_7C_{H^+}^n$. These relationships yield the steady state thickness of the barrier layer and the passive current density as

$$L_{ss} = \frac{1}{\varepsilon}\left[1-\alpha-\frac{\alpha\alpha_d}{\alpha_3}\left(\frac{\Gamma}{\chi}-1\right)\right]V + \frac{1}{\varepsilon}\left\{\frac{2.303n}{\alpha_3\chi\gamma}-\beta\left[\frac{\alpha_d}{\alpha_3}\left(\frac{\Gamma}{\chi}-1\right)+1\right]\right\}pH+\frac{1}{\alpha_3\chi K}ln\left(\frac{k_3^0}{k_7^0}\right) \tag{109}$$

and

$$I_{ss} = \Gamma F\left[k_2^0 e^{\alpha_2\alpha\gamma V}e^{-\alpha_2\alpha\varepsilon\gamma L_{ss}} + k_4^0 e^{\alpha_4\alpha\gamma V}e^{\alpha\beta\gamma pH} + k_7^0 e^{\alpha_7\alpha(\Gamma-\chi)\gamma V}e^{\beta\alpha(\Gamma-\chi)\gamma pH}\cdot C_{H^+}^n\right] \tag{110}$$

respectively. For those cases where no change in oxidation state occurs upon ejection of a cation from the barrier layer ($\Gamma = \chi$), Eq. (109) reduces to the simpler form

$$L_{ss} = \left(\frac{1-\alpha}{\varepsilon}\right)V+\frac{1}{\varepsilon}\left[\frac{2.303n}{\alpha_3\chi\gamma}-\beta\right]pH+\frac{1}{\alpha_3\chi K}ln\left(\frac{k_3^0}{k_7^0}\right) \tag{111}$$

where the rate constants are defined in Tables 4.4.4 and 4.4.5.

The dependencies of the rate constants on voltage and barrier layer thickness were derived using the partial charge development principle (D. Macdonald [1986]), which allows the exponents a_i and b_i to be expressed in terms of the extent of charge development, α_i, in the transition state, compared with that in the initial state. Thus, if α_i is close to zero, the transition state resembles the initial state in its charge distribution, while if α_i is close to one the charge distribution resembles that of the final state shown in Table 4.4.4. By way of illustration, the first reaction in Figure 4.4.29 is written as

$$m+V_M^{\chi'} \rightarrow \left[(1-\alpha_1)m+(1-\alpha_1)V_M^{\chi'}+\alpha_1 M_M+\alpha_1 v_m+\alpha_1\chi e'\right]^{\neq} \rightarrow M_M+v_m+\chi e'$$

where the entity in square brackets is the transition state (activated complex). Writing down the change in the electrochemical potential, $\tilde{\mu} = \mu^0 + RTln(a) + zF\phi$, where μ^0 is the standard chemical potential, a is the activity, z is the species charge number, and ϕ is the electrostatic potential, leads to the first entry in Table 4.4.4.

Equations (109) and (111) account for the ubiquitous observation that the thickness of the passive film (barrier layer) varies linearly with the applied voltage. This relationship is so reliable experimentally that the anodizing community quotes film thickness in terms of the anodizing voltage. The anodizing constant is equal to $(1-\alpha)/\varepsilon$ and noting that $0 < \alpha < 1$ and that $1 \times 106 < \varepsilon < 5 \times 106$ we find that the anodizing constant is 2.5 nm/V, by choosing $\alpha = 0.5$ and $\varepsilon = 2 \times 106\,V/cm$. This value is in good agreement with that determined experimentally for many metals (2.0–2.6 nm/V), mostly the valve metals (Young [1961]).

The fluxes of the defects in the barrier layer are written as

$$J_c = -D_c \frac{\partial C_c}{\partial x} + \chi K D_c C_c \tag{112}$$

$$J_i = -D_i \frac{\partial C_i}{\partial x} - \chi K D_i C_i \tag{113}$$

and

$$J_o = -D_o \frac{\partial C_o}{\partial x} - 2 K D_o C_o \tag{114}$$

for cation vacancies, metal interstitials, and oxygen vacancies, respectively, where D and C are diffusivity and concentration, $K = \varepsilon F/RT$, $R = 8.314\,\mathrm{J mol^{-1} K^{-1}}$ is the gas constant, T is the absolute temperature, and ε is the electric field strength, which is considered to be a constant according to the PDM (Chao *et al.* [1981], D. Macdonald [1999]). The continuity equations, $\partial C/\partial t = -\nabla \cdot J$, then become

$$\frac{\partial C_c}{\partial t} = D_c \frac{\partial^2 C_c}{\partial x^2} - \chi D_c K \frac{\partial C_c}{\partial x} \tag{115}$$

$$\frac{\partial C_i}{\partial t} = D_i \frac{\partial^2 C_i}{\partial x^2} + \chi D_i K \frac{\partial C_i}{\partial x} \tag{116}$$

and

$$\frac{\partial C_o}{\partial t} = D_i \frac{\partial^2 C_o}{\partial x^2} + 2 D_o K \frac{\partial C_o}{\partial x} \tag{117}$$

The initial conditions ($t = 0$) are $C_c(x) = C_c^0(x)$, $C_i(x) = C_i^0(x)$, and $C_o(x) = C_o^0(x)$. The boundary conditions are readily stated as: at $x = 0$ (the film/solution interface): $J_c(0) = k_4^0 e^{a_4 V}$ (k_4^0 has units of mol/cm^2.s), $J_i(0) = -k_4^0 e^{a_4 V} C_i(0)$ (k_4^0 is in cm/s); $J_o(0) = -(\chi/2) k_7^0 e^{a_7 V} C_o(0)$ (k_7^0 is in cm/s); at $x = L$ (the metal/film interface): $J_c(L) = k_1^0 e^{a_1 V} e^{-b_1 L} C_c(L)$ (k_1^0 in cm/s), $J_i(L) = -k_2^0 e^{a_2 V} e^{-b_2 L}$ (k_2^0 is in mol/s/cm^2); $J_o(L) = -(\chi/2) k_3^0 e^{a_3 V} e^{-b_3 L}$ (k_3^0 is in mol/s/cm^2).

Writing again Eqs (107) and (108) as

$$\frac{dL}{dt} = -\Omega k_3^0 e^{a_3 V} e^{-b_3 L} - \Omega k_7^0 e^{a_7 V} C_{H^+}^n \tag{118}$$

and

$$\begin{aligned} I = F\big[&\chi k_1^0 e^{a_1 V} e^{-b_1 L} C_c(L) + \chi k_2^0 e^{a_2 V} e^{-b_2 L} + \chi k_3^0 e^{a_3 V} e^{-b_3 L} + (\Gamma - \chi) k_4^0 e^{a_4 V}\big] \\ &+ (\Gamma - \chi) k_5^0 e^{a_5 V} C_i(0) + (\Gamma - \chi) k_7^0 e^{a_7 V} C_{H^+}^n \,\mathrm{J} \end{aligned} \tag{119}$$

we note that the current density is a function of V, L, $C_i(0)$, and $C_c(L)$. Accordingly, for any arbitrary changes δV, δL, $\delta C_i(0)$, and $\delta C_c(0)$

$$\begin{aligned} \delta I = &\left(\frac{\partial I}{\partial V}\right)_{L, C_i(0), C_c(L)} \delta V + \left(\frac{\partial I}{\partial L}\right)_{V, C_i(0), C_c(L)} \delta L \\ &+ \left(\frac{\partial I}{\partial C_i(0)}\right)_{V, L, C_c(L)} \delta C_i(0) + \left(\frac{\partial I}{\partial C_c(L)}\right)_{V, L, C_i(L)} \delta C_c(L) \end{aligned} \tag{120}$$

In EIS, the variations are sinusoidal in form, so that $\delta V = \delta V e^{j\omega t}$; $\delta L = \delta L e^{j\omega t}$; $\delta C_i(0) = \Delta C_i(0)e^{j\omega t}$, and $\delta C_c(L) = \Delta C_c(L)e^{j\omega t}$, where ΔX is the amplitude of X at the frequency $\omega = 0$.

From Eq. (16), the Faradic admittance is defined as

$$Y_f = \frac{\delta I}{\delta V} = I^V + I^L \frac{\Delta L}{\Delta V} + I^{C_i(0)} \frac{\Delta C_i(0)}{\Delta V} + I^{C_c(L)} \frac{\Delta C_c(L)}{\Delta V} \tag{121}$$

where, $I^V = \left(\frac{\partial I}{\partial V}\right)_{L,C_i(0),C_c(L)}$; $I^L = \left(\frac{\partial I}{\partial L}\right)_{V,C_i(0),C_c(L)}$; $I^{C_i(0)} = \left(\frac{\partial I}{\partial C_i(0)}\right)_{L,V,C_c(L)}$, and

$I^{C_c(L)} = \left(\frac{\partial I}{\partial C_c(L)}\right)_{L,V,C_i(0)}$. These differentials are readily evaluated from Eq. (119).

We now return to Eq. (118) and determine the response of $\frac{dL}{dt}$ to δV, δL, $\delta C_i(0)$, and $\delta C_c(L)$ by taking the total differential. Thus, the relaxation in film thickness is given as:

$$\frac{d}{dt}(\delta L) = \Omega k_3^0 a_3 e^{a_3 V} e^{-b_3 L} \delta V - \Omega k_3^0 a_3 e^{a_3 V} e^{-b_3 L} \delta L - \Omega k_7^0 a_7 e^{a_7 V} C_{H^+}^n \delta V \tag{122}$$

Noting that $\frac{d}{dt}(\delta L) = j\omega \Delta L e^{j\omega t}$, it follows that

$$j\omega \Delta L e^{j\omega t} = \Omega\left(k_3^0 a_3 e^{a_3 V} e^{-b_3 L} - k_7^0 a_7 e^{a_7 V} C_{H^+}^n\right)\Delta V e^{j\omega t} - \Omega k_3^0 b_3 e^{a_3 V} e^{-b_3 L} \Delta L e^{j\omega t} \tag{123}$$

and hence

$$\frac{\Delta L}{\Delta V} = \frac{\Omega\left(k_3^0 a_3 e^{a_3 V} e^{-b_3 L} - k_7^0 a_7 e^{a_7 V} C_{H^+}^n\right)}{\Omega k_3^0 b_3 e^{a_3 V} e^{-b_3 L} + j\omega} \quad \text{or} \quad \frac{\Delta L}{\Delta V} = \frac{\Phi_3}{1 + j\omega\tau_3} \tag{124}$$

Where

$$\Phi_3 = \left(\frac{a_3}{b_3}\right) - \left(\frac{k_7^0}{k_3^0}\right)\left(\frac{a_7}{b_3}\right) e^{(a_7 - a_3)V_e b3 L C_{H^+}^n} \tag{125}$$

and

$$\tau_3 = \frac{1}{\Omega k_3^0 b_3 e^{a_3 V} e^{-b_3 L}} \tag{126}$$

For the relaxation of cation vacancies at the metal/film interface, we write the rate of annihilation of the vacancies at the m/bl interface as $dC_c(L)/dt = -k_1 C_c(L) = -k_1^0 e^{a_1 V} e^{-b_1 L} C_c(L)$. Thus, taking the total differential yields

$$\frac{d}{dt}[\delta C_c(L)] = -k_1^0 e^{a_1 V} e^{-b_1 L}[a_1 C_c(L)\delta V - b_1 C_c(L)\delta L + \delta C_c(L)] \tag{127}$$

Substitution of $\delta X = \Delta X e^{j\omega t}$, where $X \equiv V, L, C_c(L)$, therefore gives

$$\frac{\Delta C_c}{\Delta V} = \frac{b_1 C_c(L)}{1 + j\omega\tau_1} \frac{\Delta L}{\Delta V} - \frac{a_1 C_c(L)}{1 + j\omega\tau_1} \tag{128}$$

and, upon substitution of Eq. (20), we obtain

$$\frac{\Delta C_c}{\Delta V} = \frac{b_1 C_c(L)}{(1+j\omega\tau_1)} \frac{\Phi_3}{(1+j\omega\tau_3)} - \frac{a_1 C_c(L)}{1+j\omega\tau_1} \tag{129}$$

where $\tau_1 = 1/k_1^0 e^{a_1 V} e^{-b_1 L}$.

For metal interstitials, the relaxation for $C_i(0)$ is determined through the mass balance at the film/solution interface $dC_i(0)/dt = -k_5 C_i(0) = -k_5^0 e^{a_5 V} C_i(0)$ (the units of k_5^0 are s^{-1}). Thus, the total differential of the rate of change of the concentration of metal interstitials at the barrier layer/outer layer (bl/ol) interface is $d[\delta C_i(0)]/dt = -k_5^0 [a_5 e^{a_5 V} C_i(0)\delta V + e^{a_5 V}\ \delta C_i(0)]$. Substituting the sinusoidal variations for δV and $\delta C_i(0)$, yields

$$j\omega \Delta C_i(0) e^{j\omega t} = -k_5^0 a_5 e^{a_5 V} C_i(0) \Delta V e^{j\omega t} - k_5^0 e^{a_5 V} \Delta C_i(0) e^{j\omega t} \tag{130}$$

and hence

$$\Delta C_i(0)\left(j\omega + k_5^0 e^{a_5 V}\right) = -k_5^0 a_5 e^{a_5 V} C_i(0) \Delta V \tag{131}$$

Therefore,

$$\frac{\Delta C_i(0)}{\Delta V} = \frac{-k_5^0 a_5 e^{a_5 V} C_i(0)}{j\omega + k_5^0 e^{a_5 V}} = \frac{a_5 c_i(0)}{1+j\omega\tau_3} \tag{132}$$

in which

$$\tau_3 = \frac{1}{k_5^0 e^{a_5 V}} \tag{133}$$

The expression for the faradic admittance is finally derived from Eq. (121) as

$$\begin{aligned} Y_f = I^V + I^L \frac{\Phi_3}{1+j\omega\tau_3} + I^{C_c(L)} \left(\frac{b_1 C_c(L)}{(1+j\omega\tau_1)} \frac{\Phi_3}{(1+j\omega\tau_3)} - \frac{a_1 C_c(L)}{1+j\omega\tau_1} \right) \\ - I^{C_i(0)} \frac{a_3 C_i(0)}{1+j\omega\tau_3} + j\omega C \end{aligned} \tag{134}$$

in which the parallel geometric capacitance $C_g = \hat{\varepsilon}\varepsilon_0/L$, where $\hat{\varepsilon}$ is the dielectric constant and $\varepsilon_0 = 8.85 \times 10^{-14}$ *F/cm* $\varepsilon_0 = 8.85 \times 10^{-14}$ F/cm is the vacuum permittivity. The second, third, and fourth terms on the right side of Eq. (134) correspond to relaxations in the barrier layer thickness, the concentration of cation vacancies at the metal/barrier layer interface, and the concentration of metal interstitials at the barrier layer/solution interface, respectively. Note that we ignore any relaxation that might occur in the concentration of hydrogen ion at the barrier layer/solution interface. Thus, this derivation assumes that the solution is pH buffered with a buffer having infinite buffer capacity.

At this point, it is necessary to comment on the forms of the equations for the relaxations in the defect concentrations. Prior work (Chao *et al.* [1982], D. Macdonald and Smedley [1990], Pensado-Rodriguez *et al.* [2001]) solved the complex form of the Nernst–Planck equation to obtain the amplitudes of the variations in the concentrations at the interfaces. This required the equating of rates and fluxes at the interfaces, resulting in terms containing $\sqrt{\omega}$ and $\sqrt{D}$ in the expression for the interfacial impedance, where ω is the angular frequency and D is the diffusivity of the defect. The present analysis uses the rate expressions at the interfaces directly, but because the rates are equal to the species fluxes then the two approaches presumably contain the same information. Importantly, the approach used here is capable of simulating pseudo-Warburg behavior in which the phase angle is greater than $\pi/4$, and is considerably simpler mathematically.

The differentials in Eq. (134) are defined as

$$I^V = \left(\frac{\partial I}{\partial V}\right)_{L,C_i(0),C_c(L)}$$
$$= F\left[a_1\chi k_1^0 e^{a_1V}e^{-b_1L}C_c(L) + a_2\chi k_2^0 e^{a_2V}e^{-b_2L} + a_3\chi k_3^0 e^{a_3V}e^{-b_3L} + a_4(\Gamma-\chi)k_4^0 e^{a_4V} + a_5(\Gamma-\chi)k_5^0 e^{a_5V}C_i(0) + a_7(\Gamma-\chi)k_7^0 e^{a_7V}C^n_{H^+}\right] \tag{135}$$

$$I^L = \left(\frac{\partial I}{\partial L}\right)_{V,C_i(0),C_c(L)} = -\chi F\left[b_1k_1^0 e^{a_1V}e^{-b_1L}C_M(L) + b_2k_2^0 e^{a_2V}e^{-b_2L} + b_3k_3^0 e^{a_3V}e^{-b_3L}\right] \tag{136}$$

$$I^{C_c(L)} = \left(\frac{\partial I}{\partial C_c(L)}\right)_{V,L,C_i(0)} = \chi F k_1^0 e^{a_1V}e^{-b_1L}$$
$$I^{C_i(0)} = \left(\frac{\partial I}{\partial C_i(0)}\right)_{V,L,C_c(L)} = (\Gamma-\chi)Fk_5^0 e^{a_5V} \tag{137}$$

In the above expressions, the parameters that appear on the right side [L, $C_c(L)$, and $C_i(0)$] are identified with the steady-state quantities. At steady state

$$\frac{d^2C_i}{dx^2} - \chi K\frac{dC_i}{dx} = 0 \tag{138}$$

$$\frac{d^2C_i}{dx^2} - \chi K\frac{dC_i}{dx} = 0 \tag{139}$$

and

$$\frac{d^2C_o}{dx^2} + 2K\frac{dC_o}{dx} = 0 \tag{140}$$

the solutions of which are

$$C_c(x) = A_c e^{\chi KX} + B_c \tag{141}$$

$$C_i(x) = A_i e^{-\chi KX} + B_i \tag{142}$$

and

$$C_o(x) = A_o e^{-2KX} + B_o \tag{143}$$

The expressions for the fluxes therefore become

$$J_c = \chi KD_cB_c \tag{144}$$

$$J_i = -\chi KD_iB_i \tag{145}$$

and

$$J_o = -2KD_o B_o \tag{146}$$

From the boundary conditions, the coefficients are obtained as

$$\begin{aligned} B_c &= k_4^0 e^{a_4 V} / \chi K D_c \\ B_i &= k_2^0 e^{a_2 V} e^{-b_2 L} / \chi K D_i \end{aligned} \tag{147}$$

and

$$B_o = \left(\frac{\chi}{2}\right) k_2^0 e^{a_2 V} e^{-b_2 L} \Big/ 2KD_o \tag{148}$$

Because the fluxes for a given species at two interfaces are equal at steady state: $k_4^0 e^{a_4 V} = -k_1^0 e^{a_1 V} e^{-b_1 L} C_c(L)$, $k_5^0 e^{a_5 V} C_i(0) = k_2^0 e^{a_2 V} e^{-b_4 L}$, and $k_6^0 e^{a_6 V} C_o(0) = k_3^0 e^{a_3 V} e^{-b_3 L}$, we obtain

$$C_c(L) = \left(\frac{k_4^0}{k_1^0}\right) e^{(a_4 - a_1)V} e^{-b_1 L} \tag{149}$$

$$C_i(0) = \left(\frac{k_2^0}{k_5^0}\right) e^{(a_2 - a_5)V} e^{-b_2 L} \tag{150}$$

and

$$C_o(0) = \left(\frac{k_3^0}{k_6^0}\right) e^{(a_3 - a_6)V} e^{-b_3 L} \tag{151}$$

The rate constants, k_1^0, k_2^0, k_3^0, $k_3'^0$, k_4^0, k_5^0, in the above equations and in Table 4.4.5 are functions of pH. The PDM postulates that the potential drop across the film/solution interface is $\phi_{f/s} = \phi_f - \phi_s = \alpha V + \beta pH + \phi_{f/s}^0$ and that across the metal/film interface is therefore $\phi_{m/f} = \phi_m - \phi_s = (1-\alpha)V - \varepsilon L - \beta pH - \phi_{f/s}^0$, in which $\alpha = d\phi_{f/s}/dV$ is the dependence of the potential drop across the film/solution interface on V (i.e. the polarizability of the bl/ol interface), $\beta = d\phi_{f/s}/dpH$ is the dependence of the potential drop across the film/solution interface on pH, $\phi_{f/s}^0$ is the value of $\phi_{f/s}$ in the standard state ($V = 0$, pH = 0); $\gamma = F/RT$, and $K = \varepsilon\gamma = \varepsilon F/RT$.

The application of reaction mechanism analysis (RMA) is illustrated with data for Alloy-22 in saturated NaCl solutions at elevated temperature (80°C). Alloy-22 has been selected as the material for constructing canisters for the disposal of high level nuclear waste (HLNW) in the Yucca Mountain repository that is now undergoing site licensing evaluation. Measured data for the steady state current density as a function of voltage for Alloy-22 (nominally 22Cr13Mo3.6Fe2.8W1.0Co0.1VbalNi) in 6.2 m NaCl (sat) + 0.001 M HCl (pH = 3) at 80°C is shown in Figure 4.4.30 (D. Macdonald *et al.* [2004]). The data were measured in the ascending and descending voltage directions, with the lack of hysteresis demonstrating that the system was at steady state.

The current/voltage regime can be divided into two regions. At voltages lower than ca. 0.7 V_{she}, the current is independent of voltage, corresponding to the passive state. At voltages greater than 0.7 V_{she}, the current increases exponentially with voltage [$\log(I_{ss})$ versus V is linear]; that is, the polarization behavior is Tafelinian in

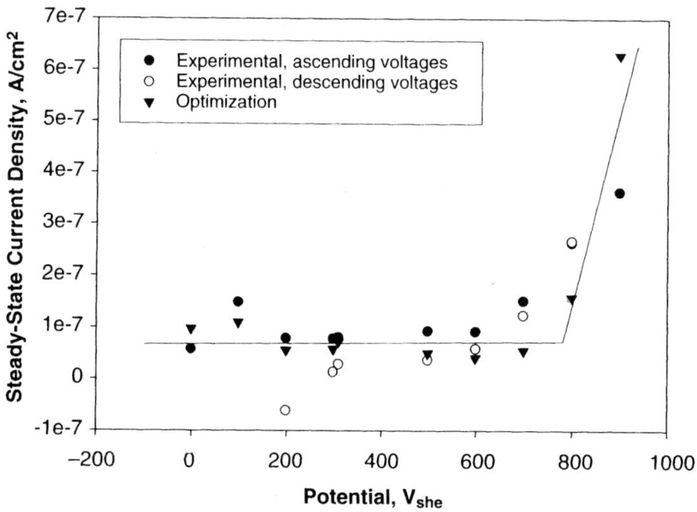

Figure 4.4.30. Steady state current density as a function of voltage for Alloy-22 in 6.2 m NaCl (sat) + 0.001 m HCl (pH = 3) at 80°C. (After D. Macdonald *et al.* [2004]).

form. Also shown in this plot are data calculated from the PDM using parameters determined from optimization of the model on experimental impedance data (see below). The excellent agreement between the calculated and experimental steady-state polarization behavior attests to the validity of the model.

The barrier layer of the passive film on Alloy-22 is known to be defective chromic oxide and Mott–Schottky analyses have demonstrated that it is n-type in electronic character (D. Macdonald *et al.* [2004]). Accordingly, the principal defects are oxygen vacancies or cation interstitials or both. The reader will note that, in the passive range (0–700 mV$_{she}$, Figure 4.4.30), the barrier layer dissolves to form Cr^{3+}, and hence no change in oxidation state occurs (i.e. $\Gamma = \chi$). Thus, the last term on the right side of Eq. (108) is voltage independent. Because the film is n-type in electronic character, we may neglect any contribution to the current from cation vacancies (which would dope the barrier layer p-type) and hence the current density reduces to

$$I_{ss} = \Gamma F \left\lfloor k_2^0 e^{\alpha_2 \alpha \gamma V} e^{-\alpha_2 \alpha \varepsilon \gamma L_{ss}} + k_7^0 C_{H^+}^n \right\rfloor \tag{152}$$

The quantity $e^{\alpha_2 \alpha \gamma V} e^{-\alpha_2 \alpha \varepsilon \gamma L_{ss}}$ is obtained by substituting Eq. (111) into Eq. (152), noting that there is no change in oxidation state upon ejection of a cation from the barrier layer or upon dissolution of the barrier layer itself. The result is a current density that is independent of voltage, as observed experimentally. Because of space limitations, the transpassive behavior will not be dealt with here. It is sufficient to note that substitution of Eq. (109) for L_{ss}, for the case where a change in oxidation state occurs upon ejection of a cation from the barrier layer or dissolution of the film, leads to a Tafel relationship, assuming one dominant defect, again in agreement with experiment.

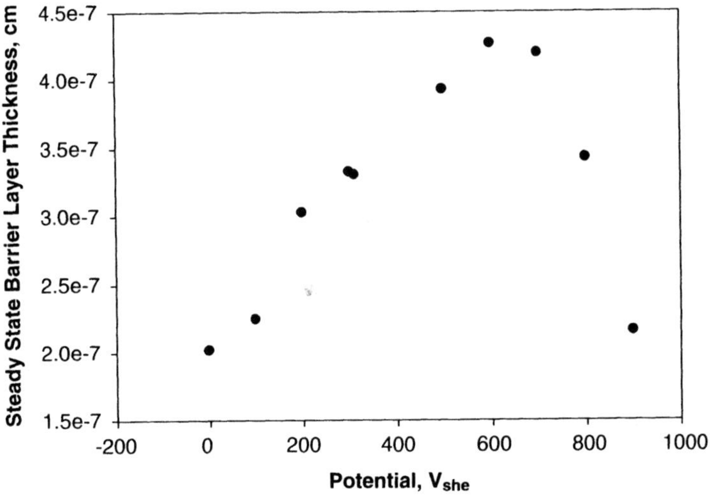

Figure 4.4.31. Steady state thickness of the barrier layer as a function of voltage for Alloy-22 in 6.2 m NaCl (sat) + 0.001 m HCl (pH = 3) at 80°C. (After D. Macdonald *et al.* [2004]).

The barrier layer thickness may be estimated from the measured capacitance at high frequency (e.g. at 5 kHz) using the "parallel plate" formula, $L_{ss} = \hat{\varepsilon}\varepsilon_0/C$, with a reasonable estimate for the dielectric constant (in this case, $\varepsilon_0 = 30$). The thickness data so calculated are summarized in Figure 4.4.31. As seen, the film thickness in the passive state ($V < 0.7\ V_{she}$) is found to increase linearly with voltage, in accordance with Eq. (111), but in the transpassive state the thickness is found to decrease with increasing voltage.

According to the point defect model (D. Macdonald [1999]), the steady state thickness of the barrier oxide layer is given by Eq. (109), which is reproduced here as

$$L_{ss} = \frac{1}{\varepsilon}\left[1 - \alpha - \frac{\alpha\alpha_5}{\alpha_2}\left(\frac{\Gamma}{\chi} - 1\right)\right]V + \frac{1}{\varepsilon}\left[\frac{2.303n}{\alpha_2\chi\gamma} - \beta - \frac{\alpha_5\beta(\Gamma - \chi)}{\alpha_2\chi}\right]pH + \frac{1}{\alpha_2\chi K}\ln\left(\frac{k_2^o}{k_5^o}\right) \tag{153}$$

where, with reference to Figure 4.4.29, Γ is the oxidation state of Cr in solution (3 or 6), χ is the oxidation state of chromium in the barrier layer (assumed to be 3), and the other parameters are as previously defined. As proposed elsewhere (D. Macdonald [1999]), Eq. (153) can be written in a compact form as

$$L_{ss} = \frac{1}{\varepsilon}\left[1 - \alpha - \frac{\alpha\alpha_5}{\alpha_2}\left(\frac{\Gamma}{\chi} - 1\right)\right](V - E_{bl}^e) \tag{154}$$

where E_{bl}^e is the equilibrium potential for the barrier layer formation reaction. In the case of Alloy-22, this reaction may be written as

$$(2+x)Cr+(3-y)H_2O=Cr_{(2+x)}O_{(3+x)}+(6-2y)H^+ +(6-2y)e^- \tag{155}$$

which accounts for the possibilities that the barrier layer is cation rich ($x > 0$, $y = 0$) or oxygen deficient ($y > 0$, $x = 0$), or both ($x > 0$, $y > 0$). Because of the lack of data on the thermodynamic properties of defective Cr(III) oxides, it is not possible to calculate E^e_{bl} in an *a priori* manner. However, in previous work (e.g. that for the formation of the passive film on Zn (D. Macdonald [1999])), E^e_{bl} has been found to equal the equilibrium potential for the formation of the stoichiometric oxide (ZnO). Inspection of Figure 4.4.31 shows that the voltage of zero thickness for the barrier layer is $-0.556\,V_{she}$ compared with $-0.789\,V_{she}$ calculated for the formation of stoichiometric Cr_2O_3 at 80°C under the prevailing conditions. This difference suggests that the defective barrier layer on Alloy-22 (see below) is considerably less stable thermodynamically than is the stoichiometric oxide. On the other hand, it may be argued that the thickness data plotted in Figure 4.4.31 contain a systematic error, because of the use of the parallel plate capacitor expression or because of the use of an inappropriate value for the dielectric constant. If this were the case, the thickness versus voltage correlation for potentials lower than $0.6\,V_{she}$ should be displaced upward by 0.37 nm, in order for the potential of zero thickness to agree with the calculated equilibrium voltage for Cr_2O_3/Cr. Indeed, as previously noted (D. Macdonald [1999]), such a procedure may be used to internally calibrate passive film thickness measurements.

Finally, for the passive range, we note that a least squares fit of L_{ss} versus voltage yields an anodizing constant of $\partial L_{ss}/\partial V = 1.6$ nm/V compared with 2.1 nm/V calculated from Eq. (154) using the values of α and ε listed in Table 4.4.6 (see below). The latter value is characteristic of passive films on a large number of metals and alloys, which yield $\partial L_{ss}/\partial V$ values ranging from 1.9 to 2.4 nm/V (Young [1961]). This would suggest that the derived L_{ss} values are only semiquantitative at best, which would not be surprising in light of the fact that they were calculated in the optimization algorithm using the parallel plate capacitor approximation. However, we note that Moffat and Latanision [1992], using *ex situ* XPS measurements of passive film thickness on chromium in 1 M H_2SO_4, found an anodizing constant of 1.5 nm/V, which is in good agreement with the value obtained in this study for passive Alloy-22.

From Eq. (154), we see that if, at some voltage, the oxidation state of the metal species being ejected from the barrier layer increases (e.g. from 3 to 6, in the case of a Cr_2O_3 barrier layer), the steady state thickness of the barrier layer is predicted to undergo a concomitant decrease with increasing voltage ($\partial L_{ss}/\partial V = -4.37$ nm/V, as is clearly shown by the data in Figure 4.4.31). This explanation of the existence of the transpassive state was presented previously (D. Macdonald [1992], D. Macdonald [1999]) and it appears to be a general phenomenon; that is, the transpassive state is induced by the oxidative emission of cations from, *and* by the oxidative dissolution of, the barrier later, with the latter resulting in a decrease in the barrier layer thickness as the voltage is increased [see Eq. (154) with $\Gamma > \chi$]. The oxidative emission of cations accounts for the change in electronic type—n-type to p-type, in the case of Alloy-22 (D. Macdonald *et al.* [2004])—in terms of a massive

Table 4.4.6. Parameter Values from Optimization of the Point Defect Model on the Experimental Impedance Data for Alloy-22 in Deaerated, Saturated NaCl (6.2 m, pH = 3) Solution at 80°C as a function of applied potential (D. Macdonald *et al.* [2004]); the parameters apply to the reduced reaction set shown in Figure 4.4.32

V vs. SHE	−0.002	0.098	0.198	0.398	0.498	0.598
R_s (ohms)	0.1	0.1	0.1	0.1	0.1	0.1
C_{dl} (F/cm^2)	3.36e–05	3.59e–05	3.11e–05	3.66e–05	4.28e–05	3.92e–05
ε (V/cm)	1.4e6	1.4e6	1.4e6	1.4e6	1.4e6	1.4e6
$\hat{\varepsilon}$	30	30	30	30	30	30
k_1^{00} (mol/cm^2.s)	1.66e–13	9.15e–14	1.22e–13	1.17e–13	1.24e–13	1.12e–13
k_2^{00} (mol/cm^2.s)	5.44e–15	5.79e–15	6.69e–15	5.07e–15	5.51e–15	4.90e–15
k_5^{00} (mol$^{0.4}$/cm$^{0.2}$.s)	1.14e–09	1.91e–09	1.29e–09	1.35e–09	1.60e–09	1.63e–09
α	0.7	0.7	0.7	0.7	0.7	0.7
α_1	0.213	0.186	0.154	0.118	0.112	0.102
α_2	0.185	0.191	0.120	0.126	0.123	0.121
n	0.6	0.6	0.6	0.6	0.6	0.6
β (V)	−0.005	−0.005	−0.005	−0.005	−0.005	−0.005
ϕ^0 (V)	−0.1	−0.1	−0.1	−0.1	−0.1	−0.1
Ω (cm^3/mol)	14.59	14.59	14.59	14.59	14.59	14.59
L_{ss} (cm)	8.64e–08	9.63e–08	1.3e–07	1.42e–07	1.68e–07	1.83e–07
I_{ss} (A/cm^2)	4.33e–08	3.36e–08	3.33e–08	5.33e–08	5.18e–08	8.82e–08

	Mean (± Standard Deviation)	Origin
R_s (uncompensated resistance)	0.1	Assumed
C_{dl} (double layer capacitance)	(3.65 ± 0.38)e–05	Second stage optimization
ε (field strength)	1.4E6	First stage optimization
$\hat{\varepsilon}$ (dielectric constant)	30	First stage optimization
k_1^{00} (rate constant)	(1.22 ± 0.22)e–13	Second stage optimization
k_2^{00} (rate constant)	(5.57 ± 0.58)e–15	Second stage optimization
k_5^{00} (rate constant)	(1.49 ± 0.26)e–09	Second stage optimization
α polarizability of bl/ol interface	0.7	First stage optimization
α_1 transfer coefficient	0.148 ± 0.041	Second stage optimization
α_2 transfer coefficient	0.144 ± 0.031	Second stage optimization
n (kinetic order of dissolution wrt [H$^+$]	0.6	First stage optimization
β dependence of pot. drop accross bl/ol inerface on pH)	−0.005	Assumed
ϕ^0 (constant)	−0.1	First stage optimization
Ω (mole volume/cation)	14.59	Calculated for Cr_2O_3

increase in the concentration of cation vacancies in the barrier layer, such that they become the dominant defect, and both cation emission and oxidative dissolution account for the sharp increase in the current as the potential is made more positive [cf. Eq. (152) with $\Gamma > \chi$]. Because the passive film on Alloy-22 is exclusively n-type in electronic character (D. Macdonald *et al.* [2004]), it was assumed that cation vacancies play no role in determining the crystallographic or electronic structure of the barrier layer, and it was also assumed that passivity was entirely a property of the barrier layer. On the basis of these observations and assumptions, the impedance data were interpreted in terms of the reduced reaction set PDM shown in Figure 4.4.32.

Typical experimental impedance spectra for the passive state on Alloy-22 are shown in Figure 4.4.33, in the form of Nyquist plots, in which the imaginary component of the impedance is plotted against the real component for decreasing frequencies. Also plotted in this figure are impedance data calculated from the model described above, after optimization of the reduced PDM (Figure 4.4.32) on the experimental data to derive values for the various model parameters (D. Macdonald *et al.* [2004]). The DataFit software (Version 7.1, http://www.curvefitting.com) was employed in this work for optimization, in order to obtain values for the transfer coefficients, α_i, and standard rate constants, k_i^{00}, for the i elementary interfacial reactions, the polarizability of the film/solution interface, α, the electric field strength (ε), and other parameters as described below. The values for the kinetic order of the dissolution reaction at the film/solution interface with respect to the [H$^+$], and for β and $\phi_{f/s}^0$ were assumed. The assumption of values for n and β is the result of the fact that we do not have any data upon which the model can be optimized with respect to pH as the independent variable.

Optimization was carried out in two stages. In the first stage, all of the parameters except R_s (resistance of the solution between the specimen and the tip of the

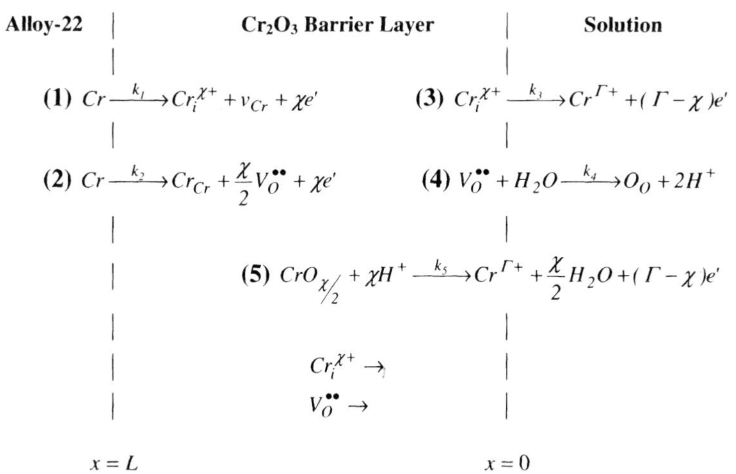

Figure 4.4.32. Reduced reaction set PDM used for interpreting the impedance of passive Alloy-22 in 6.2 m NaCl + 0.001 m HCl (pH = 3) at 80°C. The defect notation is as given in Figure 4.4.29.

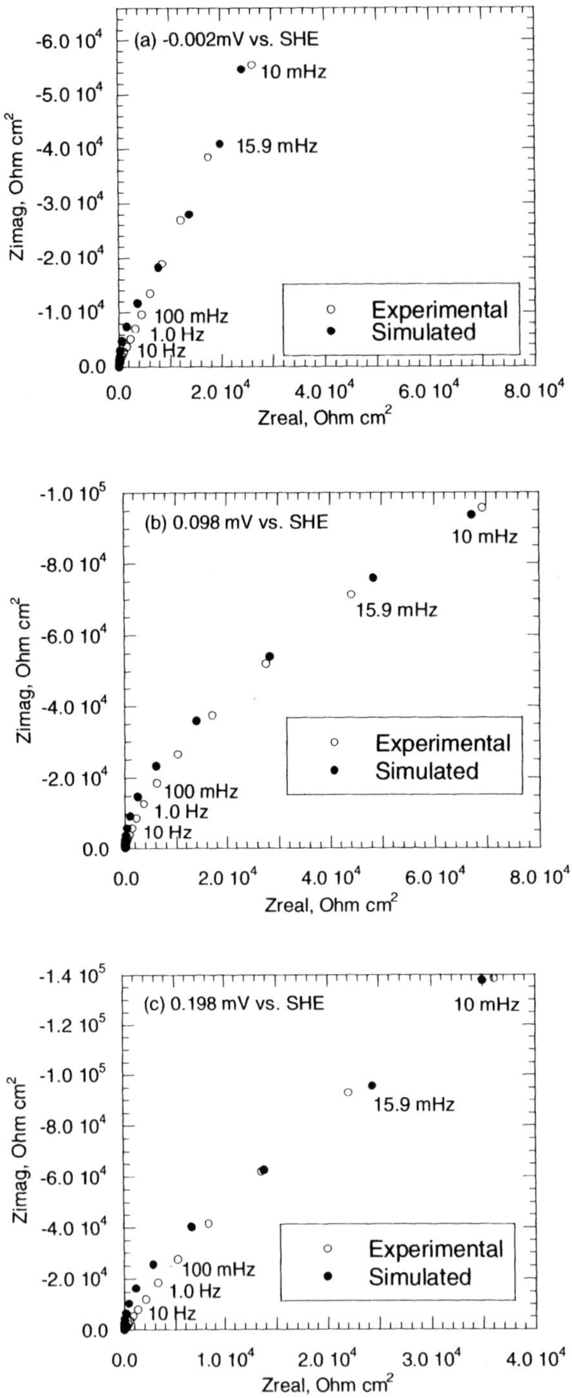

Figure 4.4.33. Nyquist impedance diagrams for Alloy-22 in deaerated, saturated NaCl (6.2 m) + 0.001 m HCl (pH = 3) solution at 80°C as a function of potential in the passive state, showing a comparison between the experimental and simulated data, with the latter being calculated from parameter values obtained by optimization of the reduced PDM (Figure 4.4.32) on the experimental data (After D. Macdonald *et al.* [2004]).

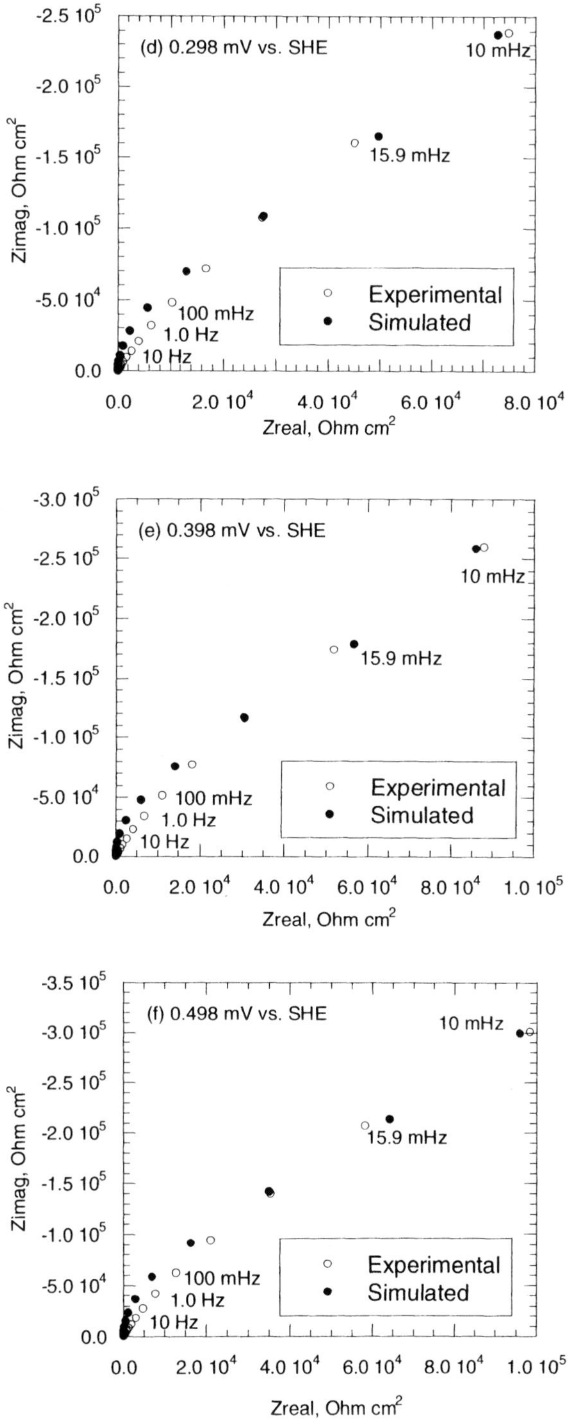

Figure 4.4.33. *Continued*

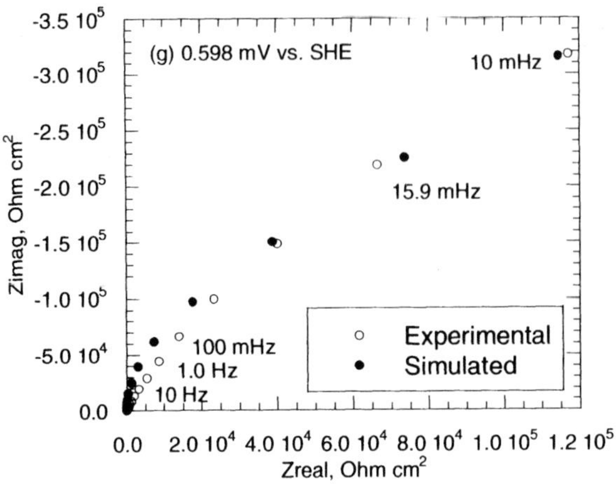

Figure 4.4.33. *Continued*

reference electrode Luggin probe), n, β, and ϕ^0_{fls} were treated as variables and the optimization procedure yielded values for them. For example, the dielectric constant was found to range from 26 to 34, without any systematic dependence on voltage, except that the two largest values of 34 were found at the highest voltages. On the other hand, Moffat and Latanision [1992], in their study of passive chromium, found that the dielectric constant of the Cr_2O_3 passive film increased systematically with increasing voltage when calculated from the XPS-derived thickness using the parallel plate capacitor expression. Likewise, the value for the electric field strength was found to vary from 0.9×10^6 V/cm to 1.9×10^6 V/cm, while that for α ranged from 0.69 to 0.71.

In order to reduce the number of variables, we averaged the values obtained in the first stage of the optimization for the dielectric constant (30), electric field strength (1.4×10^6 V/cm), and α (0.7), and treated these parameters as being constants in the final optimization stage. In performing the optimization, we also assumed that the geometric capacitance of the oxide film (C_g) and the double layer capacitance on the solution side of the film/solution interface (C_{dl}) are connected in series and are in parallel with the Faradaic impedance due to the charge transfer reactions at the interfaces. Both C_g and C_{dl} were initially treated as variables, but the choosing of a dielectric constant in the second stage of the optimization essentially fixed the geometric capacitance (C_g) to a value that is consistent with the film thickness (i.e. $C_g = \hat{\varepsilon}\varepsilon_0/L_{ss}$). Values for the fundamental model parameters for passive Alloy-22 in saturated NaCl brine at 80°C, pH = 3, are presented in Table 4.4.6 and comparisons between the experimental and simulated impedance data are displayed in Figure 4.4.33.

The correlation between experimental impedance data and simulated data is acceptable for the Nyquist plots shown, in most cases, and is similar to that reported in other studies (D. Macdonald and Smedley [1990], Pensado-Rodriguez *et al.*

Table 4.4.7. Reduced Parameter Value Set Used to Simulate the Effect of Dissolution Rate on the Impedance of Alloy-22 in 6.2 m NaCl + 0.001 m HCl (pH = 3) at 80°C

Parameter	Value	Dimensions
C_{dl} (double layer capacitance)	30	μF
R_s (uncompensated resistance)	4.0	Ohms
ε (electric field strength)	1.4e6	V/cm
$\hat{\varepsilon}$ (dielectric constant)	30	
k_1^{00}	1.0e–14	mol/cm^2.s
k_2^{00}	1.0e–14	mol/cm^2.s
α	0.70	
α_1	0.05	
α_2	0.20	
n	0.60	
β	–0.50e–3	V
ϕ_0	–0.10	V

[2001], Bojinov *et al.* [2000], Bojinov *et al.* [2002], Betova *et al.* [2004], Bojinov *et al.* [2003], Betova *et al.* [2002]). Thus, the impedance model based on the PDM provides a reasonable account of the experimental data, although other models may be equally successful. In any event, the parameter values obtained by optimization of the model on impedance data that span the entire passive range are found not to depend on voltage in any systematic manner, which, in itself, is a test of the viability of the model. The parameter values have been averaged to yield a single set for passive Alloy-22 (shown in Table 4.4.7, last column). It is these data that were used to estimate the steady-state passive current densities shown in Figure 4.4.30 (identified as "optimization").

During the analysis of the impedance data for Alloy-22, it became evident that the impedance locus depended in a very sensitive manner on the magnitude of the rate constant for the film dissolution reaction at the barrier layer/solution interface. Using parameter values that are typical for Alloy-22 (Table 4.4.7), Nyquist plots of the impedance of Alloy-22 in 6.2 m NaCl + 0.001 m HCl (pH = 3) at 80°C, together with the experimental data for a voltage of 0.398 V_{she}, are shown in Figure 4.4.33. The data are plotted with equal scales on the two axes, which is the accepted convention for Nyquist plots.

The data plotted on Figure 4.4.34 are re-plotted in Figures 4.4.35 and 4.4.36 in the Bode plane, as log(|Z|) versus log(frequency) (Figure 4.4.35) and as the phase angle versus log(frequency) (Figure 4.4.36). For low values of the standard rate constant for barrier layer dissolution, the system is almost entirely capacitive, as shown by the vertical locus in Figure 4.4.34, by the linear plot of log(|Z|) versus log(frequency) of slope equal one in Figure 4.4.35, and by the phase angle of $\pi/2$ over most of the frequency range in Figure 4.4.36. However, as the standard rate constant for barrier layer dissolution is increased, significant deviations occur from capacitive

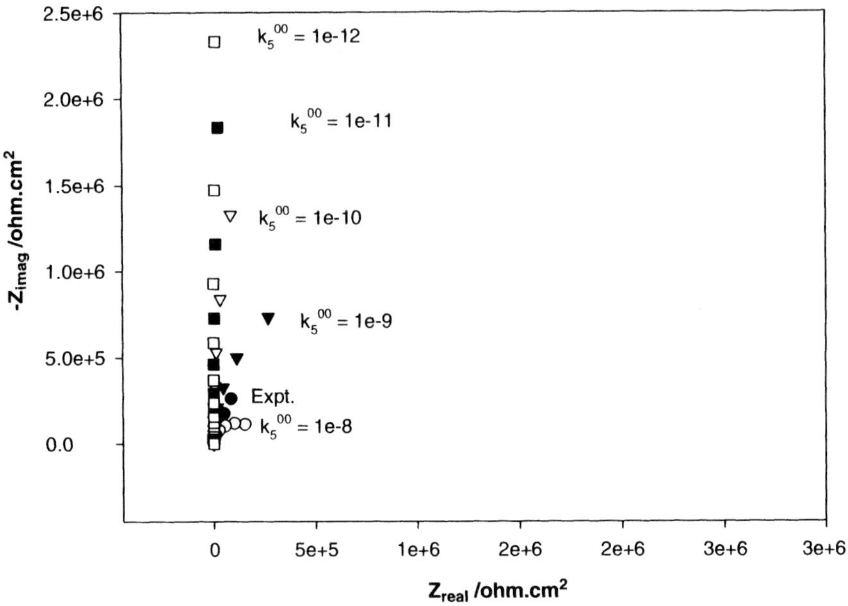

Figure 4.4.34. Bode impedance magnitude plots for Alloy-22 in 6.2 m NaCl + 0.001 m HCl (pH = 3) at 80°C as a function of the standard rate constant for the dissolution of the passive film, calculated using the data listed in Table 4.4.6. Some experimental data are included for comparison. These data gave rise to the parameter values listed in Table 4.4.7 as determined by optimization of the PDM on the experimental data. The standard rate constant, k_5^{00} has units of $mol^{0.4}/cm^{0.2}.s$, corresponding to a kinetic order for film dissolution with respect to H^+ of 0.6.

behavior, manifest in Figure 4.4.34 by the departure of the locus from the vertical, in Figure 4.4.35 by a slope of less than one, and in Figure 4.4.36 of a phase angle being less than $\pi/2$, particularly at low frequencies, where the impact of dissolution is most manifest. Importantly, as the rate constant increases, the magnitude of the impedance decreases, corresponding to a thinning of the barrier layer. These simulations attest to the importance of barrier layer dissolution in determining the impedance, and hence the corrosion resistance, of Alloy-22.

4.4.7.4 Bilayer Passive Films

As noted earlier in this chapter, passive films generally comprise bilayer structures with a defective barrier layer adjacent to the metal and a porous outer layer adjacent to the solution. An example of such a case is the passive film that forms on lithium in contact with an alkaline solution (Pensado-Rodriguez *et al.* [1999a, 1999b, 2001]. In this particular case, the authors concluded, from a thermodynamic analysis (Pensado-Rodriguez *et al.* [1999a, 1999b]), that the barrier layer was the defective hydride, $Li_{1-x}H_{1-y}$, in which the principal defects are hydrogen vacancies ($V_H^{\bullet}$) and lithium vacancies (V_{Li}'). The same thermodynamic analysis (Pensado-Rodriguez

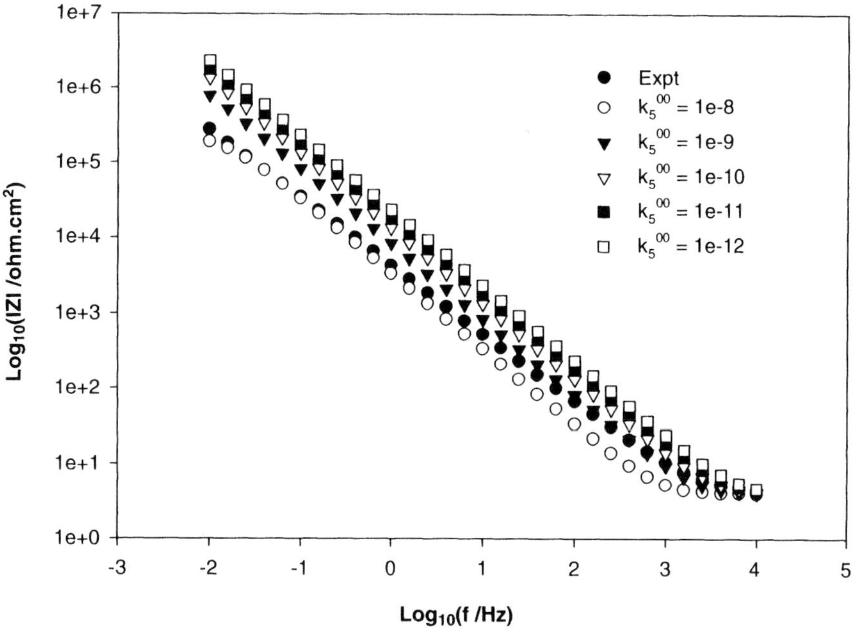

Figure 4.4.35. Bode impedance magnitude plots for Alloy-22 in 6.2 m NaCl + 0.001 m HCl (pH = 3) at 80°C as a function of the standard rate constant for the dissolution of the passive film, calculated using the data listed in Table 4.4.6. Some experimental data are included for comparison. These data gave rise to the parameter values listed in Table 4.4.7 as determined by optimization of the PDM on the experimental data. The standard rate constant, k_5^{00} has units of $mol^{0.4}/cm^{0.2}.s$, corresponding to a kinetic order for film dissolution with respect to H^+ of 0.6.

et al. [1999a]) indicated that the outer layer could only be LiOH, and, because it forms by a precipitation process, it is assumed to be nondefective but porous. The point defect model for this system is shown in Figure 4.4.37.

An important feature of this model is the inclusion of the direct cathodic evolution of hydrogen [Reaction (6), Figure 4.4.37], such that, at open circuit, the total current flowing in the system is zero. As before, the lattice conservative Reaction (1) describes the annihilation of cation vacancies at the m/bl interface, while cation vacancies are generated at the bl/ol interface by ejection of cations, as described by the lattice conservative process, Reaction (3). Thus, cation vacancies move from the bl/ol interface, where they are generated, to the m/bl interface, where they are annihilated. Likewise, hydride vacancies are generated in a lattice nonconservative process, as described by Reaction (2), and are annihilated in the lattice conservative Reaction (4). The lithium hydride barrier layer is converted into the LiOH outer layer by reaction with water, as described by the lattice nonconservative Reaction (5). This reaction also produces hydrogen. Thus, the thickness of the lithium hydride barrier layer reflects a balance between the rates of the two lattice nonconservative processes, Reactions (2) and (5), in a manner that is analogous to the oxide barrier

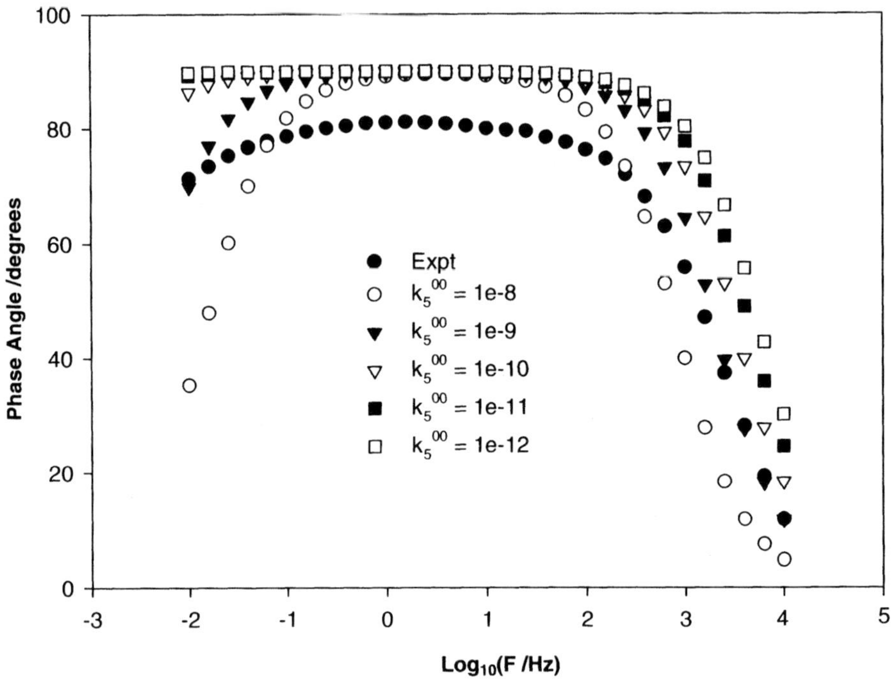

Figure 4.4.36. Bode phase angle plots of the impedance of Alloy-22 in 6.2 m NaCl + 0.001 m HCl (pH = 3) at 80°C as a function of the standard rate constant for the dissolution of the passive film, calculated using the data listed in Table 4.4.6. Some experimental data are included for comparison. These data gave rise to the parameter values listed in Table 4.4.7 as determined by optimization of the PDM on the experimental data. The standard rate constant, k_5^{00} has units of $mol^{0.4}/cm^{0.2}.s$, corresponding to a kinetic order for film dissolution with respect to H^+ of 0.6.

layer case (Figure 4.4.29). Because the outer layer is electronically an insulator, reduction of water is considered to occur at the bl/ol interface, as described by Reaction (6). Finally, the LiOH outer layer dissolves at the ol/solution interface, as described by Reaction (7) at a rate that depends upon the degree of saturation of the solution.

An important feature of this model is the invocation of the Gibbs–Thompson equation to describe the dependence of the Gibbs' energy of LiOH (s) in the pores of the outer layer on the curvature of the internal pore surface. This analysis shows that the pore radius cannot go below a critical value, because below this critical radius LiOH precipitation is no longer spontaneous. It is for this reason that the pores remain open, but, because the concentration of Li^+ in the pores is determined by the rates of Reactions (3) and (8), the critical radius is potential dependent, such that the radius decreases as the voltage is made more positive (Pensado-Rodriguez [1999]).

Using mathematical techniques similar to those described previously in this chapter and elsewhere (Chao *et al.* [1982], D. Macdonald and Smedley [1990]), the

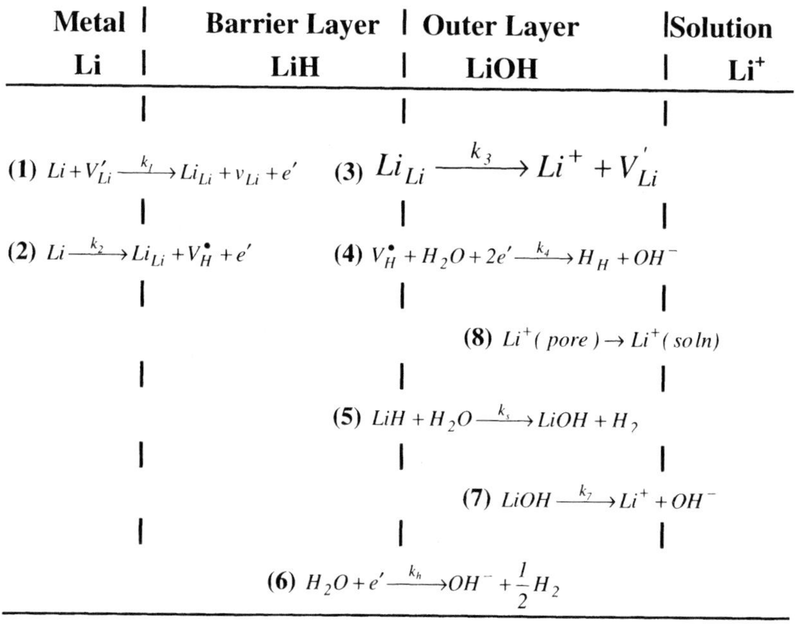

Figure 4.4.37. Interfacial reactions leading to the generation and annihilation of point defects within the bilayer passive film on lithium. In this figure, $V_H^{\bullet}$ and H_H represents vacant and occupied positions on the hydrogen sublattice of the barrier layer.

admittance of the system, Y, is the sum of the faradaic admittance and $i\omega C$, where C is the geometric capacitance of the passive film (reasonably approximated by a parallel plate capacitor model); i.e

$$Y = i\omega C + \frac{D_1 + D_2(1+i)\sqrt{\omega}}{1 + A_1(1+i)\sqrt{\omega}} = \frac{CA_1(1+i)\omega^{3/2} + i\omega C + D_2(1+i)\sqrt{\omega} + D_1}{1 + A_1(1+i)\sqrt{\omega}} \tag{156}$$

Furthermore, consideration of the solution resistance, R_s, yields the interfacial impedance as

$$Z = \frac{(1+i)\sqrt{\omega} + B_1}{i(1+i)C\omega^{3/2} + B_1 iC\omega + B_2(1+i)\sqrt{\omega} + B_3} + R_s \tag{157}$$

where B_1, B_2 and B_3 are

$$B_1 = \frac{1}{A_1} = \frac{\sqrt{2}J_{V_\infty}}{\alpha_3\gamma\alpha c_W^{ss}\big|_{BOI}\sqrt{D_W^{ss}}} \tag{158}$$

$$B_2 = \frac{D_2}{A_1} = \frac{\sqrt{2}FJ_{V_\infty}\theta^{ss}k_s^o}{c_o\sqrt{D_W^{ss}}}\left[\frac{4\left(1 - \frac{c_{V'_{Li}}^{MBI}k_1^{o'}}{c_o k_2^{o'}} + \frac{k_h^{ss}}{k_s^{o'}}\right)A_1}{\xi_2 + 2\alpha_3\gamma\alpha V_o} + \frac{c_o}{k_s^o}\sqrt{\frac{D_W^{ss}}{2}}\right] \tag{159}$$

and

$$B_3 = \frac{D_1}{A_1} = \frac{\sqrt{2}FJ_{V_\infty}\theta^{ss}k_s^o}{c_o\sqrt{D_W^{ss}}}\left[\frac{4\left(1-\frac{c_{V_{Li}}^{MBI}k_1^{o'}}{c_ok_2^{o'}}+\frac{k_h^{ss}}{k_s^o}\right)}{\xi_2+2\alpha_3\gamma\alpha V_o}+\frac{c_{V_{Li}}^{MBI}k_1^{o'}}{c_ok_2^{o'}}\right] \tag{160}$$

and the rate constants, $k_i^{0'}$, and transfer coefficients, α_i, refer to the reactions shown in Figure 4.4.37. The parameter θ^{SS} corresponds to the porosity of the outer layer in the steady state and D_W^{SS} is the diffusivity of water in the pores. The other parameters are defined in the original publication (Pensado-Rodrigueuz [2001]).

Optimization of the model on experimental impedance data for lithium in 12 M KOH (Figure 4.4.38) and in 12 M KOH + 0.58 M sucrose (Figure 4.4.39) produced

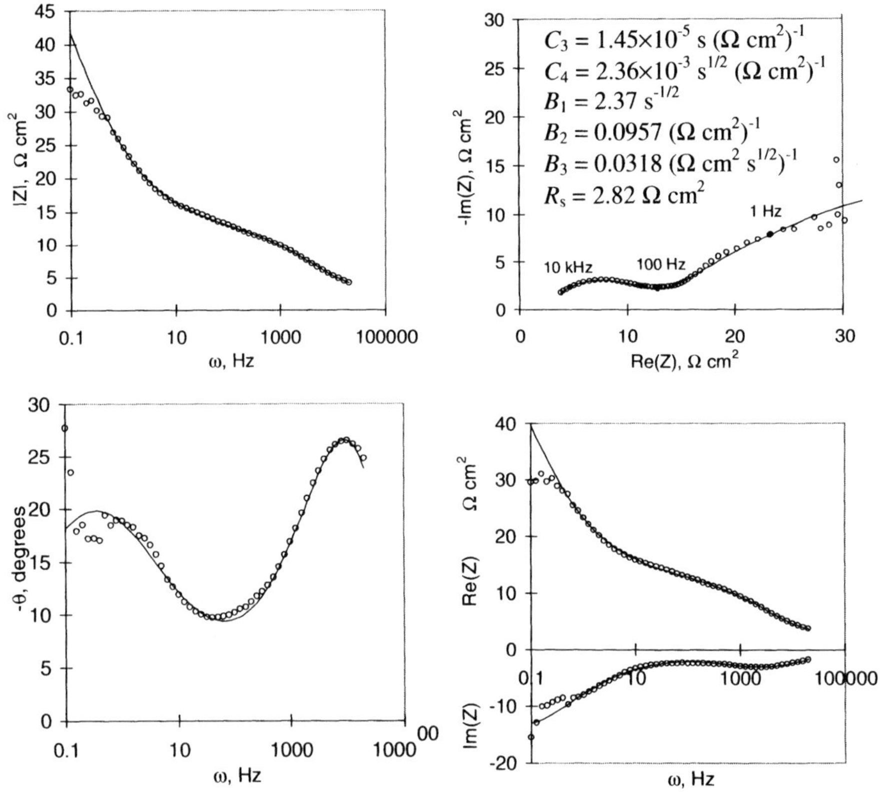

Figure 4.4.38. Nyquist and Bode plots of impedance data for lithium polarized at $-2.56\,V_{she}$ in a 12 M KOH electrolyte. Experimental data are represented by circles. Points at three frequencies are highlighted in the Nyquist plots. The optimization parameters, used in the computation of the solid lines, are indicated in the figure. (After Pensado-Rodriguez *et al.* [2001]).

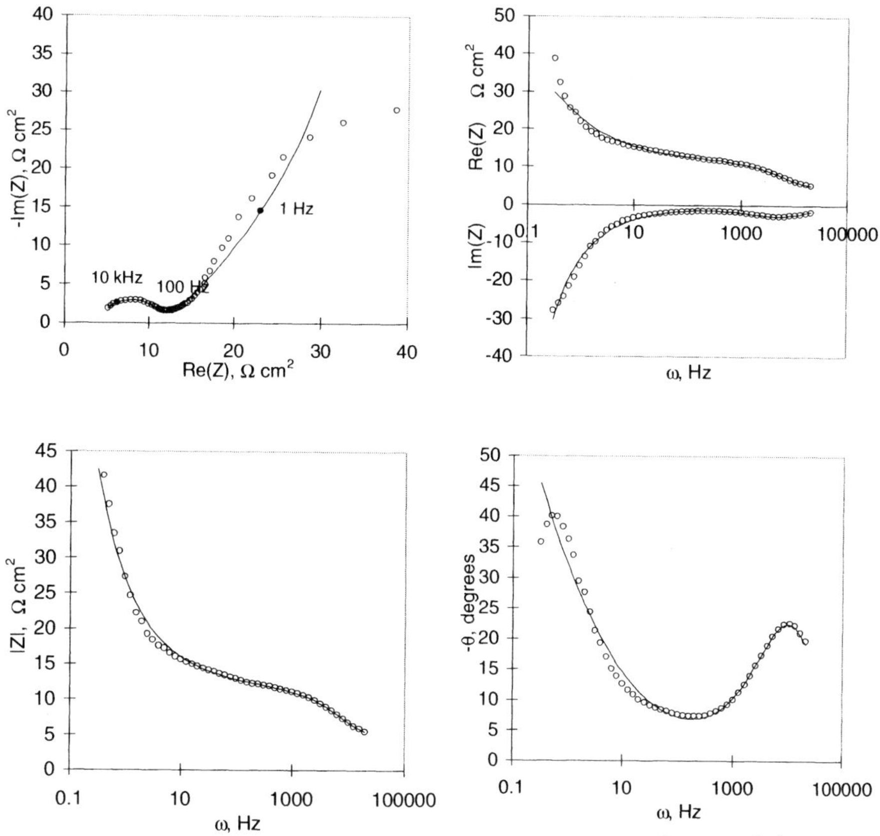

$C_3 = 1.89\times10^{-5}$ s $(\Omega\ \text{cm}^2)^{-1}$
$C_4 = 1.15\times10^{-3}$ s$^{1/2}$ $(\Omega\ \text{cm}^2)^{-1}$
$B_1 = 2.75$ s$^{-1/2}$
$B_2 = 0.13$ $(\Omega\ \text{cm}^2)^{-1}$
$B_3 = -3\times10^{2}$ $(\Omega\ \text{cm}^2\ \text{s}^{1/2})^{-1}$
$R_s = 4.38$ Ω cm^2

Figure 4.4.39. Nyquist and Bode plots of impedance data for lithium polarized at $-2.56\,V_{she}$ in a 12 M KOH + 0.58 M sucrose electrolyte. Experimental data are represented by circles. Points at three frequencies are highlighted in the Nyquist plot. The optimization parameters, used in the computation of the solid lines, are indicated in the figure. (After Pensado-Rodriguez *et al.* [2001]).

values for the various parameters, as shown in the Figures. Sucrose was explored in this study as an inhibitor of hydrogen evolution. As seen, the model effectively reproduces the experimental impedance data, with only minor changes in the parameter values. These differences have been attributed to the presence of sucrose in the solution.

To the author's knowledge, the Li/LiOH + sucrose system is the most sophisticated RMA performed to date on a passive system. Although numerical values could be determined only for groups of parameters, it is evident that measurement of the impedance as a function of all of the independent variables of the system would be necessary for a complete analysis from which values for the individual parameters would be obtained. This is the normal evolution of scientific enquiry and it is expected that RMA will follow this same, well-trodden path.

4.4.8 Equivalent Circuit Analysis

Digby D. Macdonald
Michael C. H. McKubre

Because electrochemical impedance techniques have their genesis in electrical engineering, great emphasis has been placed in the past on identifying an equivalent "electrical circuit" for the interface. Although this exercise is useful, in that the equivalent circuit is capable of mimicking the behavior of the system, the circuits that often are adopted are too simplistic to be of any interpretive value. Thus, the impedance data plotted in Figure 4.4.9, for example, cannot be interpreted in terms of a simple RC circuit of the type shown in Figure 4.4.40, if for no other reason that the circuit does not delineate the partial anodic and cathodic reactions. Also, this circuit strictly yields a semicircle in the complex plane that is centered on the real axis, whereas the locus of the experimental data is clearly neither semicircular nor centered on the abscissa. Nevertheless, an equivalent circuit of the type shown in Figure 4.4.40 is frequently assumed, and values are often calculated for the various components. It is important to realize, however, that the choice of an "incorrect" equivalent circuit has no consequence for the measured polarization resistance or corrosion rate, since the resistance $|Z|_{\omega\to 0} - |Z|_{\omega\to\infty}$ is independent of the form of the reactive component.

The shape of the impedance locus plotted in Figure 4.4.9 suggests that a more appropriate equivalent circuit is an electrical transmission line of the type shown in Figure 4.4.41. Although extensive use has been made of transmission lines in the past (de Levie [1964, 1965, 1967], McKubre [1976], Atlung and Jacobsen [1976], Park and Macdonald [1983], Lenhart *et al.* [1984]) to model rough surfaces, porous electrodes, corroding interfaces, polarization in soils and clays, and coated surfaces, they have not been universally embraced by corrosion scientists and electrochemists. The particular transmission line shown in Figure 4.4.41 was developed by Park and

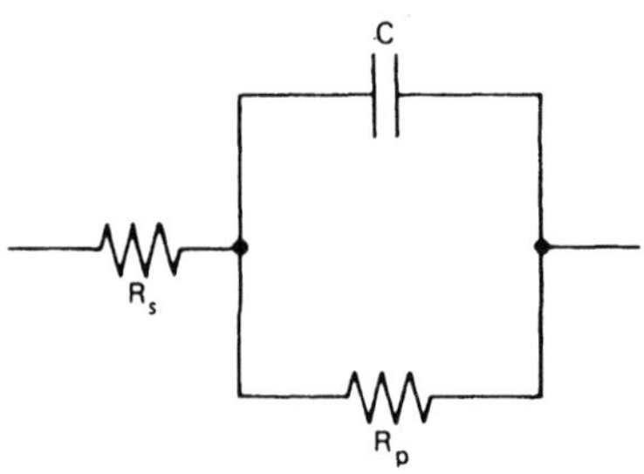

Figure 4.4.40. Simple electrical equivalent circuit for a corroding interface. Note that this circuit does not delineate the partial anodic or cathodic reactions.

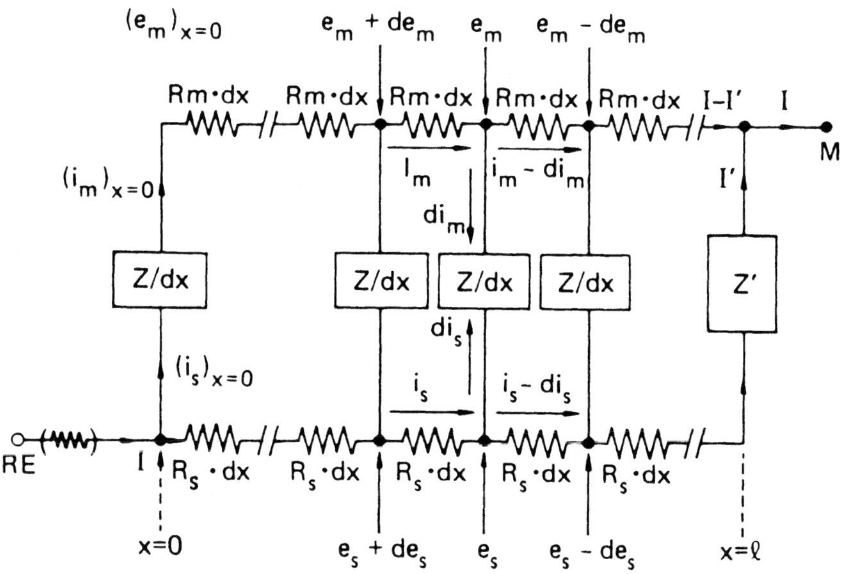

Figure 4.4.41. Discretized form of the transmission line model. e_m and e_s are the potentials in the magnetite and solution phases, respectively. Here i_m and i_s are the currents in the magnetite and solution phases, respectively; I and I' are the total current and the current flowing across the metal–solution interface and base of the pore, respectively; RE and M designate the reference electrode and metal (working electrode) locations, respectively. (Reprinted with permission from J . R. Park and D. D. Macdonald, Impedance Studies of the Growth of Porous Magnetite Films on Carbon Steel in High Temperature Aqueous Systems, *Corros. Sci.* **23**, 295 [1983]. Copyright 1983, Pergamon Journals Ltd.)

Macdonald [1983] and by Lenhart, Macdonald, and Pound [1984] to model porous magnetite films on carbon steel in high-temperature chloride solutions and to describe the degradation of porous $Ni(OH)_2$—NiOOH battery electrodes in alkaline solution upon cyclic charging and discharging. We will discuss the first case in some detail, because it is a good example of how a transmission line can provide a physical picture of the processes that occur with a complex corrosion reaction.

The work of Park and Macdonald [1983] was performed in an attempt to understand the phenomenon of "denting" corrosion in pressurized water reactor (PWR) steam generators (Garnsey [1979]). Briefly, this phenomenon, which is represented schematically in Figure 4.4.42, occurs because of the rapid growth of magnetite in the crevice between an Inconel 600 steam generator tube and the carbon steel support plate. The growing crystalline magnetite can deform or "dent" the tube and may result in leakage from the primary (radioactive) circuit to the secondary (nonradioactive) side via stress corrosion cracks that nucleate and grow in the highly stressed regions of the tubes. Previous work by Potter and Mann [1965], among others (see Park [1983]), had indicated that the growth of magnetite on carbon steel in simulated tube–support plate crevice environments is kinetically linear; that is, the growing film offers no protection to the underlying metal. However, this

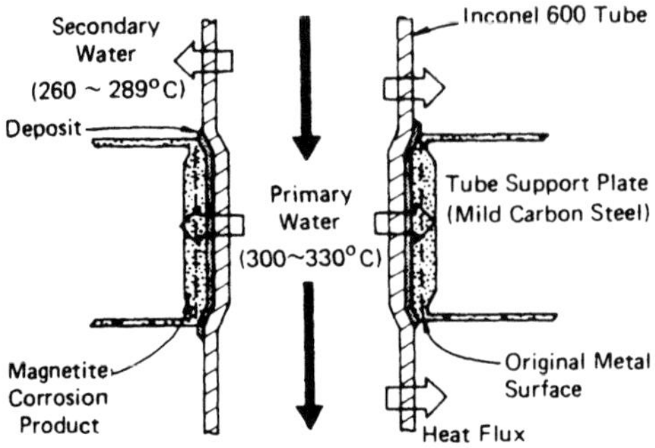

Figure 4.4.42. Schematic representation of denting corrosion in PWR steam generators.

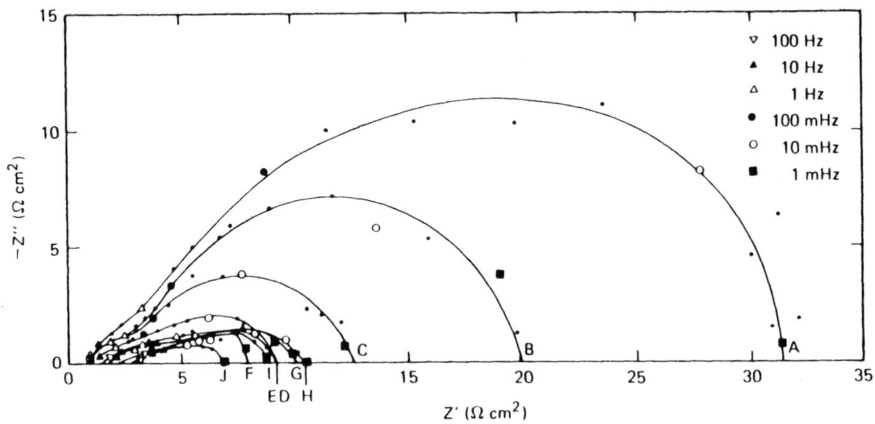

Figure 4.4.43. Impedance diagrams at the corrosion potential as a function of time for carbon steel exposed to a 0.997 m NaCl + 0.001 m $FeCl_3$ + 3500 ppb O_2 solution at 250°C. A = 11 h, B = 22 h, C = 35 h, D = 46 h, E = 58 h, F = 70 h, G = 86.5 h, H = 112 hr, I = 136 h, J = 216 h.

conclusion was based on *ex situ* analyses of weight loss and film growth, neither of which yield real-time estimates of the rate of corrosion in the environment of interest.

In the work conducted by Park and Macdonald [1983], impedance spectroscopy was used to measure the polarization resistance of corroding carbon steel in a variety of chloride (1M)-containing solutions at temperatures from 200 to 270°C. Typical complex plane impedance spectra for one such system as a function of time are shown in Figure 4.4.43. In all cases (different exposure times), the impedance spectra are characteristic of a system that can be represented by an electrical transmission line. Also of interest is the observation (Park [1983]) that the inverse of the polarization resistance, which is proportional to the instantaneous corrosion rate, increases with increasing exposure time. Accordingly, the corrosion process is kinetically autocatalytic, rather than being linear, as was previously reported. This autocatalytic behavior has been found for a wide range of solutions that simulate crevice environments, particularly those containing reducible cations, such as Cu^{2+}, Ni^{2+}, and Fe^{3+}. The large amount of experimental polarization resistance data generated in the study showed that these reducible cations greatly accelerate the corrosion rate and yield autocatalytic behavior, with the effect lying in the order $Cu^{2+} > Ni^{2+} > Fe^{3+}$. The enhanced rates were found to be far greater than could be accounted for by hydrolysis (to form H^+) alone, and it was concluded that reduction of the foreign cations was largely responsible for the rapid corrosion of the steels. However, the autocatalytic nature of the reaction was also partly attributed to the buildup of chloride in the porous film. This buildup occurs for the reason of electrical neutrality, in response to the spatial separation of the anodic (iron dissolution at the bottom of the pores) and cathodic (H^+ or cation reduction at the magnetite–solution interface) processes.

The model that was adopted to account for the fast growth of porous magnetite on carbon steel is shown schematically in Figure 4.4.44, and an ideal single pore is depicted in Figure 4.4.45. This single pore can be represented electrically by the continuous transmission line, as in Figure 4.4.46, and by the discretized transmission line shown in Figure 4.4.41, where Z is the impedance per unit length down the pore wall and Z' is the impedance of the metal–solution interface at the bottom of the pore. The impedance of a single pore is readily derived using Kirchhoff's equations to yield

$$Z_p = \frac{R_m R_s l}{R_m + R_s} + \frac{2\gamma^{1/2} R_m R_s + \gamma^{1/2}(R_m^2 + R_s^2)C + \delta R_s^2 S}{\gamma^{1/2}(R_m + R_s)(\gamma^{1/2} S + \delta C)} \tag{161}$$

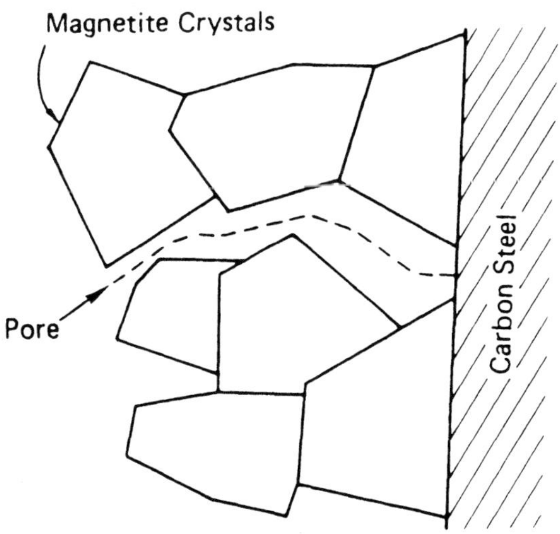

Figure 4.4.44. Schematic of fast growth of magnetite in acidic chloride solutions at elevated temperatures. (Reprinted with permission from J. R. Park and D. D. Macdonald, Impedance Studies of the Growth of Porous Magnetite Films on Carbon Steel in High Temperature Aqueous Systems, *Corros. Sci.* **23**, 295 [1983]. Copyright 1983, Pergamon Journals Ltd.)

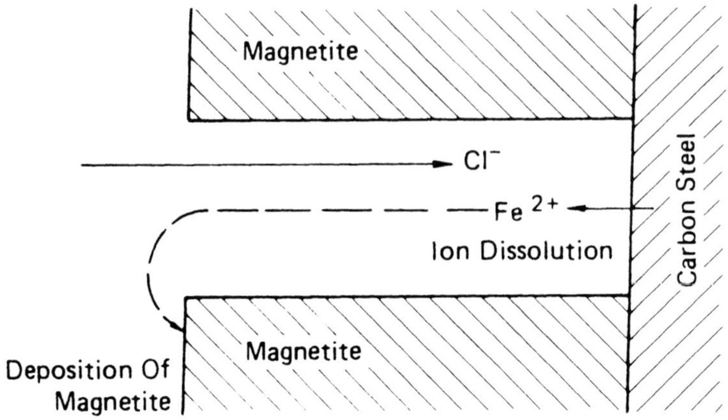

Figure 4.4.45. Ideal single-pore model for the growth of porous magnetite film in acidic chloride solutions at high temperature. (Reprinted with permission from J. R. Park and D. D. Macdonald, Impedance Studies of the Growth of Porous Magnetite Films on Carbon Steel in High Temperature Aqueous Systems, *Corros. Sci.* **23**, 295 [1983]. Copyright 1983, Pergamon Journals Ltd.)

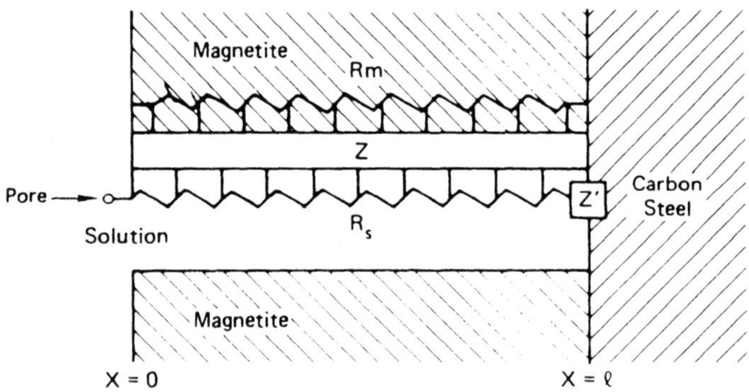

Figure 4.4.46. Transmission line model for a single one-dimensional pore: R_m = magnetite resistance per unit length of a pore, R_s = solution resistance unit length of a pore (Ω/cm), Z = pore wall–solution interface impedance for unit length of a pore (Ω/cm), Z' = metal–solution interface impedance at the base of a pore (Ω), X = distance from the mouth of the pore along the pore axis (cm). (Reprinted with permission from J. R. Park and D. D. Macdonald, Impedance Studies of the Growth of Porous Magnetite Films on Carbon Steel in High Temperature Aqueous Systems, *Corros. Sci.* **23**, 295 [1983]. Copyright 1983, Pergamon Journals Ltd.)

where

$$\gamma = \frac{R_m + R_s}{Z} \tag{162}$$

$$\delta = \frac{R_m + R_s}{Z'} \tag{163}$$

$$C = \cosh(\gamma^{1/2} l) \tag{164}$$

$$S = \sinh(\gamma^{1/2} l) \tag{165}$$

Here R_m is the resistance of the magnetite per unit length of the pore, R_s is the corresponding quantity for the solution, and l is the length of the pore. For an oxide consisting of n independent parallel pores, the total impedance of the film then becomes

$$Z_T = Z_p / n \tag{166}$$

To calculate the total impedance, it is necessary to assume models for the pore wall–solution and pore base–solution interfacial impedances (Z and Z', respectively). In the simplest case, both interfaces are assumed to exist under charge transfer control, in which case Z and Z' can be represented by the equivalent circuit shown in Figure 4.4.40. Typical impedance spectra calculated for various values of the resistance at the pore base are shown in Figure 4.4.47. These theoretical curves generally exhibit the features displayed by the experimental data (Figure 4.4.43). Similar calculations have been carried out for porous films involving mass transfer control within the pores, and impedance spectra that compare well with experiment have again been generated (Park and Macdonald [1983], Park [1983]).

4.4.8.1 Coating

An effective means of reducing the rate of corrosion of a metal is to protect the surface with a coating. However, coatings generally are not impervious to water or even ions, so that corrosion reactions still proceed at the metal–coating interface,

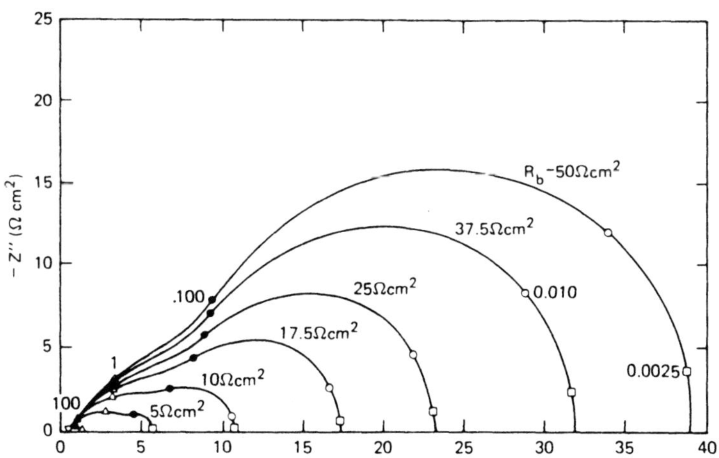

Figure 4.4.47. Calculated impedance spectra for a porous magnetite film on carbon steel in high-temperature aqueous sodium chloride solution as a function of the resistance at the base of the pore. N (no. of pores/cm^2) = 10^4, ρ (solution) = 18.5 Ω-cm, ρ (Fe_3O_4) = 116 Ω-cm, θ (film porosity) = 0.01, L (film thickness) = 0.025 cm, R_W (pore wall–solution resistance) = 10^3 Ω-cm^2, C_w (capacitance at pore wall–solution interface) = 2×10^{-2}/cm^2, c_b (capacitance at pore base) = 7×10^{-3} F/cm^2.

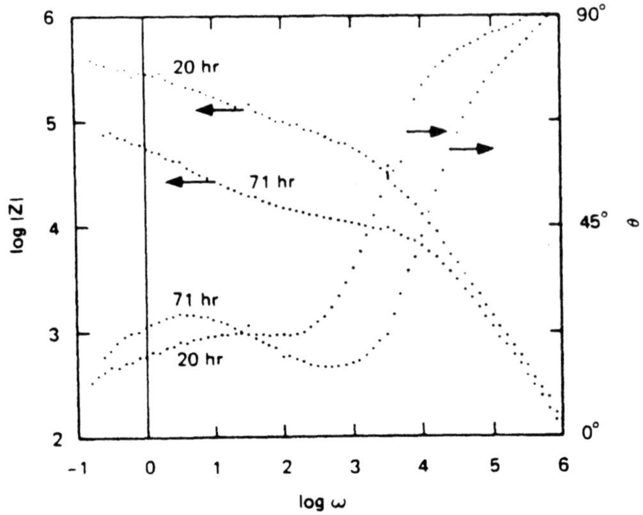

Figure 4.4.48. Bode plots for phosphated and coated 1010 carbon steel in 0.5 N NaCl solution. (From F. Mansfield, M. W. Kendig, and S. Tsai, Evaluation of Organic Coating Metal Systems by AC Impedance Techniques. Reprinted with permission from *Corrosion*, **38**, 482 [1982], NACE, Houston, TX.)

albeit at a low rate. Because the rate of corrosion depends upon the transport of corrosive species through the film, in addition to the reactions occurring at the metal surface, it is evident that the overall process of attack is a very complicated one. It is therefore not surprising that coated metals have been studied using impedance spectroscopy.

Perhaps one of the more extensive studies of this type is that reported by Mansfeld, Kendig, and Tsai [1982]. These workers evaluated polybutadiene coatings on carbon steel and aluminum alloys that had been subjected to different surface treatments, including phosphating for the steel and exposure to a conversion coating in the case of the aluminum alloys. Typical Bode plots ($\log|Z|$ and θ vs. $\log \omega$) for 1010 carbon steel that had been degreased, phosphated, and then coated with polybutadiene ($8 \pm 2\,\mu$m thick), are shown in Figure 4.4.48 for two times after exposure to 0.5 N NaCl solution at ambient temperature. The low-frequency behavior of the phase angle ($\sim\pi/8$) suggests the existence of a transport process through the coating that is best described in terms of a transmission line (see earlier discussion in this section). However, the authors chose to analyze their data in terms of a classical Randles-type equivalent electrical circuit containing a normal semiinfinite Warburg diffusional impedance element. Randles plots were then used to evaluate Warburg coefficients (σ), and polarization resistances (R_p) were calculated directly from the impedance magnitude at $\omega \to 0$ and $\omega \to \infty$ [Eq. 4]. The R_p and σ parameters are plotted in Figure 4.4.49 as a function of time for a variety of surface pretreatments (including phosphating), as listed in Table 4.4.8. Both parameters are observed to decrease with time after an initial increase, at least in the case of R_p. The simulta-

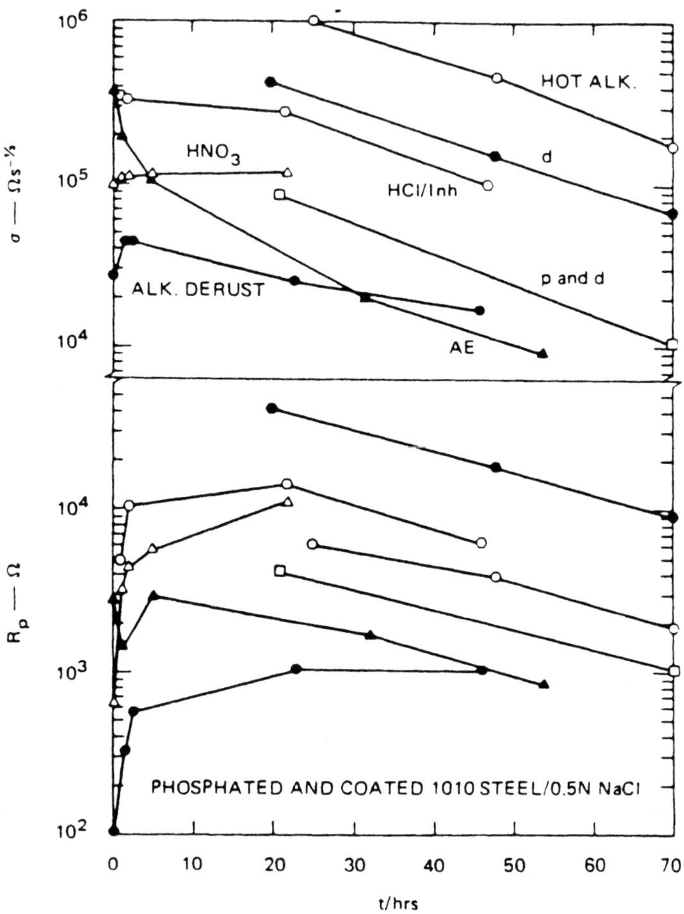

Figure 4.4.49. Time dependence of resistance R_p and Warburg coefficient σ for different surface pretreatment procedures applied to 1010 carbon steel. (From F. Mansfield, M. W. Kendig, and S. Tsai, Evaluation of Organic Coating Metal Systems by AC Impedance Techniques. Reprinted with permission from *Corrosion*, **38**, 482 [1982], NACE, Houston, TX.)

neous decrease in R_p and σ was rationalized in terms of a model involving the development of ionically conducting paths through the polybutadiene coating as the corrosion reaction proceeds at the metal-coating interface.

4.4.9 Other Impedance Techniques

In keeping with the theme established at the beginning of this section that impedance techniques can be used to analyze many cause-and-effect phenomena, a number of other transfer functions have been defined. Three such functions, the electro-

Table 4.4.8. Surface Pretreatment Procedures Used by Mansfeld *et al.* [1982] Prior to Applying Polybutadiene Coatings to 1010 Carbon Steel

Pretreatment	Procedure
Degrease (d)	Trichloroethylene, 15 min, TT-C-490, method II
Polish (p)	Wet polish, 600-grit SiC
p + d	
d + hot alkaline cleaning	5 min, 100°C, TT-C-490, method III
d + alkaline derusting	10 min, 50°C, TT-C-490, method V, type III
p + d + HCl/Inh[a]	5 min, RT, 42 v/o HCl + 5.8 g/l 2-butyne-1,4-diol
d + anodic etch (AE)	2 min, RT, 30 v/o H_2SO_4, 0.1 A/cm^2 30 min, RT
d + conc. HNO_3	
Zinc phosphate coating	6.4 g/l ZnO, 14.9 g/l H_3PO_4, 4.1 g/l HNO_3, 95°C, 30 min

[a] Inhibitor, see Mansfeld *et al.* [1982].

chemical hydrodynamic impedance (EHI), fracture transfer function (FTF), and the electrochemical mechanical impedance (EMI), are discussed briefly below.

4.4.9.1 Electrochemical Hydrodynamic Impedance (EHI)

In the EHI technique (Deslouis *et al.* [1980, 1982], Bonnel *et al.* [1983]), the rate of mass transport of reactants to, or products from, an electrochemical interface is modulated (sine wave), and the response current or potential is monitored at constant potential or constant current, respectively, depending upon whether the transfer function is determined potentiostatically or galvanostatically. Experimentally, the transfer function is most easily evaluated using a rotating-disk electrode, because of the ease with which the rotational velocity can be modulated. For a fast redox reaction occurring at the surface of a disk, the potentiostatic and galvanostatic electrochemical hydrodynamic admittances are given as follows:

$$Y_{HD}^{I} = \left(\frac{\delta I}{\delta \Omega}\right)_V = -\frac{3}{2\Omega_0}\exp(-0.26pj)\cdot\frac{C_\infty - C_0}{1.288}\cdot\frac{3^{1/3}}{\delta}\cdot\frac{K_1(s)}{Ai(s)}\cdot D\cdot\frac{Z_D}{Z} \tag{167}$$

$$Y_{HD}^{V} = \left(\frac{\delta V}{\delta \Omega}\right)_I = -\frac{3}{2\Omega_0}\exp(-0.26pj)\cdot\frac{C_\infty - C_0}{1.288}\cdot\frac{K_1(s)}{Ai'(s)}\cdot R_T(k_f - k_b) \tag{168}$$

where Ω_0 is the mean angular velocity of the disk, $p = \omega/\Omega_0$, ω is the modulation frequency, $Ai(s)$ and $Ai'(s)$ are the Airy function of the first kind and its first derivative, $s = 1.56jpSc^{1/3}$, Sc (Schmidt number) $= v/D$, v is the kinematic viscosity, D is the diffusivity, $j = \sqrt{-1}$, Z_D is the diffusion impedance, R_1 = the charge transfer resistance, k_f and k_b are the forward and backward rate constants for the redox reaction, Z is the overall impedance, and $K_1(s)$ is given by

$$K_1(s) = -\sum_0^\infty \frac{s^n}{n_!}\int_0^\infty \frac{\delta^n Ai}{\delta \xi^n}\left[\xi_2 \exp\left(-\frac{\xi^3}{6}\right) + \xi^4 \rho_0(\xi)\exp\left(\frac{\xi^3}{6}\right)\right]d\xi \tag{169}$$

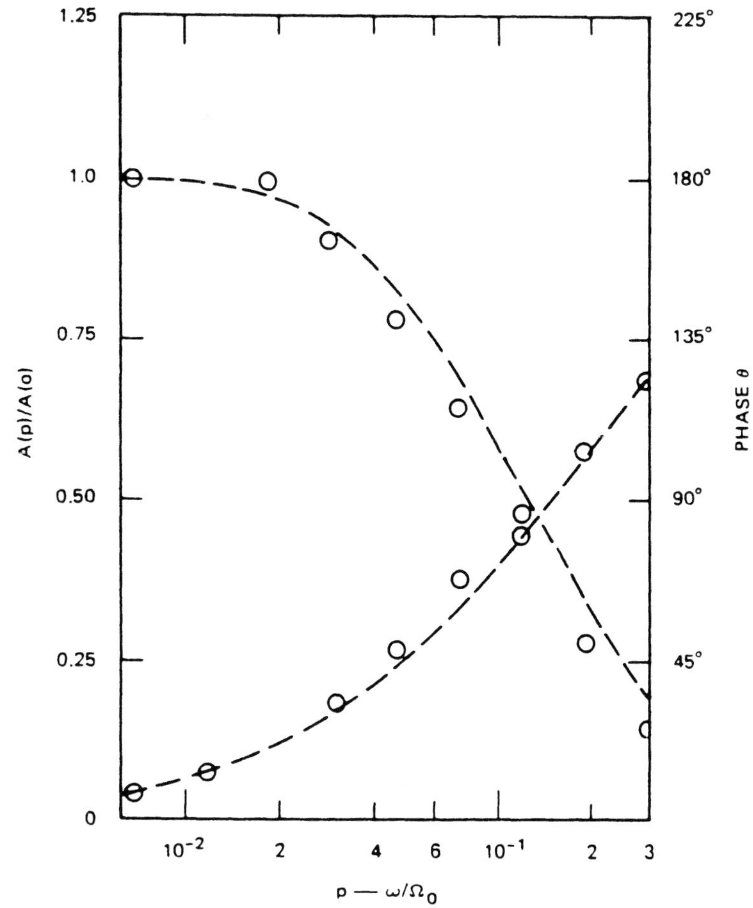

Figure 4.4.50. Bode plots of the reduced EHI modulus and phase angle as a function of reduced frequency (ω/Ω_0) for carbon steel in 3% NaCl solution. The EHI was measured in the galvanostatic mode [Eq. (168)] at a mean rotational velocity (Ω_0) of 1000 rpm and at $I = 0$ (open-circuit conditions). (After Bonnel *et al.* [1983]).

The application of EHI analysis to a corroding system has been described by Bonnel *et al.* [1983] in their study of the corrosion of carbon steel in neutral chloride solutions. An example of their data is shown in Figure 4.4.50 for carbon steel soon after immersion into air-saturated 3% NaCl solution. The steel surface is believed to be free of solid corrosion products under these conditions. The best fit of Eq. (168) to the experimental data shown in Figure 4.4.50 (broken line) is found for $Sc = 615$ and $D(O_2) = 1.63 \times 10^{-5}\,cm^2/s$. The value for the diffusivity of oxygen was found to be consistent with the steady state current–voltage curve obtained for this system. A significant advantage claimed for the EHI technique is that it may be used to detect the presence of a porous corrosion product film on a corroding surface (Bonnel *et al.* [1983]), but in general the analysis is complicated, and the method has yet to demonstrate any clear advantages over more conventional techniques for examining coupled mass transfer-charge-transfer phenomena.

4.4.9.2 Fracture Transfer Function (FTF)

Chung and Macdonald [1981] defined a transfer function for the propagation of a crack through an elastic–plastic material under transient loading conditions. Designating the stress intensity as $K_I(t)$ and the instantaneous crack velocity as $da(t)/dt$ (both time-dependent quantities), the fracture transfer function may be defined as

$$\bar{H}_F(s) = \frac{£[K_I - K_{ISCC}]}{£[da(t)/dt]} \tag{170}$$

where K_{ISCC} is the critical stress intensity for slow crack growth, and $\mathcal{L}$ designates the Laplace transform. The transfer function $\bar{H}_F(s)$ is a function of the Laplace frequency s.

Application of this analysis to transient crack growth data for AISI 4340 steel in 3% NaCl solution (Chung [1983]) is shown in Figure 4.4.51. It is seen that the transfer function $\bar{H}_F(s)$ increases linearly with the Laplace frequency s. Accordingly, the electrical analog for crack propagation is simply a resistance and inductance in series (Figure 4.4.52). However, the data shown in Figure 4.4.51 also reveal that the inductance, but not the resistance, is a function of the applied stress intensity; that is

$$\bar{H}_F(s) = R + sL(K_I) \tag{171}$$

It is clear, therefore, that the system is nonlinear, so that $\bar{H}_F(s)$ is not expected to obey the Kramers–Kronig transforms.

The parameter R provides a measure of the resistance to crack propagation, such that $1/R$ is proportional to the steady state crack velocity. The R values obtained from the intercept at $s = 0$ indicate that the steady state crack velocity is independent of stress intensity.

The inductance contained in the equivalent circuit shown in Figure 4.4.52 corresponds to an energy adsorption process, presumably due to the plastic deformation of the steel matrix in front of the crack tip. The size of the plastically deformed region, and hence the amount of energy deposited in the matrix in front of the crack tip, is known to increase with increasing K_I (Chung [1983]), and this accounts for the observed increase in $L(K_I)$.

4.4.9.3 Electrochemical Mechanical Impedance

Instead of the purely mechanical case discussed above, it is also possible to define an impedance in terms of the electrochemical response of a fractured specimen to a mechanical input that causes crack propagation.

One case in which this analysis can be usefully applied is that of the electrochemically assisted (stress corrosion) cracking of metals submerged in an electrolyte and subjected to a cyclic mechanical load or stress, corresponding to vibrational

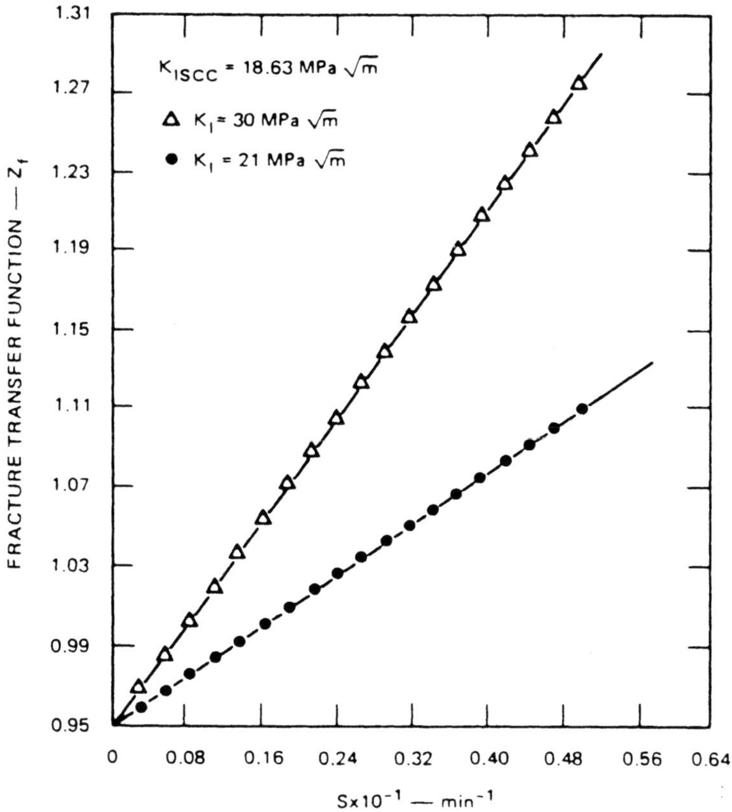

Figure 4.4.51. Fracture transfer function vs. Laplace frequency.

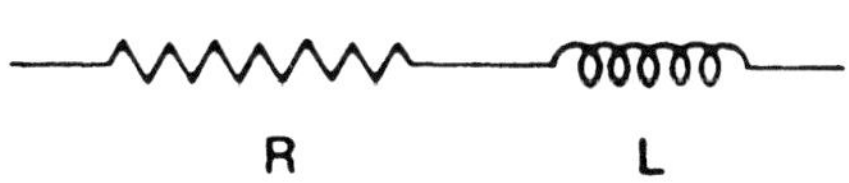

Figure 4.4.52. Series RL equivalent circuit used to represent the growth of a stress corrosion crack through AISI 4340 steel in 3% NaCl solution.

effects (Eiselstein, McKubre, and Caligiuri [1983, 1985]). With some simplification, crack growth due to cyclical film rupture at the growing crack tip, followed by metal dissolution, can be considered as described below.

Figure 4.4.53 shows schematically the case of a crack growing into a metal under constant current but cyclic load conditions. The loading conditions can be controlled to achieve a sinusoidal variation in the crack opening displacement D:

$$D = \overline{D} + \tilde{d}\sin(\omega t) \tag{172}$$

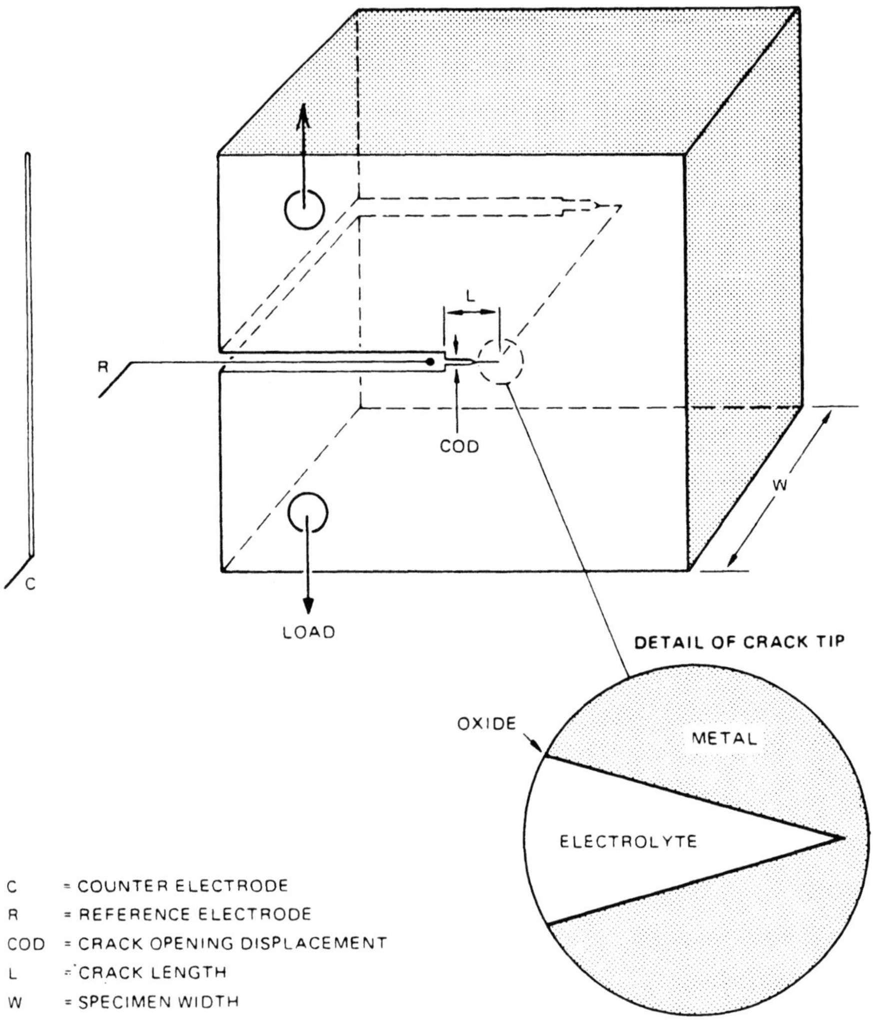

Figure 4.4.53. CT specimen geometry for determining the electrochemical/mechanical impedance for the propagation of a crack.

where $\overline{D}$ is the mean crack opening and $\tilde{d}$ the amplitude of the superimposed sinusoidal perturbation. Both $\overline{D}$ and $\tilde{d}$ will increase with crack length. To a first approximation, the area exposed at the crack tip can be considered to be proportional to the crack opening angle,

$$A = \gamma W \tan(D/L) \tag{173}$$

$$A \approx \gamma W(D/L) = \frac{\gamma W}{L}[\overline{D} + \tilde{d}\sin(\omega t)] \tag{174}$$

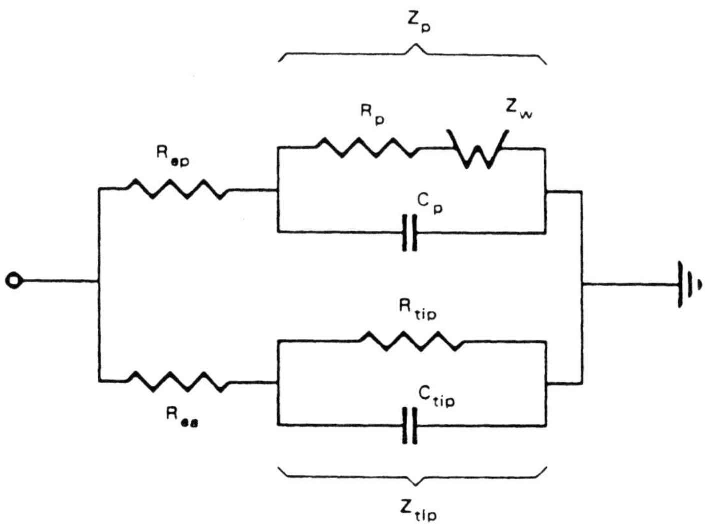

R_{ep} = ELECTROLYTE RESISTANCE TO CRACK WALLS
R_p = CHARGE TRANSFER RESISTANCE AT PASSIVE WALLS
C_p = DOUBLE LAYER CAPACITANCE AT PASSIVE OXIDE
Z_W = DIFFUSIONAL IMPEDANCE AT PASSIVE WALLS
Z_p = INTERFACIAL IMPEDANCE AT PASSIVE WALLS
R_{ea} = ELECTROLYTE RESISTANCE TO CRACK TIP
R_{tip} = CHARGE TRANSFER RESISTANCE AT EXPOSED METAL
C_{tip} = DOUBLE LAYER CAPACITANCE OF EXPOSED METAL
Z_{tip} = INTERFACIAL IMPEDANCE AT CRACK TIP

Figure 4.4.54. Simplified equivalent circuit for a stress corrosion crack.

where γ is a proportionality constant determined by the mechanical properties of the metal and the oxide film (Eiselstein *et al.* [1985]) and W is the specimen width.

Figure 4.4.54 shows the approximate[1] equivalent circuit for the inside of the crack, with the impedance of the oxide-passivated walls appearing in parallel with the impedance element due to dissolution of exposed metal at the crack tip. The variation in crack opening displacement, described by Eq. (172), results in a sinusoidal perturbation at the exposed tip area such that

$$C_{\text{tip}} = C^0_{\text{tip}} A \approx \frac{C^0_{\text{tip}} \gamma W}{L} [\bar{D} + \tilde{d} \sin(\omega t)] \tag{175}$$

[1] Because of the potential distribution along the crack, this equivalent circuit should more properly be represented as a nonuniform, finite transmission line for the oxide wall impedance, with the crack tip as a terminating impedance. Such a case is shown in Figure 4.4.46.

$$R_{\text{tip}} = \frac{R_{\text{tip}}^{0}}{A} \approx \frac{R_{\text{tip}}^{0} L}{\gamma W[\overline{D} + \tilde{d}\sin(\omega t)]} \tag{176}$$

Without going through the algebra, for the equivalent circuit shown in Figure 4.4.54 we can obtain an expression for the electrochemical–mechanical impedance (Z_{em}) for this system, defined as the ratio of the ac voltage that appears at the reference electrode to the crack opening displacement due to the sinusoidal load variation under dc galvanostatic conditions:

$$Z_{\text{em}} = \frac{\tilde{V}_{\text{ref}}}{\tilde{d}} = \overline{V}_{\text{tip}}[R_{e,p} + Z_{p}]\left[\frac{1}{R_{\text{tip}}^{0}} + j\omega C_{\text{tip}}^{0}\right]\frac{\gamma W}{L} \tag{177}$$

Under optimal conditions, Z_{em}, measured over a range of frequencies, can be used to deconvolve the equivalent circuit parameters in a manner analogous to that of electrochemical impedance analysis. The advantages of this method, in the example given, are that measurements can be made under dynamic load conditions and that the perturbation is imposed at the point of principal interest, the crack tip.

Figure 4.4.55 presents the real vs. imaginary components of the electrochemical–mechanical impedance response measured for an HY80 steel specimen of geometry shown in Figure 4.4.53 immersed in 3.5 wt % NaCl. Due to the form of Eq. (177), these plots are somewhat more complex than a conventional Nyquist plot. Nevertheless, these data are amenable to standard methods of electrical

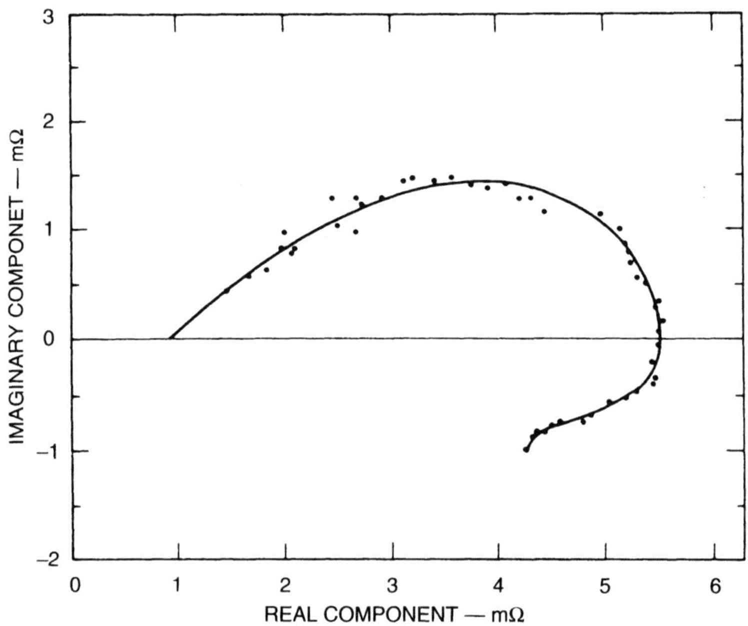

Figure 4.4.55. Complex plane plot of the electrochemical–mechanical impedance for the propagation of a crack through HY80 steel in 3.5% NaCl solution at 25°C under sinusoidal loading conditions.

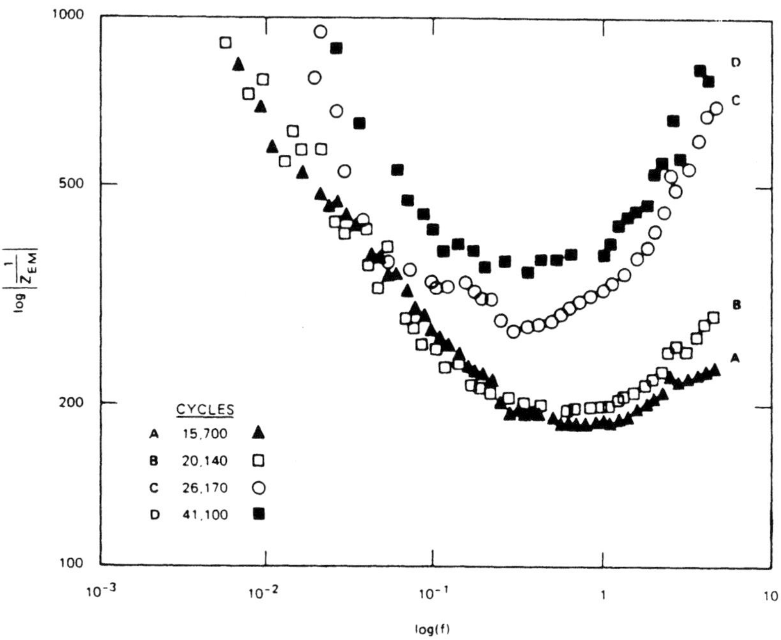

Figure 4.4.56. Log ($|1/Z_{em}|$) vs. log(frequency) for an HY80 specimen of the geometry shown in Figure 4.4.51, exposed to 3.5 wt% NaCl at 25°C.

analysis. We observe that the processes at the growing crack tip dominate at low frequencies, and the properties of greatest interest, R_{tip} and C_{tip}, can be deconvolved at limitingly low frequencies.

Figure 4.4.56 shows the logarithm of the electrochemical–mechanical admittance ($Y_{em} - 1/Z_{em}$) plotted vs. log frequency, for a typical data set, as a function of the number of mechanical load cycles experienced by the specimen during crack growth. These data show some evolution in the crack tip parameters (the low-frequency, descending portion of the curves), but considerably more changes appear in the crack wall impedance, reflecting the increased area with increasing crack length. Because of the potential distribution along the crack, this equivalent circuit should more properly be represented as a nonuniform, finite transmission line for the oxide wall impedance, with the crack tip as a terminating impedance. Such a case is shown in Figure 4.4.46.

This section was written by Michael C. H. McKubre and Digby D. Macdonald.

4.5 ELECTROCHEMICAL POWER SOURCES

4.5.1 Special Aspects of Impedance Modeling of Power Sources

Evgenij Barsoukov

4.5.1.1 Intrinsic Relation Between Impedance Properties and Power Sources Performance

It is widely recognized that apart from such characteristics as voltage and discharge capacity, internal energy losses constitute an important property of electrochemical power sources. There are multiple reasons for such losses. Usually both or one electrode of the power source is a porous layer of active material affixed on a surface of the current collector and in contact with a liquid or solid electrolyte. Electrodes are separated by a layer of electrolyte and, in some cases, an ionic conductive separator. If we trace the path of electric charge during charging or discharging, we will observe that it travels through an electronic pathway from one particle of active material to another, therefore experiencing ohmic resistance. To pass from an electronic to an ionic conductive medium, charge has to cause electrochemical reaction at the surface of particles, which is an activated process and is therefore associated with energy loss described by the Butler–Volmer equation. For small voltages the current–voltage relationship at this boundary can be characterized using a charge-transfer resistance R_{ct}. If charge passes for a prolonged time, the surface concentration of active materials decreases and has to be replenished by diffusion. The associated relationship with voltage is complex, but it can be said in general that diffusion-limited supply appears to an external observer as an additional time-dependent resistive term. After passing the boundary, charge than travels through an ionic pathway in the electrolyte filling the pores and between electrodes. The voltage–current relationship there is also similar to ohmic at usual currents, while becoming more complex at high currents.

These multiple resistive components will all degrade performance of the power source. Impedance spectroscopy helps in two different ways. First, it has the unique ability to separate different resistive contributions, using only measurements at power sources terminals. This allows researchers to identify the largest hindrance to high power capability of a power source and to investigate quantitatively how different steps help to eliminate it. Second, quantifying different components allows one to model power source current/voltage/time behavior. Older models of power sources assumed that all internal current hindrances could be represented by a single resistor in series with the voltage of the power source. For this approach no distinguishing of different components is needed, only their sum is observed. However, limitations of such a model become apparent when the load on a power source is variable, as is now the case in most applications. The reason for failure of the simple model is the transient nature of the voltage response of power source to applied load. Basically, voltage does not drop immediately by an $I \cdot R$ value, where R is the sum of all resistive contributions, but drops by a time-dependent value which approaches

I·R only at infinite time. The only way to describe power capability of a power source at variable load is therefore to measure parameters describing all resistive and capacitive contributions to discharge hindrance and to use them in a physically relevant model to calculate the desired current/voltage/time relationship. Impedance spectroscopy does exactly this—it allows one both to test if a certain model is appropriate and to obtain parameters of this model for the particular power source.

4.5.1.2 Linear Time-Domain Modeling Based on Impedance Models, Laplace Transform

Electrochemical systems are in general nonlinear which often makes analytical treatment of their kinetics prohibitively complex. Fortunately, in many cases where small voltage changes are involved, a linear approximation, equivalent to representing the electrochemical system as an electric network made of ideal resistors, capacitors and inductors, can be employed. The voltage response of resistances is not only linear but also time independent, and it follows Ohms law, $i = V/R$, where R is not only independent of i and V but also independent of time. However, the current/voltage dependence of such elements as capacitors and inductors is time dependent and expressed by differential equations

$$i(t) = [\mathrm{d}v(t)/\mathrm{d}t]C, v = [\mathrm{d}i(t)/\mathrm{d}t]L \tag{1}$$

The reason why power sources exhibit transient voltage response and therefore deviate from Ohm's law is the presence of capacitive elements in parallel with resistive ones, for example, a double-layer capacitance in parallel with R_{ct} and pseudocapacitance associated with diffusion process in parallel with diffusion resistance.

Fortunately, there is an indirect way to apply a treatment similar to Ohm's law to time-dependent systems. Linear differential equations have a very useful property to be transformed into linear equations, e.g. to a form similar to that of Ohm's law, when subjected to Laplace transform, given in Eq. (2).

$$F(s) = \int_0^{\infty} e^{-s \cdot t} \cdot f(t)dt \tag{2}$$

Here s is a complex frequency in form of $\delta + \omega \cdot i$. For example, a differential equation describing the behavior of a capacitor C: $i(t) = \mathrm{d}v(t)/\mathrm{d}t \cdot C$, is transformed to $\mathcal{L}[i(t)] = I(s)$ where $I(s) = C \cdot s \cdot V(s) - s \cdot v(0)$. I would like to emphasize here that due to advances in symbolic computing both direct and inverse Laplace transform are no longer the laborious table-searching procedures they used to be. Analytical software such as Maple or Mathematica can yield the Laplace transform (if it exists) of a function of almost any complexity in a matter of seconds, which makes this technique very convenient to use for any researcher. So, I encourage the reader not to be intimidated by Eq. (2), get hold of some appropriate software, and think Laplace transform as something easily done.

Now, as can be seen from the result above, the relationship between voltage and current in the Laplace domain does not involve differential equations and is therefore much easier to solve for complex systems. Moreover, equations for a capacitor and an inductor can be formally represented in the same general form as that for resistors. Indeed, assuming that initial voltage on capacitor $v(0)$ is zero, we can rearrange the above equation as

$$I(s) = V(s)/Z(s) \text{ where } Z(s) = 1/(C \cdot s). \tag{3}$$

Similarly, for an inductor $Z(s) = s \cdot L$. It can be seen that Eq. (3) has the same form as Ohm's law and the quantity $Z(s)$ plays the same role as resistance in Ohm's law. That allows us to treat the voltage/current relationship for any network of capacitors, resistors and inductors in the Laplace domain as if they all were resistors, only substituting the quantity $Z(s)$ instead of resistance for each element. For example the Laplace current for a serially connected resistor R and a capacitor C will be $I(s) = V(s)/(Z(s) + R)$ which can be rearranged as

$$I(s) = V(s)/(1/C \cdot s + R) \tag{4}$$

Now, to find the current/voltage relationship in the time domain, we only need to substitute $V(s)$ for the particular excitation of interest and carry out the inverse Laplace transform for the entire equation. However, first we need to know how the particular excitation we want to apply looks in the Laplace domain. Here is an example of how to do it for a voltage step. In the time domain it can be expressed as $v(t) = \text{Heaviside}(t) \cdot V_1$ where the function Heaviside(t) takes a value 1 if $t > 0$ and 0 otherwise. V_1 is the value of voltage applied during the step. The Laplace transform of this function, $V(s) = \mathscr{L}[\text{Heaviside}(t) \cdot V_1] = V_1/s$. To find the Laplace current of the above mentioned network to pulse excitation, we substitute this $V(s)$ into Eq. (4) and perform inverse Laplace transformation of the resulting equation. We obtain

$$i(t) = V_1 \cdot \frac{exp\left(\frac{-t}{R \cdot C}\right)}{R} \tag{5}$$

You might note that neither impedance nor ac response was mentioned yet. The Laplace transform is handling a non-periodic excitation (such as pulse); therefore the expression for $Z(s)$ is not identical to the impedance function of the same circuit. However, response to periodic excitation is related in an elegant way to response to arbitrary excitation expressed in the Laplace domain. In Section 1.3 is described how time-domain response to periodic excitation can be "folded" into the frequency domain by using Fourier transform where imaginary frequency is equal to the inverse period of the periodic excitation. Differential equations relating capacitor and inductor currents and voltages (Eq. (1)) are transformed to the frequency domain as $I(\omega \cdot i) = C \cdot \omega \cdot i \cdot V(\omega \cdot i)$ and $I(\omega \cdot i) = V(\omega \cdot i)/(L \cdot \omega \cdot i)$ which allows one, like in the case of the Laplace transform, to present it in a form similar to Ohm's law: $I(\omega \cdot i) = V(\omega \cdot i)/Z(\omega \cdot i)$ where the quantity $Z(\omega \cdot i)$ is called impedance and is given as $Z(\omega \cdot i) = 1/(C \cdot \omega \cdot i)$ for a capacitor and $Z(\omega \cdot i) = L \cdot \omega \cdot i$ for an inductor.

By comparing $Z(\omega \cdot i)$ with the quantity $Z(s)$ in Eq. (3), obtained from the Laplace transform, we can see that they become identical if the real part, δ, in $s = \delta + \omega \cdot i$ is set to zero. Substituting instead of the complex $s = \delta + \omega \cdot i$ a purely imaginary frequency $s = \omega \cdot i$ we convert all expressions in the Laplace domain, which can be obtained, for example, by solving differential equations for a particular distributed system, as in Section 2.1.3, into impedance equations. Alternatively, all parameters of impedance functions (such as C and L in the above example) obtained by fit of an experimental impedance spectrum can be substituted into Laplace-domain equations. Time domain response to arbitrary (non-periodic) excitation can then be obtained by inverse Laplace transform as described earlier in this chapter.

4.5.1.3 Expressing Model Parameters in Electrical Terms, Limiting Resistances and Capacitances of Distributed Elements

In Section 2.2.2 is described how to express impedance functions for distributed systems in the form of a dimensionless function multiplied by a resistance factor. Dealing with dimensionless quantities simplifies theorctical analysis of particular impedance responses and is useful for comparing response of systems with different dimensions. However, properties of electrochemical power sources are often considered in the context of a certain electrical device, and this is convenient when optimized parameters themselves have dimensions commonly used with electric elements, e.g. Ohm and Farad. In some cases, such usage has become common in electrochemistry, as is the case when expressing the impedance of the electrochemical double layer as charge transfer resistor [R_{rc}] in parallel with double layer capacitance [C_{cl}], as shown in Section 1.3.2. By looking at the impedance plot in Figure 1.3.2 it is possible to estimate the approximate value of R_{ct} as the diameter of the semicircle in the complex plot. Moreover, if we would have an electrochemical power source with such an impedance (which is a good approximation with some fuel cells) we could immediately proceed from impedance analysis to modeling of this power source performance in the actual device, by assigning the impedance model parameter R_{ct} as the internal resistance of the device.

Unfortunately, not all impedance models are as simple as the above, mostly because multiple steps govern charge transfer kinetics and because electrochemical systems are distributed rather then discrete. In order to evaluate the relative importance of the contributions to the dc resistance from different elements, we need to know how to characterize in electrical terms processes such as diffusion, which are commonly described in electrochemistry by specific physical quantities such as the diffusion coefficient. To achieve this, an impedance function for a particular process may be analyzed to find its limiting form at infinitely low and infinitely high frequencies. $Lim\ \{\underline{\omega} \to 0, Re[Z(\omega)]\}$ will be the equivalent of the dc resistance for this particular process. The limiting series capacitance of a power source may be obtained by analyzing $Lim\ \{\underline{\omega} \to 0, -1/\omega \cdot Im[Z(\omega)]\}$. Further, the impedance function may be rearranged in terms of these limiting quantities so that they can be directly obtained from fitting the experimental impedance spectrum.

We will exemplify this approach for the finite-length-diffusion case. The impedance function for this case is described in Section 2.1.3, Eq. (135) for a reflective boundary and Eq. (136) for a transmissive one. The first case corresponds to diffusion into a layer of electrolyte or active material with limited thickness, often used to analyze the impedance of battery materials. The second case applies to diffusion through a membrane, important in fuel-cell analysis. Before finding the limiting resistance, we need to make sure that impedance equation is not in normalized form, e.g. produces unit Ohm. Considering that *dE/dc* is the voltage change with volumetric concentration, (volt · mole/m^3) we need to divide it by electrode area *S* so that Eqs (135) and (136) involve the total impedance in Ohms. Correspondingly modified equations are given below:

$$Z(\omega) = \frac{dE}{dc} \cdot \frac{1}{z \cdot F \cdot S} \cdot \frac{coth\left(\sqrt{\frac{i \cdot \omega}{D}} \cdot l^2\right)}{\sqrt{i \cdot \omega \cdot D}} \tag{6}$$

$$Z(\omega) = \frac{dE}{dc} \cdot \frac{1}{z \cdot F \cdot S} \cdot \frac{\tanh\left(\sqrt{\frac{i \cdot \omega}{D}} \cdot l^2\right)}{\sqrt{i \cdot \omega \cdot D}} \tag{7}$$

Here D is the diffusion coefficient in m^2/sec; z is the charge of diffusing ions; and l is the diffusion length.

On finding the real part from Eq. (6) and taking $Lim\ \{\underline{\omega} \to 0,\ Re[Z(\omega)]\}$, we obtain the limiting resistance

$$R_d = \frac{dE}{dc} \cdot \frac{\ell}{D \cdot S \cdot F \cdot z} \cdot \frac{1}{3} \tag{8}$$

The imaginary part of the impedance approaches infinity at zero frequency since its dependence on frequency approaches that of a series capacitor. Considering that the impedance of a such a capacitor is $Z(\omega) = 1/\mathrm{i} \cdot \omega \cdot C$ and $Im[Z(\omega)] = -1/\omega \cdot C$, we can find the capacity knowing the imaginary part of the impedance to be $C = -1/\omega \cdot Im[Z(\omega)]$. To find the limiting capacity for reflective diffusion, we use this relation at the zero frequency limit, $C = Lim\{\underline{\omega} \to 0, -1/\omega \cdot Im[Z(\omega)]\}$, where $Z(\omega)$ is given by Eq. (6). The resulting capacity is shown below:

$$C_d = \frac{l \cdot S \cdot F \cdot z}{\frac{dE}{dc}} \tag{9}$$

Figure 4.5.1 compares the impedance spectrum of reflective finite-length diffusion with the spectrum of its limiting equivalent circuit, a limiting resistance in series with a limiting capacitance. It can be seen that these responses approach each other as the frequency decreases.

The limiting equivalent circuit may be directly applied in analysis of power source performance because the direct-current response, which corresponds to the zero-frequency limit of the impedance spectrum, is of most importance. Conversion of Eqs (8) and (9) allow one to find the ohmic characteristics of a given power source at direct current. They also allow one to obtain a close initial guess for impedance fitting of the equivalent circuit by direct observation of the impedance plot. R_d will

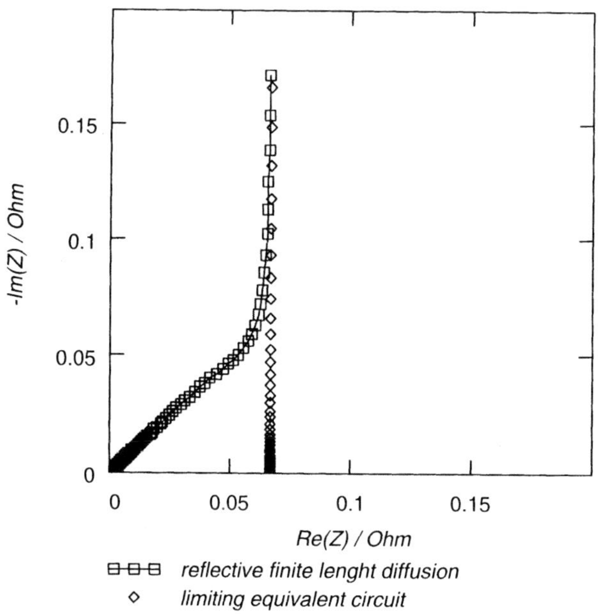

Figure 4.5.1. Reflective finite-length diffusion impedance spectrum.

be the value on the right side of the Re-axis, and C_d is found for the lowest frequency point using the relation described above. The values found can be used to find D and dE/dc or used as initial guesses for fitting. Alternatively, Eq. (6) can be rearranged in terms of limiting parameters, as shown below:

$$Z(\omega) = \sqrt{\frac{3Rd}{Cd \cdot i \cdot \omega}} \cdot tanh\left(\sqrt{3 \cdot Rd \cdot C_d \cdot i \cdot \omega}\right) \tag{10}$$

Fitting experimental response to an equation expressed in electrical terms has the additional advantage that the errors in all parameters have electrical units, thus allowing one to evaluate their significance when used to model a particular electrical device. In this way it can be easily seen which parameter becomes poorly defined in the fit from the point of view of its effect on the overall electric response.

If a system has more then one kinetic step, for example a charge transfer reaction is followed by diffusion, treated first by Ho *et al.* [1980], the contributions of each kinetic step can be compared if all impedance equations are expressed in terms of their limiting resistances and capacitances. In this example the equivalent circuit of a charge transfer reaction can be represented as R_{ct} in parallel with double layer capacitance C_{dl} and its admittance as the sum of the admittances of the parallel elements. The limiting resistance of this equivalent circuit will be R_{ct}. The overall impedance of charge transfer resistance followed by diffusion is given by summing up impedances of these two serially connected equivalent circuits and considering the admittance of parallel capacitor:

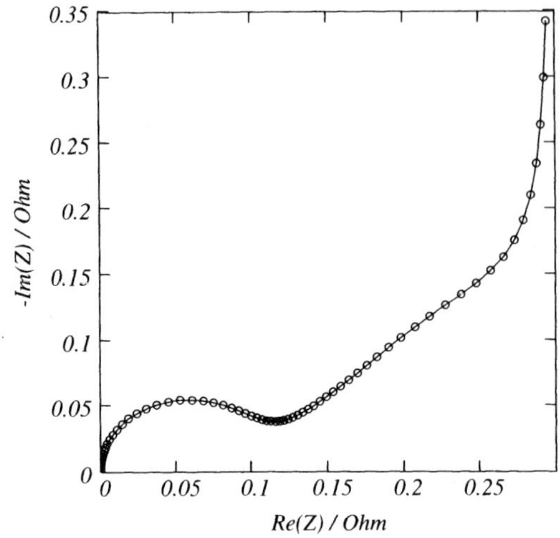

Figure 4.5.2. Diffusion impedance in series with surface impedance.

$$Z(\omega) = \frac{1}{\dfrac{1}{R_{ct} + \sqrt{\dfrac{3 \cdot R_d}{C_d \cdot i \cdot \omega}} \cdot coth(\sqrt{3 \cdot R_d \cdot C_d \cdot i \cdot \omega})} + i \cdot \omega \cdot C_{dl}} \qquad (11)$$

The impedance spectrum of such a system is shown in Figure 4.5.2.

It can be seen that the semicircle due to the charge transfer on electrode/electrolyte boundary is followed by a 45° line, characteristic for diffusion, which changes to a vertical line due to the limited diffusion length and reflective boundary. The effective series resistance observed by applying dc current to this circuit for a prolonged time is just a sum of R_{ct} and R_d. The relative "importance" of charge transfer and diffusion for the performance of this particular power source is simply quantified as the ratio between R_{ct} and R_d; in the above example it is 0.1 Ohm and 0.2 Ohm. Thus for this case, the diffusion is twice as important for the dc performance than is the charge transfer.

4.5.1.4 Discretization of Distributed Elements, Augmenting Equivalent Circuits

A description of the electrochemical kinetics of power sources requires treatment of two different kinds of processes. The first, intensive, can be thought of as localized, occurring in a specific volume, that is negligible compared to volume of the entire system, such as charge transfer or double-layer capacitance. These processes are described by ordinary differential equations and their equivalent circuits consist of basic building blocks representing losses and storage—resistors and capacitors.

Once a system is represented as an equivalent circuit, additional complications can be easily taken into account. For example nonlinearity can be introduced by making a resistor value dependent on current. Numerical solution for networks of nonlinear capacitors and resistors are straightforward with standard theoretical treatment and free computational tools (such as SPICE, Hageman [1993]) widely available.

Processes included in the second class are extensive and so are distributed over a large volume, which can include the entire test system. Diffusion is one example of such a process. Because parameters describing such processes are different depending on position (for example concentration changes with distance from an electrode), two coordinates—time and space—are needed for description and differential equations in partial derivatives have to be employed. These can be analytically solved using the Laplace transform in many cases; however, sometimes only numerical solutions are possible, particularly if the system parameters are inhomogeneous or the system is nonlinear. One of the ways to obtain such numerical solutions involves subdividing the spatially distributed system into small "compartments," each of them thought of as separate localized processes. This approach bridges the second type of systems with the first, because each of the localized compartments can be represented as an equivalent circuit and the entire distributed process can be described as the response of a network of two or three-dimensionally connected equivalent circuits of localized processes.

Section 2.1.3 shows how the process of diffusion in one dimension can be represented by a chain of resistors and capacitors, and Section 2.1.6 shows how porous electrodes can be represented by a similar network. While this approach is valid even for a distributed process with no boundary (like diffusion into infinite space), discretization is even more important for the case where a distributed process is limited in space. In this case, a finite number of discrete elements can describe the system to arbitrary precision, and can be used for numerical calculations, as treated in next chapter, even if no analytical solution is possible. Another convenience of discretizing a distributed process is the resulting ability to add additional "subprocesses" directly to the equivalent circuit rather than starting the derivation by formulating a new differential equation. For example the equivalent network representing electric response of a pore is given in Figure 4.5.3.

Here ρ is the specific conductivity of the electrolyte in a pore (wall conductivity assumed infinite), and c is the specific capacity of the walls per unit length of the pore. Because we formulated the problem as electric network, we abstract from

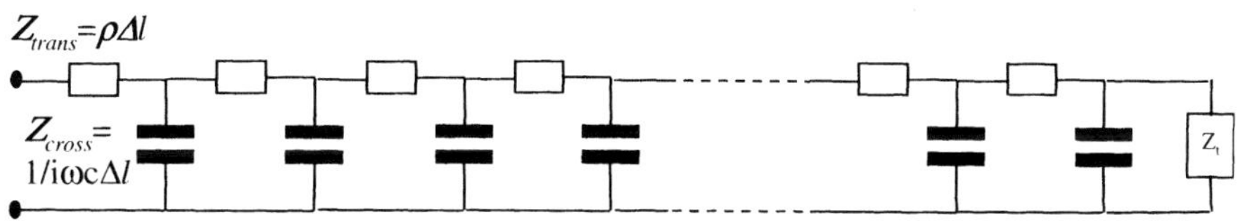

Figure 4.5.3. Discretization of a transmission-line model of a porous electrode or diffusion process.

underlying physical meaning and seek a solution for input impedance in terms of complex specific impedances per unit length. Instead of ρ we will have the complex function $Z_{trans}(w)$, and instead of specific capacitance $c\Delta l$ we have specific surface impedance $Z_{cross}(w)$. The boundary condition at the side where a pore contacts the bulk solution can be considered as the termination impedance of the network, Z_t. The solution for the input impedance of a chain-connected network is found in the theory of electric networks (Zinke and Vlcek [1987]) as

$$Z(w) = Z_0(w) \cdot \frac{1 + \dfrac{\dfrac{Z_t(w)}{Z_0(w)} - 1}{\dfrac{Z_t(w)}{Z_0(w)} + 1} \cdot e^{-2\cdot\gamma(w)\cdot l}}{1 - \dfrac{\dfrac{Z_t(w)}{Z_0(w)} - 1}{\dfrac{Z_t(w)}{Z_0(w)} + 1} \cdot e^{-2\cdot\gamma(w)\cdot l}} \tag{12}$$

Here Z_0 is commonly called *the characteristic impedance* of the transmission line, and γ is its *propagation coefficient* defined as

$$Z_0 \equiv \sqrt{Z_{trans} \cdot Z_{cross}}$$
$$\gamma \equiv \sqrt{\frac{Z_{trans}}{Z_{cross}}} \tag{13}$$

This bulky equation reduces to simple and familiar equations for situations where the terminating impedance is infinite (when pores are terminated by a wall) or when it is zero, as in the case of conduction through a membrane. Note that these equations for conduction through a porous layer are mathematically identical to those for diffusion, given in Chapter 2.1.3.1

$$Z(w) = Z_0 \cdot coth(l \cdot \gamma \cdot (w)) \tag{14}$$

$$Z(w) = Z_0 \cdot tanh(l \cdot \gamma \cdot (w)) \tag{15}$$

Formulating the problem in a discretized way allows us to extend it effortlessly to more complicated cases. Let's assume, for example, that in addition to double-layer capacitance we will have an electrochemical reaction on the pore surface, as would be the case in a battery or fuel-cell electrode. The equivalent circuit for the pore surface now will involve a capacitor in parallel with a charge transfer resistance, R_{ct}; and the surface impedance Z_{cross} will be given as follows:

$$Z_{cross} = \frac{1}{\left(\dfrac{1}{R_{ct}} + C \cdot \omega \cdot li\right)} \tag{16}$$

To obtain the impedance of a porous electrode with a reaction on its surface we only need to substitute the above expression into Eqs. (13, 14). This relation will then be valid only at higher frequencies where diffusion can be neglected.

To extend the validity to lower frequencies, we can consider solid-state diffusion of reacted species into the bulk of the particles of the active material. The diffusion length into a particle will be limited by the particle size, and finite-length reflective Warburg impedance can be used for its description. We simply need to add a finite length Warburg impedance to R_{ct} in Eq. (16) to obtain the impedance of pore surface Z_{cross}, which was earlier derived in Eq. (11). Again, to obtain impedance for a porous electrode with reaction and diffusion on the pore surface we just need to substitute a new Z_{cross} into Eq. (14). This convenient approach, of augmenting equivalent circuits with increasing complexity, has been successfully applied to the analysis of conducting polymers (Paasch *et al.* [1993], Fletcher [1992], Popkirov *et al.* [1997]) but is also applicable for batteries (Barsoukov *et al.* [1998, 1999]) and supercapacitors (Conway and Pell [2002]).

4.5.1.5 Nonlinear Time-Domain Modeling of Power Sources Based on Impedance Models

In Section 4.5.1.2 we discussed time-domain modeling based on an equivalent circuit used in impedance analysis. However, this is only possible for the case of small excitations, where all parameters can be assumed independent of the magnitude of the excitation (system is linear). This is generally not the case over the operational range of power sources, for example in the course of gradual discharge of a battery from full to empty its impedance spectrum is changing significantly, as shown in Figure 4.5.4 based on data from Barsoukov *et al.* [1999].

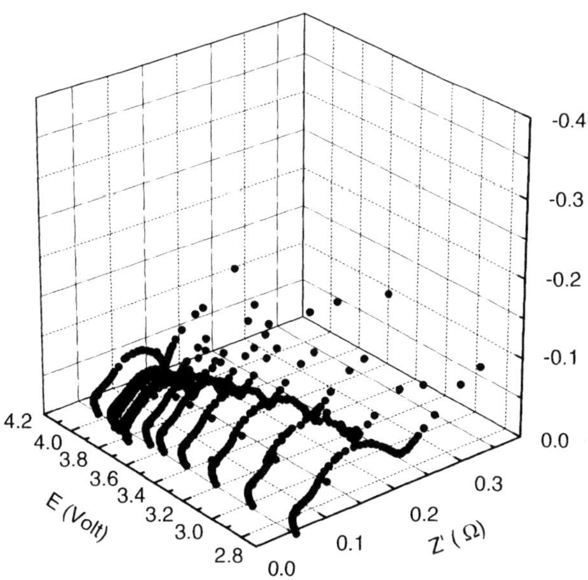

Figure 4.5.4. Impedance spectra of Sony 18650 Li-ion battery dependent on open circuit voltage. Spectra are from 1 kHz to 1 mHz (frequency decreases from left to right).

Because of such large changes, any linear model would fail to describe the entire discharge process. The same conclusion applies in a lesser degree to supercapacitors, whose operational range of voltages is much smaller but still enough to cause significant change of electrochemical properties.

Another type of nonlinearity is dependence on load. The Butler–Volmer equation predicts linearity of voltage/current dependence only within a voltage change below 20 mV. Battery response to current can stay linear at much higher overvoltages, however, because most of the voltage drop in batteries is due to resistive (conductivity of active material and electrolyte) and diffusion contributions, which are both linear regardless of load. However, at high loads the battery response also becomes nonlinear and the effective impedance decreases with increasing load (Mauracher and Karden [1997]).

Yet another type of nonlinearity observed in distributed system is due to the differences of the material state dependent on the distance from the electrode. This occurs at high discharge rates in rechargeable batteries and results in "material blocking" or "starvation" of charge carriers. Areas near to electrode have higher "states of discharge" than less accessible areas, resulting in a decrease of conductivity and increase of diffusion hindrance in these areas, thus preventing still available material from reacting. This inhomogenity effect makes the apparent battery impedance at high discharge rate increase compared with lower rates. Note that Butler–Volmer and "blocking" effects change apparent impedance in opposite directions, so only complete nonlinear modeling can indicate which one will prevail in a particular material at a particular rate.

Fortunately, the formulization of an electrochemical model as a discretized equivalent circuit allows one to perform nonlinear modeling considering all these factors. Resistances and capacitances have to be made dependent on the state of charge of the battery as well as other factors which can influence it, such as discharge rate and temperature. The dependence of parameters on the charge state can be obtained by fitting multiple impedance spectra measured at different states of charge. Because "state of charge" is not an electric quantity that can be used in simulation of electric circuit behavior, parameter change can be represented as a function of open circuit voltage instead. Open circuit voltage is itself one of the simulation parameters, and corresponds to the voltage at the largest series capacitor in a particular "chain" of the distributed circuit. In the case of a simple system with the equivalent circuit shown in Figure 4.5.3, the nonlinear circuit for time-domain modeling, where the number of discrete elements is chosen as N = 8, is given in Figure 4.5.5.

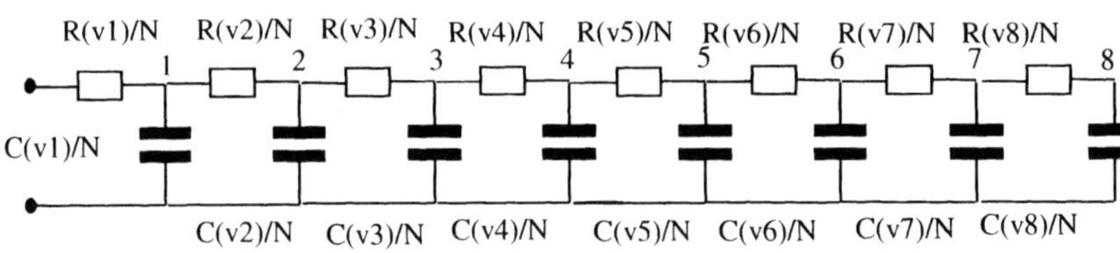

Figure 4.5.5. Discretized transmission line for non-linear time domain simulation.

Here functions $R(v)$ and $C(v)$ can be obtained by piecewise-linear interpolation of the dependence of R and C parameters obtained by fitting the experimental spectra at different voltages (such as in Figure 4.5.4) to the impedance function in Eq. (10). Any other suitable smooth interpolation can be used. The impedance function has to be expressed in terms of electric parameters, as described in Section 4.5.1.3. For use in a discretized equivalent circuit, the values obtained from the fit have to be divided or multiplied by the number of chains, depending on the series or parallel position of the electric element. So, for series resistors it has to be divided, and for parallel, multiplied. It should be considered that the low-frequency limit of $Re(Z)$, used as a fitting parameter in the equation, is not always a simple sum of the discrete elements that constitute a transmission line. In particular, in Eq. (10) the R_d is 1/3 of the specific resistance multiplied by the transmission line length, as can be seen from Eq. (8). Therefore resistance of single chain shown in Fig 4.55 will be R_d*3/N.

V1–V8 are voltages at the nodes with the given number. Note that this nonlinear equivalent circuit covers two types of nonlinearity—the one due to changing state of charge (as voltage reflects state of charge) and the one due to inhomogeneity, because state of charge for the calculation of parameters in each chain is not assumed to be the same everywhere along the transmission line, but is calculated at every node as a result of simulation. The nonlinearity due to the Butler–Volmer type of relation can also be introduced as, for example, dependence of R_{ct} value on voltage drop across it, shown in Eq. (17).

$$R_{ct}(V_{R_{ct}}) = \frac{V_{R_{ct}} \cdot R_{ct}}{K \cdot \left[e^{\frac{\beta \cdot V_{R_{ct}}}{K}} - e^{-(1-\beta) \cdot \frac{V_{R_{ct}}}{K}} \right]} \tag{17}$$

Here V_{Rct} is the voltage across resistor; R_{ct} is the linearized value obtained from the fit of impedance spectra; K is RT/F and β is the transfer coefficient that can be assumed to have a value of 0.5. A more exact treatment allows one to derive a similar relationship using the Frumkin isotherm (Levi and Aurbach [1999]).

Methods to calculate time dependent current/voltage response for circuits whose parameters depend on voltage or current are well developed in theoretical electronics and free open source tools for numerical calculations are available, such as general electronic circuit simulator SPICE (Hageman [1993]).

4.5.1.6 Special Kinds of Impedance Measurement Possible with Power Sources—Passive Load Excitation and Load Interrupt

Impedance measurement techniques for most electrochemical systems are covered in detail in Chapter 3. One specific aspect of measurement on power sources arises from the special importance of low frequencies down to 1 mHz. Because of the very long time needed to collect each single low frequency point, the system might change significantly in the course of measurement if frequencies are measured sequentially. Spectrum measurement time can be shortened up to order of magnitude by applying all frequencies of interest simultaneously, and analyzing system response by FFT (Popkirov and Schindler [1992]). As an added bonus, nonlinearity of the response can be detected if additional frequencies appear in the response spectrum (Popkirov [1994]). It is also possible to qualitatively discriminate between nonlinear, not stationary and noncausal distortions by analyzing response to wide-band excitation signal as shown by Schoukens *et al.* [2001]. Another problem of impedance measurement on power sources is associated with using an external energy source to excite a voltage or current response of the system. Because electrochemical power sources are usually of very low resistance and always have voltage bias, an external power source needs quite a high power capability to achieve a response with an acceptable signal-to-noise ratio. For large power sources, such as cranking lead acid batteries, there are no commercially available potentiostats/galvanostats available to support the needed excitation power. Fortunately, electrochemical power sources offer a unique opportunity to simplify the test set-up, because they *are* energy sources themselves. Their impedance can be revealed by applying a varying load across their terminals and observing the transient voltage change. When the load variation is periodic, impedance can be found as $Z(w) = V^{\sim}(w)/I^{\sim}(w)$ where $V^{\sim}$ and $I^{\sim}$ are Fourier transforms of measured current and voltage. If excitation is nonperiodic, impedance can be found as $Z(s) = V(s)/I(s)$ where V and I are Laplace transforms of voltage and current, and only the imaginary part of s is substituted into the equation. There can be many different variations of this approach, reviewed in particular in Barsoukov *et al.* [2002]. In many cases, current measurement is not needed, as the applied load is known, so a test set-up may consist of a digital voltmeter with data logging capability at a high rate, and a low-pass filter set to $\frac{1}{2}$ of the sampling frequency.

The simplest method is "constant load". It is accomplished by instantaneously connecting the terminals of a power source resting at a voltage V_0 to a resistor with a resistance value R which will cause approximately a 20–40 mV voltage drop across the power source. The resistor should have a low thermal coefficient (such as constantah based) and appropriate power rating and cooling so that no extensive heating occurs. Voltage has to be sampled with a sampling interval at least twice shorter than the inverse of the maximum frequency and the duration of sampling should be more, or equal to, the inverse of the minimum frequency needed to be measured (Nyquist theorem). Sampling points need not be equidistant in time. To save memory, and even improve noise rejection, sampling with a logarithmically increasing time between samples can be used. It is also necessary to use low-pass filtering at the frequency equal to half of the inverse of sampling interval to prevent aliasing.

To obtain impedance spectra, acquired arrays $v[i]$, $t[i]$ are used to estimate parameters of a differentiable interpolation function $v(t)$. These parameter values are then substituted into Laplace transform of this function, $V(s)$. To achieve maximum possible noise rejection, it is useful to employ a physically meaningful interpolation function for the particular system under test and to fit $v[i]$, $t[i]$ data to the function to find its parameters. For most electric systems, such a function would be a sum of exponentials, as proposed to be use in such conversion by Macdonald [1993]. In this case $v(t)$ and its Laplace transform $V(s)$ are shown below.

$$v(t) = \sum_{i=0}^{n} k_i \cdot e^{\frac{-t}{\tau_i}} \tag{18}$$

$$V(s)=\sum_{i=0}^{n}\frac{k_i}{s+\frac{1}{\tau_i}} \tag{19}$$

The use of differentiable smoothing splines, or even linear piecewise interpolation in conjunction with some noise-reduction method, can also be practical if the nature of the system is not known. Once parameters of $V(s)$ are found by a fit of data to $v(t)$, the impedance spectrum in the frequency range defined by the sampling interval and duration of sampling can be obtained using Eq. (20). Fit can be performed either by linear regression with fixed values of time constants τ_i (for example logarithmically distributed from duration of sampling step to the duration of sampling period), or using nonlinear fit where both number and values of time constants are optimized to achieve statistically best description of data (Provencher [1976]).

$$Z(s)=\frac{V_0}{s}\cdot\frac{R}{V(s)}-R \tag{20}$$

Another similar method suitable for power sources testing is based on "current interrupt". If the load is applied for a long time, so that it can be assumed that the system under test has stabilized and no longer changes in the time interval similar to the necessary measurement duration, the load is disconnected and sampling starts in the instant of disconnection. Acquired data values $v[i]$, $t[i]$ are again used in a fit to a selected carrier function $v(t)$ to obtain its parameters which are then substituted into the equation of this function's Laplace transform to obtain $V(s)$. The impedance function is then obtained as shown below.

$$Z(s)=\frac{s}{I}\cdot V(s)+\frac{1}{s\cdot C_{ser}} \tag{21}$$

Note that this method is not sensitive to series capacitance so it has to be determined from the voltage slope immediately prior to measurement as $I/dv/dt$.

This method has the disadvantage that load application time prior to measurement is fixed. This can be inconvenient if a measurement has to be performed in a particular instant of time based on some other criteria, for example we might want to limit maximum voltage drop between two measurements during the battery discharge. Variation of this method allows one to use excitation periods of arbitrary length, even much shorter ones than the subsequent sampling period. Eq. (19) should be modified in this case to correct for arbitrary excitation length pulse t_{pulse} as shown in Eq. (22), and then substituted into Eq. (21) to find impedance, omitting the addition of serial capacitance which is already taken care of by correction.

$$V(s)=\sum_{i=0}^{n}\frac{k_i}{\left(s+\frac{1}{\tau_i}\right)\cdot\left(e^{\frac{t_{pulse}}{\tau_i}}-1\right)} \tag{22}$$

For this method it is preferable to use an electronic load with constant current rather then a resistive load, to make sure that current is unchanged during the excitation pulse.

4.5.2 Batteries

Evgenij Barsoukov

4.5.2.1 Generic Approach to Battery Impedance Modeling

Despite the large variety of existing battery chemistries, the basic design of batteries shares several common elements. All batteries have porous electrodes consisting of particles of energy-storing material and conductive additive, held together by polymeric binder. They also include electrolyte and a separator, preventing electric contact between opposite electrodes. Instead of focusing on particular chemical compositions, a general approach to battery kinetics can be based on analyzing the impedance of the above common components. This not only allows one to create a working kinetic model of the battery as a whole but also allows the evaluation of the effect of each component on material power capability and even a comparison between different chemistries.

Figure 4.5.6 shows the kinetic steps that are common for most batteries, originally proposed for the Li-ion battery in Barsoukov *et al.* [2000]. These include (from the right) electronic conduction through the particles and ionic conduction through the electrolyte in cavities between particles. On the surface of each particle, charge transfer involves the resistance of an insulating layer and activated electron transfer resistance on the electronic/ionic conduction boundary. Further, ions have to diffuse into the bulk of particles via solid-state diffusion. Other subsequent processes, such as the formation of new crystalline structures, can also become limiting kinetic steps at frequencies below 1 mHz. Not indicated on the figure are conduction through the electrolyte and porous separator, and conduction through wires, which both dominate high-frequency behavior above 10 kHz. The inductance of wires and electrode winding becomes a factor at even higher frequencies. In subsequent sections, we will have a closer look at the impedance of all these components.

Impedance of Electrolyte and Membrane. The theoretical background of impedance spectroscopy applied to liquid and solid electrolytes is covered in detail in Sections 2.1.2 and 2.1.3. In order to achieve high conductivity, batteries always use a large excess of conductive salts in liquid electrolytes. This allows them to be classified as "supported electrolytes" and allows the use of a simplified theoretical approach that assumes an absence of electric field in the bulk electrolyte and disregards the migration term in Eq. (78) of that chapter. In practice, this means that we have to consider only the diffusion contribution to electrolyte impedance, while conduction and relaxation effects of bulk electrolyte are shifted to the high frequency region and therefore appear as a series resistance contribution in impedance measurements in practical batteries. In order to measure electrolyte conduction and relaxation impedance excluding electrode effects, the four electrode arrangement can be used, where two reference electrodes are placed in close proximity to two other electrodes that are unable to exchange ions with the electrolyte (such as Pt) in order to exclude diffusion contributions. Voltage response should then be measured differentially between the reference electrodes.

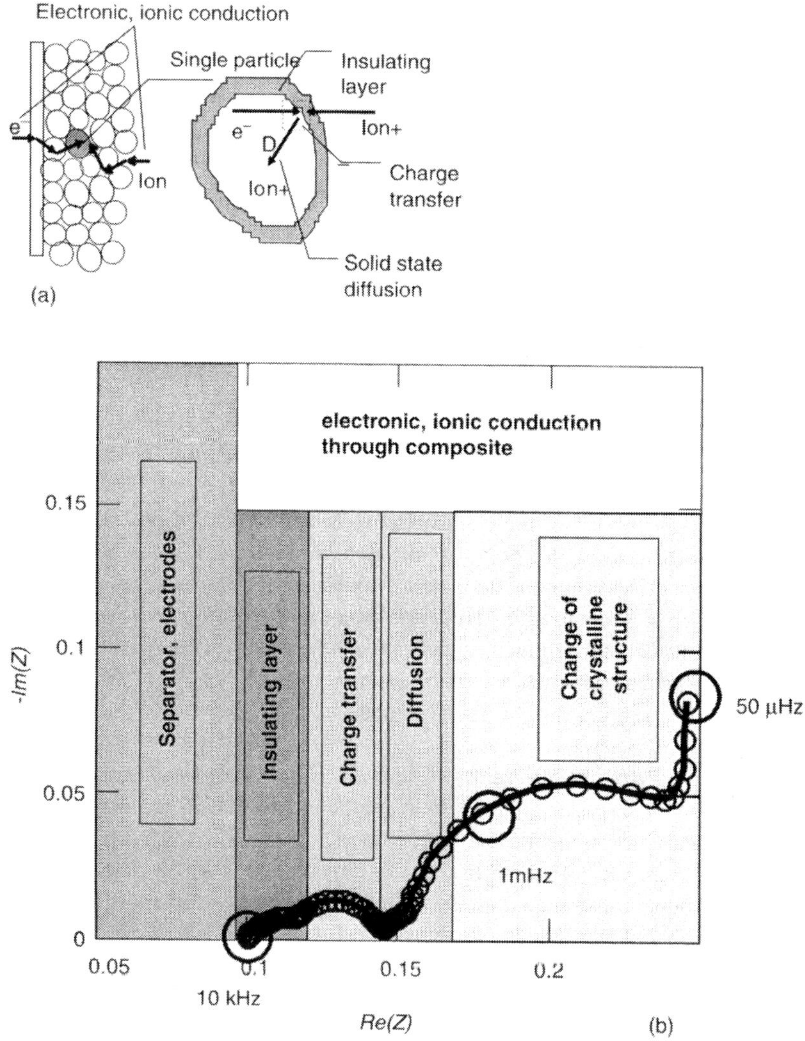

Figure 4.5.6. (*a*) Kinetic steps common in most batteries; (*b*) typical impedance spectra of intercalation material.

In the case of solid electrolytes, that are currently being tested but have not yet been used in commercial batteries, the situation is more complicated as they cannot be treated as being supported, and all relaxation, migration, and diffusion effects can overlap in the same frequency region. See Section 2.1.3 for the treatment of this case.

The diffusion impedance of a bulk electrolyte can be described by a finite length Warburg impedance with transmissive boundary (Eq. (7)). A transmissive boundary is appropriate because an ion produced at the cathode is consumed at the anode and *vice versa* during battery electrochemical processes. A more precise treatment, using

Eq. (12), is one where electrodes and diffusion in the electrolyte between electrodes are represented by means of an ion-producing electrode in series with a transmission line, terminated with an electrode that consumes diffusing ions. In this case, the terminating impedance Z_t will be the impedance of this electrode. An alternative treatment of diffusion with imperfect boundary conditions is given in Franceschetti *et al.* [1991].

However, because of the high concentration of the ions taking part in a charge/discharge reaction and the very thin electrolyte layer used in practical batteries, this impedance is negligible for most liquid electrolytes applied in practice, and its influence will only appear in the high frequency region or, most commonly, as a *series resistance contribution.* On the other hand, in the case of gel electrolytes, the Warburg impedance of the electrolyte layer can be well resolved, especially if a thicker layer of electrolyte is used for research purposes (Qian *et al.* [2002]). The impedance spectrum of polymer and gel electrolytes appears as a depressed semicircle. Typical analysis of the spectrum in order to determine specific conductivity is given in Section 2.1.2.3, as devised by Cole and Cole [1941].

An important contribution of the electrolyte to overall battery impedance is that associated with electrolyte in the pores of the active material. Because the cross section of pores is much less then the geometric dimensions of electrodes, even high-conductivity electrolyte leads to considerable resistance along the pores. Pores also have a significant surface capacitance, which makes the effect of the impedance of the electrolyte in pores appear over a wide frequency range. This effect will be discussed in detail in the section on porous electrodes.

Impedance of the Particle Surface. Electrochemical reactions, formation of a double layer, and adsorption all contribute to the particle surface impedance. Adsorbed species can diffuse inside the bulk of active material in batteries, except for the case of metal anodes. Because of this, the impedance of the particle surface will involve a Warburg impedance element rather than a simple capacitor to represent adsorption (a model for metals can, however, use a capacitor instead of Z_w). The Randles [1947] equivalent circuit in Figure 4.5.7 encompasses all these steps. Note that in batteries, active materials are not dissolved in the electrolyte but are

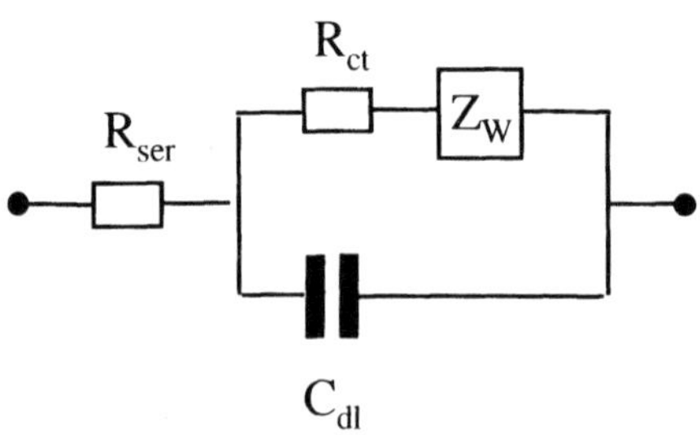

Figure 4.5.7. Randles equivalent circuit. R_{ser} represents the series resistance, R_{ct} the charge transfer resistance of the electrochemical reaction, and Z_w solid state diffusion and other subsequent reactions.

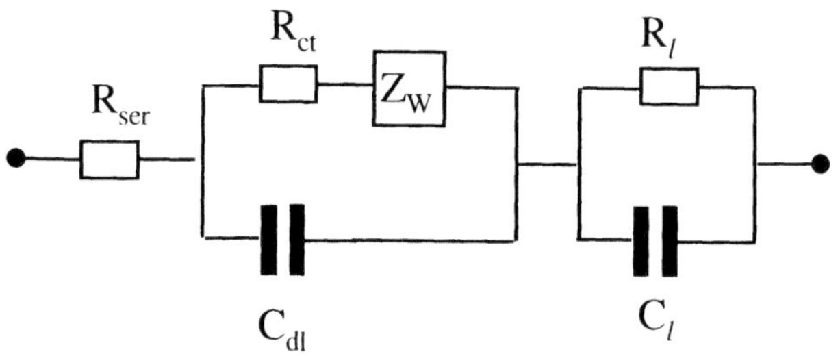

Figure 4.5.8. Equivalent circuit of particle surface impedance taking into account a passivating layer.

typically solid; therefore Z_w involves solid state diffusion, as proposed by Ho *et al.* [1980], in contrast to diffusion in electrolyte originally discussed by Randles. The diffusion component will be discussed more in a later section on this subject.

On the surface of Li-ion anodes, the formation of an ionically conductive passivating layer, or SEI, has been observed (E. Peled [1979, soloced to Rist]). Other battery electrodes can also build up similar inactive layers due to different degradation reactions. Passivating layers, apart from their critical importance for preventing irreversible reactions in the battery, have measurable effects on the discharge kinetics and therefore can be analyzed by impedance spectroscopy. The growth of such layers has been investigated by Barsoukov *et al.* [1998]. A simple model, treating surface impedance as the equivalent circuit in Figure 4.5.8, allowed the estimation of the dielectric constant, thickness, and ionic resistance of the layer.

In order to evaluate specific electrochemical characteristics, such as exchange current density $i_0 = R \cdot T/n \cdot F \cdot R_{ct} \cdot S$, estimation of the particles' surface area S is necessary. Note that the experimental estimation of the surface boundary between electronically and ionically conductive media in a composite material consisting of multiple particles has not been successful to date. BET surface area estimation usually tends to severely overestimate this surface area. It correlates well with irreversible capacity loss during first intercalation (Iijima *et al.* [1995]) but not with the surface impedance of materials (Aurbach *et al.* [2001]), which indicates that loosely electrically connected microparticles make a major contribution to BET surface area but not to the electrochemically active area. In order to estimate only the area that is electrically accessible and also to use the same value of S for both diffusion and surface kinetics, it is common to use a summary geometric area of the particles of the active material as an estimate for surface area. See further details in the "bringing it all together" section.

Diffusion. While an electron transfer reaction only occurs on a particle surface, most of the electrochemically active material resides in the bulk of the particles. Many electrochemically active substances are not suitable as battery materials, despite their promising specific capacity and voltage, because their crystals are too dense and so only the surface is active. An open crystalline structure, allowing ions to enter and leave the crystal, is a common characteristic of materials used in most modern batteries. Different from primary battery materials, such as MnO_2, rechargeable battery materials should be also be able to maintain and protect their crystalline structure from a change to dense form (collapse) and from dissolution. Cathode material in a Ni–Cd battery, both anode and cathode in Ni–MH and Li-ion one, follow this principle. The process of ion movement inside the crystals is controlled by solid state diffusion. The above considerations make quantitative characterization of this process critical for determining suitability and improving the kinetics of battery materials.

Ions diffuse toward the center of particles, therefore their diffusion path is limited, e.g. the boundary condition is reflective. Impedance analysis of the finite length diffusion for different electrode geometries and boundary conditions is summarized by Jacobsen and West [1995]. Particle geometries occurring in battery materials are thin plate (planar), spherical, and cylindrical. Below are equations for corresponding geometries, modified so that parameters are expressed in electrical terms.

For planar geometry, see Eq. (10).

For spherical geometry:

$$Z(s)=\frac{tanh(\sqrt{3\cdot C_d\cdot R_d\cdot s})}{\sqrt{\frac{3\cdot C_d\cdot s}{R_d}}-\frac{1}{R}\cdot tanh(\sqrt{3\cdot C_d\cdot R_d\cdot s})} \tag{23}$$

Transformation from electrical unit parameters to electrochemical ones is given as

$$\frac{dE}{dc}=\frac{F\cdot n}{C_d}\cdot\frac{4}{3}\cdot\pi\cdot r^3 \qquad D=\frac{r^2}{3\cdot C_d\cdot R_d} \tag{24}$$

For cylindrical geometry:

$$Z(s)=\frac{I_0(\sqrt{2\cdot R_d\cdot C_d\cdot s})}{(\sqrt{2\cdot R_d\cdot C_d\cdot s})\cdot I_1(\sqrt{2\cdot R_d\cdot C_d\cdot s})}\cdot R_d \tag{25}$$

Here $I_0(x)$ and $I_1(x)$ are Bessel-functions of the first kind, with 0 and 1 order correspondingly. Transformation from electrical unit parameters to electrochemical ones is given for this case as:

$$\frac{dE}{dc}=\frac{F\cdot n\cdot\pi\cdot r^2\cdot h}{C_d} \qquad D=\frac{r^2}{2\cdot C_d\cdot R_d} \tag{26}$$

Here r is the radius and h the length of the cylindrical rod. Equations (10) and (23)–(25) describe the impedance of a single particle. Obviously, a composite electrode consists of a huge number of such particles that are not in parallel or in series in the electrical sense, but are distributed as a chain network along the thickness of the composite. The way that impedances of single particles are combined into impedance of a whole electrode will be described in the "bringing it all together" section.

Diffusion in solid state materials when the activities of diffusing species are not concentration independent does not follow Fick's law. The diffusion coefficient has to be modified by a *thermodynamic enhancement factor* taking into account the activity change with concentration to bring the diffusion equation to semi-Fick's form. The modified diffusion coefficient, $D_c = D_0 L$ is called the *chemical diffusion coefficient* and L is the *enhancement factor*, see details in Section 2.1.3.2. All solid state diffusion coefficients treated in this chapter are chemical diffusion coefficients even if not explicitly stated.

The dE/dc dependence on state of charge of battery materials is also significantly different from Nernstian but is often well described by the Frumkin isotherm which takes into account attractive or repulsive interactions of adsorbed species, as reviewed by Levi and Aurbach [1999]. The actual dE/dc in the case of any particular material can be obtained by discharge/relaxation experiments, and knowledge of its value can significantly assist quantitative analysis of impedance spectra, as will be shown in the section on battery-specific improvement in impedance spectra fitting.

It is important to note that even if the experimental dE/dc dependence is used in Eqs (11), (21)–(24), the estimated diffusion coefficient will still be the *chemical diffusion* coefficient. However, Levi and Aurbach [1999] have shown that analysis of experimental dE/dX (where X is degree of occupation of intercalation sites) allows one to obtain a concentration-independent diffusion coefficient through calculation of the enhancement factor explicitly, as shown below

$$L = \frac{e}{k \cdot T} \cdot X \cdot \frac{dE}{dX} \tag{27}$$

or, in terms of the Frumkin isotherm:

$$L = \frac{1 + g \cdot (1 - X) \cdot X}{1 - X} \tag{28}$$

Here g is measure of interaction between intercalation sites in the Frumkin isotherm, and X is the degree of occupation of intercalation sites. X can be expressed from the charge Q passed during intercalation as $X = Q/Q_{max}$, where Q_{max} is the maximum charge stored by the particular intercalation phase. In many cases, such as intercalation into graphite, multiple phases are present and therefore the entire intercalation region must be subdivided into subregions for analysis.

Other deviations from Fick's diffusion relationships can be caused by the kinetics of new phase formation, which appears as a secondary reaction reducing the concentration of the diffusing species. This effect not only alters the apparent diffusion coefficient but actually changes the shape of the impedance spectrum, introducing deviation from $\sqrt{\omega}$ dependence at low frequencies below 1 mHz. The impedance spectra of Li-ion intercalation anodes exhibiting this effect were analyzed by Barsoukov *et al.* [2000]. It is convenient to express the needed modification of the impedance equations at the equivalent circuit level, as in Figure 4.5.3. In parallel with a diffusion pseudocapacitance, $c\Delta l$, we can add a resistor R_p in series with the capacitor C_p, which represents an additional activated-concentration-reducing process due to new phase formation along the diffusion patch. This will modify Eqs (11), (21) and (24) so that the diffusion pseudocapacitor admittance C_d s is replaced by the admittance of the new parallel circuit, $1/(R_p + 1/C_p s) + C_d\, s$, as for planar diffusion:

$$Z(s) = \sqrt{\frac{R_d}{\dfrac{1}{R_p + \dfrac{1}{C_p \cdot s}} + C_d \cdot s}} \cdot tanh\left[\sqrt{R_d \cdot \left(\frac{1}{R_p + \dfrac{1}{C_p \cdot s}} + C_d \cdot s\right)}\right] \tag{29}$$

A detailed treatment for spherical particles of cathode intercalation material is given by Barsoukov *et al.* [2003].

Porous Electrode. Most battery electrodes comprise an open structure consisting of small particles compressed together, as shown in Figure 4.5.6. This structure does not have well-defined pores (such as cylindrical) but rather an irregular network of interconnected space between particles filled with electrolyte. The absence of well-defined pores complicates *ab initio* deduction of electrolyte impedance; however, the frequency dependence of the impedance of porous materials is well described by the ladder-network approach originally proposed by de Levie [1963] for cylindrical pores. See also Section 2.1.6 for treatment of various geometries of porous electrodes.

Different from porous metallic electrodes, where metal-particle resistance is often low and can be neglected, for battery electrodes both the electronic resistance of particles and the ionic resistances of the electrolyte in pores are important. A theoretical treatment of the impedance of porous electrodes using ladder-networks with both ionic and electronic conductivity is given by Paasch *et al.* [1993] by direct solution of the relevant differential equations and by Fletcher [1992] using the formalism developed for chain-connected two-ports. Both approaches allow one to modify the equations in order to consider arbitrary impedance functions of particle surface. Unfortunately, analysis of actual experimental data from electrode/electrolyte impedance spectra measurements by nonlinear fitting does not allow one to distinguish between contributions of the ionic and electronic components or even to estimate the two separate conductivities. The effect of the presence of the two conductivities is manifested mostly in a shift of the high-frequency limit of the impedance spectrum, where the imaginary part goes to zero. However, in an actual experimental set-up this limit (series resistance) can make an unknown contribution and obscure the effect of the two separate conductivities. A combination of electrode–electrode and electrolyte–electrolyte measurement across a porous electrode can still allow one to measure separate conductivities (see Popkirov *et al.* [1995], Wang *et al.* [2001]), but for the usual electrode–electrolyte set-up it is only possible to obtain a combined ionic–electronic conductivity for the porous electrode.

Despite the inability to distinguish between ionic and electronic contributions, information about the combined specific resistance of a porous electrode is often useful because it defines the battery-impedance dependence on electrode thickness, as investigated by Barsoukov *et al.* [1999a]. Since porous electrode resistance is a dominant contributor to the total electrode impedance, it is also critical for battery modeling, as shown by Doyle and Newman [1995]. Such information can be obtained by analyzing experimental data using a transmission line model as in Figure 4.5.3, which results in Eq. (14) when reflective boundary conditions are assumed. For a porous electrode, Z_{trans} will represent the combined ionic and electronic specific resistance, ρ_m, and Z_{cross} will be the specific impedance of particles themselves, as defined by the charge transfer and diffusion processes described in previous chapters.

Bringing it all Together. As shown in the previous chapter, the transmission line model may be used to analyze the impedance spectra of porous battery electrodes. A simple substitution of the expressions for surface and impedance diffusion discussed above, scaled accordingly to particle size, for Z_{cross} transforms Eq. (14) to an equation for entire battery electrode impedance. Such a complete model of battery electrode behavior, valid in the wide frequency range from 10 kHz down to 100 μHz, has been applied to the Li-ion electrode situation by Barsoukov *et al.* for flake-type particles [2000] and spherical particles [2003]. The model can also be used without modification for other intercalation-type battery electrodes, particularly $Ni(OH)_2O$ cathodes used in NiMH and NiCd batteries, and metal hydride anodes. The equivalent circuit for this model is basically the same as that in Figure 4.5.3, except for Z_{cross} being replaced by the surface and diffusion impedance, as shown in Figure 4.5.9.

The frequency dependence of the impedance of this equivalent circuit is obtained by replacing Z_{cross} in Eq. (14) by the impedances of R_{ct}, C_{dl} and the Z_d circuit and is given by Eq. (30)

$$Z(s) = \sqrt{\frac{R_m}{s \cdot C_{dl} + \dfrac{1}{R_{ct} + Z_d(s)}}} \cdot coth\left[\sqrt{R_m \cdot \left[s \cdot C_{dl} + \frac{1}{R_{ct} + Z_d(s)}\right]}\right] \tag{31}$$

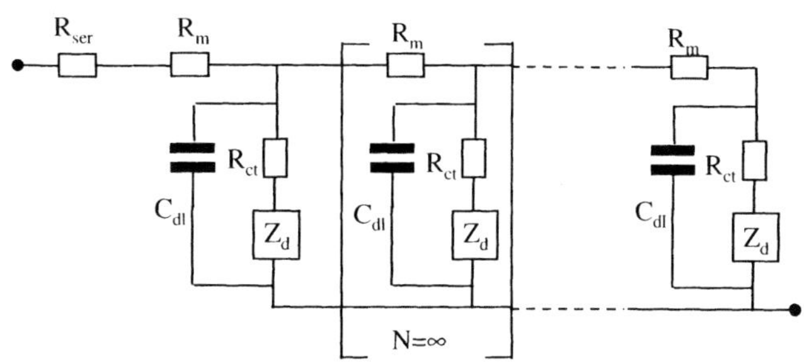

Figure 4.5.9. Equivalent circuit of a battery insertion electrode. Here R_m is the distributed resistance of the transmission line representing electronic and ionic resistance of the layer of active material; the charge transfer resistance and passivation layer resistance of the particle interface R_{ct}; and the double layer capacitance C_{dl}. Z_d is the impedance of diffusion and charge storage processes inside the particles.

For spherical particles, the diffusion impedance $Z_d(s)$ is given by Eq. (22). To include the kinetics of new crystalline phase formation, this equation must be modified by replacing the admittance of the pseudocapacitance of the diffusion process, $C_d \cdot s$, with the admittance of a circuit including both nonactivated charge storage, C_d, and the activated charge storage due to new phase formation: C_p and R_p. This results in the modified equation given below:

$$Z_d(s) = \frac{tanh\left[\sqrt{3 \cdot R_d \cdot \left(s \cdot C_d + \dfrac{1}{R_p + \dfrac{1}{C_P \cdot s}}\right)}\right]}{\sqrt{\dfrac{3 \cdot \left(s \cdot C_d + \dfrac{1}{R_p + \dfrac{1}{C_p \cdot s}}\right)}{R_d}} - \dfrac{1}{R_d} \cdot tanh\left[\sqrt{3 \cdot Rd \cdot \left(s \cdot C_d + \dfrac{1}{R_p + \dfrac{1}{C_p \cdot s}}\right)}\right]} \tag{32}$$

Experimental data and fits by the above functions (Barsoukov *et al.* [2003]) is shown in Figure 4.5.10. The spectrum of Li-intercalation cathodes has a depressed semicircle in the high frequency region, whose size and frequency dependence is determined by R_{ct} and C_{dl}. Combined alectronic and ionic resistance of the porous material, R_m contributes to its "stretching" along the X-axis. In the lower frequency

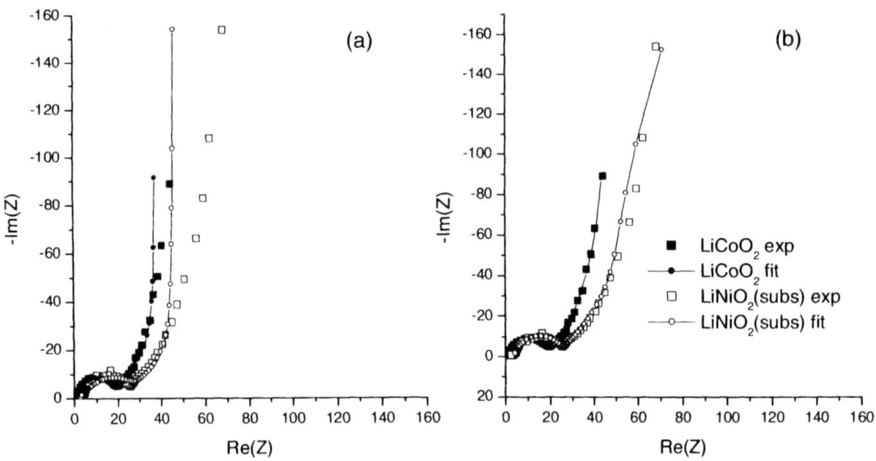

Figure 4.5.10. Experimental impedance spectra of Li-cobalt oxide (RC-41) and substituted Li-nickel oxide (FL-2) cathodes in the fully charged state and in the frequency range from 10 kHz to 200 μHz obtained by multifrequency FFT impedance measurement. Solid lines indicate the fit by Eqns (29) and (21) (*a*) and the fit with inclusion of the kinetics of phase formation by Eqns (29) and (30) is presented in part (*b*).

region between 0.1 Hz and 1 mHz the 45° inclination, characteristic for diffusion processes, can be seen. It then starts to approach a vertical line typical for finite-length reflective diffusion. However, this line never becomes perfectly vertical, even at 200 μHz. It can be seen that including the kinetics of phase formation as in Eq. (31) considerably improves the fit in the low-frequency region where such kinetics significantly influence the spectrum.

Equations (31) and (32) can be used to analyze impedance spectra without knowledge of structural electrode parameters (thickness, density, etc). However, we need this information in order to transform the ohmic parameters obtained by a fit into specific electrochemical parameters. In particular, this information can be used to calculate the effective surface area of the particles. Particles used in practical batteries can usually be treated either as thin plates (Levi and Aurbach [1997]) or as pseudospherical in shape (Barsoukov [2003]), and have a narrow size distribution due to sieving. Particle size values are provided by material manufacturers. The number of particles in a given volume can be estimated from the ratio of their crystallographic density of particles, σ_p, to the density of the composite-electrode film, σ. This allows one to calculate the electrochemically active surface area for a composite electrode for thin-plate particles as $S = xAd\sigma/[l\sigma_p]$ and for spherical particles as $S = 3xAd\sigma/[r\sigma_p]$. Here x is the fraction of active material in the composite; A is the geometric area of the electrode; d is the thickness of the composite electrode; σ is the density of the composite electrode; σ_p is the true density of particles; and l and r are the thickness of the plate and radius of spherical particles, respectively.

Using knowledge of the particle area S, the charge transfer resistance, expressed as the ohmic parameter R_{ct} (Ω), can be used to obtain the specific charge resistance R^s_{ct} [$\Omega\,\mathrm{m}^2$] as $R_{ct} = R^s_{ct}/S$. Correspondingly, C_{dl} (Farad) can be transformed into the specific double-layer capacitance C^s_{dl} (Farad/m^2) as $C_{dl} = C^s_{dl}S$. The combined ionic and electronic resistance of mesoporous material R_m (Ω) is given by $\rho d/A$ where ρ ($\Omega\,$m) is the specific resistance of the active material (sum of ionic and electronic resistances), A is the geometric surface area of the electrode (m^2), and d is the thickness of the layer of active material (m).

Estimates of the diffusion parameters R_d and capacitance C_d obtained from data fitting can be used to calculate the chemical diffusion coefficient D (cm^2 sec^{-1}), if the particle radius is known, as $D = r^2/(3C_dR_d)$. The diffusion pseudocapacitance C_d is related to the Emf-relation of the material so dE/dc (volt·cm^3/mole) can be obtained from C_d as $dE/dc = 4Fz\pi r^3 S/(3C_d)$. All these specific parameters can be used as sample-independent characteristics of a particular intercalation material and allow one to predict the change of impedance spectra (and battery performance) with changing active layer particle size or thickness (Barsoukov *et al.* [2000]).

Battery-specific Improvements Possible in Impedance Spectra Fitting. Ultra low-frequency (below 1 mHz) behavior of batteries is of principal importance for evaluation of their performance, as batteries are mostly used in devices with constant or slightly variable load. Unfortunately, low-frequency impedance measurements take a long time and can be often distorted by system change during the measurement as well as by voltage drift inherent in the components of the measurement system. The good news is that part of the information defining the impedance spectrum at low frequencies can be obtained from dc measurement and integrated with impedance-spectrum data analysis. All battery equivalent circuits discussed in the previous chapter, and particularly those defining finite length diffusion, can be simplified at frequencies approaching zero to a limiting equivalent circuit comprising a resistor in series with capacitor (exemplified by Figure 4.5.1). Ho *et al.* [1980] have shown that the value of this limiting capacitor at different states of charge of the battery can be obtained by differentiation of the open-circuit voltage dependence on passed charge, $C_s = 1/(dE/dq)$. The $E(q)$ dependence can be obtained either by discharging the battery at a very low rate, such as 1/20 C rate (full discharge in 20 h), by applying a low-rate voltage ramp or by applying current pulses followed by relaxation periods with variable duration terminated when dV/dt is below a predefined value (1 μV/sec or less).

The capacity information obtained can be used to increase confidence in the parameters obtained by fitting a finite-length diffusion model to the impedance data measured at a given depth of discharge (DOD). For example the capacity obtained by differentiation of the above $V(dod)$ dependence at $dod = 0$ (as in Figure 4.5.11*b*) can be assigned to C_d in Eq. (10), and used as fixed parameter in the fit of the impedance spectrum measured at this DOD. In this case, one's confidence in the estimation of other low-frequency parameters (R_d) will be significantly increased. R_d cannot be determined at all if the frequency range of the spectrum is not sufficient to include the "bend" point from 45° to vertical dependence in the complex plot. However, when C_d information is available, R_d determination become possible even if only the 45° part of the spectrum is available. This approach can be repeated with C_s values at other DODs by analyzing a series of impedance spectra measured at different discharge levels of the battery.

Knowledge of the limiting capacitance is also useful in the analysis of more complex models, where several capacitive elements occur in parallel (such as in Eq. (31)). At the zero-frequency limit this circuit will be simplified to one series resistor and a series capacitor, which will be a sum of all parallel capacitors which have no resistors in parallel with them. In this case, the sum of all parallel capacitive elements, can be fixed in the fit as equal to C_s, again reducing the degrees of freedom of the fit and improving the estimated standard deviations of all the parameters. An obvious way to do this would be to define one parallel capacitance through another as $C_1 = C_s - C_2$, and fit only C_2. However, the above definition can result in C_1 becoming negative although an impedance fitting program such as LEVM allows restricting the fit parameters to physically meaningful positive values. A better method of estimating the summary capacitance is that of fitting the ratio between the parameters ($r = C_1/C_2$), while expressing the parameters themselves as $C_1 = C_s r/(1 + r)$ and $C_2 = C_s/1 + r$. This way, both C_1 and C_2 are guaranteed to remain positive with their sum equal to C_s. The same method can be used to limit the sum of any number of parallel capacitances.

Parameter Correlation for Quality Control, State and Health Monitoring. Apart for using impedance spectroscopy to analyze specific kinetic proper-

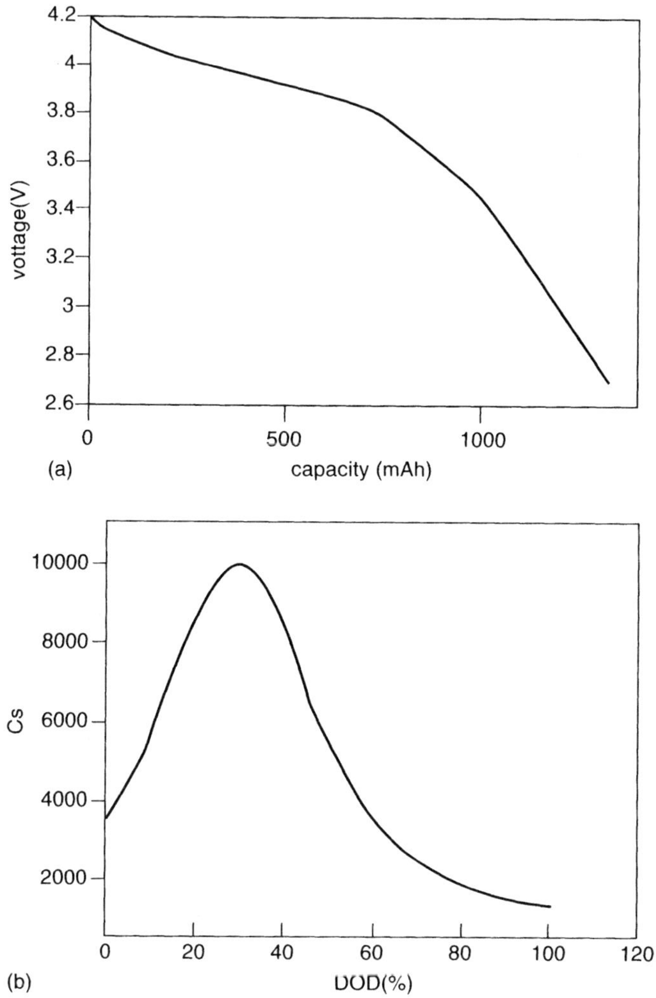

Figure 4.5.11. (*a*) Voltage dependence on discharge capacity of Sony Li-ion battery. (*b*) Limiting capacitance, C_s (Farad) calculated by differentiation the dependence in (*a*) vs. depth of discharge (DOD).

ties of battery materials, it can be used to evaluate certain aggregate behavioral properties of a battery such as state of charge, degree of degradation, and production faults. Once a wide frequency range impedance spectrum is analyzed in the boundaries of the relevant model, it provides a set of parameter estimates that reflect all aspects of battery operation. In the ideal case where the model captures all aspects of the impedance spectra, the set of model parameter values represents the optimum data compression and noise reduction possible for the spectral data. For this reason it is much better to use parameters of the model rather then raw impedance data for correlation with target properties.

Then, since battery kinetics is fully determined by the estimated parameters, in most cases some of these parameters would have significant correlation with a particular property of the battery we are interested in. It is sometimes possible to decide which parameter would be likely to have the best correlation with a particular property based on *a priori* knowledge of the mechanism of that particular property. For example charge transfer resistance R_{ct} is likely to be correlated with the state of charge, as suggested by the Butler–Volmer equation. An overview of approaches for determining the charge state by impedance correlation is given by Rodrigues *et al.* [2000]. However, in more complex cases statistical analysis will be needed to find parameters with best correlation (Yoon *et al.* [2001]). In some cases, no single parameter will exhibit significant correlation, but the combination of several parameters analyzed using multivariant regression techniques such as locally-weighted regression (Cleverland and Grosse [1990]) will show useful predictive ability for a particular property. A review of multivariant statistical methods applied to battery research has been published by Hagan and Fellowes [2003], and a particular application to Li-ion batteries appears in Kim *et al.* [1999].

The choice of a model to analyze spectral data should involve a compromise between the complete capture of all kinetic features and the lowest possible standard deviation for parameters that show promising correlation. Indeed, the standard deviations of parameters, as estimated from the spectrum fit, define the lower limit of uncertainty of the target property correlated with them. For example if R_{ct} is correlated with state of charge (SOC) of a battery linearly as SOC = $R_{ct}^{*}\,a + b$, the uncertainty of the SOC estimation will be at least $\delta R_{ct}^{*}a$ where δR_{ct} is the absolute value of the standard deviation from the impedance fit. Of course, correlation itself is never perfect, so additional error terms will be introduced from a and b. To minimize the standard deviation of critical parameters, the model should be as simple as possible even if it leads to a worse fit in frequency areas weakly influenced by parameters of interest. These frequency areas should preferably be excluded from the fit in order to prevent distortion of the parameters of interest and increase of their uncertainties. However, if making a model more physically relevant leads to a better fit in spectral areas defining the parameters of interest, a more complex model might actually decrease the standard deviations of the relevant parameters, despite having more parameters in the fit. For example, the Li-ion battery impedance spectrum from 1 kHz to 0.1 Hz in Figure 4.5.12 appears as two depressed semicircles in a complex plot. It can be fitted to two serially connected parallel RC elements representing surface insulating layer R_l and C_l, charge transfer resistance R_{ct} and double layer capacitance C_{dl}, as in Figure 4.5.8.

The relative standard deviations for R_{ct} and C_{dl} are then 1.8% and 3.8% respectively. However, it can be seen from Figure 4.5.12*a* that this fit does not adequately reflect the depression of the semicircles. Recognition that the electrode is porous suggests using a transmission line model as in Figure 4.5.3 where Z_{trans} will be the impedance of the particle surface as in Figure 4.5.8. This model is more complex since it adds the additional parameter—resistance of the porous layer, ρ. It can be seen from Figure 4.5.12*b* that it leads to a better fit of the experimental data. Despite the increased complexity of the model, the relative standard deviations of the param-

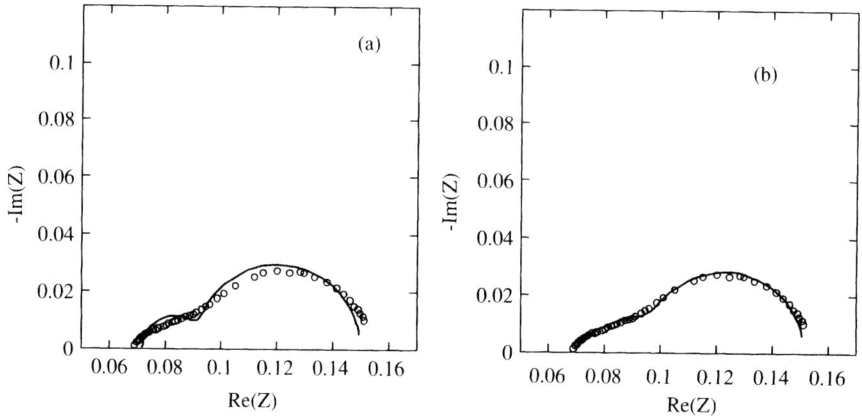

Figure 4.5.12. Impedance spectra of Li-ion battery in fully charged state from 1 kHz to 0.1 Hz; (*a*) fit with 2RC model and (*b*) fit with distributed 2RC model.

eters of interest, R_{ct} and C_{dl}, decreased to 1.7% and 2.8%. Moreover, the more physically relevant model is more likely to yield parameter estimates that better reflect the nature of observed process and therefore involve a better correlation with the target properties. In addition, the additional information represented by the parameter ρ can also be used for correlation.

4.5.2.2 Lead Acid Batteries

While an exhaustive overview of aspects of impedance investigation of a particular battery chemistry is outside the scope of this work, the present generic approach may be exemplified for each particular chemistry by looking separately at the impedance of the electrolyte, the material surface, effects of diffusion and porosity, and the boundary effects, combined in one model. Subsequently, the model can be simplified for purposes of correlation or discretized and made nonlinear if time-domain simulation is required.

The electrolyte in a lead acid battery is a highly conductive, concentrated H_2SO_4 solution. Its ionic resistance is always negligible and makes a small contribution to the series resistance of the battery which depends on state of charge. Unique to lead acid batteries is the fact that the charge reaction produces and discharge captures SO_4^{2-} ions, and therefore the electrolyte concentration is changing during charge and discharge. This involves reflective boundary conditions for diffusion through the electrolyte (Mauracher *et al.* [1997]), resulting in pseudocapacitive behavior conforming to Eq. (10), which is, however, apparent only at very low frequencies, below 100 μHz. A typical simulated impedance spectrum of a lead acid battery is given in Figure 4.5.13.

The impedance of a Pb/$PbSO_4$ anode surface can be represented quite well by the Randles equivalent circuit, as in Figure 4.5.7. It appears as a small semicircle at

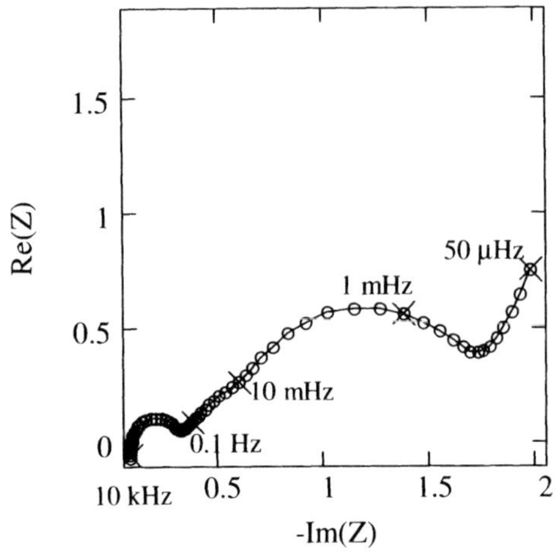

Figure 4.5.13. Typical impedance spectrum of a lead acid battery from 10 kHz to 50 μHz. Simulation is based on data in Mauracher *et al.* [1997].

the high-frequency side of the battery impedance spectrum (Armstrong and Bladen [1976]) and constitutes a rather small contribution to the overall impedance.

Cathode material, on the other hand, dominates the impedance spectrum and involves a depressed semicircle ranging from 1 Hz to 1 mHz. This semicircle may also be interpreted as caused by diffusion through a porous layer (Mauracher *et al.* [1997]); however a greater than five times change of this impedance with bias current, observed by the above authors, and weak dependence on temperature, demonstrated by Hawkins and Barling [1995], both indicate a strong contribution of faradaic charge transfer resistance in this part of the spectrum. Very high pseudo-capacitance of this material (Maya and Penazzi [1985]) indicates the presence of an activated adsorption mechanism typical of supercapacitors rather than that involving simple double layer capacitance. These results explain the significant nonlinear dependence of the associated resistance on current. It is also known that porous PbO_2 is formed during charging and has high conductivity for a metal oxide, and yet much lower conductivity than a metal, which explains the high impedance of the cathode. To take into consideration the combined specific resistance of porous material and electrolyte in pores, as well as the surface impedance of particles, the transmission line model of Figure 4.5.3 can be used, where Z_{cross} will be the impedance of the surface of the PbO_2 particles. An equivalent circuit taking into account activated adsorption, C_{ad}, is shown in Figure 4.5.14.

The function describing the frequency dependence of such an impedance element is equivalent to that of diffusion in Eq. (28), but the meaning of the parameters must be redefined as $R_d = R_m$, $R_p = R_{ct}$, $C_d = C_{dl}$, and $C_p = C_{ad}$. A complete equivalent circuit of the lead acid battery will involve serially connected elements described in previous paragraphs for the electrolyte and anode and cathode, with the addition of series resistance and inductance.

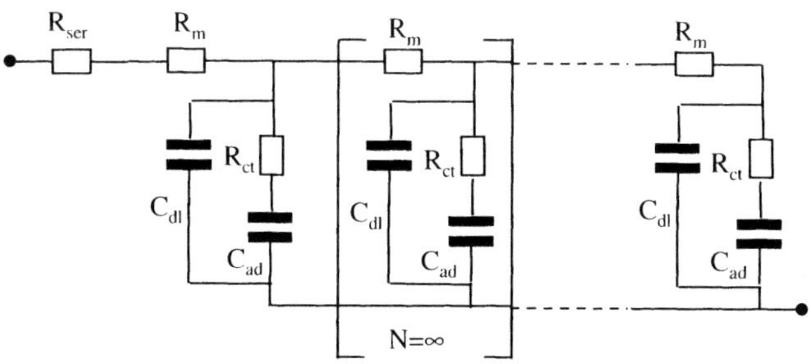

Figure 4.5.14. Porous electrode with activated adsorption on particles' surface.

The dependence on state of charge of a lead acid battery has been investigated by Hawkins and Barling [1995]. Discharge results in an impedance increase at most frequencies, but especially in the low frequency region dominated by the PbO_2 porous layer. A significant increase in impedance takes place also on overcharge (floating) the cells, probably due to passivation of the positive electrode. Increased impedance starts already at 83% SOC, as shown by Salkind *et al.* [2001]. The use of impedance data for determining the state of charge of lead acid batteries is reviewed by Rodrigues *et al.* [2000].

4.5.2.3 Nickel Cadmium Batteries

The electrolyte and separator in Ni–Cd batteries do not contribute significantly to overall impedance due to the use in such batteries of highly conductive, concentrated KOH/LiOH solution. The effect of electrolyte and current collector resistance adds to the serial resistance of the cell. Typical impedance spectra of a Ni–Cd battery is shown in Figure 4.5.15.

It can be seen that the impedance decreases from the fully charged state until about 75% state of charge and then stays mostly unchanged until almost end of discharge, where it slightly increases again. Detailed investigation of state of charge dependence of anode and cathode separately was performed by Hammouche *et al.* [2004].

The anode of a Ni–Cd battery typically consists of a mix of Cd and CdO powders with the addition of a conductive additive (acetylene black). The impedance of the anode-particle surface is determined by the activated adsorption of OH^- anions first on the metal surface, with subsequent conversion into $Cd(OH)_2$ and hydrated CdO layers (Duhirel *et al.* [1992])). Reaction products are also present in a partly dissolved $Cd(OH)_3^-$ state. The activated adsorption mechanism of the anode reaction, as well as porous structure of the electrode, makes it appropriate to use for its analysis the equivalent circuit shown in Figure 4.5.14. It was shown by Xiong *et al.* [1996], by separate impedance measurements on the anode and cathode, that most of the impedance decrease during discharge is due to the anode, as the initial formation of a $Cd(OH)^-_{ad}$ monolayer is the rate limiting step of the reaction. The

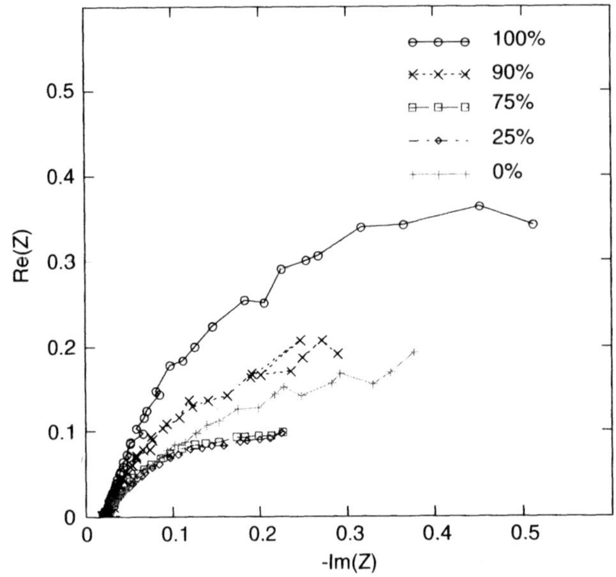

Figure 4.5.15. Impedance spectra of a cylindrical AA-size Ni–Cd battery at states of charge (SOC) of 100, 90, 75, 25 and 0%, measured using a multiwave FFT impedance spectrometer covering the frequency range from 1 kHz to 1 mHz.

consumption of Cd also increases the surface roughness and therefore the active surface area, contributing to the impedance decrease. However, the slight increase of impedance near the end of the discharge may be due to exhaustion of active cathode material in NiO(OH) form, which indicates that the impedance of the cathode is a significant part of the overall impedance.

The cathode of a modern Ni–Cd battery consists of controlled particle size spherical $NiO(OH)_2$ particles, mixed with a conductive additive (Zn or acetylene black) and binder and pressed onto a Ni-foam current collector. Nickel hydroxide cathode kinetics is determined by a solid state proton insertion reaction (Huggins *et al.* [1994]). Its impedance can therefore be treated as that of intercalation material, e.g. considering H^+ diffusion toward the center of solid-state particles and specific conductivity of the porous material itself. The porous nature of the electrode can be accounted for by using the transmission line model (D.D. Macdonald *et al.* [1990]). The equivalent circuit considering both diffusion within particles and layer porosity is given in Figure 4.5.9. Using the diffusion equations derived for spherical boundary conditions, as in Eq. (30), appears most appropriate.

Because the Ni–Cd battery impedance spectrum does not show a gradual change with state of charge, single parameter correlation in order to determine SOC appears unfeasible (Reid [1990]). However, multivariate correlation has been successfully used (Rodrigues *et al.* [2000]), and the frequency where *Im*(*Z*) changes sign from negative to positive was found by Hammouche *et al.* [2004] to be a good predictor of SOC. On the other hand, Ni–Cd battery degradation with cycling is accompanied by a significant impedance increase, especially at low frequencies (Haak *et al.* [1984]). Impedance spectra have therefore been widely used for estimating the state of health of Ni–Cd batteries. The reason for the impedance increase at the cathode side was found by D. D. Macdonald [1990] to involve a decrease of the pore size and an increase of interparticle resistance of the porous layer.

4.5.2.4 Nickel Metal-hydride Batteries

The electrolyte and cathode material of Ni–MH batteries are identical to those discussed in the Ni–Cd section. The impedance contributions of the anode and cathode are typically similar in magnitude, as shown by Reid [1990]. Most of the differences between these battery types are due to the metal-hydride anode, using several different alloy types. Most common alloys have been investigated by impedance spectroscopy: type AB_5 ($LaNi_{4.7}Al_{0.3}$) by Kuriyama [1996], and Cheng [1999]; more modern ($MmNi_{3.6}Co_{0.7}Al_{0.3}Mn_{0.4}$) by Li *et al.* [2004]); type AB_2 (for example $ZrCr_mFe_nCo_pV_o$ where m + n + p + o = 2) (Rajalakshmi *et al.* [2001]); Mg_2Ni type (for example $Mg_{1.95}Y_{0.05}Ni_{0.92}Al_{0.08}$) investigated by Cui and Luo [2000]; and other types, especially those containing vanadium. The AB_5 type currently exhibits the best cycle ability and is most commonly used in commercial batteries, while others show higher capacity and are being developed to improve cycle ability.

Typically, the anode consists of small particles of hydride-building alloy held together by a binder and conductive additive (acetylene black) and pressed onto a Ni-foam current collector. The impedance of the particle surface is determined by the charge transfer resistance of hydrogen reduction, double layer capacitance, and the impedance of subsequent solid-state diffusion into the bulk of the particle. To take into account electronic resistance between the particles and ionic resistance of electrolyte in pores, as well as the impedance of the particle surface, we can use the transmission line model of Figure 4.5.9. Because particle shape is best approximated as a sphere, diffusion with spherical boundary conditions, as in Eq. (30) can be used for Z_d.

The dependence of a Ni–MH battery on its state of charge is mostly defined by an increase of anode impedance with discharge, except for the fully discharged state where increasing cathode impedance becomes pronounced. Some typical impedance spectra measured using the FFT multisine method after repeated discharge/relaxation at different depths of discharge (DOD) on a Hitachi AA-size Ni–MH battery are shown in Figure 4.5.16 as a perspective complex plot.

A gradual impedance increase in the areas associated with charge transfer and diffusion behavior is apparent, while the series resistance mostly associated with the electrolyte and separators remains unchanged. Bundy *et al.* [1998]) used multivariant correlation in order to predict the state of charge based on impedance spectra for a large set of commercial Ni–MH cells and found that despite cell-to-cell variations it was possible to achieve a 7% average error in prediction if a few abnormal cells were removed. Hammouche *et al.* [2004] found that the frequency where $Im(Z)$ changes sign from negative to positive has a good correlation with SOC.

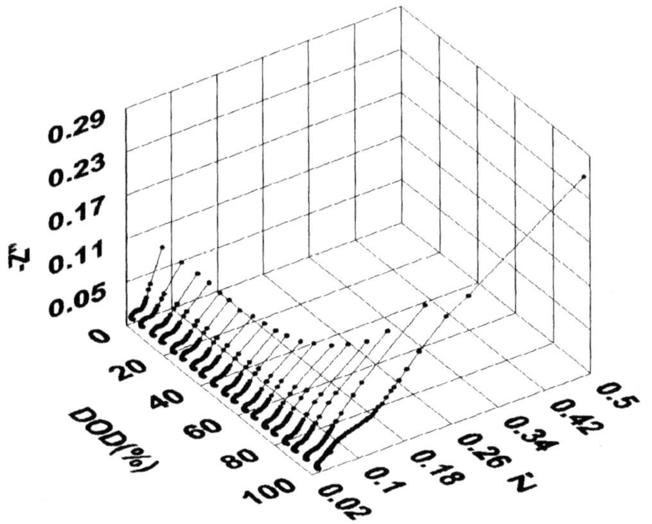

Figure 4.5.16. Impedance spectra measured using the FFT multisine method after repeated discharge/relaxation at different DOD on a Hitachi AA Ni–MH. The frequency range extends from 1 kHz to 1 mHz.

The impedance of a metal hydride anode typically decreases during the first few charge–discharge cycles, apparently because of an increase of the active surface. A similar increase can be produced by pretreatment with a base and reducing agent (Cheng *et al.* [1999]). In the case of battery deterioration with subsequent cycling, both series resistance increase and overall "stretching" of the impedance spectra can indicate depletion of the electrolyte both in the separator and in the pores of the anode and cathode material. Replenishment of the electrolyte decreases the impedance to its original state. Further degradation is associated with corrosion and fracturing of anode material which also results in an increase of interparticle resistance and eventually to irreversible loss of active surface (Cheng *et al.* [1998]).

4.5.2.5 Li-ion Batteries

The Li-ion battery is the most actively investigated battery of modern times. Its basic design consists of a cathode and anode with an open crystalline structure capable of repeatedly intercalating Li-ions, and a high-stability electrolyte and separator. Research of Li-metal anode-based batteries largely overlaps with Li-ion battery research. A Li-ion battery uses an aprotic (no mobile hydrogen atoms) electrolyte, quite different from other batteries. The need for such electrolyte arises because of the high cell voltage of the battery (4.2 V) which exceeds the water stability region both on the cathode and on the anode side. Electrolytes used can be subdivided into inorganic solid (Bates *et al.* [2000]), solid polymer (Armand [1983], Aihara *et al.* [2003]), plasticized (Rajendran *et al.* [2001]), gelled (Krok *et al.* [1999], Saito *et al.*

[2001]), and liquid (Blomgren [2003]). All of these electrolytes have been the subject of investigations using impedance spectroscopy, however only liquid and, to some extent, gel electrolytes are currently of widespread commercial use because of their superior conductivity.

The impedance spectrum of polymer and gel electrolyte appears as a depressed semicircle in the frequency region between 100 kHz and 0.1 Hz, which can be analyzed using the Cole and Cole [1941] approach, as described in Section 2.1.2.3. Typically, polymeric, plasticized, and gel Li-ion conductors show abnormally low conductivity as compared to that expected from self-diffusion coefficients calculated using other methods such as PMFG-NMR (Clericuzio *et al.* [1995]). In addition to the usual attribution of this effect to ion association, the incomplete removal of the electrode impedance effect during analysis can contribute to an apparent increase in the electrolyte resistance.

Impedance is typically measured in a two-electrode configuration where the electrolyte is compressed between two blocking (steel, platinum) or nonblocking Li-electrodes (Qian *et al.* [2002]). Analysis of electrolyte impedance in the presence of electrode impedance is complicated and usually assumes that the electrolyte is responsible for the highest frequency region of the spectrum, about 1 kHz. To improve confidence in the conductivity estimation, measurements with several layer thicknesses should be performed. To remove the effect of the electrode impedance in a test setup, four-electrode measurements have also been proposed (Bruce *et al.* [1988]). Typically, two pseudoreference electrodes made of Li-foil strips are pressed through a cavity in the middle of circular main electrodes to the surface of the polymer electrolyte under test.

Because of the inherent reactivity of the electrolyte with intercalated Li, the choice of solvent/conductive salt is based on a compromise between its conductivity over a wide temperature range and its ability to build a passivating layer on the intercalating host surface. Initially, a propylene carbonate (PC)/ethylene carbonate (EC) combination was widely used because of the high conductivity of the first and the good passivating properties of the second. Presently, more complicated mixtures of multiple solvents (Plichta *et al.* [2001]) and passivation enhancing additives such as vinylene carbonate (Aurbach *et al.* [2002]) are used to improve cycle ability and low temperature performance. Impedance investigations of solvent conductivity dependent on temperature and conductive salt (Jow *et al.* [2003]) are common and can use the standard treatment for conductivity calculation (Section 2.1.2.3).

The separator in a Li-ion battery is typically a thin (15 μm) microporous polypropylene film. It prevents the electrodes from shorting directly or through Li microdendrite growth on overcharge, and it also serves as a thermal shut-down safety device. When heated above 150°C (for example due to an internal short in a cell) the separator melts and its pores close, thus preventing current flow and thermal runaway. It is common to investigate the shut-down behavior of separators by measurement of cell impedance at selected frequencies, such as 1 kHz, dependent on temperature (Uchida [2003]).

Separate impedance investigation of electrodes is possible using a three-electrode configuration, where the electrode of interest (anode or cathode) is

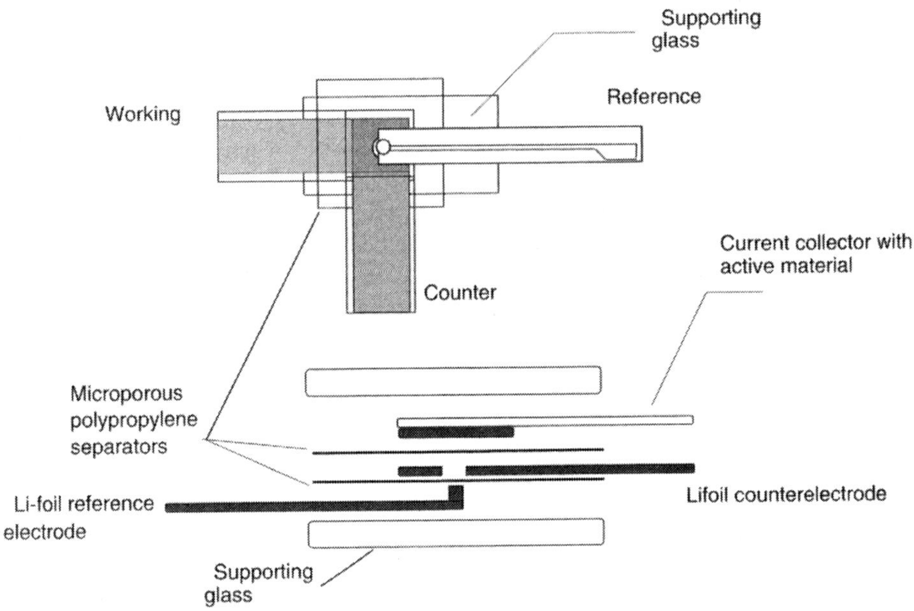

Figure 4.5.17. Three-electrode arrangement for impedance measurement on Li-intercalation electrode in battery-like conditions.

typically paired with a Li-metal counter electrode and a Li-metal reference electrode. The electrodes are sandwiched with a separator, compressed together in a holding arrangement and immersed in electrolyte. An example of such arrangement, used by Barsoukov *et al.* (1999a), is shown in Figure 4.5.17.

In this case, electrodes are compressed between two glass plates with a spring. Placement of the reference electrode through a hole in the counter electrode provides the most symmetric position, important for measurement reproducibility and for preventing inductive artifacts often associated with a conductive path to the reference electrode coinciding with that of the current flow between the main electrodes. Detailed discussion of three-electrode measurement is given in Dolle *et al.* [2001].

Anode. Most common intercalation anodes are based on natural graphite and hard carbon graphitized to different degrees, with the degree of graphitization significantly influencing the capacity, kinetics, and first cycle irreversible capacity loss (Dahn *et al.* [1993]). Popular are preshaped carbons such as MCMB with spherical particles and cylindrical carbon fibers (Sawai *et al.* [1994]). A well-defined particle surface allows one to reduce the specific area of an electrode and therefore the self-discharge and first cycle capacity loss. The surface impedance of a carbon anode is determined by two factors—the resistance and capacitance of the passivating layer and the charge transfer resistance of the intercalation reaction, followed by solid state diffusion toward the center of the particles, which can be formalized in terms of the equivalent circuit shown in Figure 4.5.18.

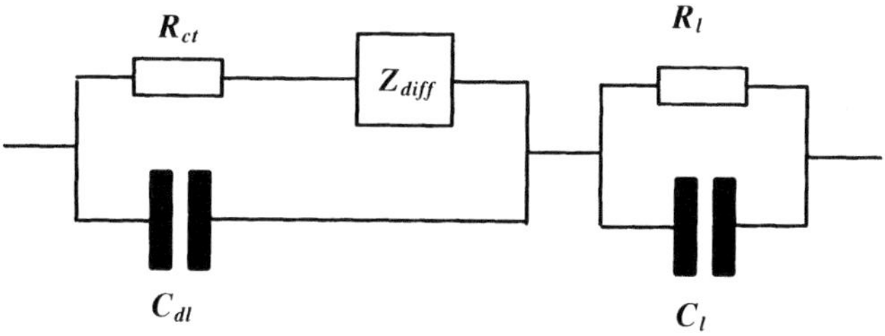

Figure 4.5.18. Equivalent circuit of the particle surface of a Li-intercalation anode.

Impedance spectroscopy was first applied to the investigation of SEI on lithium metal by Thevenin [1985]. For porous electrodes, both factors have been investigated on MCMB by studying EIS during the growth of the passivation layer Barsoukov *et al.* [1998]. Typical impedance spectra measured during the growth of the passivation layer are shown in Figure 4.5.19.

It can be seen that a high-frequency semicircle corresponding to the passivation layer grows noticeably at room temperature and is almost unchanged at low temperature, indicating a different character of the formed layer. Quantitative analysis of the surface characteristics is only possible by using a model that considers the porosity of the active layer because the influence of the ionic and electronic resistance of the active layer is important in all frequency ranges. The transmission line model in Figure 4.5.3, where impedance, Z_{cross}, is given by the equivalent circuit in Figure 4.5.18, can be used to analyze the above impedance spectra.

The diffusion impedance Z_{diff} should be expressed differently depending on the predominant shape of the particles: Eq. (10) for planar and Eq. (22) for spherical particles. Planar particles are rarely found in commercial batteries but ultrathin carbon films can be a good model system for studying intercalation kinetics, as demonstrated by Levi and Aurbach [1997]. At very low frequencies the impedance spectrum starts to deviate from the shape expected due to diffusion alone. It was suggested by Barsoukov *et al.* [1999a] that new-phase-formation kinetics starts to influence the solid-state concentration of the free Li^+ ions. Modified diffusion equations that take this effect into consideration have been derived for the planar (Barsoukov [1999b]) and the spherical case (Barsoukov [2003]), Eq. (31). A correct description of phase formation effects allows one to investigate processes with long time constants critical for the dc performance of intercalation materials. See also Section 4.5.2.1, Bringing it All Together, for more details. A dependence of carbon-anode impedance on state of charge has been reported by Barsoukov [1999a] and a typical spectrum development is shown on the perspective complex plot in Figure 4.5.20.

It can be seen that changes in the charge transfer resistance (high-frequency depressed semicircle) and in the chemical diffusion coefficients defining the low frequency dependence involve Li intercalation phase changes. The resistance R_{ct} and diffusion hindrance increase in areas where change from one phase to the next occurs. The value of impedance spectra for correlation with state of charge is problematic for the Li-ion battery because impedance increases and decreases in the

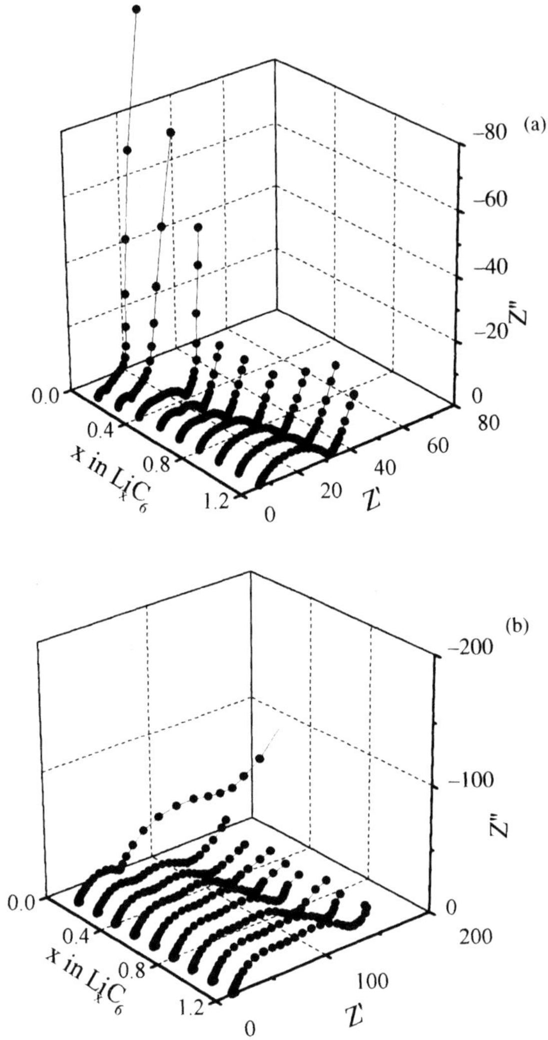

Figure 4.5.19. Impedance spectra measured using a multisine FFT impedance spectrometer in the range from 10 kHz to 1 mHz in a three-electrode configuration on a hard-carbon anode during first charge at room temperature (*a*) and at −20°C (*b*).

course of discharge several times. Only multivariant correlation using information from multiple frequencies or multiple model parameters can allow confident estimation (Yoon *et al.* [2001]).

A few other materials have been investigated as lithium intercalation anodes—in particular, amorphous Sn, SnO, and SnO_2. Impedance analysis can be performed using models similar to those for with carbon-based electrodes (Hong *et al.* [1999]). The increase in the specific resistance of a porous electrode due to the growth of

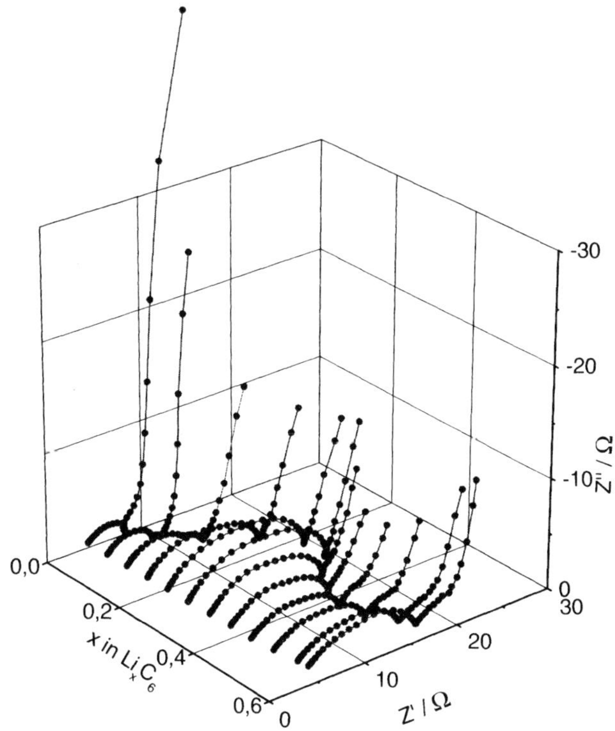

Figure 4.5.20. Impedance spectra measured using a multisine FFT impedance spectrometer in the range from 10 kHz to 1 mHz in a three-electrode configuration on a hard-carbon anode during discharge/relaxation sequences at different states of charge.

Sn-particle size and the resulting electrical disconnection of particles in line with an increase of the percolation threshold for larger particle size was found to be the main reason for fast capacity fade. Doping elements and manufacturing of glass-like materials allowed improvement of the originally poor cycle ability in addition to their excellent capacity (Miyasaka [1998]). Unfortunately, the champion of this idea, the Fuji Corporation, dropped the project for financial reasons, and since then commercial interest in novel, but hardly economical when compared to carbons anode materials, remains low. A still promising application for these materials involves all-solid RF-sputtering of deposited thin film batteries (Strømme [1996]) because of their ease of deposition compared to carbon.

Cathode. $LiCoO_2$ (Ozawa [1994]) has been a cathode of choice since the inception of the Li-ion battery, with $LiMn_2O_4$ (Kim and Pyun [2002]) being investigated because of its fast kinetics and higher capacity; nevertheless it has not been of wide commercial use due to fast degradation above 60°C (Amatucci *et al.* [1999]). Research is continuing on finding doping ions that stabilize spinel structure (Sun *et al.* [1999]). More recently, $LiNiO_2$ and its derivatives replacing part of Ni for Co (especially $LiNi_{0.8}Co_{0.2}O_2$) improved its thermal stability and are starting to find application due to higher voltage and capacity. A new class of phosphate materials, particularly $LiFePO_4$ (Prosini [2002]) and $Li_3V_2(PO_4)_3$ (Saïd *et al.* [2003]) involving a simple synthetic route including carbonization to improve electronic conductivity, has recently led to a promising alternative material with similar capacity and price but highly increased thermal run-away temperature.

Typical impedance spectra for $LiCoO_2$ and $LiNiO_2$ cathodes are shown in Figure 4.5.10. The surface impedance of Li-intercalation cathode materials is similar to that of anodes and is dominated by surface film impedance and by charge transfer resistance. Because of the relatively low conductivity of the cathode material, 5–15% of conductive additive (carbon black) is added to the material. Still the high material resistance is a major contributor to the overall electrode impedance and can be treated by the transmission line model similar to that discussed for anodes. Capacity fading is also largely attributed to an increase of material resistance due to the growth of insulating layers on particle surfaces and the resulting particle disconnection (Aurbach *et al.* [2002]). Solid state diffusion in intercalation cathodes can only be analyzed as part of an overall impedance model because the material resistance of the porous electrode can lead to a 45° angle in the complex plot which can be mistaken for diffusion effect. Nonlinear fitting to a model which includes both diffusion and material resistance factors (Section 4.5.2.1) allows one to separate their effects because the diffusion effect is expressed only in the low-frequency area but material resistance influences all frequency regions.

The dependence of wide-frequency-range impedance spectra of $LiCoO_2$ on state of charge has been investigated by Barsoukov *et al.* [2000]. Typical impedance spectra are shown in the perspective complex plot of Figure 4.5.21. It can be seen that although most changes occur in the charge transfer resistance of the material, changes in diffusion hindrance are also significant.

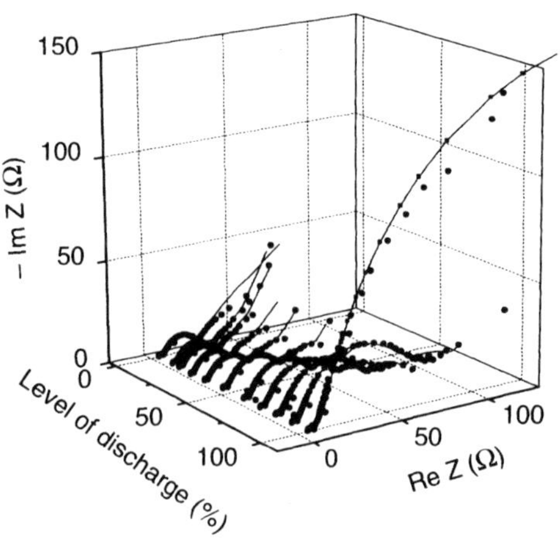

Figure 4.5.21. Impedance spectra measured at various levels of discharge on $LiCoO_2$ based composite cathode in the frequency range 10 kHz to 1 mHz (markers), and the results of non-linear fits to the model in Eqs. 31, 32.

4.5.3 Impedance Behavior of Electrochemical Supercapacitors and Porous Electrodes

Brian Conway

4.5.3.1 Introduction

The history of the concept of the double layer at interfaces with liquid electrolyte solutions or ionic solids, and at colloids, goes back to early work on colloids and electrode interfaces in the 19th century and through the 20th, but is beyond the scope of present article. However, general reference sources are Adam [1941], Adamson [1982], and Conway [1999].

Double layers of charges, negative and positive, separated by a solvent layer of some 0.3 nm in thickness, arise at electrode and colloid interfaces with electrolyte solutions. Such a configuration of charges, e.g. electrons or electron-deficiency at electrode interfaces and conjugate ionic charges in a solution in contact with the electrode, has the nature of a "parallel plate" capacitor (model of von Helmholtz [1853]). The capacitance, C, of such a distribution of charges is given in classical electrostatics by

$$C = A\varepsilon\varepsilon_o/d \tag{33}$$

for an area A cm^{-2}, with plates separated by a distance d with a dielectric between the plates having dielectric permittivity ε, with ε_o being the permittivity of free space, equal to 8.84×10^{-14} Fcm^{-1}. For a distance $d = 0.3$ nm and an effective dielectric permittivity of the interphasial solution layer at an electrode equal to ca. 6–10 for water at a charged interface, C per cm^2 would have a large value, in the order of 20 μF cm^{-2}, as found, for example, at Hg or Au electrodes in aqueous H_2SO_4 or $HClO_4$ solutions.

The main factor of principal significance is that, per cm^2, interfacial double-layer capacitance values are very large in comparison with that of hardware capacitors of the same, 1 cm^2 area where values are on the order of picofarads. Interfacial double-layer capacitance values are on the order of 20 μF cm^{-2}, a figure that arises on account of the very small separation of charges across an electrode or colloid interface, on the order of 0.3 nm.

For comparison, a hardware vacuum ($\varepsilon = 1$) capacitor having $d = 0.01$ cm, say, would have to have an area some 10^5 times larger to provide a specific capacitance comparable with that, per cm^2, of the double-layer capacitor. Even with bulk water as the dielectric, the area would need to be some 10^3 times larger! Interfacial double-layers thus provide high capacitance density and correspondingly large specific charge-storage densities.

Hence, for double-layers established at high specific-area materials, such as powdered or fibrous carbons, or metals, or their oxides, very high specific capacitance per gram can be attained. This is the principle that has underlain the development of electrochemical capacitors, now often referred to as "supercapacitors" (Conway [1999]) or "ultracapacitors".

Electrochemical capacitors are a relatively recent development in the technology of electric charge storage and delivery devices, though an early original patent

(Becker [1958]) describes charge storage by the double-layer capacitance of high specific-area carbon material. The basis of such a device of significance for practical development is that a variety of carbon powder, fiber, and aerogel materials can be prepared or obtained with specific real areas up to 1000~2000 $m^2 g^{-1}$. At electrode interfaces, including those of various carbon materials, an interfacial double-layer capacitance arises (Becker [1958], Parsons [1954], Stern [1924], Grahame [1947]) having values in the range 10 to 30 $\mu F cm^{-2}$, typically 15~25 $\mu F cm^{-2}$ for Au, Hg and other metals. Thus, for an electrode made from C powder or fibers having real accessible area of 1000 $m^2 g^{-1}$, an overall electrode capacitance of 200 $F g^{-1}$ will arise for an interface having a specific double-layer capacitance of, say, 20 $\mu F cm^{-2}$ (Grahame [1947]).

Note (cf. Conway [1999]) that a practical electrochemical double-layer capacitor device always requires, of course, a two-electrode configuration per cell, with one electrode interface being charged against the other by opposite directions of polarization, as in a single battery cell. Each electrode has, however, its own characteristic impedance spectrum which is not necessarily the same for the two interfaces due to asymmetry of cation/anion adsorption (Grahame [1947]). Each of the individual electrode impedances can be evaluated in a suitable two-, or three-electrode configuration, or the overall cell impedance can be evaluated in a two-terminal configuration; then the capacitances of the two interfaces are measured reciprocally in series, the overall capacitance then being formally half that of each electrode in the symmetrical case, but practically less than such a value.

Contrary to what was naively believed in the early stages of development of electrochemical capacitors, that is very high power-densities would be achieved in their use because of their capacitative nature (supposedly like that of Figure 4.5.22*a*), electrochemical supercapacitors do have some intrinsic limitations on their power delivery. This arises for two reasons: (a) due to the usual "iR" loss when a significant equivalent (ESR) or real series resistance exists in the electrical configuration of the capacitance element; and (b) on account of the internal distributed resistance in multiple series/parallel (Figure 4.5.22) connections (de Levie [1963, 1964], Keiser *et al.* [1976]) representing an high-area porous carbon, double-layer electrochemical capacitance. This causes a dispersion of measured capacitance with increasing frequency in ac impedance spectroscopy, due to an attenuation of amplitude of the addressing alternating voltage down the pores of a porous electrode matrix that has been referred to as the "penetration effect" (de Levie [1963,1964]) or "porous-electrode factor". Note that the "penetration effect" applies, in principle, to any form of modulation signal such as linear-sweep voltammetry, potential-step chronoamperometry as well as to AV modulation. Further details of this matter are treated later in this section.

The origin of this kind of behavior is fundamental to the understanding of the difference of capacitance response to a modulating signal when the capacitance is developed in a porous-electrode matrix in comparison with that at a solid, plane electrode of the same (e.g. carbon) material for which a circuit of type 1(b) applies. It can be understood in terms of the hierarchy of equivalent circuits (illustrated in Figure 4.5.22), representing the electrical behavior of a pure capacitance (vacuum

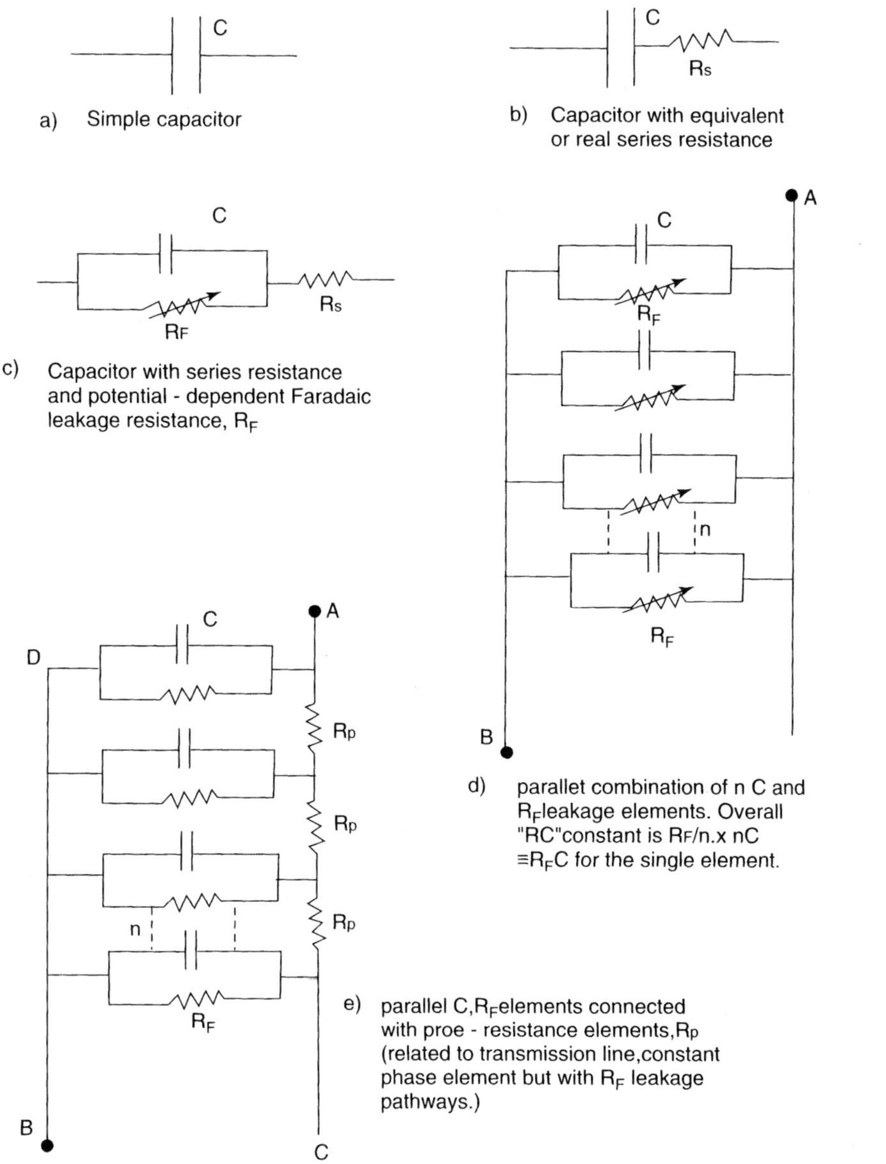

Figure 4.5.22. Hierarchy of equivalent circuits for capacitor/resistor combinations, including that for a porous electrode. (*a*) simple capacitor; (*b*) capacitor with equivalent or real series resistance; (*c*) capacitor with series resistance and potential-dependent Faradaic leakage resistance, R_p; (*d*) parallel combination of C with leakage resistances; (*e*) series/parallel ladder circuit for a porous electrode leading to transmission-line behavior.

"dielectric") without series or parallel resistance elements, through to the complex circuit for porous electrodes (Figure 4.5.22*e*).

In terms of the relationship between voltage V, across the plates of any capacitor and the stored charge density q per cm^2, the capacitance, C, is given by the fundamental equation in electrostatics:

$$C = q/V \tag{34}$$

or, differentially,

$$C = \mathrm{d}q/\mathrm{d}V \tag{35}$$

when the capacity, C, is not constant with charging V, as is often experimentally found. q and V are fundamental and conjugate variables in any study of capacitative devices, especially by means of modulation procedures such as impedance spectroscopy (see definition in Chapters 1 and 2 and in Lasia [1999]) (modulation of voltage by a sinusoidal alternating voltage, AV, at various frequencies, ω) and linear-sweep voltammetry (linear modulation of voltage at various sweep-rates, $\pm \mathrm{d}V/\mathrm{d}t$). These two complementary procedures have provided major instrumental methods for the study of electrochemical systems, especially for examination of the behavior of the double-layer and its capacitance at metal electrodes and semiconductors.

4.5.3.2 The Time Factor in Capacitance Charge or Discharge

The "RC" Time Constant. A capacitance either stores incoming charge or delivers charge. The rate of charge delivery or acceptance (discharging or charging processes) is usually time dependent except for a pure capacitance, like a capacitor in vacuum, connected to a source of, or sink for, charge by connections of zero resistance or inductance. In any capacitor device having a real material dielectric, the capacitance is effectively connected in series with a real or an *equivalent series resistance*, the ESR. The latter can arise when the polarization of the dielectric upon charging or discharging is time dependent and thus behaves in a way equivalent to a capacitance in a series with a resistance, since the charging, that is admission of charge, is displaced with respect to potential in the range of overall applied potential, so that the capacitance is apparently "dispersed". This effect is not to be confused with the –90° phase angle between current and voltage inherent in the charging of any pure capacitance.

The time scale of admission or delivery of charge to or from a capacitance in series with a real or equivalent resistance, R, is characterized by the time constant RC, the product of C and R. The behavior of such an RC combination can be treated in terms of a generalized *reduced* time scale, t/RC. The quantity t/RC enters into any time-based response of a capacitor to change of voltage or admission or delivery of charge, and appears in ac impedance studies as a reciprocal time variable, the frequency f (Hertz) or the circular frequency ω, $= 2\pi f$. If RC is small, the response time scale is correspondingly short while, if RC is large, it is long. Systems with large RC require low frequency to be significantly AV modulated while for smaller RCs they can be modulated by and respond to higher f values.

The "time factor" in charging and discharging capacitors is a prime factor in their applications as it determines attainable *rates* of those processes. This is of special importance in optimizing performance of capacitor devices such as: double-layer supercapacitors (Conway [1999]) and pseudocapacitors designed for load-leveling; hybridizing with battery or fuel-cell systems in electric vehicles; and as hybrid electrode systems (faradaic combined with non-faradaic, capacitative electrodes). In this respect, the "time factor" therefore determines the power levels at which an electrochemical capacitor device can be operated and consequently the forms of Ragone plots (Conway [1999], Ragone [1968], Pell and Conway [1996]) representing operating energy densities as a function of achievable power densities on discharge or recharge.

Charging of a Capacitance Through a Series or Equivalent-series Resistance; Potentiostatic Case. In this case, a constant voltage step is applied between the outer end of the series resistance, R_s, and the further terminal of the capacitor (Figure 4.5.22*b*). In this circuit, constant V becomes distributed across R_s and C as $iR_s + V_c$ (= V) where i is the time-dependent response charge current. A time-dependent potential, V_c, develops across C and the charging kinetics follows as:

$$C.\mathrm{d}(V - iR_s)/\mathrm{d}t = i_t \tag{36}$$

Since, for a potentiostatic step, the overall V is constant, then

$$-R_s C.\mathrm{d}i_t/\mathrm{d}t = i_t \tag{37}$$

or

$$\mathrm{d}i_t/i_t = \mathrm{d}\ln i_t = -\frac{1}{R_s C}\cdot \mathrm{d}t \tag{38}$$

that is, upon integration, noting that $t = 0$, i is the initial current i_i,

$$i_t = i_i \exp-[t/R_s C] \tag{39}$$

so that the charging current, i_t, decreases exponentially in time, scaled by $1/R_sC$. Contrast the Eq. 41 below for discharge across a resistance, R_p, in parallel. Initially, all the applied, constant V operates across R_s ($V_C = 0$) so that, from Ohm's law, the initial current is simply V/R_s; then Eq. (39) can be conveniently written as

$$i_t = \frac{V}{R_s}\exp-[t/R_s C] \tag{40}$$

Discharge of a Capacitor into a Parallel Load Resistance, R_p. A complementary and simple way of evaluating C is from the time-dependence of voltage across C when C is allowed to be discharged through a constant, parallel resistance, R_p.

Then the discharge current is $C\ \mathrm{d}V/\mathrm{d}t = I$ and

$$i = V/R_p \tag{41}$$

Then

$$\frac{dV}{V} = dt/R_pC \equiv d\ln V = \frac{1}{R_pC}\cdot dt \tag{42}$$

and integrating, taking V_t as the potential across the capacitor at time t and V_i as the initial voltage at time $t = 0$, the time dependence of V is found as a "first-order" type of decay relation:

$$V_t/V_i = \exp-[t/R_pC] \tag{43}$$

t/R_pC is a characteristic reduced time scale for the discharge where R_pC is the system's time constant. Experimental evaluation of V_t/V_i or $\ln V_t/V_i$ as $f(t)$ using Eq. (43) then enables C to be determined for a given R. V_t is nowadays easily and accurately determined by means of a digital oscilloscope or some other digital-recording computer arrangement.

Impedance of the Series and Parallel RC Circuits. In the simplest analysis, the circuits I and II represent the electrical configurations encountered in measurement and interpretation of experimental impedance spectroscopic behavior of the double-layer at electrode interfaces. They provide the basis for analysis of impedance spectra of more complex "*RC*" networks that arise in representation of supercapacitor behavior.

(i) Series *RC* circuit: analogous to two resistors in series, the respective impedances Z_R and Z_C of a series *RC* circuit add up linearly:

$$Z_{RC} = Z_R + Z_C \tag{44}$$

However, Z_C is frequency dependent and has to be written as an imaginary quantity, introducing the imaginary factor $j = \sqrt{-1}$. Then, since $Z_R = R_s$ and $Z_C = 1/j\omega C = -j/\omega C$ at frequency ω,

$$Z_{RC} = R_S - j/\omega C \tag{45}$$

which has obvious complementary real and imaginary impedance components, the capacitative one always being dependent on ω. Note, for $\omega \to \infty$ $(1/\omega \to 0)$ or for very large C, $Z_{RC} \to R_S$. This is commonly identified on a Nyquist complex-plane plot as the intercept on the Z_R axis ($\equiv Z'$ in complex-plane plot notation) as $\omega \to \infty$, $Z_C \to 0$ (Figure 4.5.23).

(ii) Parallel *RC* circuit: here Z_R and Z_C are in a parallel configuration, so they add reciprocally:

$$\frac{1}{Z_{RC}} = \frac{1}{Z_R} + \frac{1}{Z_C} \tag{46}$$

Introducing the frequency dependence of Z_C and $Z_R \equiv R_p$, Eq. (46) becomes

$$\frac{1}{Z_{RC}} = \frac{1}{R_p} + j\omega C \tag{47}$$

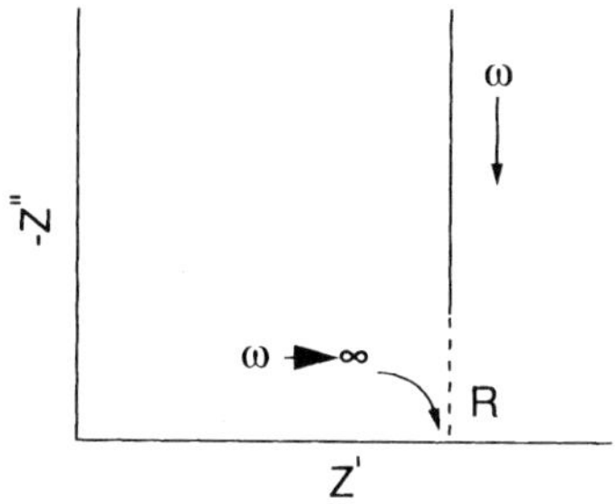

Figure 4.5.23. Complex-plane or Nyquist plot for the impedance spectrum of a simple series *RC* circuit.

$$= \frac{1 + j\omega R_p C}{R_p} \tag{48}$$

In order to separate the real and imaginary components of impedance in this equation, numerator and denominator are multiplied by the rationalizing complex conjugate of $1 + j\omega R_p C$, viz $j\omega R_p C - 1$, giving

$$\frac{1}{Z_{RC}} = \frac{(j\omega R_p C + 1)(j\omega R_p - 1)}{R_p(j\omega\omega_p C - 1)} \tag{49}$$

$$= -\frac{\omega^2 R_p C^2 + 1}{R_p(j\omega R_p C - 1)} \tag{50}$$

so that

$$Z_{RC} = -\frac{j\omega R_p^2 C - R_p}{\omega^2 R_p^2 C^2 + 1} \tag{51}$$

$$= -\frac{j\omega R_p^2 C}{\omega^2 R_p^2 C^2 + 1} + \frac{R_p}{\omega^2 R_p^2 C^? + 1} \tag{52}$$

the first term of which is the imaginary part of Z_{RC} and the second the real part which are thus separated as a $f(\omega)$ as Z'' and Z', respectively, which can be plotted against each other in a Nyquist diagram (see Section 4.5.3.3) for various values of ω, Figure 4.5.24.

In qualitatively considering the impedance behavior of series or parallel *RC* circuits, it is always useful to deduce the limiting cases that arise for $\omega \to 0$ or $\omega \to \infty$. Then, respectively, $Z_C \to \infty$ and $Z_C \to 0$. The behavior for these cases is usually obvious, bearing in mind $Z_C = 1/j\omega C$.

4.5.3.3 Nyquist (or Argand) Complex-Plane Plots for Representation of Impedance Behavior

A very convenient and informative way of representing the impedance behavior of "*RC*" circuits is by plotting the imaginary (Z'') versus the real (Z') components of

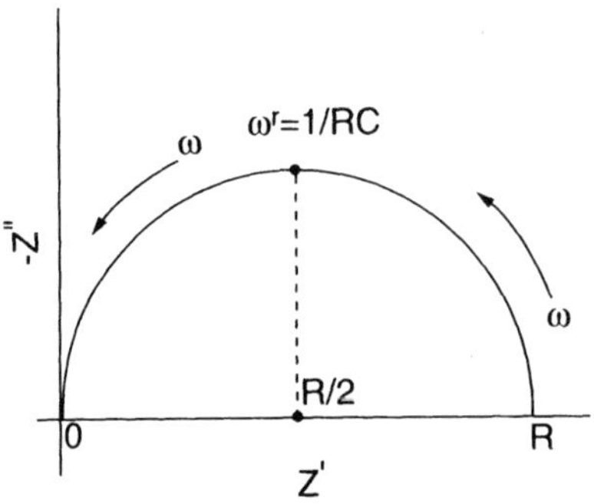

Figure 4.5.24. Complex-plane or Nyquist plot for the impedance spectrum of a simple parallel *RC* circuit, showing top-point characteristic frequency $\omega = 1/RC$.

the impedance vector (*Z*) for a range of frequencies, ω, in a so-called Nyquist or Argand diagram drawn in the "complex plane". This procedure originated in the works of Cole and Cole [1941] on the behavior of dielectrics where real (ε') and imaginary (ε'') components of the dielectric constant (ε) of a material medium can be distinguished and measured over a range of frequencies (Böttcher [1952]). The Cole–Cole plot for a lossy dielectric is in the form of a semicircle, intercepting the axis of real (ε') values at two critical points (Böttcher [1952]). The dispersion of dielectric constant with frequency arises from a phase-difference between the dielectric displacement vector, **D**, and the field vector, **E**, causing the displacement of the electric polarization, **P**. The phase-difference is due to three factors: (a) electric resistivity, leading to a response component in-phase with **E**; (b) relaxation effects in permanent dipole orientation in the dielectric medium associated with the dipole polarization component of **P**; and (c) high-frequency resonance effects arising from electronic and atomic polarization (Böttcher [1952]).

Application of the above type of analysis to electrode processes, including study of interfacial capacitance, forms the basis of so-called impedance spectroscopy, now commonly applied to electrochemical systems, though originating (cf. Cole and Cole [1941]) in evaluation of the properties of dielectrics (Böttcher [1952]), Smyth [1938], Frohlich [1949]) over wide ranges of frequencies up to mega-, and gigaHz values, interpreted in terms of processes involving "dielectric loss" and absorption of the energy of the modulation signal. Dielectric loss in capacitors is in most cases related to the dissipative energies associated with the requirement of periodic (at the frequency f) orientation and reorientation of permanent and induced dipoles in the molecular-dielectric material, and their mutual interactions in that material, especially in the case of liquids.

For electrochemical systems, an important development in application of impedance spectroscopy was the use of complex-plane diagrams by the Sluyters [1964] who showed that the real and imaginary components of measurable impedance at an electrode interface could also be plotted out, like ε'' vs ε', in the complex-plane giving semicircle diagrams and related figures, depending on conditions. Such

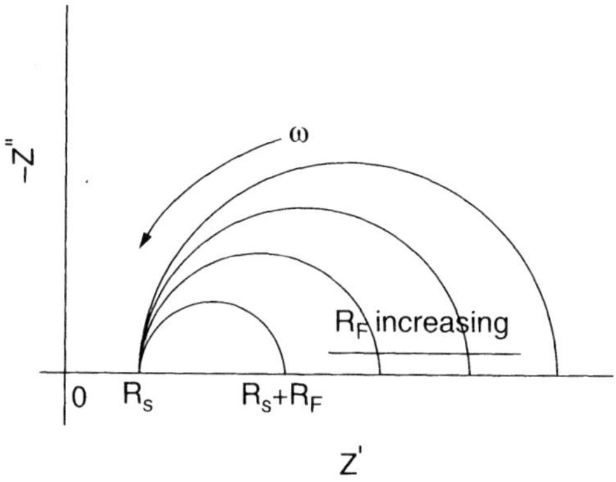

Figure 4.5.25. Complex-plane or Nyquist plots for the impedance spectrum of a double-layer capcitance in parallel with a potential-dependent (f_r) Faradaic resistance, Z_f'.

plots arise in electrochemistry from: (a) the double-layer capacitance coupled with (b) a Faradaic reaction resistance (Figure 4.5.25) related to the exchange current density, j_o, of the electrode process; and (c) the series resistance of the solution in contact with the test electrode. Other factors can be: (d) presence of a pseudocapacitance, C_ϕ, related to potential dependence of coverage by electrosorbed intermediates in the electrode reaction (Conway and Gileadi [1962], Eucken and Weblus [1951]); and (e) possible involvement of diffusion control in the kinetics of the electrode process, represented by Warburg impedance elements (Randles [1947,1952]) dependent on concentration of reactive species and frequency, giving, in fact, a square-root relation in ω with a 45° phase angle at all frequencies.

In the general case of real (Z') and imaginary (Z'') components of Z involving C and R elements we can write the ω-dependence of Z'' and Z' according, for example, to the equivalent circuits treated above. It is found (Sluyters-Rehbach and Sluyters [1964]) that

$$(Z'')^2 - RZ' + (Z')^2 = 0 \tag{53}$$

which can be written in the form

$$(Z'')^2 + (Z' - R/2)^2 = (R/2)^2 \tag{54}$$

which is the equation for a circle when Z'' is plotted against Z' in the complex plane for various values of the experimental AV frequency values. The radius of the resulting (semi) circle is $R/2$ and the plot has intercepts of 0 (for zero series resistance) and R (the Faradaic resistance related to j_o) on the Z' axis, i.e. for $Z'' = 0$. Note that usually Z'' values are negative arising from the double-layer capacitance impedance, equal to $1/j\omega C$, $\equiv -j/\omega C$.

In the complex-plane representation of the impedance behavior of a parallel RC circuit, it is convenient to identify the maximum (so-called "top point") in the semicircular plot which is at a critical frequency $\omega = 1/RC$, the reciprocal of the time constant for the response of the circuit. A similar situation arises for a series RC

circuit if the AV response is plotted in the admittance, Y'', Y' plane. Figure 4.5.25 shows the behavior of a double-layer in parallel with a potential-dependent Faradaic resistance.

The above paragraph gives, in brief, the analysis first published by the Sluyters [1964] for complex-plane analysis of the impedance spectrum of an electrode process at an interface exhibiting a double-layer capacitance; hence its great importance as a basis for examining the impedance spectroscopy of supercapacitor systems, particularly those based (Conway [1999]) on double-layer capacitance (Grahame [1947]).

Historically, it should be noted that double-layer capacitance measurements were first reliably made (at Hg) by Bowden and Rideal [1928] using the dc charging-current method and by Proskurnin and Frumkin [1935] by means of ac modulation. Randles [1947,1952] pioneered the examination of impedance of an electrode process (e.g. redox reactions), using phase-sensitive detector instrumentation to record the frequency dependence of the separated real and imaginary components of Z. Under diffusion-control, the $\omega^{1/2}$ dependence of Z'' and Z' were found due to the Warburg impedance element.

4.5.3.4 Bode Plots of Impedance Parameters for Capacitors

A useful procedure, complementary to plotting real and imaginary components of impedance (Section 4.5.3.3) in the complex-plane, is the use of so-called Bode plots (Lasia [1999]). In such representations, the overall impedance vector, Z, in rotating vector diagrams, is related to the corresponding phase-angle, ϕ. Z is related to its real and imaginary components through the Pythagoras relation

$$Z = \sqrt{(Z'')^2 + (Z')^2} \tag{55}$$

and each component is $Z \sin \omega t$ or $Z \cos \omega t$.

The phase angle is the angle between the directions of the real and imaginary components of Z at a given frequency, ω. For a pure capacitor, ϕ is –90° while for a resistor it is zero. For an *RC* circuit, it will have some values in between, depending on ω. For a capacitor, in terms of a rotor diagram, the voltage, V_C, lags behind the response current by 90°.

Bode plots represent Z and ϕ plotted versus ω, usually over a wide range of ω values, on a logarithmic scale. An advantage of Bode plots, but not to the exclusion of Nyquist plots, is that the impedance behavior at high frequencies is shown with equal weight, along the plot, to that at low frequencies whereas, in Nyquist plots, high-frequency data tend to become bunched together towards the $\omega \to \infty$ intercept on the Z' axis. In the phase-angle plot, an approach to pure capacitative behavior at low ω can usually be identified by $\phi \to -90°$. Deviations from such limiting behavior indicate that the AV response is not purely capacitative so that the behavior has to be modeled in terms of a so-called *constant phase element*, CPE (see Section 4.5.3.8). Examples of Bode plots for Z and ϕ are shown in Figure 4.5.26*a* and *b*, respectively, for interfacial double-layer behavior at an Au electrode in aq. 0.5 M H_2SO_4 (Germain *et al.* [2004]).

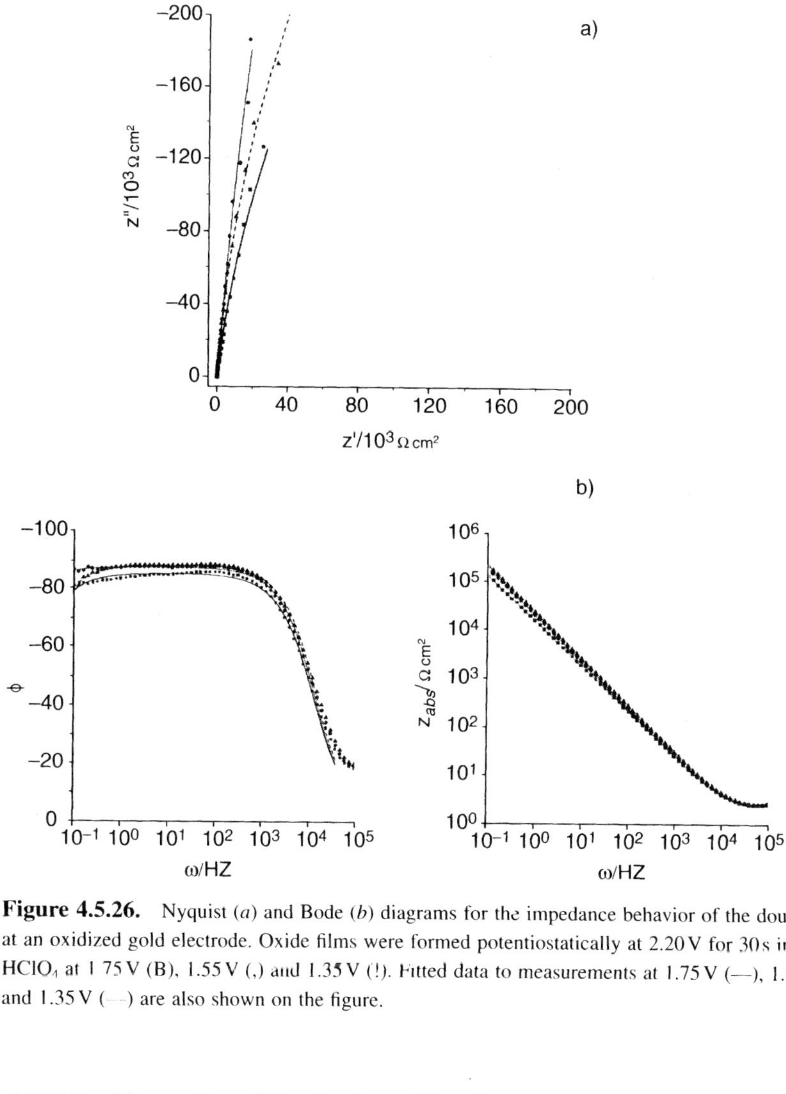

Figure 4.5.26. Nyquist (*a*) and Bode (*b*) diagrams for the impedance behavior of the double layer at an oxidized gold electrode. Oxide films were formed potentiostatically at 2.20 V for 30 s in 0.5 M $HClO_4$ at 1.75 V (B), 1.55 V (,) and 1.35 V (!). Fitted data to measurements at 1.75 V (—), 1.55 V (. . .) and 1.35 V (- -) are also shown on the figure.

4.5.3.5 Hierarchy of Equivalent Circuits and Representation of Electrochemical Capacitor Behavior

The representation of impedance spectra of electrochemical systems in terms of the ac response behavior of equivalent circuits has become a common practice in the field. A physically reasonable circuit is chosen and values of its *C*, *R* and *L* components are selected to give the best fit to the observed frequency spectral response over a wide range of frequencies. It should be noted, however, that a particular, selected circuit model may not uniquely represent the impedance behavior of the system; nevertheless, the circuit modeling approach is widely used, aided by special computer programs.

An alternative, but related procedure, is to make a purely analytical treatment of impedance of an electrochemical system, as was usefully carried out by Armstrong and Henderson [1972].

While treatment of the impedance spectrum of a unit capacitor is simple, if not trivial, when the problem of response of double-layer capacitance at the interface of an high-specific-area matrix, for example of carbon powder, felt or aerogel, is to be evaluated, the problem becomes very complex owing to the distributed series-parallel *RC* configurations throughout the electrode matrix, not to mention some usual contribution of 5 ~ 10% from Faradaic pseudocapacitance due to reactive functionalities at the extended carbon surface.

A seminal treatment of the electrical behavior of a porous electrode was first given in detail in a series of classic papers by de Levie [1963,1964], with important extensions to pores of various specific shapes by Keiser *et al.* [1976]. These works provided the essential bases for much of the basic work on electrochemical supercapacitors and the special effects that arise when a series/parallel, distributed network of *C* and *R* elements is involved.

However, it is useful first to illustrate the complexity of such systems by showing a hierarchy of equivalent *RC* circuits that can be developed, from which an ultimate model of a porous-electrode *CR* device can be constructed. Note that the behavior of an "electrochemical capacitor" device is, electrically, far from that of a pure capacitor in its ac response spectrum to AV modulation; this is primarily due to the complexity of the distributed, internal "connections" within the matrix, associated with resistivity of electrolyte channels and the intrinsic resistance of the *C* microparticles or fibrils and their interparticle contact resistances which usually depend on the pressure applied during fabrication of electrode structures.

A hierarchy of such *CR* equivalent circuits, from that for a simple capacitor element, through to that (Figure 4.5.22*e*) of a porous, electrochemical capacitor electrode, was shown in Figure 4.5.22. The final circuit of the series shown is that in Figure 4.5.22*e*; its characteristics are that: (a) the external capacitor element is directly accessible for input or delivery of charge (except for a possible ESR), i.e. it has a small time constant and good localized power density; and (b) that inner surface regions have double-layer capacitance elements that are accessible only through progressively increasing cumulative resistance down the pores, i.e. further into the three-dimensional porous structure of the electrode. Of course, the real electrical complexity of a porous electrode will depend on the pore-size and particle-size distributions of the carbon material, a difficult situation to model in a reliable way.

However, in various papers by Miller [1972], he has been able to model the impedance spectrum and power-density (related to the distribution of *RC* time-constant values) of practical carbon double-layer capacitor devices by a five-element ladder circuit of *C* and *R* elements as illustrated in Figure 4.5.27. The resulting impedance behavior is exemplified in Figure 4.5.28.

Pell *et al.* [1999] examined the behavior of a five-element circuit made up with series/parallel connections of hardware capacitor and resistor elements. They were able to directly demonstrate the nonuniformity of charging of such a network in time

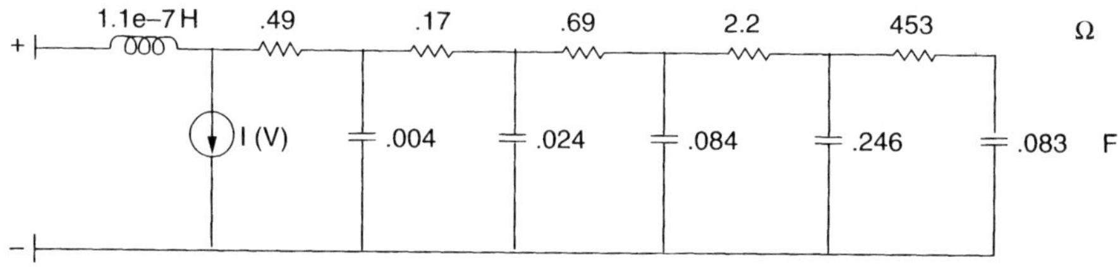

Figure 4.5.27. Five-element ladder RC circuit modeling the impedance behavior of a porous, electrochemical capacitor, designed for memory back-up (from Miller [1972]).

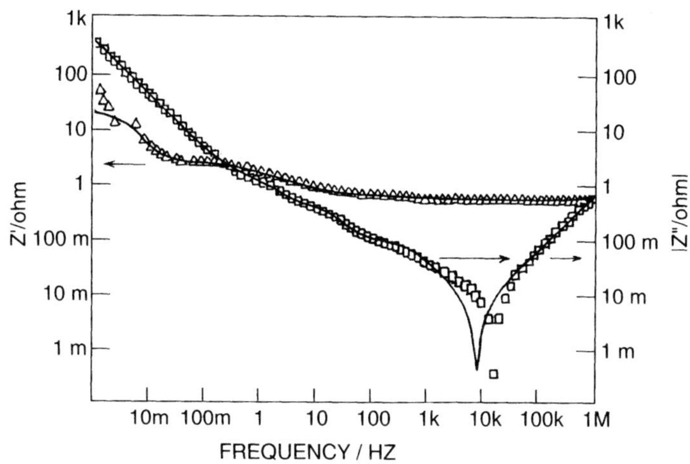

Figure 4.5.28. Impedance behavior of the five-element circuit of Figure 4.5.27 (from Miller [1972]).

(corresponding to the signal "penetration effect" of de Levie [1963,1964]) and also an interesting internal redistribution of charging of the capacitor elements in time, following a single voltage pulse of short duration. Being an hardware model circuit, potentials at all connections down the ladder circuit could be directly measured as a function of time during the charging pulse and during subsequent open-circuit relaxation (Figure 4.5.29).

Perhaps the most interesting and practically significant aspect of the impedance spectrum of a ladder *CR* circuit, like that shown in Figure 4.5.22*e*, is that it exhibits a linear relation between its $-Z''$ and Z' components with a –45° phase angle at all frequencies. This is the characteristic behavior of a transmission line and was experimentally demonstrated by de Levie [1963,1964] using a model porous electrode made up, as a brush, from a large number of fine wires in parallel configuration, constrained within a cylindrical tube. This model has anticylindrical, parallel pores and demonstrates (de Levie [1963,1964]), experimentally, a –45° phase-angle behavior over an appreciable range of frequencies (Figure 4.5.30). Pell and Conway [2001] tested a similar brush electrode of Ag or Au wires, and demonstrated a related

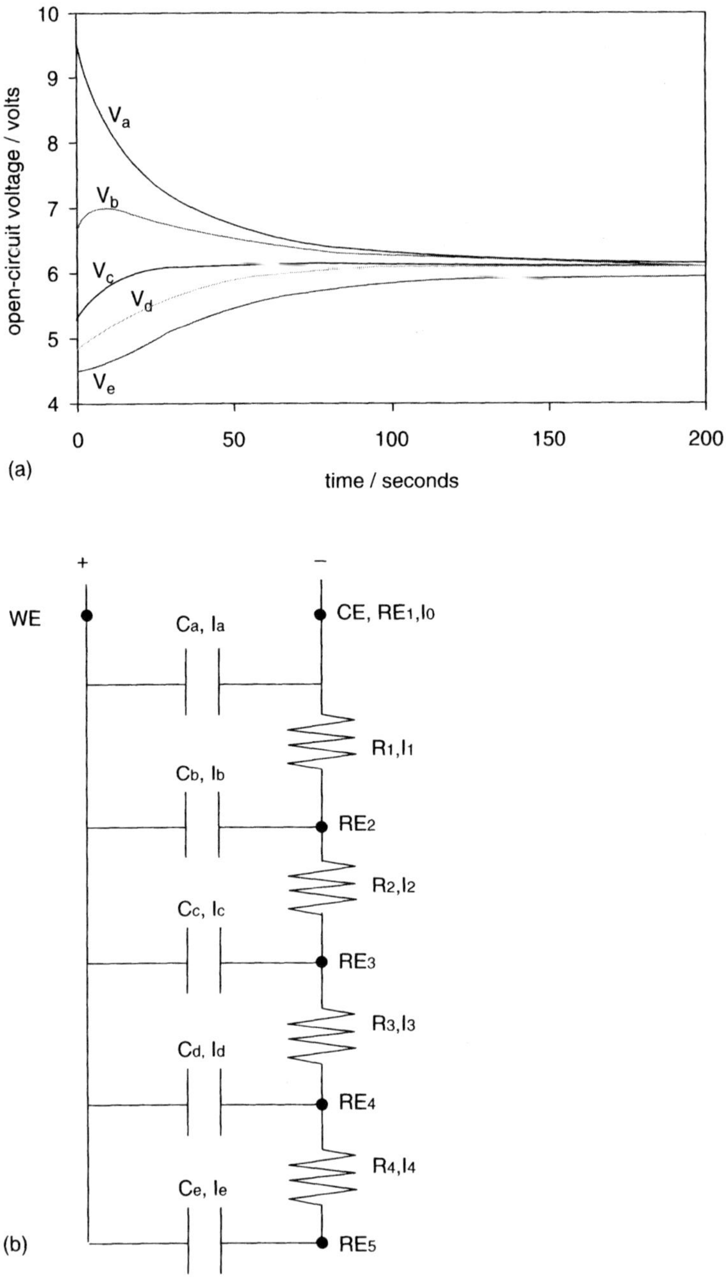

Figure 4.5.29. (*a*) Time-dependent, open-circuit potential (V) behavior of a five-element hardware *RC* ladder circuit following charging (from Pell *et al.* [1999]). (*b*) Hardware ladder circuit of capacitors, C_a, C_b, etc. and resistors R_1, R_2, etc. Component currents, I, as indicated. WE = "working electrode", CE = "counter electrode".

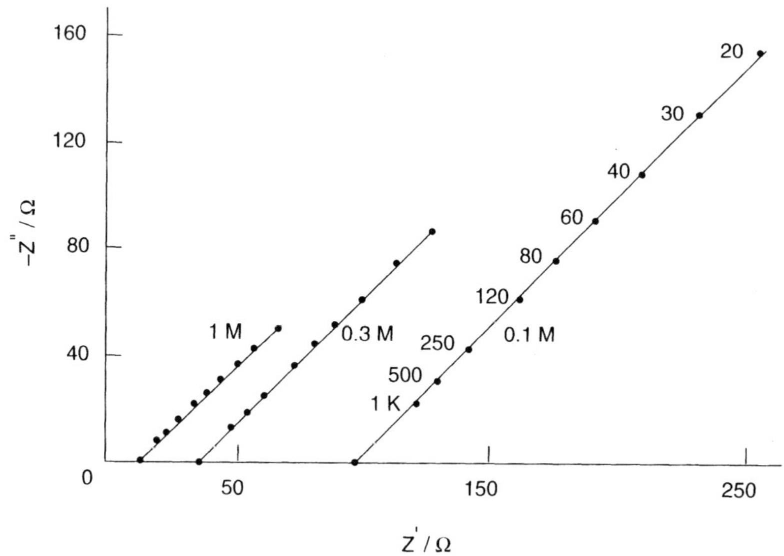

Figure 4.5.30. Impedance behavior of the de Levie brush electrode in the complex plane as a function of frequency for three electrolyte (KCl) concentrations (from de Levie [1963, 1964]).

dispersion of the cyclic voltammetry response at various increasing sweep-rates, not observed at single wires at normalized identical current densities (Figure 4.5.31). Thus, as is to be expected, the signal penetration effect and consequent attenuation of signal amplitude and current response down pores is well demonstrated by the brush-electrode model both in its ac impedance and cyclic voltammetric behaviors.

These types of experiments demonstrate the fundamentally different types of signal response behavior in the impedance of a model or real porous matrix electrode from that of a plane electrode of the same material. Practically, a double-layer type of electrochemical capacitor, based on an high specific-area carbon material, therefore exhibits a *power spectrum* so that, unlike a plane electrode double-layer capacitance, it exhibits high-power performance only for a small fraction of the chargeable area when addressed by short-period pulses or by elevated frequency sinusoidal signals while a bigger fraction of the total pore area (inner regions of the electrode matrix) becomes available for charging under longer time pulse modulation or by low-frequency AV. In the dc limit, all of the potential interfacial double-layer capacitance is ultimately available for charging or discharging.

This situation is the origin of the early misapprehension that double-layer supercapacitor devices were capable of operation at very high power densities, like that for a regular capacitor. Because of differences of surface accesability, it is only some fraction of the rated dc capacitance that is available for rapid charging, therefore at high power densities. This situation is of great importance for performance specifications which have to include data on the operating power spectrum of the device, which will also usually depend on temperature since internal electrolytic resistance is rather sensitive to changes of ambient temperature.

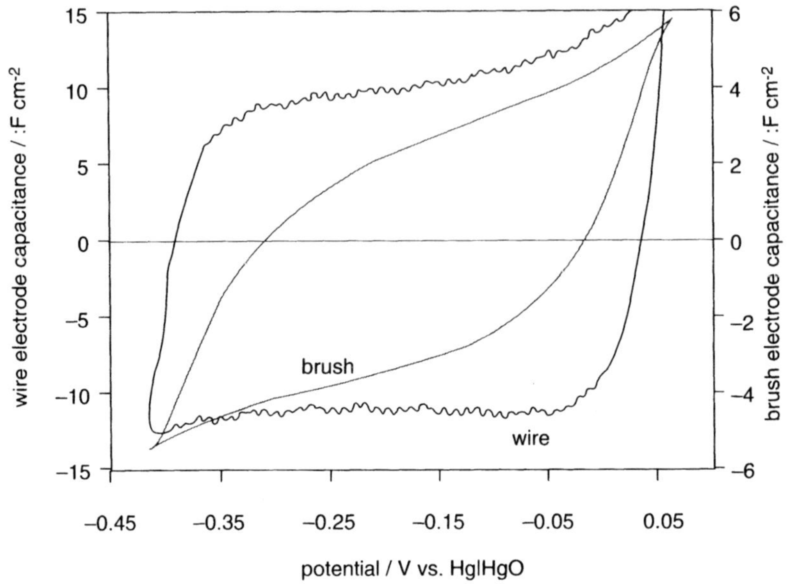

Figure 4.5.31. Cyclic voltammogram for a gold-wire brush electrode at 20 mV s^{-1} in comparison with that for a single gold wire electrode on a normalized identical current-response scale (from Pell and Conway [2001]).

The impedance spectra of porous carbon supercapacitor electrodes have a characteristic inflected form as shown typically in Figure 4.5.32. Three regions can be distinguished in this type of complex-plane plot. At relatively high frequencies, a linear relation between $-Z''$ and Z', initially arises having a near 45° slope (for $-Z''$ plotted with the same scale sensitivity as that of Z') which is characteristic (at such frequencies) of transmission-line behavior, in contrast to a Warburg 45° phase-angle plot which arises only at low frequencies when a diffusion-controlled process is determining the impedance spectrum. Note that the high-frequency 45° line extrapolates (first region) with respect to $-Z'' \rightarrow 0$, to an intercept on the Z' axis that gives the resistance, R_s (ESR) in series with the capacitor device. In well-conducting solutions, ESR values for double-layer supercapacitor electrodes can be as low as fractions of milliohms in well-designed devices.

Progressing to lower frequencies (second region) in Figure 4.5.32, a knee or smooth inflection arises, going into an (ideal) vertical, third region of the complex-plane plot in which almost pure capacitative behavior is manifested ($-Z''$ simply increasing according to $1/j\omega C$). Extrapolating the vertical section of the plot in Figure 4.5.32 to zero $-Z''$ gives another intercept on the Z' axis equal to $R_s + \Omega$, where Ω represents an effective internal resistance of the distributed RC network of the porous electrode. Ω is often appreciably larger than the ESR, R_s, and determines, with R_s, the time response and power-spectrum of the distributed RC network and thus practical aspects of device performance. Ω can have values up to tens of mΩ

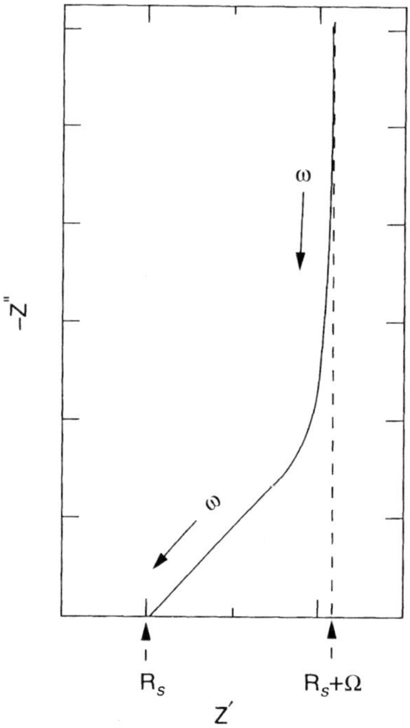

Figure 4.5.32. Characteristic form of a complex-plane impedance plot for a porous capacitor electrode with series resistance R_s (intercept at $\omega \to \infty$) and cumulative distributed resistance, Ω.

depending on the type of porous electrode material and the electrolyte used in the device. The shape of Figure 4.5.32 was reproduced in the calculated complex-plane plots of Keiser *et al.* [1976] for various values of the form-factor representing different shapes of pores (Figure 4.5.33).

4.5.3.6 Impedance and Voltammetry Behavior of Brush Electrode Models of Porous Electrodes

A brush of many thin wires, oriented in a parallel way and constrained to be in lateral contact, provides a useful and geometrically well-defined model of a porous electrode. Of course, the model pores are linear and of anticylindrical cross section while pores in a real powder electrode, or an electrode made from unoriented carbon fibers, are of random size and geometry.

De Levie [1963,1964] was the first, it seems, to examine the frequency response of a wire-brush electrode of the above kind and provide the background mathematical treatment required to interpret the experimental behavior. The important aspect of the observed impedance spectroscopy over a wide range of frequencies was that transmission-line (Figure 4.5.22*e*) behavior resulted, as illustrated in Figure 4.5.30. The characteristic behavior is a linear complex-plane plot having a 45° phase-angle

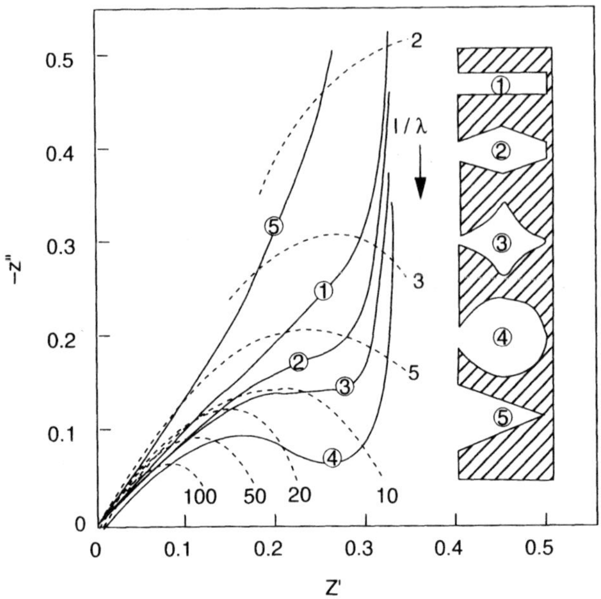

Figure 4.5.33. Impedance behavior for a series of pore shapes, as shown, in a porous electrode. l is pore length and λ is a form factor (see original literature from Keiser *et al.* [1976]).

over all the experimental range of frequencies of the addressing AV signal. As noted earlier, this behavior is the same as that of a transmission line where its resistance increases down its length accompanying increasing self-capacitance. The ladder equivalent-circuit for a porous electrode, behaving as a transmission line, was shown in Figure 4.5.22*e*.

Pell and Conway [2001] carried out linear-sweep voltammetry studies at various sweep-rates on brush electrodes made from multiple wires of Ag and Au in an inactive electrolyte in order to relate the effects to those of increasing frequency in the de Levie experiments. The purpose of this work was to compare the cyclic voltammetry current response behavior of a brush electrode made from 50 linear and parallel wires (de Levie [1963, 1964]) with that of a free *single* wire of the same metal (Ag) and length, over the same range of modulation sweep rates. For the purpose of the comparison, the response current profile (for a given sweep rate) was normalized to that of the single wire by dividing by the number (50) of wires in the brush. The resulting comparison is shown in Figure 4.5.31 in which it is clearly seen that the normalized response-current profile for a brush electrode is appreciably distorted from the (expected) more rectangular profile observable for the double-layer capacitance at the wire. The distortion has the form, typical in cyclic voltammetry, for the influence of a series iR-drop. In the AV modulation experiment of the de Levie [1963,1964], the distortion effect of progressive iR-drop down the pores of the brush electrode is manifested by transmission-line behavior with a 45° phase angle between imaginary and real components of the impedance at all frequencies result-

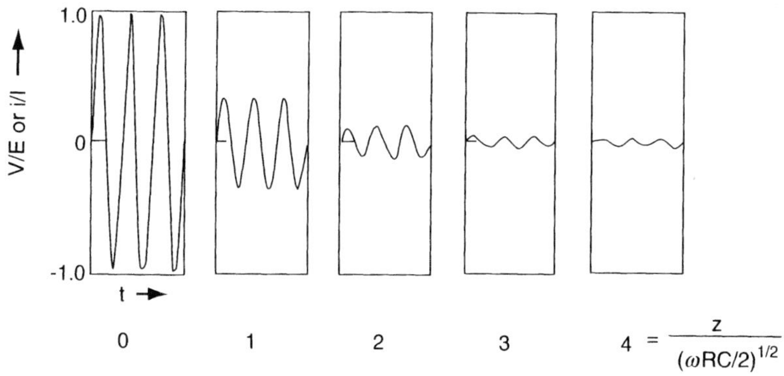

Figure 4.5.34. Attenuation of a sinusoidal signal down a pore due to the de Levie signal penetration effect. Dispersions shown for five values of the reduced distance, z, down the pore (see original literature from de Levie [1963, 1964]).

ing in a linear relation in the complex-plane plot (Figure 4.5.30). This figure shows plots for three concentrations of the supporting electrolyte, KCl, with intercepts on the Z' axis corresponding to their solution resistances.

The "porous-electrode" effects, both in AV and linear-sweep voltammetry, arise from attenuation of the amplitude of modulation signals down pores having finite resistance and hence progressively increasing iR-drop down the pores. This was referred to by de Levie [1963, 1964] in terms of attenuated "penetration" of the signal down pores, as illustrated for sinusoidal modulation in Figure 4.5.34 due to de Levie signal penetration effect [1963,1964] in which a reduced scale is used. Closely similar attenuation of the addressing signal amplitude arises in linear-sweep modulation of potential as applied to the brush electrodes in the experiments of Pell and Conway [2001].

A general schematic, Bode-plot type of representation of dispersion of phase angle with frequency for porous-electrode impedance is shown in Figure 4.5.35, according to de Levie [1963,1964]. Note the 45° phase-angle values that arise due to pore exhaustion at low log ω and the phase angle of 45° for double-layer capacitance at sufficiently high log ω.

At this point, it is of interest to relate the current responses that arise in modulation of the double-layer capacitance by liner-sweep voltammetry (cyclic voltammetry) at a sweep-rate $s = \pm dV/dt$ comparatively to that by alternating voltage. For voltammetry applied at a capacitance, C, the response current, i, is given by

$$i = C\mathrm{d}V/\mathrm{d}t \equiv Cs \tag{56}$$

where C may itself be a function of potential V as arises at most metal electrodes and high-specific-area carbon materials. The derived C is i/s and has the nature of a differential capacitance (Grahame [1947]), and s is constant over the selected range of potentials applied with constant *linear variation* in time. In the case of AV modulation, the addressing signal is of course $V = V_o \sin \omega t$ where V_o is the

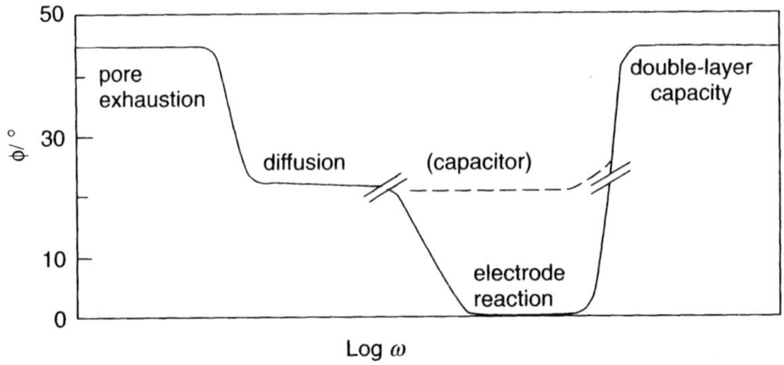

Figure 4.5.35. Illustrating dispersion of phase angle of impedance of a porous electrode over a wide range of frequencies, ω, covering various dispersion mechanisms, as indicated, with increasing ω (from de Levie [1963, 1964]).

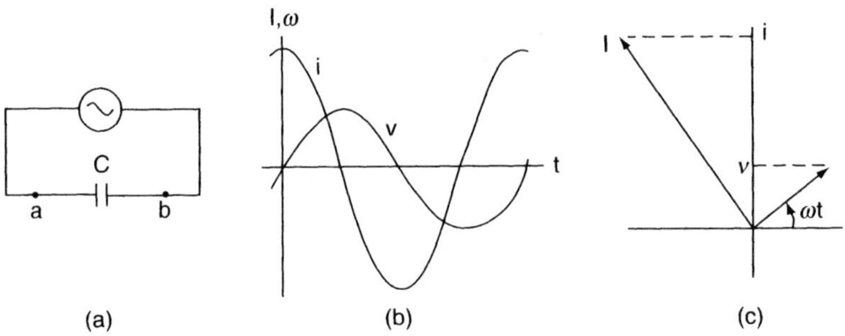

Figure 4.5.36. Phase relation between current and voltage for a sinusoidal voltage signal addressed to a simple capacitance element, C, phase angle −90°. (a) sinusoidal voltage source; (b) current, i, vs. voltage, V, relation as a function of time, t; (c) vector diagram for phase relation between i and V.

amplitude of the sinusoidal signal. Since, for any capacitance element, the response current to modulation is that given by Eq. (56), then

$$i(\mathrm{AV}) = C\mathrm{d}(V_o \sin \omega t)/\mathrm{d}t \tag{57}$$

$$= CV_o\omega \cos \omega t \tag{58}$$

so that i(AV) is out of phase with V(AV) by −90° (Figure 4.5.36). The time-dependent modulation variable is here the frequency ω which is the analog (Schultze and Koppitz, [1976]) of sweep-rate s, and V_o is the analog of the potential range of voltage sweep in cyclic voltammetry. Note, of course, that unlike in voltammetry where s is constant (in any one direction of sweep), the sign and magnitude of $\pm\mathrm{d}V/\mathrm{d}t$ is continuously varying in AV modulation through the half-cycle from 0 to maxima at $\cos \omega t = 1$, i.e. twice in the *complete* cycle at 0 and π in the 2π angular cycle.

Note that impedance measurements can be made either by recording current fluctuation induced by alternating (sinusoidal) voltage (AV) or, vice versa, by recording voltage fluctuations induced by alternating current (ac).

4.5.3.7 *Impedance Behavior of Supercapacitors Based on Pseudocapacitance*

Nature and Origins of Pseudocapacitance. Formally, pseudocapacitance of an electrode is recognized when the accepted or delivered charge, q, is Faradaic and is a function of potential, V. Then there is a derivative dq/dV, which is a capacitance quantity and can be measured as such, except that the charge originates from a *Faradaic* electron-transfer process involving a potential-dependent change of oxidation state and sometimes of chemical phase of the surface of the electrode material. In contrast, double-layer capacitor devices involve mainly *non-Faradaic* accumulation of charge difference across an interface that is electrostatic (Grahame [1947]), rather than electrochemical, in origin.

The concept of pseudocapacitance had arisen early in electrochemistry, being associated with potential-dependence of chemisorption of H atoms, e.g. at Pt (Conway and Gileadi [1962]), and of anions such as the halide ions which are adsorbed at a variety of electrode metals, with partial charge transfer (electrosorption valency), including Hg, (Schultze and Koppitz [1976]).

A different type of pseudocapacitance arises at the interfaces of oxide films at transition metals, e.g. RuO_2, IrO_2, Co_3O_4, WO_3. It originates from oxidation/reduction couples that are active in the near-surface regions and at actual surfaces of such oxide materials. The (pseudo-) capacitative behavior of RuO_2 films was first recognized by Trasatti and Buzzanca [1971] and led to the invention of the ruthenium oxide supercapacitor device by Conway, Kozlowska and Birss in collaboration with Continental Group Inc. in 1975 and following years. In the RuO_2 case, and for related oxides, as above, the charge storage arises from quasi-solid-state redox reactions in the oxide surface regions, probably involving several oxidation states of Ru. Further details are given in (b) below.

Since such surface redox reactions usually involve Faradaic charge transfer, the specific pseudocapacitances that can be manifested are some five to ten times larger than the specific double-layer capacitance (always present and significant at all electrode interfaces) at the same electrode. Hence such pseudocapacitor systems are of importance for practical development as has been the case with RuO_2, but mainly for military applications owing to the high cost of that material.

Pseudocapacitance can arise in several distinguishable ways that can be summarized as follows:

(a) from potential-dependence of surface coverage by Faradaically deposited (i.e. by electrosorption) atomic or radical species, e.g. H, OH or O, at electrode surfaces such as Pt, Rh, Pd, Ru, and Ni. The electrochemical formation of monolayers of H or OH and O requires specific charges of ca. $220\,\mu C\,cm^{-2}$ for H at polycrystalline Pt or $440\,\mu C\,cm^{-2}$ for O species which are passed over potential ranges of ca. 0.35 to 0.5 V, respectively. This gives average capacitances of ca. $700\,\mu F\,cm^{-2}$ for H or ca. $800\,\mu F\,cm^{-2}$ for O species, e.g. at Pt or Au. For H, however, the electrosorption takes place in several peaks, as measured by means of cyclic voltammetry (Will and Knorr [1960]) at polycrystalline Pt and in different specific ways at single-crystal principal-index planes. These adsorption pseudocapacitance values are on the order of 10~30 times the double-layer capacity at Pt over the oxide-free and H-free surfaces of Pt. Note, however, that any pseudocapacitance is in parallel configuration to a double-layer capacitance that is always significant at any electrode/solution interface, though it is usually much smaller. See Section 4.5.3.7. for the equivalent circuit.

The general theory of pseudocapacitance of underpotential deposited H and metal adatoms was given in detail in the paper of the present author, with Gileadi, in 1962, and also for cases (Gileadi and Conway [1964]) where a Faradaically deposited species, e.g. H in the hydrogen-evolution reaction, is an intermediate in an overall Faradaic reaction such as H_2 evolution proceeding at various overpotentials above the reversible potential. Related papers have been published by Laviron [1972] for electroactive adsorbed small organic molecules which also give rise to pseudocapacitance.

(b) From potential dependence of the state of oxidation/reduction of redox-electroactive oxide films, as e.g. at Ru, as RuO_2 and at other transition-metal oxides, as noted above. The development of redox pseudocapacitance follows formally the same equations as for pseudocapacitance of (Langmuir) chemisorption processes (see Table 4.5.1 for comparative equations). The same equations apply to the notional pseudocapacitance arising in a Nernst-type redox reaction in solution, e.g. the $[Fe(CN)_6]^{4-}/[Fe(CN)_6]^{3-}$ couple at an inert Au electrode where state of oxidation in the couple (≡ charge) is a thermodynamic function of potential.

Tilak *et al.* [1992] made useful calculations of the impedance of a redox pseudocapacitance, C_p, (e.g. as for RuO_2) for a planar electrode in comparison with that for a semiinfinite and an infinite pore in a porous electrode. The results were plotted as Nyquist diagrams (e.g. see Figures 17.14 and

Table 4.5.1. Correlation of Types of Systems Giving Rise to Pseudocapacitance with Application to Impedance of Supercapacitors

System Type	Essential Relations
a) Redox system:	$E = E_o + \frac{RT}{zF} \ln \Re/(1 - \Re)$
Ox + ze ⇌ Red	$\Re = [Ox/([Ox]) + [Red])$
b) Intercalation system:	$E = E_o + \frac{RT}{zF} \ln X/(1 - X)$
Li^+ into "MA_2"	X = occupancy fraction of layer lattice sites (e.g. for Li^+ in TiS_2)
c) Underpotential deposition:	$E = E_o + \frac{RT}{zF} \ln \theta/(1 - \theta)$
$M^{z+} + S + ze \rightleftharpoons S.M$ (S ≡ surface lattice sites)	θ = 2-dimensional site occupancy fraction

Note: (b) and (c) can be regarded as mixing of occupied (X or θ) sites with unoccupied sites, $(1 - \theta)$ or $(1 - X)$, respectively. Also $\theta + (1 - \theta) = 1$ or $X + (1 - X) = 1$.

17.15 in Adamson [1982]). Their derived results were based on the following equation for the impedance in the Laplace form:

$$Z_{(p)} = \frac{1}{\sigma A}\left[\frac{\coth ml}{m}\right] \tag{59}$$

where

$$m = \left[\frac{1}{\sigma'}\left\{pC_{dl} + \frac{pC_\phi}{pC_\phi R_f^1 + 1}\right\}\right]^{\frac{1}{2}} \tag{60}$$

and, in the latter equation, $p = aj + b$ is the complex frequency in Laplace domain, $\sigma' S = \sigma A$, S is the surface area per unit length of a pore of length l, σ is the specific conductance, A the cross-sectional area, R_f the faradaic resistance related to the reciprocal of the exchange current-density, j_0, of the charging process for C_ϕ and R_f^1 is a faradaic resistance modified by a factor related to electrosorption valencies (Schultze and Koppitz, [1976]), Lorenz and Salie [1961]) of the surface redox species, "Ox" and "Red" (see Tilak *et al.* [1992]).

(c) From the potential-dependence of extents, X, of intercalation of Li into Li-battery layer-oxide or sulfide hosts where pseudocapacitance arises (like for the case of two-dimensional chemisorption [UPD] referred to above) because of potential dependence of extents of quasi-two-dimensional Li^+ ion intercalation into layer-lattice host electrode materials. Again, large pseudocapacitances can arise but, for such systems, the distinction between "battery" and "supercapacitor" behavior becomes nebulous (see Conway [1991] for discussion of this transition situation).

(d) A fourth type of pseudocapacitance behavior arises in the charging and discharging of conducting polymers such as polypyrrole, polythiophene, polyaniline and their derivatives as reported in much literature (see Conway [1999], Rudge, Raistrick, *et al.* [1994], Rudge, Davy, *et al.* [1964], D'Aprano and Leclerc [1993], Gholamian *et al.* [1987] for example). The charging process can be regarded as a Lewis acid/base oxidation reaction (on positive charge), the extents of which change with potential giving rise to a large pseudocapacitance like that for other surface redox systems. Because the polymer chains acquire electronic conductivity due to conjugation upon charging and then behave as molecular wires, their electrochemical charging behavior has been referred to as equivalent to that of a one-dimensional double-layer with induction of charge analogous to doping of a semiconductor material.

Such conducting-polymer materials give excellent cyclic voltammograms, reversible over ca. 0.8 V and, in the case of poly-p-fluorophenyl thiophene (Rudge, Davy, *et al.* [1964]) both positive *and* negative reversible charging can be achieved over a range of ca. 3.0 V. These materials are relatively cheap to synthesize but have a cycle life of only several thousands of cycles (compare RuO_2 devices which have cycle lives of >ca. 10^5!). The lifetime of these conducting polymer systems is restricted by organic dissociation processes that originate from generation of the positively-charged sites on the polymer chains under repetitive charge/discharge cycling.

Note that generation of conducting-polymer matrices for electrochemical devices is best carried out on a high-specific-area carbon fiber support material. This gives rise to good capacitance density and hence electrical charge-storage density (cf. Rudge, Raistrick, *et al.* [1994], Rudge, Davy, *et al.* [1964],), as well as good electrolytic conductivity inside the matrix.

As with porous carbon double-layer capacitor devices, all the above pseudocapacitance systems are subject to the same *RC* time-scale restrictions on charging rates arising from the de Levie "porous electrode" effects which give rise to dispersion of capacitances with increase of charging rate or with respect to AV modulation rate (ω) in studies of their impedance spectroscopy.

Frequency-Dependence of Pseudocapacitance, and Equivalent Circuits. Supercapacitors based on pseudocapacitance suffer "non-ideality" of their capacitative behavior, like that of double-layer capacitors, but for two reasons: (a) from the porous-electrode effect associated with a distributed resistance/capacitance networks in their equivalent circuits and (b) because the pseudocapacitance is charged or discharged through a series potential-dependent Faradaic reaction resistance, subject to Tafel polarization (cf. Gileadi and Conway [1964]), as illustrated in the equivalent-circuit diagram of Figure 4.5.37. Additionally, in the case of adsorption capacitance of anions (Frumkin and Melik-Gaykazyan [1951]), a Warburg diffusional element (Randles [1947,1952]) (also in series with the pseudocapacitance element) may be required in the case of fast adsorption kinetics, e.g. as studied by Kerner and Pajkossy [2002].

The Faradaic resistance is associated with finite rates of charging the pseudocapacitance and arises from slowness of the electrochemical changes (e.g. of chemical redox state of transition-metal oxides or of deposition of adsorbed H at, say, Pt).

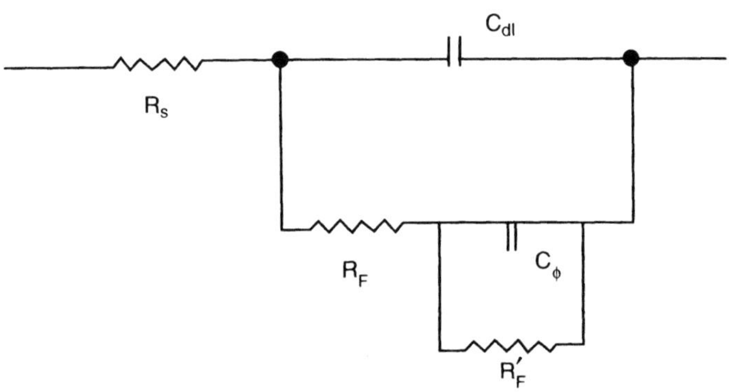

Figure 4.5.37. Equivalent circuit for a pseudocapacitance in parallel with a double-layer capacitance through a Faradaic resistance, R_F, and a possible leakage resistance, R'_F. R_s is equivalent series resistance.

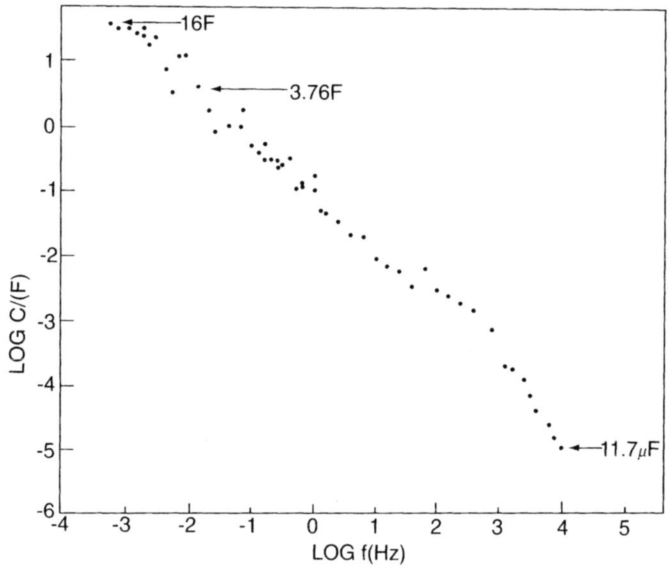

Figure 4.5.38. Broad range of dispersion of pseudocapacitance of a porous RuO_2 electrode with frequency (from Wojtowicz and Conway, unpublished; see Conway [1999]).

Hence there can be a dispersion of the pseudocapacitance with frequency of a modulation signal or of duration of a potentiostatic pulse. Figure 4.5.38 shows the broad range of dispersion of the redox pseudocapacitance of a RuO_2 electrode in aqueous H_2SO_4 at 298 K with AV frequency.

Since the pseudocapacitance element in pseudocapacitor charge-storage devices is always in parallel (through the Faradaic resistance, Figure 4.5.38) with an interfacial double-layer capacitance, the latter is also subject to the "non-ideality effects" to be discussed in Section 4.5.3.8 associated with electrode interface inhomogeneity and anion adsorption effects. However, the apparent dispersion associated with finite rates of charging the pseudocapacitance element will usually be much more apparent than that of the double-layer itself represented by a CPE (Section 4.5.3.8III). In the case of pseudocapacitance associated with H electrosorption, Frumkin, Dolin and Ershler [1940] treated the behavior of Pt electrodes in terms of heterogeneity by considering the electrosorptive charging as taking place through parallel reaction channels having distributed rate constants. The resulting behavior is manifested as a dispersion of measured overall capacitance with frequency.

In the case of pseudocapacitance associated with anion chemisorption (so-called "specific adsorption"), as referred to in Section 4.5.3.8, it can be fundamentally impossible to make a separation between the non-Faradaic, electrostatic double-layer capacitance and the parallel Faradaic pseudocapacitnce, so that an equivalent circuit such as that in Figure 4.5.37 would be inapplicable. This aspect of interfacial capacitance behavior was treated theoretically by Delahay [1966], as mentioned in Section 4.5.3.8 below.

4.5.3.8 Deviations of Double-layer Capacitance from Ideal Behavior: Representation by a Constant-phase Element (CPE)

Contrary to the situation at Hg, which provides the ideal smooth electrode surface (notwithstanding thermal fluctuations on the sub-nanoscale), impedance spectroscopy studies on the double-layer capacitance at solid electrodes, usually (but depending on conditions), show deviations from ideal behavior. Such effects are manifested as a frequency dependence of the capacitance which, on a Nyquist plot, are exhibited by deviations from the expected vertical linear plot of $-Z''$ vs. Z' corresponding to frequency dependence of $-Z''$ being simply $1/j\omega C$ (Figure 4.5.32). Such plots, for solid electrodes, show typically a significant slope of <90° from the Z' axis, i.e. a systematic deviation from a Z'-independent vertical line; in some cases this deviation shows some curvature away from the $-Z''$ axis in the Nyquist plot. Note that, in earlier years, for example in the work on double-layer capacitance at Hg in various electrolytes by Grahame [1947], only a single modulation frequency of 1000 Hz was usually employed. This is satisfactory for Hg where little or no dispersion of capacitance arises with frequency but, at other electrodes, a range of frequencies is desirable to characterize the impedance behavior, especially at solid electrodes.

The deviations referred to here and in the literature (Pajkossy [1994], Brug *et al.* [1984]) are mainly of interest in fundamental aspects of electrode interfacial behavior and do not materially influence the performance specifications (and their interpretation) of electrochemical capacitors.

Impedance behavior, corresponding to anomalous capacitance dispersion requiring representation by a CPE, has been examined with respect to a specific form of irregularity or roughness, namely a surface having fractal, that is self-similar geometry on various scales, and was treated by Mulder *et al.* [1990] and by Pajkossy [1991].

One of the first observations that ideal capacitance behavior, like that at Hg, is not observed at solid electrodes was given in an analysis by Brug *et al.* [1984]. The anomalous dispersion behavior of capacitance in such cases is manifested as a sloping line, that is not vertical, in the complex plane plot for the impedance of a series *RC* situation and a related depression of Nyquist-plot semicircles for a Faradaic process involving a reaction-kinetic resistance in parallel with the double-layer capacitance element. The deviation from ideal capacitance behavior corresponds to a frequency-dependent phase-angle and the nonideal behavior can be represented by a so-called constant phase element, CPE, that is a complex impedance having the special property that its phase angle is independent of frequency. Such a characteristic behavior seems to have been noted for the first time by Fricke [1932], a long time ago.

This capacitance dispersion at solid electrodes depends strongly on the state of the electrode's surface, e.g. its roughness and degree of polycrystallinity, and also, importantly, on anion adsorption. The deviations from ideal capacitative behavior can be empirically represented by the so-called "constant phase element" (CPE) having the following dependence on frequency and containing the double-layer capacitance quantity, T:

$$Z(\text{CPE}) = 1/T(j\omega)^{\varphi} \tag{61}$$

from which the double-layer capacitance (C_{dl}) itself can be extracted as

$$C_{dl} = \sqrt[\varphi]{\frac{T}{R_s^{(\varphi-1)}}} \tag{62}$$

The parameter φ is related to the phase angle ϕ by $\phi = \varphi(\pi/2)$. The exponent, φ, is <1, usually in the range 0.9 to 0.99 depending on the electrode material, its surface preparation and whether or not adsorbable anions are present in the electrolyte. $\phi \rightarrow 1$ obviously corresponds to true capacitance behavior.

Significantly, Brug *et al.* [1984] noted that the observable nonideal double-layer capacitance, represented by a CPE, is due to inhomogeneity of the electrode surface in such a way that only the value of the capacitance is distributed *along* the electrode surface. Then, it is concluded that the equivalent circuit for the above situation, with the interface in contact with an homogeneous electrolyte solution, should be represented by a parallel *ladder network* of series *C–R* elements but not of the kind proposed by de Levie [1963,1964] for a three-dimensional porous electrode having progressively increasing resistance down pores connected cumulatively with *C* elements (circuit *e* in Figure 4.5.22).

The significance of the CPE element and its properties have been definitively and thoroughly discussed in the important papers of Pajkossy [1994,1991] and of Brug *et al.* [1984]. The smoother and cleaner the electrode, the closer is the parameter φ to unity.

The deviations from ideal capacitative behavior of the double layer at solid electrodes also become apparent in the impedance spectroscopy behavior of electrode reactions at such electrodes where a Faradaic resistance is coupled with the capacitance; then, instead of regular semicircular complex-plane plots, the Nyquist plots are arcs of semicircles, i.e. with centers depressed below the Z' axis.

Physical interpretation of the significance of a required CPE in the response of an *RC* circuit has been controversial, as explained in Pajkossy [1994]. Theoretical attempts to rationalize capacitance dispersion, as empirically represented by a CPE element can be divided into two groups: (a) the commonest explanation, originating from Frumkin's works, is that the dispersion effect originates from micro-(or nano-?) scopic roughness of the electrode metal's surface. Current-density inhomogeneities then arise across the surface so that effects of solution bulk resistance and interfacial capacitance become mutually obscured leading to inability of the Z' and Z'' components of impedance to be separated. Thus, owing to various forms of roughness, the effective local solution resistance varies along the surface. A related effect in (a) is due to surface inhomogeneity arising from polycrystallinity and resultant faceting.

A second possible explanation (b) is that the capacitance dispersion is entirely of interfacial origin, not involving bulk-solution resistance. This effect is associated with anion adsorption, probably coupled with complexity of anion adsorption kinetics (cf. Frumkin and Melik-Gaykazyan [1951], Pajkossy [1994]), at a roughened surface. This conclusion arises because it is found that at a deliberately roughened

Pt surface, as studied by Pajkossy [1994], almost ideally capacitative behavior can, in fact, be observed; this is obviously a critical result indicating that it is not inseparable coupling between solution resistance and capacitance at a roughened (Pt) electrode surface (Pajkossy [1994]) that is the origin of dispersion effects. This led (Pajkossy [1994]) to the conclusion that it is ion (anion) adsorption that plays a crucial role in capacitance dispersion, because of frequency-dependent adsorption pseudocapacitance associated with anion chemisorption and associated kinetics of that process (Pajkossy [1994], Pajkossy *et al.* [1996]).

In regard to electrode-roughness effects, it is interesting that Borisova and Ershler [1950] found that capacitance dispersion could be noticeably decreased if the electrode material, Tl, Cd or Pb, had been melted and allowed to cool down into the form of a droplet prior to measurements. Thus, Conway *et al.* [1956] found that lead melted in H_2 and allowed to form a drop in a thin glass bulb appeared to be specularly reflective, like Hg.

The results of Scheider [1975] at Au in Cl^- solutions had the form of CPE behavior but were attributed to roughness due to the *R–C* coupling at the surface. However, it seems that the Cl^- adsorption effect was dominant because of coupling to the roughness factor.

It is to be noted (cf. Pajkossy [1994], Brug *et al.* [1984]) that the roughness-factor effect is really a limiting case of the de Levie porous electrode effect. The impedance spectrum for such systems can have a part that can be approximately modeled by a CPE having the φ parameter equal to 0.5.

From the results of Pajkossy [1994] at Pt in the presence of Cl^- ion, it seems clear that dispersion of the double-layer capacitance is a direct result of Cl^- anion adsorption effects. Similar conclusions arise in the case of Cl^- ion-containing solutions at Au (Pajkossy [1994], an Germain *et al.* [2004]). Also to be noted is the fact that in the oxide region at Pt, no anomalous dispersion behavior arises despite the possibility of roughness of the oxide film itself. In the presence of extensively formed, anodic oxide films at Pt or Au, Cl^- ion chemisorption (as well as that of HSO_4^- ion) is much diminished, although at low O coverages the surface oxidation of Pt and Au is competitively inhibited (Conway and Novak [1981]).

Interestingly, the capacitance in Cl^--containing solutions is smaller than in SO_4^{2-} solutions alone and smaller, also, than that for the oxide-film region (Pajkossy [1994]). This is not explicable (cf. Pajkossy [1994]) in terms of simple specific adsorption effects since theories predict an increase, rather than a decrease, of capacitance. In these cases, it seems that the double-layer and adsorption pseudocapacitance are coupled in an inseparable way as argued by Delahay [1966] and as pointed out in the recent studies by Germain *et al.* [2004] in our laboratory in work at Au (cf. Cahan *et al.* [1991]).

In a paper by Kerner and Pajkossy [2002], anion adsorption rates have been measured by impedance spectroscopy at Au (111), for SO_4^{2-}, Cl^-, Br^- and I^-. Cl^- ion adsorption is very fast, so the equivalent circuit for the process not only has to include a Faradaic pseudocapacitance and its corresponding reaction resistance, but also a Warburg element for the anion diffusion. The interfacial capacitance is then plotted in the Nyquist plane as real vs. imaginary capacitance components, C' and

C'', having the form of a semicircle or depressed semicircle, depending on whether (relatively) the diffusion is fast or slow (Pajkossy *et al.* [1996]).

An important conclusion from the paper by Brug *et al.* [1984] is that involvement of a CPE at solid electrodes used for studies of the impedance of Faradaic reactions can severely influence the frequency dispersion of interfacial admittance, leading to large errors in the determined Faradaic rate parameters. However, those authors note that it is feasible to account for the CPE effect correctly and to check the results of impedance analysis with respect to their internal consistency. The latter can be checked by a Kramers–Kronig analysis (cf. Lasia [1999]) which requires, however, detailed frequency-response data. Their approach was supported by experimental impedance studies on proton reduction at single and polycrystalline Au electrodes and on reduction of tris-oxalato-Fe(III) (Brug *et al.* [1984]).

Acknowledgment

Brian E. Conway wishes to thank Dr. W.G. Pell, Senior Research Associate in his group, for many useful discussions on impedance spectroscopy and voltammetry of electrochemical capacitors, as published in detail in journal papers elsewhere. Acknowledgement is also made to the Natural Sciences and Engineering Research Council of Canada for support of that work, which has led to the present contribution.

4.5.4 Fuel Cells

Nobert Wagner

4.5.4.1 Introduction

General Aspects and Classification of Fuel Cell Systems. Despite their relatively long history (discovered by C. F. Schönbein and invented by W. R. Grove in 1839, as reported by Bossel [2000]) fuel cells have attracted much interest during the past 35 years. The need for an efficient, non-polluting power source for vehicles in urban environments, emphasized by legislative initiatives, has resulted in increased attention to the option of fuel-cell-powered vehicles of high efficiency and low emissions. Recently, fuel cells have also been used in a variety of new applications like portable devices (e.g. radio communications, mobiles, lap-top, etc.), residential applications (combined heat and power generation) and other transport applications (e.g. submarines, ships, rail guided vehicles, etc.).

Fuel cells can continuously convert the chemical energy of a fuel (hydrogen, methanol, methane, etc.) and an oxidant into electrical energy at up to 70% efficiency with very low pollutant emissions. Depending on the type of electrolyte used in a fuel cell, one can distinguish five main types:

- alkaline fuel cell (AFC)
- polymer electrolyte membrane fuel cell (PEFC or PEMFC)—also called solid polymer fuel cell (SPFC)

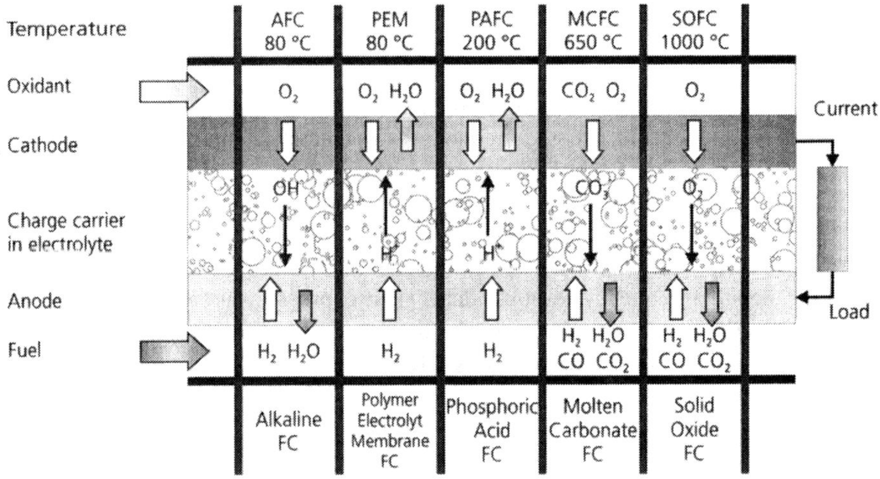

Figure 4.5.39. Schematics of main types of fuel cell.

- phosphoric acid fuel cell (PAFC)
- molten carbonate fuel cell (MCFC)
- solid oxide fuel cell (SOFC)

Schematics of these fuel cells are shown in Figure 4.5.39. Both electrodes, cathode and anode, of a fuel cell are porous electrodes in order to increase the reaction area and thereby the current generated. The key technical problems common to all types of fuel cell are the optimal design and stability of these porous electrodes and as well as the necessity to increase the three-phase interface between the electrolyte, the gaseous (liquid) reactant and the electronically conducting catalyst in the electrode. Other research and development challenges are the investigation of the reaction mechanism and kinetics at each electrode/electrolyte interface, determination of degradation (poisoning) mechanism, production of cheap and efficient electrodes with low catalyst loading, development of suitable catalysts, long term stability, etc. In order to improve the efficiency of the fuel cell systematically, a better understanding of the electrochemical reactions and mass transport in the fuel cell is essential. Moreover, quality control and understanding of degradation require new non-destructive methods and a better understanding of experimental results based on modeling and simulation. However, EIS is such an *in situ* investigation method and is increasingly applied in fuel cell research and development.

Detailed descriptions of fuel cell technology are extensively presented in several recent monographs by Appleby and Foulkes [1993], Kordesch and Simader [1996], Vielstich *et al.* [2003], Kurzweil [2003], Blumen and Mugerwa [1993]; a brief description of the main types will be given in Sections 4.5.4.2 to 4.5.4.4.

Electrochemical Characterization of Fuel Cells—Correlation Between EIS and Current/Voltage Characteristic of Fuel Cells. The performance of a fuel cell depends not only on electrochemical properties of the electrode/electrolyte interfaces and electrolyte, but also on the construction of its components such as the gas leads and the gas distributor. One way to characterize the performance of a fuel cell consists of measuring a steady state current/potential curve. From such a graph, one obtains information about the entire fuel cell comprising the sum of the electrochemical behavior of the electrode/electrolyte interfaces, conductivity of the electrolyte, the influences of the gas supply and the electrical contacts between the individual components.

In order to determine current/potential curves, one either scans the whole potential range continuously, beginning from the open-circuit potential (OCP) to full load, with a constant voltage scan rate lying between 0.1 and 10 mV/s and records the resulting current, or one gradually alters the potential, and the steady state current is recorded after a definite time period. One problem with this method is the criteria by which one defines the steady state. Either a distinct time period is fixed after one guesses the genesis of the stationarity (second to minute range) or one allows the current to alter over a distinct range in a time unit that should not exceed (for example 2 mA in 120 s).

Depending on the criterion used, one gets distinctly different results. If the boundary conditions during the measurement of a steady state current/potential curve are not known, a comparable statement is not possible. Furthermore, one must consider that a current/potential curve normally is measured beginning from the open circuit potential or full load and then the potential increases or decreases. One switches from a continuous operation point of 500 mV or 700 mV for instance, to a starting potential of about 1000 mV or 100 mV. Hence it follows that the whole system is suddenly altered after the jump, especially the moisture content of the electrolyte and electrodes. This conditioning potential, different from the potential at which the measurement is made, has a decisive influence on the current/potential curve and the performance of the fuel cell.

Another method to characterize the performance of fuel cells is electrochemical impedance spectroscopy (EIS). EIS was initially applied by Graham [1947] and later by Delahay [1965] to study the double layer capacitance and was also applied in ac polarography to separate the double layer charging current from the total current, and it is now applied to characterize electrode processes and complex interfaces—a review is given by Lasia [1999]. The analysis of the system response yields information about the reactivity and structure of the interface, also about the electrochemical reactions and mass transport limitations taking place there. Its importance recently increased due the modern computer controlled devices and the corresponding analytical software. With the help of the EIS, it is possible to take measurements without all the external influences, directly "*in situ*" in a running cell, up to a current of 100 A (for example using the external electronic load EL101 from Zahner-elektrik). Even after the measurement, the cell can work at the same operation point. The ac response of the fuel cell provides additional information on the investigated system. By the analysis of the impedance spectra one tries to describe the system-specific physical parameters of the electrochemical set up with an equivalent circuit. Ideally, it is then possible to separate the individual voltage losses of the single components, namely of the anode, cathode, electrolyte and perhaps even to identify diffusion processes. Such results can allow one to draw

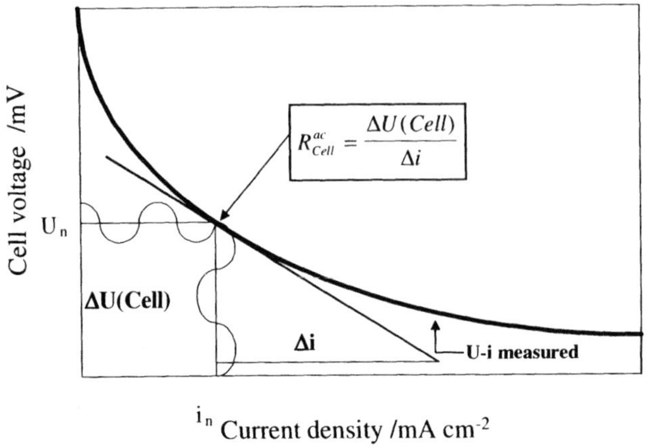

Figure 4.5.40. Schematically representation of the correlation between impedance of fuel cell and current/voltage curve, after Wagner [2002].

conclusions about the limiting processes which determine the performance of the fuel cell.

This powerful technique was first used by Mund [1975] to obtain information about the rate process in the porous electrodes of an AFC containing Raney nickel or porous silver and is widely used for the investigation of the reaction kinetic and diffusion phenomena. Springer *et al.* [1996] used a combination of models and experiments to describe the effect of different operation points and cell components on the impedance spectrum and to attribute the lower catalytic activity of the cathode, the conductivity of the catalytic layer and the decrease of the conductivity of the electrolyte to the low humidity of the air stream. Furthermore, the individual reaction steps (for example the oxygen reduction reaction (ORR) and the hydrogen oxidation reaction (HOR)) in a fuel cell still were investigated. Arico *et al.* [1996] as well as Holze and Vielstich [1984] used this technique to investigate the ORR in H_2SO_4 in addition to neutral and alkaline environments and concluded that the cathodic overpotential is primarily due to the slow diffusion of oxygen and its reaction products. With measurements over the entire performance range of the hydrogen/oxygen ratio, one tries to separate the single contributions to the performance loss of a fuel cell during load or even at OCP.

Theory of the Conversion of Resistance to an Overpotential. The correlation between impedance measurements and current/potential curve was first described by Wagner [2002] and is given schematically in Figure 4.5.40 and Figure 4.5.41. The polarization resistance of the cell (R_{Cell}) measured at U_n (index n = counter) corresponds to the tangent to the current/potential curve at that potential (U_n). The polarization resistance of the cell is the impedance at frequencies near 0 Hz where only ohmic parts attract attention. To obtain the polarization resistance of the cell, one has to extrapolate the simulated impedance (model) at very low fre-

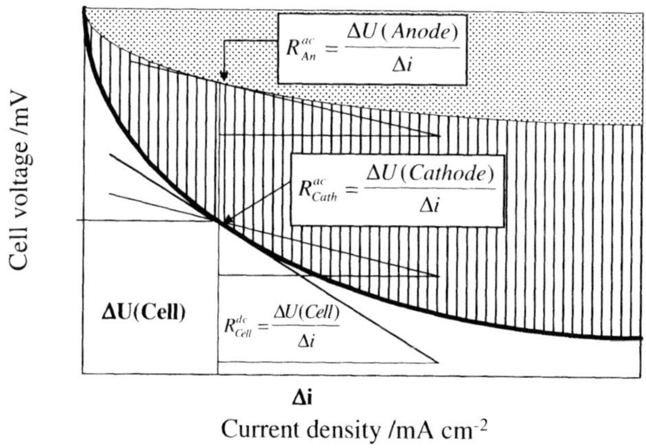

Figure 4.5.41. Schematically representation of the correlation between polarization resistances (anode, cathode and cell) of fuel cell and current/voltage curve.

quency (e.g. 1 nHz) or sum up the individual resistances, obtained after fitting the measured spectra with an equivalent circuit.

Assuming that the current/potential curve can be expressed by an Eq. (63) of second order and the resistance is defined by Eq. (64), then the parameters a_n, b_n and c_n from Eq. (63) are given by the Eqs (64)–(67).

$$U_n = a_n I_n^2 + b_n I_n + c_n \tag{63}$$

$$R_n = \left.\frac{\partial U}{\partial I}\right|_n \tag{64}$$

with:

$$a_n = \frac{R_{n+1} - R_n}{2(I_{n+1} - I_n)} \tag{65}$$

$$b_n = R_{n+1} - 2a_n I_{n+1} \tag{66}$$

$$c_n = U_{n-1} - a_n I_{n-1}^2 - b_n I_{n-1} \tag{67}$$

This method was applied successfully to determine the different individual contributions—representing different reaction steps—to the overall overvoltage in the case of a PEFC; an example is given in Section 4.5.4.3.

Theory of Time Resolved Electrochemical Impedance Spectroscopy (TREIS). A prerequisite for the development and the improvement of fuel cells is the knowledge of the mechanistic processes that take place during operation. The understanding of the kinetic behavior of the fuel cells requires the variation of different experimental parameters. Often, the variation of distinct parameters causes situations where steady state conditions are no longer fulfilled. In practice, EIS analysis often suffers from the fact that the steady state condition is violated due to time instability of the examined systems. While an EIS measurement is running, the examined system should not change its dynamic behavior. Unfortunately, the violation of steady state conditions complicates the evaluation of experimentally obtained impedance spectra because all relevant physical models for the interpretation of the data are based on steady state conditions. The time- and frequency-dependent relationship between current and voltage of an electrochemical system is often called "two-pole impedance transfer function" (TTF). All properties, which are influencing the current–voltage dependence must be stable in this time interval. Otherwise the TTF will be falsified in a way that is equivalent to a violation of causality.

Nevertheless it is possible to investigate such "drifting" systems and to obtain relevant data for the development of fuel cells using EIS. Enhanced numerical procedures are required to compensate or to eliminate the drift effects of systems with states that change with time.

Causal Transfer Functions. Valid transfer functions H in accordance with the causality rule are complex functions of the angular frequency ω and may be separated into real and imaginary components $R(\omega)$ and $X(\omega)$:

$$H(j\omega) = R(\omega) + jX(\omega) \tag{68}$$

If the transfer function H is in accordance with the causality rule, the components R and X are no longer independent of each other. Causality in the meaning of system theory forces couplings between the real and imaginary part, which are known as Kramers–Kronig relations (KKT) or Hilbert relations (HT), for details see Section 3.1.2.9 (The use of Kramers–Kronig Transforms).

These relations offer the possibility to examine measured transfer functions (impedance spectra) on errors caused by time instability or time drift. However, KK-checking techniques have fundamental problems in their application to practical measurements. Therefore many attempts have been made to overcome these limitations by means of different interpolation procedures. An attempt is the Z-HIT approximation, applied by Schiller *et al.* [2001], Agarwal *et al.* [1995], and Ehm *et al.* [2000], an approximation formula for the calculation of the impedance modulus course from the phase angle by integration.

During operation of fuel cells different time-dependent effects can occur, depending on the type of fuel cell and experimental conditions like gas flow rates (flooding of the electrode, changing of the gas composition), temperature (corrosion or sintering), impurities in the hydrogen (CO) leading to poisoning of the anode in the case of PEFC, degradation of the electrodes during long time of operation etc.

One has to assume that the system changes its state not only between two measurements but also during the recording of a single spectrum. The latter fact causes problems for the evaluation of the spectrum, because the recording of an impedance spectrum with one frequency after each other requires a finite time, while the measurement at high frequencies requires less time than the measurement at low frequencies. Due to the fact that the recording of a single spectrum in the frequency range, for example from 10 kHz to 50 mHz requires about 20 minutes, the influence of the changed state on the measured spectrum is not negligible. For this reason, one

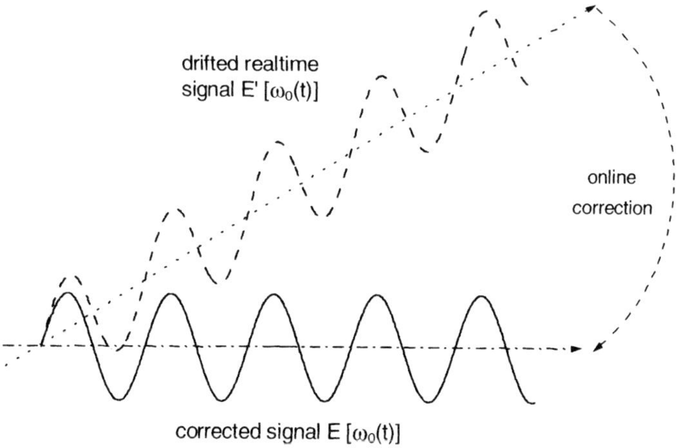

Figure 4.5.42. Schematic representation of real time-drift compensation during impedance spectrum measurement.

of the fundamental prerequisites for the evaluation of impedance measurements is violated. Nevertheless, it is possible to reconstruct "quasi steady state" (and therefore "quasi causal") spectra from drift affected impedance data using *improved evaluation techniques* which are denoted as the real-time drift compensation, the time course interpolation and the Z-HIT refinement. These techniques were applied successfully to the interpretation of time dependent impedance spectra of a fuel cell which exhibits non-steady-state behavior.

Real Time-Drift Compensation. During recording an impedance spectrum, a potential drift due to changing of the state of the system can occur. The magnitude of drift is frequency dependent and should be compensated by software during the measurement as exemplified in Figure 4.5.42.

Time-Course Interpolation. Recording an impedance spectrum one frequency after each other requires a finite time, while the measurement at high frequencies requires less time than the measurement at low frequencies. As a consequence, the system at the start is in another state than at the end if the system changes the state during the impedance measurement. It is impossible to eliminate the drift which is caused by the finite measuring time, performing only a single spectrum. In particular, the data recorded at low frequencies are affected and a mathematical procedure has to be applied to check the data with respect of causality and linearity (Z-HIT) to avoid erroneous interpretations resulting from a fit of the drift-affected data.

According to the idea of Savova-Stoynov and Stoynov (Savova-Stoynov and Stoynov [1985, 1992], Stoynov [1990]), recording of a series of impedance measurements at distinct time intervals offers the possibility to eliminate the drift and therefore to reconstruct an impedance spectrum which is acquired in an "infinite" short time. The time course interpolation procedure is schematically depicted in

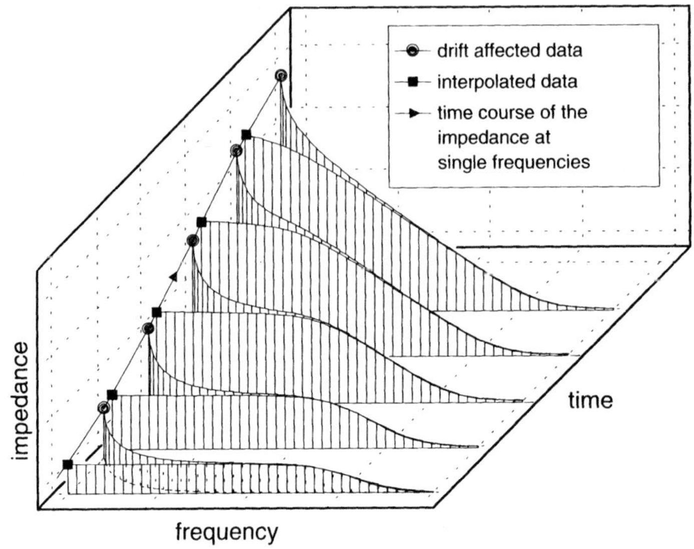

Figure 4.5.43. Schematic representation of time-course interpolation.

Figure 4.5.43 and outlined in the following. In Figure 4.5.43, the primarily recorded impedance data as well as the reconstructed set of different measurements of the series are plotted against the frequency. Additionally, the elapsed time of the experiment is involved as a third parameter. As mentioned above, the measurement at lower frequencies requires a longer time for the registration and therefore the measured curve is shifted to the back along the time axes. It should be noted, that the absolute value of the shift is frequency dependent because the acquisition time for each measured frequency is different. As indicated in Figure 4.5.43 for the lowest frequency, a single impedance spectrum can be reconstructed by interpolating the impedance value from the time course of the series using an appropriate smoothing function at the time of the start of this distinct measurement. The interpolation procedure is repeated for each measured frequency resulting in a data set where the effect of finite measurement time is significantly reduced or even eliminated for each recorded spectrum of the series.

Applying all three techniques to time drifting systems, quasi-steady-state impedance spectra at defined times can be obtained, namely time resolved electrochemical impedance spectra (TREIS). In Figure 4.5.44, for illustration, an EIS recorded at the PEFC with flooded cathode, after 4450 s of "dead end" operation at 2 A, is given (for details see 4.5.4.3).

Cylindrical Porous Electrode Model. In electrocatalysis, and in catalysis generally, there is a great interest in increasing the real surface area, especially the electrochemical active area of electrodes. In such cases, porous electrodes are used. First investigations of porous electrodes with EIS were applied by de Levie (for details see Section 2.1.6—Rough and Porous Electrodes), presenting a model describing

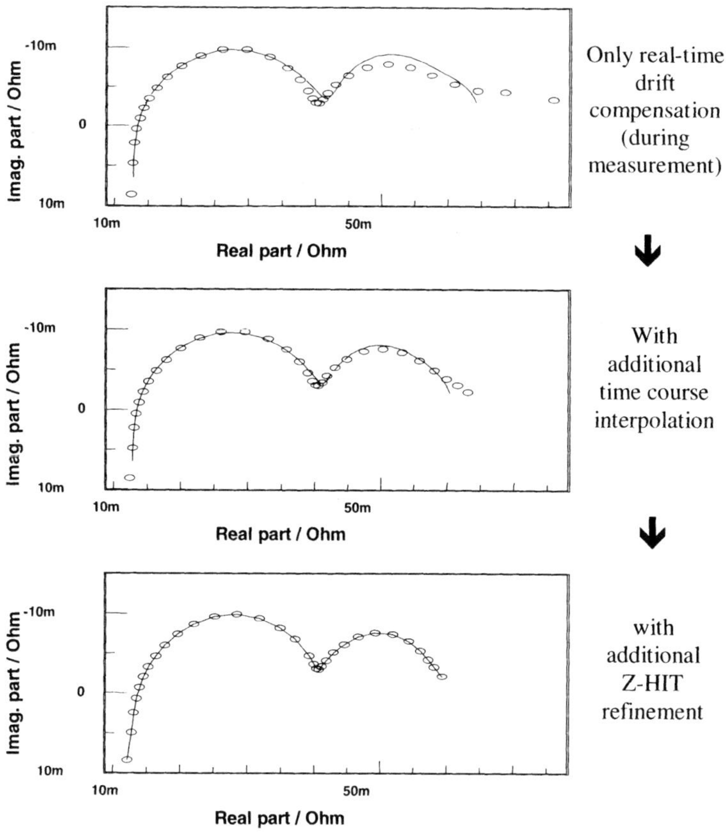

Figure 4.5.44. Nyquist plot of a representative example of the series, recorded during flooding the cathode of a PEFC, after 4450 s of "dead end" galvanostatically operation at 2 A.

the pores of a porous electrode as essentially circular cylindrical channels of uniform diameter and of semiinfinite length. Based on ac impedance measurements, applied for the investigation of porous lead electrodes in sulfuric acid, Göhr *et al.* [1983] and Göhr [1997] proposed another model of the porous electrode. This model, schematically represented in Figure 4.5.45, is integrated in the simulation and evaluation program Thales (Zahner Elektrik, Kronach, Germany). In this model the cylindrical pore is considered as a transmission line made of a large number of infinitesimally thin sections, with impedance elements of pore's ground surface (Z_n), pores electrolyte (Z_p), pore's wall surface (Z_q), porous layer (Z_s) and surface layer (Z_o). The basic equations for the impedance of a porous electrode are given in Table 4.5.2.

A review, describing gas diffusion electrodes and flooded electrodes, is given by Szpak [1991]. The great number of very interesting publications, regarding ac impedance measurements on porous electrodes and model electrodes shows the importance and wide applications of the method, from porous layers to fuel cell electrodes.

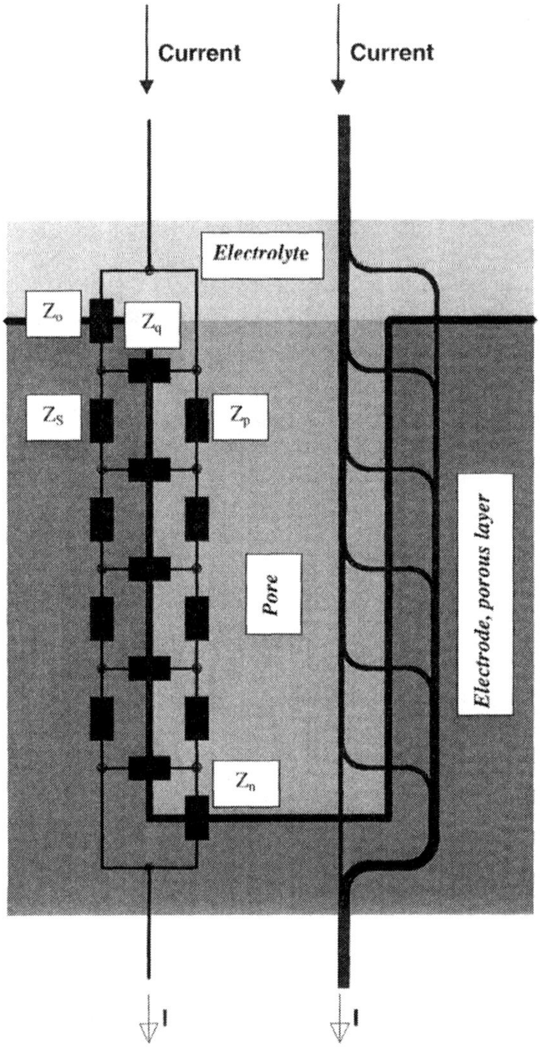

Figure 4.5.45. One-dimensional network distributed homogeneous cylindrical porous electrode model after Göhr [1997].

Impedance Spectra with Inductive Behavior at Low Frequencies

Relaxation Impedance. Based on the concept of impedance elements, Göhr [1986] described the Faradaic impedances as connections of impedance elements each of which is associated with a single process. One of such an impedance element is the relaxation impedance, describing the surface relaxation of the interface and explaining the development of the pseudoinductive behavior in the low frequency range (frequency $< 3\,\text{Hz}$) in the impedance spectra of the fuel cell. This behavior was first found by Müller *et al.* [1999] during poisoning the anode of a PEFC with

Table 4.5.2. Basic Equations for the Impedance Z of a Porous Electrode with Cylindrical Pores, After Göhr [1997]

$\mathbf{Z}^* = \sqrt{(Z_p + Z_s) \cdot Z_q}$	$\mathbf{Z}\# = \dfrac{Z_p \cdot Z_s}{(Z_p + Z_s)}$	
$\mathbf{C} = \cosh\left(\dfrac{Z_p + Z_s}{Z^*}\right)$	$\mathbf{S} = \sinh\left(\dfrac{Z_p + Z_s}{Z^*}\right)$	
$\mathbf{p} = \dfrac{Z_p}{Z_p + Z_s}$	$\mathbf{q}_o = \dfrac{Z^*}{Z_o}$	$\mathbf{v} = \dfrac{Z_p + Z_s}{Z^*}$
$\mathbf{q_n} = \dfrac{Z^*}{Z_n}$	$\mathbf{s} = \dfrac{Z_s}{Z_p + Z_s} = 1 - \mathrm{p}$	

$$\mathbf{Z} = Z\# + Z^* \cdot \frac{C + (1 + C) \cdot 2 \cdot p \cdot s + S \cdot (p^2 \cdot q_n + s^2 \cdot q_o)}{S \cdot (1 + q_n \cdot q_o) + C \cdot (q_n + q_o)}$$

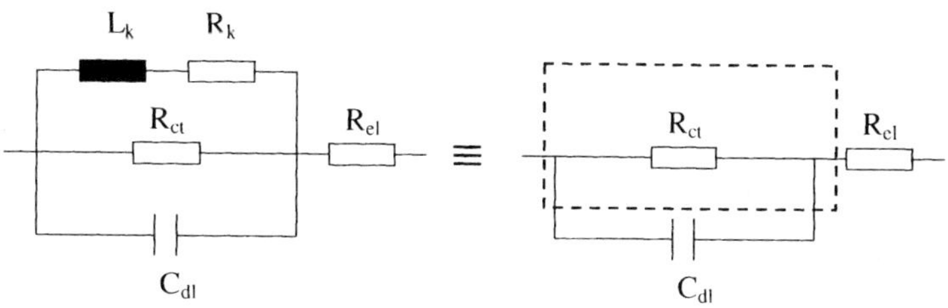

Figure 4.5.46. Equivalent circuit with relaxation impedance (Z_k).

a mixture of H_2/CO and will be discussed in detail in Section 4.5.4.3. The surface relaxation impedance represents a Faradaic impedance (Z_F) at non-equilibrium potential with a potential-dependent transfer reaction rate: $k = k(\varepsilon)$ and its time dependent relaxation according to Eqs (69) and (70):

$$Z_F = \frac{R_{ct}}{1 + {R_{ct}}/{Z_K}} \tag{69}$$

$$Z_K = \frac{1 + j\omega\tau_K}{I_F \cdot {d\ln k}/{d\varepsilon}}$$

$$\text{with} \quad R_K = \frac{1}{I_F \cdot {d\ln k}/{d\varepsilon}} \quad \text{and} \quad X_K = j\omega\tau_K \cdot R_K = j\omega \cdot L_K \tag{70}$$

In Eq. (69) R_{ct} denotes the charge transfer resistance, Z_K is defined as the relaxation impedance and is schematically shown as series combination of R_k and L_k as well as the box surrounding R_{ct} in Figure 4.5.46, according to Eq. (70), where I_F denotes the Faraday current, τ_K the time constant of relaxation and the expression $d\ln k/d\varepsilon$ is the first derivative of the logarithm of the reciprocal relaxation time constant ($k = 1/\tau_K$) against the potential ε. According to its frequency dependence, Z_K can be split-up into the relaxation-resistance R_K and the relaxation inductivity X_K, with the pseudoinductance $L_K = \tau_K \cdot R_K$ and which is proportional to the relaxation time constant τ_K.

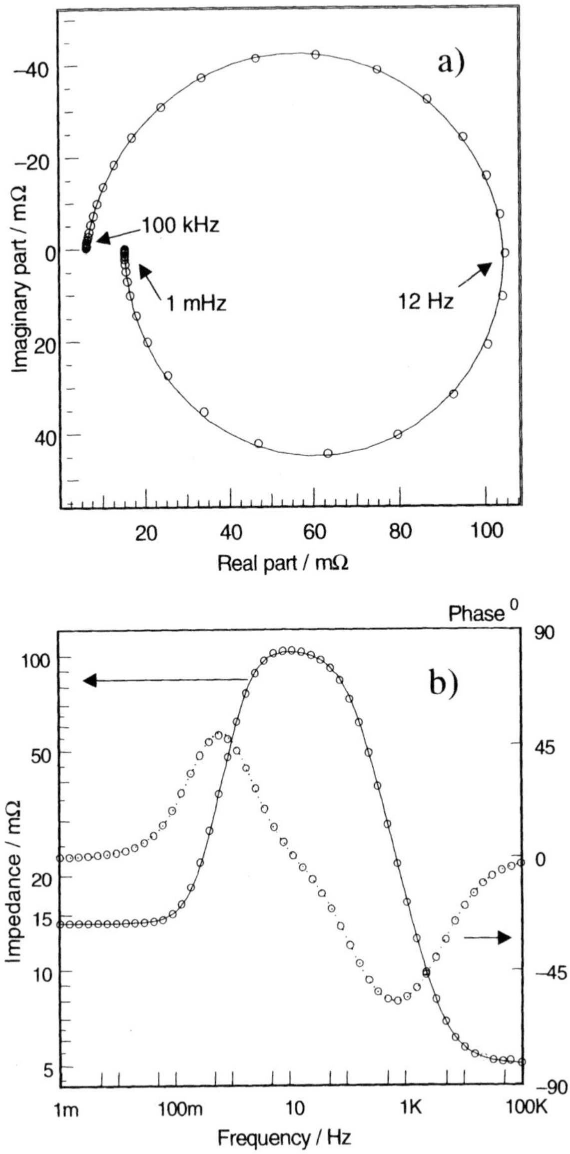

Figure 4.5.47. Impedance spectrum simulated with equivalent circuit from Fig. 4.5.46 $\tau_K = 1\,s$, $R_k = 10\,m\Omega$, $R_{ct} = 100\,m\Omega$, $C_{dl} = 10\,mF$ and $R_{el} = 5\,m\Omega$, represented as (*a*) Nyquist plot and as (*b*) Bode plot.

For illustration, in Figure 4.5.47 a simulated impedance spectra, presented as a Nyquist plot (*a*) and a Bode plot (*b*) with an inductive behavior in the low frequency range is shown, which looks similar to the impedance spectra measured at PEFC with H_2 + 100 ppm CO as fuel gas.

Inductive behavior in the low frequency range of impedance spectra was also observed in systems in which a reaction consisting of two successive electron transfer steps proceeding by an adsorbed intermediate species:

$$\text{reactant}(s) \pm e^- \Leftrightarrow \text{adsorbate} \quad \text{step 1} \tag{71}$$

$$(\text{reactant}) + \text{absorbate} \pm e^- \Leftrightarrow \text{product}(s) \quad \text{step 2} \tag{72}$$

This mechanism was proposed by Diard *et al.* [1992] to evaluate EIS for cathodic production of hydrogen by the metallic electrodeposition reaction and anodic oxidation reaction of metals (e.g. iron) close to their corrosion potentials. A similar mechanism was found by Rcrolle and Wiart [1996] studying the oxygen evolution during zinc electrowinning.

4.5.4.2 Alkaline Fuel Cells (AFC)

Electrochemical Impedance Studies of AFC Cathodes. A review of the state of the art of alkaline fuel cell is given by McLean *et al.* [2002]. The electrolyte in AFC is a concentrated KOH solution, 25–50 wt% when operated at 120°C or below and in older AFC systems, operated at 250°C a 85 w/o KOH solution was used. The AFC was first fuel cell developed to technical maturity, based on the fundamental research of Bacon [1952] and has been employed extensively in the NASA space programs: Apollo program (1960–1965) and space shuttle program (since 1981).

However, for technical use of AFC, the long-term behavior of AFC components is important, especially that of the electrodes. Nickel can be used for the hydrogen oxidation reaction (catalyst in the anode) and on the cathode silver can be used as catalyst (see next section), no expensive noble metal (platinum) is necessary, because the oxygen reduction reaction kinetics are more rapid in alkaline electrolytes than in acids and the alkaline electrochemical environment in AFC is less corrosive compared to acid fuel cell conditions. Both catalysts and electrolyte represents a big cost advantage. The advantages of AFC are not restricted only to the cheaper components, as shown by Gülzow [1996].

The life time of fuel cells is a decisive factor for their commercialization. Therefore, the degradation of fuel cell components is under increased interest and investigation. For alkaline fuel cells, several studies of degradation processes exist: Gülzow *et al.* [2002a], Kiros and Schwartz [2000], Rahman *et al.* [1997] and Wagner *et al.* [1996], they are all focused on the electrodes, because the electrolyte in AFCs can be easily exchanged and regenerated. In the case of using a silver gas diffusion electrode (GDE) in a complete alkaline fuel cell, the solubility of silver can lead in certain circumstances to a transfer of silver to the anode and the resulting plating out feature may damage the catalysts (usually Raney-Nickel) of the anode. On open circuit standing, the oxidation of silver may occur. In particular, carbon-containing cathodes suffer from oxidation at higher temperatures. Shut down and restarting procedures can lead to damages too.

Another important feature of AFC porous electrodes is the control of the location and extent of the three-phase boundary. This can be achieved by adding a hydrophobing agent such as PTFE (powder or suspension). Alternatively, a dual-porosity structure or, as proposed by Wagner and Kaz [2002], a dual-composition structure can be used.

The Oxygen Reduction Reaction (ORR). Oxygen reduction electrocatalysis is of special importance for fuel cells, metal–air batteries and industrial chlorine–alkali electrolysis as well. Silver is a known electrocatalyst for the reduction of oxygen in alkaline fuel cell cathodes. To enhance its catalytic activity, silver is used in a form with high specific area, such as porous electrodes where the silver particles are dispersed in the porous matrix. The porous PTFE-bonded gas-diffusion electrodes can be prepared by a cold rolling process, first described by Winsel [1985]. The electrodes consist of the electrocatalytic powder (silver, Raney-silver, silver-covered PTFE particles or silver oxide), the organic binding agent (PTFE), pore forming material (e.g. $NaHCO_3$) and a metal wire gauze to stabilize mechanically the electrode and to collect the current. In gas diffusion electrodes, hydrophobic and hydrophilic pore systems are required. In AFC the hydrophilic system allows the penetration of the electrolyte into the electrode and the transport of the ions to or from the reaction zone; the hydrophobic pore system is required for the transport of the oxygen to the reaction zone. In addition, to give the mechanical stability, the PTFE in the electrodes forms a hydrophobic pore system, whereby the spider web of PTFE fibers in and on the electrodes are formed during preparation. Details of electrode preparation, activation (silver oxide reduction) and experimental set-up are given by Gülzow *et al.* [2003].

Making use of the fact that the electrolyte is a liquid, the use of a reference electrode (Hg/HgO) is possible and the two electrodes of a fuel cell, anode and cathode, can be investigated independently from each other, in half-cell configuration. Since the kinetics of the oxygen reduction reaction is much slower than the kinetics of the hydrogen oxidation reaction the performance loss of the fuel cell will be determined mainly by the cathode and therefore the ORR was intensively investigated with different steady state and potential step techniques on smooth and porous electrodes. The reduction process has been found to follow two reaction pathways: the direct $4e^-$ path and the $2e^-$ path (generally referred to as the sequential path) involving the formation of peroxide as an intermediate product (Hoare [1968]). To elucidate the reaction mechanism and to obtain the kinetic parameters for oxygen reduction on PTFE-bonded gas-diffusion silver electrodes in alkaline solution the ac impedance method was applied by several investigators: Mund [1975b], Holze and Vielstich [1984].

In order to analyze the kinetics of the oxygen reduction reaction on silver gas diffusion electrodes, EIS were measured at different potentials The measured impedance spectra are represented as Bode plots, where the logarithm of impedance magnitude $|Z|$ and the phase-shift (α) are plotted vs. the logarithm of the frequency (f). Impedance spectra have been measured at 18°C in the potential range from OCP (91 mV) to −230 mV vs. Hg/HgO (Figure 4.5.48 and Figure 4.5.49). These potential depending spectra can be classified into two different groups, with a transition potential range between 40 mV and 50 mV. The first group of impedance spectra exhibits

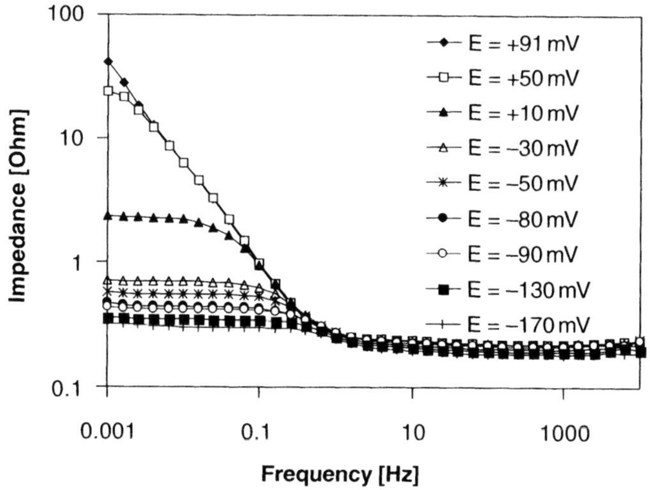

Figure 4.5.48. Bode diagram (log|Z|vs. log f) of the impedance spectra measured in the potential range from OCP (91 mV) to −170 mV during ORR on a Ag GDE in 25 wt% KOH at 18°C, p_{O2} = 1 bar.

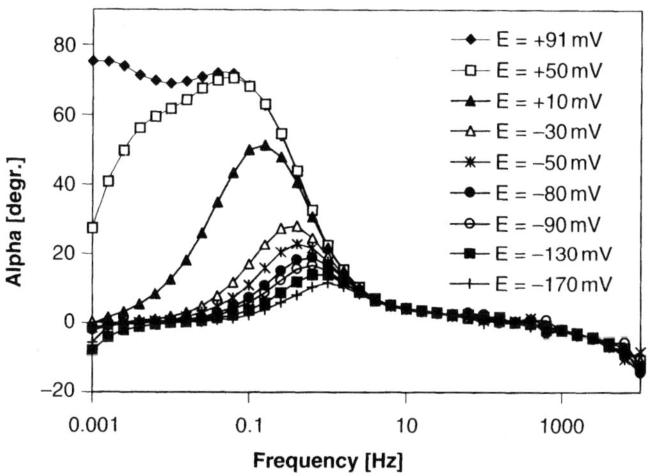

Figure 4.5.49. Bode diagram (phase-shift vs. log f) of the impedance spectra measured in the potential range from OCP to −170 mV during ORR on a Ag GDE in 25 wt% KOH at 18°C, p_{O2} = 1 bar.

two maxima in the phase-shift (two time constants), one maximum at low frequencies ($f < 10$ mHz) and an other one in the frequency range 100 mHz $< f <$ 1 Hz. The low frequency phase-shift maximum disappears at potentials more cathodic than 50 mV while the other phase-shift maximum moves toward higher frequencies when the cathodic polarization is increased. The second group of impedance spectra shows only one phase-shift maximum (one time constant).

The impedance magnitude at the low frequency end of the measured frequency range decreases with increasing cathodic polarization. The impedance magnitude at frequencies $f > 100\,\mathrm{Hz}$ (R_{el}) is constant in the potential range from OCP to $-90\,\mathrm{mV}$ and at more cathodic potentials the electrolyte resistance gets smaller.

In order to evaluate the measured impedance spectra and to obtain kinetic data of the oxygen reduction reaction, the reaction steps can be translated into an appropriate equivalent circuit which contains various impedance elements representing the involved reaction steps. These elements are generally represented as ohmic, capacitive or inductive components with particular dependencies of their complex impedance upon the frequency of the ac signal. The particular linking of these impedance elements described by an equivalent circuit is based upon the relationship between the processes represented by these elements. Subsequently occurring steps are represented by a series connection of the elements while steps occurring simultaneously are represented by a connection in parallel. In the case of porous electrodes the connection of the elements is more complicated.

The main features of the impedance measurements can be explained using the cylindrical-pore model (see Section 4.5.4.1, Figure 4.5.45) proposed by Göhr [1997] where the impedance elements of the pore's wall surface (Z_q) are the double layer capacity (C_{dl}), charge transfer resistance (R_{ct}) and impedance elements of oxygen adsorption or heterogeneous surface reaction ($R_{ad/het}$ and $C_{ad/het}$). The modified equivalent circuit is given in Figure 4.5.50.

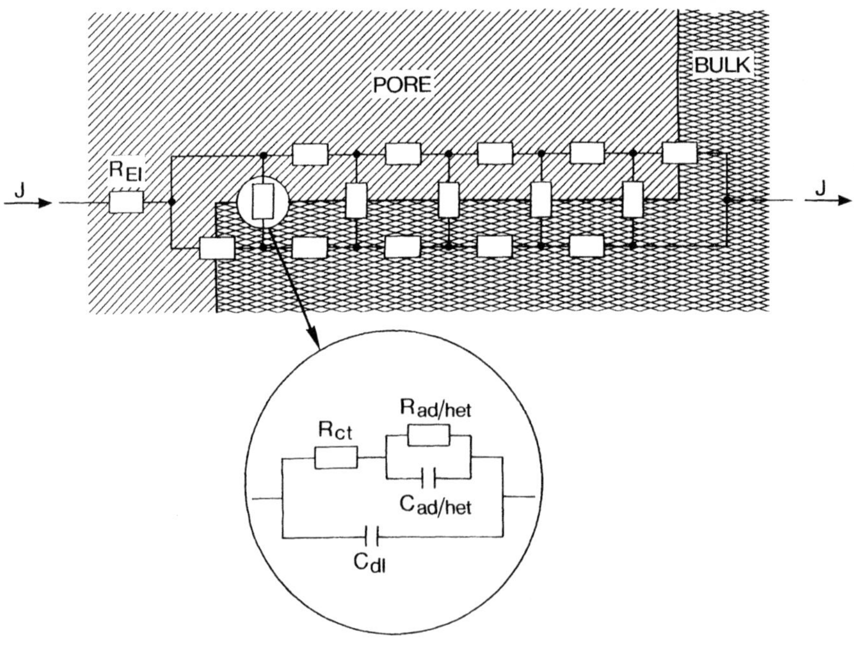

Figure 4.5.50. Equivalent circuits for the pore's wall surface.

The adsorption/heterogeneous impedance ($Z_{ad/het}$) is given according to Pilla [1970]:

$$Z_{ad/het} = RT(k - \omega i)/N^2 F^2 c_S A(k^2 + \omega^2) \tag{73}$$

with A = electrode surface, k = first-order reaction rate constant, F = Faraday constant, c_s = surface concentration of reacting species, i = imaginary quantity and $\omega = 2\pi f$.

As shown by Holze *et al.* [1986], this can be separated into an ohmic and a capacitive part connected in parallel:

$$R_{ad/het} = RT/(n^2 F^2 c_s A k) \tag{74}$$

$$C_{ad/het} = n^2 F^2 c_s A/(RT) \tag{75}$$

The numerical values of the impedance parameters, as a function of the applied potential obtained with the complex nonlinear least-square fit program, using for the pore's wall surface the equivalent circuit from Figure 4.5.50, are given in Table 4.5.3 with the corresponding accuracy of the fit.

There is a good agreement between experiment and the results of the fit. The standard deviation of the phase-shift, which is more sensitive to the fitting parameters, was found to be in the range of 0.1 and 0.3 degree.

The most potential sensitive impedance elements are $R_{ad/het}$ and R_{ct}. The double layer capacity (C_{dl}) is approximately constant in the measured potential range and the high value (170 mF/cm^2) indicates a high electrochemical active surface. Assuming that the double layer capacity for polished silver electrodes measured at the same experimental conditions is about 40 μF/cm^2 according to Wagner *et al.* [1996] a surface factor ($C_{dl\text{-}porous}/C_{dl\text{-}polished}$) of 4250 is obtained.

Table 4.5.3. Numerical Values of the Impedance Elements as a Function of the Applied Potential

	Impedance Elements of Pore's Wall Surface (Z_q)					
Potential (mV)	$R_{ad/het}$ (Ωcm^2)	$C_{ad/het}$ (mF/cm^2)	R_{ct} (Ωcm^2)	C_{dl} (mF/cm^2)	Z_p (mΩ)	R_{el} (mΩcm^2)
91	1734 ± 72	251 ± 6	115 ± 5	166 ± 2	46 ± 4	1313 ± 62
50	105 ± 3	435 ± 18	79 ± 16	163 ± 4	45 ± 4	1309 ± 64
10	2.4 ± 0.5	492 ± 24	10 ± 1	179 ± 2	49 ± 4	1307 ± 56
−30	0.5 ± 0.0	590 ± 100	2.3 ± 1	167 ± 1	45 ± 6	1298 ± 77
−50	0.4 ± 0.1	605 ± 106	1.6 ± 0.1	169 ± 5	43 ± 11	1300 ± 148
−70	0.3 ± 0.1	624 ± 135	1.2 ± 0.1	168 ± 8	45 ± 7	1285 ± 37
−90	0.3 ± 0.1	639 ± 102	1 ± 0.1	171 ± 4	42 ± 4	1264 ± 95
−80	0.3 ± 0.1	283 ± 94	1 ± 0.1	159 ± 5	50 ± 9	1194 ± 68
−130	0.2 ± 0.2	303 ± 25	0.6 ± 0.3	162 ± 8	49 ± 3	1134 ± 38
−170	0.2 ± 0.2	312 ± 212	0.5 ± 0.5	162 ± 25	48 ± 5	1080 ± 24
−200	0.1 ± 0.2	321 ± 31	0.3 ± 0.1	167 ± 10	30 ± 2	1007 ± 99

Long Term Investigation of AFC Electrodes During Oxygen Reduction. In order to analyze in detail the degradation (aging) process of silver electrodes during oxygen reduction, electrochemical impedance spectroscopy (EIS) has been performed by Wagner *et al.* [2004a]. The PTFE-bonded gas diffusion silver electrodes has been operated galvanostatically at 100 mA cm^{-2} over 1400 h and impedance spectra were recorded at the open circuit potential, daily at the beginning and after 2 weeks of operation in 48 and 72 h intervals, respectively. In the impedance spectra, represented as Bode plots (Figure 4.5.51) or as Nyquist plots (Figure 4.5.52), a change during operation at 100 mA cm^{-2} is observed. During the measurement period of each impedance spectrum at OCP the electrode state stays constant; the electrode degrades only during operation at 100 mA cm^{-2}. First spectrum (0 h), recorded prior to the beginning of the long-term operation, differ from the EIS recorded after loading. During load, at first, the electrode is activated (second spectrum recorded after 24 h) and a slight change with operation time can be observed.

The main features of the impedance measurements can be explained using the cylindrical-pore model proposed by Göhr [1997]. Taking into account only the pore electrolyte resistance (R_{por}) and the pore wall impedance, then the porous electrode model and pore impedance can be simplified. The spectra can be simulated by an equivalent circuit (Figure 4.5.53) consisting of a series combination of uncompensated electrolyte resistance (R_{el}) and pore impedance (see transmission-line model from Figure 4.5.50). The impedance of the pores wall surface is given by the charge transfer resistance (R_{ct}), Warburg impedance (W) and double layer capacity (C_{dl}).

The most time sensitive impedance elements are the double layer capacity, the Warburg impedance (W) and the charge transfer resistance. The Warburg impedance is related to a diffusion process, as derived in section 2.2.2.2 (Distributed Circuit

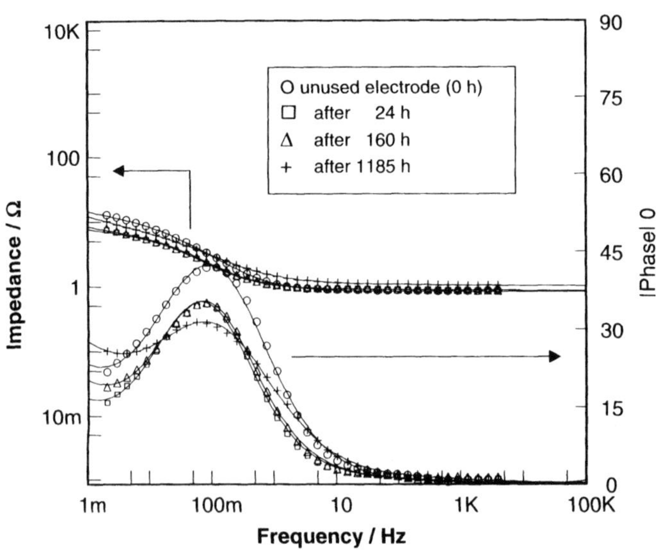

Figure 4.5.51. Bode diagram of the impedance spectra of the silver GDE measured at OCP, 70°C, in 30 wt% KOH during oxygen reduction at different times of operation at 100 mA cm^{-2}.

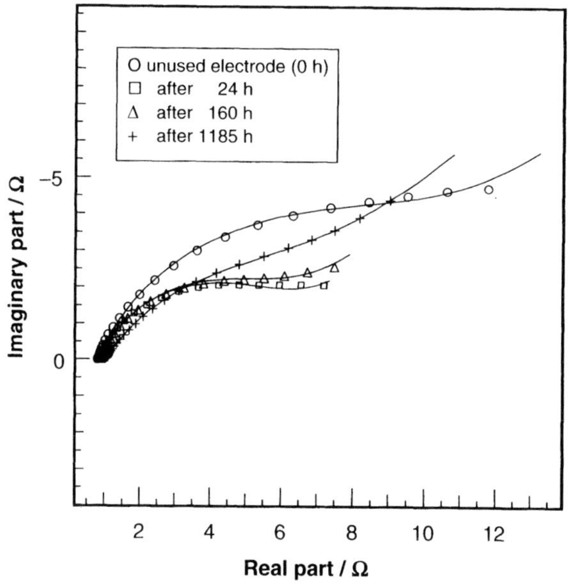

Figure 4.5.52. Nyquist diagram of the impedance spectra of the silver GDE measured at OCP, 70°C, in 30 wt% KOH during oxygen reduction at different times of operation at 100 mA cm^{-2}.

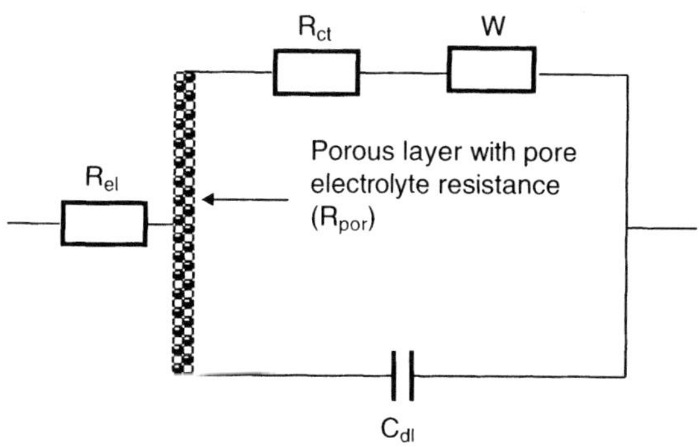

Figure 4.5.53. Equivalent circuit (porous electrode model) for the evaluation of the impedance spectra of the silver GDE measured during ORR at different times of operation at 100 mA cm^{-2}.

Elements). From the time dependency of the impedance parameters, the aging effects can be related to the decrease of the double layer capacity (Figure 4.5.54), a change of the porous structure and catalyst activity (Figure 4.5.55) and an increase of the Warburg impedance (Figure 4.5.56). The decrease of the capacity is given by a decrease of the electrochemically active surface. The increase of the Warburg impedance indicates that the diffusion of reacting gas is hindered. In the Nyquist diagram, one can observe the infinite diffusion as a straight line with a slope of 1 (real part = imaginary part) in the low frequency range of the spectra. In the Bode diagram, due to the logarithmic scale of the impedance, the diffusion can not be seen so clearly.

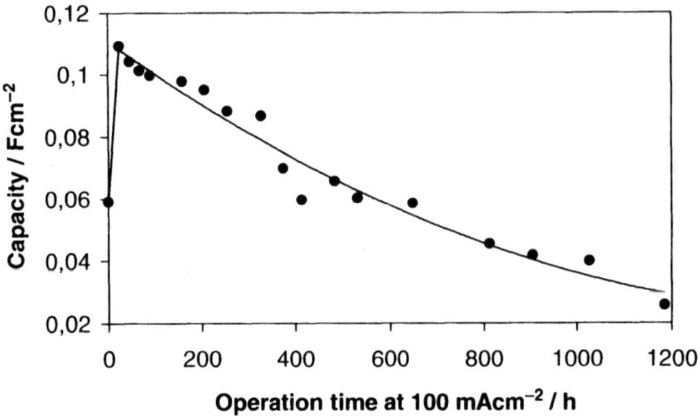

Figure 4.5.54. Time dependency of double layer capacity (C_{dl}) obtained with a least square fit of the measured impedance spectra at open circuit potential.

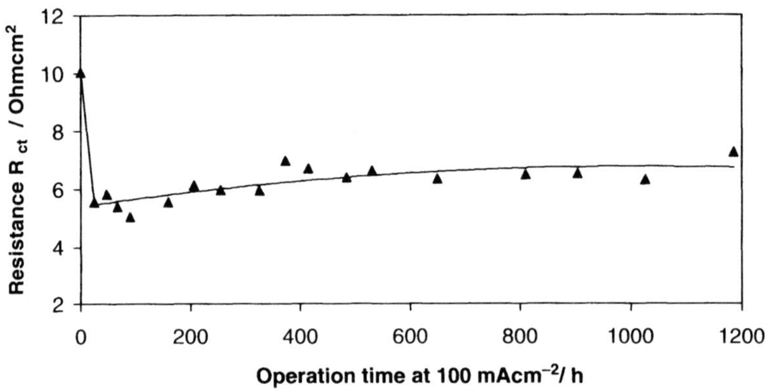

Figure 4.5.55. Time course of charge transfer resistance R_{ct} obtained with a least square fit of the measured impedance spectra at open circuit potential.

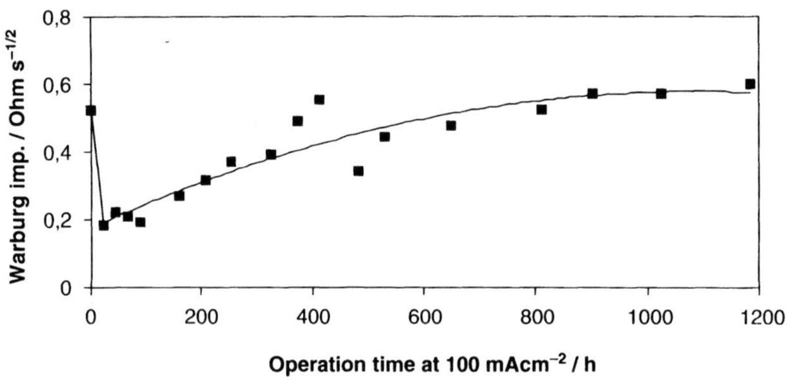

Figure 4.5.56. Time course of Warburg impedance obtained with a least square fit of the measured impedance spectra at open circuit potential.

The aging effect is also observed by scanning electron microscopy (SEM) and x-ray photoelectron spectroscopy (XPS). Consequently, the combination of electrochemical and physical methods is very helpful for the study of degradation processes as shown by Wagner *et al.* [2004a].

The Hydrogen Oxidation Reaction (HOR). The kinetic of hydrogen reduction/oxidation reaction was first investigated by Gerischer and Mehl [1955] using an ac bridge. First ac impedance measuements on porous Nickel (Raney-Nickel) electrodes during HOR in alkaline solution were reported by Mund [1975b] and Mund and v. Sturm [1975a]. They proposed an equivalent circuit identical with the equivalent circuit shown in Figure 4.5.50, only $R_{ad/het}$ (indicated by R_2) and $C_{ad/het}$ (indicated by C_2) have a different meaning. They found for $\omega \rightarrow 0$

$$R_2 = \alpha R_{ct} \coth \alpha \tag{76}$$

$$C_2 = (R_{ct}b)^{-1}\{1 + (\alpha - 3\sinh\alpha\cosh\alpha)/(2\alpha\cosh^2\alpha)\} \tag{77}$$

With

$$\alpha = 1(b/D)^{1/2} \tag{78}$$

$$b = 2i_0/nF(r_p c_{v0} + 2c_{s0}) \tag{79}$$

Where l represent the pore length (cm), D the effective diffusion coefficient (cm2 s^{-1}), i_0 is the exchange current density (A cm^{-2}) of the charge transfer reaction, r_p is the pore radius (cm), c_{v0} the saturation concentration of H_2 in the solution (mol cm^{-3}) and c_{s0} the saturation concentration of H_2 on the electrode surface (mol cm^{-2}).

Latest publications regarding hydrogen oxidation reaction in alkaline solution are given by Lasia [1998] and a remarkable overview by Conway and Tilak [2002].

4.5.4.3 Polymer Electrolyte Fuel Cells (PEFC)

Electrochemical Impedance Studies of Polymer Electrolyte Fuel Cells (PEFC). Due to the low temperature of operation, CO_2 tolerance by the electrolyte, combination of high power density with high energy conversion efficiency and modular construction, PEFCs are considered as one of the most promising options for powering future cars and small combined power units. As shown by Gottesfeld and Zawodzinski [1997] during the last 10–15 years a lot of key barriers for the development of PEFC for terrestrial applications have been successfully overcome. However, there is still need for further research and development effort—an important goal is to increase the efficiency and to produce efficient, cheap electrodes and membrane-electrode assemblies (MEA) with low noble metal catalyst loading, high CO tolerance and long life time. In practice, the research and development efforts are focused on the development of nanodispersed catalysts, non-noble metal catalyst, production techniques of MEAs with low noble metal catalyst loading and suitable for mass production, as proposed by Bevers *et al.* [1998] and Gülzow *et al.* [1999], development of high temperature membranes to enhance the kinetics and to lower the CO influence (up to 160°C) and integration of the MEA into a fuel cell system with humidifier, gas conditioner, bipolar plate, fuel storage systems, etc.

The main difference between the AFC and PAFC is the gas-tight solid polymer electrolyte membrane, a solid proton exchange membrane which has as its main function the transport of protons from anode to cathode. To investigate the physical and electrochemical origins of the performance loss in PEFC—operated at different conditions like high current densities, fuel composition (neat H_2, H_2 + 100 ppm CO, H_2O), flow rates, temperature, air or pure oxygen, etc.—electrochemical impedance studies on different PEFC systems with different electrodes and membranes were performed, as mentioned in Section 4.5.4.1. First impedance measurements and interpretation of EIS performed to characterize PEFC were reported by Srinivasan *et al.* [1988], Fletcher [1992], Wilson *et al.* [1993] and Poltarzewski *et al.* [1992]. With increasing research and development effort to improve the PEFC performance and availability of suitable instrumentation the number of publications has increased.

Determination of Performance Losses in PEFC. EIS Measured at Different Cell Voltage (Current Load). Applying a classical three-electrode cell with one reference electrode is extremely difficult for the investigation of electrochemical systems with solid electrolytes. Therefore, first the anode and cathode transfer functions at open circuit potential (OCP) can be determined independently, without a reference electrode, using a symmetrical gas supply of hydrogen and oxygen at the two electrodes of the cell. Thus, cathode and anode impedance at OCP can be determined directly with two independent experiments, as published by Wagner [2002] and shown in Figure 4.5.57. as a Bode diagram. From the charge transfer resistance R_{ct} at open cell voltage an apparent exchange current density i_0 according to $R_{ct} = RT/zFi_0$ can be determined.

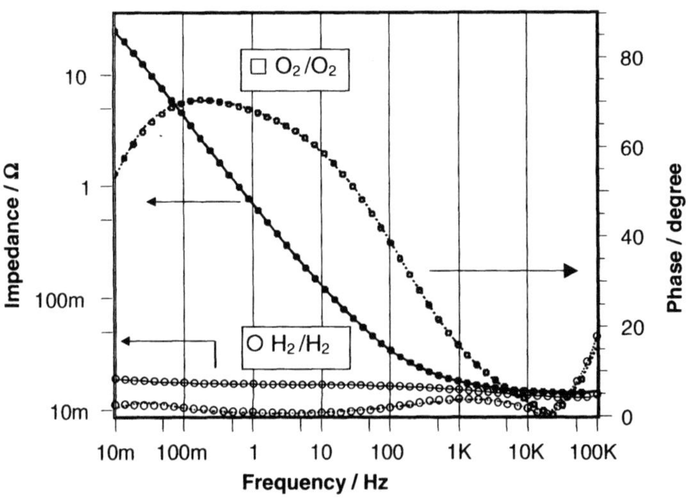

Figure 4.5.57. Bode diagram of the measured impedance spectra, PEFC (Nafion-impregnated electrodes) at 80°C, at pH_2 and pO_2 = 2 bar absolute, symmetrical gas supply, at open cell potential.

The preliminary impedance spectrum (Figure 4.5.57) measured at open cell voltage with symmetrical gas supply (H_2/H_2 and O_2/O_2) at 80°C with impregnated (1 mg cm^{-2} Nafion suspension, Aldrich Chemie) E-TEK-electrodes (20% Pt/C, 0.4 mg cm^{-2}, thickness 200 μm, geometric surface of 23 cm^2), have shown that the impedance in the low frequency range is much higher when operated symmetrically with oxygen (symmetrical cathode arrangement) than the impedance when the fuel cell is operated symmetrically with hydrogen (symmetrical anode arrangement).

Furthermore, by varying other experimental conditions such as current load, temperature, gas composition, and as recently shown by Andreaus *et al.* [2002] hydrogen humidification and membrane thickness, measured cell impedance can be split into anode impedance, cathode impedance and electrolyte resistance, without using reference electrodes. These results were used to derive appropriate equivalent circuits for the analysis of impedance spectra measured on fuel cells operating with H_2/O_2, H_2/air and H_2 + 100 ppm CO/O_2. The variation of the experimental conditions is also a useful method to confirm the accuracy of the equivalent circuit.

For illustration, in Figure 4.5.58 impedance spectra measured over the whole performance range of a PEFC from open circuit to full load were presented. The nature of the impedance spectra varied with cell potential, between open circuit potential (1024 mV) and 840 mV an exponential potential dependency of the impedance due to the high activation energy of the oxygen reduction reaction is observed in the low frequency range ($f < 1$ Hz).

With further increasing current density, an increase in the cell impedance is observed (Figure 4.5.59*a*) and an additional phase-shift maximum at 75 mHz occurs.

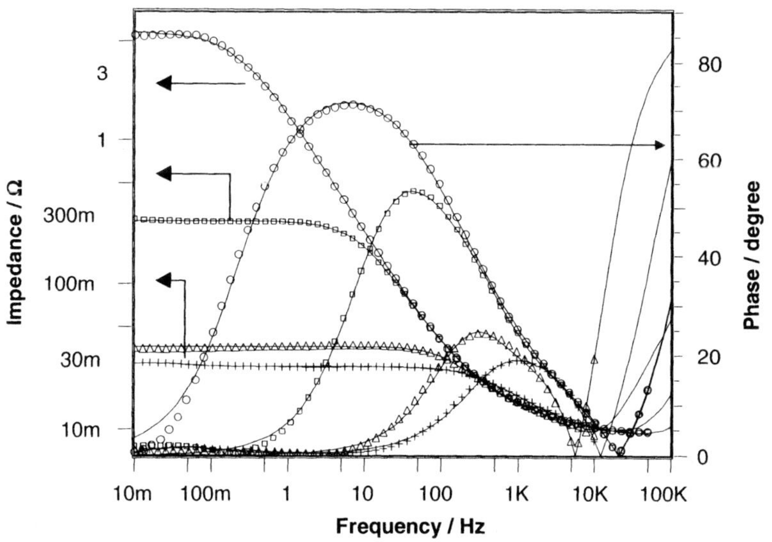

Figure 4.5.58. Bode diagram of the measured impedance spectra, PEFC (hot-pressed, Nafion-impregnated E-TEK electrodes) at 80°C, at different cell voltages: (○) OCV (1024 mV), (□) 900 mV, (Δ) 800 mV, (+) 700 mV.

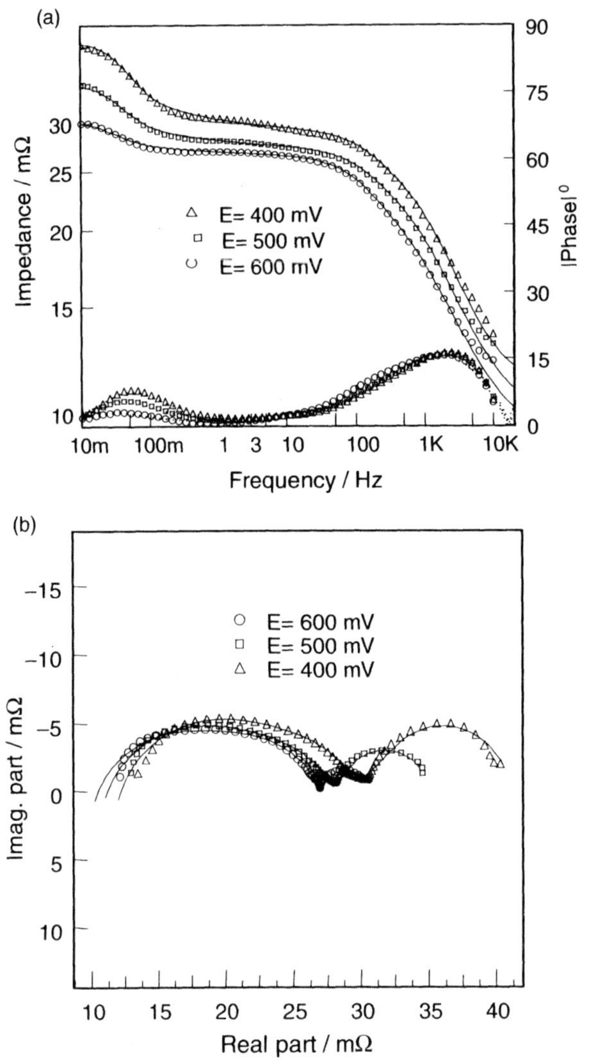

Figure 4.5.59. (*a*) Bode and (*b*) Nyquist diagram of the measured impedance spectra, PEFC (hot-pressed, Nafion-impregnated E-TEK electrodes) at 80°C, at different cell voltages: (○) 600 mV), (□) 500 mV, (Δ) 400 mV.

From this increase in the cell impedance, one can deduce that at higher load of the cell an additional overvoltage attracts attention. This additional process is related to a mass transport process (diffusion). The impedance related to the diffusion is usually found in the lowest frequency region. Therefore, the impedance has to be measured over a wide frequency range, down to 10 mHz, or even lower.

To identify and separate the different diffusion processes, it is useful to represent the measured impedance spectra also as Nyquist diagram (Figure 4.5.59*b*). In

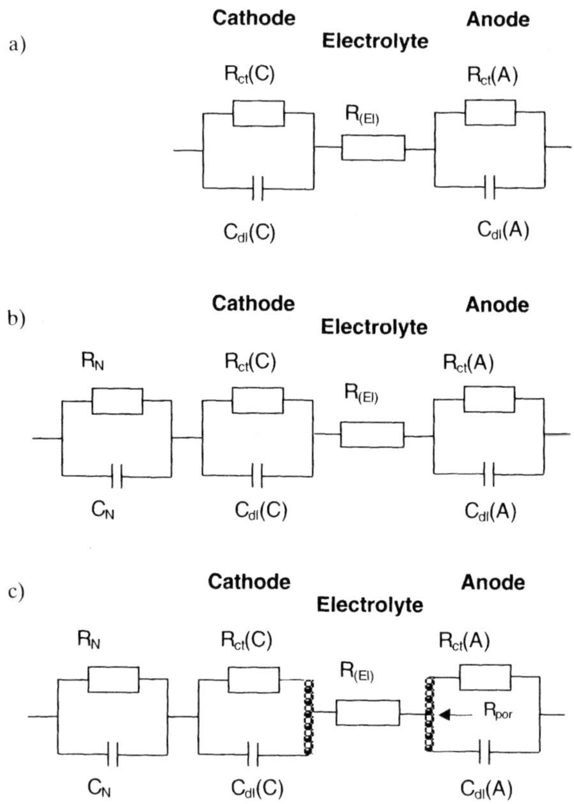

Figure 4.5.60. (*a*) Equivalent circuit (EC) of the PEFC with O_2/H_2 gas supply, for low current densities. (*b*) Equivalent circuit of the PEFC with O_2/H_2 gas supply, for high current densities with an additional diffusion step (Nernst-impedance), (*c*) Equivalent circuit of the PEFC with O_2/H_2 gas supply with Nernst-impedance and porous electrode model.

the Nyquist diagram, one can observe the finite diffusion (finite length diffusion) as an additional loop at the lowest part of the frequency range and the infinite diffusion as a straight line with a slope of 1 (real part = imaginary part). In the Bode diagram (Figure 4.5.59*a*) the difference between the two kinds of diffusion cannot be seen so clearly due to the logarithmically scale. In general, the Bode plot provides a clearer description of the electrochemical system's frequency-dependent behavior than the Nyquist plot, in which frequency values are implicit.

Starting with EIS measured with symmetrical gas supply, an equivalent circuit for the complete PEFC can be applied for the simulation of the measured impedance spectra of the PEFC. Besides a series resistance (electrolyte or membrane resistance R_{el}), the equivalent circuit (Figure 4.5.60*b*) contains three time constants of parallel *R*/*C*. In the simulation the capacitance (*C*) was replaced by CPE (CPE = constant phase element) due to the porous structure of the electrodes. The cathode can be described using a time constant for the charge transfer through the double

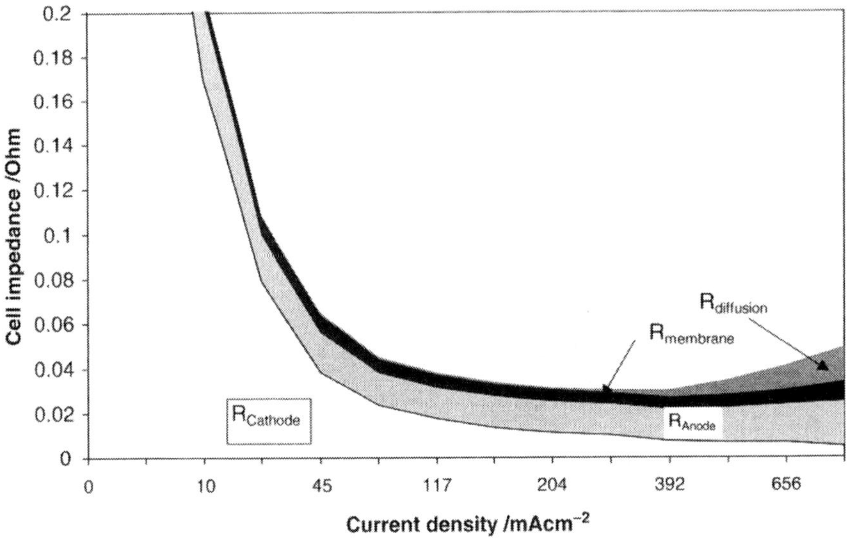

Figure 4.5.61. Current density dependency of the resistances after evaluation of the impedance spectra evaluated with EC from Figure 4.5.60.

layer ($R_{ct(C)}/CPE_{dl(C)}$, the exponent of the CPE is around 0.85, for an exponent of 1 the CPE is equal with the capacitance), a time constant for the finite diffusion of water with a Nernst-impedance like behavior ($R_{(N)}/CPE_{(N)}$, the exponent of the CPE is around 0.95) and finally the time constant of the anode ($R_{ct(A)}/CPE_{dl(A)}$, the exponent of the CPE is around 0.80). Anode and cathode can be also simulated with the porous electrode model (Figure 4.5.60*c*).

Applying Eqs 63–67 and using the current density dependency of the resistances (R_N, $R_{ct,A}$, $R_{(el)}$ and $R_{ct,C}$) represented in Figure 4.5.61, gained from the simulation with the equivalent circuit from Figure 4.5.60*c*, the individual performance losses (overpotentials) can be determined (Figure 4.5.62). At low current densities the cell overpotential is given mainly by the cathodic overpotential. At higher current densities ($i > 400\,\mathrm{mA\,cm^{-2}}$) an additional diffusion overpotential becomes noticeable. The increase in the anodic overpotential with increasing current density can be explained by assuming a partially dry out of the interface membrane/anode with simultaneous increasing of the water content depending electrolyte resistance inside the pore ($Z_p \equiv R_{pore}$)—applicable only in the case of impregnated electrodes or electrodes with electrolyte powder inside the electrode and can be minimized by appropriate water management. Therefore, a further refinement of the equivalent circuit is useful and can be made if one uses the porous electrode model, presented in Section 4.5.4.1. If one assumes that the impedance of the porous electrode is determined only by the resistance (R_{pore}) of the electrolyte inside the pore and the pore's wall surface (R_{ct}) then one can simplify the expression for the impedance of the porous electrode after Göhr [1997]:

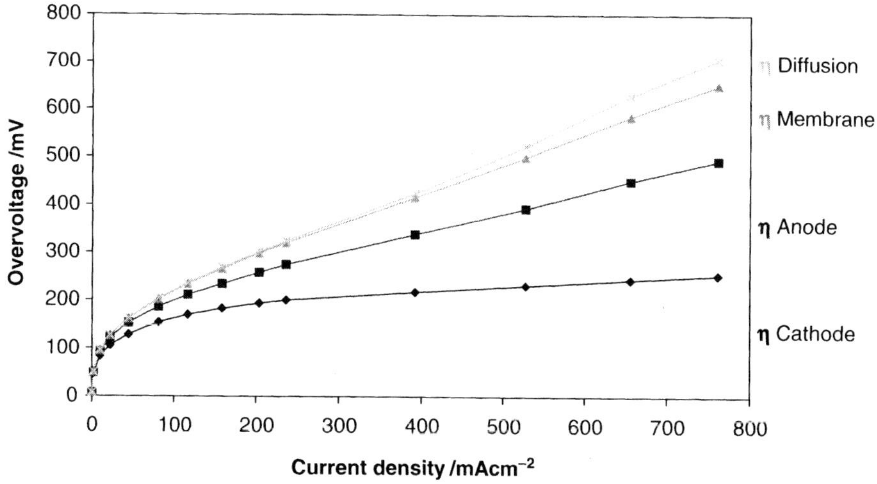

Figure 4.5.62. Cell voltage and individual performance losses of the PEFC (hot-pressed, Nafion-impregnated E-TEK electrodes) at 80°C in function of current density, calculated by integration of the individual resistances at different current densities.

$$R_p = \frac{(R_{pore} \cdot R_{ct})^{\frac{1}{2}}}{\tanh\left\{\left(\frac{R_{pore}}{R_{ct}}\right)^{\frac{1}{2}}\right\}} \tag{80}$$

The impedance of the anode can be split up, according Eq. 80 and the current density dependency is shown in Figure 4.5.63.

To validate the proposed equivalent circuit other, additional EIS has been performed, for example flooding the cathode in the "dead end" mode of operation of the cathode. The fuel cell was operated over 8 h in the galvanostatically mode of operation at constant load, so that one can assume that the impedance change with time can be attributed to the change of and in the cathode, as shown by Schiller *et al.* [2001].

EIS Measured During CO Poisoning of the Pt and Pt/Ru Anodes in a PEFC. As mentioned in the introduction, due to the high energy conversion rate and the harmless emission products, polymer electrolyte fuel cells (PEFC) are receiving more and more attention, especially for powering electric vehicles. The highest performance is achieved with hydrogen (H_2) which is the preferred fuel for low-temperature fuel cells. However, H_2 has several limitations. The storage systems for liquid or compressed H_2 are heavy and bulky. Furthermore, H_2 refueling is costly and takes time. An additional obstacle is the lack of an infrastructure to distribute H_2 to the consumer, as shown for example by Cleghorn *et al.* [1997].

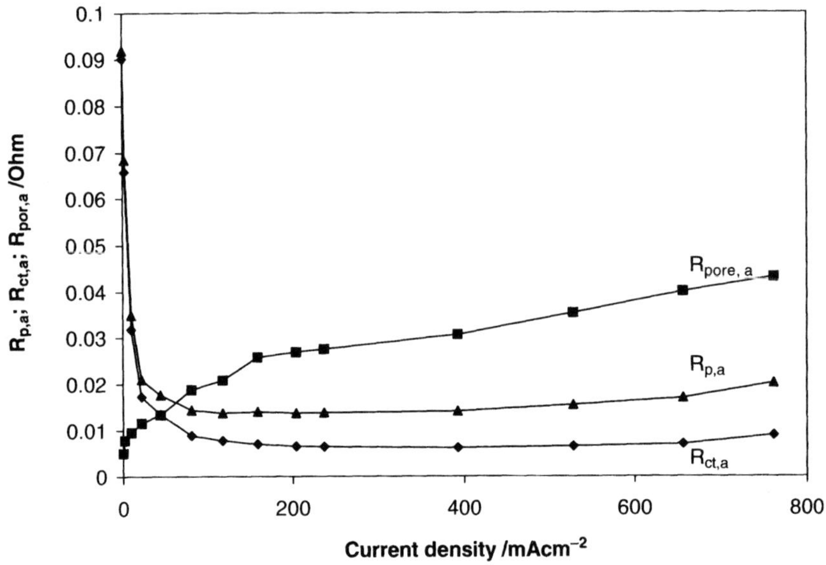

Figure 4.5.63. Porous anode resistance (▲ $R_{p,a}$), charge transfer resistance (◆ $R_{ct,a}$) and electrolyte resistance (■ $R_{pore,a}$) in function of the current density.

An alternative to the use of H_2 as fuel is methanol, which is a liquid fuel and easy to handle. This can be directly transformed to electrical current in a DMFC (direct methanol fuel cell). The DMFC allows a simple system design. However, presently achieved performance data of DMFC is not satisfactory and material costs are too high. As another alternative, methanol or hydrocarbons (e.g. natural gas, biogas) can be transformed to hydrogen on board the electric vehicle by a reformation reaction. This allows use of the H2-PEFC cell, which has a higher level of development. The reformate feed gas may contain up to 2.5% carbon monoxide (CO) by volume, which can be reduced to about 50 ppm CO using a selective oxidizer (Wilkinson *et al.* [1997]).

The performance of platinum which is known as one of the most effective catalysts for the hydrogen oxidation in polymer membrane fuel cells, is influenced by even traces of carbon monoxide: compared with the use of pure hydrogen, the maximum power density is more than halved in the presence of only 5 ppm carbon monoxide. One possible explanation for the decrease of the fuel cell performance is that the carbon monoxide blocks or limits the active sites of the platinum catalyst due to adsorption, which leads to an inhibition of the hydrogen oxidation reaction. In the last two decades, intensive work has been devoted to finding electrocatalysts that are tolerant to CO in hydrogen at operating temperatures below 100°C. This system has been subject to numerous studies using electrochemical methods like potentiostatic measurement, potentiodynamic measurement and stripping voltammetry. Detailed kinetic studies were performed, for example by Vogel *et al.* [1975], Schmidt *et al.* [1998] and Koper *et al.* [2001].

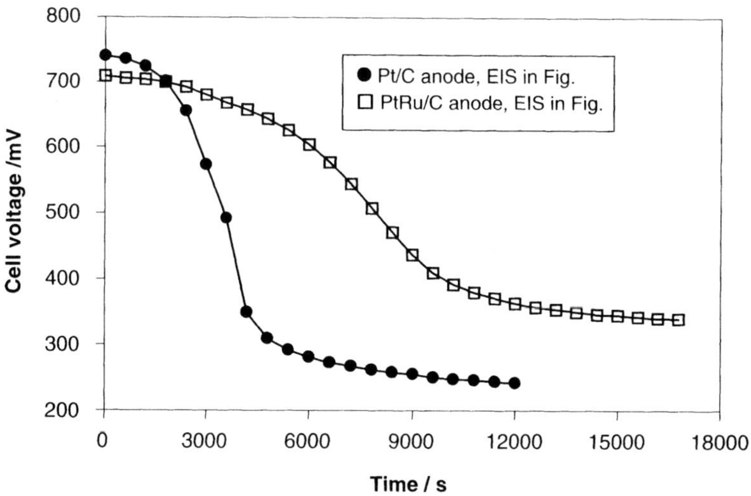

Figure 4.5.64. Evolution of cell voltage during CO poisoning of Pt/C (•) and PtRu/C (□) anodes in galvanostatic mode of fuel cell operation at 217 mA cm^{-2}, 80°C.

The first *in situ* impedance measurements were reported by Müller *et al.* [1999], who reported the time dependence of impedance spectra in galvanostatic conditions for CO poisoning Pt anodes. Other publications, with detailed kinetic data were published by Ciureanu and Wang [1999, 2000], Wang *et al.* [2001], Leng *et al.* [2002], Kim *et al.* [2001a, b] and recently by Wagner and Gülzow [2004b].

For the development of improved electrocatalysts which are less sensitive with respect to the presence of carbon monoxide, a mechanistically understanding of the poisoning process of the anode is desirable. The progressive poisoning with carbon monoxide of a fuel cell can be monitored using time resolved electrochemical impedance spectroscopy (TREIS).

As shown in detail by Wagner and Schulze [2003], the CO poisoning of the Pt and PtRu anode during galvanostatic mode of operation with H_2 + 100 ppm CO causes a change of the state of the fuel cell which is reflected in a decrease of the cell voltage during operation (Figure 4.5.64) and in time dependency the recorded impedance spectra. Besides, an increase of the total impedance of the fuel cell, in the case of galvanostatic mode of operation, the occurrence and the increase of a pseudoinductive behavior at frequencies lower than 3 Hz is observed. It is useful to operate the fuel cell in the galvanostatic mode of operation because, at constant current density, the impedance of the cathode and the membrane resistance can be assumed to be constant during the course of impedance measurements and the changes in the impedance spectra during poisoning the anode with CO can be attributed exclusively to the impedance of the anode.

Representative time-dependent impedance spectra of the series, i.e. time resolved impedance spectra, are depicted in Figure 4.5.65 for the fuel cell with Pt anode and in Figure 4.5.66 for the fuel cell with PtRu anode as Nyquist plots,

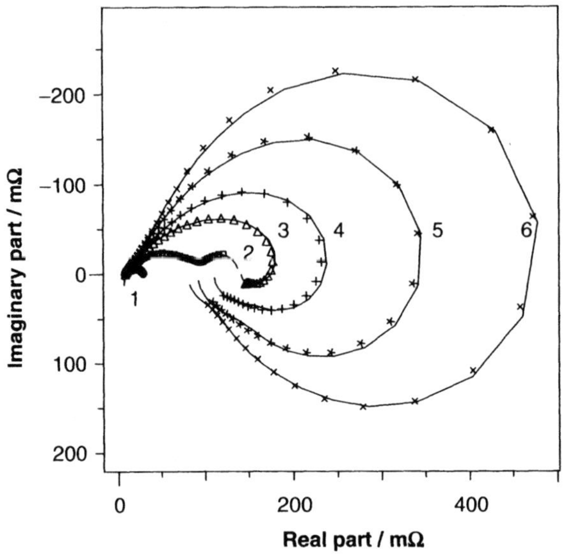

Figure 4.5.65. Nyquist plot of EIS measured at: (1) 0 s, (2) 3601 s, (3) 5402 s, (4) 7204 s, (5) 9605 s and (6) 11404 s during CO poisoning of Pt/C anode during galvanostatic mode of operation with H_2 + 100 ppm CO (anode), at 5 A (cell surface = 23 cm^2), 80°C and oxygen (cathode).

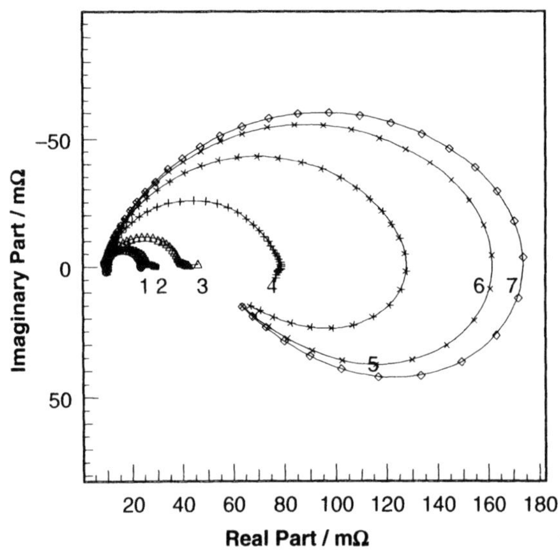

Figure 4.5.66. Nyquist plot of EIS measured at: (1) 0 s, (2) 3000 s, (3) 6000 s, (4) 9000 s, (5) 12000 s, (6) 15000 s and (7) 16800 s during CO poisoning of PtRu/C anode during galvanostatic mode of operation with H_2 + 100 ppm CO (anode), at 5 A (cell surface = 23 cm^2), 80°C and oxygen (cathode).

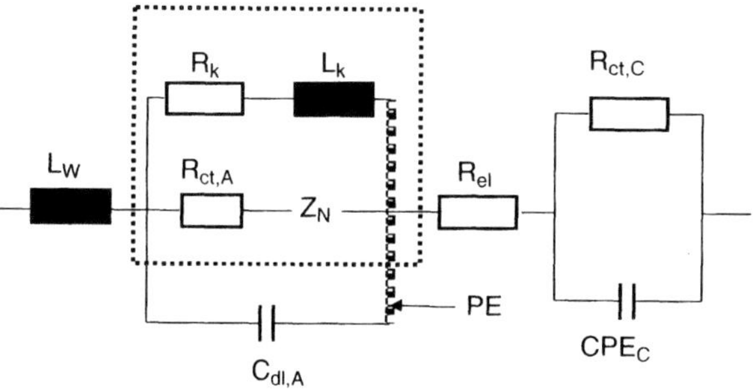

Figure 4.5.67. Equivalent circuit used for evaluation of interpolated (time dependent) impedance spectra during hydrogen oxidation and CO poisoning of the anode and oxygen reduction at the fuel cell cathode.

whereby the experimental data are represented by symbols and the solid lines in the figures represent the modeled curves after fitting the experimental data with the equivalent circuit from Figure 4.5.67.

In series to both half cells, the resistance of the membrane itself—denoted as the electrolyte resistance (R_{el})—as well as a parasitic wiring inductance (L_w) due to the mutual induction effect has to be taken into account. The impedance of the cathodic half cell (oxygen reduction) is approximated using a charge transfer resistance ($R_{ct,C}$) in parallel to a constant phase element (CPE_C). This simple equivalent circuit describes the partial impedance of the cathodic half cell with sufficient accuracy. In contrast, the impedance of the anode (hydrogen oxidation) is more complicated, due to the carbon monoxide poisoning and the appearance of an inductive loop in the low frequency range of the impedance spectra. The impedance of the anode is modeled using a porous electrode (PE) in series to a double layer capacity ($C_{dl,A}$), which is in parallel to the Faradaic impedance (Z_F), which contains a surface relaxation impedance Z_K (see Section 4.5.4.1) parallel to the charge transfer resistance ($R_{ct,A}$) in series with a finite diffusion impedance element (Nernst-impedance, Z_N), so that the expression for the Faradaic impedance from Eq. (69) was extended and is given by Eq (81):

$$Z_F = \frac{R_{ct} + Z_N}{1 + R_{ct}/Z_K} \tag{81}$$

On the basis of this model and the equivalent circuit shown in Figure 4.5.67, the changes and differences, depending on the used anode in the fuel cell (Pt/C or PtRu/C) in the impedance spectra during the experiment, are dominated by the changes of the charge transfer resistance of the anode ($R_{ct,A}$), the surface relaxation impedance (R_K, τ_K) and the finite diffusion impedance (Z_N).

The evolution of the charge transfer resistance of the anodes is shown in Figure 4.5.68 as a function of the elapsed time. The Nernst impedance (Z_N) represented in Figure 4.5.69 shows the greatest difference between the two anodes. The Nernst

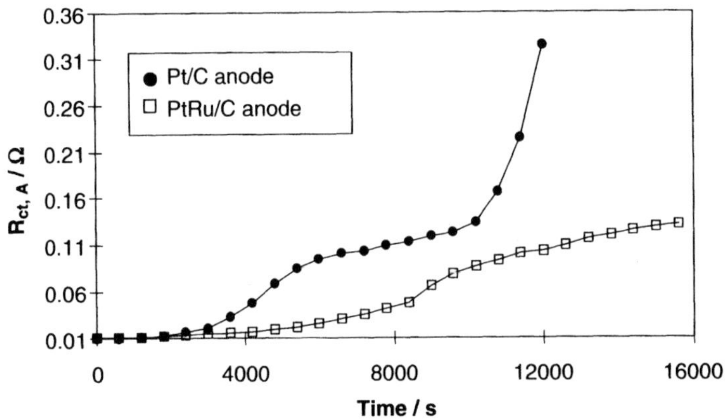

Figure 4.5.68. Time elapsed of the charge transfer resistance of the (•) Pt/C and (□) PtRu/C anodes after evaluation of the time dependent impedance spectra with the equivalent circuit from Figure 4.5.67.

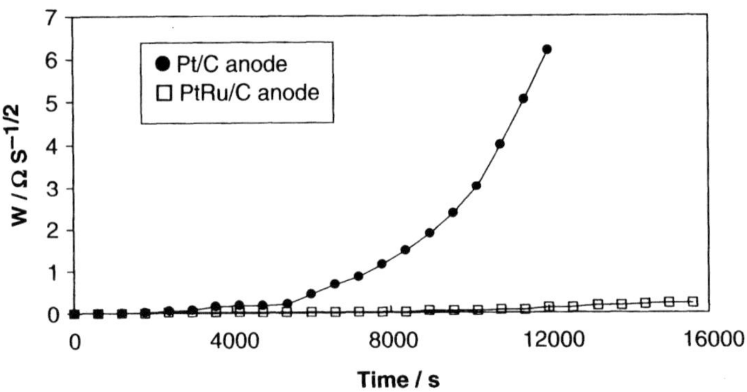

Figure 4.5.69. Time elapsed of the Warburg parameter W from the Nernst impedance (Eq. 82) for the (•) Pt/C and (□) PtRu/C anodes after evaluation of the time dependent impedance spectra with the equivalent circuit from Figure 4.5.67.

impedance Z_N contains two parameters: the Warburg parameter W ("diffusion resistance", at low frequencies) and a diffusion time constant (k_N), determined by the constant of diffusion (D_k) and diffusion layer thickness (d_N). The Nernst impedance is calculated by Eq. (82):

$$Z_N = \frac{W}{\sqrt{j\omega}} \tanh \sqrt{\frac{j\omega}{k_N}} \tag{82}$$

with

$$K_N = D_k \big/ d_N^2$$

Electrochemical Impedance Studies of Direct Methanol Fuel Cells (DMFC). The direct methanol fuel cell (DMFC) has been investigated for more than 50 years and has recently been discussed as an interesting option for a fuel-cell-based mobile power supply system; a historical overview is given by McNicol *et al.* [1999] and Lamy *et al.* [2002] discussed recent advances in the development of direct alcohol fuel cells (DAFC). However, methanol is the most promising organic fuel; other alcohols have also been considered for use in a fuel cell, but until now the most advanced system is the DMFC. The DMFC uses liquid methanol/water solution as fuel which is, compared to hydrogen, easy to handle and to distribute. But the DMFC has some particular disadvantages, like low electrocatalytic activity of the catalyst regarding methanol oxidation, anode poisoning by strongly adsorbed intermediates and methanol cross-over through the proton exchange membrane, leading to limited performances of the DMFC. Currently, research is focused on overcoming these problems by developing new membranes (Jörissen [2002]) applicable at higher temperatures (Savadogo [2004]), with low methanol cross over and high ionic conductivity, to find new electrocatalysts for the methanol oxidation (Burstein *et al.* [1997], Wasmus and Küver [1999]) or methanol-tolerant oxygen reduction catalysts (Reeve *et al.* [2000]), or to increase the electrocatalytic activity of the catalysts using new production steps (Lizcano-Valbuena [2003]), and finally increasing the utilization degree of the catalyst by improving the coating technique during MEA production (Gülzow *et al.* [2002]).

Until now there have been only a few EIS studies published; the first studies of DMFC have been reported by Müller and Urban [1998]. They separated the anode and cathode impedance using different fuel cell configurations. The impedance of the anode was measured by eliminating the contribution on the cathode. This was achieved by operating the cathode as a dynamic hydrogen electrode (DHE) without oxygen, so that instead of oxygen reduction, protons are reduced and hydrogen was evolved. DMFC cathode impedance spectra were obtained after subtracting the anode impedance from the complete cell impedance. For the modeling of the Faradaic impedance of DMFC anodes, Müller *et al.* [1999] used the same equivalent circuit as proposed by Göhr [1986] and used by Wagner *et al.* [1999] for modeling the PEFC anode during CO poisoning. The model (equivalent circuit) and notations used by Müller *et al.* [1999] was first derived by Harrington and Conway [1987] for the formulation of the kinetic of hydrogen evolution reaction and is shown in Figure 4.5.70.

Assuming for the methanol oxidation a reaction mechanism with intermediate adsorbed CO:

$$CH_3OH \rightarrow (CO)ads \rightarrow CO_2$$

then the Faradaic impedance (Z_F) can be expressed by:

$$Z_F = \frac{j\omega + C}{A + B} = \left(\frac{1}{R_\infty} + \frac{1}{R_0 + j\omega L_S} \right)^{-1} \tag{83}$$

where

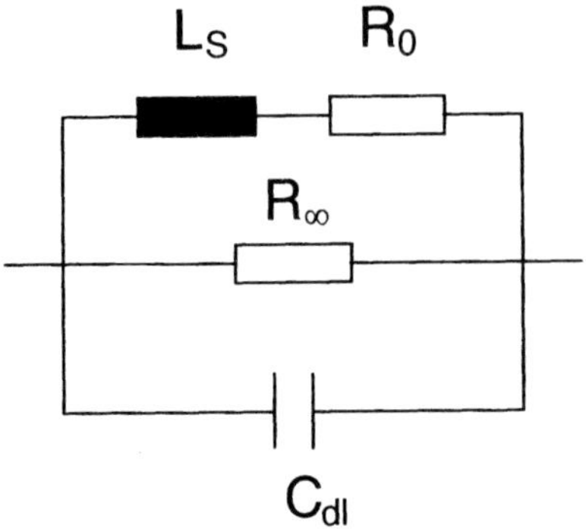

Figure 4.5.70. Equivalent circuit for modeling the Faradaic impedance of DMFC anodes (for single-adsorbate mechanisms without diffusion control, proposed by Harrington and Conway [1987].

$$R_{\ni} = 1/A;\ R_0 = C/B \text{ and } L_S = 1/B$$

The parameters A, B and C are defined by Müller *et al.* [1999] by:

$$A = F\left(\frac{\partial r_e}{\partial E}\right)_\Theta;\ B = \frac{F^2}{q_{CO}}\left(\frac{\partial r_{CO}}{\partial \Theta}\right)_E\left(\frac{\partial r_{CO}}{\partial E}\right)_\Theta;\ C = \frac{-F}{q_{CO}}\left(\frac{\partial r_{CO}}{\partial \Theta}\right)_E$$

where r_e is the net rate of production of electrons, r_{CO} the net rate of $(CO)_{ads}$, Θ the fractional surface coverage of CO and q_{CO} is the charge required for adsorption of CO to complete surface coverage.

Recently, Liu *et al.* [2003] studied the impedance behavior of methanol oxidation on a Pt–Ru anode with different electrolyte (Nafion) content in the anode and found also an inductive loop in the low frequency range of the EIS.

4.5.4.4 Solid Oxide Fuel Cells (SOFC)

As pictured in Figure 4.5.39, the Solid Oxide Fuel Cell (SOFC) uses a ceramic (solid oxide) electrolyte. The ionic conductivity (O^{2-} ion conduction) requirement for the ceramic electrolyte necessitates high operating temperatures (600°C to 1000°C). Such high operating temperatures promote fast reaction kinetics, even with non-noble metal electrocatalysts and allow reforming of hydrocarbon (CO, CH_4, gasoline, etc.) fuels within the fuel cell. But high operating temperature makes great demands on materials and requires very careful selection of components with matching thermomechanical properties. A comprehensive treatise on both the fundamental and technological aspects of SOFC is given by Minh and Takahashi [1995].

Electrochemical impedance measurements were, at first, used mainly to determine the conductivity of SOFC components (electrolyte and electrode materials). According to Verkerk and Burggraaf [1983] the equivalent circuit describing the Pt/Yttria-stabilized ZrO_2 (YSZ) interface consists of a parallel combination of the charge transfer resistance and a Warburg impedance related to the diffusion of oxygen atoms on the electrode surface, in series with a second Warburg impedance and the double layer capacity in parallel. The same equivalent circuit was used by Nicoloso *et al.* [1988] for the study of the effect of impurities on YSZ conductivity. Given the low conductivity of YSZ, research has been focused on finding new materials with higher conductivity, even at lower temperatures, as reported by Van Herle *et al.* [1994], or to use anode-supported SOFCs with thin gas-tight electrolyte films by dc sputtering, as reported by Wanzenberg *et al.* [2003], or by vacuum plasma spraying as applied by Schiller *et al.* [2000]. As shown by Wanzenberg *et al.* [2003] and reported by Bossel [1993] and Fleig and Maier [1996], the electrode contact area have some influence on impedance (conductivity) measurements. Boukamp *et al.* [1991] investigated the kinetic of the oxygen reduction reaction—the cathodic reaction in the SOFC, which is also important for the determination of the oxygen concentration in oxygen gas sensors (car exhaust).

EIS Measured at SOFC at Different Experimental Conditions—At Different Current Densities. EIS was applied by van Heuveln [1993] to determine the electrical losses in the SOFC, further refinements have been made by Richter [1997] to represent the porous structure of the electrode using the porous electrode model proposed by Göhr [1997].

Electrochemical impedance spectra of single SOFCs were reported by Wagner *et al.* [1998]. They investigated SOFC with cermet anode (ZrO_2–30% Ni, 30 μm thick) produced by vacuum plasma spraying at DLR (details are given by Lang *et al.* [2002]) deposited onto 640 μm thick YSZ electrolyte substrates (ZrO_2–8 mol % Y_2O_3, Kerafol GmbH, Germany), and screen printed layers of (La,Sr)MnO_3 (40 μm thick, Siemens AG, Germany) was used as cathode.

Due to the high operating temperature of the SOFC, the charge transfer resistance of the anode and the cathode are in the same order of magnitude. Therefore the contribution of the anode cannot be neglected. Even at OCV (1234 mV) and at lower cell voltage (higher current density), in addition to the charge transfer resistances (R_{ct}) and double layer capacities (C_{dl}) of the electrodes, a phase-shift maximum in the low frequency range of the spectra (at 10 Hz) with a Nernst-impedance like behavior is observed, as indicated in the impedance spectra presented in Figure 4.5.71*a* as a Bode plot and Figure 4.5.71*b* as a Nyquist plot. The relaxation at 10 Hz is attributed to gas diffusion along the gas supply channels on the anode side caused by inefficient flushing, similar to the mechanism proposed by Geyer *et al.* [1997]. Primdahl and Mogensen [1997, 1998] also found this behavior. They call it "gas conversion impedance" which is related to the change of atmosphere composition over the anode or "non uniform diffusion impedance", as denoted by Van Herle *et al.* [1994].

The same equivalent circuit used to evaluate the PEFC impedance spectra (Figure 4.5.60) can be applied for the simulation of the measured impedance spectra of the SOFC. The time constant of the cathode ($R_{ct(C)}/C_{dl(C)}$) is given by the charge transfer through the double layer of the cathode. The time constants due to anode

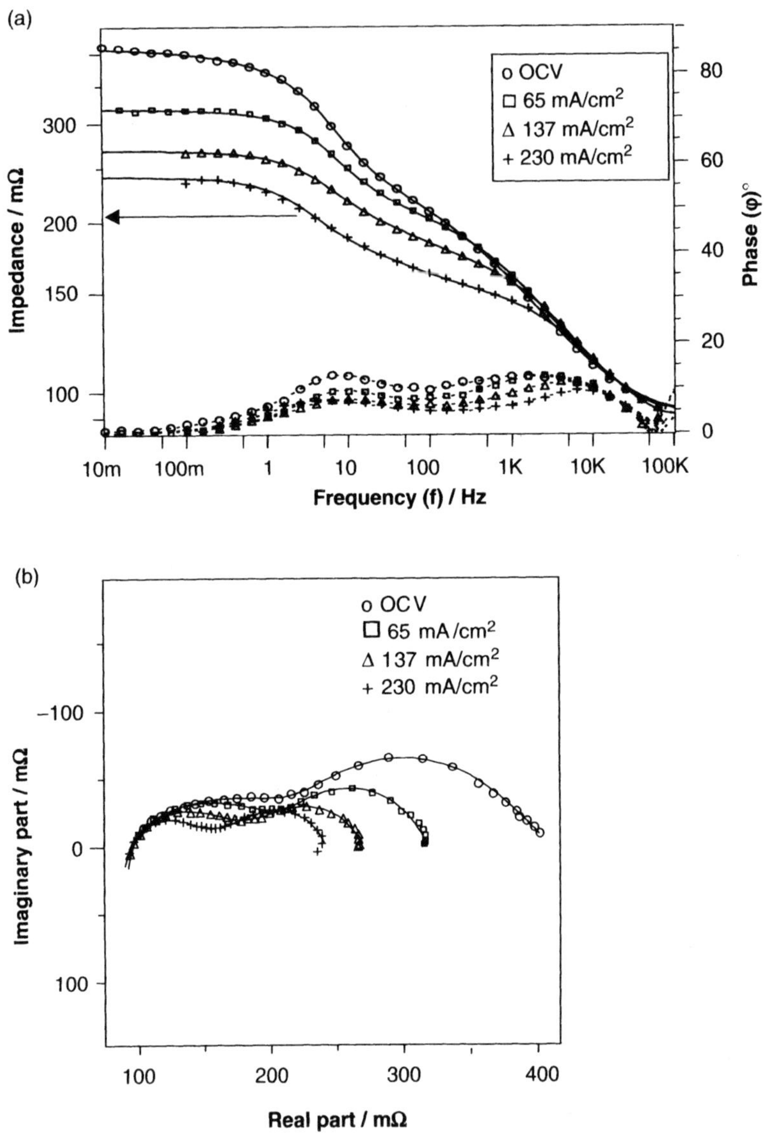

Figure 4.5.71. (*a*) Bode plot and (*b*) Nyquist plot of impedance spectra, SOFC-cell (ZrO_2–Ni/YSZ/(La,Sr)MnO_3, geometric surface 5 cm^2) measured at 950°C and different current densities.

and cathode respectively are determined by variation of the gas supply, as shown by Lang *et al.* [2002]. From a least square fit one obtains a set of impedance parameters. Using these impedance parameters, as shown in the PEFC section (Eqs (63) to (67)), one can calculate the voltage losses at the components of a SOFC. In the case of a linear current/voltage characteristic, these voltage losses can also can be cal-

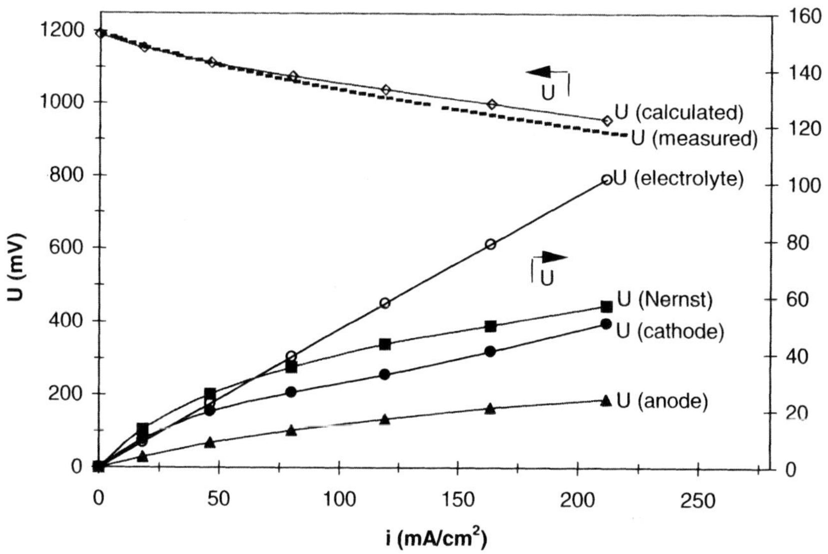

Figure 4.5.72. Cell voltage (measured and calculated) and calculated voltage losses at the SOFC (Ni-cermet/YSZ/(La,Sr)MnO_3) measured at 950°C at different current densities.

culated from the product of the impedance parameters magnitude ($R_{ct(A)}$, $R_{ct(C)}$, $R_{(N)}$ and R_{El}) and the current densities (Figure 4.5.72). The calculated cell voltage is in good agreement with the cell voltage, which was measured in a independent second experiment with an external load.

EIS Measured in SOFC—Anode Supplied with Dry and Moistened Hydrogen. The water vapor content of the fuel gas fed to the SOFC-anode has a strong effect on the electrochemical characteristics of the cell. Even using "dry" hydrogen a small amount of water is present in the hydrogen due to water production at exchange current density (i_0). In the case of insufficiently sealed cells, chemically-produced water also reduces the OCV below 1000 mV. Therefore the value of the OCV can be used as an indication of quality, for example the sealing of the cell. In Figure 4.5.73 impedance spectra measured at open circuit voltage with dry hydrogen and with humidified hydrogen, obtained after passage of hydrogen through a bubbling bottle at room temperature, are shown.

In order to explain the water vapor influence, one has to refer to the following equations. According to Geyer *et al.* [1997], the dependence of the anodic charge transfer resistance ($R_{ct,A}$ in the equivalent circuit used) on the mole fraction of water vapour (X) in the gas can be explained by assuming a Butler–Volmer kinetic where the exchange current density (i_0) is proportional to powers of the concentrations of the reactants (hydrogen and water vapour):

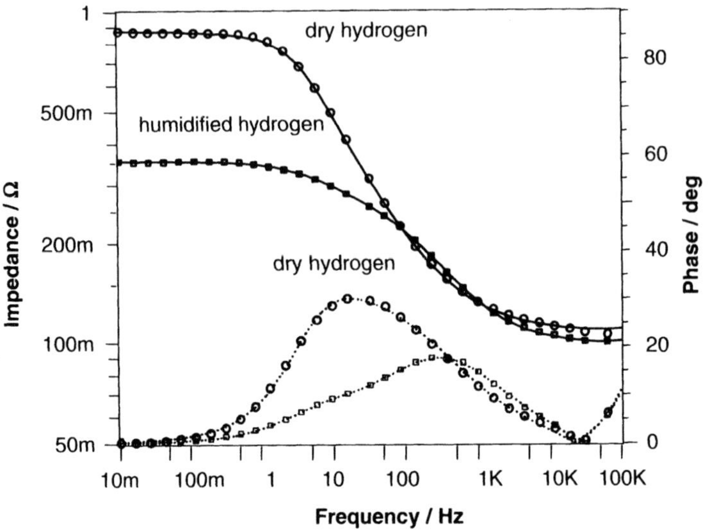

Figure 4.5.73. Bode plot of the measured impedance spectra, SOFC (ZrO_2-Ni/YSZ/(La,Sr)MnO_3) at 950°C, measured at different OCV- with dry hydrogen (○) at ••

$$RT/2Fi_0 = R_{ct,A}(X) \propto \frac{1}{X^{\alpha}(1-X)^{(1-\alpha)}} \tag{84}$$

where α is the charge transfer coefficient, R the gas constant and T the absolute temperature.

Another effect due to the presence of water vapor in the fuel gas is observed in the open cell voltage. The observed differences of the open cell voltage, depending on the water vapor content in hydrogen, can be explained by the Nernst equation. The gas diffusion overvoltage (η_N) is given by the Nernst-equation:

$$\eta_N = \frac{RT}{2F} \ln \frac{X}{(1-X)\sqrt{p_{O_2}}} \tag{85}$$

This diffusion overvoltage leads to the Nernst impedance. Operating the fuel cell with oxygen ($p_{O_2} = 1$), no diffusion overvoltage for $X = 0.5$ occurs. Therefore, the open cell voltage, given by Eq. (86), is very sensitive to the water vapor content in the fuel gas and can be used to determine the water vapor in the fuel gas quantitatively.

$$OCV = E_{0,th} + |\eta_N| \tag{86}$$

where $E_{0,th}$ is the standard cell potential at operating temperature. $E_{0,th}$ can be calculated using Eq. (87).

$$\Delta_f G_T^0 = -zFE_{0,th} \tag{87}$$

where $\Delta_f G_T^0$ is the standard molar Gibbs energy change of formation of water vapor at the temperature T in kJ/mol and z the number of transferred during the overall reaction. From Lide [1997] for T = 950°C (1223 K) we found $\Delta_f G_T^0 = -180.152$ kJ/mol. Using Eq. (87) one calculates $E_{0,th} = 934\,\text{mV}$.

The mole fraction of water vapor (X) in the gas can be calculated with the semi-empirical Antoine equation, as shown by Baehr [1988]:

$$\log p_{H_2O}^{sat} = A - \frac{B}{C + (T - 273{,}15)} \tag{88}$$

where $A = 8.0732991$, $B = 1656.39$, $C = 226.86$ and $p_{H_2O}^{sat}$ in mbar.

$$X = p_{H_2O}^{sat} / p \tag{89}$$

Using Eq. (88) and (89), for $T = 298.15\,\text{K}$ and $p = 1014\,\text{mbar}$ the calculated mole fraction of water vapor content (X) reaches 3 mol%. Inserting $X = 0.03$ in Eq. (85) and combining it with Eq. (86) one obtains an open circuit voltage of 1117 mV. This is in very good agreement with the measured OCV of 1114 mV.

Temperature Dependency of EIS Measured in SOFC at Open Cell Voltage. In order to calculate the activation energy of the oxygen reduction reaction at the cathode, hydrogen oxidation reaction at the anode and ionic conduction

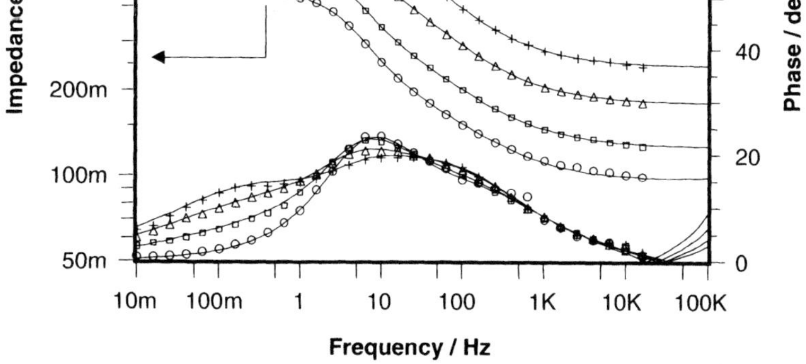

Figure 4.5.74. (*a*) Bode plot and (*b*) Nyquist plot of impedance spectra, SOFC-cell (ZrO_2–Ni/YSZ/(La,Sr)MnO_3, geometric surface 5 cm^2) measured at different cell temperatures.

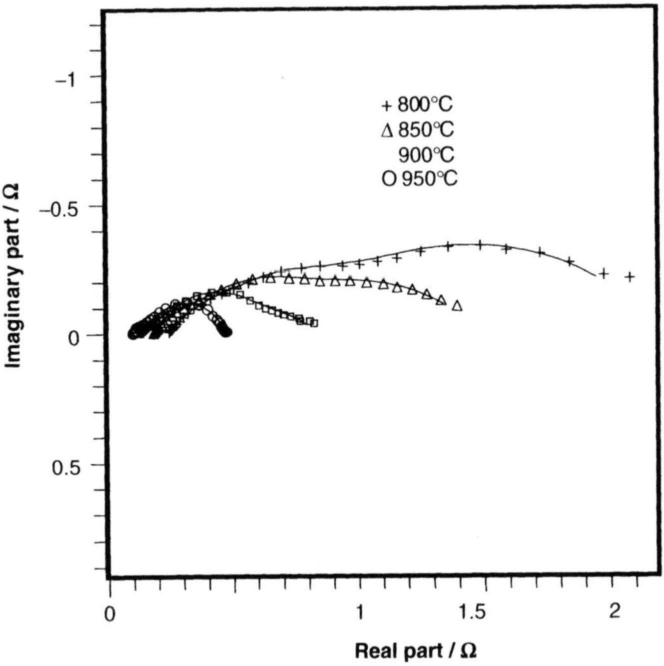

Figure 4.5.74. *Continued*

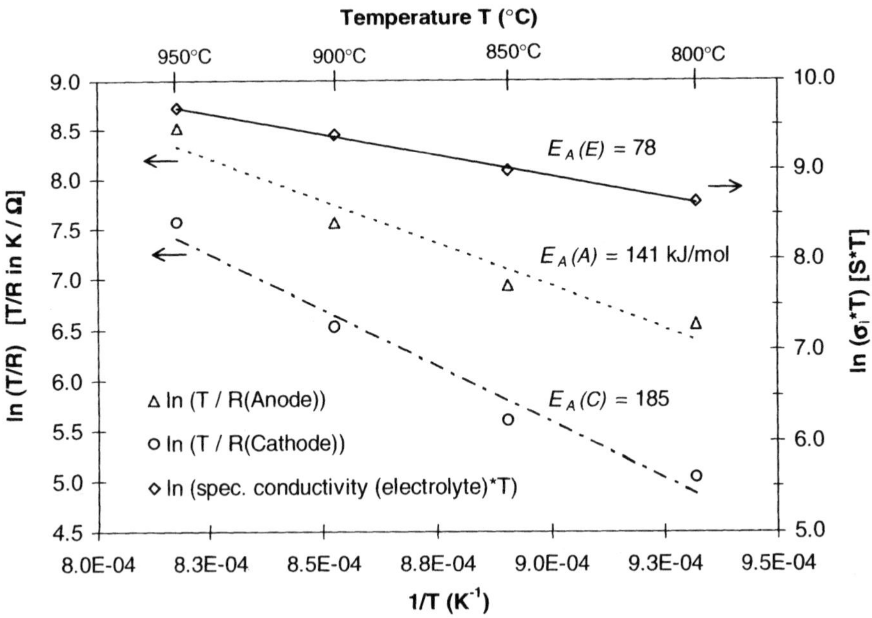

Figure 4.5.75. Arrhenius plot for the determination of activation energies of SOFC processes, after evaluation of EIS measured at different temperatures at OCV.

(O^{2-}) in the ceramic (solid oxide) electrolyte impedance spectra at OCV and different temperature, have to be measured. From the evaluation of the impedance spectra measured at different temperatures (Figure 4.5.74*a* and *b*) by applying the proposed equivalent circuit, a set of resistances (charge transfer and electrolyte resistance) are calculated and using an Arrhenius approach, the activation energy for each process can be calculated, as shown in Figure 4.5.75.

Appendix

Abbreviations and Definitions of Models

BNN Barton, Nakajima, Namikawa response, Section 4.2, Eq. (5)
CK0 The combination of a capacitance in parallel with the K0 response model
CK1 The combination of a capacitance in parallel with the K1 response model
CK1S The CK1 model with a CPE, the SCPE, in series with it
CK1EL The CK1 model with a four-parameter electrode-effect model in series
CMF The corrected modulus formalism response model
CNLS Complex nonlinear least squares
CPE The constant-phase distributed response element
CSD Conductive-system dispersion
DE Distributed circuit element
DEC Conductive-system DE
DED Dielectric-system DE
DRT Distribution of relaxation times
DSD Dielectric-system dispersion
EL An electrode-response model that includes the SCPE
EM Effective medium
K0 The Kohlrausch frequency response model directly associated with stretched-exponential temporal response
K1 The Kohlrausch frequency response model derived from the K0 model; see Section 4.2, Eq. (1) with $\varepsilon_Z = \varepsilon_{C1\infty}$, Eq. (3), and Eq. (4). Some composite models are the CK0, CK1, PK1, CK0S, CK1S, EMK1, CK1EL, CPK1. Parallel elements appear on the left side of K0 or K1, and series ones on the right. C denotes a parallel capacitance or dielectric constant.
NCL Nearly constant loss: $\varepsilon''(\omega)$ nearly independent of frequency over a finite range
PCPE A CPE in parallel with other response; denoted in a composite model by P
OMF The original modulus formalism response model; Section 4.2, Eq. (1) with $\varepsilon_Z = \varepsilon_\infty$, Eq. (2), and Eq. (3)
SCPE A CPE in series with other response; denoted in a composite model by S
ZARC Another name for the ZC response model
ZC A complex-power-law response model; Cole–Cole response at the impedance level

Impedance Spectroscopy, Second Edition, edited by Evgenij Barsoukov and J. Ross Macdonald
ISBN 0-471-64749-7

References

H. Abe, M. Esashi, and T. Matsuo [1979] ISFET's Using Inorganic Gate Thin Films, *IEEE Trans. Electron Devices* **ED-26**, 1939–1944.

P. Abelard and J. F. Baumard [1982] Study of the d.c. and a.c. Electrical Properties of an Yttria-Stabilized Zirconia Single Crystal [$(ZrO_2)_{0.88}$–$(Y_2O_3)_{0.12}$], *Phys. Rev.* **B26**, 1005–1017.

[1984] Dielectric Relaxation in Alkali Silicate Glasses: A New Interpretation, *Solid State Ionics* **14**, 61–65.

M. Abramowitz and I. A. Stegun [1965] *Handbook of Mathematical Functions*, NBS Publication 55, p. 376.

N. K. Adam [1941] *The Physics and Chemistry of Surfaces*, 3rd ed., Oxford University Press, London.

A. W. Adamson [1982] *Physical Chemistry of Surfaces*, 4th ed., John Wiley and Sons, New York.

K. El Adham and A. Hammou [1983] "Grain Boundary Effect" on Ceria-Based Solid Solutions, *Solid State Ionics* **9–10**, 905–912.

S. B. Adler [2002] Reference Electrode Placement in Thin Solid Electrolytes, *J. Electrochem. Soc.* **149**, E166–172.

P. Agarwal, M. E. Orazem, and L. H. Garcia-Rubio [1995] Application of Measurement Models to Impedance Spectroscopy. III. Evaluation of Consistency with the Kramers-Kronig Relations, *J. Electrochem. Soc.* **142**, 4159–4168.

Y. Aihara, J. Kuratomi, T. Bando, T. Iguchi, H. Yoshida, T. Ono, and K. Kuwana [2003] Investigation on Solvent-Free Solid Polymer Electrolytes for Advanced Lithium Batteries and their Performance, *Journal of Power Sources* **114**, 96–104.

S. Alexander, J. Bernasconi, W. R. Schneider, and R. Orbach [1981] Excitation Dynamics in Random One-Dimensional Systems, *Rev. Mod. Phys.* **53**, 175–198.

D. P. Almond and A. R. West [1981] Measurement of Mechanical and Electrical Relaxations in Beta-Alumina, *Solid State Ionics* **3/4**, 73–77.

[1983a] Mobile Ion Concentrations in Solid Electrolytes from an Analysis of AC Conductivity, *Solid State Ionics* **9–10**, 277–282.

[1983b] Impedance and Modulus Spectroscopy of "Real" Dispersive Conductors, *Solid State Ionics* **11**, 57–64.

D. P. Almond, G. K. Duncan, and A. R. West [1983] The Determination of Hopping Rates and Carrier Concentrations in Ionic Conductors by a New Analysis of ac Conductivity, *Solid State Ionics* **8**, 159–164.

D. P. Almond, C. C. Hunter, and A. R. West

[1984] The Extraction of Ionic Conductivities and Hopping Rates from A. C. Conductivity Data, *J. Mat. Sci.* **19**, 3236–3248.

D. P. Almond, A. R. West, and R. J. Grant

[1982] Temperature Dependence of the A. C. Conductivity of Na Beta-Aluminum, *Solid State Comm.* **44**, 1277–1280.

G. Amatucci, A. Du Pasquier, A. Blyr, T. Zheng, and J.-M. Tarascon [1999] The

Impedance Spectroscopy, Second Edition, edited by Evgenij Barsoukov and J. Ross Macdonald
ISBN 0-471-64749-7

Elevated Temperature Performance of the LiMn2O4/C System: Failure and Solutions, *Electrochim. Acta* **45**, 255–271.

B. Andreaus, A. J. McEvoy, and G. G. Scherer [2002] Analysis of Performance Losses in Polymer Electrolyte Fuel Cells at High Current Densities by Impedance Spectroscopy, *Electrochim. Acta* **47**, 2223–2229.

A. Appleby and F. R. Foulkes [1993] *Fuel Cell Handbook*, Krieger Publishing Comp., Malabar FL, USA.

W. I. Archer and R. D. Armstrong [1980] The Application of A. C. Impedance Methods to Solid Electrolytes, *Electrochemistry*, Chemical Society Specialist Periodical Reports **7**, 157–202.

A. Arico, P. Creti, H. Kim, R. Mantegna, N. Giordano, and V. Antonucci [1996] Analysis of the Electrochemical Characteristics of a Direct Methanol Fuel Cell Based on a Pt-Ru/C Anode Catalyst, *J. Electrochem. Soc.* **143**, 3950–3959.

M. Armand [1983] Polymer Solid Electrolytes—an Overview, *Solid State Ionics* **9–10**, 745–754.

R. D. Armstrong [1974] The Metal Solid Electrolyte Interphase, *J. Electroanal. Chem.* **52**, 413–419.

R. D. Armstrong and K. L. Bladen [1976] The Anodic Dissolution of Lead in Oxygenated and Deoxygenated Sulphuric Acid Solutions, *J. Appl. Electrochem.* **7**, 345.

R. D. Armstrong and K. Edmondson [1973] The Impedance of Metals in the Passive and Transpassive Regions, *Electrochim. Acta* **18**, 937–943.

R. D. Armstrong and M. Henderson [1972] Impedance Plane Display of a Reaction with an Adsorbed Intermediate, *J. Electroanal. Chem.* **39**, 81–90.

R. D. Armstrong and R. Mason [1973] Double Layer Capacity Mesurement Involving Solid Electrolytes, *J. Electroanal. Chem.* **41**, 231–241.

R. D. Armstrong, M. F. Bell, and A. A. Metcalfe [1977] The Anodic Dissolution of Molybdenum in Alkaline Solutions. Electrochemical Measurements, *J. Electroanal. Chem.* **84**, 61.

R. D. Armstrong, M. F. Bell, and A. A. Metcalfe [1978] The AC Impedance of Complex Electrochemical Reactions, *Electrochemistry*, Chemical Society Specialist Periodical Reports **6**, 98–127.

R. D. Armstrong, T. Dickinson, and J. Turner [1973] Impedance of the Sodium β-alumina Interphase, *J. Electroanal. Chem.* **44**, 157–167.

R. D. Armstrong, T. Dickinson, and P. M. Willis [1974] The A. C. Impedance of Powdered and Sintered Solid Ionic Conductors, *J. Electroanal. Chem.* **53**, 389–405.

R. D. Armstrong, W. P. Race, and H. R. Thirsk [1968] Determination of Electrode Impedance over an Extended Frequency Range by Alternating-Current Bridge Methods, *Electrochim. Acta* **13**, 215.

F. Artuso, F. Bonino, F. Decker, A. Lourenco, and E. Masetti [2002] Study of Lithium Diffusion in RF Sputtered Nickel-Vanadium Mixed Oxides Thin Films. *Electrochim. Acta* **47**, 2231–2238.

M. Atanasov, C. Daul, J.-L. Barras, L. Benco, and E. Deiss [1999] Polarizable Continuum Model for Lithium Interface Transitions Between a Liquid Electrolyte and an Intercalation Electrode. *Solid State Ionics* **121**, 165–174.

S. Atlung and T. Jacobsen [1976] On the AC-Impedance of Electroactive Powers: γ-Manganese Dioxide, *Electrochim. Acta* **21**, 575–584.

D. Aurbach, K. Gamolsky, B. Markovsky, Y. Gofer, M. Schmidt, and U. Heider [2002] On the use of Vinylene Carbonate (VC) as an Additive to Electrolyte Solutions for Li-ion Batteries, *Electrochim. Acta* **47**, 1423–1439.

D. Aurbach, J. S. Gnanaraj, M. D. Levi, E. A. Levi, J. E. Fischer, and A. Claye [2001] On the Correlation Among Surface Chemistry, 3D Structure, Morphology, Electrochemical and Impedance Behavior of Various Lithiated Carbon Electrodes, *Journal of Power Sources* **97–98**, 92–96.

D. Aurbach, B. Markovsky, E. Rodkin, Y. S. Levi, H.-J. Cohen, D. H. Kim, and M. Schmidt [2002] On the Capacity Fading of LiCoO2 Intercalation Electrodes: the Effect of Cycling, Storage, Temperature, and Surface Film Forming Additives, A., *Electrochim. Acta* **47**, 4291–4306.

E. Avendano [2002] *Electrochromic Properties of Nickel Based Oxide Films*, Licentiate Thesis, UPTEC 02 009, Uppsala University.

E. Avendano, A. Azens, J. Isidorsson, R. Karmhag, G. A. Niklasson, and C. G. Granqvist [2003] Optimized Nickel-Oxide-Based Electrochromic Thin Films, *Solid State Ionics* **165**, 169–173.

A. Azens and C. G. Granqvist [2003] Electrochromic Smart Windows: Energy Efficiency and Device Aspects, *J. Solid State Electrochem.* **7**, 64–68.

U. Bach, D. Lupo, P. Comte, J. E. Moser, F. Weissortel, J. Salbeck, H. Spreitzer, and M. Grätzel [1998] Solid-State Dye-Sensitized Mesoporous TiO_2 Solar Cells with High Photon-to-Current Conversion Efficiencies, *Nature* **395**, 583–585.

F. T. Bacon [1952] Improvements Relating to Galvanic Cells and Batteries, *Patent GB 667298.*

S. P. S. Badwal [1983] New Electrode Materials for Low Temperature Oxygen Sensors, *J. Electroanal. Chem.* **146**, 425–429.

[1984] Kinetics of the Oxygen Transfer Reaction at the $(Cu_{0.5}Sc_{0.5})O_{2\pm x}$/YSZ Interface by Impedance Spectroscopy, *J. Electroanal. Chem.* **161**, 75–91.

S. P. S. Badwal and H. J. de Bruin [1978] Polarization at the $(U_{0.7}Y_{0.3})O_{2+x}$/YSZ Interface by Impedance Dispersion Analysis, *Aust. J. Chem.* **31**, 2337–2347.

S. P. S. Badwal, M. J. Bannister, and M. J. Murray [1984] Non-Stoichiometric Oxide Electrodes for Solid State Electrochemical Devices, *J. Electroanal. Chem.* **168**, 363–382.

J.-S. Bae and S.-I. Pyun [1995] Electrochemical Lithium Intercalation Reaction of Anodic Vanadium Oxide Film. *J. Alloys Compounds* **217**, 52–58.

H. D. Baehr [1988] *Thermodynamik*, Springer Verlag, p. 212.

P. R. Barabash and R. S. C. Cobbold [1982] Dependence on Interface State properties of Electrolyte: Silicon Dioxide-Silicon Structures, *IEEE Trans. Elect. Devices* **ED-29**, 102–108.

A. J. Bard and L. R. Faulkner [1980] *Electrochemical Methods: Fundamentals and Applications*, Wiley, New York.

G. Barker [1958] Square-Wave Polarography. III. Single Drop Polarography, *Anal. Chim. Acta* **18**, 118.

[1969] Noise Connected with Electrode Processes, *J. Electroanal. Chem.* **21**, 127.

G. Barker, R. Faircloth, and A. Gardner [1958] Use of Faradaic Rectification for the Study of Rapid Electrode Processes, IV. Theoretical Aspects of Square Wave Polarography, *Nature* **181**, 247–248.

E. Barsoukov, J. H. Kim, J. Hun Kim, C. O. Yoon, and H. Lee [1998] Effect of Low-Temperature Conditions on Passive Layer Growth on Li Intercalation Materials: In situ Impedance Study, *J. Electrochem. Soc.* **145**, 2711–2717.

E. Barsoukov, J. H. Kim, J. Hun Kim, C. O. Yoon, and H. Lee [1999a] Kinetics of Lithium Intercalation into Carbon Anodes: in situ Impedance Investigation of Thickness and Potential Dependence, *Solid State Ionics* **116**, 249–261.

E. Barsoukov, J. H. Kim, C. O. Yoon, and H. Lee [1999b] Universal Battery Parameterization to Yield a Non-Linear Equivalent Circuit Valid for Battery Simulation at Arbitrary Load, *J. Power Sources* **83**, 61–70.

E. Barsoukov, J. H. Kim, D. H. Kim, K. S. Hwang, C. O. Yoon, and H. Lee [2000] Parametric Analysis using Impedance Spectroscopy: Relationship between Material Properties and Battery Performance, *J. New Materials for Electrochem. Sys.* **3**, 301–308.

E. Barsoukov, S. H. Ryu, and H. Lee [2002] A Novel Impedance Spectrometer Based on Carrier Function Laplace Transform of the Response to Arbitrary Excitation, *J. Electroanal. Chem.* **536**, 109–122.

E. Barsoukov, D. H. Kim, H.-S. Lee, H. Lee, M. Yakovleva, Y. Gao, and J. F. Engel [2003] Comparison of Kinetic Properties of LiCoO2 and LiTi0.05Mg0.05Ni0.7Co0.2O2 by Impedance Spectroscopy, *Solid State Ionics* **161**, 19.

J. L. Barton [1966] La Relaxation Diélectrique de Quelques Verres Ternaries Silice Oxyde Alcalin Oxyde Alcalino-Terreux, *Verres et Refr.* **20**, 324–328.

H. Bässler [1993] Charge Transport in Disordered Organic Photoconductors, *Phys. Stat. Sol. B* **175**, 15–56.

J. B. Bates, N. J. Dudney, B. Neudecker, A. Ueda, and C. D. Evans [2000] Thin-Film Lithium and Lithium-Ion Batteries, *Solid State Ionics* **135**, 33–45.

H. H. Bauer [1964] Theory of Faradaic Distortion: Equation for the 2nd-Harmonic Current, *Australian J. Chem.* **17**, 715.

J. E. Bauerle [1969] Study of Solid Electrolyte Polarization by a Complex Admittance Method, *J. Phys. Chem. Solids* **30**, 2657–2670.

[1976] Correlation between Impedance, Microstructure, and Compositions of Calcia-Stabilized Zirconia, *Electrochim. Acta* **21**, 303–310.

H. E. Becker [1957]

U.S. Patent No. 2,800,616 (7,23,57), to GM Co.

N. M. Beekmans and L. Heyne [1976] Correlation between Impedance, Microstructure, and Compositions of Calcia-Stabilized Zirconia, *Electrochim. Acta* **21**, 303–310.

J. G. Berberian and R. H. Cole [1969] Null Instrument for Three- and Four-Terminal Admittance Measurements at Ultra-low Frequencies, *Rev. Sci. Instrum.* **40**, 811.

P. Bergveld [1970] Development of an Ion-Sensitive Solid State Device for Neurophysiological Measurements, *IEEE Trans. Biomed. Eng.* **BME-17**, 70–71.

P. Bergveld and N. F. De Rooij [1981] History of Chemically Sensitive Semiconductor Devices, *Sensors and Actuators*, **1**, 5–15.

L. E. A. Berlouis

An Impedance Study of Flow-Through Porous Electrodes, Ph.D. Thesis, the University, Southampton, England.

H. Bernard [1981] *Microstructure et Conductivité de la Zircone Stabilisée Frittée*, Report CEA-R-5090, Commissariat a l'Energie Atomique, CEN-Saclay, France.

J. Bernasconi, H. U. Beyeler, S. Strassler, and S. Alexander [1979] Anomalous Frequency-Dependent Conductivity in Disordered One-Dimensional System, *Phys. Rev. Lett.* **42**, 819–822.

U. Bertocci [1979] AC Induced Corrosion: The Effect of an Alternating Voltage on Electrodes under Charge Transfer Control, *Corrosion* **35**, 211–215.

U. Bertocci and J. L. Mullen [1981] Effect of Large Voltage Modulations on Electrodes under Charge Transfer Control, in *Electrochemical Corrosion Testing*, ed.

F. Mansfeld and U. Bertocci, American Society of Testing and Materials, San Francisco, pp. 365–380.

I. Betova, M. Bojinov, and T. Tzvetkoff [2004] Oxidative Dissolution and Anion-Assisted Solubilization in the Transpassive State of Nickel-Chromium Alloys, *Electrochim. Acta* **49**, 2295–2306.

I. Betova, M. Bojinov, P. Kinnunen, T. Laitinen, P. Pohjanne, and T. Saario [2002] Mechanism of Transpassive Dissolution of Nickel-based Alloys Studied by Impedance Spectroscopy and Rotating Ring-disc Voltammetry, *Electrochim. Acta* **47**, 2093–2107.

D. Bevers, N. Wagner, and M. von Bradke [1998] Innovative Production Procedure for Low Cost PEFC Electrodes and Electrode Membrane Structures, *Int. J. Hydrogen Energy* **23**, 57–63.

P. Bindra, M. Fleischmann, J. W. Oldham, and D. Singleton [1973] Nucleation, *Faraday Discuss. Chem. Soc.* **56**, 180.

R. L. Birke [1971] Operational Admittance of an Electrode Process where an *a priori* Dependence of Rate Constant on Potential Is Not Assumed, *J. Electroanal. Chem.* **33**, 201–207.

J. B. Birks

Heywood Press, London, pp. 69–114.

J. Bisquert, G. Garcia-Belmonte, F. Bueno, E. Longo, and L. O. S. Bulhöes [1998] Impedance of Constant Phase Element (CPE)-Blocked Diffusion in Film Electrodes, *J. Electroanal. Chem.* **452**, 229–234.

K. L. Bladen [1977] The Anodic Dissolution of Lead in Oxygenated and Deoxygenated Sulphuric Acid Solutions, *J. Appl. Electrochem.* **7**, 345.

G. Blanc, C. Gabrielli, and N. Keddam [1975] Measurement of the Electrochemical Noise by a Cross Correlation Method, *Electrochim. Acta* **20**, 687.

G. E. Blomgren [2003] Liquid Electrolytes for Lithium and Lithium-Ion Batteries, *Journal of Power Sources* **119–121**, 326–329.

L. J. Blumen and M. N. Mugerwa (ed.) [1993] *Fuel Cell Systems*, Plenum Press, New York.

R. Boddy [1965] The Structure of the Semiconductor-Electrolyte Interface, *J. Electroanal. Chem.* **10**, 199–244.

H. W. Bode [1945] *Network Analysis and Feedback Amplifier Design*, Van Nostrand, New York, Ch. 14.

E. V. Bohn [1963] *The Transform Analysis of Linear Systems*, Addison-Wesley, Reading, Mass.

O. Bohnke and Cl. Bohnke [1988] Comparative Study of the Electrochromic Properties of WO_3 Thin Films, *Displays* **9**, 199–206.

Cl. Bohnke and O. Bohnke [1990] Impedance Analysis of Amorphous WO_3 Thin Films in Hydrated $LiClO_4$-Propylene Carbonate Electrolytes, *Solid State Ionics* **39**, 195–204.

Cl. Bohnke, O. Bohnke, and B. Vuillemin [1993] Constant Phase Angle Behavior of SnO_2/WO_3 Thin Film Electrodes in Anhydrous $LiClO_4$-Propylene Carbonate Electrolyte, *Electrochim. Acta* **38**, 1935–1940.

M. Bojinov, I. Betova, M. Bojinov, T. Latinen, K. Makela, P. Pohjanne, and T. Saario [2002] The Transpassive Dissolution Mechanism of Highly Alloyed Stainless Steels I. Experimental Results and Modelling Procedure, *Corros. Sci.* **44**, 2675–2697.

M. Bojinov, S. Cattarin, M. Musiani, and B. Tribollet [2003] Evidence of Coupling between Film Growth and Metal Dissolution in Passivation Processes, *Electrochim. Acta* **48**, 4107–4117.

M. Bojinov, G. Fabricius, T. Laitinen, T. Saario, and G. Sundholm [2000] Coupling between Ionic Defect Structure and Electronic Conduction in Passive Films on Iron, Chromium and Iron-Chromium Alloys, *Electrochim. Acta* **45**, 2029–2048.

N. Bonanos [1989] High Oxide Ion Conductivity in Bismuth Uranate, Bi_2UO_6, *Mat. Res. Bull.* **24**, 1531–1540.

N. Bonanos and E. P. Butler [1985] Ionic Conductivity of Monoclinic and Tetragonal Yttria-Zirconia Single Crystals, *J. Mat. Sci. Lett.* **4**, 561–564.

N. Bonanos and E. Lilley [1980] Ionic Conductivity of Suzuki Phases, *Solid State Ionics* **1**, 223–230.

[1981] Conductivity Relaxations in Single Crystals of Sodium Chloride Containing Suzuki Phase Precipitates, *J. Phys. Chem. Solids* **42**, 943–952.

N. Bonanos, P. Holtappels, and M. J. Jørgensen [2002] Electrochemical Evaluation Of Functionally Graded SOFC Cathodes, *Proceedings of 5th European SOFC Forum, Science and Engineering of SOFC*, 01–05 July 2002, Luzern, Switzerland.

N. Bonanos, R. K. Slotwinski, B. C. H. Steele, and E. P. Butler [1984a] Electrical Conductivity/Microstructural Relationships in Aged CaO and CaO + MgO Partially Stabilized Zirconia, *J. Mat. Sci.* **19**, 785–793.

[1984b] High Ionic Conductivity in Polycrystalline Tetragonal Y_2O_3-ZrO_2, *J. Mat. Sci. Lett.* **3**, 245–248.

A. Bonnel, F. Dabosi, C. Deslouis, M. Duprat, M. Keddam, and B. Tribollet

[1983] Corrosion Study of a Carbon Steel in Neutral Chloride Solution by Impedance Techniques, *J. Electrochem. Soc.* **130**, 753–761.

P. Bordewijk [1975] Defect-Diffusion Models of Dielectric Relaxation, *Chem. Phys. Lett.* **32**, 592–596.

T. I. Borisova and B. V. Ershler [1950] Determination of the Zero Voltage Points of Solid Metals from Measurements of the Capacity of the Double Layer, *Zh. Fiz. Khim.* **24**, 337–344.

U. G. Bossel [1993] Effects of Current Collection, Contacting and Wiring on PEN Characterization, *Proceedings of Solid Oxide Workshop*, Chexbres, Switzerland, 59–61.

U. G. Bossel [2000] *The Birth of the Fuel Cell*, The European Fuel Cell Forum.

C. J. F. Böttcher [1952] *Theory of Electric Polarization*, Elesevier Publishing Co., Amsterdam.

C. J. F. Böttcher and P. Bordewijk [1978] *Theory of Electric Polarization*, Vol. II, *Dielectrics in Time-Dependent Fields*, Elsevier, Amsterdam.

P. H. Bottelberghs and G. H. J. Broers [1976] Interfacial Impedance Behavior of Polished and Paint Platinum Electrodes at Na_2WO_4-Na_2MoO_4 Solid Electrolytes, *J. Electroanal. Chem.* **67**, 155–167.

B. A. Boukamp [1984] A Microcomputer Based System for Frequency Dependent Impedance/Admittance Measurements, *Solid State Ionics* **11**, 339–346.

[1989] Equivcrt, Univ. of Twente, NL.

B. A. Boukamp, B. A. van Hassel, and A. J. Burggraaf [1991] The use of Impedance Spectroscopy and NLLS-Fit Data Analysis in the Study of Oxygen Electrode Kinetics, *Proc.-Electrochem. Soc. (Proc. Symp. High Temp. Electrode Mater. Charact.)* 44–55.

L. Bousse [1982] Ph.D. Thesis, Twente University of Technology, the Netherlands.

L. Bousse and P. Bergveld [1983] On the Impedance of the Silicon Dioxide/Electrolyte Interface, *J. Electroanal. Chem.* **152**, 25–39.

F. P. Bowden and E. K. Rideal [1928] The Electrolytic Behavior of Thin Films. II. The Areas of Catalytically Active Surfaces, *Proc. Roy. Soc., London* **A120**, 80–89.

A. D. Brailsford and D. K. Hohnke [1983] The Electrical Characterization of Ceramic Oxides, *Solid State Ionics* **11**, 133–142.

B. Breyer and H. H. Bauer [1964] *Alternating Current Polarography and Tensammetry*, Interscience, New York.

R. Brown, D. E. Smith, and G. Booman [1968] Operational Amplifier Circuits Employing Positive Feedback for IR Compensation. I. Theoretical Analysis of Stability and Bandpass Characteristics, *Anal. Chem.* **40**, 1411.

P. G. Bruce [1987] Electrical Measurements on Polymer Electrolytes, in *Polymer Electrolyte Reviews I*, ed. J. R. MacCallum and C. A. Vincent, Elsevier, London.

P. G. Bruce and A. R. West [1983] The A. C. Conductivity of Polycrystalline LISICON $Li_{2+2x}Zn_{1-x}GeO_4$ and a Model for Intergranular Constriction Resistances, *J. Electrochem. Soc.* **130**, 662–669.

P. G. Bruce, J. Evans, and C. A. Vincent [1988] Conductivity and Transference Number Measurements on Polymer Electrolytes, *Solid State Ionics* **28–30**, 918–922.

P. G. Bruce, A. R. West, and D. P. Almond [1982] A New Analysis of ac Conductivity Data in Single Crystal Beta-Alumina, *Solid State Ionics* **7**, 57–60.

D. Bruggeman [1935] Berechnung Verschiedener Physikalischer Konstanten von Heterogenen Substanzen, I. Dielektrizitätskonstanten und Leitfähigkeiten der Mischkörper aus Isotropen Substanzen, *Annalen der Physik* Series **5 24**, 636–679.

R. Bube [1992] *Electrons in Solids: an Introductory Survey*, 3rd ed, Academic Press.

R. P. Buck [1980] Impedance Diagrams Applied to the Investigation of Ion-Selective Electrodes, *Hungarian Sci. Instr.* **49**, 7–23.

[1982] The Impedance Method Applied to the Investigation of Ion-Selective Electrodes, *Ion-Selective Electrode Rev.* **4**, 3–74.

A. Bunde, P. Maass, and M. D. Ingram [1991] Diffusion Limited Percolation: A Model for Transport in Ionic Glasses, *Ber. Bunsenges. Phys. Chem.* **95**, 977–983.

K. Bundy, M. Karlsson, G. Lindbergh, and A. Lundqvist [1998] An Electrochemical Impedance Spectroscopy Method for Prediction of the State of Charge of a Nickel-Metal Hydride Battery at Open Circuit and During Discharge, *J. Power Sources* **72**, 118–125.

A. J. Burggraaf, M. van Hemert, D. Scholten, and A. J. A. Winnubst [1985] Chemical Composition of Oxidic Interfaces in Relation with Electric and Electrochemical Properties, *Mater. Sci. Monogr.* **28B** (Reactivity of Solids), 797–802.

G. T. Burstein, C. J. Barnett, A. R. Kucernak, and K. R. Williams [1997] Aspects of the Anodic Oxidation of Methanol, *Cat. Today* **38**, 425–437.

E. P. Butler and N. Bonanos [1985] The Characterization of Zirconia Engineering Ceramics by a.c. Impedance Spectroscopy, *Mat. Sci. Eng.* **71**, 49–56.

N. Cabrera and N. F. Mott [1948] Theory of the Oxidation of Metals, *Report Progr. Phys.* **12**, 163–184.

B. D. Cahan and C.-T. Chen [1982] The Nature of the Passive Film on Iron. II. AC Impedance Studies, *J. Electrochem. Soc.* **129**, 474–480.

B. D. Cahan, H. M. Villulas, and E. B. Yeager [1991] The Effects of Trace Anions on the Voltammetry of Single Crystal Gold Surfaces, *J. Electroanal. Chem.* **306**, 213–238.

L. M. Callow, J. A. Richardson, and J. L. Dawson [1976] Corrosion Monitoring Using Polarization Resistance Measurements, Parts I and II, *Brit. Corrosion J.* **11**, 123–132.

R. Calvert [1948] A New Technique in Bridge Measurements, *Electron. Eng.* **20**, 28–29.

H. S. Carslaw and J. C. Jaeger [1959] *Conduction of Heat in Solids*, 2nd Ed., Clarendon Press, Oxford.

C. W. Carter [1925] Impedance of Networks Containing Resistances and Two Reactances, *Bell Sys. Tech. J.* **4**, 387–401.

M. Casciola and D. Fabiani [1983] Ionic Conduction and Dielectric Properties of Anydrous Alkali Metal Salt Forms of Zirconium Phosphase, *Solid State Ionics* **11**, 31–38.

E. B. Castro [1994] Analysis of the Impedance Response of Passive Iron, *Electrochim. Acta* **39**, 2117–2123.

H. Chang and G. Jaffé [1952] Polarization in Electrolytic Solutions, Part I, Theory, *J. Chem. Phys.* **20**, 1071–1077.

C.-Y. Chao, L. F. Lin, and D. D. Macdonald [1981] A Point Defect Model for Anodic Passive Films. I. Film Growth Kinetics, *J. Electrochem. Soc.* **128**, 1187–1194.
[1982] A Point Defect Model for Anodic Passive Films. III. Impedance Response, *J. Electrochem. Soc.* **129**, 1874–1879.

S. Cheng, J. Zhang, H. Liu, Y.-J. Leng, A. Yuan, and C. Cao [1998] Study of Early Cycling Deterioration of a Ni/MH Battery by Electrochemical Impedance Spectroscopy, *J. Power Sources* **74**, 155–157.

S. Cheng, J. Zhanga, M. Zhaob, and C. Caoa [1999] Electrochemical Impedance Spectroscopy Study of Ni/MH batteries *J. Alloys Compounds* **293–295**, 814–820.

D.-T. Chin and S. Venkatesh [1979] A Study of Alternating Voltage Modulation on the Polarization of Mild Steel, *J. Electrochem. Soc.* **126**, 1908–1913.

H. Chung [1983] *Transient Crack Growth in AISI 4340 Steel in Sodium Chloride Solution*, Ph.D. Thesis, Ohio State University.

H. Chung and D. D. Macdonald [1981] *Corrosion and Corrosion Cracking of Materials for Water Cooled Reactors*, Report to the Electric Power Research Institute, Palo Alto, CA, Project RP1166–1.

M. Ciureanu and H. Wang [1999] Electrochemical Impedance Study of Electrode-Membrane Assemblies in PEM Fuel Cells I. Electro-oxidation of H_2 and H_2/CO Mixtures on Pt-Based Gas-Diffusion Electrodes, *J. Electrochem. Soc.* **146**, 4031–4040.

M. Ciureanu and H. Wang [2000] Electrochemical Impedance Study of Anode CO-Poisoning in PEM Fuel Cells, *J. New Mat. Electrochem. Systems* **3**, 107–119.

S. J. C. Cleghorn, T. E. Springer, M. S. Wilson, C. Zawodzinski, T. A. Zawodzinski, and S. Gottesfeld [1997] PEM Fuel Cells for Transportation and Stationary Power Generation Applications, *Int. J. of Hydrogen Energy* **22**, 1137–1144.

M. Clericuzio, W. O. Parker, Jr., M. Soprani, and M. Andrei [1995] Ionic Diffusivity and Conductivity of Plasticized Polymer Electrolytes: PMFG-NMR and Complex Impedance Studies, *Solid State Ionics* **82**, 179–192.

W. Cleveland and E. Grosse [1990] *Fortran Routines for Locally-Weighted Regression, LOESS Program Source and Documentation*, www.netlib.org.

J. H. Clint [1992] *Surfactant Aggregation*, Blackie & Son, Glasgow.

K. S. Cole [1972] *Membranes, Ions, and Impulses*, University of California Press, Berkeley.

K. S. Cole and R. H. Cole [1941] Dispersion and Absorption in Dielectrics. I. Alternating-Current Characteristics, *J. Chem. Phys.* **9**, 341–351.

M. Colloms [1983] *Computer Controlled Testing and Instrumentation*, Pentech Press, London.

P. Colonomos and R. G. Gordon [1949] Bounded Error Analyses of Experimental

Distributions of Relaxation Times, *J. Chem. Phys.* **71**, 1159–1166.

B. E. Conway [1991] Transition from "Supercapacitor" to "Battery" Behavior in Electrochemical Energy Storage. *J. Electrochem. Soc.* **138**, 1539–1548.

B. E. Conway [1999] *Electrochemical Supercapacitors: Scientific Fundamentals and Technological Applications*, Kluwer Academic, Plenum Publishing Co., New York.

B. E. Conway and E. Gileadi [1962] Kinetic Theory of Pseudo-Capacitance and Electrode Reactions at Appreciable Surface Coverage, *Trans. Faraday Soc.* **58**, 2493–2509.

B. E. Conway and D. M. Novak [1981] Competitive Adsorption and State of Charge of Halide-Ions in Monolayer Oxide Film Growth-Processes at Pt Anodes, *J. Chem. Soc., Faraday Trans.* **I. 77**, 2341–2359.

B. E. Conway and W. G. Pell [2002] Power Limitations of Supercapacitor Operation Associated with Resistance and Capacitance Distribution in Porous Electrode Devices, *Journal of Power Sources* **105**, 169–181.

B. E. Conway and B. V. Tilak [2002] Interfacial Processes Involving Electrocatalytic Evolution and Oxidation of H_2, and the role of Chemisorbed H, *Electrochim. Acta* **47**, 3571–3594.

B. E. Conway, J. O'M. Bockris, and W. Mehl [1956] Preparation and Maintenance of Spherical Metal Electrodes in Hydrogen, *J. Sci. Instr.* **33**, 400.

B. E. Conway, H. A. Kozlowska, and E. Gileadi [1965] Significance of Nonsteady State a.c. and d.c. Measurements in Electrochemical Adsorption Kinetics. Applications to Galvanostatic and Voltage Sweep Methods, *J. Electrochem. Soc.* **112**, 341–349.

B. E. Conway, K. Tellefsen, and S. Marshall [1984] Lattice Models, Interactions, and Solvent Effects in Electrochemical Monolayer Formation, in *The Chemistry and Physics of Electrocatalysis*, ed. J. D. E. McIntyre, M. J. Weaver, and E. B. Yaeger, The Electrochemical Society Inc., Pennington, N.J. 84–12, pp. 15–33.

J. W. Cooley and J. W. Tukey [1965] An Algorithm for the Machine Computation of Complex Fourier Series, *Math. Comp.* **19**, 297.

S. Cordoba-Torresi, A. Hugot-Le Goff, and H. Takenouti [1990] Electrochromism in Metal Oxide Films Studied by Raman Spectroscopy and A.C. Techniques: Charge Insertion Mechanism, *Proc. SPIE* **1272**, 152–161.

P. Costamagna, P. Costa, and V. Antonucci [1998] Micro-Modelling of Solid Oxide Fuel Cell Electrodes, *Electrochim. Acta* **43**, 375–394.

R. T. Coverdale, E. J. Garboczi, and H. M. Jennings [1995] An Improved Model for Simulating Impedance Spectroscopy, *Computational Materials Science* **3**, 465.

E. J. Crain [1970] *Laplace and Fourier Transforms for Electrical Engineers*, Holt, Rinehart and Winston, New York.

S. C. Creason and D. E. Smith [1972] Fourier Transform Faradaic Admittance Measurements: Demonstration of the Applicability of Random and Pseudo-Random Noise as Applied Potential Signals, *J. Electroanal. Chem.* **36**, 1.

[1973] Fourier Transform Faradic Admittance Measurements: On the Use of High-Precision Data for Characterization of Very Rapid Electrode Process Kinetic Parameters, *J. Electroanal. Chem.* **47**, 9.

N. Cui and J. L. Luo [2000] An AC Impedance Study of Self-Discharge Mechanism of Nickel–Metal Hydride (Ni–MH) Battery Using Mg2Ni-Type Hydrogen Storage Alloy Anode, *Electrochim. Acta* **45**, 3973–3981.

J. R. Dahn, A. K. Sleigh, Hang Shi, J. N. Reimers, Q. Zhong, and B. M. Way [1993] Dependence of the Electrochemical Intercalation of Lithium in Carbons on the Crystal Structure of the Carbon, *Electrochim. Acta* **38**, 1179–1191.

G. D'Aprano and M. Leclerc [1993] Steric and Electronic Effects in Methyl And Methoxy Substituted Polyanilines, *J. Electroanal. Chem.* **351**, 145–148.

L. S. Darken [1948] Diffusion, Mobility, and their Interrelation through Free Energy in Binary Metallic Systems, *Trans. AIME* **175**, 184–201.

D. W. Davidson [1961] Dielectric Relaxation in Liquids. I. The Representation of Relaxation Behavior, *Can. J. Chem.* **39**, 571–594.

D. W. Davidson and R. H. Cole [1951] Dielectric Relaxation in Glycerol, Propylene Glycol, and *n*-propanol, *J. Chem. Phys.* **19**, 1484–1490.

S. K. Deb [1969] A novel Electrophotographic System. *Appl. Opt. Suppl.* **3**, 192–195.

C. de Boor [1978] *A Practical Guide to Splines*, Springer-Verlag, New York, p. 249.

P. Debye [1929] *Polar Molecules*, Chemical Catalogue Company, New York, p. 94.

P. Debye and H. Falkenhagen [1928] Dispersion of the Conductivity and Dielectric Constants of Strong Electrolytes, *Phys. Z.* **29**, 121–132, 401–426.

F. Decker, S. Passerini, R. Pileggi, and B. Scrosati [1992] The Electrochromic Process in Non-Stoichiometric Nickel Oxide Thin Film Electrodes. *Electrochim. Acta* **37**, 1033–1038.

L. C. De Jonghe [1979] Grain Boundaries and Ionic Conduction in Sodium Beta-Alumina. *J. Mat. Sci.* **14**, 33–48.

P. Delahay [1954] *New Instrumental Methods in Electrochemistry*, Interscience, New York.

P. Delahay [1965] *Double Layer and Electrode Kinetics*, Wiley-Interscience, New York.

P. Delahay [1966] The Theory of the Electrolytic Double-Layer, *J. Phys. Chem.* **70**, 2373–2379.

P. Delahay, M. Senda, and C. H. Weis [1960] Faradaic Rectification and Electrode Processes, *J. Phys. Chem.* **64**, 960.

C. Deslouis, C. Gabrielli, Ph. Sainte-Rose Franchine, and B. Tribollet [1980] Relationship between the Electrochemical Impedance and the Electrohydrodynamical Impedance Measured using a Rotating Disk Electrode, *J. Electroanal. Chem.* **107**, 193–195.

[1982] Electrohydrodynamical Impedance on a Rotating Disk Electrode, *J. Electrochem. Soc.* **129**, 107–118.

J. Devay and L. Meszaros [1980] Study of the Rate of Corrosion of Metals by a Faradaic Distortion Method. II. Effect of the Ohmic Resistance and Capacity of the Double Layer on the Measurement of the Harmonic Components of the Faradaic Currents, *Acta Chim. Acad. Sci. Hung.* **104**, 311.

J. P. Diard and B. LeGorrec [1979] Use of a New Generator with Negative Impedance for Studying the Anodic Behaviour of Zinc in an Alkaline Medium, *Electroanal. Chem.* **103**, 363–374.

J. P. Diard, B. Le Gorrec, and C. Montella [1992] Calculation, Simulation and Interpretation of Electrochemical Impedances. Part 3. Conditions for Observations of Low Frequency Inductive Diagrams for a Two-Step Electron Transfer Reaction with an Adsorbed Intermediate Species, *J. Electroanal. Chem.* **326**, 13–36.

J. L. Diot, J. Joseph, J. R. Matin, and P. Clechet [1985] pH Dependence of the Si/SiO_2 Interface State Density for EOS Systems, *Electroanal. Chem.* **193**, 75–88.

K. Doblhofer and A. A. Pilla [1971] LaPlace Plane Analysis of the Faradaic and Non-Faradaic Impedance of the Mercury Electrode, *J. Electroanal. Chem.* **39**, 91.

M. Dolle, F. Orsini, A. Gozdz, and S. Tarascon [2001] Development of Reliable Three-Electrode Impedance Measurements in Plastic Li-Ion Batteries, *J. Electrochem. Soc.*, **148**, A851–A857.

K. Donovan and T. Kreouzis [2000] Deconvolution of Time of Flight Photocurrent Transioents and Electronic Response Time in Columnar Crystals, *J. Appl. Phys.* **88**, 918–923.

M. Doyle and J. Newman [1995] Modeling the Performance of Rechargeable Lithium-Based Cells: Design Correlations for Limiting Cases, *Journal of Power Sources*, **54**, 46–51.

J. Drennan and E. P. Butler [1982] Does Alumina Act as a Grain Boundary Scavenger in Zirconia? *J. Am. Ceram. Soc.* **65(11)**, Comm. C-194–195.

Y. Duhirel, B. Beden, J.-M. Leger, and C. Lamy [1992] Passivation du Cadmium en Milieu Basique: Etude Mecanistique par Voltammetrie Cyclique, *Electrochim. Acta* **37/4**, 665–671.

S. S. Dukhin and V. N. Shilov [1974] *Dielectric Phenomena and the Double Layer in Disperse Systems and Polyelectrolytes* (English Translation), Keter, Jerusalem.

E. C. Dutoit, R. L. Van Meirhaeghe, F. Cardon, and W. P. Gomes [1975] Investigation on the Frequency-Dependence of the Impedance of the Nearly Ideally Polarizable Semiconductor Electrodes CdSe, CdS, and TiO_2, *Ber. Bunsen Gesell.* **79**, 1206–1213.

J. C. Dyre [1988] The Random Free Energy Barrier Model for ac Conduction in Disordered Solids, *J. Appl. Phys.* **64**, 2456–2468.

J. C. Dyre and T. B. Schrøder [2000] Universality of ac Conduction in Disordered Solids, *Rev. Mod. Phys.* **72**, 873–892.

W. Ehm, H. Göhr, R. Kaus, B. Röseler, and C. A. Schiller [2000] The Evaluation of Electrochemical Impedance Spectra using a Modified Logarithmic Hilbert Transform, *ACH-Models in Chemistry* **137**, 145–157.

L. E. Eiselstein, M. C. H. McKubre, and R. D. Caligiuri [1983] *Prediction of Crack Growth in Aqueous Environments*, First Annual Report by SRI to the Office of Naval Research, N00014–82-K-0343.

[1985] *Electrical and Mechanical Impedance Studies of Crack Growth*, Final Report by SRI to the Office of Naval Research.

H. Engstrom and J. C. Wang [1980] Automatic Multifrequency Measurements of the Complex Impedance of Fast Ion Conductors, *Solid State Ionics* **1**, 441–459.

I. Epelboin and M. Keddam [1970] Faradaic Impedances: Diffusion Impedance and Reaction Impedance, *J. Electrochem. Soc.* **117**, 1052–1056.

I. Epelboin, C. Gabrielli, M. Keddam, and H. Tanenouti [1975] A Model of the Anodic Behaviour of Iron in Sulphoric Acid Medium, *Electrochim. Acta* **20**, 913–916.

I. Epelboin, C. Gabrielli, M. Keddam, J.-C. Lestrade, and H. Takenouti [1972] Passivation of Iron in Sulfuric Acid Medium, *J. Electrochem. Soc.* **119**, 1631–1637.

A. Eucken and B. Weblus [1951] Adsorption of Hydrogen on Platinum According to Measurements of the Capacity of Polarization, *Zeit. Elektrochem.* **55**, 114–119.

F. G. Fabregat-Santiago, J. Garcia-Belmonte, E. Bisquert, N. S. Ferriols, P. R. Bueno, A. J. S. Longo, and S. Castro-Garcia [2001] Dynamic Processes in the Coloration of WO_3 by Lithium Insertion. *J. Electrochem. Soc.* **148**, E302–E309.

J. Fleig [2000] The Influence of Non-Ideal Microstructures on the Analysis of Grain Boundary Impedances, *Solid State Ionics* **131**, 117–127.

[2002] The Grain Boundary Impedance of Random Microstructures, Numerical Simulations and Implications for the Analysis of Experimental Data, *Solid State Ionics* **150**, 181–193.

J. Fleig and J. Maier [1996] The Impedance of Imperfect Electrode Contacts on Solid Electrolytes, *Solid State Ionics* **85**, 17–24.

[1998] A Finite Element Study of the Grain Boundary Impedance of Different Microstructures, *J. Electrochem. Soc.* **145**, 2081–2089.

S. Fletcher [1992] An Electrical Model Circuit that Reproduces the Behaviour of Conducting Polymer Electrodes in Electrolyte Solutions, *J. Electroanal. Chem.* **337**, 127–145.

D. R. Franceschetti [1981] Small-Signal A. C. Response Theory for Systems Exhibiting Impurity Oxygen Transport, *Solid State Ionics* **5**, 613–616.

[1982] *Physics of Electrochemical Systems*, Lecture Notes, State University, Utrecht, Netherlands, January, pp. 1–165.

[1984] Small-Signal AC Response Theory for Systems Exhibiting Diffusion and Trapping of an Electroactive Species, *J. Electroanal. Chem.* **178**, 1–9.

D. R. Franceschetti and J. R. Macdonald [1977] Electrode Kinetics, Equivalent Circuits, and System Characterization: Small-Signal Conditions, *J. Electroanal. Chem.* **82**, 271–301.

[1979a] Numerical Analysis of Electrical Response: Statics and Dynamics of Space-Charge Regions at Blocking Electrodes, *J. Appl. Phys.* **50**, 291–302.

[1979b] Numerical Analysis of Electrical Response: Biased Small-Signal A. C. Response for Systems with One or Two Blocking Electrodes, *J. Electroanal. Chem.* **100**, 583–605.

[1979c] Diffusion of Neutral and Charged Species under Small-Signal A. C. Conditions, *J. Electroanal. Chem.* **101**, 307–316.

[1980] Coupling of Electrode Kinetics and Space Charge Dynamics in Unsupported Systems, in *Proc. Third Symposium on Electrode Processes*, ed. S. Bruckenstein, J. D. McIntyre, B. Miller, and E. Yeager, the Electrochemical Society 1980 Proceedings **80–3**, 94–114.

[1982] Small-Signal A-C Response Theory for Electrochromic Thin Films, *J. Electrochem. Soc.* **129**, 1754–1756.

D. R. Franceschetti and P. C. Shipe [1984] Bulk Conductivity and Polarization of Ionic Crystals Exhibiting Defect Exchange between Inequivalent Sites, *Solid State Ionics* **11**, 285–291.

D. R. Franceschetti, J. R. Macdonald, and R. P. Buck [1991] Interpretation of Finite-Length-Warburg-Type Impedances in Supported and Unsupported Electrochemical Cells with Kinetically Reversible Electrodes, *J. Electrochem. Soc.* **138**, 1368–1371.

A. D. Franklin [1975] Electrode Effects in the Measurement of Ionic Conductivity, *J. Am. Ceram. Soc.* **58**, 465–473.

A. D. Franklin and H. J. de Bruin [1983] The Fourier Analysis of Impedance Spectra of Electroded Solid Electrolytes, *Phys. Stat. Sol. (a)* **75**, 647–656.

G. Frenning and M. Strømme [2001] Theoretical Derivation of the Isothermal Transient Ionic Current in an Ion Conductor: Migration, Diffusion and Space-Charge Effects, *J. Appl. Phys.* **90**, 5570–5575.

G. Frenning, A. K. Jonsson, A. L. Larsson, and M. Strømme [2003] Theoretical Investigation of Ion Conduction in Three-Layered Ion-Conductor Systems: Derivation of the Isothermal Transient Ionic Current and Frequency-Dependent Impedance, *J. Appl. Phys.* **94**, 1629–1638.

G. Frenning, M. Nilsson, J. Westlinder, G. A. Niklasson, and M. Strømme Mattsson [2001] Dielectric and Li Transport Properties of Electron Conducting and Non-Conducting Sputtered Amorphous Ta2O5 Films, *Electrochim. Acta* **46**, 2041–2046.

R. J. Friauf [1954] Polarization Effects in the Ionic Conductivity of Silver Bromide, *J. Chem. Phys.* **22**, 1329–1338.

H. Fricke [1932] The Theory of Electrolytic Polarization, *Philos. Mag.* **14**, 310–318.

[1953] The Maxwell-Wagner Dispersion in a Suspension of Ellipsoids, *J. Phys. chem.* **57**, 934–937.

K. Friestad, G. Hagen, S. Panero, and B. Scrosati [1997] Transmission Spectra and Electrochemical Impedance Analysis of Polyaniline, Prussian Blue and Tungsten Oxide with Incorporated Nickel in Electrochromic Windows, *Electrochem. Soc. Proc.* **96**, 353–368.

H. Frölich [1949] *Theory of Dielectrics*, Oxford University Press, Oxford.

A. Fromhold and E. L. Cook [1967] Diffusion Currents in Large Electric Fields for Discrete Lattices, *J. Appl. Phys.* **38**, 1546–1553.

A. N. Frumkin and V. I. Melik-Gaykazyan [1951] Determination of the Kinetics of Adsorption of Organic Substances by a.-c. Measurements of the Capacity and the Conductivity at the Boundary: Electrode-Solution, *Dokl. Acad. Nauk* **77**, 855–858.

A. N. Frumkin, P. Dolin, and B. V. Ershler [1940a] Kinetics of Processes on the Platinum Electrode. II. The Rate of Discharge of H-Ions and the Rate of the Over-All Process of Hydrogen Evolution on Platinum, *Acta Physicochimica URSS* **13**, 779–792.

[1940b] Kinetics of Processes on the Platinum Electrode. III. The Influence of the Diffusion of Molecular Hydrogen on the Capacity of the Platinum Electrode, *Acta Physicochimica URSS* **13**, 793–802.

K. Funke [1986] Debye-Hückel-Type Relaxation Processes in Solid Ionic Conductors: The Model, *Solid State Ionics* **18–19**, 183–190.

[1998] A Modified Jump Relaxation Model for Fragile Glass-Forming Ionic Melts, *Z. Phys. Chem.* **206**, 101–116.

K. Funke and R. D. Banhatti [2004] Modelling Frequency-Dependent Conductivities and Permittivities in the Framework of the Migration Concept, submitted to *Solid State Ionics*.

K. Funke, S. Bruckner, C. Cramer, and D. Wilmer [2002] Dynamics of Mobile Ions in Solid Electrolytes—Conductivity Spectra and the Concept of Mismatch and Relaxation, *J. Non-Cryst. Solids* **307–310**, 921–929.

R. M. Fuoss and J. G. Kirkwood [1941] Electrical Properties of Solids. VIII. Dipole Moments in Polyvinyl Chloride-Diphenyl Systems, *J. Amer. Chem. Soc.* **63**, 385–394.

C. Gabrielli [1981] *Identification of Electrochemical Processes by Frequency Response Analysis*, Monograph Reference 004/83, Solartron Instrumentation Group, Farnsborough, England.

C. Gabrielli and M. Keddam [1974] Recent Progress in the Measurement of Electrochemical Impedances in a Sinusoidal System, *Electrochim. Acta* **19**, 355.

C. Gabrielli, M. Keddam, H. Perrot, and R. Torresi [1994] Lithium Insertion in WO_3 Studied by Simultaneous Measurements of Impedance, Electrogravimetric and Electro-Optical Transfer Functions. *J. Electroanal. Chem.* **378**, 85–92.

R. Garnsey [1979] Boiler Corrosion and the Requirement for Feed- and Boiler-Water Chemistry Control in Nuclear Steam Generators, *Nucl. Energy* **18**, 117.

W. Gärtner [1959] Depletion-layer photoeffects in semiconductors, *Phys. Rev.* **116**, 84–87.

R. C. Garvie, R. H. Hannink, and R. T. Pascoe [1975] Ceramic Steel? *Nature* **258**, 703–704.

Y. Gefen, A. Aharony, and S. Alexander [1983] Anomalous Diffusion on Percolating Clusters, *Phys. Rev. Lett.* **50**, 77–80.

H. Gerischer [1951] Alternating-current polarization of electrodes with a potential determining step for equilibrium potential, *Z. physik chem.*, **198**, 268–313.

H. Gerischer and W. Mehl [1955] Zum Mechanismus der Kathodischen Wasserstoffabscheidung an Quecksilber, Silber und Kupfer, *Z. Elektrochem.* **59**, 1049–1060.

P. S. Germain, W. G. Pell, and B. E. Conway [2004] Evaluation and Origins of the Difference Between Double-Layer Capacitance Behaviour at Au-Metal and Oxidized Au Surfaces, *Electrochim. Acta* **49**, 1775–1788.

J. Geyer, H. Kohlmüller, H. Landes, and R. Stübner [1997] Investigations into the Kinetics of the Ni-YSZ-Cermet-Anode of a Solid Oxide Fuel Cell, *Proc. of the 5th Int. Symp. on SOFC, Proc.* **97–40**, pp.585–594, The Electrochem. Soc.

M. Gholamian, J. Sundaram, and A. Q. Contractor [1987] Oxidation of Formic Acid at Polyaniline-Coated and Modified-Polyaniline-Coated Electrodes, *Langmuir* **3**, 741–744.

E. Gileadi and B. E. Conway [1963] Kinetic Theory of Adsorption of Intermediates in Electrochemical Catalysis. *J. Chem. Phys.* **39**, 3420–3430.

J. S. Gill, L. M. Callow, and J. D. Scantlebury [1983] Corrosion Measurements Derived from Small Perturbation Nonlinearity. I. Harmonic Analysis, *Corrosion* **39**, 61–66.

S. H. Glarum [1960] Dielectric Relaxation of Isoamyl Bromide, *J. Chem. Phys.* **33**, 639–643.

S. H. Glarum and J. H. Marshall [1980] The AC Response of Iridium Oxide Films, *J. Electrochem. Soc.* **127**, 1467–1474.

H. Göhr [1997] Impedance Modelling of Porous Electrodes, *Electrochemical Applications* **1**, 2–3 (http://www.zahner.de).

H. Göhr and C. A. Schiller [1986] Faraday-Impedanz als Verknüpfung von Impedanzelementen, *Z. Phys. Chemie Neue Folge* **148**, 105–124.

H. Göhr, J. Söllner, and H. Weinzierl [1983] Kinetic Properties of Smooth and Porous Lead/Lead Sulfate Electrodes in Sulfuric Acid during Charging and Discharging, *Proc. 34th ISE Meeting, Erlangen, Poster 0715.*

S. Goldman [1950] *Transformation Calculus and Electrical Transients*, Prentice-Hall, Englewood Cliffs, N.J.

A. Gorenstein, F. Decker, W. Estrada, C. Esteves, A. Andersson, S. Passerini, S. Pantaloni, and B. Scrosati [1990] Electrochromic NiO_xH_y Hydrated Thin Films: Cyclic Voltammetry and ac Impedance Spectroscopy in Aqueous Electrolyte. *J. Electroanal. Chem.* **277**, 277–290.

S. Gottesfeld and T. A. Zawodzinski [1997] Polymer Electrolyte Fuel Cells, in *Advances in Electrochemical Science and Engineering* Vol. **5**, ed. R. C. Alkire, H. Gerischer, D. M. Kolb, and C. W. Tobias, Wiley-VCH, Weinheim, pp. 196–301.

S. Gottesfeld, J. D. E. McIntyre, G. Beni, and J. L. Shay [1978] Electrochromism in Anodic Iridium Oxide Films. *Appl. Phys. Lett.* **33**, 208–210.

D. C. Grahame [1947] The Electrical Double Layer and the Theory Of Electro-Capillarity, *Chem. Rev.* **41**, 441–501.

[1952] Mathematical Theory of the Faradaic Admittance, *J. Electrochem. Soc.* **99**, 370c–385c.

C. G. Granqvist [1995] *Handbook of Inorganic Electrochromic Materials*, Elsevier, Amsterdam.

[2000] Electrochromic Tungsten Oxide Films: Review of Progress 1993–1998. *Solar Energy Mater. Solar Cells* **60**, 201–262.

M. Grätzel [2000] Properties and Applications of Nanocrystalline Electronic Junctions, in *Handbook of Nanostructured Materials and Nanotechnology*, Vol. **3**: *Electrical Properties*, H. S. Nalwa (Ed.), Academic Press.

M. Green [1992] *Solar Cells: Operating Principles, Technology, and System Applications*, Englewood Cliffs, Prentice-Hall.

S. Greenbaum, J. J. Wilson, M. C. Wintersgill, and J. J. Fontanelli [1990] Dielectric Relaxation Studies on Polymer Electrolytes, in *Second Intern. Symp. On Polymer Electrolytes*, B. Scrosati (Ed.), Elsevier, London.

G. Greeuw, and B. J. Hoenders [1984] Theoretical Solution of the Transient Current Equation for Mobile Ions in a Dielectric Film Under the Influence of a Constant Electric Field, *J. Appl. Phys.* **55**, 3371–3375.

G. Greeuw and J. F. Verway [1984] The Mobility of Na^+, Li^+, and K^+ Ions in Thermally Grown SiO_2 Films. *J. Appl. Phys.* **56**, 2218–2224.

A. Guerrero, E. P. Butler, and P. L. Pratt [1979] The Suzuki Phase in $NaCl:Cd^{2+}$ and $NaCl:Mg^{2+}$, *J. Physique Colloque* **7**, C6/363–366.

E. Gülzow [1996] Alkaline Fuel Cells: a Critical View, *J. Power Sources* **61**, 99–104.

E. Gülzow, M. Schulze, N. Wagner, T. Kaz, A. Schneider, and R. Reissner [1999] New Dry Preparation Technique for Membrane Electrode Assemblies for PEM Fuel Cells, *Fuel Cell Bulletin* **15**, 8–12.

E. Gülzow, M. Schulze, and G. Steinhilber [2002a] Investigations of the Degradation of Different Nickel Anode Types for Alkaline Fuel Cells (AFC), *J. Power Sources* **106**, 126–135.

E. Gülzow, R. Reissner, S. Weisshaar, and T. Kaz [2002b] Progress in DMFC Development Using the Dry Spraying Preparation Technique, *Fuel Cells* **3**, 48–51.

E. Gülzow, N. Wagner, and M. Schulze [2003] Preparation of Gas Diffusion Electrodes with Silver Catalysts for Alkaline Fuel Cells, *Fuel Cells—From Fundamentals to Systems* **3**, 67–72.

T. K. Gupta, J. H. Bechtold, R. C. Kuznicki, L. H. Kadoff, and B. R. Rossing [1977] Stabilization of Tetragonal Phase in Polycrystalline Zirconia, *J. Mat. Sci.* **12**, 2421–2426.

R. Haak and D. Tench [1984] Electrochemical Photocapacitance Spectroscopy Method for Characterization of Deep Levels and Interface States in Semiconductor Materials, *J. Electrochem. Soc.* **131**, 275–283.

R. Haak, C. Ogden, and D. Tench [1982] Electrochemical Photocapacitance Spectroscopy: A New Method for Characterization of Deep Levels in Semiconductors, *J. Electrochem. Soc.* **129**, 891–893. [1984] Degradation in Nickel-Cadmium Cells Studied by Impedance Measurements, *J. Power Sources* **12**, 289–303.

P. Hagan and D. Fellowes [2003] Multivariate Statistical Methods in Battery Research *J Power Sources*, **122**, 77–84.

S. Hageman [1993] Berkeley SPICE, EDV 38/22 117.

A. Hagfeldt and M. Grätzel [1995] Light-Induced Redox Reactions in Nanocrystalline Systems, *Chem. Rev.* **95**, 49–68.

B. Hague [1957] *AC Bridge Methods*, Pitman, London.

A. Hammouche, E. Karden, and R. W. De Doncker [2004] Monitoring State-of-Charge of Ni–MH and Ni–Cd Batteries using Impedance Spectroscopy, *Journal of Power Sources* **127**, 105–111.

D. A. Harrington and B. E. Conway [1987] ac Impedance of Faradaic Reactions Involving Electrosorbed Intermediates—I. Kinetic Theory, *Electrochim. Acta* **32**, 1703–1712.

J. W. Hartwell [1971] A Procedure for Implementing the Fast Fourier Transform on Small Computers, *IBM J. Res. Develop.* **15**, 355–363.

J. M. Hawkins and L. O. Barling [1995] Some Aspects of Battery Impedance Characteristics, *Proc. IEEE INTELEC'95*.

H. Helmholtz [1853] Ueber einige Gesetze der Vertheilung elektrischer Ströme in körperlichen Leitern mit Anwendung auf die theirisch-elektrischen Versuche, *Ann. Phys. (Leipzig)* **89**, 211–233.

L. Heyne [1983] Interfacial Effects in Mass Transport in Ionic Solids, in *Mass Transport in Solids*, ed. F. Beniere and C. R. A. Catlow, Plenum Press, New York, pp. 425–456.

K. Hladky, L. M. Callow, and J. L. Dawson [1980] Corrosion Rates from Impedance Measurements: An Introduction, *Brit. Corrosion J.* **13**, 20.

C. Ho [1980] *Application of AC Techniques to Kinetic Studies of Electrochemical Systems*, Ph.D. thesis, Stanford University.

C. Ho, I. D. Raistrick, and R. A. Huggins [1980] Application of AC Techniques to Study of Lithium Diffusion in Tungsten Trioxide Thin Films, *J. Electrochem. Soc.* **127**, 343–350.

J. P. Hoare [1968] *The Electrochemistry of Oxygen*, Interscience, NY.

R. R. Adzic, F. C. Anson, and K. Kinoshita (Es), [1996] *Proc. Symp. On Oxygen Electrochemistry*, The Electrochemical Society, Pennington, NJ.

I. M. Hodge, M. D. Ingram, and A. R. West [1975] A New Method for Analyzing the a.c. Behavior of Polycrystalline Solid Electrolytes, *J. Electroanal. Chem.* **58**, 429–432.

[1976] Impedance and Modulus Spectroscopy of Polycrystalline Solid Electrolytes, *J. Electroanal. Chem.* **74**, 125–143.

P. Holtappels and C. Bagger [2002] Fabrication and Performance of Advanced multi-Layer SOFC Cathodes, *J. Europ. Ceram. Soc.* **22**, 41–48.

P. Holtappels, M. J. Jørgensen, S. Primdahl, M. Mogensen, and C. Bagger [1998] Electrochemical Performance and Structure of Composite $(La_{0.85}Sr_{0.15})_{0.9}MnO_3$/ YSZ cathodes, In *Proceedings of the 3rd European SOFC Forum 2–5 June*, France, Ed. P. Stevens, 1998, p. 311.

K. Holub, G. Tessari, and P. Delahay [1967] Electrode Impedance without *a priori* Separation of Double-Layer Charging and Faradaic Process, *J. Phys. Chem.* **71**, 2612–2618.

R. Holze [1983] *Impedanzmessungen an porösen Elektroden*, Thesis, Bonn.

R. Holze and W. Vielstich [1984] The Kinetics of Oxygen Reduction at Porous Teflon-Bonded Fuel Cell Electrodes, *J. Electrochem. Soc.* **131**, 2298–2303.

R. Holze, I. Vogel, and W. Vielstich [1986] New Oxygen Cathodes for Fuel Cells with Organic Fuels, *J. Electroanal. Chem.* **210**, 277–286.

A. Hooper [1977] A Study of the Electrical Properties of Single Crystal and Polycrystalline β-Alumina Using Complex Plane Analysis, *J. Phys. D. Appl. Phys.* **10**, 1487–1497.

G. Hsieh, S. J. Ford, T. O. Mason, and L. R. Pederson [1996] Experimental Limitations in Impedance Spectroscopy:

Part I, simulation of reference electrode artifacts in three-point measurements, *Solid State Ionics* **91**, 191–201.

G. Hsieh, T. O. Mason, and L. R. Pederson [1996] Experimental Limitations in Impedance Spectroscopy: Part II, electrode artifacts in three-point measurements on Pt/YSZ *Solid State Ionics* **91**, 203–212.

G. Hsieh, T. O. Mason, E. J. Garboczi, and L. R. Pederson [1997] Experimental Limitations in Impedance Spectroscopy: Part III, Effect of reference electrode geometry/position, *Solid State Ionics* **96**, 153–172.

R. A. Huggins, H. Prinz, M. Wohlfahrt-Mehrens, L. Jörissen, and W. Witschel [1994] Proton Insertion Reactions in Layered Transition Metal Oxides, *Solid State Ionics*, **70–71**, 417–424

B. D. Hughes, M. F. Shlesinger, and E. W. Montroll [1981] Random Walks with Self Similar Clusters, *Proc. Nat. Acad. Sci.* **78**, 3287–3291.

A. Hunt [1992] Transport in Ionic Conducting Glasses: II. Scaling relations and approximate power law behavior, *J. Phys. Condens. Matter* **4**, 5371–5381.

I. Iami and P. Delahay [1962] Faradaic Rectification and Electrode Processes. III. Experimental Methods for High Frequencies and Application to the Discharge of Mercurous Ion, *J. Phys. Chem.* **66**, 1108–1113.

N. Ibl [1983a] Fundamentals of Transport Phenomena, in *Comprehensive Treatise of Electrochemistry*, Vol. 6, ed. E. Yaeger, J. O'M. Bockris, B. E. Conway, and S. Sarangapani, Plenum Press, New York, pp. 1–63.

[1983b] Current Distribution, in *Comprehensive Treatise of Electrochemistry*, Vol. **6**, *Electrodics: Transport*, ed. E. Yaeger, J. O'M. Bockris, B. E. Conway, and S. Sarangapani, Plenum Press, New York, p. 239.

T. Iijima, K. Suzuki, and Y. Matsuda [1995] Electrodic characteristics of various carbon materials for lithium rechargeable batteries, *Synthetic Metals*, **731**, 9–20.

T. Jacobsen and K. West [1995] Diffusion Impedance in Planar, Cylindrical and Spherical Symmetry, *Electrochim. Acta* **40**/2, 255–262.

G. Jaffé [1952] Theory of Conductivity of Semiconductors, *Phys. Rev.* **85**, 354–363.

H. Jain and S. Krishnaswami [1998] Composition Dependence of Frequency Power Law of Ionic Conductivity of Glasses, *Solid State Ionics* **105**, 129–137.

J. Jamnik [2003] Impedance Spectroscopy of Mixed Conductors with Semi-Blocking Electrodes, *Solid State Physics* **157**, 19–28.

J. Jamnik and J. Maier [2001] Generalised Equivalent Circuits for Mass and Charge Transport: Chemical Capacitance and its Applications, *Phys. Chem. Chem. Phys.* **3**, 1668–1678.

J. Janata and R. J. Huber, (eds) [1985] *Solid State Chemical Sensors*, Academic Press, New York.

W. Jenseit, O. Böhme, F. U. Leidich, and H. Wendt [1993] Impedance Spectroscop: a Method for in situ Characterization of Experimental Fuel Cells, *Electrochim. Acta*, **38**, 2115–2120.

H. B. Johnson and W. L. Worrell [1982] Lithium and Sodium Intercalated Dichalcogenides: Properties and Electrode Applications, *Synthetic Metals* **4**, 225–248.

W. B. Johnson, N. J. Tolar, G. R. Miller, and I. B. Cutler [1969] Electrical and Mechanical Relaxation in CaF_2 Doped with NaF, *J. Phys. Chem. Solids* **30**, 31–42.

A. K. Jonscher [1974] Hopping Losses in Polarisable Dielectric Media. *Nature* **250**, 191–193. [1975a] Physical Basis of Dielectric Loss, *Nature* **253**, 717–719.

[1975b] The Interpretation of Non-Ideal Dielectric Admittance and Impedance Diagrams, *Phys. Stat. Sol.* (a) **32**, 665–676.

[1975c] New Interpretation of Dielectric Loss Peaks, *Nature* **256**, 566–568.

[1977] The "Universal" Dielectric Response, *Nature* **267**, 673–679.

[1980] The Universal Dielectric Responsc: A Review of Data and their New Interpretation, *Physics of Thin Films* **11**, 205–317.

[1983] *Dielectric Relaxation in Solids*, Chelsea Dielectrics Press, London.

A. K. Jonscher and J.-M. Reau [1978] Analysis of the Complex Impedance data for beta-PbF2, *J. Mat. Sci.* **13**, 563–570.

A. K. Jonsson, M. Furlani, and G. A. Niklasson [2004] Isothermal Transient Ionic Current Study of Laminated Electrochromic Devices for Smart Window Applications. *Solar Energy Mater. Solar Cells*, **84**, 361–367.

A. K. Jonscher, M. Strømme Mattsson, and G. A. Niklasson [1999] Li Intercalation in Zirconium Dioxide Films. *Defect and Diffusion Forum*, **177–178**, 51–58.

A. K. Jonsson, M. Strømme Mattsson, and G. A. Niklasson [1999] Intercalation in airconium dioxide films. *Defect and Diffusion Forum*, **177–178**, 51–58.

M. J. Jørgensen and M. Mogensen [2001] Impedance of Solid Oxide Fuel Cell LSM/YSZ Composite Cathodes, *J. Electrochem. Soc.* **148**, A433–A442.

L. Jörissen, V. Gogel, J. Kerres, and J. Garche [2002] New Membranes for Direct Methanol Fuel Cells, *J. Power Sources* **105**, 267–273.

T. R. Jow, M. S. Dinga, K. Xu, S. S. Zhang, J. L. Allen, K. Amine, and G. L. Henriksen [2003] Nonaqueous Electrolytes for Wide-Temperature-Range Operation of Li-ion Cells, *Journal of Power Sources* **119–121**, 343–348.

M. Juhl, S. Primdahl, C. Manon, and M. Mogensen [1996] Performance/ Structure Correlations for Composite SOFC Cathodes, *J. Power Sources* **61**, 173.

T. Katsube, M. Hara, T. Yaji, S. Kobayashi, K. Suzuki, and Y. Nakagawa [1986] Electrical Characterization of a Complemental Electrochromic Device. *Japan Display '86*. 376–379.

T. Kaz and N. Wagner [2002] Method for Production of Multi-Layer Electrode or Electrode Assembly and Gaseous Diffusion Electrode, *US Patent, US 2002150812 A1*.

M. Keddam, J.-F. Lizee, C. Pallotta, and H. J. Takenouti [1984] Electrochemical Behaviour of Passive Iron in Acid Medium, *J. Electrochem. Soc.* **131**, 2016–2024.

M. Keddam, O. R. Mattos, and H. J. Takenouti [1981] Reaction Model for Iron Dissolution Studied by Electrode Impedance. Determination of the Reaction Model, *J. Electrochem. Soc.* **128**, 257–266; **128**, 266–274.

H. Keiser, K. D. Beccu, and M. A. Gutjahr [1976] Estimation of the Pore Structure of Porous Electrodes by Impedance Measurements, *Electrochim. Acta* **21**, 539–543.

H. Keister, K. D. Beccu, and M. A. Gutjahr [1976] Abschatzung der Porenstrukter Poroser Elektroden aus Impedanzmessung, *Electrochim. Acta* **21**, 539–543.

T. Kenjo, S. Osawa, and K. Fujikawa [1991] High Temperature Air Cathodes Containing Ion Conductive Oxides, *J. Electrochem. Soc.* **138**, 349–355.

Z. Kerner and T. Pajkossy [2002] Measurement of Adsorption Rates of Anions on Au(111) Electrodes by Impedance Spectroscopy, *Electrochim. Acta* **47**, 2055–2063.

D. H. Kim, S. H. Ryu , and H. S. Lee [1999] *Method for Evaluating Capacity of Secondary Battery using Mathematical Calculation of Specific Resistance Components of Equivalent Circuit Model Fitted from Impedance Spectrum*, US Patent Application 20040006440.

J.-D. H. Kim, Y.-I. Park, K. Kobayashi, and M. Nagai [2001a] Effect of CO gas and Anode-Metal Loading on H_2 Oxidation in Proton Exchange Membrane Fuel Cell, *J. Power Sources* **103**, 127–133.

J.-D. H. Kim, Y.-I. Park, K. Kobayashi, M. Nagai, and M. Kunimatsu [2001b] Characterization of CO Tolerance of PEMFC by ac Impedance Spectroscopy, *Solid State Ionics* **140**, 313–325.

S. W. Kim and S. I. Pyun [2002] Analysis of Cell Impedance Measured on the LiMn2O4 Film Electrode by PITT and EIS with Monte Carlo Simulation, *J. Electroanal. Chem.* **528**, 114–120.

S. W. Kim, J. Fleig, and J. Maier [2003] Space Charge Conduction: Simple Analytical Solutions for Ionic and Mixed Conductors and Application to Nanocrystalline Ceria, *Phys. Chem. Chem. Phys.* **5**, 2268.

T. Kim, J. J. Tryk, D. A. Amemiya, K. Hashimoto, and A. Fujishima [1997] Color Impedance and Electrochemical Impedance Studies of WO_3 Thin Films: H^+ and Li^+ Transport, *J. Electroanal. Chem.* **435**, 31–38.

M. Kimura, T. Osaki, and S. Toshima [1975] Interfacial Capacity between a Silver Bromide Solid Electrolyte and a Blocking Electrode, *Bull. Chem. Soc. Japan* **48**, 830–834.

R. Kirchheim [1987] Growth Kinetics of Passive Films, *Electrochim. Acta* **32**, 1619–1629.

J. G. Kirkwood and R. M. Fuoss [1941] Anomalous Dispersion and Dielectric Loss in Polar Polymers, *J. Chem. Phys.* **9**, 329–340.

Y. Kiros and S. Schwartz [2000] Long-Term Hydrogen Oxidation Catalysts in Alkaline Fuel Cells, *J. Power Sources* **87**, 101–105.

M. Kleitz and J. H. Kennedy [1979] Resolution of Multicomponent Impedance Diagrams, in *Fast Ion Transport in Solids*, ed. P. Vashishta, J. N. Mundy, and G. K. Shenoy, North-Holland, Amsterdam, pp. 185–188.

M. Kleitz, H. Bernard, E. Fernandez, and E. J. L. Schouler [1981] Impedance Spectroscopy and Electrical Resistance Measurements on Stabilized Zirconia, in *Advances in Ceramics*, Vol. 3, ed. A. H. Heuer and L. W. Hobbs, American Ceramic Society, Columbus, Ohio, pp. 310–336.

K. Kobayashi, H. Kayajima, and T. Masaki [1981] Phase Change and Mechanical Properties of Zirconia-Yttria Solid Electrolyte after Aging, *Solid State Ionics* **3/4**, 489–493.

R. Kohlrausch [1854] Theorie des Elektrischen Rückstandes in der Leidener Flasche, *Pogg. Ann. der Phys. und Chemie* **91**, 179–214.

M. T. M. Koper, T. J. Schmidt, N. M. Markovic, and P. N. Ross [2001] Potential Oscillations and S-Shaped Polarization Curve in the Continuous Electro-oxidation of CO on Platinum Single-crystal Electrodes, *J. Phys. Chem. B* **105**, 8381–8386.

A. K. Kordesch and G. Simader [1996] *Fuel Cells and their Application*, WileyVCH, Weinheim.

A. Kornyshev, W. Schmickler, and M. Vorotyntsev [1982] Nonlocal Electrostatic Approach to the Problem of a Double Layer at a Metal-Electrolyte Interface, *Phys. Rev.* **B25**, 5244–5256.

H. A. Kramers [1929] Theory of Dispersion in the X-Ray Region, *Physik. Z.* **30**, 52.

F. Krok, J. R. Dygas, B. Misztal-Faraj, Z. Florjaczyk, and W. Bzducha [1999] Impedance and Polarisation Studies of New Lithium Polyelectrolyte Gels,

Journal of Power Sources **81–82**, 766–771.

R. de L. Kronig [1926] The Theory of Dispersion of X-Rays, *J. Opt. Soc. Am.* **12**, 547.

F. Kruger [1903] *Z. Physik. Chem.*, **45**, 1.

N. Kuriyama, T. Sakai, H. Miyamura, H. Tanaka, I. Uehara, F. Meli, and L. Schlapbach [1996] Surface Treatment of a LaNi5-Type Metal-Hydride Electrode with an Alkaline Solution Dissolving Cobalt(II) Hydroxide, *J. Alloys Compounds* **238**, 128–140.

P. Kurzweil [2003] *Brennstoffzellentechnik*, Vieweg Verlag, Wiesbaden, Germany.

R. B. Laibowitz and Y. Gefen [1984] Dynamic Scaling near the Percolation Threshold in Thin Au Films, *Phys. Rev. Lett.* **53**, 380–383.

A. C. M. Lampert [2003] Large-Area Smart Glass and Integrated Photovoltaics. *Solar Energy Mater. Solar Cells* **76**, 489–499.

C. Lamy, A. Lima, V. LeRhun, F. Delime, C. Coutanceau, and J.-M. Léger [2002] Recent Advances in the Development of Direct Alcohol Fuel Cells (DAFC), *J. Power Sources* **105**, 283–296.

R. Landauer [1978] Electrical Conductivity in Inhomogeneous Media, in *Electrical Transport and Optical Properties of Inhomogeneous Media*, ed. J. C. Garland and D. B. Tanner, American Institute of Physics, N.Y, pp. 2–43.

M. Lang, T. Franco, G. Schiller, and N. Wagner [2002] Electrochemical Characterization of Vacuum Plasma Sprayed Thin-Film Solid Oxide Fuel Cells (SOFC) for Reduced Operating Temperatures, *J. Appl. Electrochem.* **32**, 871–874.

V. Lanteri, A. H. Heuer, and T. E. Mitchell [1983] Tetragonal Precipitation in the System Y_2O_3-ZrO_2, in *Proceedings of the 41st Annual Meeting of the Electron Microscopy Society of America*, ed. G. W. Bailey, San Francisco Press, San Francisco, pp. 58–59.

S. Lànyi [1975] Polarization of Ionic Crystals with Incompletely Blocking Electrodes, *J. Phys. Chem. Solids* **36**, 775–781.

A. Lasia [1998] Hydrogen Evolution/Oxidation Reaction on Porous Electrodes, *J. Electroanal. Chem.* **454**, 117–123.

[1999] Electrochemical Impedance Spectroscopy and it's Applications, in *Modern Aspects of Electrochemistry*, Vol. 32, ed. B. E. Conway, J. Bockris, and R. E. White, Kluwer Academic/Plenum Publishers, New York, pp. 143–248.

I. R. Lauks and J. N. Zemel [1979] The Silicon Nitride/Silicon Ion Sensitive Semiconductor Electrode, *IEEE Trans. Electron Devices*, **ED-26**, 1959–1964.

E. Laviron [1972] Effect of the Adsorption of the Depolarizing Agent or an Electrochemical Reaction Product on Polarogrpahic Currents. XVI. Theoretical Voltammetric Study with Linear Voltage Variation, of the Surface Reaction, *J. Electroanal. Chem.* **34**, 463–473.

C. G. Law and J. J. S. Newman [1979] A Model for the Anodic Dissolution of Iron in Sulfuric Acid, *J. Electrochem. Soc.* **126**, 2150–2155.

S.-H. Lee, H. M. Cheong, C. E. Tracy, R. Mascarenhas, Pitts, G. Jorgensen, and S. K. Deb [2000] Alternating Current Impedance and Raman Spectroscopic Study of Electrochromic a-WO_3 Films, *Appl. Phys. Lett.* **76**, 3908–3910.

W. K. Lee, J. F. Liu, and A. S. Nowick [1991] Limiting Behavior of ac Conductivity in Ionically Conducting Crystals and Glasses: A New Universality, *Phys. Rev. Lett.* **67**, 1559–1561.

A. Le Mehaute [1984] Transfer Processes in Fractal Media, *J. Statistical Physics* **36**, 665–676.

A. Le Mehaute and G. Crepy [1983] Introduction to Transfer and Motion in

Fractal Media: The Geometry of Kinetics, *Solid State Ionics* **9–10**, 17–30.

Y.-J. Leng, X. Wang, and I-M. Hsing [2002] Assessment of CO-Tolerance for Different Pt-Alloy Anode Catalysts in a Polymer Electrolyte Fuel Cell Using ac Impedance Spectroscopy, *J. Electroanal. Chem.* **528**, 145–152.

S. J. Lenhart, D. D. Macdonald, and B. G. Pound [1984] Restructuring of Porous Nickel Electrodes, *Proc. 19th IECEC Conference*, San Francisco, p. 875.

C. León, A. Rivera, A. Varez, J. Sanz, J. Santamaria, and K. L. Ngai [2001] Origin of Constant Loss in Ionic Conductors, *Phys. Rev. Lett.* **86**, 1279–1282.

C. León, J. Santamaria, M. A. Paris, J. Sanz, J. Ibarra, and A. Varez [1998] Non-Debye Conductivity Relaxation in the non-Arrhenius $Li_{0.5}La_{0.5}TiO_3$ Fast Ionic Conductor. A Nuclear Magnetic Resonance and Complex Impedance Study, *J. Non-Cryst. Solids* **235–237**, 753–760.

F. Leroy, P. Gareil, and R. Rosset [1982] Gas and Ion Selective Field Effect Transistor Electrodes, *Analysis* **10**, 351–366.

V. Levi and D. Aurbach [1997] Simultaneous Measurements and Modeling of the Electrochemical Impedance and the Cyclic Voltammetric Characteristics of Graphite Electrodes Doped with Lithium *J. Phys. Chem. B* **101**, 4630–4640.

[1999] Frumkin Intercalation Isotherm—a Tool for the Description of Lithium Insertion into Host Materials: a Review, *Electrochim. Acta* **45**, 167–185.

R. de Levie [1963] Porous Electrodes in Electrolyte Solutions, I, *Electrochim. Acta* **8**, 751.

[1964] On Porous Electrodes in Electrolyte Solutions, II, *Electrochim. Acta* **9**, 1231–1245.

[1965] The Influence of Surface Roughness of Solid Electrodes on Electrochemical Measurements, *Electrochim. Acta* **10**, 113–130.

[1967] Electrochemical Response of Porous and Rough Electrodes, in *Advances in Electrochemistry and Electrochemical Engineering*, Vol. 6, ed. P. Delahay and C. W. Tobias, Interscience, New York, pp. 329–397.

R. de Levie and D. Vukadin [1975] Dipicrylamine Transport across Ultrathin Phosphatidylethanolamine Membrane, *J. Electroanal. Chem.* **62**, 95–109.

L. M. Levinson [1981] Grain Boundary Phenomena in Electronic Ceramics, in *Advances in Ceramics*, Vol. 1, ed. L. M. Levinson, American Ceramic Society, Columbus, Ohio.

H. Li, X. Huang, and L. Chen, [1999] Electrochemical Impedance Spectroscopy Study of SnO and nano-SnO Anodes in Lithium Rechargeable Batteries, *Journal of Power Sources* **81–82**, 340–345.

S. Lia, G. L. Panb, X. P. Gao, J. Q. Qua, F. Wua, and D. Y. Songa [2004] The Electrochemical Properties of MmNi3.6Co0.7Al0.3Mn0.4 Alloy Modified with Carbon Nanomaterials by Ball Milling, *Journal of Alloys and Compounds*, **364**, 250–256.

R. Liang, B. G. Pound, and D. D. Macdonald [1984] Impedance Analysis of Passive Films on Single Crystal Nickel, *Ext. Abst. Electrochem. Soc.* **84–2**, 356.

D. R. Lide (ed.) [1997] *CRC Handbook of Chemistry and Physics*, 77th Edition, CRC Press.

E. Lilley and J. E. Strutt [1979] Bulk and Grain Boundary Ionic Conductivity in Polycrystalline β''-Alumina, *Phys. Stat. Sol.* (a) **54**, 639–650.

L. F. Lin, C.-Y. Chao, and D. D. Macdonald [1981] A Point Defect Model for Anodic Passive Films, II, Chemical Breakdown, *J. Electrochem. Soc.* **128**, 1194–1198.

C. P. Lindsey and G. D. Patterson [1980] Detailed Comparison of the Williams-Watts and Cole-Davidson Functions, *J. Chem. Phys.* **73**, 3348–3357.

S. H. Liu [1985] Fractal Model for the ac Response of a Rough Interface, *Phys. Rev. Lett.* **55**, 529–532.

Y.-C. Liu, X.-P. Qiu, W.-T. Zhu, and G.-S. Wu [2003] Impedance Studies on Mesocarbon Microbeads Supported Pt-Ru Catalytic Anode, *J. Power Sources* **114**, 10–14.

W. H. Lizcano-Valbuena V. A. Paganin, C. A. P. Leite, F. Galembeck, and E. R. Gonzalez [2003] Catalysts for DMFC: Relation Between Morphology and Electrochemical Performance, *Electrochim. Acta* **48**, 3869–3878.

J. Llopis and F. Colom [1958] Study of the Impedance of a Platinum Electrode Acting as Anode, *Proc. 8th meeting of the CITCE*, Madrid, 1956, Butterworths, London, pp. 414–427.

W. Lorenz and G. Salié [1961a] Reaction Steps of the Electrochemical Phase-Boundary Reaction, *Zeit. Phys. Chem N. F.* **29**, 390–396.

[1961b] Potential Dependence and Charge Transition Coefficients of the Tl/Tl+ Reaction, *Zeit. Phys. Chem, N. F.* **29**, 408–412.

D. D. Macdonald [1977] *Transient Techniques in Electrochemistry*, Plenum Press, New York.

[1978a] An Impedance Interpretation of Small Amplitude Cyclic Voltammetry. I. Theoretical Analysis for a Resistive-Capacitive System, *J. Electrochem. Soc.* **125**, 1443–1449.

[1978b] An Impedance Interpretation of Small Amplitude Cyclic Voltammetry. II. Theoretical Analysis of Systems that Exhibit Pseudoinductive Behavior. *J. Electrochem. Soc.* **125**, 1977–1981.

[1992] The Point Defect Model for the Passive State, *J. Electrochem. Soc.* **139**, 3434–3449.

[1999] Passivity: The Key to Our Metals-Based Civilization, *Pure Appl. Chem.* **71**, 951–986.

D. D. Macdonald and M. C. H. McKubre [1981] Electrochemical Impedance Techniques in Corrosion Science, in *Electrochemical Corrosion Testing*, ed. Florian Mansfield and Ugo Bertocci, STP 727, ASTM, Philadelphia.

[1982] Impedance Measurements in Electrochemical Systems, in *Modern Aspects of Electrochemistry*, Vol. 14, ed. J. O'M. Bockris, B. E. Conway, and R. E. White, Plenum Press, New York.

D. D. Macdonald and S. I. Smedley [1990] An Electrochemical Impedance Analysis of Passive Films on Ni(111) in Phosphate Buffer Solutions, *Electrochim. Acta* **35**, 1949–1956.

D. D. Macdonald and M. Urquidi-Macdonald. [1985] Application of Kramers-Kronig Transforms to the Analysis of Electrochemical Impedance Data. I. Polarization Resistance, *J. Electrochem. Soc.* **132**, 2316–2319.

[1986] Distribution Functions for the Breakdown of Passive Films, *Electrochim. Acta* **31**, 1079–1086.

D. D. Macdonald, A. Sun, N. Priyantha, and P. Jayaweera [2004] An Electrochemical Impedance Study of Alloy-22 in NaCl Brine at Elevated Temperature: II. Reaction Mechanism Analysis, *J. Electroanal. Chem.*, in press.

J. R. Macdonald [1953] Theory of a.c. Space-Charge Polarization Effects in Photoconductors, Semiconductors, and Electrolytes, *Phys. Rev.* **92**, 4–17.

[1954] Static Space-Charge Effects in the Diffuse Double Layer, *J. Chem. Phys.* **22**, 1317–1322.

[1955] Capacitance and Conductance Effects in Photoconducting Alkali Halide Crystals, *J. Chem. Phys.* **23**, 275–295.

[1962] Restriction on the Form of Relaxation-Time Distribution Functions for a Thermally Activated Process, *J. Chem. Phys.* **36**, 345–349.

[1963] Transient and Temperature Response of a Distributed, Thermally Activated Process, *J. Appl. Phys.* **34**, 538–552.

[1971a] Electrical Response of Materials Containing Space Charge with Discharge at the Electrodes, *J. Chem. Phys.* **54**, 2026–2050.

[1971b] The Impedance of a Galvanic Cell with Two Plane-Parallel Electrodes at a Short Distance, *J. Electroanal. Chem.* **32**, 317–328.

[1974a] Binary Electrolyte Small-Signal Frequency Response, *J. Electroanal. Chem.* **53**, 1–55.

[1974b] Simplified Impedance/ Frequency-Response Results for Intrinsically Conducting Solids and Liquids, *J. Chem. Phys.* **61**, 3977–3996.

[1975] Binary Electrolyte Small-Signal Frequency Response, *J. Electroanal. Chem.* **53**, 1–55.

[1976a] Complex Rate Constant for an Electrochemical System Involving an Adsorbed Intermediate, *J. Electroanal. Chem.* **70**, 17–26.

[1976b] Interpretation of AC Impedance Measurements in Solids, in *Superionic Conductors*, ed. G. D. Mahan and W. L. Roth, Plenum Press, New York, pp. 81–97.

[1984] Note on the Parameterization of the Constant-Phase Admittance Element, *Solid State Ionics* **13**, 147–149.

[1985a] Ambiguity in Models of Small-Signal Response of Dielectric and Conducting Systems and a New Umbrella Model, *Bull. Am. Phys. Soc.* **30**, 587.

[1985b] New Aspects of Some Small-Signal ac Frequency Response Functions, *Solid State Ionics* **15**, 159–161.

[1985c] Frequency Response of Unified Dielectric and Conductive Systems Involving an Exponential Distribution of Activation Energies, *J. Appl. Phys.* **58**, 1955–1970.

[1985d] Generalizations of "Universal Dielectric Response" and a General Distribution-of-Activation-Energies Model for Dielectric and Conducting Systems, *J. Appl. Phys.* **58**, 1971–1978.

[1987] Relaxation is Systems with Exponential or Gaussian Distributions of Activation Energies, *J. Appl. Phys.* **61**, 700–713.

[1991] Precision of Impedance Spectroscopy Estimates of Bulk, Reaction Rate, and Diffusion Parameters, *J. Electroanal. Chem.* **307**, 1–11.

[1993] Some New Directions in Impedance Spectroscopy Data Analysis, *Electrochim. Acta* **38**, 1883–1890.

[1995] Exact and Approximate Nonlinear Least Squares Inversion of Dielectric Relaxation Spectra, *J. Chem. Phys.* **102**, 6241–6250.

[1996] Analysis of Dispersed, Conducting-System Frequency-Response Data, *J. Non-Cryst. Solids* **197**, 83–110.

[1997a] Accurate Fitting of Immittance Spectroscopy Frequency-Response Data using the Stretched Exponential Model, *J. Non-Cryst. Solids* **212**, 95–116.

[1997b] Limiting Electrical Response of Conductive and Dielectric Systems, Stretched-Exponential Behavior, and Discrimination Between Fitting Models, *J. Appl. Phys.* **82**, 3962–3971.

[1998] The Ngai Coupling Model of Relaxation: Generalizations, Alternatives, and their use in the Analysis of non-Arrhenius Conductivity in Glassy Fast-Ionic Materials, *J. Appl. Phys.* **84**, 812–827.

[1999a] Dispersed Electrical-Relaxation: Discrimination between Conductive and Dielectric Properties, *Braz. J. of Physics* **29**, 332–346. http://www.physics.unc.edu/~macd.

[1999b] Critical Examination of the Mismatch-and Relaxation Frequency-Response Model for Dispersive Materials, *Solid State Ionics* **124**, 1–19.

[2000a] Comparison of Parametric and Nonparametric Methods for the Analysis and Inversion of Immittance Data: Critique of Earlier Work, *J. Comput. Phys.* **157**, 280–301.

[2000b] Comparison of the Universal Dynamic Response Power-Law Fitting Model for Conducting Systems with Superior Alternative Models, *Solid State Ionics* **133**, 79–97.

[2000c] On Relaxation-Spectrum Estimation for Decades of Data: Accuracy and Sampling-Localization Considerations, *Inv. Problems* **16**, 1561–1583.

[2001a] Nearly Constant Loss or Constant Loss in Ionically Conducting Glasses: A Physically Realizable Approach, *J. Chem. Phys.* **115**, 6192–6199.

[2001b] Assessing Some Sacred Cows: Strengths and Weaknesses of Some Dispersive Relaxation-Response Models for Thermally Activated Conductive Systems, *4th International Discussion Meeting on Relaxations in Complex Systems*, Crete, Greece, 18 June.

[2001c] Scaling and Modeling in the Analysis of Dispersive Relaxation of Ionic Materials, *J. Appl. Phys.* **90**, 153–161.

[2002a] New Model for Nearly Constant Dielectric Loss in Conductive Systems: Temperature and Concentration Dependencies, *J. Chem. Phys.* **116**, 3401–3409.

[2002b] Discrimination Between Series and Parallel Fitting Models for Nearly Constant Loss Effects in Dispersive Ionic Conductors, *J. Non-Cryst. Solids* **307–310**, 913–920.

[2002c] Some Alternate Models for Nearly Constant Loss in Conductive Systems, *Phys. Rev. B* **66**, 064305.

[2002d] Resolution of Conflicting Views Concerning Frequency-Response Models for Conducting Materials with Dispersive Relaxation, and Isomorphism of Macroscopic and Microscopic Models, *Solid State Ionics* **150**, 263–279.

[2003a] Comparison and Evaluation of Several Models for Fitting the Frequency Response of Dispersive Systems, *J. Chem. Phys.* **118**, 3258–3267.

[2003b] Effective-Medium Model for Nearly Constant Loss in Ionic Conductors, *J. Appl. Phys.* **94**, 558–565.

[2004] On two Incompatible Models for Dispersion in Ionic Conductors, *J. Appl. Phys.* **95**, 1849–1853.

[2005a] Immittance Spectroscopy: Models, Data Fitting, and Analysis, *Solid State Ionics*, to be published.

[2005b] Universality, the Barton, Nakajima, and Namikawa relation, and Scaling for Dispersive Ionic Materials, submitted to *Phys. Rev. B.*

J. R. Macdonald and C. A. Barlow, Jr. [1963] Relaxation, Retardation, and Superposition, *Rev. Mod. Phys.* **35**, 940–946.

J. R. Macdonald and M. K. Brachman [1956] Linear System Integral Transform Relations, *Rev. Mod. Phys.* **28**, 393–422.

J. R. Macdonald and G. B. Cook [1984] Analysis of Impedance Data for Single Crystal Na Beta-Alumina at Low Temperatures, *J. Electroanal. Chem.* **168**, 335–354. (Also presented at a meeting of the Southeastern Section of the American Physical Society, November 1983, and Sixth Australian Electrochemistry Conference, Geelong, February, 1984.)

[1985] Reply to Comments by Almond and West on Na β-Alumina Immittance Data Analysis, *J. Electroanal. Chem.* **193**, 57–74.

J. R. Macdonald and D. R. Franceschetti [1978] Theory of Small-Signal A-C Response of Solids and Liquids with

Recombining Mobile Charge, *J. Chem. Phys.* **68**, 1614–1637.

[1979a] Compact and Diffuse Double Layer Interaction in Unsupported System Small-Signal Response, *J. Electroanal. Chem.* **99**, 283–298.

[1979b] A Method for Estimating Impedance Parameters for Electrochemical Systems That Exhibit Pseudoinductance, *J. Electrochem. Soc.* **126**, 1082–1083.

J. R. Macdonald and J. A. Garber [1977] Analysis of Impedance and Admittance Data for Solids and Liquids, *J. Electrochem. Soc.* **124**, 1022–1030.

J. R. Macdonald and C. A. Hull [1984] Pseudo Reaction Rate in the AC Response of an Electrolytic Cell, *J. Electroanal. Chem.* **165**, 9–20.

J. R. Macdonald and R. L. Hurt [1986] Analysis of Dielectric or Conductive System Frequency Response Data Using the Williams-Watts Function, *J. Chem. Phys.* **84**, 496–502.

J. R. Macdonald and S. W. Kenkel [1985] Comparison of Two Recent Approaches towards a Unified Theory of the Electrical Double Layer, *Electrochim. Acta* **30**, 823–826.

J. R. Macdonald and J. C. Phillips [2005] Topological Derivation of Shape Exponents for Stretched Exponential Relaxation, *J. Chem. Phys.* *122*, 074510.

J. R. Macdonald and Potter, L. D. Jr. [1987] A Flexible Procedure for Analyzing Impedance Spectroscopy Results: Description and Illustrations, Solid State Ionics **24**, 61–79. (The current MS-DOS and Windows versions of the LEVM CNLS fitting and inversion program are freely available for downloading from http://www.physics.unc.edu/~macd.)

J. R. Macdonald, D. R. Franceschetti, and A. P. Lehnen [1980] Interfacial Space Charge and Capacitance in Ionic Crystals: Intrinsic Conductors, *J. Chem. Phys.* **73**, 5272–5293.

J. R. Macdonald, D. R. Franceschetti, and R. Meaudre [1977] Electrical Response of Materials with Recombining Space Charge, *J. Phys. C: Solid State Phys.* **10**, 1459–1471.

J. R. Macdonald, A. Hooper, and A. P. Lehnen [1982] Analysis of Hydrogen-Doped Lithium Nitride Admittance Data, *Solid State Ionics* **6**, 65–77.

J. R. Macdonald, J. Schoonman, and A. P. Lehnen [1981] Three Dimensional Perspective Plotting and Fitting of Immittance Data, *Solid State Ionics* **5**, 137–140.

[1982] The Applicability and Power of Complex Nonlinear Least Squares for the Analyses of Impedance and Admittance Data, *J. Electroanal. Chem.* **131**, 77–95.

J. R. Macdonald and Solartron Inc. [2003] LEVMW Immitance Spectra Fitting Program for MS Windows, http://www.solartronanalytical.com/downloads/software.html

P. B. Macedo, C. T. Moynihan, and R. Bose [1972] The Role of Ionic Diffusion in Polarization in Vitreous Ionic Conductors, *Physics Chem. of Glasses* **13**, 171–179.

P. B. Macedo, R. Bose, V. Provenzano, and T. A. Litowitz [1972] In *Amorphous Materials*, ed. R. W. Douglas and B. Ellis, Interscience, London, p. 251.

J. Maier [1986] On the Conductivity of Polycrystalline Materials, *Ber. Bunsenges Phys. Chem.* **90**, 26–33.

[1996] Interface Effects in Mixed Conductors, in *High Temperature Electrochemistry, Ceramics and Metals, Proc. 17th Risø Intl. Symposium on Materials Science*, ed. F. W. Poulsen, N. Bonanos, S. Linderoth, M. Mogensen, and B. Zachau-Christiansen, Risø National Laboratory, Denmark, Sept. 2–6, 1996, pp. 678–676.

[2004] *Physical Chemistry of Ionic Materials: Ions and Electrons in Solids*, Wiley.

M. Maja and N.Penazzi [1985] Ac Behaviour of Lead Dioxide Electrode: Antimony Effect., *Electrochim. Acta* **30**, 773—778.

B. Mandelbrot [1983] *The Fractal Geometry of Nature*, Freeman, New York.

F. Mansfeld [1976] Electrochemical Monitoring of Atmospheric Corrosion Phenomena, *Adv. Corr. Sci. Tech.* **6**, 163.

F. Mansfeld, M. W. Kendig, and S. Tsai [1982] Evaluation of Organic Coating Metal Systems by AC Impedance Techniques, *Corrosion* **38**, 478–485.

S. W. Martin and C. A. Angell [1986] D. C. and A. C. Conductivity in Wide Composition Range Li_2O-P_2O_5 Glasses, *J. Non-Cryst. Solids* **83**, 185–207.

N. Matsui [1981] Complex-Impedance Analysis for the Development of Zirconia Oxygen Sensors, *Solid State Ionics* **3/4**, 525–529.

A. Matsumoto and K. Higasi [1962] Dielectric Relaxation of Nonrigid Molecules at Lower Temperature, *J. Chem. Phys.* **36**, 1776–1780.

P. Mauracher and E. Karden [1997] Dynamic Modelling of Lead/Acid Batteries using Impedance Spectroscopy for Parameter Identification, *J. Power Sources* **67**, 69–84.

J. C. Maxwell [1881] *A Treatise on Electricity and Magnetism*, 2nd Ed., Clarendon Press, Oxford.

J. C. Maxwell-Garnett [1904] Colours in Metal Glasses and Metal Films, *Trans. Royal Soc.* **CCIII**, 385–420.

J. F. McCann and S. P. S. Badwal [1982] Equivalent Circuit Analysis of the Impedance Response of Semiconductor/Electrolyte/Counterelectri de Cells, *J. Electrochem. Soc.* **129**, 551–559.

N. G. McCrum, B. E. Read, and G. Williams [1976] *Anelastic and Dielectric Effects in Polymeric Solids*, Wiley, London, see p. 111.

M. C. H. McKubre [1976] *An Impedance Study of the Membrane Polarization Effect in Simulated Rock Systems*, Ph.D. thesis, Victoria University, Wellington, N.Z.

[1981] The Use of Harmonic Analysis to Determine Corrosion Rates, Paper number 135 in the *160th meeting of the Electrochemical Society*, Denver, Colorado.

[1983] *The Electrochemical Measurement of Corrosion Rates in Cathodically Protected Systems*, EPRI Report CS-2858 on Project 1689–7, SRI International, Menlo Park, Calif.

[1985] Unpublished data.

M. C. H. McKubre and G. J. Hills [1979] A Digitally Demodulated Synchronous Detector for Fast, High Precision Impedance Measurement, unpublished work.

M. C. H. McKubre and D. D. Macdonald [1984] Electronic Instrumentation for Electrochemical Studies, in *A Comprehensive Treatise of Electrochemistry*, ed. J. O'M. Bockris, B. E. Conway, and E. Yeager, Plenum Press, New York, pp. 1–98.

M. C. H. McKubre and H. B. Sierra-Alcazar [1985] *Sulfur Side Supporting Studies for the Sodium Sulfur Battery*, Interim Report on EPRI Contract 128–2, SRI International, Menlo Park, Calif.

M. C. H. McKubre and B. C. Syrett [1986] Harmonic Impedance Spectroscopy for the Determination of Corrosion Rates in Cathodically Protected Systems, in *Corrosion Monitoring in Industrial Plants Using Nondestructive Testing and Electrochemical Methods*, ASTM STP 908, ed. G. C. Moran and P. Labine, American Society for Testing and Materials, Philadelphia, pp. 433–458.

D. S. McLachlan, M. Blaszkiewicz and R. E. Newman [1990] Electrical Resistivity of Composites, *J. Am. Ceram. Soc.* **73**, 2187–2203.

R. E. Meredith and C. W. Tobias [1962] Conduction in Heterogeneous Systems, in *Advances in Electrochemistry and Electrochemical Engineering*, ed. C. W. Tobias, Vol. 2, Interscience, New York, p. 15.

L. Meszaros and J. Devay [1982] Study of the Rate of Corrosion by a Faradaic Distortion Method, IV, Application of the Method for the Case of Reversible Charge Transfer Reaction, *Acta Chim. Acad. Sciences*, Hungary, **109**, 241–244.

A. Metcalfe [1977] The Anodic Dissolution of Silver into Silver Rubidium Iodide-Impedance Measurements, *J. Electroanal. Chem.* **88**, 187–192.

N. Q. Minh and T. Takahashi [1995] *Science and Technology of Ceramic Fuel Cells*, Elsevier, Amsterdam.

S. P. Mitoff [1968] Properties Calculations for Heterogeneous Systems, in *Advances in Materials Research*, Vol. 3, ed. H. Herman, Wiley-Interscience, New York, pp. 305–329.

T. Miyasaka [1998] United States Patent 6,416,902.

J. Mizusaki, K. Amano, S. Yamuchi, and K. Fueki [1983] Response and Electrode Reaction of Zirconia Oxygen Gas Sensor, in *Analytical Chemistry Symposium Series*, Vol. 17, Elsevier, New York, pp. 279–284.

T. P. Moffat and R. M. Latanision [1992] Electrochemical and X-Ray Photoelectron Spectroscopy Study of the Passive State of Chromium, *J. Electrochem. Soc.* **139**, 1869–1879.

P. M. S. Monk [1999] Charge Movement through Electrochromic Thin Film Tungsten Trioxide. *Crit. Rev. Solid State Mater. Sci.* **24**, 193–226.

E. W. Montroll and J. T. Bendler [1984] On the Levy (or Stable) Distribution and the Williams-Watts Model of Dielectric Relaxation, *J. Statistical Physics* **34**, 129–162.

E. W. Montroll and M. F. Shlesinger [1984] On the Wonderful World of Random Walks, in *Non-Equilibrium Phenomena. II. From Statistics to Hydrodynamics*, ed. J. L. Lebowitz and E. W. Montroll, Elsevier, New York, pp. 1–121.

E. W. Montroll and G. H. Weiss [1965] Random Walks on Lattices, II, *J. Math. Phys.* **6**, 167–181.

C. I. Mooring and H. L. Kies [1977] AC Voltammetry at Large Amplitudes: A Simplified Theoretical Approach, *J. Electroanal. Chem.* **78**, 219.

C. T. Morse [1974] A Computer Controlled Apparatus for Measuring AC Properties of Materials over the Frequency Range 10^{-5} to 10^{+5} Hz, *J. Phys.* **E7**, 657–662.

N. F. Mott [1939] Copper-Cuprous Oxide Photocells, *Proc. Royal Soc.* (London) **A171**, 281–285.

C. T. Moynihan, L. P. Boesch, and N. L. Laberge [1973] Decay Function for the Electric Field Relaxation in Vitreous Ionic Conductors, *Physics and Chem. Glasses* **14**, 122–125.

J. T. Mueller and P. M. Urban [1998] Characterization of Direct Methanol Fuel Cells by ac Impedance Spectroscopy, *J. Power Sources* **75**, 139–143.

W. H. Mulder, J. H. Sluyters, T. Pajkossy, and L. Nyikos [1990] Tafel Current at Fractal Electrodes. Connection with Admittance Spectra, *J. Electroanal. Chem.* **285**, 103–115.

B. Müller, N. Wagner, and W. Schnurnberger [1999] Change of Electrochemical Impedance Spectra (EIS) with Time during CO-Poisoning of the Pt-Anode in a Membrane Fuel Cell, *in Proton Conducting Membrane Fuel Cells II*, ed. S. Gottesfeld and T. F. Fuller, *Electrochem. Soc. Proc.* **98–27**, 187–199.

J. T. Müller, P. M. Urban, and W. F. Hölderich [1999] Impedance Studies on Direct Methanol Fuel Cell Anodes, *J. Power Sources* **84**, 157–160.

K. Mund [1975] Impedanzmessungen an Porösen Gestützten Raney-Nickel-Elektroden für Alkalische Brennstoffzellen, *Siemens Forsch.- und Entwickl.- Ber.* **4**, 68.

K. Mund and F. v. Sturm [1975] Degree of Utilization and Specific Effective Surface Area of Electrocatalysts in Porous Electrodes, *Electrochim. Acta* **20**, 463–467.

H. Näfe [1984] Ionic Conductivity of ThO_2- and ZrO_2-based Electrolytes between 300 and 2000 K, *Solid State Ionics* **13**, 255–263.

J. Nagai, T. Kamimori, and M. Mizuhanshi [1986] Electrochromism in Amorphous Lithium Tungsten Oxide Films. *Solar Energy Mater.* **13**, 279–295.

T. Nakajima [1972] Correlations Between Electrical Conduction and Dielectric Polarization in Inorganic Glasses, in 1971 Annual Report, *Conference on Electrical Insulation and Dielectric Phenomena* (National Academy of Sciences, Washington, DC), 168–176.

H. Namikawa [1975] Characterization of the Diffusion Process in Oxide Glasses Based on the Correlation between Electric Conduction and Dielectric Relaxation, *J. Non-Cryst. Solids* **18**, 173–195.

J. S. Newman [1973] *Electrochemical Systems*, Prentice-Hall, Englewood Cliffs, N.J.

K. L. Ngai [1979] Universality of Low-frequency Fluctuation, Dissipation, and Relaxation Properties of Condensed Matter. I. Comments, *Solid State Phys.* **9**, 127–140.

[1980] Universality of Low-frequency Fluctuation, Dissipation, and Relaxation Properties of Condensed Matter. II. Comments, *Solid State Phys.* **9**, 141–155.

[1998] Evidence of Interaction between Oxygen Ions from Conductivity Relaxation and Quasielastic Light Scattering Data of Yttria-Stabilized Zirconia, *Phil. Mag. B* **77**, 187–195.

[1999] Properties of the Constant Loss in Ionically Conducting Glasses, Melts, and Crystals, *J. Chem. Phys.* **110**, 10576–10584.

K. L. Ngai and C. León [2002] Cage Decay, Near Constant Loss, and Crossover to Cooperative Ion Motion in Ionic Conductors: Insight from Experimental Data, *Phys. Rev. B* **66**, 064308.

K. L. Ngai and C. T. White [1979] Frequency Dependence of Dielectric Loss in Condensed Matter, *Phys. Rev.* **B20**, 2475–2486.

K. L. Ngai, A. K. Jonscher, and C. T. White [1979] On the Origin of the Universal Dielectric Response in Condensed Matter, *Nature* **277**, 185–189.

E. H. Nicollian and J. R. Brew [1982] *MOS (Metal Oxide Semiconductor Technology) Physics and Technology*, Wiley, New York.

E. H. Nicollian and A. Goetzberger [1967] The Si-SiO_2 Interface: Electrical Properties as Determined by the Metal-Insulator-Silicon Conductance Technique, *Bell Syst. Tech. J.* **46**, 1055–1133.

N. Nicoloso, W. Kernler, B. Leibold, A. Löbert, and W. Weppner [1988] Electrical Response of Oxygen Sensing TiO_2 Surfaces and Fractal Pt/YSZ-Interfaces, *Solid State Ionics* **28–30**, 1637–1643.

I. M. Novoseleskii, N. N. Gudina, and Yu. I. Fetistov [1972] Identical Equivalent Impedance Circuits, *Sov. Electrochem* **8**, 546–548.

A. S. Nowick, A. V. Vaysleyb, and W. Liu [1998] Identification of Distinctive Regimes of Behavior in the ac Electrical Response of Glasses, *Solid State Ionics* **105**, 121–128.

J. O'Brien, N. A. Gordon, H. Mathew, and B. P. Hichwa [1999] Electrochromic Coatings-Applications and Manufacturing Issues, *Thin Solid Films* **345**, 312–318.

T. Odagaki and M. Lax [1980] AC Hopping Conductivity of a One-dimensional Bond-Percolation Model. *Phys. Rev. Lett.* **10**, 847–850.

H. Oheda [1981] Phase-shift Analysis of Modulated Photocurrents: its Application to the Determination of the Energetic Distribution of Gap States, *J. Appl. Phys.* **52**, 6693–6700.

K. B. Oldham [1957] Faradaic Rectification: Theory and Application to the Hg^{++}/Hg Electrode, *Trans. Faraday Soc.* **53**, 80–90.

L. Onsager [1926] Theory of Electrolytes, I, *Phys. Z.* **27**, 388–392.

[1927] Theory of Electrolytes, II, *Phys. Z.* **28**, 277–298.

B. O'Regan and M. Grätzel [1991] A low-cost, High-Efficiency Solar Cell Based on Dye-Sensitized Colloidal TiO_2 Films, *Nature* **353**, 737–740.

K. Ozawa [1994] Lithium-ion Rechargeable Batteries with LiCoO2 and Carbon Electrodes: the LiCoO2/C System, *Solid State Ionics* **69**, 212–221.

G. Paasch, K. Micka, and P. Gersdorf [1993] Theory of the Electrochemical Impedance of Macrohomogeneous Porous Electrodes, *Electrochim. Acta* **38**, 2653–2662.

T. Pajkossy [1991] Electrochemistry at Fractal Surfaces, *J. Electroanal. Chem.* **300**, 1–11.

[1994] Impedance of Rough Capacitive Electrodes, *J. Electroanal. Chem.* **364**, 111–124.

G. J. Brug. A. Van den Eeden, M. Sluyters-Rehbach, and J. H. Sluyters [1984] The Analysis of Electrode Impedances Complicated by the Presence of a Constant Phase Element, *J. Electroanal.Chem.* **176**, 275–295.

T. Pajkossy, T. Wandlowski, and Kolb D. M. [1996] Impedance Aspects of Anion Adsorption on Gold Single Crystal Electrodes, *J. Electroanal. Chem.* **414**, 209–220.

J. R. Park [1983] *The Mechanism of the Fast Growth of Magnetite on Carbon Steel under PWR Crevice Conditions*, Ph.D. thesis, Ohio State University.

J. R. Park and D. D. Macdonald [1983] Impedance Studies of the Growth of Porous Magnetite Films on Carbon Steel in High Temperature Aqueous systems, *Corros. Sci.* **23**, 295.

R. Parsons [1954] in *Modern Aspects of Electrochemistry*, Vol. 1, ed. J.O'M. Bockris and B. E. Conway, Butterworths, London.

W. G. Pell and B. E. Conway [1996] Quantitative Modeling of Factors Determining Ragone Plots for Batteries and Electrochemical Capacitors, *J. Power Sources* **63**, 255–266.

[2001] Voltammetry at a de Levie Brush Electrode as a Model for Electrochemical Supercapacitor Behavior, *J. Electroanal. Chem.* **500**, 121–133.

W. G. Pell, B. E. Conway, W. A. Adams, and J. de Oliveria [1999] Electrochemical Efficiency in Multiple Discharge/Recharge Cycling of Supercapacitors in Hybrid EV Applications, *J. Power Sources* **80**, 134–141.

O. Pensado-Rodriguez, M. Urquidi-Macdonald, and D. D. Macdonald [1999a] Electrochemical Behavior of Lithium in Alkaline Aqueous Electrolytes. I. Thermodynamics, *J. Electrochem. Soc.* **146**, 1318–1325.

O. Pensado-Rodriguez, J. Flores, M. Urquidi-Macdonald, and D. D. Macdonald [1999b] Electrochemical Behavior of Lithium in Alkaline Aqueous Electrolytes. II. Point Defect Model, *J. Electrochem. Soc.* **146**, 1326–1335.

[2001] The Electrochemistry of Lithium in Alkaline Aqueous Electrolytes. Part III:

Impedance Model, *J. Electrochem. Soc.* **148**, B386–B399.

L. Peter [1990] Dynamic Aspects of Semiconductor Photoelectrochemistry, *Chem. Rev.* **90**, 753–769.

G. Pfister and H. Scher [1978] Dispersive (Non-Gaussian) Transient Transport in Disordered Solids, *Adv. Phys.* **27**, 747–798.

J. C. Phillips [1994] Microscopic theory of the Kohlrausch constant β_k, *J. Non-Cryst. Solids* **172–174**, 98–103.

A. A. Pilla [1970] Transient Impedance Technique for the Study of Electrode Kinetics: Application to Potentiostatic Methods, *J. Electrochem. Soc.* **117**, 467.

[1972] Electrochemistry, in *Computers in Chemistry and Instrumentation*, Vol. 2, ed. J. S. Mattson, H. B. Mark, Jr., and H. C. Macdonald, Jr., Marcel Dekker, New York, p. 139.

[1975] Introduction to Laplace Plane Analysis, *Information Chemistry*, ed. S. Fujiwara and H. B. Mark, Japan Science Society, pp. 181–193.

E. C. Potter and G. M. W. Mann [1965] The Fast Linear Growth of Magnetite on Mild Steel in High Temperature Aqueous Conditions, *Corros. J.* **1**, 26.

E. J. Plichta, M. Hendrickson, R. Thompson, G. Au, W. K. Behl, M. C. Smart, B. V. Ratnakumar, and S. Surampudi [2001] Development of Low Temperature Li-ion Electrolytes for NASA and DoD Applications, *J. Power Sources* **94**, 160–162.

Z. Poltarzewski, P. Staiti, V. Alderucci, W. Wieczorek, and N. Giordano [1992] Nafion Distribution in Gas Diffusion Electrodes for Solid-Polymer-Electrolyte-Fuel-Cell Applications, *J. Electrochem. Soc.* **139**, 761–765.

G. Popkirov and R. N. Schindler, [1992] A new Impedance Spectrometer for the Investigation of Electrochemical Systems, *Rev. Scientific Instruments* **63**, 5366–5372.

[1994] A New Approach to the Problem of "Good" and "Bad" Impedance Data in Electrochemical Impedance Spectroscopy, *Electrochim. Acta* **39**, 2025.

G. Popkirov, E. Barsoukov, and R. N. Schindler [1997] Investigation of Conducting Polymer Electrodes by Impedance Spectroscopy during Electropolymerization under Galvanostatic Conditions, *J. Electroanal. Chem.* **425**, 209–216.

[1995] Electrochemical Impedance Spectroscopy of Twin Working Electrodes Bridged with Conducting Polymer Layer, *Electrochim. Acta* **40**, 1857–1862.

M. Porto, P. Maass, M. Meyer, A. Bunde, and W. Dieterich [2000] Hopping Transport in the Presence of Site Energy Disorder: Temperature and Concentration Scaling of Conductivity Spectra, *Phys. Rev. B* **61**, 6057–6062.

B. G. Pound and D. D. Macdonald [1985] *Development of Noise Impedance Techniques for the Measurement of Corrosion Rates*, Final Report by SRI International on Sohio Contract NT-MEM-2219.

R. W. Powers [1984] The Separability of Inter- and Intragranular Resistivities in Sodium Beta-Alumina Type Ceramics, *J. Mat. Sci.* **19**, 753–760.

R. W. Powers and S. P. Mitoff [1975] Analysis of the Impedance of Polycrystalline Beta-alumina, *J. Electrochem. Soc.* **122**, 226–231.

S. Primdahl and P. V. Hendriksen [1996] Pitfalls in Solid Electrode Characterisation, in *High Temperature Electrochemistry, Ceramics and Metals, Proc. 17th Risø Intl. Symposium on Materials Science*, ed. F. W. Poulsen, N. Bonanos, S. Linderoth, M. Mogensen, and B. Zachau-Christiansen, Risø National Laboratory, Denmark, Sept. 2–6, 1996, pp. 403–410.

S. Primdahl and M. Mogensen [1998] Gas Conversion Impedance: A Test Geometry Effect in Characterization of Solid Oxide Fuel Cell Anodes, *J. Electrochem. Soc.* **145**, 2431–2438.

P. P. Prosini , M. Lisi, D. Zanec, and M. Pasquali [2002] Determination of the Chemical Diffusion Coefficient of Lithium in LiFePO4, *Solid State Ionics* **148**, 45–51.

M. Proskurnin and A. N. Frumkin [1935] Determination of the Capacity of the Electric Double Layer, *Trans. Faraday Soc.* **31**, 110–115.

S. W. Provencher [1976] An Eigenfunction Expansion Method for the Analysis of Exponential Decay Curves. *J. Chem. Phys.* **64**, 2772, http://s-provencher.com/pages/discrete.shtml

S.-I. Pyun, and J.-S. Bae [1996] Lithium Ion Transport in r.f.-Magnetron Sputtered WO_3 Film as a Function of Lithium Content. *J. Alloys Compounds* **245**, L1–L4.

S.-I. Pyun, Kim, D.-J., and Bae, J.-S. [1996] Hydrogen Transport Through r.-f. Magnetron Sputtered Amorphous and Crystalline WO_3 Films. *J. Alloys Compounds* **244**, 19–22.

X. Qian, N. Gu, Z. Cheng, X. Yang, E. Wang, and S. Dong [2002] Plasticizer Effect on the Ionic Conductivity of PEO-Based Polymer Electrolyte, *Materials Chemistry and Physics* **74**, 98–103.

D. V. Ragone [1968] in *Proc. Soc. Automotive Engineers Conference on Review of Battery Systems for Electric Vehicles*, Soc. of Automotive Engineers, Warrendale, Pa.

S.-U. Rahman, M. A. Al-Saleh, A. S. Al-Zakri, and S. Gultekin [1997] Preparation of Raney–Ni Gas Diffusion Electrode by Filtration Method for Alkaline Fuel Cells, *J. Appl. Electrochem.* **27**, 215–220.

I. D. Raistrick [1983] Lithium Insertion Reactions in Tungsten and Vanadium Oxide Bronzes, *Solid State Ionics* **9/10**, 425–430.

I. D. Raistrick, C. Ho, and R. A. Huggins [1976] Ionic Conductivity of Some Lithium Silicates and Aluminosilicates, *J. Electrochem. Soc.* **123**, 1469–1476.

I. D. Raistrick, C. Ho, Y.-W. Hu, and R. A. Huggins [1977] Ionic Conductivity and Electrode Effects on β-PbF_2, *J. Electroanal. Chem.* **77**, 319–337.

N. Rajalakshmi, K. S. Dhathathreyana, and S. Ramaprabhub [2001] Electrochemical Investigations of ZrCrmFenCopVo (m + n + o + p = 2) Electrode for Ni–MH Battery Applications, *Int. J. Hydrogen Energy* **26**, 1097–1102.

S. Rajendran, R. Kannan, and O. Mahendran [2001] Study on Li ion Conduction Behaviour of the Plasticized Polymer Electrolytes Based on Poly Acrylonitrile, *Materials Letters* **48**, 331–335.

D. O. Raleigh [1976] The Electrochemical Double Layer in Solid Electrolytes, in *Electrode Processes in Solid State Ionics*, ed. M. Kleitz and J. Dupuy, Reidel, Dordrecht/Boston, pp. 119–147.

A. C. Ramamurthy and S. K. Rangarajan [1977] Fournier Polarography Revisited, *J. Electroanal. Chem.* **77**, 267–286.

J.-P. Randin and R.Viennet [1982] Proton Diffusion in Tungsten Trioxide Thin Films, *J. Electrochem. Soc.* **129**, 2349–2354.

J. E. B. Randles [1947] Kinetics of Rapid Electrode Reactions, *Disc. Faraday Soc.* **1**, 11–19.

[1952] Kinetics of Rapid Electrode Reactions. II. Rate constants and activation energies of electrode reactions, *Trans. Faraday Soc.* **48**, 828–832.

J. E. B. Randles and K. W. Somerton [1952] Kinetics of Rapid Electrode

Reactions, Parts 3 and 4, *Trans. Farad. Soc.* **48**, 937–950, 951–955.

S. K. Rangarajan [1974] A Simplified Approach to Linear Electrochemical Systems. I. The Formalism; II. Phenomenological Coupling; III. The Hierarchy of Special Cases; IV. Electron Transfer through Adsorbed Modes, *J. Electroanal. Chem.* **55**, 297–327, 329–335, 337–361, 363–374.

[1975] Non Linear Relaxation Methods. III. Current Controlled Perturbations, *J. Electroanal. Chem.* **62**, 31–41.

G. P. Rao and A. K. Mishra [1977] Some AC Techniques to Evaluate the Kinetics of Corrosion Kinetics, *J. Electroanal. Chem.* **77**, 121.

D. Ravaine and J.-L. Souquet [1973] Application du Trace des Diagrammes D'Impedance Complexe de la Determination de la Conductivité Electrique des Verres Silice-Oxyde Alcalin. *Compt. Rend. Acad. Sci.* (Paris) **277C**, 489–492.

R. W. Reeve, P. A. Christensen, A. J. Dickinson, A. Hamnett, and K. Scott [2000] Methanol-Tolerant Oxygen Reduction Catalysts Based on Transition Metal Sufides and their Application to the study of Methanol Permeation, *Electrochim. Acta*, **45**, 4237–4250.

L. E. Reichl [1980] *A Modern Course in Statistical Physics*, University of Texas Press, Austin, Tex. Reticon Corporation

[1977] *Fast Fourier Transform Module: Product Guide*, Reticon Model R5601, Sunnyvale, Calif.

M. A. Reid [1990] Impedance Studies of Nickel/Cadmium and Nickel/Hydrogen Cells using the Cell Case as a Reference Electrode, *J. Power Sources* **29**, 467–476.

G. Reinhardt and W. Göpel [1998] Electrode Reactions at Solid Electrolytes: Finite Difference Calculations to Describe Geometric and Electrical Properties of Planar Devices, Proceedings of the Electrochemical Society 97-24 (Ionic and Mixed Conducting Ceramics), 610–630.

C. Rerolle and R. Wiart [1996] Kinetics of Oxygen Evolution on Pb and Pb-Ag Anodes during Zinc Electrowinning, *Electrochim. Acta* **41**, 1063–1069.

E. Rhoderick and R. H. Williams [1988] *Metal-Semiconductor Contacts*, Oxford Science Publications, 2nd edn.

F. Richter [1997] Impedance Measurements under High Current for Development and Quality Control of Solid Oxide Fuel Cells (SOFCs), *Electrochemical Applications* **1**, 3–7 (available from http://www.zahner.de).

A. Rivera, C. León, J. Sanz, J. Santamaria, C. T. Moynihan, and K. L . Ngai [2002] Crossover of Near-Constant Loss to Ion Hopping Relaxation in Ionically Conducting Materials: Experimental Evidences and Theoretical Interpretation, *J. Non-Cryst. Solids* **305**, 88–95.

S. Rodrigues, N. Munichandraiah, and A. K. Shukla [2000] A Review of State-of-Charge Indication of Batteries by Means of a.c. Impedance Measurements, *Journal of Power Sources* **87**, 12–20.

B. Roling, C. Martiny, and S. Murugavel [2001] Ionic Conduction in Glass: New Information on the Interrelation Between the "Jonscher Behavior" and the "Nearly Constant-Loss Behavior" from Broadband Conductivity Spectra, *Phys. Rev. Lett.* **87**, 085901-1-085901-4.

A. Roos, D. R. Franceschetti, and J. Schoonman [1984] The Small Signal AC Response of $La_{1-x}Ba_xF_{3-x}$ Solid Solutions, *Solid State Ionics* **12**, 485–491.

P. R. Roth [1970] Digital Fourier Analysis. Some of the Theoretical and Practical Aspects of Measurements Involving Fourier Analysis by Digital Instrumentation, *Hewlett Packard J.* **21**, 2.

J. Rouxel [1979] Recent Progress in Intercalation Chemistry: Alkali Metal in Chalcogenide Hosts, *Rev. Inorg. Chem.* **1**, 245–279.

A. Rudge, J. Davey, I. D. Raistrick, and S. Gottesfeld [1994] Conducting Polymers as Active Materials in Electrochemical Capacitors, *J. Power Sources* **47**, 89–107.

A. Rudge, I. D. Raistrick, S. Gottesfeld, and J. P. Ferraris [1994] A study of the Electrochemical Properties of Conducting Polymers for Application in Electrochemical Capacitors, *Electrochim. Acta* **39**, 273–287.

M. Y. Saïd, J. Barker, H. Huang, J. L. Swoyer, and G. Adamson [2003] Performance Characteristics of Lithium Vanadium Phosphate as a Cathode Material for Lithium-Ion Batteries, *Journal of Power Sources* **119–121**, 266–272.

Y. Saito, H. Kataoka, T. Sakai, and S. Deki [2001] Conduction Prsoperties of Lithium Gel Electrolytes Investigated by Impedance Spectroscopy and Pulsed-Field Gradient NMR with Electric Field, *Electrochim. Acta* **46**, 1747–1751.

A. Salkind, T. Atwater, P. Singh, S. Nalaturyk, S. Damodar, C. Fennie Jr., and D. Raisner [2001] Dynamic Characterization of Small Lead-Acid Cells, *J. Power Sources* **96**, 151–159.

R. P. Sallen and E. L. Key [1955] *IEEE Trans.*, CT-2, 74; D. Hilberman, *IEEE Trans.* **CT-15**, 431 (1968).

J. R. Sandifer and R. P. Buck [1974] Impedance Characteristics of Ion Selective Glass Electrodes, *J. Electroanal. Chem.* **56**, 385–398.

N. Sato [1978] in *Passivity of Metals*, ed. R. P. Frankenthal and J. Kruger, The Electrochemical Society, Pennington, NJ, p. 29.

N. Sato and M. Cohen [1964] Kinetics of Anodic Oxidation of Iron in Neutral Solution, *J. Electrochem. Soc.* **111**, 512–522.

O. Savadogo [2004] Emerging Membranes for Electrochemical Systems Part II. High Temperature Composite Membranes for Polymer Electrolyte Fuel Cells, *J. Power Sources* **127**, 135–161.

B. Savova-Stoynov and Z. Stoynov [1985] Computer Analysis of Non-Stationary Impedance Data, Proc. Symp., *Computer Aided Acquisition and Analysis of Corrosion Data*, ed. M. W. Kendig, U. Bertocci, and J. E. Strutt, Electrochem. Soc. Inc., Pennington, Vol. 85–3, 152–158.

[1992] Four-Dimensional Estimation of the Instantaneous Impedance, *Electrochim. Acta* **37**, 2353–2355.

G. Schiller, R. H. Henne, M. Lang, R. Ruckdäschel, and S. Schaper [2000] Development of Vacuum Plasma Sprayed Thin-Film SOFC for Reduced Operating Temperature, *Fuel Cells Bulletin* **21**, 7–12.

C. A. Schiller, F. Richter, E. Gülzow, and N. Wagner [2001] Validation and Evaluation of Electrochemical Impedance Spectra of Systems with States that Change with Time, *J. Phys. Chem. Chem. Phys.* **3**, 374–378.

K. Sawai, Y. Iwakoshi, and T. Ohzuku [1994] Carbon Materials for Lithium-ion (shuttlecock) Cells, *Solid State Ionics* **69**, 273.

D. F. Schanne and E. Ruiz-Ceretti [1978] *Impedance Measurements in Biological Cells*, Wiley, New York.

W. Scheider [1975] Theory of the Frequency Dispersion of Electrode Polarization: Topology of Networks with Fractional Power Frequency Dependence, *J. Phys. Chem.* **79**, 127–136.

H. Scher and E. W. Montroll [1975] Anomalous Transit-Time Dispersion in Amorphous Solids, *Phys. Rev. B* **12**, 2455–2477.

H. Scher and M. Lax [1973a] Stochastic Transport in Disordered Solids. I. Theory, *Phys. Rev.* **B7**, 4491–4502.

[1973b] Stochastic Transport in Disordered Solids. II. Impurity Conduction, *Phys. Rev.* **B7**, 4502–4519.

M. F. Schlesinger [1984] Williams-Watts Dielectric Relaxation: A Fractal Time Stochastic Process, *J. Statist. Phys.* **36**, 639–648.

M. F. Schlesinger and E. W. Montroll [1984] On the Williams-Watts Function of Dielectric Relaxation, *Proc. Nat. Acad. Sci.* **81**, 1280–1283.

T. J. Schmidt, H. A. Gasteiger, G. D. Stäb, P. M. Urban, D. M. Kolb and R. J. Behm [1998] Characterization of High-Surface-Area Electrocatalysts Using a Rotating Disk Electrode Configuration, *J. Electrochem. Soc.* **145**, 2354–2358.

W. Schottky [1939] "Zur Halbeitertheorie der Sperrschiet- und Spitzengleichrichter, *Z. Phys.* **113**, 367–414.

[1942] Vereinfachte und erweterte Theorie der Randschichte gleichrichter, *Z. Phys.* **118**, 539–592.

J. Schoukens, R. Pintelon, Y. Rolain, and T. Dobrowiecki [2001] Frequency Response Function Measurements in the Presence of Nonlinear Distortions, *Automatica* **37**, 939–946.

E. J. L. Schouler [1979] *Etude des Cellules a Oxyde Electrolyte Solide par la Methode des Impédances Complexes: Applications à la Mesure Precise de la Conductivité et a l'Etude de la Réaction d'Electrode à Oxygène*, Ph.D. thesis, Institut National Polytechnique de Grenoble.

E. J. L. Schouler, G. Giroud, and M. Kleitz [1973] Application selon Bauerle du Tracé des Diagrammes d'Admittance Complexe en Électrochimie des Solides, *J. Chim. Phys.* **70**, 1309–1316.

E. J. L. Schouler, M. Kleitz, E. Forest, E. Fernandez, and P. Fabry [1981] Overpotential of H_2-H_2O/YSZ Electrodes in Steam Electrolyzers, *Solid State Ionics* **5**, 559–562.

E. J. L. Schouler, N. Mesbahi, and G. Vitter [1983] In Situ Study of the Sintering Process of Yttria Stabilized Zirconia by Impedance Spectroscopy, *Solid State Ionics* **9**, **10**, 989–996.

J. Schrama [1957] *On the Phenomenological Theory of Linear Relaxation Processes*, Ph.D. thesis, University of Leiden, Netherlands.

D. Schroder [1998] *Semiconductor Material and Device Characterization*, John Wiley & Sons, 2nd edn.

J. W. Schultze and F. D. Koppitz [1976a] Bond Formation in Electrosorbates. I. Correlation Between the Electrosorption Valency and Pauling's Electronegativity for Aqueous Solutions, *Electrochim. Acta* **21**, 327–336.

[1976b] Bond Formation in Electrosorbates. II. Electrosorption and Double Layer Properties in Nonaqueous Solvents, *Electrochim. Acta* **21**, 337–343.

H. J. Schütt and E. Gerdes [1992] Space-charge Relaxation in Ionically Conducting Oxide Glasses. I. Model and frequency response, *J. Non Cryst. Solids* **144**, 1–13.

H. G. Scott [1975] Phase Relationships in the Zirconia-Yttria System, *J. Mat. Sci.* **10**, 1527–1535.

B. Scrosati [1981] Non Aqueous Lithium Cells, *Electrochim. Acta* **26**, 1559–1567.

M. Seralathan and R. de Levie [1987] On the Electrode Admittance of Complex Electrode Reactions, in *Electro-analytical Chemistry*, ed. A. J. Bard, Marcel Dekker, New York.

N. Sharma, M. Deepa, and S. A Agnihotry [2002] Impedance Studies of Li Inserted Sol-Gel Derived WO_3 Films, *Solid State Ionics* **152–153**, 873–875.

W.-M. Shen, W. Siripala, and M. Tomkiewicz [1986] Electrolyte Electroreflectance Study of Surface Optimization of *n*-$CuInSe_2$ in Photoelectrochemical Solar Cells, *J. Electrochem. Soc.* **133**, 107–111.

W.-M. Shen, M. Tomkiewicz, and D. Cahen [1986] Impedance Study of Surface Optimization of n-$CuInSe_2$ in Photoelectrochemical Solar Cells, *J. Electrochem. Soc.* **133**, 112–116.

D. L. Sidebottom [1999] Universal Approach for Scaling the ac Conductivity in Ionic Glasses, *Phys. Rev. Lett.* **82**, 3653–3656.

H. B. Sierra-Alcazar [1976] *The Electrodeposition of Zinc*, Ph.D. thesis, University of Newcastle-upon-Tyne.

H. B. Sierra-Alcazar, A. N. Fleming, and J. A. Harrison [1977] A Rapid On-Line Method for Electrochemical Investigations, *Surface Technol.* **6**, 61.

A. Sihvola [1999] *Electromagnetic Mixing Formulas and Applications*, IEE Electromagnetic Waves Series 47, ed. P. J. B. Clarricoats and E. V. Jull, The Institute of Electrical Engineers, London, UK.

S. Silverman, G. Cragnolino, and D. D. Macdonald [1982] An Ellipsometric Investigation of Passive Films Formed on Fe-25 Ni-xCr Alloys in Boric Buffer Solution, *J. Electrochem. Soc.* **129**, 2419–2424.

W. M. Siu and R. S. C. Cobbald [1979] Basic Properties of the Electrolyte-Silicon Dioxide-Silicon System: Physical and Theoretical Aspects, *IEEE Trans. Electron Devices*, **ED-26**, 1805–1815.

R. K. Slotwinski, N. Bonanos, and E. P. Butler [1985] Electrical Properties of $MgO + Y_2O_3$ and $CaO + Y_2O_3$ Partially Stabilized Zirconias, *J. Mat. Sci. Lett.* **4**, 641–644.

J. H. Sluyters [1960] On the Impedance of Galvanic Cells, I, *Rec. Trav. Chim.* **79**, 1092–1100.

[1963] On the Impedance of Galvanic Cells. V. The Impedance of a Galvanic Cell with Two Plane Parallel Electrodes at a Short Distance, *Rec. Trav. Chim.* **82**, 100–109.

J. H. Sluyters and J. J. C. Oomen [1960] On the Impedance of Galvanic Cells. II. Experimental Verification, *Rec. Trav. Chim.* **79**, 1101–1110.

M. Sluyters-Rehbach and J. H. Sluyters [1964a] On the Impedance of Galvanic Cells. IX. The double layer capacity of the mercury-M HClO4 interface at high direct current potentials, *Rec. Trav. Chim. Pays Bas* **83**, 217–222.

[1964b] Impedance of Galvanic Cells. X. The impedence of the H+/H2(Hg)-electrode in M perchloric acid, *Rec. Trav. Chim. Pays Bas* **83**, 581–586.

[1964c] The Impedance of Galvanic Cells. XI. The impedance of the Hg-Hg2++ electrode in 1M and 0.1M HClO4, *Rec. Trav. Chim. Pays Bas* **83**, 967–975.

[1970] Sine Wave Methods in the Study of Electrode Processes, in *Electroanalytical Chemistry*, Vol. 4, ed. A. J. Bard, Marcel Dekker, New York, pp. 1–127.

[1984] AC Techniques, in *Comprehensive Treatise of Electrochemistry*, Vol. 9, ed. E. Yaeger, J. O'M. Bockris, B. E. Conway, and S. Sarangapani, Plenum Press, New York.

D. E. Smith [1966] AC Polarography and Related Techniques: Theory and Practice, in *Electroanalytical Chemistry*, Vol. 1, ed. A. J. Bard, Marcel Dekker, New York, pp. 1–155.

[1971] Computers in Chemical Instrumentation: Application of On-Line Digital Computers to Chemical Instrumentation, *CRC Crit. Rev. Anal. Chem.* **2**, 248.

[1976] The Acquisition of Electrochemical Response Spectra by On-Line Fast Fourier Transform: Data Processing in Electrochemistry, *Anal. Chem.* **48**, 221A.

P. H. Smith [1939] Transmission Line Calculator, *Electronics* **12**, 29–31.

[1944] An Improved Transmission Line Calculator, *Electronics* **17**, 130–133, 318–325.

W. Smyrl [1985a] Digital Impedance for Faradaic Analysis. I. Introduction to Digital Signal Analysis and Impedance Measurements for Electrochemical and Corrosion systems, *J. Electrochem. Soc.* **132**, 1551–1555.

[1985b] Digital Impedance for Faradaic Analysis. II. Electrodissolution of Cu in HCl, *J. Electrochem. Soc.* **132**, 1555–1562.

W. Smyrl and L. L. Stephenson [1985] Digital Impedance for Faradaic Analysis. III. Copper Corrosion in Oxygenated 0.1 N HCl, *J. Electrochem. Soc.* **132**, 1563–1567.

C. P. Smyth [1938] *Dielectric Behavior and Structure*, McGraw Hill, New York.

C. Solartron Instrumentation Group [1977] *1170 Series Frequency Response Analyzers: Operation Manual*, Farnborough, Hampshire, England.

L. Solymar and D. Walsh [1999] *Electrical Properties of Materials*, Oxford Science Publications, 6th ed.

P. R. Sorensen and T. Jacobsen [1982] Conductivity Charge Transfer and Transport Number: An Investigation of the Polymer Electrolyte LiSCN-Poly (Ethylene Oxide), *Electrochim. Acta* **27**, 1671–1675.

T. E. Springer, T. A. Zawodzinski, M. S. Wilson, and S. Gottesfeld [1996] Characterisation of Polymer Electrolyte Fuel Cells Using AC Impedance Spectroscopy, *J. Electrochem. Soc.* **143**, 587–599.

S. Srinivasan, E. A. Ticianelli, C. R. Derouin, and A. Redondo [1988] Advances in Solid Polymer Electrolyte Fuel Cell Technology with Low Platinum Loading Electrodes, *J. Power Sources* **22**, 359–375.

U. Staudt and G. Schön [1981] Automatic Spectroscopy of Solids in the Low Frequency Range, *Solid State Ionics* **2**, 175–183.

B. C. H. Steele [1976] High Temperature Fuel Cells and Electrolysers, in *Electrode Processes in Solid State Ionics*, ed. M. Kleitz and J. Dupuy, Reidel Dordrecht/Boston, pp. 368–384.

B. C. H. Steele and E. P. Butler [1985] Electrical Behaviour of Grain Boundaries in Zirconia and Related Ceramic Electrolytes, *Brit. Ceram. Proc.* **36**, 45–55.

B. C. H. Steele, J. Drennan, R. K. Slotwinski, N. Bonanos, and E. P. Butler [1981] Factors Affecting the Performance of Oxygen Monitors, in *Advances in Ceramics*, Vol. 3, ed. A. H. Heuer and L. W. Hobbs, American Ceramic Society, Columbus, Ohio, pp. 286–309.

O. Stern [1924] The Theory of the Electrolytic Double-Layer, *Zeit. Elektrochem.* **30**, 508–516.

M. Stern and A. L. Geary [1957] Electrochemical Polarization. I. Shape of Polarization Curves, *J. Electrochem. Soc.* **104**, 56.

Z. Stoynov [1990] Impedance Modelling and Data Processing: Structural and Parametrical Estimation, *Electrochim. Acta*, **35**, 1493–1499.

M. Strømme, A. Gutarra, G. A. Niklasson, and C. G. Granqvist [1996b] Impedance Spectroscopy on Lithiated Ti Oxide and Ti Oxyfluoride Thin Films, *J. Appl. Phys.* **79**, 3749–3757.

M. Strømme, J. Isidorsson, G. A. Niklasson, and C. G. Granqvist [1996] Impedance Studies on Li Insertion Electrodes of Sn Oxide and Oxyfluoride, *Journal of Applied Physics* **80**, 233–241.

M. Strømme Mattsson [2000] Li Insertion into WO_3: Introduction of a New Electrochemical Analysis Method and Comparison with Impedance Spectroscopy and the Galvanostatic Intermittent Titration Technique, *Solid State Ionics* **131**, 261–273.

M. Strømme Mattsson and G. A. Niklasson [1999] Isothermal Transient Ionic Current as a Characterization Technique for Ion Transport in Ta_2O_5, *J. Appl. Phys.* **85**, 8199–8204.

M. Strømme Mattsson, G. A. Niklasson, and C. G. Granqvist [1996a] Fractal Dimension of Li Insertion Electrodes Studied by Diffusion-Controlled Voltammetry and Impedance Spectroscopy, *Phys. Rev. B* **54**, 2968–2971.

M. Strømme Mattsson, J. Isidorsson, and T. Lindström [1999b] A Theoretical Analysis of Chronoamperometric Electrochemical Ion Intercalation, *J. Electrochem. Soc.* **146**, 2613–2615.

M. Strømme Mattsson, G. A. Niklasson, and C. G. Granqvist [1996c] Li Diffusion in Ti Oxyfluoride Films: Thermal Activation Energy and Jump Length Derived from Impedance Spectroscopy, *J. Appl. Phys.* **80**, 2169–2174.

M. Strømme Mattsson, G. A. Niklasson, and C. G. Granqvist [1997] Diffusion of Li, Na and K in Fluorinated Ti Dioxide Films: Applicability of the Anderson-Stuart Model, *J. Appl. Phys.* **81**, 2167–2172.

M. Strømme Mattsson, G. A. Niklasson, K. Forsgren, and A. Hårsta [1999a] A Frequency Response and Transient Current Study of β-Ta_2O_5: Methods of Estimating the Dielectric Constant, dc Conductivity and Ion Mobility, *J. Appl. Phys.* **85**, 2185–2191.

J. E. Strutt, M. W. Weightman, and E. Lilley [1976] A Versatile Cell for the Measurement of Ionic Thermo–Currents (ITC), *J. Phys.* **E9**, 683–685.

E. C. Subbarao [1981] Zirconia: An Overview, in *Advances in Ceramics*, Vol. 3, ed. A. H. Heuer and L. W. Hobbs, American Ceramic Society, Columbus, Ohio, pp. 1–24.

E. C. Subbarao and H. S. Maiti [1984] Solid Electrolytes with Oxygen Ion Conductivity, *Solid State Ionics* **11**, 317–338.

Y.-K. Sun, D.-W. Kim, and Y.-M. Choi [1999] Synthesis and Characterization of Spinel LiMn2-xNixO4 for Lithium/Polymer Battery Applications, *Journal of Power Sources* **79**, 231–237.

S. Sunde [1996a] Monte Carlo Simulations of Conductivity of Composite Electrodes for Solid Oxide Fuel Cells, *J. Electrochem. Soc.* **143**, 1123–1132.

[1996b] Monte Carlo Simulations of Polarisation Resistance of Composite Electrodes for Solid Oxide Fuel Cells, *J. Electrochem. Soc.* **143**, 1930–1939.

[1997] Calculations of Impedance of Composite Anodes for Solid Oxide Fuel Cells, *Electrochim. Acta* **42**, 2637–2648.

[2000] Simulations of Composite Electrodes in Fuel Cells, *J. Electroceramics* **5**, 153–182.

J. S. E. M. Svensson and C. G. Granqvist [1986] Electrochromic Hydrated Nickel Oxide Coatings for Energy Efficient Windows: Optical Properties and Coloration Mechanism, *Appl. Phys. Lett.* **49**, 1566–1568.

B. C. Syrett and D. D. Macdonald [1979] The Validity of Electrochemical Methods for Measuring Corrosion Rates of Copper-Nickel Alloys in Seawater, *Corrosion* **35**, 505–508.

S. M. Sze [1981] *Physics of Semiconductor Devices*, John Wiley and Sons, 2nd ed.

[1985] *Semiconductor Devices, Physics, and Technology*, Wiley, New York.

S. Szpak [1991] Experimental Simulation of Porous Electrodes, in *Techniques for Characterization of Electrodes and Electrochemical Processes*, ed. Varma and J. R. Selman, Wiley-Interscience, New York, pp. 677–716.

I. Takashi, K. Suzuki, and Y. Matsuda [1995] Electrodic Characteristics of Various Carbon Materials for Lithium Rechargeable Batteries, *Synthetic Metals* **73**, 9–20.

B. Tanner [1995] *Introduction to the Physics of Electrons in Solids*, Cambridge University Press.

J. Thevenin [1985] Passivating Films on Lithium Electrodes. An Approach by Means of Electrode Impedance Spectroscopy, *Journal of Power Sources* **14**, 45–52.

B. V. Tilak, C. P. Chen, and S. K. Rangarajan [1992] A Model to Characterize the Impedance of Electrochemical Capacitors arising from Reactions of the Type Oad + n e- .dblharw. Rad, *J. Electroanal. Chem.* **324**, 405–414.

[1993] A Model to Characterize the Impedance of Electrochemical Capacitors arising from Reactions of the type Oad + n e- .dblharw.r Rad. (Erratum to document cited in CA117(2):15999r) *J. Electroanal. Chem.* **356**, 319–321.

M. Tomkiewicz [1979] Relaxation Spectrum Analysis of Semiconductor-Electrolyte Interface-TiO_2, *J. Electrochem. Soc.* **126**, 2220–2225.

Y.-T. Tsai and D. W. Whitmore [1982] Nonlinear Least-Squares Analysis of Complex Impedance and Admittance Data, *Solid State Ionics* **7**, 129–139.

E. Tuncer, S. M. Gubański, and B. Nettelblad [2000] Dielectric Relaxation in Dielectric Mixtures: Application of the Finite Element Method and its Comparison with Dielectric Mixture Formulas, *J. Appl. Phys.* **89**, 8092–8100.

E. Tuncer, S. M. Gubański, B. Nettelblad [2001] Dielectric relaxation in dielectric mixtures: Application of the finite element method and its comparison with dielectric mixture formulas, *J. Appl. Phys.* **89** (12), 8092–8100.

E. Tuncer, Y. V. Serdyuk, and S. M. Gubański [2002] *IEEE Transactions on Dielectrics and Electrical Insulation*, Vol. 9, No. 5, 809–828.

V. A. Tyagai and G. Y. Kolbasov [1972] Use of Kramers-Kronig Relations for the Analysis of Low Frequency Impedance, *Elektrokhmiya* **8**, 59.

I. Uchida, H. Ishikawa, M. Mohamedi, and M. Umeda [2003] AC-Impedance Measurements during Thermal Runaway Process in Several Lithium/Polymer Batteries, *Journal of Power Sources* **119–121**, 821–825.

H. H. Uhlig [1978] in *Passivity of Metals*, ed. R. P. Frankenthal and J. Kruger, The Electrochemical Society, Pennington, NJ, p. 1.

M. Urquidi-Macdonald, S. Real, and D. D. Macdonald [1986] Application of Kramers-Kronig Transforms in the Analysis of Electrochemical Impedance Data, II, Transformations in the Complex Plane, *J. Electrochem. Soc.* **133**, 2018–2024.

L. K. H. van Beek [1965] Dielectric Behaviour of Heterogeneous System, in *Progress in Dielectrics*, Vol. 7, ed. J. B. Birks, Heywood Press, London, pp. 69–114.

T. van Dijk and A. J. Burggraaf [1981] Grain Boundary Effects on Ionic Conductivity in Ceramic $Gd_xZr_{1-x}O_{2-(x/2)}$ Solid Solutions, *Phys. Stat. Sol.* (a) **63**, 229–240.

R. L. Van Meirhaeghe, E. C. Dutoit, F. Cardon, and W. P. Gomes [1976] On the Application of the Kramers-Kronig Relations to Problems Concerning the Frequency Dependence of Electrode Impedance, *Electrochim. Acta* **21**, 39–43.

J. van Herle, A. J. McEvoy, and K. R. Thampi [1994] Conductivity Measurements of Various Yttria-Stabilized Zirconia Samples, *J. Materials Science* **29**, 3691–3701.

F. H. van Heuveln [1993] Some Remarks Considering the Determination of Electrical Losses in Fuel Cell Systems, *Proceedings of Solid Oxide Workshop*, Chexbres, Switzerland, pp. 47–52.

W. van Weperen, B. P. M. Lenting, E. J. Bijrank, and H. W. der Hartog [1977]

Effect of the Ce^{3+} Concentration on the Reorientation of Dipoles in SrF_2:Ce^{3+}, *Phys. Rev.* **B16**, 1953–2958.

R. Vannier, O. Théry, C. Kinowski, M. Huvé, C. van Tendeloo, E. Stuart, and F. Abraham [1999] Zr Substituted Bismuth Uranate, *J. Mater. Chem.* **9**, 435–443.

M. J. Verkerk and A. J. Burggraaf [1983] Oxygen Transfer on Substituted ZrO_2, Bi_2O_3, and CeO_2 Electrolytes with Platinum Electrodes, *J. Electrochem. Soc.* **130**, 78–84.

M. J. Verkerk, B. J. Middlehuis, and A. J. Burggraaf [1982] Effect of Grain Boundaries on the Conductivity of High-Purity Zirconium Oxide-Yttrium Oxide Ceramics, *Solid State Ionics* **6**, 159–170.

E. J. W. Verwey [1935] Electrolytic Conduction of a Solid Insulator at High Fields: The Formation of the Anodic Oxide Film on Aluminium. *Physica* **2**, 1059–1063.

M. Veszelei, M. Strømme Mattsson, L. Kullman, A. Azens, and C. G. Granqvist [1999] Zr-Ce Oxides as Candidates for Optically Passive Counter Electrodes, *Solar Energy Mater. Solar Cells* **56**, 223–230.

K. J. Vetter [1967] *Electrochemical Kinetics*, Academic Press, New York.

K. J. Vetter and F. Gorn [1973] Kinetics of Layer Formation and Corrosion Processes of Passive Iron in Acid Solutions, *Electrochim. Acta* **18**, 321–326.

W. Vielstich, A. Lamm and H. Gasteiger (ed.) [2003] *Handbook of Fuel Cells—Fundamentals, Technology, Applications*, John Wiley & Sons.

W. Vogel, J. Lundquist, P. Ross, and P. Stonehart [1975] Reaction Pathways and Poisons–II. The Rate Controlling Step for Electrochemical Oxidation of Hydrogen on Pt in Acid and Poisoning of the Reaction by CO, *Electrochim. Acta* **20**, 79–93.

J. Volger [1960] Dielectric Properties of Solids in Relation to Imperfections, in *Progress in Semiconductors*, Vol. 4, ed. A. F. Gibson, F. A. Kröger, and R. E. Burgess, Heywood, London, pp. 205–236.

C. Wagner [1933] Uber die Natur des Elektrischen Leitvermogens von a Silbersulfide, *Z. Phys. Chemie* **B21**, 42–52.

[1953] Investigation on Silver Sulfide, *J. Chem. Phys.* **21**, 1819–1827.

K. W. Wagner [1914] Explanation of the Dielectric Fatigue Phenomenon on the Basis of Maxwell's Concept, in *Arkiv für Electrotechnik*, ed. H. Schering, Springer-Verlag, Berlin.

N. Wagner [2002] Characterization of Membrane Electrode Assemblies in Polymer Electrolyte Fuel Cells using a.c. Impedance Spectroscopy, *J. Appl. Electrochem.*, **32**, 859–863.

N. Wagner and E. Gülzow [2004] Change of Electrochemical Impedance Spectra (EIS) with Time during CO-poisoning of the Pt-anode in a Membrane Fuel Cell, *J. Power Sources* **127**, 341–347.

N. Wagner and M. Schulze [2003] Change of Electrochemical Impedance Spectra (EIS) During CO-poisoning of the Pt and Pt-Ru-Anodes in a Membrane Fuel Cell (PEFC), *Electrochim. Acta* **48**, 3899–3907.

N. Wagner, E. Gülzow, M. Schulze, and W. Schnurnberger [1996] Gas Diffusion Electrodes for Alkaline Fuel Cells, in *Solar Hydrogen Energy, German-Saudi Joint Program on Solar Hydrogen Production and Utilization Phase II 1992–1995*, ed. H. Steeb and H. Aba Oud, Stuttgart.

N. Wagner, M. Schulze, and E. Gülzow [2004] Long Term Investigations of Silver Cathodes for Alkaline Fuel Cells, *J. Power Sources*, **127**, 264–272.

N. Wagner, W. Schnurnberger, B. Müller, and M. Lang [1998] Electrochemical

Impedance Spectra of Solid-Oxide Fuel Cells and Polymer Membrane Fuel Cells, *Electrochim. Acta* **43**, 3785–3793.

C. Wang, A. J. Appleby, and F. E. Little [2001] Comparison of the Electrochemical Impedance Spectroscopy Characteristics of Insertion Electrode Materials Used in Secondary Metal Hybride and Lithium-Ion Electrodes, *J. Electrochem. Soc.* **148**, A762–A767.

M. Wang and J. M. Bell [1999] The Kinetic Behaviour of Ion Injection in WO_3 Based Films Produced by Sputter and Sol-Gel Deposition: Part II. Diffusion coefficients. *Solar Energy Mater. Solar Cells* **8**, 411–429.

X. Wang, I.-M Hsing, Y.-J. Leng, and P.-L. Yue [2001] Model Interpretation of Electrochemical Impedance Spectroscopy and Polarization Behavior of H_2/CO Mixture Oxidation in Polymer Electrolyte Fuel Cells, *Electrochim. Acta* **46**, 4397–4405.

E. Wanzenberg, F. Tietz, D. Kek, P. Panjan, and D. Stöver [2003] Influence of Electrode Contacts on Conductivity Measurements of Thin YSZ Electrolyte Films and the Impact on Solid Oxide Fuel Cells, *Solid State Ionics* **164**, 121–129.

K. E. D. Wapenaar and J. Schoonman [1981] Small Signal AC Response of $Ba_{1-x}La_xF_{2+x}$ Crystals, *Solid State Ionics* **2**, 253–263.

K. E. D. Wapenaar, H. G. Koekkoek, and J. van Turnhout [1982] Low-Temperature Ionic Conductivity and Dielectric Relaxation Phenomena in Fluorite-Type Solid Solutions, *Solid State Ionics* **7**, 225–242.

E. Warburg [1899] Uber das Verhalten Sogenannter Unpolarisierbarer Electroden Gegen Wechselstrom, *Ann. Phys. Chem.* **67**, 493–499.

R. M. Waser [1989] Electrochemical Boundary Conditions for Resistance Degradation of Doped Alkaline Earth Titanates, *J. Am. Ceramic Soc.* **72**, 2234.

S. Wasmus and A. Küver [1999] Methanol Oxidation and Direct Methanol Fuel Cells: a Selective Review, *J. Electroanal. Chem.* **461**, 14–31.

M. Watanabe, K. Rikukawa, K. Sanui, and N. Ogata [1985] Evaluation of Ionic Mobility and Transference Number in a Polymeric Solid Electrolyte by Isothermal Transient Ionic Current Method. *J. Appl. Phys.* **58**, 736–740.

G. H. Weiss and R. J. Rubin [1983] Random Walks: Theory and Selected Applications, *Adv. Chem. Phys.* **52**, 363–505.

W. Weppner and R. A. Huggins [1977] Determination of the Kinetic Parameters of Mixed Conducting Electrodes and Application to the System Li_3Sb, *J. Electrochem. Soc.* **124**, 1569–1578.

A. R. West [1983] Private communication to J. R. Macdonald.

D. White, S. Ramdas, J. L. Hutchinson, and P. D. Billyard [1989] Surface Profile Imaging of a Bismuth Uranium Oxide, Bi_2UO_6, *Ultramicroscopy* **31**, 124–131.

S. Whittingham [1978] Chemistry of Intercalation Compounds: Metal Guests in Chalcogenide Hosts, *Prog. Solid State Chem.* **12**, 41–49.

M. S. Whittingham and R. A. Huggins [1971] Measurement of Sodium Ion Transport in β-Alumina Using Reversible Solid Electrodes, *J. Chem. Phys.* **54**, 414–416.

D. P. Wilkinson and D. Thompsett [1997] Materials and Approaches for CO and CO_2 Tolerance for Polymer Electrolyte Fuel Cells, in *Proceedings of the Second National Symposium on New Materials for Fuel Cell and Modern Battery Systems*, ed. O. Savadogo and P. R. Roberge, Montreal (Quebec) Canada, p. 268.

F. G. Will and C. A. Knorr [1960] Investigation of Formation and Removal of Hydrogen and Oxygen Coverage on Platinum by a New, Nonstationary Method, *Zeit. Elektrochem.* **64**, 258–259.

G. Williams and D. C. Watts [1970] Non-Symmetrical Dielectric Relaxation Behavior Arising from a Simple Empirical Decay Function, *Trans. Faraday Soc.* **66**, 80–85.

G. Williams, D. C. Watts, S. B. Dev, and A. M. North [1970] Further Considerations of Non-Symmetrical Dielectric Relaxation Behavior Arising from a Simple Empirical Decay Functions, *Trans. Farad. Soc.* **67**, 1323–1335.

M. S. Wilson, F. H. Garzon, K. E. Sickafus, and S. Gottesfeld [1993] Surface Area Loss of Supported Platinum in Polymer Electrolyte Fuel Cells, *J. Electrochem. Soc.* **140**, 2872–2877.

J. M. Wimmer, H. C. Graham, and N. M. Tallan [1974] Microstructural and Polyphase Effects, in *Electrical Conduction in Ceramics and Glasses*, Part B, ed. N. M. Tallan, Marcel Dekker, New York, pp. 619–652.

J. Winkler, P. V. Hendriksen, N. Bonanos, and M. Mogensen [1998] Geometric Requirements of Solid Electrolyte Cells with a Reference Electrode, *J. Electrochem. Soc.* **145**, 1184–1192.

A. Winsel [1985] *Poröse Gaselektroden*, German Patent, DE 3342969.

T. Wong and M. Brodwin [1980] Dielectric Relaxation Phenomena Associated with Hopping Ionic Conduction, *Solid State Comm.* **36**, 503–508.

Y. G. Wu, G. M. Wu, X. Y Ni, and X. Wu [2000] Ion Transport in Electrochromic Nickel Oxide Thin Films, *Solar Energy Mater. Solar Cells* **63**, 217–226.

Z. Wu and M. Liu [1997] Modelling of Ambipolar Transport Properties of Composite Mixed Ionic-Electronic Conductors *Solid State Ionics* **93**, 65–84.

X. Y. Xiong, H. V. Poorten, and M. Crappe [1996] Impedance Parameters of Ni/Cd Batteries—Individual Electrode Characteristics, *Electrochim. Acta*, **41**, 1267–1275.

C. Yoon, Y. Barsukov, and J. H. Kim [2001] *Method of and Apparatus for Measuring Battery Capacity by Impedance Spectrum Analysis*, United States Patent 6,208, 147.

N. Yoshiike, M. Ayusawa, and S. Kondo [1984] Electrochemical Properties of WO_3 xH_2O. III. Complex Plane Analysis of the Film on SnO_2. *J. Electrochem. Soc.* **131**, 2600–2605.

L. Young [1961] *Anodic Oxide Films*, Academic Press, New York, NY.

T. Zeuthen [1978] Potentials and Small-Signal Impedance of Platinum Micro-Electrodes in Vivo and Vitro, *Med. Biol. Eng. Comput.* **16**, 483–488.

L. Zhang and D. D. Macdonald [1998] On the Transport of Point Defects in Passive Films, *Electrochim. Acta* **43**, 679–691.

O. Zinke and A. Vlcek [1987] *Lehrbuch der Hochfrequencztechnik*, v.1,3, Springer-Verlag, pp. 60–64.

Index

Impedance Spectroscopy, Second Edition, edited by Evgenij Barsoukov and J. Ross Macdonald
ISBN 0-471-64749-7